Mathematical Methods in the Earth and Environmental Sciences

The Earth and environmental sciences are becoming progressively more quantitative due to the increased use of mathematical models and new data analysis techniques. This accessible introduction presents an overview of the mathematical methods essential for understanding the Earth's processes, providing an invaluable resource for students and early career researchers who may have missed (or forgotten) the mathematics they need to succeed as scientists. Topics build gently from basic methods such as calculus to more advanced techniques including linear algebra and differential equations. The practical applications of the mathematical methods to a variety of topics are discussed, ranging from atmospheric science and oceanography to biogeochemistry and geophysics. Including over 530 exercises and end-of-chapter problems, as well as additional computer codes in Python and MATLAB, this book supports readers in applying appropriate analytical or computational methods to solving real research questions.

Adrian Burd is an associate professor at the Department of Marine Sciences at the University of Georgia. As a marine scientist, he applies mathematical tools to understand marine systems, including the carbon cycle in the oceans, the health of seagrass and salt marshes, and the fate of oil spills. His work has taken him around the globe, from the heat of Laguna Madre and Florida Bay to the cold climes of Antarctica.

Mathematical Methods in the Earth and Environmental Sciences

ADRIAN BURD
University of Georgia

CAMBRIDGE
UNIVERSITY PRESS

CAMBRIDGE
UNIVERSITY PRESS

Shaftesbury Road, Cambridge CB2 8EA, United Kingdom

One Liberty Plaza, 20th Floor, New York, NY 10006, USA

477 Williamstown Road, Port Melbourne, VIC 3207, Australia

314–321, 3rd Floor, Plot 3, Splendor Forum, Jasola District Centre, New Delhi – 110025, India

103 Penang Road, #05–06/07, Visioncrest Commercial, Singapore 238467

Cambridge University Press is part of Cambridge University Press & Assessment, a department of the University of Cambridge.

We share the University's mission to contribute to society through the pursuit of education, learning and research at the highest international levels of excellence.

www.cambridge.org
Information on this title: www.cambridge.org/9781107117488

DOI: 10.1017/9781316338636

First published 2019

A catalogue record for this publication is available from the British Library

Library of Congress Cataloging-in-Publication data
Names: Burd, Adrian, author.
Title: Mathematical methods in the earth and environmental sciences / Adrian Burd (University of Georgia).
Description: Cambridge ; New York, NY : Cambridge University Press, 2019. | Includes bibliographical references and index.
Identifiers: LCCN 2018041841 | ISBN 9781107117488 (hardback)
Subjects: LCSH: Earth sciences–Mathematics. | Earth sciences–Mathematical models. | Environmental sciences–Mathematics. | Environmental sciences–Mathematical models. | Research–Statistical methods.
Classification: LCC QE33.2.M3 .B87 2019 | DDC 550.1/51–dc23
LC record available at https://lccn.loc.gov/2018041841

ISBN 978-1-107-11748-8 Hardback

Additional resources for this publication available at www.cambridge.org/burd

Contents

Preface

The Earth and environmental sciences, like all scientific disciplines, are rooted in observational studies. However, recent years have seen an increasing demand for researchers who are also comfortable with the language of mathematics. Mathematical and computer models are now commonplace research tools as the power of desktop computers has increased and the public release of computer codes has made modeling more accessible to those who previously may have hesitated at using a computer model in their research. In addition, the increasing availability of large regional and global multivariate data sets has stimulated the use of new data analysis techniques requiring an understanding of the mathematics that underlies them if they are not to be used as black boxes. Consequently, there is an increasing need for students and researchers who are comfortable manipulating mathematical expressions and who can understand the mathematics underlying the analysis methods that they use.

Many students enter the Earth and environmental sciences with diverse academic backgrounds. As a result, they often find themselves unprepared for the level of mathematics that they need for their coursework or research. This is often because there can be gaps of several years between when students first encounter mathematical techniques such as calculus and when they finally end up using those techniques on a regular basis. Others may never have been exposed to the mathematical tools in the first place. This book is aimed primarily at those students and researchers who find themselves in such situations and need a gentle reminder or introduction to the mathematical methods they need.

Many students, and dare I say even some experienced researchers, are either intimidated by mathematics or find it of little value. I have found an informal approach, providing context for the uses of mathematics, makes things less intimidating. Some may disagree, and I too find myself tempted to veer into the beauty and technical aspects of mathematics from time to time as my background in theoretical physics and astronomy come to the fore. However, in this book I have resisted this and tried to stick to a practical and informal style at the expense of cutting a few corners from time to time. It is my hope that teachers who see the need for more rigor can use this book as a framework to introduce students to the concepts and techniques they need, and then backfill the rigor within the classroom.

To the student who is using this book for self-study, you should work through all the derivations and equations in the text. At the beginning of the book, you will find that most of the steps in derivations are presented. However, as your skills and understanding develop as you work through the text, it is assumed that you will be able to fill in missing steps. The exercises and problems are there to help you refine and develop your understanding and intuition, so you should attempt as many as you can. It is also a good idea to have access to another text, because developing an understanding for a new topic is often a function of point of view. I have provided suggestions for further reading at the end of each chapter, and I hope that this will give ideas.

This book is not meant to be an exhaustive exploration, but rather an introduction to give the reader the tools and techniques they need to successfully do their science. Indeed, large books have been written on the subjects of individual chapters in this book. Some students in certain disciplines (such as geophysics) will require more rigorous detail and coverage of topics that are not included in the book. Some teachers may find their favorite topics missing, and I hope they will understand that given the constraints on the size of the text not everything could be covered. The lists of further reading at the end of each chapter give some advice for the reader on where they can find more information if needed. However, I hope that this book will provide a practical and accessible foundation to the tools most will need. My aim is to get students thinking mathematically.

The material for this book has come primarily from three courses that I teach or co-teach at the University of Georgia. Quantitative Methods in Marine Science is a graduate-level course designed to do precisely what this book is aimed to do, introduce new graduate students to the mathematical techniques they will need for their research and other courses. This is an intense, one-semester course and covers the material in Chapters 1 through 5 and the early parts of Chapter 6. I teach a more advanced graduate-level course, Modeling Marine Systems, that covers material from Chapters 6 to 11, but mostly concentrates on material for solving ordinary and partial differential equations. This course has been taken by students from a wide range of disciplines including marine science, geology, agriculture, and other environmental sciences, and it has been these students who have prompted me to explore the mathematical needs of students in related disciplines. Finally, Mathematics and Climate is a course for both undergraduate and graduate students that I co-teach with Professor Malcolm Adams from the University of Georgia Mathematics Department. It involves the application of dynamical systems to understanding climate. This course is also taken by students from many disciplines including mathematics, geology, geography, and economics.

The exercises and problems are an essential component of the book, especially if you are using it for self-study. As others have said before, mathematics is not a spectator sport, and one needs to practice solving problems using the techniques one learns. Most of the exercises embedded within each chapter are short and designed to practice a technique or to develop understanding through solving a small problem. The problem sets at the end of each chapter are generally more involved, some involve practicing techniques, others use those techniques to solve problems relevant to the Earth and environmental sciences. Supplemental problems are also available.

Computers play an integral part in many of the mathematical techniques introduced here, and throughout the book you will come across the symbol 🖵 in the margin. This indicates that there are supplemental computer codes available that are relevant to the material being discussed. These are not aimed to teach programming, but rather the application of techniques and are available from the Cambridge University Press website (www.cambridge.org/burd) for this book as well as the author's Github site (`https://github.com/BurdLab`).

Acknowledgments

Writing a book like this cannot be achieved in isolation, and there are many people whom I need to thank for their advice, time, support, and the ability to draw on some of their material. I would first and foremost like to thank the many students who have endured my courses over the years. Their passion, comments, and questions over the years have helped to shape my teaching, research, and the contents and style of this book. I would also like to thank those who, over the years, have had the patience to formally and informally teach me both science and pedagogy.

I would like to especially thank George Jackson and Ellen Toby for their lasting friendship and inspiration. George has been a patient and invaluable sounding board for many of my scientific and pedagogical ideas over the years. I would like to thank Mark Denny, without whose encouragement and inspiration I would not have embarked on this adventure. Malcolm Adams, through his own example, has helped me improve my teaching of mathematical topics and I thank him for allowing me to draw on material from the course we co-teach for this book. Several people have read and commented on different parts of this book, and I would like to thank Dr. Mark Denny, Dr. George Jackson, Dr. Malcolm Adams, Dr. Anusha Dissanayake, Dr. Sylvia Schaefer, and Chandler Countryman for their valuable comments and suggestions. It goes without saying that I alone am responsible for the contents of the book, and none of them should be blamed for any remaining errors (see Problem 5.9).

I would like to thank Dinesh Singh Negi for help with LaTeX macros, and the Stackexchange community (https://tex.stackexchange.com) for repeated help and advice, particularly with developing tikz code for the figures of the book. Their generosity of time and spirit are a wonderful example of what can be achieved by a community sharing knowledge and expertise. I would also like to express my gratitude and utmost appreciation to Susan Francis, Sarah Lambert, Cheryl Hutty and the whole team at Cambridge University Press. It has been a long journey from a suggestion made at a conference in Gothenburg to the finished manuscript, and I thank them for their patience, wisdom, and advice; without them, this book would not exist.

Lastly, I would like to like to thank my family and friends for their love and support, and especially my wife, Sylvia, who has had to live with the creation of this book over the last few years.

1 Estimation and Dimensional Analysis

How large a crater does an asteroid make when it impacts the Earth? How much does sea level change as global temperature changes? What is the average distance between bacterial cells in the ocean? Simple questions such as these frequently give us insight into more complicated ones, such as how often do large asteroids collide with the Earth, and can bacteria communicate with each other in the oceans? These are complicated questions, and to get accurate answers often involves using complicated computer simulations. However, by simplifying the problem we can often get a good estimate of the answer and a better understanding of what factors are important to the problem. This improved understanding can then help guide a more detailed analysis of the problem. Two techniques we can use to simplify complicated problems and gain intuition about them are *back-of-the-envelope calculations* and *dimensional analysis*.

Back-of-the-envelope calculations are quick, rough-and-ready estimates that help us get a feeling for the magnitudes of quantities in a problem.[1] Instead of trying to get an exact, quantitative solution to a problem, we aim to get an answer that is within, say, a factor of 10 (i.e., within an order of magnitude) of the exact one. To do this we make grand assumptions and gross approximations, all the time keeping in mind how much of an error we might be introducing. Back-of-the-envelope calculations also help us to understand which variables and processes are important in a problem and which ones we can ignore because, quantitatively, they make only a small contribution to the final answer.

Dimensional analysis is another useful tool we can use to simplify a problem and understand its structure. Unlike back-of-the-envelope calculations, which provide us with a quantitative feeling for a problem, dimensional analysis helps us reduce the number of variables we have to consider by examining the structure of the problem. We will rely on both techniques throughout this book.

1.1 Making Estimates on the Back of the Envelope

One of the first steps we have to take when tackling a scientific question is to understand it. What are the variables we need to consider? What equations do we need? Are there

[1] The myth of back-of-the-envelope calculations is that one should need a piece of paper no bigger than the size of the back of an envelope to do them. In reality, one sometimes needs a little more than that. However, the name conjures the right spirit, to use intuition and approximations to make the calculation as simple as you can, but not too simple!

assumptions we can use that will make the problem easier? Can we make an initial, rough estimate of the answer? This sort of understanding is needed whether we are tackling a complicated research problem, or a problem in a textbook. When we first start working on a new problem, we might feel unsure of how to proceed to a solution, particularly if the problem is in an area we are unfamiliar with. Our initial impulse is often to list all the variables and processes we think might be important and see if something leaps out at us. Back-of-the-envelope calculations can help us reduce this list by determining which variables and processes play quantitatively important roles in the problem.

To make good back-of-the-envelope calculations we need to be comfortable making good estimates of numbers. A good estimate is one that is likely within an order of magnitude of the actual value. We might wonder how we know this if we do not know the actual value. We do not. Like a painter who roughly sketches a scene, trying different arrangements and perspectives before undertaking the actual painting, or a writer trying different outlines before writing a book, we use back-of-the-envelope calculations to help us build a broad understanding of the problem we are tackling. We want to learn which variables and processes might be important for a more detailed investigation. For that we need good quantitative estimates. Estimating that the Earth is 2 km in diameter, or that a microbial cell is 1 m in diameter, will definitely lead us into trouble. But estimating that the diameter of the Earth is 12000 km, or that a microbial cell is 1 μm in diameter is acceptable. An actual microbial cell may be 2 μm in diameter, but this is only a factor of 2 different from our estimate. A more accurate value for the equatorial diameter of the Earth is 12756.28 km (Henderson and Henderson, 2009), so our estimate is only 6% off from the original and is far easier to remember. The idea is to develop a feeling for the magnitude of numbers, to build an intuition for the sizes of objects and rates of processes.

How accurate do we need our estimates to be? We may be tempted to give our answers to many decimal places or significant figures, but we should resist this because we are making only rough estimates. For example, using our estimate for the diameter of the Earth, we can estimate its surface area using $A = \pi d^2 \approx 3 \times (12 \times 10^6)^2 \approx 4.4 \times 10^{14}$ m^2 (a more accurate value is 5×10^{14} m^2, so our estimate is about 12% lower than the accurate value, good enough for a back-of-the-envelope calculation). Doing the calculation on a calculator yielded $A = 4.523889 \times 10^{14}$ m^2, but all the digits after the first or second significant figure are meaningless because we used an estimate of the diameter that differed from an accurate value by 6%. Keep in mind that the aim of a back-of-the-envelope calculation is to obtain a rough estimate, not a highly precise one, and a good rule of thumb is to keep only the first two or three significant figures when making an estimate—this also reduces the number of digits you have to write down and so minimizes the chances of copying a number incorrectly.

Our first back-of-the-envelope calculations will demonstrate how they can help us visualize the scales and magnitudes of quantities in a problem. In science, we frequently come across numbers that are either much larger or much smaller than those we experience in our daily lives. This can make them hard to visualize or think about clearly. For example, in the oceans bacteria are responsible for much of the natural cycling of elements such as carbon and nitrogen, and bacterial abundances in the surface waters are typically 10^5–10^6 cells cm^{-3}. But does this mean that the cells are crowded in the water and almost touching each other? Or are they well separated? Having a good feeling or intuition for this helps us understand processes such as the ability of bacteria to take up nutrients, or to detect chemical signals that indicate the presence of food. We will return to this problem a bit later.

One simple technique that can help us visualize very large or small numbers is to compare them with similar quantities that we might be more familiar with. As an example, let us think about visualizing the Gulf Stream, which is a large, surface current in the North Atlantic Ocean that transports water and heat northward from the subtropics to more temperate latitudes. The transport of water in the Gulf Stream increases from approximately 3×10^7 m^3 s^{-1} near Florida, to approximately 1.5×10^8 m^3 s^{-1} near Newfoundland (Henderson and Henderson, 2009). These numbers are large, and it is hard to visualize a flow of hundreds of millions of cubic meters per second; we are probably not even used to visualizing volumes of water in units of cubic meters.

To put the flow of the Gulf Stream in perspective, we can compare it with something more familiar, but what should we choose? We experience the flow of water from a tap (or faucet) whenever we wash our hands, so we have an intuitive feeling for that. The idea is then to think, "How many taps would have to be turned on to obtain a total flow equivalent to that of the Gulf Stream?" However, the flow from a single tap is too small to make a meaningful comparison—we would end up with numbers as large as the ones we had trouble visualizing in the first place. Comparing the flow of the Gulf Stream to something that is larger and that we have seen for ourselves might make more sense. One possibility is to use the flow of a large river, such as the Amazon, instead of a tap. This has the advantage of having a much larger flow rate than a tap, and we stand a good chance of having seen a large river personally, or in movies, so we can visualize what it is like.

Exercise 1.1.1 What is the typical flow speed of a medium to large river? This question is intentionally vague to encourage you to use your experience. When you walk by a large river, is it flowing faster than your walking speed? Would you have to sprint to keep up, or could you amble along at a leisurely walking pace? You then have to ask how fast you walk!

Exercise 1.1.2 Taking the average width near the river mouth to be 20 km, and the average depth of 10 m, use your answer from Exercise 1.1.1 to estimate the discharge (in m^3 s^{-1}) of the Amazon River. Compare your answer with the number given in Table 1.1. If your answer is more than an order of magnitude different from that in the table, determine which of your estimated numbers could be improved.

Now that we have an estimate for the discharge of the Amazon River, we can compare it with the flow of the Gulf Stream. By simple comparison, the flow of the Gulf Stream is between 150 and 750 Amazon Rivers, or approximately between 2000 and 9000 Mississippi Rivers,[2] while the Amazon itself is equivalent to more than 10 Mississippi Rivers. In making this calculation we have effectively come up with our own "unit"—one Amazon River's worth of flow—for visualizing the transport of water on the scales

[2] The flow of water in the Gulf Stream is 50 times greater than the combined discharge of all the rivers that flow into the Atlantic Ocean.

Table 1.1 Approximate average discharge (volume rate of flow) of some major rivers in the world.

River	Average discharge ($m^3\ s^{-1}$)
Amazon	2.0×10^5
Congo	4.0×10^4
Ganges/Brahmaputra	3.8×10^4
Orinoco	3.6×10^4
Yangtze	3.0×10^4
Mississippi	1.7×10^4

Henderson and Henderson (2009).

of ocean currents. To do so, we came up with a quantity that is a few orders of magnitude different from the one we are interested in—it would be inappropriate to use the same scale for a small stream, for example. The point is, we can come to grips with quantities that are far larger or smaller than those we experience every day by comparing them with things that are more familiar to us.

Exercise 1.1.3 Estimate the amount of milk (or your favorite drink) you drink in a week, and use this to estimate how long would it take you to consume 1 m^3 of it.

Exercise 1.1.4 What is the volume of water in a standard Olympic-sized swimming pool (50 m long, 25 m wide, and 3 m in depth)?

Exercise 1.1.5 How long would it take you to fly a distance equivalent to the diameter of the Earth?

Exercise 1.1.6 How many times could the Earth fit in the distance between the Earth and the Moon, and between the Earth and the Sun?

Exercise 1.1.7 Rainfall rates in a hurricane can be as high as 3.5 cm per hour within 56 km of the center of the hurricane. If that rainfall rate occurred uniformly over a circle of radius 56 km for 1 hour, how many Olympic-sized swimming pools would this fill?

Back-of-the-envelope estimates frequently involve more detailed calculations, but we need to always keep in mind that we are seeking an estimate, an answer that is likely accurate to within a factor of about ten. To do this, we sometimes need to know good estimates to certain numbers (e.g., the diameter of the Earth) and we need to learn how to make judicious approximations. Let us look at a more involved example.

Example 1.1 The average concentration of gold in seawater is approximately 100 fmol kg^{-1} of seawater (Falkner and Edmond, 1990; Henderson and Henderson, 2009). This is a very small number, but we can visualize it by recasting this number in terms of something more familiar. For example, if a gold ring contains 4 g of gold, how many rings could one make using all of the gold in the world's oceans?[3]

[3] The extraction of gold from seawater has actually been put forward as a serious business proposition several times.

We rarely come across a femto (see Appendix A) of anything in our daily lives, so it is hard to visualize what 100 fmol (i.e., 100×10^{-15} moles) of gold looks like. Instead of thinking about such a low concentration, we might ask what is the mass of gold in 1 kg of seawater. This raises another question: what does 1 kg of seawater look like? The density of seawater is approximately[4] 1000 kg m^{-3}, so 1 kg of water occupies 10^{-3} $m^3 = 1$ L, or the equivalent of a milk carton. The atomic weight of gold is 197 g mol^{-1}, so 100 fmol kg^{-1} is the same as

$$100 \times 10^{-15} \text{ mol} \times 197 \text{ g mol}^{-1} \approx 100 \times 10^{-15} \times 200 \text{ g} = 2 \times 10^{-11} \text{ g L}^{-1}.$$

Notice that we have approximated 197 by 200 to make our numbers easy, and which incurs an error of only 1.5%. To calculate the total amount of gold in the oceans, we need to know the total volume of the world's oceans. Knowing the radius r of the Earth (≈ 6000 km) we can calculate its surface area: $4\pi r^2 \approx 12 \times (6 \times 10^3)^2 \approx 12 \times 36 \times 10^6 = 432 \times 10^6$ km^2, assuming $\pi \approx 3$. The average depth of the oceans is 4 km and they cover approximately 70% of the Earth's surface (Henderson and Henderson, 2009), so we can calculate the volume of the oceans; $\approx 1.2 \times 10^9$ km^3 or $\approx 1.2 \times 10^{18}$ m^3 or $\approx 1.2 \times 10^{21}$ L. Now, we estimated earlier that 1 L of seawater contained about 2×10^{-11} g of gold, so the total amount of gold in the oceans is approximately 2×10^{10} g of gold, enough to make about 5×10^9 rings.

Let us return to the problem of the bacteria in the ocean that we described earlier in this section. Typical abundances of bacteria in seawater are 10^6 cm^{-3}. How can we determine if the cells are crowded together or not? One approach is to think of the distance between the cells in units of the typical size of a cell. First, we have to estimate a typical distance between cells. One way to do this is to assume that the cells are uniformly distributed in the 1 cm^3, so that the typical distance between them will be

$$l = \left(\frac{1}{10^6 \text{ cm}^{-3}} \right)^{1/3} = 10^{-2} \text{ cm} = 100 \text{ µm}.$$

A typical diameter of a bacterial cell is about 1 µm, so this means that we could fit 100 bacterial cells between each bacterium. From the perspective of an individual bacterium, that is quite a low density of cells and has implications for the mechanisms that bacteria use to detect chemical signals and survive in the oceans.

The real power of the back-of-the-envelope calculation appears when we want to obtain quick, approximate answers to complicated problems. This can be useful if we want to know whether or not a problem is worth pursuing in more detail, or whether it is a small (though possibly interesting) effect in the big scheme of things.

Example 1.2 Between 1900 and 2010, Greenland lost an estimated 9×10^{12} tonnes of ice. We might wonder how much of this ice contributed to global sea level rise. To figure this out, we can estimate the rise in global sea level if all this melting ice contributed to sea level rise. First, we need to determine the volume of ice that has been lost. We can use the

[4] The density of seawater varies with temperature, salt content (i.e., salinity), and pressure. The average density of seawater at the surface is 1025 kg m^{-3}. So our estimate introduces an error of approximately 2%.

fact that 1 kg of water occupies a volume of 1 L—bearing in mind our approximation from Example 1.1. So, 1 m^3 of water has a mass of 1 tonne and 9×10^{12} tonnes of ice occupies 9×10^{12} m^3, or 9×10^3 km^3.

To obtain the rise in sea level this would cause, we need to make some simplifications about the shape of the oceans. As we move offshore, the depth of the ocean generally increases relatively slowly until we reach what is called the *shelf break*, where the depth increases more rapidly from an average of about 130 m down to the *abyssal plain* at a depth of about 4000 m. The shallow coastal regions make up less than 10% of the total area of the oceans. So, we can approximate an ocean basin as being a straight-walled container with sides 4000 m tall. We will also assume that the melting ice gets uniformly distributed throughout all the world's oceans, so we can combine them into a single ocean. To get the change in sea level height, we simply divide the volume added to the oceans from the melting ice by the total surface area of the oceans. We have already estimated that the surface area of the Earth is about 4.4×10^{14} m^2, so knowing that the oceans cover approximately 70% of the Earth's surface, we can estimate the area of the oceans as approximately 3×10^7 km^2 to get approximately 30 mm.

It is always a good idea to perform a "sanity check" after doing such a calculation, just to make sure that our approximations are reasonable. Over the twentieth century, global sea levels rose approximately 19 cm (Jevrejeva et al., 2008), so we estimate that about 15% of this came from Greenland losing ice.

Exercise 1.1.8 The Greenland ice sheet contains approximately 2.8×10^6 km^3 of ice. Estimate the mass of this ice sheet and compare it with the 9×10^{12} tonnes that was lost between 1900 and 2010. Estimate the rise in sea level if *all* of the Greenland ice sheet were to melt and flow into the oceans.

Solving back-of-the-envelope calculations can often involve many steps, and sometimes we get stuck and cannot readily see what the next step in the calculation should be. One tactic to use to get unstuck is to examine the units of the quantities we need to calculate and see if that provides enough information to move ahead. To illustrate this, consider the following question: atmospheric carbon dioxide concentrations are increasing and values are often given in units of parts per million (ppm). But at the same time, we hear that humans emit several gigatonnes (10^{15} tonnes) of carbon into the atmosphere per year (Le Quéré et al., 2016). How many gigatonnes of carbon emitted yields a 1 ppm change in atmospheric CO_2 concentration?

The first sticking point we have here is one of units: what is meant by parts per million? Parts per million by mass? By volume? This is quite an abused notation, and we have to be careful that we understand how it is being used in the context of the question. In atmospheric sciences, these units are really mole-fractions—that is, 1 ppm is really shorthand for "1 mole of specific stuff for every million moles of all the stuff combined." It just so happens that for gases, actually for ideal gases, the mole-fraction is the same as the volume fraction (ppmv) because of the ideal gas laws,[5] and atmospheric gases at room temperature and surface pressures behave almost like ideal gases.

[5] An ideal gas is an idealized gas of particles that only interact through collisions, with no forces of attraction or repulsion between them, and the collisions are "perfectly elastic," which means that none of the kinetic energy

Because 1 ppmv is a mole-fraction, we need to know how many moles of gas there are in total in the atmosphere in order to know how many moles of CO_2 are present. We could calculate this if we knew the molecular weight (grams per mole) of air and the total mass of the atmosphere. To tackle the first part we need to know the composition of air (approximately 79% N_2 and 21% O_2) and the molecular weights of the components of air (the molecular weight of N_2 is 28 and that of O_2 is 32). If we assume that the atmosphere is well mixed so that the composition of the air is everywhere the same, then the molecular weight of air is approximately

$$\left(0.79\frac{\text{mol N}_2}{\text{mol air}} \times 28\frac{\text{g N}_2}{\text{mol N}_2}\right) + \left(0.21\frac{\text{mol O}_2}{\text{mol air}} \times 32\frac{\text{g O}_2}{\text{mol O}_2}\right)$$
$$= 22.12\frac{\text{g N}_2}{\text{mol air}} + 6.72\frac{\text{g O}_2}{\text{mol air}}$$
$$= 28.84\frac{\text{g}}{\text{mol air}},$$

or about 29 g per mole.

Next, we need to estimate the total mass or total number of moles of gas in the atmosphere. Calculating the volume or mass of the atmosphere is difficult—the concentration of gases is not uniform with height, and where does the atmosphere end? But we might be able to find a way to estimate the mass of the atmosphere by listing what we know about it. In this way we can see if there are any quantities we know that have units containing mass. We know an average surface temperature, but it is hard to see how knowing something with units of temperature will help us calculate a mass. We need to estimate a mass, so we should try and list relevant variables that have the units of weight or force in it them.[6] How about pressure? Pressure is defined as a force per unit area, and Newton's laws tell us that force is mass multiplied by acceleration. Atmospheric pressure at the surface of the Earth is 1.01×10^5 Pa (N m^{-2}). To get the total force of the mass of the whole atmosphere, we need to estimate the surface area of the Earth, which is about 510×10^6 km^2, and we can look up the acceleration due to gravity (9.81 N kg^{-1}). So, the mass of the atmosphere is the atmospheric pressure at the surface multiplied by the surface area of the Earth and divided by the acceleration due to gravity,

$$\left(1.01 \times 10^5 \times \frac{1\ \text{kg}}{9.81\ \text{N}}\right) \times (510 \times 10^6\ \text{km}^2) \times \left(\frac{1000\ \text{m}}{1\ \text{km}}\right)^2 \approx 5.2 \times 10^{18}\ \text{kg}.$$

Combining this with the average molecular weight we estimated earlier, the number of moles of gas in the atmosphere is

$$\left(\frac{5.2 \times 10^{21}\ \text{g}}{29\ \text{g mol}^{-1}}\right) \approx 1.8 \times 10^{20}\ \text{moles}.$$

Our next step is to determine how much of this is in the form of CO_2. Because 1 ppm is 1 part in 10^6, 1 ppm CO_2 in the atmosphere is 1.8×10^{14} moles CO_2. There is 1 carbon atom in each CO_2 molecule, so 1 ppm of CO_2 corresponds to 1.8×10^{14} moles of

of the particles motion is converted to other forms of energy. The ideal gas law implies that equal volumes of any ideal gas held at the same temperature and pressure contain the same number of molecules.

6 Recall that Newton's laws tell us that a force is a mass multiplied by an acceleration.

carbon contained in CO_2 molecules in the atmosphere. The end result is that the mass of 1 ppm of C is

$$(1.8 \times 10^{14}\,\text{moles}) \times \left(12\frac{\text{g}}{\text{mole}}\right) \sim 2 \times 10^{15}\,\text{gC} = 2\,\text{Pg C} = 2\,\text{Gt C}.$$

So 1 ppm of CO_2 corresponds to $\approx$ 2 Gt C. Knowing this allows us to quickly convert between the two sets of units when we see them in articles and research papers. It also allows us to ask other interesting questions, such as what is the contribution of fossil fuel burning to the rise in atmospheric CO_2 (see Problem 1.15)?

Back-of-the-envelope calculations can also be useful in determining spatial and temporal scales over which different processes are important. Many processes relevant to the Earth and environmental sciences have characteristic scales that determine how fast they occur and over what distances they work. For example, typical wind speeds over most of the United States vary between 4 and 5 m s^{-1}, but can be greater than 10 m s^{-1}. Open ocean surface currents have typical speeds of 0.1–2 m s^{-1}. So, we might expect the transport of gaseous pollutants in the atmosphere to be approximately 2–40 times faster than the transport of dissolved pollutants in the surface ocean.

Diffusion is an important process in both air and water, and we will meet it often in this book. Diffusion has the effect of smoothing out differences in concentration and is characterized by a quantity called the *diffusion coefficient* ($\mathcal{D}$) which, analogous to a velocity, is a measure of how fast diffusion can spread material. However, whereas velocity is a length divided by a time, the diffusion coefficient is a length squared divided by a time—we can think of it as the square of the distance a particle diffuses divided by the time it takes to diffuse that distance (Berg, 1993). This difference has important consequences for the distances and times over which diffusion is an important process. For example, the diffusion of a small molecule in air is roughly 10^{-5} m^2 s^{-1}, whereas in water it is $\approx 10^{-9}$ m^2 s^{-1} (Denny, 1993). We can use this to estimate the time (t) it will take a small molecule to diffuse a given distance (l), say 1 cm, in air and water:

$$t_{\text{air}} \sim \frac{l^2}{\mathcal{D}_{\text{air}}} = \frac{10^{-4}}{10^{-5}} = 10\,\text{s} \quad \text{and} \quad t_{\text{water}} \sim \frac{l^2}{\mathcal{D}_{\text{water}}} = \frac{10^{-4}}{10^{-9}} = 10^5\,\text{s} \sim 1\,\text{day}.$$

So, diffusion is a far slower process in water than in air, all other things being equal. What is more, because the diffusion coefficient is characterized by a length squared, it takes relatively longer to diffuse further distances. For example, to diffuse 10 cm takes 10^3 seconds (approximately 15 minutes) in air, and 10^7 seconds (about 116 days) in water. So, knowing something about the units of the diffusion coefficient and its value allowed us to estimate these diffusion times.[7]

Exercise 1.1.9 Estimate the time it takes a small molecule to diffuse a distance of 1 μm, 10 mm, 1 m, and 10 m in both air and water.

Exercise 1.1.10 Estimate the surface area of the Earth, the total surface area occupied by oceans, the total surface area occupied by land, and the total volume of the oceans.

[7] This is a calculation that is quick and easy to do, and can often be used to impress friends, family, and colleagues.

Exercise 1.1.11 Given that the average concentration of the salt in the oceans is 35 ppt, estimate the total mass of salt in the oceans and compare that to the mass of humanity on planet Earth.

1.2 Scaling

The phenomena we want to understand and explain in the Earth and environmental sciences cover a large range of spatial and temporal scales. At the smallest scales we might want to understand the processes of microbial interactions and how they affect biogeochemistry, or the nucleation of raindrops in the atmosphere. At the opposite end, the largest spatial scales encompass the planet, or large fractions of it. Consequently, it is useful to know if there are some general, unifying frameworks that allow us to understand how the importance of certain processes changes with scale. This is where *scaling arguments* become important.

There are generally two types of scaling that occur, *isometric*, or geometric, scaling and *allometric* scaling. Isometric scaling describes situations where the variables scale geometrically: for example, if you double the length of the side of a cube, the new surface area will be four times the old surface area, and the new volume will be eight times the old volume. In other words, the *shape* of the object stays the same, even though the size has increased. This geometric scaling can help us to understand how the importance of many processes changes with scale. For example, a microbial cell takes up nutrients through the surface of the cell, so all other things being equal, a cell B with twice the diameter of cell A should be able to take up nutrients four times faster than cell A. However, a cell's metabolic rate (a measure of how fast it uses energy) depends on its volume—the larger the cell, the more of it there is that has to be kept going. So, our geometric scaling argument implies that it should be harder for larger cells to obtain sufficient nutrients to support their energy needs than smaller cells, all other things being equal. What is more, this will vary by cell size according to the ratio of the cell surface area to its volume, and if the cells are spherical

$$\frac{\text{area}}{\text{volume}} = \frac{6}{\text{diameter}}.$$

Not all objects have a simple geometry like a sphere, so we might wonder what we use for a typical length scale when the object we are studying is not a sphere. Generally, most objects will have some characteristic length scale that is relevant to the problem and that we can choose to use. For example, if we are interested in relating maximum running speed to body length, we might choose stride length as a measure of length and relate this to body size. We could choose another variable such as leg length, and we could develop a scaling relationship using it, but leg length by itself is not necessarily a good indicator of running speed. A cheetah is about 60–90 cm feet tall at the shoulder, but has a running stride of several meters in length, much longer than a human.

A more fundamental question is whether or not geometric scaling always works. Galileo recognized that geometric scaling arguments often fail, even though data show that a scaling relationship still exists. These relationships, where a scaling relationship is not isometric, are *allometric*. If you watch a King Kong movie, you will see that Kong is just a geometrically scaled ape. However, if you carefully compare a mouse with an elephant, you will notice that the legs of a mouse seem thinner compared to their body size than a simple geometric scaling would suggest. This implies that larger animals require disproportionately thicker legs to support themselves.

Now, given that it is harder to break a thick branch than a thin twig, we might suspect that the diameter of the leg determines how easy it is to break it. The material that animal bone is made from is pretty similar between animals, so we expect the strength to be similar between animals. But what do we mean by "strength"? In this case we mean the strength of the bone to withstand fracture and buckling from just bearing the weight of the animal; in other words, a static measure of strength. The mechanical strength of a cylindrical shape (a good approximation to the shape of a leg bone) is proportional to its cross-sectional area (A)—cylinders with larger diameters are harder to buckle than those with smaller diameters and the same length. So, we expect that a heavier animal would need to have thicker bones, and hence thicker legs, to support its weight, so that $A \propto M$, where M is the animal's mass (if we double the body mass, we might expect to have to double the strength of the bone by doubling its cross-sectional area. The length (l) of a bone should scale with the size (L) of the animal,[8] so $l \propto L \propto M^{1/3}$, where we have assumed that the mass is proportional to the animal's volume. With the length and area of the bone, we can get a scaling for the bone mass, and so an estimate for the animal's skeletal mass (m), $m \propto A \times l \propto M \times M^{1/3} \propto M^{4/3}$, or $m = aM^{4/3}$ where a is a constant. If we take logarithms of this expression, we obtain the equation of a straight line $\log(m) = \log(a) + (4/3)\log(M)$, with a slope of 4/3. If we plot data for various bird species, for example, we find a slope closer to 1.0 than 1.33 (Figure 1.1). This is interesting because Figure 1.1 indicates that there is indeed a nice scaling relationship between skeletal mass and total body mass for birds, but it is not quite the relationship we expected from our geometric scaling argument. This tells us that something else is going on, and our assumptions are incorrect, so this is an example of an allometric scaling. In this case, there are several possibilities. One is that bone size is not determined by the ability of the bone to bear the animal's weight when standing still, but rather bone size is related to the ability of the bone to withstand dynamic processes such as walking (Prothero, 2015). Another possibility is that the structure of bones in a large bird is in some way different from that in smaller birds. So, comparing our scaling argument to data has revealed the assumptions behind our geometric simple scaling to be incorrect and has presented us with some interesting questions.

As another example, consider the scaling of river basins. River networks are formed from small streams that merge into larger rivers that themselves merge into still larger rivers until the final, large river discharges into the oceans (Figure 1.2). We can use two lengths to characterize the shape of the river basin, the length (L) and the average width (W). The area of the river basin is then $A \approx LW$. Observations of river basins show that

$$W \propto L^H, \qquad \text{where} \qquad \frac{1}{2} \lesssim H \lesssim 1$$

[8] This implies that a larger animal has more difficulty supporting its weight than a smaller one, and argues that an animal like King Kong could not exist; its bones would break when it moved. An excellent, if somewhat gruesome, description of this effect is given in Haldane (1945).

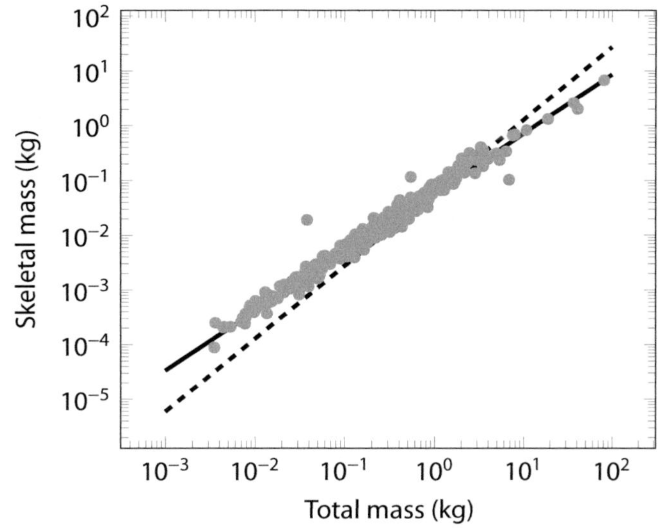

Figure 1.1 A plot of skeletal mass against body mass for 270 species of birds. The solid line has an equation, skeletal mass $= 0.059 \times$ (body mass)$^{1.082}$, whereas the dashed line has an equation, skeletal mass $= 0.059 \times$ (body mass)$^{1.333}$. The data are taken from Martin-Silverstone et al. (2015).

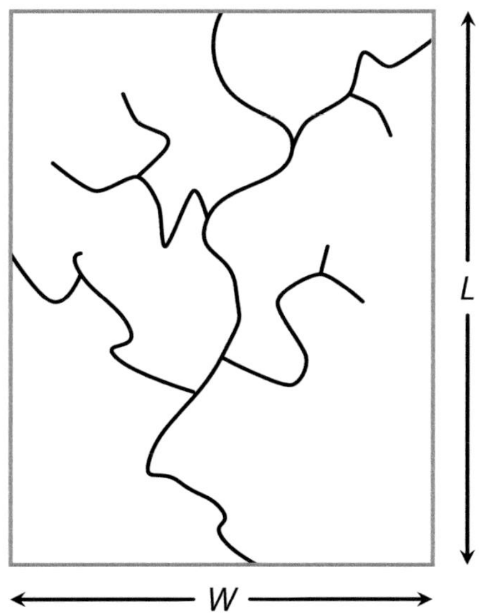

Figure 1.2 A river basin of length L and width W.

so that we have the scaling relationship $A \sim L^{1+H}$. We can use the aspect ratio (α) to characterize the shape of the basin

$$\alpha = \frac{W}{L} \propto L^{H-1} = A^{-\frac{1-H}{1+H}}.$$

Because $H \lesssim 1$ we have that $(1 - H)/(1 + H) > 0$ and so the exponent of A is always negative. This means that river basins with a large area tend to be long with relatively small widths, whereas small river basins tend to have widths that are more comparable to their lengths. A particularly useful model of river networks has $H = 1/2$ in which case $A \sim L^{3/2}$ or $L \sim A^{2/3}$, so that the length of the longest river in the network scales as the area to the power of 2/3. This is known as Hack's law (Hack, 1957; Dodds and Rothman, 2000) and can be used to help determine the physical processes leading to the formation and morphology of river networks.

Scaling arguments provide a means for generating unifying principles for how quantities that interest us change as sizes and lengths change. Simple isometric arguments do not always work, but when they fail they can spur us to come up with relationships that do explain the observed data. Allometric scaling relationships are useful for defining the properties of organisms and have been used in large-scale ecosystem models (Follows and Dutkiewicz, 2011) and explaining how processes scale with size (Peters, 1983; Schmidt-Nielsen, 1984; Niklas, 1994).

Exercise 1.2.1 Do you think that the abundance of a species will scale isometrically or allometrically with the area surveyed? What about the number of species?

1.3 Dimensional Analysis

Dimensional analysis is a very powerful technique that arises from a very simple idea: the requirement that the mathematical expressions we use in our equations should not depend on the units we use for the variables in those expressions. For example, the formula for the volume of a sphere should be the same whether we decide to measure the diameter of the sphere in units of meters, feet, inches, or any other unit of length. This seems obvious, but supplemented with an understanding of the factors that are important in a problem, dimensional analysis provides us with a tool to start developing equations for phenomena that interest us. An immediate consequence of this is that the dimensions of terms on the left- and right-hand side of an equality must be the same; this is called the *principle of dimensional homogeneity*. So for example, with Newton's law that the force acting on a body is the product of the mass of the body and its acceleration ($F = ma$), the dimensions of force must have the same dimensions (and units) as the dimensions of mass multiplied by the dimensions of acceleration. This simple requirement has profound implications (Barenblatt, 1996) and, as we will see shortly, restricts the form of the mathematical functions we can use in developing our equations.

What are the consequences of requiring our equations to be dimensionally homogeneous? Firstly, it implies that we cannot add or subtract quantities that have different dimensions (i.e., we cannot add apples to tomatoes). Why? We can calculate a velocity v from the distance (d) travelled and the time (t) it takes to travel that distance using the equation $v = d/t$. If $d = 1$ m and $t = 1$ s then $v = 1$ m s^{-1}. Let us now measure d in different units, say centimeters. Changing units is basically the same as multiplying d by a constant factor $a = 100$. The velocity in the new units is then $u = (a \times d)/t = av$. In other words, 1 m s^{-1} is the same velocity as 100 cm s^{-1}. What happens if we try to use the formula $v = d + (1/t)$ instead. Changing the units for d in the same way again produces $u = (a \times d) + 1/t \neq av$. So, the only way we can change units and keep our equation the same is if we only add and subtract quantities with the same units.

What happens if we calculate an area or volume but change the units of length? The area of a square with sides of length L is $A = L^2$. Changing the units of length involves multiplying L by a factor a, so the area in the new units is $\tilde{A} = (aL)^2 = a^2L^2 = a^2A$. Similarly, the volume (V) of a cube becomes $\tilde{V} = (aL)^3 = a^3V$. A function $f(x)$ such that $f(ax) = a^k f(x)$ is called a *homogeneous* function, where the constant k is called the *order* of the homogeneous function—such functions play a crucial role in what follows.

What we have shown suggests that a mathematical function that represents a measurable quantity has to be a homogeneous function, or be formed by adding and subtracting homogeneous functions of the same order.[9] For example, we could not have a function $F(L) = L^2 + L^3$ because changing the units of L as before would give $F(aL) = a^2L^2 + a^3L^3 \neq a^k F(L)$, so we have tried to add an area to a volume which does not make sense. Similarly, we cannot have a function $G(L) = \sin(L)$ because $G(aL) = \sin(aL) \neq a\sin(L)$. The only way that the functions $F(L)$ and $G(L)$ could represent measurable quantities is if L were dimensionless. For example, L could be the ratio of two lengths, and so be dimensionless. This tells us that the arguments of functions such as sine, cosine, tangent, logarithm, and exponential have to be dimensionless.

Exercise 1.3.1 Show that $\ln(x)$, e^x, and $\cos(x)$ are not homogeneous functions.

Before we go any further, we should pause to consider the difference between the dimension of a quantity and its units. A dimension is a measurable quantity: a length, a mass, a time, etc. In the SI system of units, there are seven base dimensions:[10] mass ([M]), length ([L]), time ([T]), electric current ([I]), thermodynamic temperature ([Θ]), an amount of a substance ([N]), and luminous intensity ([J]). Other measurable quantities are derived from these dimensions. Some quantities, such as π, are dimensionless, and by convention these have a dimension of 1 (i.e., $[\pi] = 1$). For example, velocity has dimensions of length per unit time ($[L][T]^{-1}$), and energy has dimensions $[M][L]^2[T]^{-2}$. While the dimension of a quantity tells us what it is, its units give the quantity a numerical value. For example, a *length* ([L]) can be measured in units of centimeters, feet, furlongs, light years, or one of many other units of length. Similarly, a velocity can be measured in units of cm s^{-1}, m d^{-1}, km h^{-1}, and so on, but although the same velocity will have different numerical values in these different systems of units, they all have the dimensions of $[L][T]^{-1}$.

We can often use the principle of dimensional homogeneity to write down the relationship between quantities that we are interested in. As a simple example, let us assume we travel at a constant velocity v and want to know how long it will take to cover a distance d. Looking at the dimensions of the quantities we know, d has dimensions of length ([L]) and velocity has dimensions of $[L][T]^{-1}$. We want an equation that combines these two

[9] This can be shown rigorously; for example, see Barenblatt (1996).
[10] The standard notation for a dimension is to include it in square brackets.

quantities to produce a result that has dimensions of time. There is only one way to do this such that the left- and right-hand sides of the equation both have dimensions of [T]:

$$t \propto \frac{d}{v}, \qquad [\mathrm{T}] = \frac{[\mathrm{L}]}{[\mathrm{L}][\mathrm{T}]^{-1}} = \frac{1}{[\mathrm{T}]^{-1}} = [\mathrm{T}].$$

Because it deals with quantities that have dimensions, the principle of dimensional homogeneity, and dimensional analysis in general, cannot tell us anything about dimensionless constants (e.g., π) that appear in an equation. It can tell us that the relationship between the diameter (d) and volume (V) of a sphere is $V \propto d^3$, but it cannot tell us that the constant of proportionality is $\pi/6$. This is why we have used a proportionality sign in the above equation. So, we can obtain a relationship between the quantities we are interested in by knowing the dimensions of those quantities and by demanding that the dimensions on either side of the equals sign are the same.

Exercise 1.3.2 Energy appears in many forms in science, from chemical energy to the energy released by an earthquake. The dimensions of energy are a force multiplied by a length. Two important forms of energy are potential energy and kinetic energy. If we change the height of an object above the ground, then the change in its potential energy is $mg\Delta h$, where m is the object's mass, g is the acceleration due to gravity, and Δh is the change in height. Show that the dimensions of $mg\Delta h$ are energy.

Exercise 1.3.3 Some bacteria are motile and move by using a flagellum, a whip-like appendage that extends from the cell. You read in an article that the power (P) required to propel a bacterium at a velocity u is given by $P = 50\mu ul$ where μ is the dynamic viscosity of the fluid (with units N s m^{-2}) and l is the length of the flagellum. Using the principle of dimensional homogeneity, determine if this formula is correct. If it is not, what is the correct formula?

Exercise 1.3.4 The orbital time of a satellite in a near circular orbit around the Earth depends on the gravitational constant G (dimensions $[\mathrm{L}]^3[\mathrm{M}]^{-1}[\mathrm{T}]^{-2}$), the mass M of the Earth, and the distance R between the center of the Earth and the satellite. Use dimensional homogeneity to show that the period of the satellite's orbit is given by

$$t_{\mathrm{orbit}} \propto \frac{R^{3/2}}{\sqrt{GM}}.$$

Exercise 1.3.5 If S has the dimensions of a concentration, what are the dimensions of K in the following formula?

$$V = V_{\mathrm{max}} \frac{S^2}{(K + S)^2}$$

The hard part about using dimensional analysis is determining what the important variables in a problem are. This requires a mixture of insight, understanding, and sometimes an educated guess or two. Now we will examine a more involved case.

Example 1.3 Let us use the principle of dimensional homogeneity to find equations for the speed of waves moving across the surface of a fluid in the following cases: (1) deep-water waves, (2) shallow-water waves, (3) capillary waves.

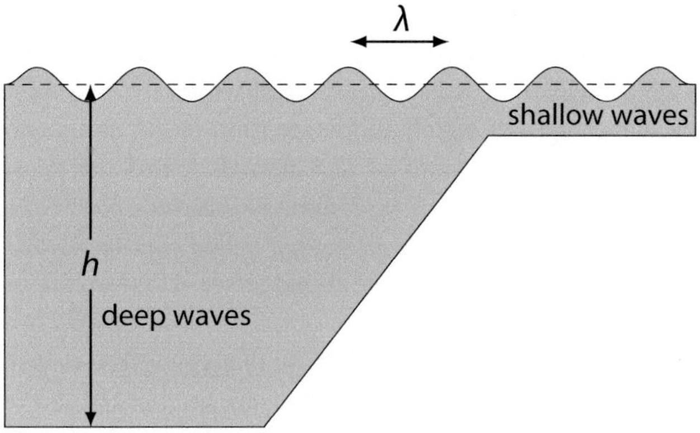

Figure 1.3 A diagram showing the meaning of wavelength (λ) and depth (h) for the dimensional analysis of wave speed.

Our first task is to list the variables that might be important (Figure 1.3).[11] In each case we are looking for a speed (c), which has dimensions of $[L][T]^{-1}$, so at least some of the remaining variables need to include dimensions [L] and [T] if the principle of dimensional homogeneity is to be satisfied. Waves can be characterized by their wavelength (λ), which has the dimension of length. We are looking for expressions for both deep- and shallow-water waves, so we might suspect that we need to consider the depth of the fluid (H). These three variables are all lengths and do not include the dimension of time, so we need to look for additional variables. If we perturb the surface of a flat, stationary fluid (by throwing a pebble into the liquid, for example), a force has to act to return the fluid surface to its previous state. We know from the physics of the situation that two likely forces are gravity and the fluid surface tension—both act to smooth out waves on the surface. We can use the gravitational acceleration (g), with dimensions $[L][T]^{-2}$, to characterize the strength of gravity.[12] Surface tension (γ) is defined as a force per unit length, and so has dimensions[13] of $[M][T]^{-2}$. Notice that we have introduced a variable that includes a dimension of mass, whereas the variable we are interested in, c, does not include mass. This means we will possibly need another variable with a dimension containing mass to cancel the one from surface tension. One possible variable is the fluid density (ρ) with dimensions $[M][L]^{-3}$.

Which of these two forces, gravity or surface tension, dominates? Are they equally important, or is one more important for certain types of wave? We might expect from experience that gravity controls waves on large spatial scales and surface tension controls waves on small spatial scales. To quantify this, we need to find a way to compare the two forces, and a standard technique for comparing two quantities is to form a ratio from them and see if it is larger or smaller than one. But a ratio of what? We need to have γ and g in the ratio because they characterize the strengths of surface tension and gravity. But g is in itself not a force—to obtain a force we need to multiply g by something that contains the dimensions of mass. It would be hard to know the mass of material in a wave, but

[11] It is *always* a good idea to draw a diagram when trying to solve a problem.

[12] The stronger the gravitational pull of a body, the greater the acceleration toward it.

[13] Newton's laws tell us that force is mass multiplied by acceleration, so the dimensions of a force must be $[M][L][T]^{-2}$. Consequently, the dimensions of a force per unit length are $[M][T]^{-2}$.

we have a better chance of knowing the density of the fluid in the wave. The ratio $\gamma/(\rho g)$ has dimensions of $[L]^2$, so $\sqrt{\gamma/(\rho g)}$ has dimensions of length. So, if we set the ratio to 1 we can calculate the length scale at which the two forces balance and thereby estimate the wavelength at which surface tension is the dominant restoring force in a wave. The surface tension of water is 72.8 mN m^{-2}, and the density of water is 10^3 kg m^{-3}, so $\sqrt{\gamma/(\rho g)} \approx 3 \times 10^{-3}\mathrm{m} = 3$ mm. So, as we expected, surface tension is important for waves with very small wavelengths (called capillary waves), whereas gravity is important for large waves such as those found at sea. Now we are in a position to see if we can find some relationships.

Deep-water waves, such as those that occur in the open ocean, occur on the surface of a fluid that is sufficiently deep that the motions of the fluid caused by the wave do not interact with the hard bottom at the base of the fluid. In this case, we can disregard H,[14] leaving λ as the only relevant length scale. Gravity provides the restoring force for these waves—which are called *gravity waves*—so surface tension can also be neglected. If we neglect surface tension, then we have only one other variable in our list that includes mass (the density, ρ), and so we should neglect this as well. We will assume for the moment that c is proportional to g raised to the power a multiplied by λ raised to the power b (based on our discussion of homogeneous functions), i.e.,

$$c_d \propto g^a \lambda^b,$$

where a and b are constants whose values we do not know yet. To find them, we will use the dimensions of the variables. If this equation is correct, then the principle of dimensional homogeneity implies that

$$\underbrace{[L][T]^{-1}}_{c} = (\underbrace{[L][T]^{-2}}_{g})^a (\underbrace{[L]}_{\lambda})^b = [L]^{a+b}[T]^{-2a}.$$

This equation can only be true if the powers of [L] and [T] on both sides of the equals sign are separately the same, so that

$$\begin{aligned} 1 &= a + b \qquad \text{by comparing [L] on both sides} \\ -1 &= -2a \qquad \text{by comparing [T] on both sides,} \end{aligned}$$

which has a solution $a = b = 1/2$. So, our final equation is

$$c_d \propto g^{1/2}\lambda^{1/2} = \sqrt{g\lambda}. \tag{1.1}$$

Unlike deep-water waves, shallow-water waves feel the influence of the hard bottom at the base of the fluid, so now the depth of the fluid comes into play instead of the wavelength (see Exercise 1.3.6 to see what happens if we include both λ and H). So, we can write

$$c_s \propto g^a H^b.$$

The corresponding equation for the dimensions is

$$\underbrace{[L][T]^{-1}}_{c} = (\underbrace{[L][T]^{-2}}_{g})^a (\underbrace{[L]}_{H})^b = [L]^{a+b}[T]^{-2a},$$

[14] The properties of the wave are not influenced by the hard ocean floor, so it does not matter how deep the ocean is.

which gives the same equations for a and b as before, so

$$c_s \propto \sqrt{gH}, \tag{1.2}$$

which is not too surprising because the only difference we made over the deep-water wave case was to substitute H for λ, both of which have the dimensions of a length.

As we have seen, the dominant restoring force for capillary waves is surface tension, not gravity. So, we can neglect g but have to include γ. This introduces the dimension of mass into the equation, so we will need to also include ρ. Following the same procedure as above, we arrive at the equation

$$c_c \propto \sqrt{\frac{\gamma}{\lambda\rho}}. \tag{1.3}$$

We have seen that $\sqrt{\gamma/(g\rho)}$ has dimensions of $[\mathrm{L}]^2$, so the quantity

$$\mathrm{Bo} = \sqrt{\frac{\lambda^2 g\rho}{\gamma}}, \tag{1.4}$$

where λ is the wavelength, is dimensionless. This dimensionless ratio is called the *Bond number* and tells us whether gravity or surface tension is the dominant restoring force for a wave of wavelength λ. As we shall see, dimensionless numbers play an important role in understanding the world.

Exercise 1.3.6 Repeat the analysis leading to Equation (1.2), but include both λ and H in the analysis so that $c \propto g^a \lambda^b H^c$. What happens when you try to solve for a, b, and c? What does this tell you about your choice of variables?

Dimensional analysis can help us if we have missed a variable in our problem. We can illustrate this using a biogeochemical problem. Phytoplankton are small (diameter ~ 1–100 μm) single-celled aquatic organisms that can photosynthesize (Denny, 2008). The biological activity of phytoplankton provides a means for the oceans to sequester carbon dioxide from the atmosphere (Denny, 2008; Williams and Follows, 2011), and photosynthesis by phytoplankton in the oceans provides roughly half of the oxygen in the atmosphere. Can we find a relationship between the metabolic rate of a phytoplankton cell and its size? Phytoplankton need both light and nutrients (such as carbon, nitrogen, and phosphorus) to grow. These cells acquire nutrients through a complicated set of processes that involves diffusion of molecules to the cell wall and subsequent transport of these molecules across the cell wall. If there are abundant nutrients in the water, then the rate at which molecules are transported across the cell wall can limit the overall rate of photosynthesis. But in conditions where nutrients are not abundant, transport of nutrients to the cell surface limits nutrient uptake. If we consider such nutrient-poor conditions, can we find a relationship between the cell size, diffusion, and metabolic rate ($\mathcal{M}$)?

The first thing we need to do is to ask what variables are potentially important to the problem:

- The problem involves diffusion, and so the diffusion coefficient ($\mathcal{D}$) must appear. Recall that the diffusion coefficient is a measure of how fast particles can travel by diffusion, and it has dimensions $[\mathrm{L}]^2[\mathrm{T}]^{-1}$.

- Cell size, for example radius r, which has dimensions [L], is the thing we are interested in.
- Metabolic rate ($\mathcal{M}$) has units of moles of oxygen per cubic meter of cell volume per second and so has dimensions $[\mathrm{N}][\mathrm{L}]^{-3}[\mathrm{T}]^{-1}$ and is a measure of the rate of consumption of resources.

With these variables we can write the equation

$$r = f(\mathcal{D}, \mathcal{M}), \tag{1.5}$$

which says the size of the cell at equilibrium is some as yet unknown function of $\mathcal{D}$ and $\mathcal{M}$. To find that equation, we assume that $r \propto \mathcal{D}^{\alpha}\mathcal{M}^{\beta}$ and balance the dimensions on each side of the equal sign:

$$\underbrace{[\mathrm{L}]}_{r} = \underbrace{([\mathrm{L}]^2[\mathrm{T}]^{-1})^{\alpha}}_{\mathcal{D}}\underbrace{([\mathrm{N}][\mathrm{L}]^{-3}[\mathrm{T}]^{-1})^{\beta}}_{\mathcal{M}}. \tag{1.6}$$

Collecting up terms, we find

$$[\mathrm{L}] = [\mathrm{L}]^{2\alpha-3\beta}[\mathrm{N}]^{\beta}[\mathrm{T}]^{-\alpha-\beta}. \tag{1.7}$$

For Equation (1.7) to be true, the dimensions of both sides of the equality have to be the same, giving us the equations

$$1 = 2\alpha - 3\beta \tag{1.8a}$$

$$0 = \beta \tag{1.8b}$$

$$0 = -\alpha - \beta. \tag{1.8c}$$

But these equations result in a contradiction: Equations (1.8b) and (1.8c) imply that $\alpha = \beta = 0$, which contradicts Equation (1.8a). We have missed something, and that is the concentration (C) of the nutrient in the water (Figure 1.4). The reason this is important is that we have made some assumptions without properly exploring their consequences. We assumed that the rate of diffusion of nutrients to the cell was the controlling rate and that this depended only on the diffusion constant. However, the rate of diffusion depends on more than just the diffusion coefficient $\mathcal{D}$, it also depends on the gradient of the concentration of whatever is diffusing. Diffusion always acts to smooth out gradients, and steeper gradients result in faster diffusion rates. So, we need to add an estimate for the gradient of nutrient concentration. This is hard to do in detail, but remember that we are trying to find a simple relationship between cell size, diffusion rate, and metabolic rate. So, let us assume that the cell is sitting in a bath of nutrients such that far from the cell the nutrient concentration (C_∞) is constant. Our assumption that the transport rate across the cell membrane is not a controlling factor implies that this rate is much faster than any other relevant rate in the calculation. We can simplify our problem by assuming that as soon as a nutrient molecule touches the cell surface, that molecule is instantaneously transported inside the cell. This means that the concentration of nutrient molecules at the cell surface is zero (they disappear inside the cell as soon as they touch it). The gradient of nutrient concentration is then determined by C_∞.

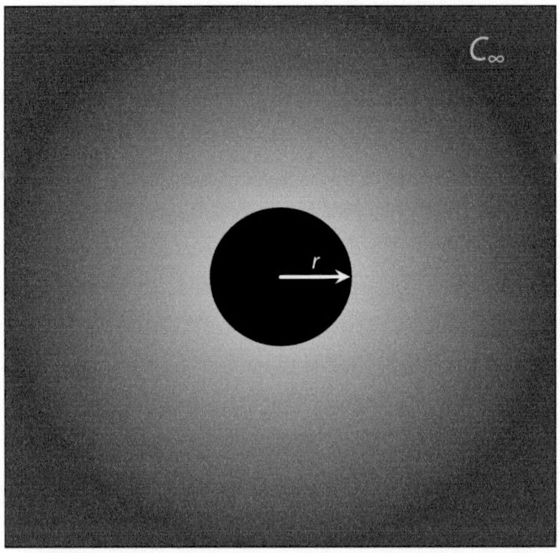

Figure 1.4 A phytoplankton cell of radius r takes up nutrients from the surrounding seawater. We assume that the nutrient concentration far from the cell is constant and has a value C_∞. The concentration of nutrient decreases toward the cell surface.

If we include the nutrient concentration, then Equation (1.5) becomes $r = f(\mathcal{D}, \mathcal{M}, C_\infty)$ and Equation (1.6) becomes

$$\underbrace{[\mathrm{L}]}_{r} = \underbrace{([\mathrm{L}]^2[\mathrm{T}]^{-1})}_{\mathcal{D}}{}^{\alpha}\underbrace{([\mathrm{N}][\mathrm{L}]^{-3}[\mathrm{T}]^{-1})}_{\mathcal{M}}{}^{\beta}\underbrace{([\mathrm{N}][\mathrm{L}]^{-3})}_{C_\infty}{}^{\gamma}. \tag{1.9}$$

Equating the powers of the different dimensions on both sides of the equation gives us

$$\alpha = \frac{1}{2} \qquad \beta = -\frac{1}{2} \qquad \gamma = \frac{1}{2},$$

and so we have the relationship

$$r \propto \sqrt{\frac{C_\infty \mathcal{D}}{\mathcal{M}}}. \tag{1.10}$$

Recall that dimensional analysis alone cannot tell us about any numerical constants that might be factors in the equation, but if we did the calculation in detail,[15] we would find that

$$r = \sqrt{\frac{3C_\infty \mathcal{D}}{\mathcal{M}}}.$$

Equation (1.10) can be rewritten as

$$\mathcal{M} \propto \frac{C_\infty \mathcal{D}}{r^2}, \tag{1.11}$$

which suggests that smaller phytoplankton cells fare better than larger ones in regions of the ocean where nutrients are in low concentration. In broad terms, this is indeed what is seen, with larger cells blooming only when there is an injection of nutrients into the surface waters, normally via some physical process such as upwelling.[16]

[15] We will do this in Chapter 10.
[16] These are generally a type of phytoplankton called diatoms.

Exercise 1.3.7 The efficiency with which streams and rivers drain a watershed is measured by the drainage density (the sum of the lengths of all the streams and rivers in the watershed divided by the total area of the watershed—a measure of how closely spaced the streams and rivers are). The channel frequency is the number of streams and rivers per unit area in the water shed. Use the principle of dimensional homogeneity to derive a relationship for channel frequency in terms of drainage density.

Exercise 1.3.8 Use the principle of dimensional homogeneity to find an expression relating the difference in pressure ($\Delta\rho$) between the center of a tornado and the air outside it, the wind speed, and the air density.

1.4 Dimensionless Ratios and the Pi Theorem

Dimensionless ratios are formed from quantities such that the numerator and denominator have the same dimensions. They play a central role in dimensional analysis and are important in fluid dynamics and many other disciplines, as we will see throughout the book. For the moment, we will use dimensionless numbers to help simplify our problems. Dimensionless ratios have a special significance because they allow us to scale phenomena that occur under very different circumstances. For example, the *Rossby number* is important for understanding fluid flows on a rotating planet and is defined by

$$R_o = \frac{V}{fL} = \frac{V^2/L}{Vf}, \tag{1.12}$$

where V is a typical value of a horizontal fluid velocity, L is a characteristic length scale, and f, the Coriolis parameter with dimensions $[\mathrm{T}]^{-1}$, is a measure of the strength of the Coriolis force resulting from the rotation of the planet and depends on latitude and the angular velocity of the planet. The Rossby number tells us whether accelerations resulting from inertial forces (V^2/L) or Coriolis forces are more important for a given flow. When $R_o \gg 1$ the effects of the rotation of the Earth on the motion of fluids can be neglected and if $R_o \ll 1$, then effects resulting from the rotation of the Earth are important. For example, large-scale (say on scales of 10^3 km) atmospheric velocities are typically 10 m s^{-1}, and at midlatitudes $f \approx 10^{-4}\ \mathrm{s}^{-1}$, giving $R_o \approx 0.1$, and such flows are approximately geostrophic.

Exercise 1.4.1 Estimate the Rossby number for flow draining a bathtub in midlatitudes.

Exercise 1.4.2 Estimate the Rossby number for large-scale ocean flows where the typical velocity is 10 cm s^{-1}.

Dimensionless ratios play a prominent role in a theorem called the Buckingham Pi theorem,[17] which is where the full power of dimensional analysis can be seen. The theorem

[17] Although this theorem is associated with the American physicist Edgar Buckingham (1867–1940), many scientists played important roles in formulating it. These include Joseph Bertrand, Lord Rayleigh, A.Vaschy,

allows us to find sets of dimensionless ratios from a given set of variables and tells us how many we can form. We can state the theorem loosely in the following way: Consider a physical variable Y that is a dimensionally homogeneous function of k other variables,

$$Y = f(x_1, x_2, \ldots, x_k),$$

with n independent dimensions (i.e., [L], [M], [T] etc.), then we can reduce the number of variables by rewriting the equation in terms of $(k - n)$ dimensionless variables formed from combinations of the original k variables. Buckingham wrote these new dimensionless variables using the Greek capital letter Π, hence the name of the theorem.

1.4.1 Application of the Buckingham Pi Theorem

To see how this works, let us look at a couple of examples. For our first example, we will follow the derivation of a famous equation from the world of turbulence (Tennekes and Lumley, 1972). A full understanding of turbulence remains an unsolved problem, but we can think of laminar and turbulent flows in the following way. To keep a fluid moving requires a continual input of energy at a rate that is equal to or greater than the rate at which energy is dissipated within the fluid by friction and viscosity. Low rates of energy input tend to produce smooth, well-ordered flows where there is very little mixing; these are called *laminar flows*. By contrast, *turbulent flows* are highly energetic, chaotic and characterized by vortices that interact with each other on many spatial scales and the energy input into the flow is transferred to smaller and smaller spatial scales through the interaction of vortices of different sizes, until eventually the viscosity of the fluid dissipates the energy as heat.

As mentioned, we do not have a full understanding of turbulence, so it is important to have simplified descriptions, and that is where dimensional analysis can help. The question we will ask is: how does the interaction of the vortices of different sizes determine the energy per unit mass (the energy density) on different spatial scales? As a measure of length we will use an inverse length (k) that has dimensions

$$[k] = \frac{1}{[L]},$$

so that larger length scales correspond to smaller values of k and vice versa. The Russian mathematician Andrej Kolmogorov[18] hypothesized that at any length scale, the flow of energy depended only on the average rate at which energy per unit mass (the energy density) is dissipated. This rate of energy density dissipation has dimensions

$$[\epsilon] = [\text{Energy}]\frac{1}{[\text{Time}]}\frac{1}{[\text{Mass}]} = \frac{[M][L]^2}{[T]^2}\frac{1}{[T]}\frac{1}{[M]} = \frac{[L]^2}{[T]^3}.$$

We would like to find a relationship between the two variables k and ϵ, and the energy density per unit wave number (called the spectral energy density), which has dimensions

$$[E(k)] = [L]^3[T]^{-2}. \tag{1.13}$$

and others. It was Buckingham's work that introduced the notation, and hence the theorem's name, we use today.

[18] Kolmogorov (1903–1987) made major discoveries in many fields of mathematics and physics, including fluid dynamics, probability, and classical mechanics.

We have three variables (k, ϵ, and $E(k)$) and two dimensions ([L] and [T]), so the Buckingham Pi theorem says that we can reformulate the problem in terms of a single dimensionless ratio (Π_1). We can write

$$\Pi_1 = E^a k^b \epsilon^c \implies [L]^{3a-b+2c}[T]^{-2a-3c} = 0,$$

and a solution is $a = 3$, $b = 5$, $c = -2$. So, the simplest dimensionless ratio we can form from these variables is

$$\Pi_1 = \frac{E^3 k^5}{\epsilon^2}.$$

Now, Π_1 is just a number with no dimensions, and we have already discovered that dimensional analysis cannot tell us the values of any proportionality constants, so we can write

$$E(k) \propto \epsilon^{2/3} k^{5/3}. \tag{1.14}$$

Experiments show that this 5/3 law holds over a wide range of length scales (Figure 1.5), but breaks down at large scales where the method of energy input (wind, stirring etc.) becomes important. At small scales, viscous dissipation becomes important and the slope of the energy spectrum steepens. In between these limits, where the energy cascades from larger to smaller length scales (the *inertial subrange*), Equation (1.14) holds well.

Exercise 1.4.3 Show that Equation (1.13) is dimensionally correct.

Exercise 1.4.4 Using the dimensions of the kinematic viscosity (ν, with dimensions $[L]^2[T]^{-1}$) and ϵ, create a variable η with dimensions of length. This is the Kolmogorov length scale and represents the smallest scales at which turbulence acts.

Our next example will be a little more complicated and involves finding the *terminal velocity* of an object falling through a fluid. This kind of phenomenon occurs in many different disciplines, from raindrops falling through the atmosphere, or sediment particles such as sand grains sinking through water. When an object falls through a fluid, it experiences the force of gravity that accelerates its downward motion as well as drag forces arising from the viscosity of the fluid that impede this motion. The falling object is said

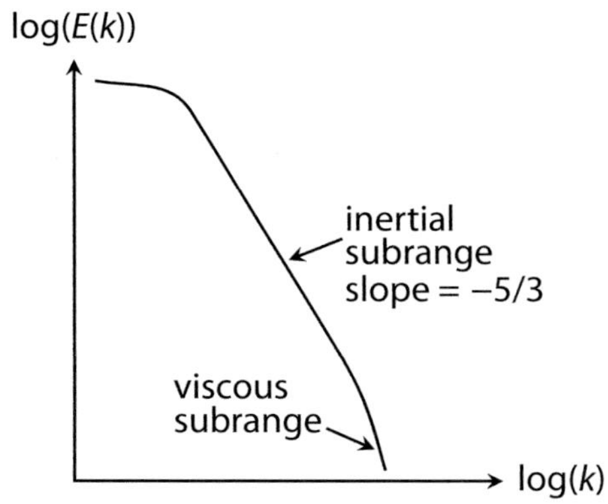

Figure 1.5 A schematic plot of the Kolmogorov turbulence spectrum.

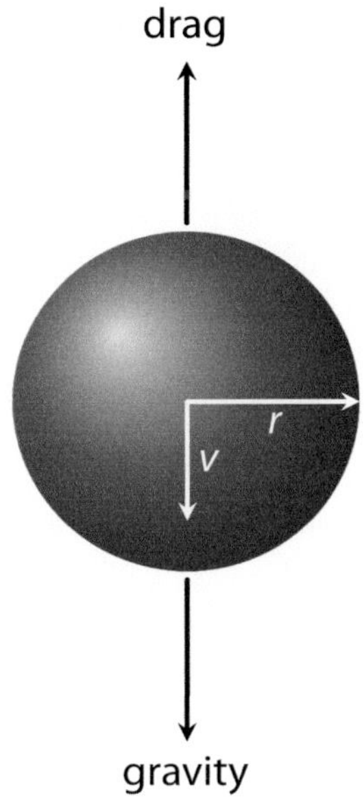

Figure 1.6 A sphere of radius r falling slowly at a velocity v under the balanced forces of gravity accelerating the sphere downward and drag, which acts in the opposite sense.

to reach terminal velocity when the drag forces balance the gravitation forces—i.e., the upward and downward force are equal. Recall that Newton's second law says the net force on an object equals the mass of the object multiplied by its acceleration. So, if the upward and downward forces on the particle are in balance, there is no net force on the body, and it will neither accelerate nor decelerate and its velocity is constant. The variables that are important in determining the terminal velocity are:

- The terminal velocity itself (v, $[L][T]^{-1}$)—this is the variable we are looking for.
- The particle size (e.g., radius r, [L])—we expect that the drag, which results from friction between the fluid and the particle surface, will vary with the surface area of the particle, and hence its radius.
- Gravity, as represented by the acceleration due to gravity (g, $[L][T]^{-2}$)—gravity provides the force accelerating the particle downward.
- The density of the particle (ρ_p, $[M][L]^{-3}$)—what we really want is the mass of the object because the force of gravity acts on the mass, not the density. However, the mass of the sphere is $(4/3)\pi r^3 \rho_p$, and we already have r in our list, so we can use density instead of mass.
- The density of the fluid (ρ_f)—we know that if a particle has the same density as water, it will not sink, so the fluid density characterizes the buoyancy forces acting on the falling particle.
- The viscosity of the fluid—this is a measure of the "stickiness" of the fluid and affects the drag force. There are two ways of representing viscosity, the *kinematic viscosity* (ν) with dimensions $[L]^2[T]^{-1}$ and the dynamic viscosity (μ) with dimensions $[M][L]^{-1}[T]^{-1}$. We could choose either of these, but we will choose to use the dynamic viscosity.

In this problem we have six variables and three dimensions, so the Buckingham Pi theorem says we need to look for three dimensionless variables. How do we find these variables? We could try guessing, but there are some more systematic methods we can use. In this example we will use the method of *repeating variables*. The Pi theorem tells us that we are looking for three dimensionless ratios, so we choose three of the six variables to appear in the calculation of each of the three Π ratios; these are the repeating variables that we will raise to unknown powers. To calculate each Π ratio we use our three repeating variables and *one* of the remaining variables, which we do not raise to a power. We are reasonably free to choose our repeating variables as we like, but our choices should satisfy the following rules of thumb:

- All of the dimensions (i.e., [M], [L], and [T]) of the problem must appear in the collection of repeating variables.
- We also should not choose the dependent variable we are interested in (in this case, v) as one of the repeated variables.

Let us choose ρ_f, r, and g as our repeating variables.

Exercise 1.4.5 Check that all of the dimensions of the falling sphere problem are represented in the chosen repeating variables ρ_f, r, and g.

Next, we write a dimensionless ratio as one of the remaining variables multiplied by $\rho_f^a r^b g^c$, where a, b, and c are constants, i.e.,

$$\Pi_1 = v\rho_f^a r^b g^c \qquad \Rightarrow 0 = [L][T]^{-1}([M][L]^{-3})^a([L])^b([L][T]^{-1})^c,$$

from which we find, using the principle of dimensional homogeneity, that $a = 0$, $b = c = -1/2$, and so our first dimensionless ratio is

$$\Pi_1 = \frac{v}{(rg)^{1/2}}.$$

Similarly, choosing $\Pi_2 = \rho_p \rho_f^a r^b g^c$ and $\Pi_3 = \mu\rho_f^a r^b g^c$ gives

$$\Pi_2 = \frac{\rho_p}{\rho_f}, \qquad \Pi_3 = \frac{\mu}{\rho_f r^{3/2} g^{1/2}}.$$

All of these Π factors are dimensionless, so we can write $\Pi_1 = Af(\Pi_2, \Pi_3)$, or

$$\frac{v}{(rg)^{1/2}} = Af\left(\frac{\rho_p}{\rho_f}, \frac{\mu}{\rho_f r^{3/2} g^{1/2}}\right), \tag{1.15}$$

where A is an unknown constant and $f(\Pi_2, \Pi_3)$ denotes an unknown function of Π_2 and Π_3.

Sometimes with dimensional analysis this is as far as we can get. We then have to rely on experimental data, plotting Π_1 against Π_2 and Π_3 to determine the shape of the unknown function. Although this may seem unsatisfactory, it is useful to note that we started with six variables and we would have had to do many experiments varying five different variables and measuring their effect on v. However, by using dimensional analysis we have reduced the problem to three variables, which is far more manageable.

In this case, however, we can improve on Equation (1.15) by rearranging it and using some intuition about the problem. Firstly, let us rearrange the equation such that only the dependent variable (v) is on the left-hand side:

$$v = A(rg)^{1/2} f\left(\frac{\rho_p}{\rho_f}, \frac{\mu}{\rho_f r^{3/2} g^{1/2}}\right).$$

We also know that, physically, when the particle has reached its terminal velocity, the gravitational and drag forces must balance, so mathematically the parameters representing these forces must be separated with one in the numerator and one in the denominator (so that their ratio is 1 when the forces are equal). The way we have written Π_3 would imply that the parameter representing the gravitational force (i.e., g) could cancel out of the equation (it appears in the numerator in the prefactor, and in the denominator Π_3). However, we can legitimately correct this problem by using $1/\Pi_3$ instead of Π_3. Because Π_3 is dimensionless, we are not affecting the dimensions of the problem. We now have two factors of $g^{1/2}$, one in the prefactor and one in the function f. For simplicity we suspect that the function should be such that these two factors appear in the final equation as g and not g^2 or $g^{2/3}$, for example. This means that $1/\Pi_3$ cannot be raised to any power or appear in a function. The simplest way to achieve this is to make it a proportionality factor, so that

$$v = Ar^{1/2} g^{1/2} f\left(\frac{\rho_p}{\rho_f}, \frac{\rho_f r^{3/2} g^{1/2}}{\mu}\right) = (Ar^{1/2} g^{1/2}) \left(\frac{\rho_f r^{3/2} g^{1/2}}{\mu}\right) \tilde{f}\left(\frac{\rho_p}{\rho_f}\right) = A\frac{\rho_f r^2 g}{\mu} \tilde{f}\left(\frac{\rho_p}{\rho_f}\right),$$

where $\tilde{f}()$ is now an unknown function of only the ratio ρ_p/ρ_f.

We can go one more step because we know that if the density of the particle equals the density of the fluid, then $v = 0$. This means that the two densities must appear in the equation as $(\rho_p - \rho_f)$. The simplest choice for $\tilde{f}$ is then (remembering that the function is a function of the ratio ρ_p/ρ_f)

$$\tilde{f} = \frac{\rho_p}{\rho_f} - 1.$$

Our final expression is therefore

$$v = Ar^2 \frac{g}{\mu} (\rho_p - \rho_f). \tag{1.16}$$

At this point we still have to determine the constant A and we would still have to determine the veracity of the equation using experiment. However, using the Buckingham Pi theorem allowed us to simplify the problem by reducing the number of variables we had to consider. When combined with our intuition about the problem, we were able to determine the form of Equation (1.16). The terminal velocity calculated from Equation (1.16) is called the *Stokes velocity*.

Exercise 1.4.6 What would happen if we had chosen slightly different variables to start with? Redo the calculation of the terminal velocity using the *kinematic viscosity* (ν), which has dimensions [L][T]$^{-2}$, instead of the dynamic viscosity.

This problem has shown that once we have a set of dimensionless ratios we can combine and manipulate them to form different dimensionless ratios (e.g., when we exchanged Π_3

for $1/\Pi_3$) without affecting the overall nature of our solution. The reason we are allowed to do this is that the ratios are dimensionless, they are just numbers without a dimension of length, or mass etc. Numerical values will change, but the relationships between variables (which is what we are interested in) will remain the same. To illustrate this, think of the ratio of the circumference (s) to the diameter (d) of a circle: $\pi = s/d$. We could equally well have defined the ratio as $\tilde{\pi} = d/s$ so that $\tilde{\pi} = 1/\pi \approx 0.31831$. The only difference is that we would have to memorize $\tilde{\pi}$ instead of π. The important thing is that the relationship between d and s is the same and dimensionally correct—s is proportional to d.

We have seen that the interpretation of the dimensionless Π ratios obtained by using the Pi theorem is not always intuitively obvious and can require some thought to put the ratios into a more transparent form. This can definitely be the case as we tackle more complicated problems. For example, let us consider an impactor, such as an asteroid, striking a planet such as Earth (Holsapple and Schmidt, 1982). The Earth is continually being bombarded by interplanetary objects. Most are small and burn up in the atmosphere, but a few have been large enough to strike Earth, sometimes with catastrophic consequences. If we were able to relate the size of the craters we see on Earth to the size of the objects hitting it, then using the observed distributions of these objects in space we could develop predictions for the frequency of catastrophic impacts and the likelihood of a catastrophic impact occurring in the future. The first thing we have to think about is what variable we want to choose to represent the size of the crater. We might think about using the diameter of the crater, but then we would also have to consider its depth. When an object strikes the Earth, it ejects material from the Earth's surface, leaving behind a crater. This requires energy, and the more energy the impact has, the greater the amount of material ejected. So, we might consider using the volume of the crater as a measure of its size. We will use the Buckingham Pi theorem to develop a relationship between the volume of the crater formed by the impact and the characteristics of the two bodies (Figure 1.7). As before, we should start by making a list of all the things we think could be relevant to the problem. First, we need to know something about the impactor, which we take to be an asteroid:

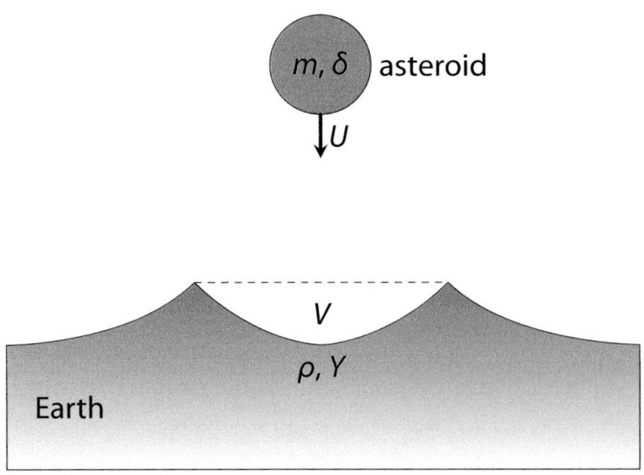

Figure 1.7 An asteroid of mass m and density ρ impacts the Earth with a velocity U, leaving a crater of volume V.

- The asteroid velocity, U with dimensions $[L][T]^{-1}$—we might suspect the faster the asteroid is moving, the larger the impact crater will be because the impactor has more kinetic energy. The kinetic energy of a body of mass m moving at a velocity v is the energy associated with its motion and is given by $E = (1/2)mv^2$.
- The mass of the asteroid, m with dimensions [M]—the more massive the asteroid, the larger the impact crater because the greater the energy involved.
- The density of the asteroid, δ with dimensions $[M][L]^{-3}$—an asteroid made up of a loose mix of ice and rock would probably leave less of an impact crater than one made of dense rock. Note that if the asteroid is spherical, then we can calculate its radius from its density and mass. We could, at this point, have chosen the radius instead of the mass or density of the asteroid.

We also need to know something about the planet being hit, the Earth for example.

- The density of the material where the impact occurs, ρ with dimensions $[M][L]^{-3}$—just like the asteroid, we expect it is harder to make a crater if the material it impacts with has a higher density.
- A measure of how easy it is to deform the Earth, characterized by Young's modulus Y with dimensions $[M][L]^{-1}[T]^{-2}$—Young's modulus is a measure of the stiffness of a material; that is, how easily it deforms. It has the same units as pressure, a force per unit area.
- The attractive gravitational force of the planet, g with dimensions $[L][T]^{-2}$—if the planet has a greater gravitational force, it will produce a greater acceleration of the asteroid toward it, producing a more energetic impact. We assume that the asteroid is much smaller than the planet, and so we disregard the gravitational attraction of the asteroid compared with that of the planet.

These variables and their dimensions are summarized in Table 1.2.

We can see that there are three independent dimensions ([M], [L], and [T]) and seven variables (U, m, δ, ρ, Y, g, and V). The Buckingham Pi theorem then tells us that we are looking for four dimensionless ratios. To find them, we will use a different approach than before. We start by writing a general Π ratio as a function of all the variables in the problem:

$$\Pi = m^{k_1} U^{k_2} \delta^{k_3} \rho^{k_4} Y^{k_5} g^{k_6} V^{k_7}. \tag{1.17}$$

Table 1.2 Quantities and dimensions used in the calculation of crater volume

Quantity	Symbol	Dimensions
Impactor velocity	U	$[L][T]^{-1}$
Impactor mass	m	$[M]$
Impactor density	δ	$[M][L]^{-3}$
Planet density	ρ	$[M][L]^{-3}$
Young's modulus	Y	Force/Area: $[M][L]^{-1}[T]^{-2}$
Gravitational acceleration	g	$[L][T]^{-2}$
Crater volume	V	$[L]^3$

If we substitute the dimensions of each of these quantities, we obtain

$$\Pi = [\mathrm{M}]^{k_1+k_3+k_4+k_5}[\mathrm{L}]^{k_2-3k_3-3k_4-k_5+k_6+3k_7}[\mathrm{T}]^{-k_2-2k_5-2k_6},$$

and because Π is dimensionless, we must have

$$\begin{aligned} 0 &= k_1 + k_3 + k_4 + k_5 \\ 0 &= k_2 - 3k_3 - 3k_4 - k_5 + k_6 + 3k_7 \\ 0 &= -k_2 - 2k_5 - 2k_6. \end{aligned}$$

Now we have a problem of solving three equations in seven unknowns. The way we do this is to use our intuition about the problem to choose four of the ks and solve for the remaining three in terms of these four—it will turn out that we can always rearrange things again at the end, so our choice at this stage does not really have much effect on our final answer. However, we want to be sensible about our choices. For example, we have two different densities in our problem, so we might not want to have both densities appearing in our four chosen variables. Let us choose k_1, k_2, k_3, and k_6 as our four variables. In terms of these variables, the remaining k values are

$$\begin{aligned} k_5 &= -\frac{1}{2}(k_2 + 2k_6) \\ k_4 &= -\left(k_1 + k_3 - \frac{1}{2}(k_2 + 2k_6)\right) \\ k_7 &= \frac{1}{3}(k_6 - 3k_1). \end{aligned}$$

But we still do not have a good solution because we do not know the values of k_1, k_2, k_3, or k_6. We have no equations to solve to obtain values for these variables, so we have to choose values for them. Although we are at liberty to choose these values arbitrarily, we might want to make sensible choices that will help us. We are going to choose values that deliberately pick out one of these four k parameters at a time. Remember, we are going to get a single dimensionless grouping (we are looking for four of them) for each set of value choices.

- If we choose $k_1 = 1$ and $k_2 = k_3 = k_6 = 0$, then we find from Equation (1.17) that

$$\Pi_1 = \frac{m}{\rho V} = \frac{\text{mass of impactor}}{\text{mass of ejecta}} \tag{1.18}$$

- Choosing $k_2 = 1$ and $k_1 = k_3 = k_6 = 0$ gives

$$\Pi_2 = \frac{U\rho^{1/2}}{Y^{1/2}} \tag{1.19}$$

- With $k_3 = 1$ and $k_1 = k_2 = k_6 = 0$ we get

$$\Pi_3 = \frac{\delta}{\rho} = \frac{\text{density of impactor}}{\text{density of planet}} \tag{1.20}$$

- Lastly, with $k_6 = 1$ and $k_1 = k_2 = k_3 = 0$ we get

$$\Pi_4 = \frac{g\rho V^{1/3}}{Y}. \tag{1.21}$$

Notice how, by choosing each of the four k parameters to have a value of 1 in turn, we have created a full set of ks that cannot be converted into each other.

Out of these Π factors, two of them are easy to interpret, and we have done this above. However, the remaining two are harder. But, if we make use of the other dimensionless factors (remember, they are dimensionless factors, just numbers), we can turn them into something easier to interpret.

For example, we can square Π_2 and use Π_3 to get the following:

$$\Pi_2^2 = \frac{U^2\rho}{Y} = \left(\frac{U^2\delta}{Y}\right)\left(\frac{\rho}{\delta}\right) = \left(\frac{U^2\delta}{Y}\right)\frac{1}{\Pi_3},$$

and since Π_3 is a dimensionless ratio (i.e., just a number), we can define a new Π ratio

$$\bar{\Pi}_2 = \frac{U^2\delta}{Y}.$$

Recalling a little physics, we can see that the numerator of $\bar{\Pi}_2$ has something to do with the kinetic energy per unit mass of the asteroid and the denominator is related to the strength of the material. Lastly, we can play around with Π_4 to find that

$$\Pi_4 = \frac{g}{U^2}\left(\frac{m}{\rho}\right)^{1/3}. \tag{1.22}$$

Exercise 1.4.7 Calculate the Π values if we had chosen each of k_1, k_2, k_3, or k_6 to have been 2 instead of 1 and compare them to those obtained above.

Exercise 1.4.8 Calculate the Π values if we had chosen $k_1 = k_2 = k_3 = k_6 = 1$, then $k_1 = k_2 = k_3 = k_6 = 2$, $k_1 = k_2 = k_3 = k_6 = 3$, and $k_1 = k_2 = k_3 = k_6 = 4$. Compare them with the Π values we obtained in the text.

Exercise 1.4.9 Derive Equation (1.22).

So, putting all of this together, we can write

$$V = \frac{m}{\rho} f\left(\frac{U^2\delta}{Y}, \frac{\delta}{\rho}, \frac{g}{U^2}\left(\frac{m}{\rho}\right)^{1/3}\right). \tag{1.23}$$

To discover the mathematical form of the unknown function, we need to do experiments and plot these variables against each other. But using the Buckingham Pi theorem has allowed us to reduce the number of variables we have to consider and provided us with a set of nice scaling relationships for the volume of the impact crater. With such a relationship we can use the measured properties of craters to infer the properties of the impactor or vice versa.

1.5 Dimensional Analysis: Some Remarks

Dimensional analysis is a very useful tool to have in our toolbox, but it does have its limitations. Primary among these is the fact that the results we obtain depend on our ability to identify the relevant and important variables in a problem. This relies on our intuition and understanding of the processes involved in the problem (e.g., using the kinetic energy in our asteroid example to determine some important variables). However, many phenomena in the natural world are very complicated and we should not hope to consider all the possible variables and parameters that might be relevant. So when we use dimensional analysis, we are implicitly making decisions as to what variables are important for understanding a given phenomenon and which ones we can safely neglect. This is where our intuition and understanding come into play—making dimensional analysis difficult, but an extremely useful tool for improving our understanding. Consequently, the relationships we derive may not always apply. For example, in our impactor problem we implicitly assumed that the asteroid was much smaller than the Earth. If the impactor was the size of the Moon, for example, then it would do a lot more damage than just leave a crater.[19]

1.6 Further Reading

There are now several interesting books that contain examples of back-of-the-envelope questions. An excellent book to start with is *Guesstimation* (Weinstein and Adam, 2008). This contains a large number of intriguing questions with detailed workings for each question. The mathematics is quite elementary. A book that covers physics and the environment and makes extensive use of back-of-the-envelope calculations is *Physics of Societal Issues* (Hafemeister, 2007). A classic text that examines a series of environmentally related problems is *Consider a Spherical Cow* (Harte, 1988). A more mathematically involved book, but one that is well worth the effort to work through, is *Mathematics in Nature* (Adam, 2003). The mathematically technical aspects of dimensional analysis are covered in the first couple of chapters of *Scaling, Self-Similarity, and Intermediate Asymptotics* (Barenblatt, 1996) and *Applied Dimensional Analysis and Modeling* (Szirtes, 2007), which contains many worked examples from a wide range of disciplines. A classic exposition of scaling relationships in biology is the essay *On Being the Right Size* by Haldane (1945); it is entertaining, thought provoking, and well worth reading. A wonderful book that covers a lot of mathematical problem-solving techniques that are used by scientists on a daily basis is *Street-Fighting Mathematics* by Mahajan (2010), but be warned, to get the most from it you will have to have studied through to at least Chapter 7 of this book. When dealing with estimations and quantitative problems in general, it is a good idea to have a reference of typical values for the variables you are interested in. For example, in this chapter we have made use of the collection of data in *The Cambridge Handbook of Earth Science Data* (Henderson and Henderson, 2009), which contains a lot of useful numbers for the Earth and environmental sciences. It is also a good idea to make your own collection that includes typical values for important parameters in your own field of research or study.

[19] One hypothesis for the formation of the Moon is that the early Earth was struck by a large impactor, approximately the size of Mars.

Problems

1.1 Differences between the physical properties of air and water lead to some interesting comparisons between the atmosphere and oceans.

1. Estimate the mass of the oceans (assuming a density of 1000 kg m^{-3} and compare it to the mass of the atmosphere (assume a density of about 1 kg m^{-3}).
2. Estimate the weight of the atmosphere per square meter of the Earth's surface, and estimate how deep you would have to dive in the oceans to experience the same weight per square meter.
3. Heat capacity is a measure of how much heat energy a substance has to absorb or emit for its temperature to change by 1 K. The specific heat capacity of air is 10^3 kJ kg^{-1} whereas that for water is 4×10^3 kJ kg^{-1}. Estimate how deep a layer of the ocean has the same heat capacity as the whole atmosphere.

1.2 If all the water in the oceans were shaped into a sphere, what would the diameter of that sphere be, and compare it to the diameter of the Earth. Name two cities that have the same distance between them as the diameter of the ocean sphere.

1.3 Oceanographers measure transport of water using a unit called the Sverdrup: 1 Sv is equivalent to a flow of water of 10^6 m^3 s^{-1}. Estimate the flow of water from a kitchen tap (or faucet) and compare this to a flow of 1 Sv.

1.4 Estimate the ratio of the acceleration due to gravity between the poles and equator given that the radius of the Earth at the poles is approximately 6357 km and that the equator is 6378 km.

1.5 Estimate the amount of CO_2 that enters the atmosphere globally from automotive transport. Assume that 1 gallon of gasoline weighs about 2.5 kg and is comprised solely of octane (C_8H_{18}) with a molecular weight of 114 and that each molecule of octane produces 8 molecules of CO_2 when it burns. (The atomic number of carbon is 12 and that of oxygen is 8.)

1.6 Estimate the rise in global sea level (in meters) if all the ice in Antarctica melted. Assume that Antarctica has an area of about 14×10^6 km^2 with an average ice thickness of 1800 m.

1.7 The total amount of precipitable water in the atmosphere is estimated as 1.24×10^{16} kg.

1. What would the depth (in centimeters) of water be if the precipitable water was spread uniformly over the whole Earth?
2. The estimated amount of total precipitation in a year over the whole of the planet is 3.96×10^{17} kg. How deep a layer is this (in centimeters) if it is spread evenly over the planet?

3. If it rained continuously, how many days would it take for the total amount of water in the atmosphere to fall out as rain? (Note, you will have to estimate a typical rainfall rate.)
4. Vaporizing a liquid requires energy. Some of the energy from the Sun goes into evaporating liquid water on the surface of the Earth. If the energy required to vaporize 1 kg of water is ~ 2.4×10^6 J kg^{-1}, what percentage of the total solar energy falling on the Earth goes to evaporating water (assume a total solar input of approximately 10^{14} kW)?

1.8 In this problem we are going to look at some detailed analysis of the atmosphere using some back-of-the-envelope calculations.

1. Given that an average wind speed in the Earth's atmosphere is 10 m s^{-1}, estimate the kinetic energy of the Earth's atmosphere. Hint: kinetic energy is the energy due to motion and is given by $E_k = (1/2)mv^2$, where m is the mass and v is the velocity.
2. Compare the value you get with the total solar energy being inputted to the Earth.
3. Now estimate the energy in an average hurricane. This can come from two processes: evaporation of water to form precipitation and the kinetic energy of the winds.

   a. Estimate the energy for evaporation.
   b. Estimate the energy from wind and compare it with that from evaporation.

1.9 In this problem we are going to track the dilution of a river plume as it enters and moves over the ocean. Being lighter than seawater, we will assume that the river plume sits on top of the ocean, and we will also assume that there is negligible mixing between the plume and the ocean below. We will also assume that the river plume is vertically homogeneous and maintains a constant thickness of 10 m. We are going to follow a parcel of water in this river plume—we can think of a parcel of water as being a given volume of water, such as 1 m^3.

1. What assumptions justify our consideration of a single parcel of water instead of the whole plume?
2. The region of the ocean that the plume is in has an average precipitation rate of 1 mm d^{-1}. What volume (in m^3) of water is added to your parcel of the plume in one day?
3. When the parcel of water enters the ocean, it has a certain concentration of dissolved inorganic carbon (DIC). Rainwater does not contain any DIC, however, so rainwater will dilute the DIC in the river water. What is the ratio of final to initial concentrations of DIC in the plume over the period of 1 day?
4. When the parcel of river water enters the ocean, it has a DIC concentration of 300 μM. Assuming that there are no chemical or biological changes affecting inorganic carbon, what is the final DIC concentration of the parcel after 100 days?

1.10 One hypothesis for the sudden extinction of the dinosaurs 65 million years ago is that the Earth was hit by an asteroid. The impact resulted in a large amount of dust being sent into the atmosphere where it blocked out sunlight and led to a colder and darker surface. Estimates are that about 20% of the mass of the asteroid was converted into dust, which ended up being spread uniformly over the Earth in a layer of 2 g cm^{-2} after settling out of the atmosphere. Astronomers tell us that a typical large asteroid has a density of approximately 2 g cm^{-3} (Carry, 2012). Estimates from impact studies suggest that a mass of material amounting to 60 times the mass of the asteroid would be blasted out to form the resulting crater and that 20% of this material would also be put into the upper atmosphere (Toon et al., 1997; Kring, 2007).

1. Estimate the total mass of dust that covered the Earth after the impact, and from that, estimate the radius of the asteroid (assume the asteroid was spherical).
2. If we assume that the dust particles were spherical, then the cross-sectional area of each particle is that of a disk. If each particle had a radius of 5×10^{-7} m and a density of 2 g cm^{-3}, calculate the combined cross-sectional area of all the particles ejected into the atmosphere.
3. In reality, the dust particles would have had a range of sizes. Repeat the calculation in 2., assuming a particle radius of 10^{-5} m.
4. Compare the two answers you got in 2. and 3. with the surface area of the Earth. Would you expect the dust particles to block out all the sunlight falling on the Earth?

1.11 From time to time a rumor will spread on the internet that an alignment of planets will tear the Earth apart or allow people to float. Given that the gravitational force between two bodies of masses M_1 and M_2 that are a distance r apart is proportional to $(M_1 M_2)/r^2$ and the tidal force is proportional to $(M_1 M_2)/r^3$, investigate the veracity of these ideas.

1.12 Between 1955 and 2010, the average temperature in the top 700 m of the Earth's oceans increased by 0.18°C. If this top part of the ocean is the only part of the ocean affected by global warming, estimate the annual rate of sea level rise over this period due to thermal expansion of water alone (the change in volume ΔV in water undergoing a change in temperature ΔT (in °C) is $\Delta V = 1.5 \times 10^{-4} V \Delta T$, where V is the starting volume of the water).

1.13 One common thought for combatting the rise in atmospheric CO_2 is to plant more plants. Plants on land take up about 120 PgC from the atmosphere through photosynthesis annually. If there were no processes to replenish atmospheric CO_2, and no human input of CO_2, how long would it take for terrestrial plants to remove all of the carbon dioxide from the atmosphere, assuming a current concentration of 400 ppm?

1.14 Oxygen was almost entirely absent from the atmosphere of the early Earth, and it was only with the evolution of photosynthesizing microbes that atmospheric oxygen became plentiful. If all photosynthesis suddenly stopped on Earth, so that there was no replenishment of oxygen, how long could humans continue to breathe? Assume that only humans consume oxygen.

1.15 How much gasoline do you have to use to produce the equivalent of 1 ppm of CO_2? Determine if this is a lot of gasoline by comparing it with some annual gasoline consumption numbers for whole countries. Assume that burning 1 L of gasoline produces approximately 2.25 kg of CO_2.

1.16 One of the important characteristics of a river drainage basin is the drainage density. This is defined as the total length of the river channels in a basin divided by the area of the drainage basin. What are the dimensions of drainage density? Drainage density depends on factors such as the local climate, the way the land is used, and the properties of the soil. Use dimensional analysis and the Buckingham Pi theorem to derive the following expression for drainage density (D):

$$D = \frac{1}{H} f\left(Q\kappa, \frac{Q\rho H}{\mu}, \frac{Q^2}{gH}\right),$$

where H is a measure of the relief of the basin ([L]), Q is called the runoff intensity ($[\mathrm{L}]^{-1}[\mathrm{T}]^{-1}$), κ is a measure of erosion ($[\mathrm{L}]^{-1}[\mathrm{T}]$), ρ is the fluid density, μ the dynamic viscosity of the fluid, and g the acceleration due to gravity.

1.17 The motion of water in the oceans is influenced by interactions with the seabed through bottom drag or friction. This is a force that slows down the current. Write down the dimensions of the drag force. One expects that the drag force depends on the water speed (u), the viscosity of the fluid μ (the more turbulent the fluid, the greater the drag), a typical length scale (l) of the flow, and lastly the density of the water (ρ). Use the Buckingham Pi theorem to derive the following expression for the drag force (F):

$$F \propto \rho l^2 u^2 f\left(\frac{\mu}{\rho l u}\right).$$

It turns out that the quantity

$$l^2 f\left(\frac{\mu}{\rho l u}\right)$$

is constant[20], which is typically denoted as C_D. Substituting this into the equation yields the standard quadratic law for frictional forces in a flow.

1.18 Use the principle of dimensional homogeneity to derive an expression for the speed of a seismic wave that depends on the axial modulus (dimensions $[\mathrm{M}][\mathrm{L}][\mathrm{T}]^{-2}$)—a measure of the incompressibility of the rock—and the density of the rock.

1.19 Heavy rainfall often results in sheets of water moving downhill. These sheets of water can result in significant erosion of bare soil. Use the Buckingham Pi theorem to show that

[20] You can intuitively justify this if you recall that a pressure is a force per unit area and realize that, since $\mu/(\rho l u)$ is dimensionless, then $f(\mu/(\rho l u))$ is also dimensionless.

$$Q = \nu\rho f\left(S_0, \frac{D}{\nu}, \frac{IL}{\nu}, \frac{\tau_c}{\tau_0}\right),$$

where Q is the mass of soil transported per unit width of the flow per unit time, D is the discharge per unit width of the flow (dimensions $[L]^2[T]^{-1}$), I is the intensity of rainfall (dimensions $[L][T]^{-1}$), L is the length of the runoff, ν is the kinematic viscosity of the fluid, and ρ is the density of the water.

1.20 Large-scale fluid motion on a rotating planet is affected by pressure gradients (e.g., winds flow from areas of high to low pressure) and the rotation of the planet, which gives rise to the Coriolis effect. If these two effects are in balance, then the resulting flow is called a geostrophic flow. Use the principle of dimensional homogeneity to show that the velocity (v) of the geostrophic flow is

$$v \propto \frac{1}{f\rho}\frac{\Delta P}{\Delta x},$$

where ΔP is the change in pressure over the distance Δx, ρ is the fluid density, and f is the Coriolis parameter, which characterizes the strength of the Coriolis force and has dimensions $[T]^{-1}$.

2 Derivatives and Integrals

Change is one of the more important aspects of the world that we study as scientists. Objects move, chemical reactions change the concentrations of important elements, populations increase and decrease. Quantities we are interested in change over time, over space, and with respect to each other. Calculus provides tools to help us understand and predict those changes. For example, how rapidly does the surface temperature of a planet change with changes in the concentration of greenhouse gases? How fast does the intensity of light decrease as you get deeper in the ocean? Measuring these rates is a critical step in understanding them, but if we can develop equations that relate rates to the quantities that influence those rates, we can make predictions that can be tested, and that is very powerful.

2.1 Derivatives, Limits, and Continuity

Differential calculus provides us with the tools to calculate rates of change. We can visualize how fast a variable (y) changes with respect to another variable (x) by plotting a graph of y against x; regions where y changes rapidly will be steeper than those where the rate of change is more gradual. However, we would like to go beyond visualizing the rate of change and develop mathematical tools that allow us to calculate it.

If y changes linearly with x then the graph will be a straight line and the rate of change of y with x is simply the slope or gradient ($\Delta y/\Delta x$) of that line and is constant for all values of x (Figure 2.1a). However, if y changes nonlinearly with x, then the slope of the curve will change as x changes, meaning that we need to be able to calculate the slope of the curve at a given point (e.g., point P in Figure 2.1b). The slope of the curve at any given point on the curve is the slope of the tangent line to the curve at that point. We know how to calculate the slope of a straight line, so we only have to find how to determine the tangent line to the curve. We can construct the tangent geometrically by first choosing two points (A_1 and A_2 in Figure 2.2) that are equidistant either side of point P and drawing the line that passes through them. This line is called a *chord*, and we can calculate its slope. We can then pick two points (B_1 and B_2) that are closer to P and do the same thing. We continue like this until the two points we choose are practically identical. The curve in Figure 2.2 is $y = x^3$, and we can see that the slopes of the different lines gradually approach that of the tangent to the curve at P. This is a geometric representation of what we do when we calculate a derivative; the derivative of a curve $y = y(x)$ tells us how the slope of the curve changes as x changes. We need to translate our geometric representation into something that allows us to write an equation.

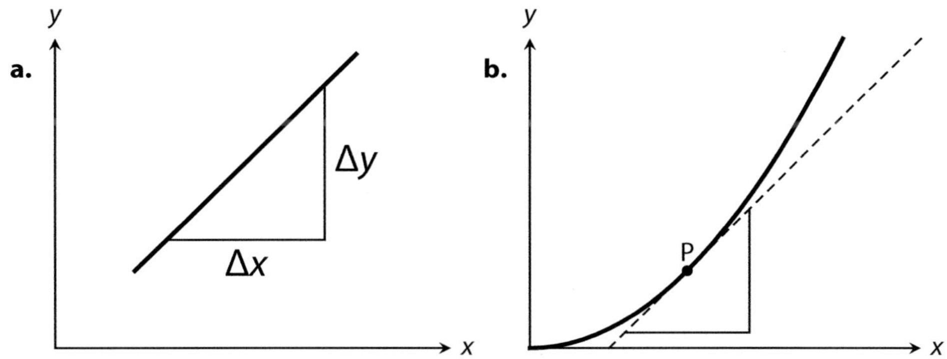

Figure 2.1 The slope of a straight line (**a.**) and curve (**b.**). In **a.**, the slope (or gradient) of a straight line is simply $\Delta y/\Delta x$ and is a constant. In **b.**, the slope of a curve at a point P is the slope of the tangent to the curve at that point and varies along the curve as the curve becomes more or less steep.

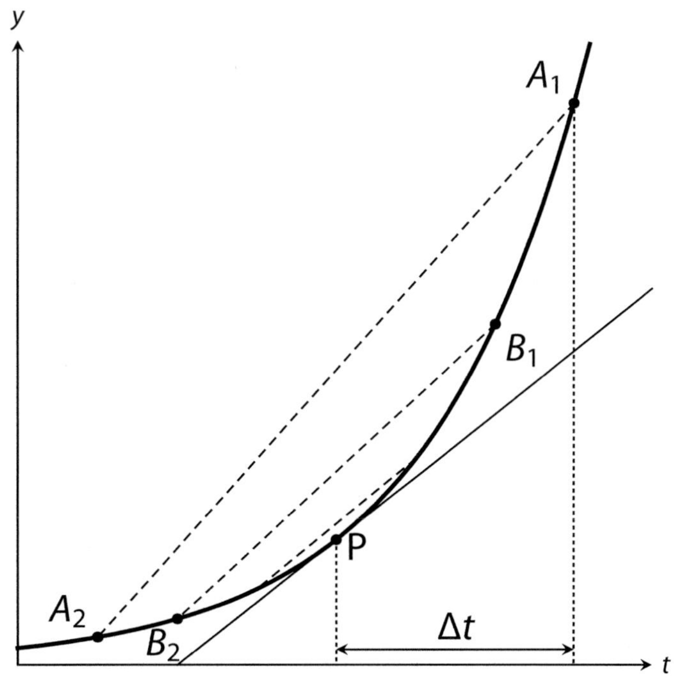

Figure 2.2 The gradient of a curve as a succession of chords. The actual tangent to the curve at the point P is shown as a solid line.

Say we want to find the derivative of the curve $y = x^3$ for any value of x. What we want to do is calculate the values of y for two given values of x that are close together and lie either side of a point x_0. Let us choose $x_{A_1} = x_0 + \Delta x$ and $x_{A_2} = x_0 - \Delta x$ as our two x values, where Δx is a small change in x. Then

$$y_{A_1} = (x_0 + \Delta x)^3 = x_0^3 + 3x_0^2\Delta x + 3x_0\Delta x^2 + \Delta x^3,$$
$$y_{A_2} = (x_0 - \Delta x)^3 = x_0^3 - 3x_0^2\Delta x + 3x_0\Delta x^2 - \Delta x^3.$$

Now, we need to find the slope of the straight line connecting the points (x_{A_1}, y_{A_1}) and (x_{A_2}, y_{A_2}),

$$\frac{\Delta y}{\Delta x} = \frac{y_{A_1} - y_{A_2}}{x_{A_1} - x_{A_2}} = \frac{x_0^3 + 3x_0^2\Delta x + 3x_0\Delta x^2 + \Delta x^3 - (x_0^3 - 3x_0^2\Delta x + 3x_0\Delta x^2 - \Delta x^3)}{(x_0 + \Delta x) - (x_0 - \Delta x)}$$
$$= \frac{6x_0^2\Delta x + 2(\Delta x)^3}{2\Delta x} = 3x_0^2 + \Delta x^2. \tag{2.1}$$

In our geometric representation of the derivative, we found the tangent by letting the points either side of x_0 get closer and closer together. That means we need to let Δx get smaller and smaller, and as $\Delta x \to 0$, terms involving Δx^2 and Δx^3 become very much smaller than those involving x_0. So, as this happens, Equation (2.1) becomes

$$\frac{y_{A_1} - y_{A_2}}{x_{A_1} - x_{A_2}} = \frac{\Delta y}{\Delta x} = 3x_0^2.$$

We did not choose x_0 to be any specific or special point, so this formula must hold for any value of x_0, and we can write the derivative of $y(x) = x^3$ as

$$\frac{dy}{dx} = 3x^2, \tag{2.2}$$

where we have used the normal nomenclature for a derivative on the left-hand side of the equation.

What we have done here is quite interesting and powerful. Formally, we have taken the limit as the difference between the x values becomes infinitesimally small. The derivative of a function $y(x)$ is written formally as

$$\frac{dy}{dx} = \lim_{\Delta x \to 0} \frac{y(x + \Delta x) - y(x)}{\Delta x}. \tag{2.3}$$

However, we need to think a little more carefully about this process because it is not as innocent as it at first appears, and if we do not take appropriate care we can end up with nonsense when we take a derivative. In Equation (2.3) we took the limit of a ratio, but it is easier to see what is happening if we take the limit of a simpler function. The concept of *taking a limit* of a function is represented graphically in Figure 2.3. The idea is intuitively simple: as we approach the point x_0 from either above ($x_0 + \Delta x > x_0$) or below ($x_0 - \Delta x < x_0$), the value of the function $y = f(x)$ tends to a value $f(x_0) = F$ as $\Delta x \to 0$. A more formal statement is given in Definition 2.1.

Definition 2.1 The function $f(x)$ tends to the limit F if for any number $\epsilon > 0$, there exists another number ($\Delta x > 0$) such that if $|x - x_0| < \Delta x$ then $|f(x) - F| < \epsilon$.

It can be difficult to see how this formal statement corresponds with our intuitive understanding of a limit.[1] Our intuitive understanding is that as the value of x approaches x_0, the

[1] We should always try and do this when we come across a formal mathematical statement such as the one in Definition 2.1.

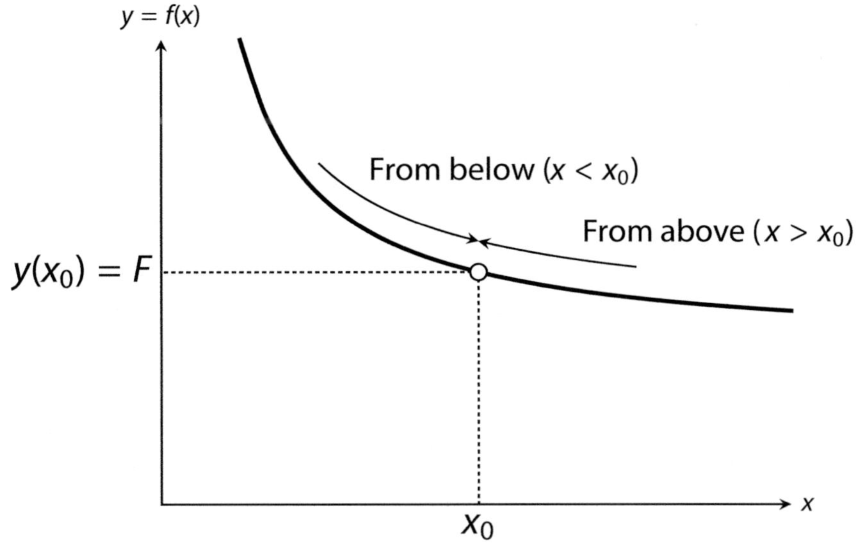

Figure 2.3 As x approaches the value x_0, the function $y = f(x)$ approaches the value $f(x_0) = F$. We can approach x_0 by decreasing x starting from $x > x_0$ (i.e., approaching from above) or by increasing the value of x starting from a point $x < x_0$ (approaching from below) and in both cases $f(x)$ approaches the values $f(x_0) = F$.

value of the function $f(x)$ will approach the value F. To see the correspondence between this and Definition 2.1, let us look at each part of the definition. The quantity $|x - x_0|$ is the absolute value of the difference between x and x_0. Because this is an absolute value it is always positive and represents the "distance" between x and x_0 irrespective of whether $x < x_0$ or $x > x_0$; in other words x can be less than x_0 and increasing toward it, or larger than x_0 and decreasing toward it. Similarly, $|f(x) - F|$ is the distance between the value $(f(x))$ of the function evaluated at x and the value of the function evaluated at x_0. As the value of Δx gets smaller and smaller, the distance between x and x_0 gets smaller and smaller. Similarly, as ϵ gets smaller, the distance between $f(x)$ and F gets smaller. Now we need to think about the main part of the definition. This says that if F is a limit of $f(x)$, then we can make the distance between $f(x)$ and F (i.e., ϵ) as small as we like, even infinitesimally small, by making the distance between x and x_0 (i.e., Δx) smaller and smaller. So in other words, Definition 2.1 is a translation of our intuitive understanding into the very concise and precise language of mathematics.

Let us see how this works in practice.

Example 2.1 As an example, let us show that

$$\lim_{x \to 2} \frac{1}{x} = \frac{1}{2}.$$

This may appear obvious, but let us prove it nonetheless—it is instructive to prove this because we have to think carefully about absolute values and inequalities. By comparison with Definition 2.1, we see that $x_0 = 2$ and $F = 1/2$. We need to show that for any value of

x such that $|x-2| < \Delta x$, there is a value of ϵ such that $|f(x) - 1/2| < \epsilon$. We can write the last inequality as

$$|f(x) - 1/2| = \left|\frac{1}{x} - \frac{1}{2}\right| = \left|\frac{2-x}{2x}\right| = \frac{|x-2|}{|2x|} < \epsilon, \tag{2.4}$$

where we have used the properties of the absolute value to write $|2-x|$ as $|x-2|$ — we did this because Definition 2.1 requires us to examine $|x-2| < \Delta x$. We know that we want Δx to be small, because we are taking a limit as $x \to 2$, so let us choose $\Delta x \leq 1$, which means that $|x-2| \leq 1$. Now, let us see what Equation (2.4) tells us about the possible values of x. To answer this, we need to think carefully about the different cases that arise from the use of the absolute values. If $\Delta x \leq 1$, then $|x-2| \leq 1$. This equation says that either $(x-2) \leq 1$ (taking the positive sign), in which case $x \leq 3$, or $-(x-2) \leq 1$ (taking the negative sign), which implies $-x \leq -1$, i.e., $x \geq 1$. So, making the choice that $\Delta x \leq 1$ imposes some constraints on the value of x, i.e., $1 \leq x \leq 3$. The other factor we have in Equation (2.4) is $1/|2x|$. We know that our choice of Δx has constrained x to lie between 1 and 3, so $1/|2x|$ must lie between 1/2 and 1/6, and Equation (2.4) tells us that

$$|f(x) - 1/2| \leq \frac{1}{6}|x-2|.$$

Definition 2.1 says that if the limit exists, we should be able to find a value for ϵ such that $|f(x) - 1/2| < \epsilon$. This means that to ensure $|f(x) - 1/2| < \epsilon$, we must have it that

$$|x-2| < 6\epsilon.$$

In other words, by choosing $\Delta x \leq 6\epsilon$ we are certain that $|(1/x)-(1/2)| < \epsilon$ for $|x-2| < \Delta x$. Therefore we have satisfied the definition of the limit, and $\lim_{x\to 2} \frac{1}{x} = \frac{1}{2}$. It is interesting to note that we do not actually have to know the value of ϵ to show that the limit exists.

We may like to think that all the limits we will come across are nice and well behaved like the one in Example 2.1, but unfortunately this is not the case. Some functions $y(x)$ do not have limits for all values of x, and this means that they cannot be differentiated at those points. For example, consider the function

$$f(x) = \frac{|x|}{x} = \text{sgn}(x), \tag{2.5}$$

which just gives us the sign of x (Figure 2.4); i.e., $\text{sgn}(x) = -1$ for $x < 0$ and $\text{sgn}(x) = +1$ for $x > 0$. If we take the limit as x approaches $x = 0$ from above (written as $\lim_{x\to 0^+}$), we find

$$\lim_{x\to 0^+} \text{sgn}(x) = +1,$$

but if we approach the same value ($x = 0$) from below, we find

$$\lim_{x\to 0^-} \text{sgn}(x) = -1.$$

This is a problem, because the value of sgn(0) cannot be both −1 and +1 at the same time. So, we have to conclude that this limit does not exist! Because we cannot take the limit at

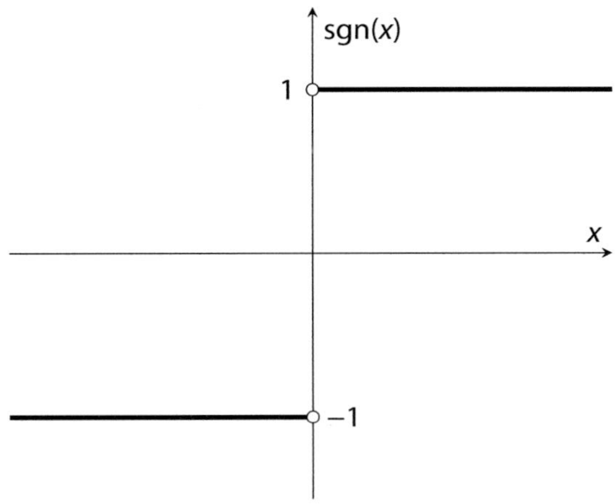

Figure 2.4 The function $y = \text{sgn}(x)$, which is discontinuous at $x = 0$.

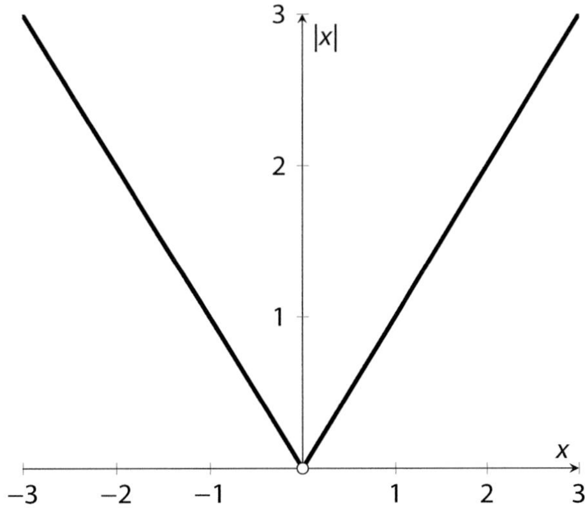

Figure 2.5 The function $|x|$, which is continuous and not smooth at $x = 0$.

this point, we cannot take the derivative of $\text{sgn}(x)$ at $x = 0$ either. Such a function is called *discontinuous*, and discontinuous functions do not have derivatives defined at the points where the discontinuity occurs, so they are *nondifferentiable*.

But this is not the only type of problem that can arise. Let us consider the function

$$f(x) = |x|$$

shown in Figure 2.5 and calculate its derivative at $x = 0$. This function is continuous at $x = 0$—the value of $f(x)$ does not jump in value—so we might think that it has a derivative there. Let us try calculating the derivative using Equation (2.3):

$$\frac{df}{dx} = \lim_{x \to 0} \frac{|x| - 0}{x - 0} = \lim_{x \to 0} \frac{|x|}{x} = \lim_{x \to 0} \text{sgn}(x). \tag{2.6}$$

But we have just shown that $\text{sgn}(x)$ is discontinuous at $x = 0$, so the limit in this equation does not exist and therefore the derivative does not exist at $x = 0$. The curve is differentiable at other points; for example, for $x < 0$ the slope of the curve is a constant (-1) and for $x > 0$ it is also a constant $(+1)$. But, as we have seen, at $x = 0$ the limit has two values depending on whether we approach $x = 0$ from below or above, so the derivative does not exist at that point. So the function $f(x) = |x|$ is continuous but nondifferentiable at $x = 0$. Functions like the one shown in Figure 2.5 with sharp corners are called *nonsmooth* functions, and they too are not differentiable at those points. Functions without sharp corners are, not surprisingly, called *smooth functions*. Some smooth functions might look like they are not smooth, but if we zoom in, the apparent corners get smoothed out. For a nonsmooth function like the one in Figure 2.5, the sharp corner will always remain no matter how much we zoom in.

Now, one might argue that nonsmooth and discontinuous functions are not useful for natural systems, but this is not strictly the case. For example, Figure 2.6 shows the density of the Earth's interior as a function of radius according to the preliminary reference Earth model (Dziewonski and Anderson, 1981). The density appears to jump at certain points, which mark the boundaries between layers of different density within the Earth. These discontinuities arise because we cannot represent the density of the Earth at arbitrarily small scales. For example, the interface between ice and water can be thought of as a boundary where density changes discontinuously. But, if instead of using density we were able to represent the number density of atoms on the scale of individual atoms, then we would not have a discontinuity at the interface, but a rapid rise in the number density of atoms as we moved from water into ice. Usually, however, we have to work at much coarser scales and deal with bulk quantities that we can easily measure. As a result, we sometimes have to deal with discontinuities. Most of the functions that we deal with in

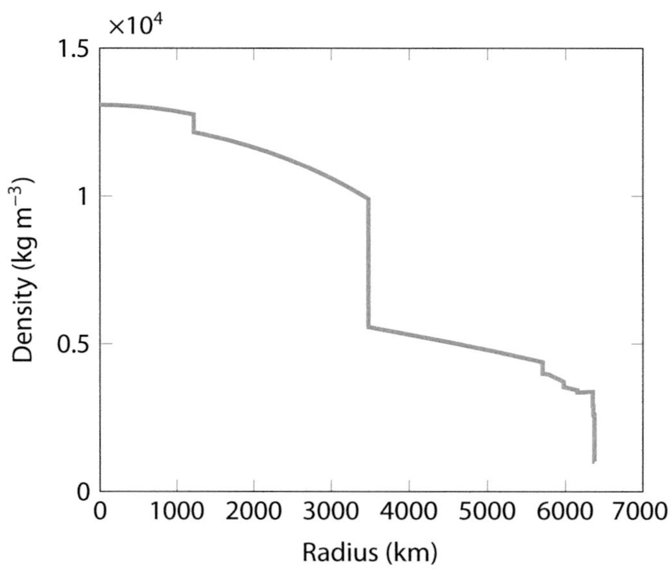

Figure 2.6 The density of the interior of the Earth as a function of radius according to the preliminary reference Earth model (Dziewonski and Anderson, 1981).

Earth and environmental sciences are smooth and continuous, but from time to time we will come across functions that are not. In these cases we need to take a little care with how we manipulate these functions, because if we do not, we will end up with nonsense, or worse!

Example 2.2 As a practical example, consider raindrops falling through the atmosphere, or river sediment particles sinking through water. Faster sinking particles can catch up and collide with slower ones, thereby creating larger raindrops or particle aggregates; this process is called *differential sedimentation*. If we consider a spherical particle of radius r_0, then the rate at which another (spherical) particle of radius r will collide with it as it sinks is given by the equation (Pruppacher and Klett, 2010)

$$K(r, r_0) = \frac{2\pi g \Delta\rho}{9\nu\rho}(r + r_0)^2 |r^2 - r_0^2|, \tag{2.7}$$

where g is the acceleration due to gravity, $\Delta\rho$ is the difference between the density of the falling particle and the surrounding fluid (e.g., air or water), ρ is the density of the surrounding fluid, and ν its viscosity. In the course of our research we may need to know whether this function is differentiable for all values of r and r_0.

Equation (2.7) tells us that particles with $r > r_0$ will catch up and collide with the particle of radius r_0, but it itself will catch up and collide with particles having $r < r_0$. However, when $r = r_0$, no collisions will occur because both particles will be falling with the same speed, and the function will have a sharp turn at that point (Figure 2.7). Just as with Equation (2.6), the derivative of this function will not exist at $r = r_0$ and the function is continuous but not smooth, so it is not differentiable at $r = r_0$.

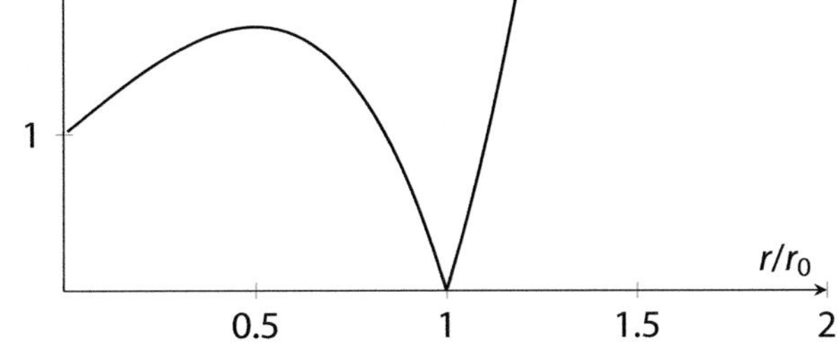

Figure 2.7 The rate of collision of two particles colliding by differential sedimentation (Equation (2.7)) showing the sharp turn in the curve when $r = r_0$.

Exercise 2.1.1 Use Equation (2.3) to show that the following functions are not differentiable at $x = 0$:

$$\text{a. } f(x) = \frac{1}{x}, \qquad \text{b. } f(x) = x\sin\left(\frac{1}{x}\right).$$

Limits have some important and useful properties that allow us to calculate the limits of complicated functions from the limits of simpler ones. If we have two functions $f(x)$ and $g(x)$ such that

$$\lim_{x\to a} f(x) = F \quad \text{and} \quad \lim_{x\to a} g(x) = G,$$

where a is a constant, then:

1. If b is a constant, the limit of $bf(x)$ is b multiplied by the limit of $f(x)$:

$$\lim_{x\to a} bf(x) = b\lim_{x\to a} f(x) = bF.$$

2. The limit of the sum (or difference) of two functions is the sum (or difference) of the limits:

$$\lim_{x\to a}(f(x) \pm g(x)) = \lim_{x\to a} f(x) \pm \lim_{x\to a} g(x) = F \pm G.$$

3. The limit of $f(x)$ multiplied by $g(x)$ is the product of their limits:

$$\lim_{x\to a}(f(x)g(x)) = \left(\lim_{x\to a} f(x)\right)\left(\lim_{x\to a} g(x)\right) = FG.$$

4. The limit of $f(x)$ divided by $g(x)$ is the ratio of their limits, so long as the limit of $g(x) \neq 0$:

$$\lim_{x\to a}\frac{f(x)}{g(x)} = \frac{\lim_{x\to a} f(x)}{\lim_{x\to a} g(x)} = \frac{F}{G}.$$

   As we will see later, we have to be a little careful with the limits of ratios of functions, particularly in cases where both $f(x)$ and $g(x)$ are either infinite or zero at the point where we are taking the limit.

Example 2.3 To see how these properties can be used, we shall calculate the limit

$$\lim_{x\to 2}\frac{3x^4}{1+\sqrt{x}}.$$

Using the rules for combining limits we get:

$$\lim_{x\to 2}\frac{3x^4}{1+\sqrt{x}} = \frac{\lim_{x\to 2} 3x^4}{\lim_{x\to 2}(1+\sqrt{x})} = \frac{3\lim_{x\to 2} x^4}{\lim_{x\to 2} 1 + \lim_{x\to 2}\sqrt{x}} = \frac{3\times 16}{1+\sqrt{2}} = \frac{32}{1+\sqrt{2}}.$$

Exercise 2.1.2 Determine if the following functions are nondifferentiable, and if so, state where and why:

$$a)\ |x+1|, \quad b)\ x|x|, \quad c)\ f(\theta) = \begin{cases} \arctan(\theta) & \theta \neq 0 \\ 1 & \theta = 0 \end{cases}.$$

Some limits require a little modification of our earlier definition. For example, how do we take the limit of $f(x)$ as $x \to \infty$? Infinity is tricky to deal with, but we can replace x in our original definition with an arbitrarily large number, say Ω, and ask that for any $x > \Omega$, the function $f(x)$ is some small distance away from the limit F. Formally, we can modify Definition 2.1 to give us Definition 2.2.

Definition 2.2 A function $f(x)$ defined on $x_0 < x < \infty$ tends to the limit F as $x \to \infty$ if there exists a number F such that for every number $\epsilon > 0$ we can find a number Ω such that $|f(x) - F| < \epsilon$ for $x > \Omega$.

Example 2.4 As an example, we can use Definition 2.2 to show that

$$\lim_{x\to\infty} \frac{1}{x} = 0.$$

According to Definition 2.2, we need to show that

$$\left|\frac{1}{x} - F\right| < \epsilon \qquad \text{for all values of} \qquad x > \Omega.$$

In this case, $F = 0$, and since we want to know the limit as $x \to \infty$, we can consider only positive values of x (this has the effect of allowing us to remove the modulus, or absolute, sign), so we now want to show that

$$\frac{1}{x} < \epsilon \qquad \text{for all values of} \qquad x > \Omega,$$

or in other words, we need to show $x > 1/\epsilon$. This looks similar to the right-hand condition above, so if we pick $\Omega = 1/\epsilon$, then

$$\frac{1}{x} < \epsilon \quad \text{for all values of } \Omega.$$

So, we have proven our result.

There are some other subtleties concerning taking limits that we have to consider. For example, let us find the following limits:

$$\lim_{x\to-1} \frac{x^2 - 1}{x + 1} \qquad \text{and} \qquad \lim_{x\to\infty} \frac{4x^2 - 1}{12x^2 + 7}. \tag{2.8}$$

In the first case, we can see an immediate problem: both the numerator and denominator are zero at $x = -1$, so we are trying to divide zero by zero. In the second case, both the numerator and the denominator are infinite at $x = \infty$, so we are trying to divide infinity by infinity. These are called *indeterminate forms* because they lead to a limit that we cannot directly evaluate. In some cases, however, we can manipulate the expressions to obtain something we can calculate. For example, we can write

$$\lim_{x\to-1} \frac{x^2 - 1}{x + 1} = \lim_{x\to-1} \frac{(x - 1)(x + 1)}{x + 1} = \lim_{x\to-1} (x - 1) = -2$$

and, for the second case,

$$\lim_{x\to\infty} \frac{4x^2 - 1}{12x^2 + 7} = \lim_{x\to\infty} \frac{4 - (1/x^2)}{12 + (7/x^2)} = \frac{4}{12} = \frac{1}{3}.$$

Indeterminate forms involve ratios of functions whose limits involve combinations of 0 and ∞ in the numerator and denominator. The two limits in Equation (2.8) are indeterminate forms of the type 0/0 and ∞/∞, but we can also have mixtures (e.g., $0/\infty$ or $\infty/0$); however, these tend to be less common.

Exercise 2.1.3 Find the limits of

$$\text{a. } \lim_{x\to 1} \frac{x^2 - 1}{x - 1}, \quad \text{b. } \lim_{x\to 0} \frac{(x + 27)^{1/3} - 3}{x}.$$

(Hint: for (b), try using a substitution $y = (x + 27)^{1/3}$).

We have seen that differentiation involves taking a limit, and sometimes that limit does not exist and we are unable to take the derivative. But we know what types of function have these problems, discontinuous and nonsmooth functions, so we know when we need to be cautious. We will now look at how we calculate derivatives of more complicated functions.

2.2 Rules for Differentiation

We can, in principle, continue to calculate derivatives in the same way that we used to obtain Equation (2.2), but it rapidly becomes cumbersome and tedious as the functions become more complicated. Instead, we make use of the properties of the derivative and rules for taking the derivative of more general functions. For example, instead of using our first-principles method to calculate the derivative of $y(x) = x^4$, $y(x) = x^5$, and so on, we can calculate the derivative of the more general equation $y(x) = ax^n$, where a and n are constants. This gives

$$\frac{dy}{dx} = \frac{d}{dx}x^n = nx^{n-1}, \tag{2.9}$$

which allows us to calculate the derivative of any power of x. A list of derivatives of other common functions is given in Appendix B.

Taking the derivative of a function is a *mathematical operation*, and it has properties that are useful for calculating the derivatives of more complicated expressions. If $f(x)$ and $g(x)$ are two differentiable functions, and a is a constant, then

1. The derivative of a function multiplied by a constant is the product of the constant with the derivative of the function

$$\frac{d}{dx}(af(x)) = a\frac{df}{dx}.$$

2. The derivative of the sum (or difference) of two differentiable functions is the sum (or difference) of the derivatives of the functions

$$\frac{d}{dx}(f(x) \pm g(x)) = \frac{df(x)}{dx} \pm \frac{dg(x)}{dx}.$$

3. The derivative of a constant is zero

$$\frac{da}{dx} = 0.$$

Example 2.5 To see how these rules can help us, let us calculate the derivative of $y(x) = 3x^2 + 4\sin(x) + 2$. Using the properties of the derivative listed above and the derivatives in Appendix B, we get

$$\frac{dy}{dx} = 3\frac{dx^2}{dx} + 4\frac{d\sin(x)}{dx} + \frac{d}{dx}(2) = 3 \times 2x + 4 \times \cos(x) + 0 = 6x + 4\cos(x).$$

These rules are useful for calculating derivatives of sums of functions, but they do not help us to differentiate functions that are formed from the product of other functions, or that are functions of other functions, such as

$$y(\theta) = \sin(\theta)\cos(\theta) \quad \text{or} \quad y(x) = x^3 \exp(4x^2 + 2x - 3).$$

To differentiate these expressions we need to make use of the *product rule* and *chain rule*.

2.2.1 Product Rule

The product rule allows us to take the derivative of a function that can be written as the product of two different functions. If $y(x) = u(x)v(x)$, then

$$\frac{dy}{dx} = \frac{d}{dx}u(x)v(x) = u(x)\frac{dv}{dx} + v(x)\frac{du}{dx}. \tag{2.10}$$

To prove this equation, we make use of Equation (2.3) and the properties of limits.

Proof Using Equation (2.3) we have

$$\begin{aligned}
\frac{d}{dx}u(x)v(x) &= \lim_{\Delta x \to 0} \frac{u(x+\Delta x)v(x+\Delta x) - u(x)v(x)}{\Delta x} \\
&= \lim_{\Delta x \to 0} \frac{u(x+\Delta x)v(x+\Delta x) - u(x)v(x) + u(x+\Delta x)v(x) - u(x+\Delta x)v(x)}{\Delta x} \\
&= \lim_{\Delta x \to 0} \frac{u(x+\Delta x)(v(x+\Delta x) - v(x)) + v(x)(u(x+\Delta x) - u(x))}{\Delta x} \\
&= \lim_{\Delta x \to 0} u(x+\Delta x)\left(\frac{v(x+\Delta x) - v(x)}{\Delta x}\right) + \lim_{\Delta x \to 0} v(x)\left(\frac{u(x+\Delta x) - u(x))}{\Delta x}\right) \\
&= \left(\lim_{\Delta x \to 0} u(x+\Delta x)\right)\left(\lim_{\Delta x \to 0} \frac{v(x+\Delta x) - v(x)}{\Delta x}\right) \\
&\quad + \left(\lim_{\Delta x \to 0} v(x)\right)\left(\lim_{\Delta x \to 0} \frac{u(x+\Delta x) - u(x))}{\Delta x}\right) \\
&= u(x)\frac{dv(x)}{dx} + v(x)\frac{du(x)}{dx}.
\end{aligned}$$

Notice that in the second line we have made use of a common trick in mathematical proofs; we have added zero $(u(x+\Delta x)v(x) - u(x+\Delta x)v(x))$ to the equation.

Example 2.6 As an example of the product rule, let us calculate the derivative of $y(\theta) = \sin(\theta)\cos(\theta)$. Comparing our function with Equation (2.10), we can set $u(\theta) = \sin(\theta)$ and $v(\theta) = \cos(\theta)$ so that

$$\frac{d}{d\theta}\sin(\theta)\cos(\theta) = \sin(\theta)\frac{d}{d\theta}\cos(\theta) + \cos(\theta)\frac{d}{d\theta}\sin(\theta),$$

and using the rules for differentiating trigonometric functions (see Appendix B) we get

$$\frac{d}{d\theta}\sin(\theta)\cos(\theta) = -\sin^2(\theta) + \cos^2(\theta).$$

Exercise 2.2.1 Use the product rule to calculate the derivatives of the following functions:

a. $f(x) = x^4\sin^2(x)$, b. $f(x) = \tan(x)\sin(x)$, c. $f(x) = x^2e^x\cos(x)$, d. $f(x) = \dfrac{\sin(x)}{x^2}$

(Hint: for (d), write $1/x^2$ as x^{-2}).

2.2.2 Chain Rule

The second important rule for taking a derivative is the *chain rule*. This is used when the function we want to find the derivative of is itself a function of something else. For example, we can think of the function $y(x) = (x+2)^3$ as being $y(u(x)) = u(x)^3$, where $u(x) = x + 2$. If $y = f(u(x))$, then the chain rule states that

$$\frac{d}{dx}f(u(x)) = \frac{df(u)}{du}\frac{du(x)}{dx}. \tag{2.11}$$

Example 2.7 We can use the chain rule to calculate the derivative of $y(x) = (x+2)^3$ with respect to x. If we let $u(x) = x + 2$, then $y = u^3$ and

$$\frac{du}{dx} = 1, \quad \text{and} \quad \frac{dy}{du} = 3u^2,$$

so, using the chain rule, we have

$$\frac{dy(u(x))}{dx} = \frac{dy}{du}\frac{du}{dx} = 3u^2 \times 1 = 3(x+2)^2.$$

Example 2.8 Let us consider an example that is a little more complicated and which requires us to use both the product rule and the chain rule. We will calculate the derivative of $y(x) = x^3\exp(4x^2 + 2x - 3)$ with respect to x. Starting with the product rule, if $y(x) = u(x)v(x)$, then we can let $u(x) = x^3$ and $v(x) = \exp(4x^2 + 2x - 3)$. The product rule then tells us that

$$\frac{dy}{dx} = x^3\frac{d}{dx}e^{4x^2+2x-3} + e^{4x^2+2x-3}\frac{dx^3}{dx} = x^3\frac{d}{dx}e^{4x^2+2x-3} + 3x^2e^{4x^2+2x-3}.$$

We can use the chain rule to calculate the remaining derivative by setting $w(x) = 4x^2 + 2x - 3$ so that

$$\frac{d}{dx}(e^{4x^2+2x-3}) = \left(\frac{d}{dw}e^w\right)\frac{dw(x)}{dx} = e^w(8x+2) = (8x+2)e^{4x^2+2x-3}.$$

Putting everything together and collecting terms we get

$$\begin{aligned}\frac{d}{dx}x^3\exp(4x^2+2x-3) &= (8x+2)\exp(4x^2+2x-3)+3x^2\exp(4x^2+2x-3)\\ &= (3x^2+8x+2)\exp(4x^2+2x-3).\end{aligned}$$

Exercise 2.2.2 Use the chain rule to calculate the derivatives of the following functions:

$$\text{a. } f(x)=(6x^3+2x-1)^3,\quad \text{b. } f(x)=\sin(2x+3),\quad \text{c.} f(x)=\mathrm{e}^{\sin(3x^2-1)}.$$

We can also combine the product rule and the chain rule to determine a formula for taking the derivative of the quotient of two functions. If $y(x)=u(x)/v(x)$, then

$$\begin{aligned}\frac{d}{dx}\frac{u(x)}{v(x)} &= \frac{d}{dx}uv^{-1} = v^{-1}\frac{du}{dx}+u\frac{d}{dx}v^{-1} = \frac{1}{v(x)}\frac{du}{dx}+u\frac{d(v^{-1})}{dv}\frac{dv}{dx}\\ &= \frac{1}{v(x)}\frac{du}{dx}-u(x)v^{-2}\frac{dv}{dx} = \frac{1}{v(x)}\frac{du}{dx}-\frac{u}{v(x)^2}\frac{dv}{dx}.\end{aligned}\tag{2.12}$$

Exercise 2.2.3 Calculate the derivatives of the following functions:

$$\text{a. } f(x)=\sin(x^2-1)\mathrm{e}^{x^2},\quad \text{b. } f(x)=\frac{\cos(3x+2)}{\sin(3x-2)},\quad \text{c. } f(x)=\frac{\mathrm{e}^{3x^2-2}}{(x^2-1)}.$$

2.2.3 Higher-Order Derivatives

As we have seen, the derivative of a function $y(t)$ with respect to t tells us how fast y changes as t changes for any value of t. If $y(t)$ represents the distance of an object from a given point and t is the time, then dy/dt is the velocity $v(t)$ of the object. Using Equation (2.3) we can write

$$v(t)=\frac{dy}{dt}=\lim_{\Delta t\to 0}\frac{y(t+\Delta t)-y(t)}{\Delta t}=\lim_{\Delta t\to 0}\frac{\Delta y}{\Delta t},$$

and because Δy is a distance and has dimensions [L], and Δt is a time and has dimensions [T], so $\Delta y/\Delta t$ has dimensions of $[\mathrm{L}][\mathrm{T}]^{-1}$ and is indeed a velocity; the mathematical process of taking the limit does not affect the dimensions of quantities.

We can now take the derivative of $v(t)$,

$$a(t)=\frac{dv}{dt}=\frac{d}{dt}\left(\frac{dy}{dt}\right)=\frac{d^2y}{dt^2}=\lim_{\Delta t\to 0}\frac{1}{\Delta t}\left(\frac{\Delta y}{\Delta t}\right).$$

This is called the *second-order derivative* of y with respect to t. Since we are thinking of y as a distance and t as being time, this has the dimensions of an acceleration. So, we have seen that the dimensions of the quantities in a derivative can help us interpret its meaning.

Exercise 2.2.4 Use the dimensions of the length y and the time t to show that the second-order derivative d^2y/dt^2 has the dimensions of an acceleration.

Exercise 2.2.5 If E represents the energy of an object and t represents time, use a dimensional argument to determine what quantity is represented by the derivative dE/dt.

We can continue in this way and take the derivative of $a(t)$, which would also be the second-order derivative of $v(t)$ and the third-order derivative of $y(t)$. There are times when a higher-order derivative of a function is zero, in which case we cannot continue to take derivatives. For example, if $y(x) = 5x^2 - 2x + 1$, then the first, second, and third derivatives of $y(x)$ are

$$\frac{dy}{dx} = 10x - 2$$
$$\frac{d^2y}{dx^2} = \frac{d}{dx}\left(\frac{dy(x)}{dx}\right) = \frac{d}{dx}(10x - 2) = 10$$
$$\frac{d^3y}{dx^3} = \frac{d}{dx}\left(\frac{d^2y}{dx^2}\right) = \frac{d}{dx}10 = 0.$$

Exercise 2.2.6 Calculate the first and second derivatives of the following functions

$$\text{a. } f(x) = e^x \sin(x), \quad \text{b. } f(x) = \frac{2x^2 - x}{x^2 - 1}, \quad \text{c. } f(x) = \sin(x)\cos(x).$$

There are different ways of writing a derivative that you will come across. Up until now we have used the standard notation, but you will also see a superscript prime being used:

$$\frac{dy(x)}{dx} = f'(x), \quad \left.\frac{dy}{dx}\right|_{x=a} = y'(a), \quad \frac{d^2y}{dx^2} = y''(x), \quad \frac{d^ny}{dx^n} = y^{(n)}(x).$$

If it is obvious what we are differentiating with respect to, then the notation is often abbreviated further so that $y''(x)$ is written as just y'' and $y^{(n)}(x)$ as just $y^{(n)}$. Sometimes, especially for derivatives with respect to time, a dot notation is used instead of a prime:

$$\frac{dy(t)}{dt} = \dot{y}(t), \quad \left.\frac{dy}{dt}\right|_{t=\tau} = \dot{y}(\tau), \quad \frac{d^2y}{dt^2} = \ddot{y}(t).$$

We have seen that knowing the dimensions of quantities can help us interpret the meaning of a derivative. But we have also seen that a first-order derivative can be interpreted geometrically as the slope of a curve. Similarly, the second derivative is the slope of the curve represented by the first derivative of $y(t)$; in other words, it is a measure of how rapidly, and in what direction, the slope of $y(t)$ is changing. If the original function $y(t)$ is gently curving upwards, then the second derivative will be positive (the slope of $y(t)$ is increasing) and small. However, if $y(t)$ rapidly gets steeper and steeper, we should expect the second derivative to be large and positive. The second derivative is therefore telling us something about the *curvature* of $y(t)$: a gentle curve will have a small second derivative, whereas a sharp curve will have a larger second derivative.

2.3 Maxima and Minima

When the derivative of a function changes sign, there is a point at which the derivative is zero. This is called a *turning point* and can be either a maximum or a minimum of the function. To find the maximum or minimum of a function $y(x)$ we calculate its derivative, set the derivative to zero, and solve the resulting equation for x. This gives the values of x where the maxima or minima occur (there could be more than one of each), and to get the corresponding values of y we just substitute these values of x back into the original equation.

Example 2.9 To find the turning points of the curve $y(x) = x^3 - 4x^2 + 6$ we first calculate the derivative $y'(x)$ and set it to zero, giving

$$y'(x) = 3x^2 - 8x = x(3x - 8) = 0.$$

This equation has two solutions, $x = 0$ and $x = 8/3$, and these are where the turning points of $y(x)$ are located on the x-axis. Substituting these values into the original equation tells us the corresponding y values. Doing this, we find the turning points are at $(x, y) = (0, 6)$ and $(x, y) = (8/3, -94/27)$.

Now we have seen how to calculate the locations of the turning points, we should determine whether a given turning point is a maximum or a minimum. There are a couple of ways to do this. If we look at Figure 2.8a, we can see that for a maximum point $(x_{\text{max}}, y_{\text{max}})$ the slope of the curve is positive for $x < x_{\text{max}}$ and negative for $x > x_{\text{max}}$. The conditions are reversed for a minimum. So, we just have to look at how the sign of the derivative changes as x changes from just smaller than x_{max} to being just a little larger than x_{max}; we do not want to pick values of x too far from x_{max} in case we jump beyond the next turning point. For example, for the turning points in Example 2.9 we can evaluate the derivative at points $x \pm 0.5$, so that

$$\left.\frac{dy}{dx}\right|_{x=-0.5} = 2.75 \qquad \left.\frac{dy}{dx}\right|_{x=0.5} = -1.25 \qquad \text{for turning point } (x, y) = (0, 6)$$
$$\left.\frac{dy}{dx}\right|_{x=2.167} = -3.25 \qquad \left.\frac{dy}{dx}\right|_{x=3.167} = 4.75 \qquad \text{for } (x, y) = (8/3, -94/27),$$

from which we see that the point $(x, y) = (0.6)$ is a maximum and the point $(x, y) = (8/3, -94/27)$ is a minimum (Figure 2.8a).

There is another way we can discover the nature of a turning point, which is sometimes easier to calculate, and that is by looking at the second derivative of the function. Recall that the second derivative is the rate of change of the slope of the original function. At a maximum, we know that the derivative changes from positive to negative as x increases, so the rate of change of the slope of the function is negative. Conversely, at a minimum of the function, the slope of the function changes from negative to positive, so it is increasing, implying that the second derivative of the function is positive. For example, the second derivative of the function used in Example 2.9 is $y''(x) = 6x - 8$. When $x = 0$, $y''(x) = -8 < 0$, so there is a maximum of the function at $x = 0$. Similarly, when $x = 8/3$, $y'' = +8 > 0$, indicating there is a minimum at this point. We can see this clearly if we plot the function used in Example 2.9, its first derivative, and second derivative (Figure 2.8b).

In addition to a maximum or minimum, a curve may also have *points of inflection*. These are places where the curve changes from being concave to convex, or vice versa. A function

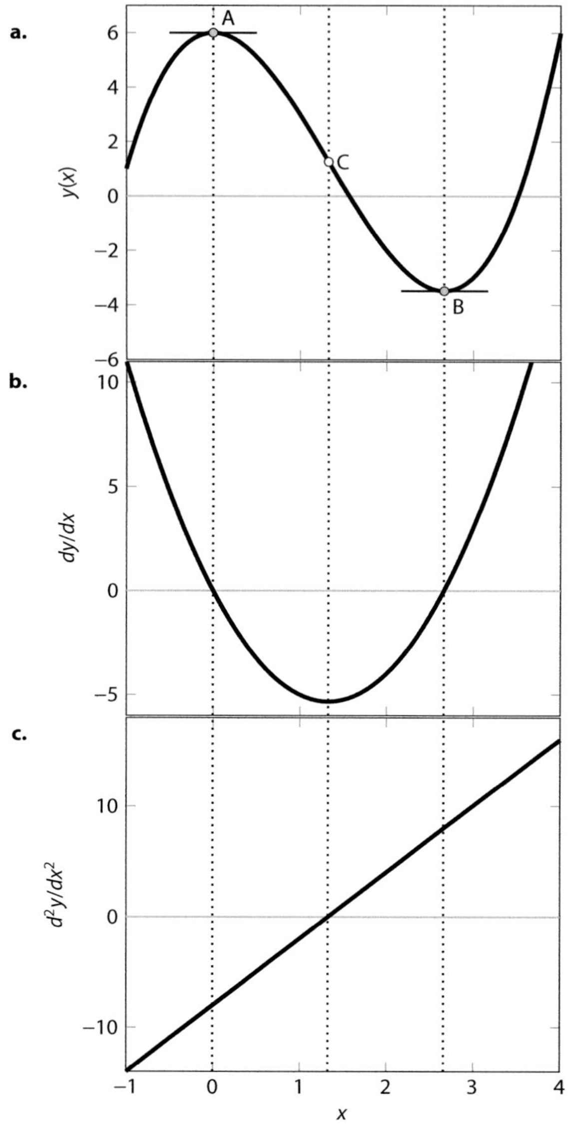

Figure 2.8 The function $y(x) = x^3 - 4x^2 + 6$ and its first and second derivative. **a.** shows the function with a local maximum at A and local minimum at B (the straight lines are tangents to the curve) and a point of inflection at C. **b.** shows that the first derivative of $y(x)$ is 0 at the maximum and minimum points, and is a minimum at the inflection point. **c.** shows that the second derivative is negative for the maximum of $y(x)$, negative for the minimum, and 0 at the point of inflection.

is *convex* if the straight line joining any two points on the curve lies either on the curve or above it (Figure 2.9). Similarly, a curve is *concave* if the line joining any two points on the curve lies either on the curve or below it. A curve can have convex portions and concave portions, and the point where the curve switches between the two is a turning point called

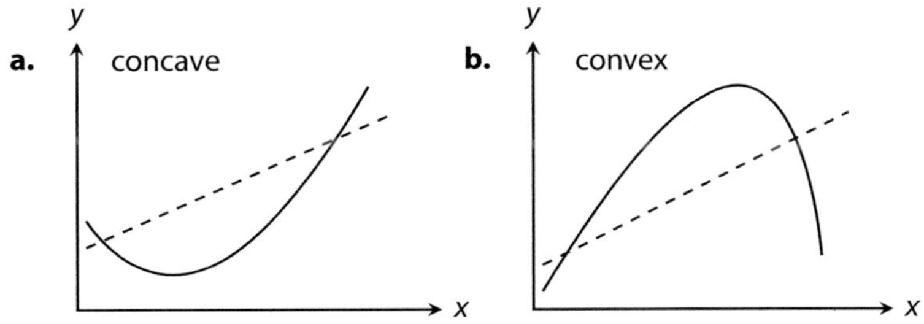

Figure 2.9 Illustrations of **a.** a concave function and **b.** a convex function.

the *point of inflection*. If we look at the function used in Example 2.9, we can see that there should be a point of inflection somewhere between the maximum and minimum of the curve. This means that the derivative should change from decreasing to increasing; in other words, the second derivative should change sign. The second derivative of the function used in Example 2.9 is

$$\frac{d^2y}{dx^2} = 6x - 8,$$

and this is zero when $x = 4/3$, which is where the point of inflection is.

Exercise 2.3.1 Find the turning points in the following functions and classify them as maxima, minima, or points of inflection:

$$\text{a. } f(x) = 3x^5 - 5x^3, \quad \text{b. } f(x) = x^3 - 12x^2 + 36x - 18.$$

Exercise 2.3.2 The air temperature during the day at a specified location is given by

$$T(t) = 21 - 10\sin\left(\frac{2\pi}{24}(t - \phi)\right),$$

where t is the time and ϕ is a phase offset. Find equations for the maximum and minimum temperatures by differentiating $T(t)$ with respect to time. If the maximum temperature occurs at 3:00 p.m., what is the value of ϕ?

2.4 Some Theorems About Derivatives

Derivatives play a large role in our science, so it is worthwhile understanding them a bit more. To begin with, there are two theorems that concern functions and their derivatives that we will find particularly useful later on. As a forewarning, some of the theorems we meet may at first sight appear to be obvious. However, there are reasons for introducing them. Firstly, though they may seem obvious, there are special cases where these theorems do not hold and we should be aware of these. Secondly, proving theorems often introduces us to techniques that can be useful in solving a wider range of problems.

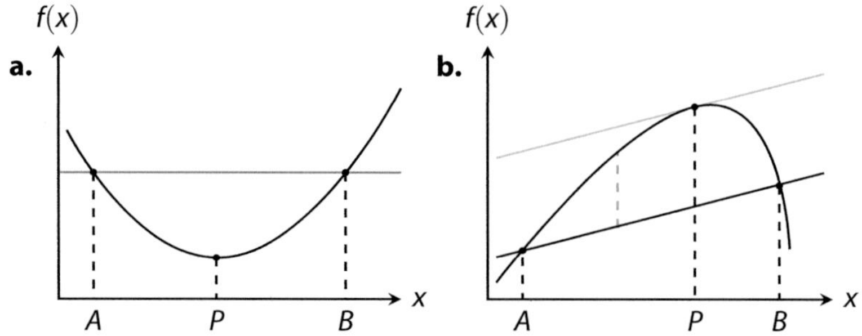

Figure 2.10 **a.** An example of Rolle's theorem. The function $f(x)$ has the same value at points A and B, and has a turning point (in this case a minimum) at point P, in between A and B. **b.** An illustration of the mean value theorem for derivatives. The average slope of the curve between points A and B is given by the line connecting the points $f(A)$ and $f(B)$ (the black line). Between points A and B there will be a point (P) at which the slope of the curve (the gray line is the tangent to the curve at point **P.** equals the slope of the line connecting points A and B.

The first theorem we will look at is called *Rolle's theorem* (Theorem 2.3).[2]

Theorem 2.3 (Rolle's Theorem) *If the function $f(x)$ is continuous between $a \leq x \leq b$ and differentiable between $a < x < b$, and $f(a) = f(b)$, then there exists a point $x = c$ such that $a < c < b$ for which $f'(c) = 0$.*

This is basically saying that if the value of a function is the same at two points ($x = a$ and $x = b$), then either the function is a constant between those points, or it has a maximum, or a minimum at some point in between (Figure 2.10a). So, why is this interesting? Let us look at a function that does not satisfy the conditions of the theorem, $f(x) = (x^2)^{1/5}$ on the interval $-1 \leq x \leq 1$ (Figure 2.11). We can see that $f(-1) = f(1) = 1$, but we know the function is not constant between these two x values. The function is continuous throughout the range of x values, but the derivative $f'(x) = (2/5)x^{-3/5}$ becomes infinite at $x = 0$, so the function is not differentiable everywhere between $x = -1$ and $x = 1$; we also cannot find a solution to the equation $f'(x) = (2/5)x^{-3/5} = 0$, which we require for there to be a maximum or minimum.

Rolle's theorem can be useful when we are looking for the roots of an equation; the roots of an equation are the solutions of $f(x) = 0$. Let us assume that $f(x)$ is continuous and differentiable for all values of x. Then, if there are two roots $x = a$ and $x = b$ $(a < b)$ to the equation, $f(a) = f(b) = 0$, so all the conditions of Rolle's theorem are satisfied. Then there has to be a value of x, call it $x = c$, such that $a \leq c \leq b$ where the derivative of $f(x)$ is zero. In other words, if the equation $f(x) = 0$ has more than one root, then it must have a maximum or minimum somewhere (i.e., $f'(x) = 0$), and this can often be easy to check.

A related but slightly less obvious theorem is the mean value theorem for derivatives (Theorem 2.4).

[2] This theorem is named after the French mathematician Michel Rolle (1652–1719), but his proof was only for functions that are polynomials.

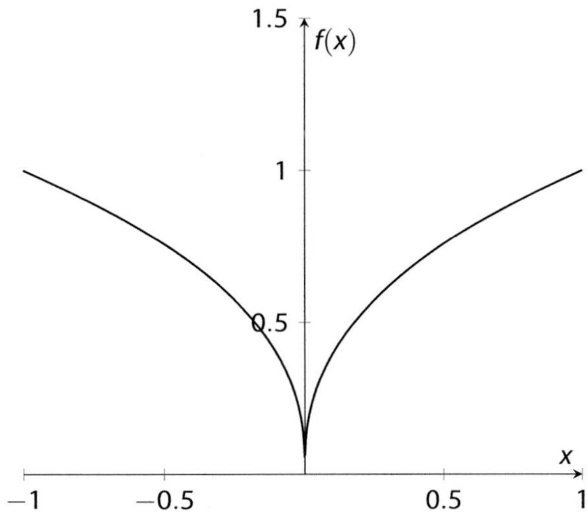

Figure 2.11 A plot of the function $f(x) = (x^2)^{1/5}$, for which Rolle's theorem does not hold.

Theorem 2.4 (Mean Value Theorem) *If the function $f(x)$ is continuous between $a \le x \le b$ and differentiable between $a < x < b$, then there is a point $x = c$ such that $a < c < b$ at which*

$$f'(c) = \frac{f(b) - f(a)}{b - a}.$$

What is Theorem 2.4 telling us? The quantity $(f(b) - f(a))/(b - a)$ is the slope of the straight line connecting points $(x, y) = (a, f(a))$ and $(x, y) = (b, f(b))$; this line is called the *secant line*. So, Theorem 2.4 is telling us that, if the function $f(x)$ is continuous and differentiable, then there is a point $x = c$ that lies between $x = a$ and $x = b$ at which the slope of the curve $y = f(x)$ is the same as the slope of the secant line connecting a and b. We can see that Rolle's theorem is a special case of the mean value theorem. It is useful to see how the proof of the mean value theorem works. We are going to use Rolle's theorem to do this, but Rolle's theorem only applies if $f(a) = f(b)$, which is not necessarily the case here. We are going to employ a useful problem-solving technique, which is to create a new function that *does* satisfy the required conditions. Our sticking point is that the secant line does not satisfy $f(a) = f(b)$, but the difference between the secant line and the function $f(x)$ (the gray, dashed line in Figure 2.10b) does satisfy these conditions. If (x, y) is a point on the secant line, then

$$\frac{y - f(a)}{x - a} = \frac{f(b) - f(a)}{b - a} \implies y = f(a) + \frac{f(b) - f(a)}{b - a}(x - a).$$

The difference between the secant line and the function $f(x)$ is then

$$g(x) = f(x) - y = f(x) - f(a) - \frac{f(b) - f(a)}{b - a}(x - a).$$

Exercise 2.4.1 Show that $g(a) = g(b) = 0$.

What is more, since $f(x)$ and $y(x)$ are both continuous and differentiable functions, all the conditions of Rolle's theorem are satisfied. Therefore, we know that there is a value $x = c$ such that $a < c < b$ and $g'(c) = 0$. We can differentiate $g(x)$ and evaluate it at $x = c$ to get

$$g'(c) = f'(c) - \frac{f(b) - f(b)}{b - a} = 0,$$

which proves Theorem 2.4.

The mean value theorem is one of the most useful theorems in calculus, because it allows us to prove other useful theorems (as we do in the next section), and it is also very useful in helping us understand the accuracy of numerical methods by placing bounds on mathematical expressions.

Example 2.10 We can use the mean value theorem to show that $|\sin(2\theta) - \sin(2\phi)| \leq 2|\theta - \phi|$. The inequality involves $\sin(2\theta)$ and $\sin(2\phi)$, so a good place to start is to look at the function $f(x) = \sin(2x)$, whose derivative is $f'(x) = 2\cos(2x)$. If we apply the mean value theorem to this function, we get

$$\frac{\sin(2\theta) - \sin(2\phi)}{\theta - \phi} = 2\cos 2\xi,$$

where ξ lies between θ and ϕ. However, we know that $|\cos(2\xi)| \leq 1$, so

$$\left| \frac{\sin(2\theta) - \sin(2\phi)}{\theta - \phi} \right| \leq 2,$$

and therefore

$$|\sin(2\theta) - \sin(2\phi)| \leq 2|\theta - \phi|.$$

This is a particular example of a more general inequality: if $f(x)$ is differentiable on $a \leq x \leq b$ and in that range, $|f'(x)| \leq M$, where M is some number, then $|f(b) - f(a)| \leq M|b - a|$.

Exercise 2.4.2 Use the mean value theorem to show that, if $\theta < \phi$, then

$$\frac{1}{1 + \phi^2} < \frac{\tan^{-1}(\phi) - \tan^{-1}(\theta)}{\phi - \theta} < \frac{1}{1 + \theta^2}.$$

2.5 Taylor's Theorem

Taylor's theorem[3] is arguably one of the most used mathematical tools in the sciences. It forms the basis for many important numerical methods and is used in analyzing the behavior of functions. Taylor's theorem will make appearances time and time again in our exploration, so it is worth spending some time examining it and understanding it.

[3] Named after the English mathematician Brook Taylor (1685–1731).

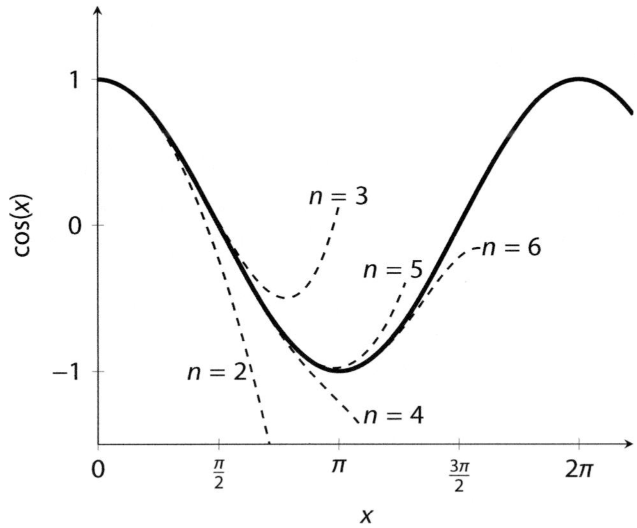

Figure 2.12 The effects of adding more terms to a Taylor series. The solid curve is the function $f(x) = \cos(x)$ and the dashed lines show Taylor series expansions about $x = 0$ containing two terms, three terms, and so on up to six terms. The more terms we add, the better the Taylor series approximates the function and for greater distances from $x = 0$. For example, the curve for $n = 3$ is a good approximation to about $x = \pi/2$, but the curve for $n = 5$ is a good approximation to $x = \pi$.

To introduce the theorem, let us imagine that we have a complicated-looking function $G(x)$ to deal with (Figure 2.12) and we would like to approximate it with a simpler one $g(x)$. There are many reasons we would want to do this: for example, it might be hard to compute the values of $G(x)$, or it might be difficult to differentiate. Polynomials are often used to approximate more complicated functions because they are easy to manipulate (i.e., add, multiply, etc.), differentiate, and as we will see, integrate. They also display a wide range of different shapes, from a straight line to curves with many turning points. The simplest polynomial is a constant ($g(x) = b_0$), which is not particularly interesting, but it does provide a baseline, or offset from zero. Next, we have first-order polynomials ($g(x) = b_0 + b_1 x$) obtained by adding a linear term (b_1 is a constant) to the constant. This is a straight line with a gradient b_1. By adding a quadratic term ($b_2 x^2$) we introduce curvature, and a cubic term ($b_3 x^3$) allows for more complicated behavior.

It is unlikely that $g(x)$ will be a good approximation to $G(x)$ for all values of x, but we might be able to make it a good approximation for some range of x values close to a particular point, say $x = a$. Let us start by looking for a polynomial in $(x - a)$, the distance from the point $x = a$. This means we want to have

$$G(x) \approx g(x) = b_0 + b_1(x - a) + b_2(x - a)^2 + b_3(x - a)^3 + \cdots , \tag{2.13}$$

where b_i ($i = 0, 1, 2, 3, \ldots$) are constants that we do not yet know. To find the values of b_i we will impose certain conditions on $g(x)$ that will define what we mean when we say that

$g(x)$ is a good approximation to $G(x)$. To start with, we would like $G(x)$ and $g(x)$ to have the same value at $x = a$. This will give us an equation for the value of b_0:

$$G(a) = g(a) = b_0 + b_1(a - a) + b_2(a - a)^2 + b_3(a - a)^3 + \cdots = b_0.$$

We also want the slopes of the two functions to be the same at $x = a$. It would not be a good approximation to have $G(a) = g(a)$ but with the slope of G increasing and that of g decreasing at $x = a$. The first derivative of the polynomial is

$$\frac{dg}{dx} = b_1 + 2b_2(x - a) + 3b_3(x - a)^2 + \cdots,$$

and setting this equal to the first derivative of $G(x)$ evaluated at $x = a$ gives

$$\left.\frac{dG}{dx}\right|_{x=a} = \left.\frac{dg}{dx}\right|_{x=a} = b_1.$$

Continuing with this line of reasoning, we would like the rate of change of the slopes of the two functions to be equal at $x = a$. That means that the second derivatives have to be equal at $x = a$, giving us

$$\frac{d^2G}{dx^2} = \frac{d^2g}{dx^2} = 2b_2 + (3 \times 2)b_3(x - a) + \cdots,$$

so that at $x = a$ we get

$$b_2 = \frac{1}{2}\left.\frac{d^2G}{dx^2}\right|_{x=a}.$$

Continuing on like this allows us to get as many b_i values that we need or have the patience to calculate. We can stop at any point we wish—for example at the $(x - a)^n$th term—and the end result will be a polynomial:

$$g(x) = G(a) + (x - a)\left.\frac{dG}{dx}\right|_{x=a} + \frac{(x - a)^2}{2!}\left.\frac{d^2G}{dx^2}\right|_{x=a} + \cdots + \frac{(x - a)^n}{n!}\left.\frac{d^nG}{dx^n}\right|_{x=a}. \quad (2.14)$$

You will sometimes see different names used for this expansion: some common ones are Taylor series, Taylor expansion, Taylor polynomial. The function $g(x)$ is a polynomial in x that approximates the function $G(x)$ in a region around $x = a$. A pertinent question to ask is, how good is this approximation? What is the effect of stopping the polynomial at the nth term? If $G(x)$ is itself a polynomial of degree n, then the $(n + 1)$th derivative of $G(x)$ is zero and the Taylor series ends at that nth term and contains a finite number of terms. In that case, $g(x)$ is a polynomial that is identical to $G(x)$. But, if $G(x)$ is not a polynomial, then we can have an infinite number of terms in the Taylor series. In this case, it makes sense to truncate the expansion at some point. However, in doing so we are approximating $G(x)$ by the polynomial $g(x)$, and if we added more terms to the series expansion, $g(x)$ would be a better approximation to $G(x)$. So, we need to determine how large an error we incur by only using a finite number of terms in the Taylor series. Taylor's theorem (Theorem 2.5) gives us an estimate for this remainder term, the difference between the polynomial approximation and the actual function.

Theorem 2.5 (Taylor's Theorem) *If $f(x)$ is any function that can be differentiated $(n+1)$ times in an interval $x = a$ to $x = a + h$, then*

$$f(x) = f(a) + f'(a)(x-a) + \frac{(x-a)^2}{2!}f''(a) + \cdots + \frac{(x-a)^n}{n!}f^{(n)}(a) + \frac{(x-a)^{n+1}}{(n+1)!}f^{(n+1)}(\xi), \quad (2.15)$$

where $a < \xi < x$.

The last term in Equation (2.15) is called the *remainder term* and represents an estimate of what is being left out by approximating the function as an nth order polynomial (see Box 2.1). If $a = 0$, then the Taylor expansion is known as a *Maclaurin series*.

Box 2.1 **Remainder Term**

To derive the remainder term in Taylor's theorem, we will look at a region around the point $x = a$, say from $x = a$ to $x = a + h$. Our original function is $f(x)$ and our approximating polynomial is $g(x)$. Let us call the remainder $R = f(x) - g(x)$. We know that at $x = a$ we have $R = f(a) - g(a) = 0$ because we have constructed $g(x)$ such that that is the case. At a distance h away from $x = a$, we have

$$R = f(a+h) - g(a+h).$$

We can try and derive an expression for R at $x = a + h$, but it is more useful to derive one for how R varies between $x = a$ and $x = a + h$. Looking at the last term in our Taylor polynomial (Equation (2.14)), we can guess that the next term, the $(n+1)$th term, would contain a factor $(x-a)^{n+1}$. So, we want a function that contains this term, but is zero at $x = a$ and has a value R at $x = a + h$. A simple possibility is (you should check this is indeed so)

$$f(x) - g(x) = R\frac{(x-a)^{n+1}}{h^{n+1}}.$$

Now we are going to doing a clever little trick. Rather than look at the function above, we are going to look at

$$F(x) = f(x) - g(x) - R\frac{(x-a)^{n+1}}{h^{n+1}}.$$

The reason for doing this is that $F(0) = F(a+h) = 0$. So now we can use Rolle's theorem to tell us that there is some point, $x = x_1$ between $x = a$ and $x = a + h$, where $F'(x) = 0$. This first derivative is

$$F'(x) = f'(x) - g'(x) - (n+1)R\frac{(x-a)^n}{h^{n+1}}.$$

But remember that we constructed our polynomial $g(x)$ such that $g'(a) = f'(a)$, so we see that $F'(a) = 0$, and we also know that there is a point $a < x_1 < a + h$ such that $F'(x) = 0$. So, we can apply Rolle's theorem again to say that there is some point (x_2) that lies between $x = a$ and $x = x_1$ such that $F''(x_2) = 0$:

$$F''(x) = f''(x) - g''(x) - n(n+1)R\frac{(x-a)^{n-1}}{h^{n+1}},$$

and using the same kind of argument, we see that there is a point (x_3) between $x = a$ and $x = x_2$ such that $F'''(x_3) = 0$. We can continue like this until we get to $F^{(n+1)}$, where things are slightly different. This is because

Equation (2.14) contains terms only up to the nth order, so the $(n+1)$th derivative of $g(x)$ is zero. Rolle's theorem tells us that $F^{(n+1)}(x_{n+1}) = 0$, so

$$F^{(n+1)}(x_{n+1}) = f^{(n+1)}(x_{n+1}) - \frac{R(n+1)!}{h^{n+1}} = 0,$$

which tells us that

$$R = \frac{h^{n+1}}{(n+1)!} f^{(n+1)}(x_{n+1}).$$

Now, we do not know the value of x_{n+1}, but that is all right; we know that it lies between $x = a$ and $x = a + h$, and we can just give it a name, call it ξ. So, the remainder term looks like

$$R = \frac{h^{n+1}}{(n+1)!} f^{(n+1)}(\xi).$$

Taylor's theorem is very useful for creating approximations to functions, and some of the more important ones are listed in Appendix B.

Example 2.11 Let us calculate the Taylor series of $y(x) = \mathrm{e}^x$ about the point $x = 0$. We obtain the Taylor expansion by taking successively higher-order derivatives and evaluating them at $x = 0$. The derivatives of $y(x) = \mathrm{e}^x$ are $y'(x) = y''(x) = \cdots = y^{(n)}(x) = \mathrm{e}^x$, which, evaluated at $x = 0$ are all equal to 1. So Taylor's theorem says that

$$y(x) = 1 + x + \frac{x^2}{2!} + \frac{x^3}{3!} + \cdots = \sum_{n=0}^{\infty} \frac{x^n}{n!}.$$

Example 2.12 The Taylor expansion of $y(x) = \mathrm{e}^{-x}$ about the point $x = 0$ is very similar to the Example 2.11, except that the derivatives now alternate in sign:

$$y(x=0) = 1, \quad y'(x=0) = -1, \quad y''(x=0) = 1, \quad y'''(x=0) = -1, \quad \text{etc.},$$

so the Taylor expansion looks like

$$y(x) = 1 - x + \frac{x^2}{2!} - \frac{x^3}{3!} + \cdots = \sum_{n=0}^{\infty} (-1)^n \frac{x^n}{n!}.$$

Example 2.13 We can often use a Taylor series that we know to calculate expansions of new functions. For example, the Taylor expansion of $y(x) = x^5\mathrm{e}^{-x}$ about $x = 0$ can be obtained from the expansion of e^{-x} by simply multiplying by x^5:

$$x^5\mathrm{e}^{-x} = x^5 \sum_{n=0}^{\infty} (-1)^n \frac{x^n}{n!} = \sum_{n=0}^{\infty} (-1)^n \frac{x^{n+5}}{n!}.$$

Notice in Example 2.13 that we used the Taylor expansion as if it were just another function, multiplying it by x^5. In general, we can treat Taylor expansions as if they were functions, and we can add, subtract, and multiply them to obtain expansions of more complicated functions.

Exercise 2.5.1 Show that the Taylor expansion of the polynomial $y(x) = 3x^3 + 2x^2 + x - 6$ about $x = a$ is exact; i.e., that there is no remainder term.

Exercise 2.5.2 Find the first three terms of the Taylor expansions of the following functions about $x = 0$:

$$\text{a. } y(x) = \sin(x), \quad \text{b. } y(x) = e^x \sin(x), \quad \text{c. } y(x) = \sin(x)\cos(x).$$

The Taylor series allows us to define an *analytic function*. These are functions that are infinitely differentiable. Remember that if we could calculate all the infinite number of terms in a Taylor series for a function $f(x)$ near $x = x_0$, then the polynomial we would get from the Taylor series would give the same values as the function. But to do this, we would need to differentiate $f(x)$ an infinite number of times. So, an analytic function can be represented exactly as an infinite Taylor series. Although polynomials do not require an infinite number of terms in their Taylor series, they can be represented exactly by a finite Taylor series, so they are analytic. As you might guess, functions that are discontinuous or not differentiable (e.g., $f(x) = |x|$) are not analytic. So, while most functions we come across in our science are analytic, some that we will need are not.

2.6 L'Hôpital's Rule

Derivatives can help us evaluate the limit of an indeterminate form. Recall from Section 2.1 that this occurs when we want to take the limit

$$\lim_{x\to x_0} \frac{f(x)}{g(x)},$$

where $\lim_{x\to x_0} f(x)$ and $\lim_{x\to x_0} g(x)$ are either 0 or ∞. In Section 2.1 we found that in some cases we could evaluate an indeterminate limit by algebraically manipulating $f(x)/g(x)$. But this is not always possible, and this is where l'Hôpital's rule[4] comes to the rescue. Informally, this rule states that if we have an indeterminate form of the type 0/0 or ∞/∞, then given certain conditions,

$$\lim_{x\to x_0} \frac{f(x)}{g(x)} = \lim_{x\to x_0} \frac{f'(x)}{g'(x)}. \tag{2.16}$$

Example 2.14 Let us use l'Hôpital's rule to evaluate the limit

$$\lim_{\theta\to 0} \frac{\sin(\theta)}{\theta},$$

[4] Named after the French mathematician Guillaume François Antoine, Marquis de l'Hôpital. Although he was not actually the first to derive it, he was the first to publish it.

which is an indeterminate form of the type 0/0 because both $\sin(\theta)$ and θ are 0 at $\theta = 0$. So, using l'Hôpital's rule, we have

$$\lim_{\theta\to 0} \frac{\sin(\theta)}{\theta} = \lim_{\theta\to 0} \frac{\cos(\theta)}{1} = \frac{1}{1} = 1.$$

Example 2.15 Sometimes we have to apply l'Hôpital's rule more than once. For example, the limit

$$\lim_{x\to\infty} \frac{e^x}{x^2}$$

is an indeterminate form of the type ∞/∞, so applying l'Hôpital's rule gives

$$\lim_{x\to\infty} \frac{e^x}{x^2} = \lim_{x\to\infty} \frac{e^x}{2x},$$

which is also an indeterminate form of the type ∞/∞. So, applying l'Hôpital's rule a second time gives

$$\lim_{x\to\infty} \frac{e^x}{x^2} = \lim_{x\to\infty} \frac{e^x}{2x} = \lim_{x\to\infty} \frac{e^x}{2} = \infty.$$

This is all very nice, but why does l'Hôpital's rule work? Why does taking the derivative allow us to calculate the limit, and why should the ratio of the derivatives have the same limit as the ratio of the original functions? To gain some insight into this, let us consider two functions $f(x)$ and $g(x)$ that are both differentiable at $x = x_0$. Taylor's theorem states that

$$f(x) = f(x_0) + (x - x_0)f'(x_0) + R_1 \quad \text{and} \quad g(x) = g(x_0) + (x - x_0)g'(x_0) + P_1, \tag{2.17}$$

where R_1 and P_1 are the remainder terms we obtain when we use the Taylor expansion only up to the first derivative. As we let x get closer and closer to x_0, both remainder terms get smaller and smaller. We know this because we have said that, for example, $f(x)$ is differentiable, so rearranging the equation gives

$$f'(x_0) = \frac{f(x) - f(x_0)}{x - x_0} - \frac{R_1}{x - x_0}.$$

But the definition of the derivative (Equation (2.3)) tells us that

$$f'(x_0) = \lim_{x\to x_0} \frac{f(x) - f(x_0)}{x - x_0},$$

so R_1 (and P_1) must get smaller faster than $(x - x_0)$ in order that the two expressions for the derivative give the same answer. Recall that this is a condition we used in deriving the Taylor series expansion. This implies that as x gets closer and closer to x_0, the function $y = f(x)$ looks more and more like its tangent line. As R_1 gets smaller and smaller, Equation (2.17) looks more and more like the equation of the tangents to $y = f(x)$ and $y = g(x)$ at $x = x_0$. If we have an indeterminate form where $f(x_0) = g(x_0) = 0$, then we can use the Taylor expansion of both functions so that near $x = x_0$

$$\frac{f(x)}{g(x)} = \frac{f(x_0) + (x - x_0)f'(x_0) + R_1}{g(x_0) + (x - x_0)g'(x_0) + P_1} = \frac{(x - x_0)f'(x_0) + R_1}{(x - x_0)g'(x_0) + P_1} = \frac{f'(x_0) + (R_1/(x - x_0))}{g'(x_0) + (P_1/(x - x_0))},$$

and because R_1 and P_1 tend to zero faster than $(x - x_0)$ as x approaches x_0, we have that

$$\lim_{x\to x_0} \frac{f(x)}{g(x)} = \lim_{x\to x_0} \frac{f'(x)}{g'(x)}.$$

This argument is not a proof of l'Hôpital's rule—for that we would have to consider the different indeterminate forms and lots of other details—but it does give us insight into how a function and its derivative behave. It is also important to realize that l'Hôpital's rule only works for indeterminate limits that have the 0/0 or ∞/∞ form.

Exercise 2.6.1 Identify whether the following limits are in either the 0/0 or ∞/∞ indeterminate forms, and use l'Hôpital's rule to evaluate them.

$$\text{a. } \lim_{x\to\infty} x\tan\left(\frac{1}{x}\right), \quad \text{b. } \lim_{\theta\to\pi/2} \frac{\cos(\theta)}{\theta-\pi/2}.$$

2.7 Using Derivatives

We motivated our brief exploration of derivatives by noting that they allow us to study rates of change—how fast is this population changing, how rapidly does density change with height in the atmosphere? The derivative is a very useful tool that can help us in many other ways.

2.7.1 Curve Sketching

We can combine our understanding of the derivative with other information to obtain a good sketch of the behavior of a curve. Let us look at a simple example. What does the function

$$y(x) = \frac{x}{x^2-4} \tag{2.18}$$

look like? To answer this, we need to follow a set of well-defined steps.

1. First, we look to see if the function is defined for all values of x. The numerator is just x, so there are no problems there. The denominator, however, is zero when $x = \pm 2$, making $y(x)$ infinite at those points. We shall deal with these points a bit later.
2. Next we try to find any maxima or minima by taking the first derivative

$$\frac{dy}{dx} = -\frac{x^2+4}{(x^2-4)^2}.$$

   For the derivative to be zero, we would need $x^2+4=0$, which implies x would have to be a complex number (Appendix C). If we restrict ourselves to real numbers, then there are no maxima or minima. What is more, the first derivative is always negative, so the function $y(x)$ is always decreasing as x increases.
3. We now take the second derivative to find any points of inflection:

$$\frac{d^2y}{dx^2} = \frac{2x(x^2+12)}{(x^2-4)^3},$$

   and the only real value of x that makes this zero is $x = 0$, so there is a point of inflection at $(x = 0, y = 0)$.

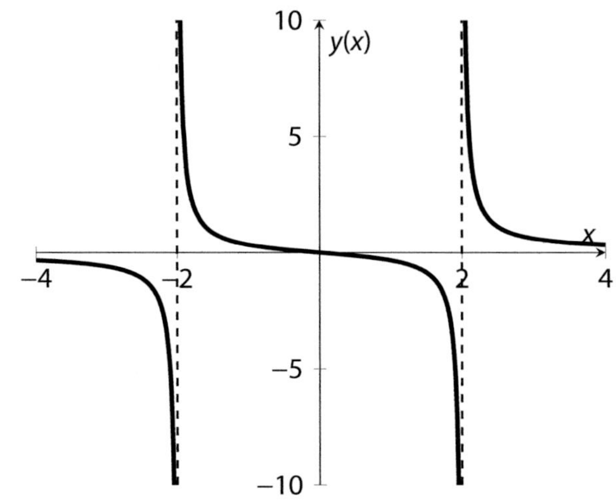

Figure 2.13 A sketch of the function given by Equation (2.18) showing the three branches of the curve separated by the two asymptotes at $x = 2$ and $x = -2$.

4. Now we need to deal with the points $x = \pm 2$. These are *vertical asymptotes*. As $x \to +2$ from above (i.e., from values greater than 2), $y(x)$ tends to $+\infty$, but as $x \to +2$ from below (i.e., from values less than 2), $y(x) \to -\infty$. Similarly, as $x \to -2$ from above, $y(x) \to +\infty$ and as $x \to -2$ from below, $y(x) \to -\infty$. This means that the function is discontinuous at $x = \pm 2$.
5. Lastly, we need to know what happens when x becomes very large. As $x \to \pm\infty$, then

$$\lim_{x\to\pm\infty} y(x) = \lim_{x\to\pm\infty} \frac{x}{x^2 - 4} = 0.$$

With all this information, we can sketch the behavior of the curve (Figure 2.13).

One might legitimately ask why it is useful to learn what a function looks like and how to sketch it when we have computer programs that will plot any function we want. There are several good reasons. The first is simply that you learn more about the behavior of the function by doing it yourself—for example, your function might have unwanted behavior at values of x that you did not plot or that were too large to plot. Related to this is the fact that learning the behavior of a function allows you to see where various approximations might be made—for example, between $x = \pm 1$, the function given by Equation (2.18) is approximately a straight line and for $|y| > 5$ the curve is approximated well by straight vertical lines at $x = \pm 2$. Lastly, knowing the behavior of different functions allows you to choose a function suitable for a given job. The following exercises illustrate some of these points.

Exercise 2.7.1 Sketch the behavior of the curve $V(C) = V_{\max}C/(K + C)$, where $C \geq 0$ and $V_{\max}$ and K are positive constants.

Exercise 2.7.2 Sketch the curve

$$y(x) = \frac{x^3 + 2x^2 - 3x + 1}{x^3}.$$

2.7.2 Newton's Method

Newton's method[5] gives us a means for solving an equation, $f(x) = 0$, using the derivative of that equation. This method is an iterative method, which means that we start with an initial guess for the solution and use it to obtain a more accurate answer, which is then used to obtain an even more accurate solution, and so on until we are satisfied with the accuracy of our answer, or exhausted. The algorithm is easy to program, so most of the tedious work can be done by a computer.

The essence of Newton's method can best be seen geometrically (Figure 2.14). Let us assume that we have an equation $f(x) = 0$ to solve and we have an initial guess (x_0) for its solution—this could be an educated guess, or we could sketch the curve of the equation and use it to make our initial guess. We can evaluate the derivative of $f(x)$ at $x = x_0$, giving us the slope of the tangent to the curve at x_0. The point where the tangent line crosses the x axis (x_1) provides a better approximation to the solution of the equation. Because the tangent is a straight line, we can write

$$y_1 = 0 = f(x_0) + \left.\frac{df}{dx}\right|_{x=x_0} (x_1 - x_0),$$

which we can rearrange to give

$$x_1 = x_0 - \frac{f(x_0)}{(df/dx)_{x=x_0}}.$$

We know everything on the right-hand side of this equation, so we can easily find the next point (x_1) in our iteration. We can evaluate the derivative of the function at this new point and go through the same procedure to get

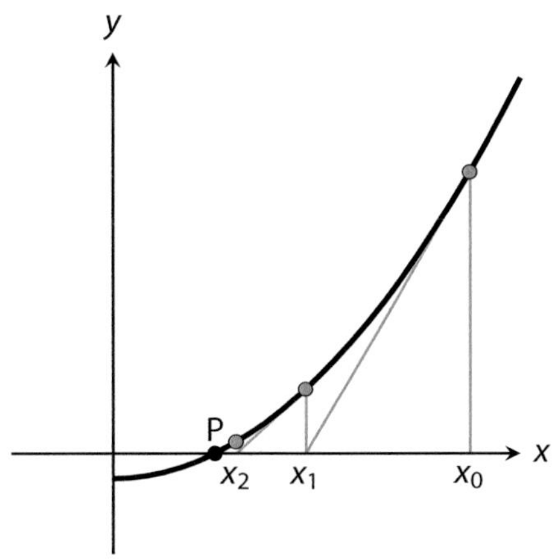

Figure 2.14 A graphical illustration of Newton's method. The root of the equation is the point P. The initial guess is $x = x_0$, from which Newton's method gives a better estimate ($x = x_1$) of the root. Using this value, Newton's method gives an even better estimate ($x = x_2$) and so on.

[5] Named after Isaac Newton (1642–1726).

$$x_2 = x_1 - \frac{f(x_1)}{(df/dx)_{x=x_1}}$$

and so on. We can see from Figure 2.14 that repeating this sequence of steps will get us closer and closer to our solution of $f(x) = 0$. In general, if we want to find the solution to an equation $f(x) = 0$ and we have an initial guess, $x = x_0$, then we can obtain better approximations (x_n) by repeatedly applying the formula

$$x_n = x_{n-1} - \frac{f(x_{n-1})}{(df/dx)_{x=x_{n-1}}} \qquad n = 1, 2, 3 \ldots \tag{2.19}$$

Example 2.16 We will use Newton's method to find the positive solution to the equation $x^2 - 3x = 2$. The first thing we have to do is put the equation in the correct form to apply the method. Newton's method works for equations of the form $f(x) = 0$, and we need to rewrite the equation as

$$f(x) = x^2 - 3x - 2 = 0.$$

The derivative of the function is $f'(x) = 3x - 3$. Choosing $x_0 = 8$ as our initial guess, then working to four decimal places,

$$\begin{aligned}
x_1 &= x_0 - \frac{f(x_0)}{(df/dx)_{x=x_0}} = 8.0 - \frac{38}{13} = 5.0769 \\
x_2 &= x_1 - \frac{f(x_1)}{(df/dx)_{x=x_1}} = 5.0769 - \frac{8.5442}{7.1538} = 3.8825 \\
x_3 &= x_2 - \frac{f(x_2)}{(df/dx)_{x=x_2}} = 3.8825 - \frac{1.4265}{4.765} = 3.5832 \\
x_4 &= x_3 - \frac{f(x_3)}{(df/dx)_{x=x_3}} = 3.5832 - \frac{0.0897}{4.1664} = 3.5617 \\
x_5 &= x_4 - \frac{f(x_4)}{(df/dx)_{x=x_4}} = 3.5617 - \frac{0.0006}{4.1234} = 3.5616.
\end{aligned}$$

The exact solution to the equation is $x = (3 \pm \sqrt{17})/2 \approx 3.56155$, which we can compare with the approximate solution $x \approx 3.5616$ we found in only five iterations. But what about the other root? This is located at $x = (3 - \sqrt{17})/2 \approx -0.56155$. To find this root we would need to start with a new initial guess such that the tangent at that point would lead us in the right direction. We can see that $f'(x) = 0$ at $x = 1$, which lies between the two roots. So any point $x > 1$ will move us toward the root $x = (3 \pm \sqrt{17})/2$, and starting at a point with $x < 1$ will move us toward $x = (3 - \sqrt{17})/2$.

Newton's method does not always work, and sometimes it can fail spectacularly. Some of these cases are explored in the problems at the end of the chapter, but in general Newton's method will fail under the following circumstances:

- If the derivative of $f(x)$ is zero, or if the function cannot be differentiated at some point near the solution you are trying to find.
- If the second derivative of the function in the neighborhood of the solution is very large, in which case the function has a strong curvature. This means that using the tangent of the curve may give a point that is not a better approximation.

- A poor choice of starting point. If we had picked $x_0 = -2$ as our starting point in Example 2.16, then we would have had problems because the negative root of the equation lies between our initial guess and our desired solution.
- Poor choices for the initial guess can also cause Newton's method to oscillate between two values and not converge on a solution.

Exercise 2.7.3 Use Newton's method to find the point where the curves $y_1(x) = 2x^2$ and $y_2(x) = e^{-x^2}$ intersect by finding the root of the equation $y_1(x) - y_2(x) = 0$.

2.8 Partial Derivatives

So far in this chapter we have only dealt with functions of one variable, $y = f(x)$. But many functions we have to deal with in science are functions of more than one variable. For example, the density of seawater is often taken to be a function of temperature and salinity. How do we take derivatives of such functions, and what do they mean?

Let us consider a function of two variables, $z = g(x, y)$. If we want to calculate its derivative, the first question we have to answer is "the derivative with respect to what?" We can look for the derivative with respect to x, or the derivative with respect to y, or a derivative with respect to variations in both x and y, making the situation quite complicated. Think back to our definition in Equation (2.3) for the derivative of a single variable function:

$$\frac{df}{dx} = \lim_{\Delta x \to 0} \frac{f(x + \Delta x) - f(x)}{\Delta x}.$$

If instead of $f(x)$ we have a function $g(x, y)$ of two variables and we want the derivative with respect to x, we do exactly the same thing as before and treat y as a constant (Figure 2.15):

$$\frac{\partial g}{\partial x} = \lim_{\Delta x \to 0} \frac{g(x + \Delta x, y) - g(x, y)}{\Delta x}. \tag{2.20}$$

That is, we move along lines of constant y calculating the rate of change of the function g with respect to x. This means that if we move to another line of constant y, the derivative with respect to x will change. Similarly, for a derivative with respect to y we hold x constant:

$$\frac{\partial g}{\partial y} = \lim_{\Delta y \to 0} \frac{g(x, y + \Delta y) - g(x, y)}{\Delta y}. \tag{2.21}$$

These derivatives are called *partial derivatives* because we are considering how the function $g(x, y)$ varies with changes in only one variable. One thing to note is the slight change in notation for a partial derivative, from d/dx to $\partial/\partial x$—the script '∂' just reminds us that we are taking a partial derivative, i.e., taking the derivative of the function with respect to one variable while holding the other variables constant. This can be generalized to functions of any number of variables.

One way to visualize the process of partial differentiation is to think about hiking in a mountain range where lines of latitude and longitude are the equivalent of the x and y

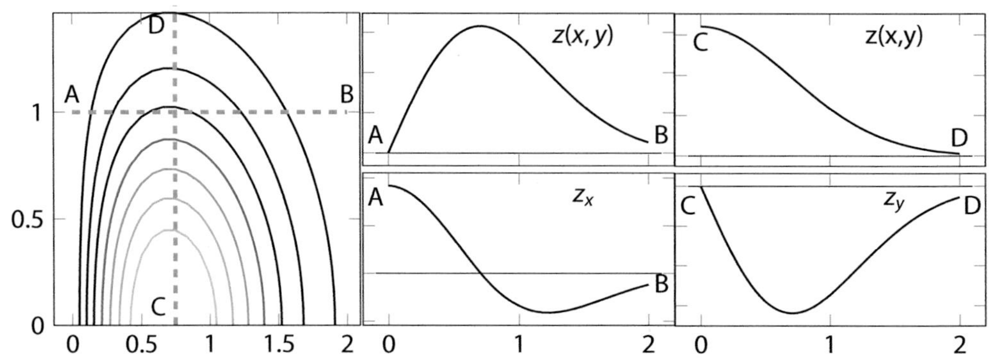

Figure 2.15 A contour plot of the function $z(x, y) = x \exp(-x^2 - y^2)$ is shown in the left-hand panel with two lines, one at a constant value of x (CD) and the other at a constant value of y (AB). The two middle panels show $z(x, y)$ along the path AB (upper panel) and z_x along the same path (lower panel)—the thin solid line is $z = 0$ in the upper panel and $z_x = 0$ in the lower one. The two right-hand panels show $z(x, y)$ and z_y along the path CD in a similar manner.

variables, and the function $g(x, y)$ gives the height of the mountain as a function of latitude and longitude. Two hikers that walk along different lines of constant latitude will traverse different paths of differing degrees of steepness; one may be on a steep slope while the other is ambling in a valley where the nearby terrain is relatively flat.

Example 2.17 Let us take the partial derivatives of the following functions with respect to each variable: a. $f(x, y) = x^2 + y^2 + z^2$, b. $g(x, y, z) = x^2 + y^2 e^{x^2}$. To take the partial derivative of $f(x, y)$ with respect to x, we treat y and z as constants to obtain

$$\frac{\partial f}{\partial x} = 2x,$$

and similarly with the derivatives with respect to y and z giving

$$\frac{\partial f}{\partial y} = 2y, \qquad \frac{\partial f}{\partial z} = 2z.$$

To take the partial derivative of $g(x, y)$ with respect to x, we treat y as a constant, giving

$$\frac{\partial g}{\partial x} = 2x + 2xy^2 e^{x^2}.$$

Similarly, we obtain the partial derivative of $g(x, y)$ with respect to y by treating x as a constant, to give

$$\frac{\partial g}{\partial y} = 2y e^{x^2}.$$

Things become a little more complicated when we take higher-order derivatives. For example, for a second-order derivative we can again take the partial derivative with respect to x or y, producing the combinations

$$\frac{\partial}{\partial x}\left(\frac{\partial f}{\partial x}\right)=\frac{\partial^2 f}{\partial x^2} \qquad \frac{\partial}{\partial y}\left(\frac{\partial f}{\partial y}\right)=\frac{\partial^2 f}{\partial y^2}$$

$$\frac{\partial}{\partial y}\left(\frac{\partial f}{\partial x}\right)=\frac{\partial^2 f}{\partial y\partial x} \qquad \frac{\partial}{\partial x}\left(\frac{\partial f}{\partial y}\right)=\frac{\partial^2 f}{\partial x\partial y}.$$

Derivatives such as

$$\frac{\partial^2 f}{\partial y\partial x} \quad \text{and} \quad \frac{\partial^2 f}{\partial x\partial y}$$

are called mixed partial derivatives, and if $f(x, y)$ is a continuous function, then

$$\frac{\partial^2 f}{\partial y\partial x}=\frac{\partial^2 f}{\partial x\partial y}. \tag{2.22}$$

Equation (2.22) is called *Clairaut's theorem*.[6]

Example 2.18 To demonstrate the validity of Clairaut's theorem, we will calculate all second-order partial derivatives of $f(x, y) = x^2y^3 + 3e^y$. First, we need to calculate the first-order partial derivatives:

$$\frac{\partial f}{\partial x}=2xy^3, \qquad \frac{\partial f}{\partial y}=3x^2y^2+3e^y.$$

Now we can calculate the four second-order derivatives:

$$\frac{\partial^2 f}{\partial x^2}=2y^3 \qquad \frac{\partial^2 f}{\partial y^2}=6x^2y+2e^y$$

$$\frac{\partial^2 f}{\partial y\partial x}=6xy^2 \qquad \frac{\partial^2 f}{\partial x\partial y}=6xy^2$$

We see that Equation (2.22) is indeed satisfied.

Exercise 2.8.1 Calculate all second first- and second-order partial derivatives of the following functions

a. $f(x, y) = \sin(2x) + \cos(2y)$, b. $f(x, y) = x^2 + y^2$.

Even though we are dealing with partial derivatives, we can still use the product rule and the chain rule to calculate derivatives. For functions of a single variable, we found that we could differentiate a function of a function $f(x(t))$ by using the chain rule. A very similar rule exists for partial derivatives. If a function $g = g(x, y)$ with $x = x(u, v)$ and $y = y(u, v)$, then

$$\frac{\partial g}{\partial x}=\frac{\partial g}{\partial u}\frac{\partial u}{\partial x}+\frac{\partial g}{\partial v}\frac{\partial v}{\partial x}, \qquad \frac{\partial g}{\partial y}=\frac{\partial g}{\partial u}\frac{\partial u}{\partial y}+\frac{\partial g}{\partial v}\frac{\partial v}{\partial y}. \tag{2.23}$$

So, if $g(u, v) = u + v$ with $u(x, y) = xy$ and $v(x, y) = x - y$, then

$$\frac{\partial g}{\partial x}=\frac{\partial g}{\partial u}\frac{\partial u}{\partial x}+\frac{\partial g}{\partial v}\frac{\partial v}{\partial x}=y+1$$

$$\frac{\partial g}{\partial y}=\frac{\partial g}{\partial u}\frac{\partial u}{\partial y}+\frac{\partial g}{\partial v}\frac{\partial v}{\partial y}=x-1.$$

[6] Named after the French mathematitican Alexis Clairaut (1713–1765), whose work helped confirm Newton's theory that the Earth was ellipsoidal in shape.

Exercise 2.8.2 Calculate all first and second partial derivatives of these functions:

$$\text{a. } f(x, y) = \sin(2x)\cos(2y), \quad \text{b. } f(x, y) = \frac{1 + x^2 e^{-x^2}}{y^2 - 1}$$

$$\text{c. } f(x, y, z) = (x^2 + y^2)e^{-(x^2+y^2+z^2)}.$$

Exercise 2.8.3 Verify that Equation (2.22) is satisfied for all pairs of second derivatives for these functions:

$$\text{a. } f(x, y, z) = xyz, \quad \text{b. } f(\theta, \phi) = \sin(\theta)\tan(\phi).$$

There are several different notations you will come across for partial derivatives.[7] For first derivatives these are

$$\frac{\partial f(x, y)}{\partial x} = \partial_x f = f_x, \quad \frac{\partial f(x, y)}{\partial y} = \partial_y f = f_y,$$

and for second derivatives

$$\frac{\partial^2 f(x, y)}{\partial x^2} = \partial_{xx} f = f_{xx}, \quad \frac{\partial^2 f}{\partial x \partial y} = \partial_{xy} f = f_{xy}, \quad \frac{\partial^2 f}{\partial y \partial x} = \partial_{yx} f = f_{yx}.$$

We can expand functions of two (or more) variables using a generalization of the Taylor polynomials (Equation (2.14)). For example, the expansion of a function $f(x, y)$ about $x = a$, $y = b$ is

$$\begin{aligned} f(x, y) &= f(a, b) + (x - a)f_x(a, b) + (y - b)f_y(a, b) \\ &\quad + \frac{1}{2!}\left((x - a)^2 f_{xx}(a, b) + 2(x - a)(y - b)f_{xy}(a, b) + (y - b)^2 f_{yy}(a, b)\right) + \cdots . \end{aligned} \tag{2.24}$$

As you can see, there are more derivatives to consider, but this is made somewhat simpler by the fact that because $f_{xy} = f_{yx}$ we can combine terms.

Things similarly become more complicated if we start looking for maxima or minima. This is because there are more possibilities to consider (Figure 2.16) when we have functions of two or more variables; there are more directions we can move in. Let us assume that a function $f(x, y)$ has a turning point at $x = a$, $y = b$. We know from dealing with functions of one variable that the first derivative is zero at a maximum or minimum. For a function of two variables, *both* first derivatives must be zero at these points for them to be either a maximum or minimum. If we expand $f(x, y)$ in a Taylor series about the turning point $(x, y) = (a, b)$, the two terms involving $f_x(a, b)$ and $f_y(a, b)$ will vanish because we are expanding the function about a maximum or minimum point. So, we can write the Taylor series as

$$\begin{aligned} \Delta f &= f(x, y) - f(a, b) \\ &= \frac{1}{2!}\left((x - a)^2 f_{xx}(a, b) + 2(x - a)(y - b)f_{xy}(a, b) + (y - b)^2 f_{yy}(a, b)\right). \end{aligned} \tag{2.25}$$

If $(x, y) = (a, b)$ is a maximum, then as we move away from that point in any direction, the function $f(x, y)$ decreases, so $\Delta f < 0$. Therefore

$$\frac{1}{2!}\left((x - a)^2 f_{xx}(a, b) + 2(x - a)(y - b)f_{xy}(a, b) + (y - b)^2 f_{yy}(a, b)\right) < 0.$$

[7] These can be very useful for saving typing!

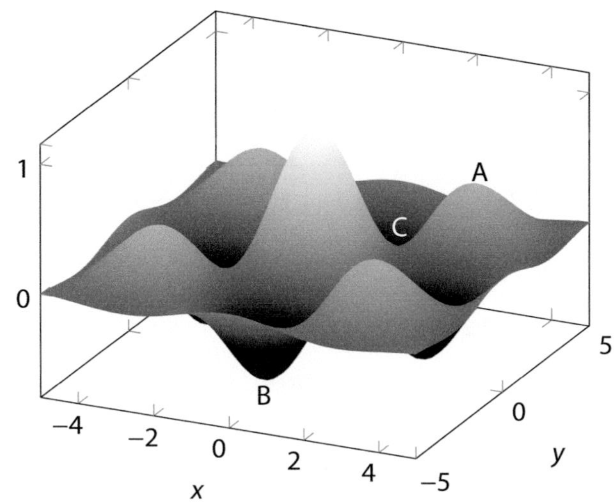

Figure 2.16 A plot of the function $f(x, y) = \cos(x)\cos(y)e^{-(x^2+y^2)}$ showing a maximum (A), a minimum (B), and a saddle (C).

But this equation holds no matter what direction we move in, so if we move along the line $y = b$ we see that $f_{xx} < 0$. Similarly, if we move along a line $x = a$, we see that $f_{yy} < 0$. These are just the conditions on the second derivative we would expect for a maximum from what we know of functions of a single variable. But for a function of two variables we have the freedom to move in directions that combine changes in x and y. Let us move in a straight line defined such that $x - a = \xi(y - b)$, where ξ is a constant number. Then, Equation (2.25) tells us that

$$\Delta f = \xi^2 f_{xx}(a, b) + 2\xi f_{xy}(a, b) + f_{yy}(a, b) < 0.$$

We can multiply this equation by $f_{xx}(a, b)$, remembering that $f_{xx} > 0$ to get

$$\begin{aligned} &\xi^2(f_{xx}(a, b))^2 + 2\xi f_{xx}(a, b)f_{xy}(a, b) + f_{xx}(a, b)f_{yy}(a, b) < 0 \\ &\implies f_{xx}(a, b)f_{yy}(a, b) - (f_{xy}(a, b))^2 > (\xi f_{xx}(a, b) + f_{xy}(a, b))^2. \end{aligned}$$

The last inequality is true for all values of ξ, so we can choose ξ such that the right-hand side of the inequality is zero. Then, because the inequality is true for any value of ξ we have

$$f_{xx}(a, b)f_{yy}(a, b) - (f_{xy}(a, b))^2 > 0.$$

The quantity on the left-hand side of the inequality,

$$H = \frac{\partial^2 f}{\partial x^2}\frac{\partial^2 f}{\partial y^2} - \left(\frac{\partial^2 f}{\partial x \partial y}\right)^2, \tag{2.26}$$

is called the *Hessian*[8] and is related to the curvature of the surface.

[8] Named after the German mathematician Ludwig Otto Hesse (1811–1874).

Exercise 2.8.4 Using similar arguments to those leading to Equation (2.26), show that if $(x, y) = (a, b)$ is a minimum, then $f_{xx} > 0$, $f_{yy} > 0$ at the point (a, b) and $H > 0$.

There is another case to consider. What happens if Δf in Equation (2.25) is zero? This means that

$$\frac{1}{2!}\left((x-a)^2 f_{xx}(a,b) + 2(x-a)(y-b)f_{x,y}(a,b) + (y-b)^2 f_{yy}(a,b)\right) = 0.$$

If we again choose a direction to move such that $x - a = \xi(y - b)$, then we get $\xi^2 f_{xx}(a,b) + 2\xi f_{x,y}(a,b) + f_{yy}(a,b) = 0$. Following similar arguments to those leading to Equation (2.26), we find that

$$f_{xx}(a,b)f_{yy}(a,b) - (f_{xy}(a,b))^2 < 0.$$

This critical point is called a saddle point, and it is a maximum in one direction (e.g., the x direction) and a minimum in the other (point C in Figure 2.16).

Example 2.19 We can use the Hessian to find and classify the stationary points of the function $g(x, y) = 2x^3 + 6xy^2 - 3y^3 - 150x$. First, we calculate the partial derivatives of the function

$$g_x = 6x^2 + 6y^2 - 150, \quad g_y = 12xy - 9y^2, \quad g_{xx} = 12x, \quad g_{yy} = 12x - 18y, \quad g_{xy} = 12y.$$

The Hessian is

$$H = g_{xx}g_{yy} - g_{xy}^2 = 36x(2x - 3y) - 144y^2.$$

To find the stationary points we set the first derivatives equal to zero, which gives the equations

$$x^2 + y^2 = 25 \quad \text{and} \quad 4xy - 3y^2 = y(4x - 3y) = 0.$$

Solving these equations for x and y gives us four stationary points

$$(x_1, y_1) = (5, 0), \quad (x_2, y_2) = (-5, 0), \quad (x_3, y_3) = (3, 4), \quad (x_4, y_4) = (-3, -4).$$

To classify the nature of each point, we need to use the second derivatives and the Hessian. At the point (x_1, y_1) we have $g_{xx} > 0$, $g_{yy} > 0$, and $H = 3600 > 0$, so (x_1, y_1) is a minimum. At (x_2, y_2), $g_{xx} < 0$, $g_{yy} < 0$ and $H > 0$, making (x_2, y_2) a maximum point. The Hessian $H < 0$ for the two remaining points, so they are saddle points.

Exercise 2.8.5 Find and classify all the stationary points of the function $g(x, y) = x^3 + y^3 - 3x - 3y$.

It is important to remember that the partial derivative involves a variation with only one variable, the others are held constant. So, if we take the partial derivative with respect to y of a function $f(x, y, z)$, we treat x and z as constants. But what do we do if we are interested in the change in $f(x, y, z)$ as all three variables change at the same time? For this, we use the *total derivative* or *total differential*

$$df(x, y, z) = \frac{\partial f}{\partial x}dx + \frac{\partial f}{\partial y}dy + \frac{\partial f}{\partial z}dz. \tag{2.27}$$

For example, say we want to know how the amount of soot changes in the atmosphere. This becomes rather complicated because there are processes such as precipitation that will remove soot from the air, but the air itself is also moving. So, if the amount of soot as a function of time (t) and position ($x(t)$, $y(t)$, $z(t)$) is $F(t, x(t), y(t), z(t))$, then

$$\frac{dF}{dt} = \frac{\partial F}{\partial t} + \frac{\partial F}{\partial x}\frac{dx}{dt} + \frac{\partial F}{\partial y}\frac{dy}{dt} + \frac{\partial F}{\partial z}\frac{dz}{dt}.$$

The first term on the right-hand side represents the rate of change of F with time by those processes (such as precipitation) not connected to the motion of the air. The other terms contain what appear to be velocities (derivatives of a position with respect to time) and represent how F changes with the motions of the air.

Example 2.20 As an example, let us calculate the total differential of $f(x, y, z) = 2xz + 3y^2z^3 + 5x + 1$. Using Equation (2.27), we have

$$df = (2z + 5)dx + 6yz^3dy + (2x + 9y^2z^2)dz.$$

Example 2.21 We can use the total derivative to calculate the derivative of $g(t, x, y, z)$ with respect to time where $g(t, x, y, z) = \mathrm{e}^{-at} + xyz + xy + z$ and $x(t) = 2\cos(t)$, $y(t) = 2\sin(t)$, and $z(t) = t$. The derivative is

$$\begin{aligned}
\frac{dg}{dt} &= \frac{\partial g}{\partial t} + \frac{\partial g}{\partial x}\frac{dx}{dt} + \frac{\partial g}{\partial y}\frac{dy}{dt} + \frac{\partial g}{\partial z}\frac{dz}{dt} \\
&= -ae^{-at} + y(z+1)\frac{dx}{dt} + x(z+1)\frac{dy}{dt} + (xy+1)\frac{dz}{dt} \\
&= -ae^{-at} - 2y(z+1)\sin(t) + 2x(z+1)\cos(t) + (xy+1) \\
&= -ae^{-at} - 4(t+1)\sin^2(t) + 4(t+1)\cos^2(t) + 4\cos(t)\sin(t) + 1 \\
&= -ae^{-at} + 4(t+1)\cos(2t) + 2\sin(2t) + 1.
\end{aligned}$$

2.9 Using Partial Derivatives

Partial derivatives will occur whenever we are interested in how a function that depends on multiple variables changes as those variables change. In this section we will explore two useful applications, and later chapters will make extensive use of partial differentiation.

2.9.1 Propagating Uncertainty

We often find we need to derive new quantities from those we measure. For example, oceanographers do not measure the density of seawater directly; instead they measure the temperature, salinity, and pressure of the seawater. To calculate the density of seawater they use the *equation of state* that relates the density to the three measured quantities.

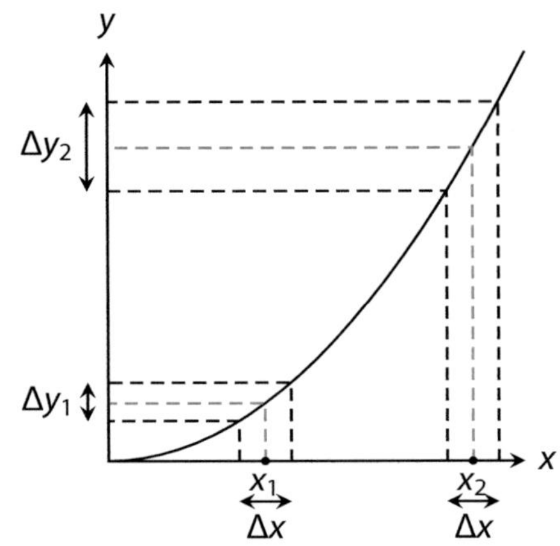

Figure 2.17 The propagation of uncertainty for a quadratic function $y = x^2$. Measurements yield two values of x (x_1 and x_2), both having the *same* uncertainty Δx. The corresponding uncertainties in the calculated values of y for the two values of x are very different because the curve is nonlinear.

However, all measured quantities have some uncertainty in them, even if that uncertainty is very small. So, how do we assign an uncertainty to the derived quantity (e.g., seawater density)[9]?

Let us start by considering an example that is easy to visualize (Figure 2.17). We make two measurements of a variable x, each having the same uncertainty Δx. Using each measurement we calculate $y = x^2$. If Δx is small, the value of y at $x - \Delta x/2$ is approximately $y - \Delta y/2 = (x - \Delta x/2)^2 = x^2 - x\Delta x + (\Delta x)^2/4$, so $\Delta y/2 \approx x\Delta x$ (where our assumption of Δx being small allows us to ignore the term in $(\Delta x)^2$) and the uncertainty in y depends on the value of x. If we make measurements of two different values of x (x_1 and x_2 in Figure 2.17) and they have the same uncertainty of $\pm\Delta x/2$, then the uncertainties in the corresponding y values will be different because the equation relating x and y is nonlinear.

It is quite easy to determine the errors for a quadratic function, and there is a general formula we can use for more complicated functions. Let us assume we are interested in the value of a variable x, where x is a function of two quantities u and v that we can measure (along with their uncertainties): $x = f(u, v)$. We can calculate a value for x by using the mean values $(\bar{u}, \bar{v})$ of the measurements $\bar{x} = f(\bar{u}, \bar{v})$, where the mean values are defined by

$$\bar{u} = \frac{1}{N}\sum_{i=1}^{N} u_i,$$

where N is the number of replicate measurements of u that were made. But what about the uncertainty in x? One measure of the uncertainty in x is the variance

$$\sigma_u^2 = \frac{1}{N}\sum_{i=1}^{N}(u_i - \bar{u})^2.$$

9 This is often called the *propagation of errors*, but I prefer the term *propagation of uncertainty* because hopefully no experimental errors have been made, one is just trying to deal with the inevitable uncertainty inherent in making a measurement.

The term in the summation is the sum of the squares of differences between each measured value and the mean. If we write

$$\Delta x = x_i - \bar{x} = (u_i - \bar{u})\frac{\partial x}{\partial u} + (v_i - \bar{v})\frac{\partial x}{\partial v}, \tag{2.28}$$

we can express the variance as

$$\begin{aligned}
\sigma_x{}^2 &= \frac{1}{N}\sum_{i=1}^{N}(x_i - \bar{x})^2 \simeq \frac{1}{N}\sum_{i=1}^{N}\left\{(u_i - \bar{u})\left(\frac{\partial x}{\partial u}\right) + (v_i - \bar{v})\left(\frac{\partial x}{\partial v}\right)\right\}^2 \\
&\simeq \frac{1}{N}\sum_{i=1}^{N}\left\{(u_i - \bar{u})^2\left(\frac{\partial x}{\partial u}\right)^2 + (v_i - \bar{v})^2\left(\frac{\partial x}{\partial v}\right)^2 + 2(u_i - \bar{u})(v_i - \bar{v})\frac{\partial x}{\partial u}\frac{\partial x}{\partial v}\right\}.
\end{aligned} \tag{2.29}$$

The first and the second terms in the brackets contain the variances of u and v, and the last term contains the *covariance* σ_{uv}, which is a measure of how much the variables u and v vary together. Using the definition of the variance we obtain the equation for the propagation of uncertainties

$$\sigma_x{}^2 \simeq \sigma_u{}^2\left(\frac{\partial x}{\partial u}\right)^2 + \sigma_v{}^2\left(\frac{\partial x}{\partial u}\right)^2 + 2\sigma_{uv}\left(\frac{\partial x}{\partial u}\right)\left(\frac{\partial x}{\partial u}\right), \tag{2.30}$$

where we have defined

$$\sigma_{uv} \equiv \frac{1}{N}\sum_{i=1}^{N}\left[(u_i - \bar{u})(v_i - \bar{v})\right]. \tag{2.31}$$

In practice it is frequently assumed that the covariance is zero, but strictly this must be justified on a case-by-case basis with careful thought and analysis.

Example 2.22 Propagation of uncertainty formulae for simple equations are easy to calculation. Let us assume that we have measurements of the variables u and v and their uncertainties σ_u and σ_v and we need to calculate the uncertainty in the following:

(a) $x = \alpha u \pm \beta v$,

(b) $x = \pm\alpha u^{\pm\beta}$,

where α and β are constants. For (a), the appropriate derivatives are

$$\left(\frac{\partial x}{\partial u}\right) = \alpha \qquad \left(\frac{\partial x}{\partial v}\right) = \pm\beta,$$

so using Equation (2.30), we get

$$\sigma_x{}^2 = \alpha^2\sigma_u{}^2 + \beta^2\sigma_v{}^2 \pm 2\alpha\beta\sigma_{uv}.$$

In (b), x is a function of only one variable (u), so we only need the derivative

$$\frac{\partial x}{\partial u} = \pm\alpha\beta u^{\pm\beta-1} = \pm b\frac{x}{u},$$

and using Equation (2.30), we get

$$\frac{\sigma_x}{x} = \beta\frac{\sigma_u}{u}.$$

Exercise 2.9.1 Use the propagation of uncertainties to find the uncertainty in x given the values of the variables u and v and their associated uncertainties σ_u^2 and σ_v^2.

$$\text{a. } x = uv, \quad \text{b. } x = u/v, \quad \text{c. } x = \alpha e^{\pm \beta u}, \quad \text{d. } x = \alpha \ln(\pm \beta u)$$

Although these formulae can be used to propagate uncertainties, it is important to appreciate the assumptions made in deriving them. Chief among these is that the uncertainties are small. What do we mean by *small*? Small with respect to what? Look back at Figure 2.17. We have assumed that Δx can be written using only first derivatives of the function (Equation (2.28)), and in doing so we have approximated a function by a straight line (the first derivative being the slope of a straight line). This is a process called *linearization*, and it results in an approximation to the curve if Δx is small. But if Δx is large enough, then even though the mean x value is located in the middle of the interval Δx, the mean value of y need not be in the middle of the range Δy. This will happen if the function $y = f(x)$ is strongly nonlinear and has a large curvature.

2.9.2 Fitting a Straight Line

A common data analysis task is to find the straight line or curve that best fits a set of observed data points. We may want to do this to provide a better understanding of the relationships between variables in a data set, or we might be trying to use the data to estimate the values of important parameters in a theoretically derived equation. This process is called *curve fitting* or *regression*, and the most common use is fitting a straight line to the points. If we have a set of N data points $((x_i, y_i),\ i = 1 \ldots N)$, then to fit a straight line to them requires finding values for a_0 and a_1 such that the straight line

$$\tilde{y} = a_0 + a_1 x \tag{2.32}$$

is the best fit to the data, where the variable $\tilde{y}$ represents the value of y that we get from using the x data values in Equation (2.32). This is an example of *linear regression* because the quantities we want to find (a_0 and a_1) appear linearly in the equation. What do we mean by the "best fit" straight line? There are several definitions we can use, but the most useful is the line that minimizes the sum of the squares of the distance in the y direction of each data point from the straight line (Figure 2.18), which is called a least-squares regression. If we had chosen just the distances, then distances of points below the line would cancel some of those above the line. This in and of itself is not a problem, but it makes finding the minimum harder.[10] We will minimize the function

$$\chi^2 = \sum_{i=1}^{N} \frac{(\tilde{y}_i - y_i)^2}{\sigma_i^2} = \sum_{i=1}^{N} \frac{(a_0 + a_1 x_i - y_i)^2}{\sigma_i^2}, \tag{2.33}$$

where σ_i is the uncertainty in the y_i measurement. Why have we divided by σ_i^2? We would like points that have a smaller measured uncertainty to count more toward the value of χ^2

[10] We could have used the absolute value of the distance in the y direction, but as we have seen in Section 2.1, the absolute value has some unfortunate mathematical properties. Consequently, the square of the distance is usually chosen so as to make our mathematical lives easier.

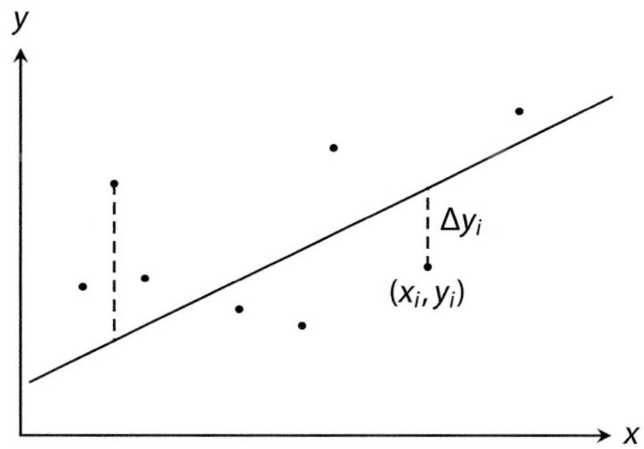

Figure 2.18 Fitting a straight line to data points. Eight data points $((x_i, y_i, i = 1\ldots8)$ are represented together with the "best fit" straight line. The least-squares technique minimizes the sum of the squares of the vertical distances (Δy_i) between each data point (x_i, y_i) and the straight line.

than those with larger measured uncertainties, so we give them more weight by dividing by σ_i^2; this is called a *weighted least-squares* fit.

We want to find the straight line (i.e., the values of a_0 and a_1) that minimizes the value of χ^2. We know how to find the minimum of a function: we take partial derivatives of χ^2 with respect to a_0 and a_1 and set the derivatives to zero. Doing so gives us

$$\frac{\partial \chi^2}{\partial a_0} = \sum_{i=1}^{N} \frac{2(a_o + a_1 x_i - y_i)}{\sigma_i^2} = 0, \quad \frac{\partial \chi^2}{\partial a_1} = \sum_{i=1}^{N} \frac{2x_i(a_o + a_1 x_i - y_i)}{\sigma_i^2} = 0. \tag{2.34}$$

These look ugly, but we can make them appear much nicer and see their structure by grouping terms together. For example, the first equation can be written

$$\sum_{i=1}^{N} \frac{(a_o + a_1 x_i - y_i)^2}{\sigma_i^2} = a_o \sum_{i=1}^{N} \frac{1}{\sigma_i^2} + a_1 \sum_{i=1}^{N} \frac{x_i}{\sigma_i^2} - \sum_{i=1}^{N} \frac{y_i}{\sigma_i^2} = S a_0 + S_x a_1 - S_y = 0,$$

where the quantities S and S_y, and S_{xy} are defined by

$$S = \sum_{i=1}^{N} \frac{1}{\sigma_i^2}, \quad S_x = \sum_{i=1}^{N} \frac{x_i}{\sigma_i^2}, \quad S_y = \sum_{i=1}^{N} \frac{y_i}{\sigma_i^2}, \quad S_{xx} = \sum_{i=1}^{N} \frac{x_i^2}{\sigma_i^2}, \quad S_{xy} = \sum_{i=1}^{N} \frac{x_i y_i}{\sigma_i^2}. \tag{2.35}$$

Making these substitutions in both equations in Equation (2.34) gives us the two equations

$$S a_0 + S_x a_1 = S_y, \qquad S_x a_0 + S_{xx} a_1 = S_{xy},$$

where each quantity S, S_x, S_y, S_{xy}, and S_{xx} is just a number that we calculate from the data, and the only things we do not know are a_0 and a_1. We have two equations in two unknowns that we can solve to give

$$a_0 = \frac{S_y S_{xx} - S_{xy} S_x}{\Delta}, \quad a_1 = \frac{S S_{xy} - S_x S_y}{\Delta}, \quad \Delta = S S_{xx} - S_x S_x. \tag{2.36}$$

So, we can directly calculate the values of a_0 and a_1 that give the best fit of a straight line to the observed data. But we can do more! A straightforward but tedious calculation using the propagation of uncertainties from Section 2.9.1 gives us the uncertainties in a_0 and a_1

$$\sigma_{a_0}^2 = \frac{S_{xx}}{\Delta}, \qquad \sigma_{a_1}^2 = \frac{S}{\Delta}. \tag{2.37}$$

These equations tell us something very interesting and very useful for the design of experiments: the magnitudes of the uncertainties in the parameters a_0 and a_1 depend only on the values of the independent variable x_i and the uncertainty in the measured values (σ_i^2), and do *not* depend on the measured values y_i. So, to design an experiment that minimizes the uncertainties in a_0 and a_1 we have to *maximize* Δ. Equation (2.36) tells us that to maximize Δ we need to maximize the difference between S_{xx} and $S_x S_x$. These quantities are sums over the x_i data values, so this is equivalent to maximizing the range of x values; the larger range of x that you cover, then the better will be your estimates of the slope and intercept.

Typically, when we perform experiments, we control the x variable (the *independent variable*) and assume that it has negligible uncertainty. We then measure replicates of the y variable (the *dependent variable*), often getting a range of values for the same value of x; a regression using data where there are uncertainties only in the y variables is called a *Type I regression*. However, it is not always possible to have negligible uncertainty in the x values, especially with measurements made in the field. In such cases, we have to modify our regression technique and use a *Type II regression* that also takes into account the uncertainties in the x values (Legendre and Legendre, 2012).

2.10 Integration

Integration is another important and useful tool for understanding processes that vary spatially and temporally. For example, the growth of phytoplankton in the ocean varies with location, time, and depth. The rate of growth of new biomass is called *primary production* and changes with available light and nutrients, both of which vary with space and time. Figure 2.19 shows hypothetical profiles of primary production at two locations in the ocean, say the tropical Pacific (A) and the North Atlantic (B). Which location has the greater primary production? We can make a meaningful comparison between these sites by dividing the depth range into smaller depth intervals, estimating the production for each curve within each of these depth intervals and summing these numbers multiplied by the size of the depth interval over the whole depth range.This will tell us the total or *integrated* production over the top 50 m at both locations. But this is also the area under each of the two curves. As another example, we may be interested in the total absorption and scattering of light as it passes through the atmosphere to the surface of the Earth. This will also be an integrated quantity that will depend on the length of the path that light has travelled through the atmosphere. Both of these examples interpret an integral as a sum—a sum of rates with depth in the ocean or atmosphere.

Integration is also closely associated with differentiation, and integrals are often called *antiderivatives*. To see why this is, let us consider a situation where we know the derivative of a function, $f(t) = dF/dt$, but we do not know the function $F(t)$ itself. How can we calculate F? Let us assume that we do know the value of the function at a single value of $t = t_0$, $F(t_0)$. We are stuck without any more information, so let us also assume the simplest

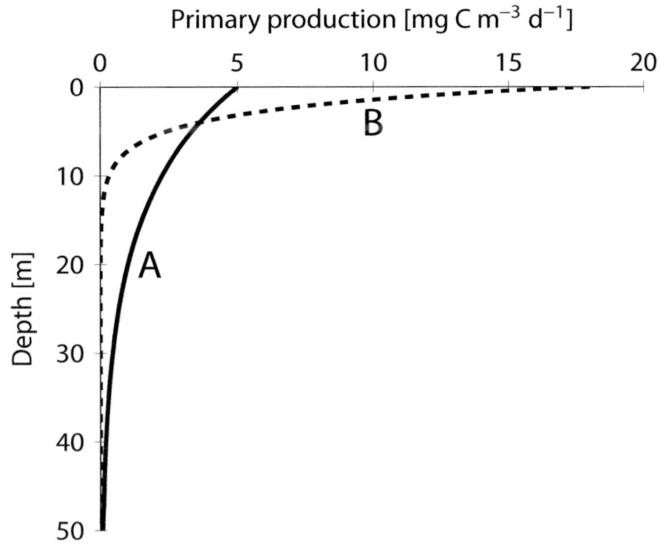

Figure 2.19 Hypothetical depth profiles of primary production in two regions of the ocean.

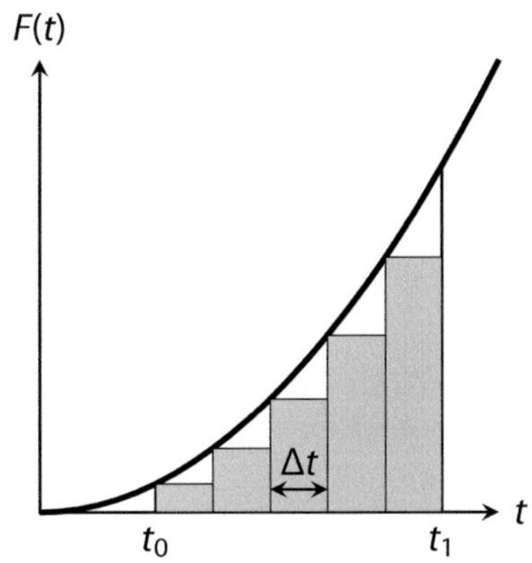

Figure 2.20 The Riemann integral showing the area under a curve as the sum of rectangular areas.

thing we can about the derivative $f(t)$—that it is constant. We know that a straight line has a constant derivative, so F must be the equation of a straight line, and we can calculate the value of F at any value of t

$$F(t) = F(t_0) + (t - t_0)f.$$

In other words, the value of $F(t)$ is the starting value $F(t_0)$, plus the derivative (dF/dt) multiplied by the interval $t - t_0$. But, in general, the derivative $f = dF/dt$ is not constant. However, we know that we can approximate a curve by a straight line (the tangent to the curve) over an interval $(t - t_0)$ if we make the interval very small. So, if we want to integrate the derivative f between $t = t_0$ and $t = t_1$, we first subdivide the interval into a large number (N) of small subintervals of size $\Delta t = (t_1 - t_0)/N$ (Figure 2.20). If we

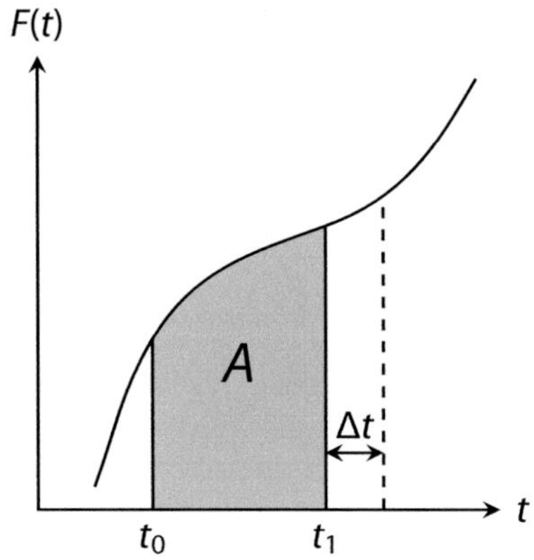

Figure 2.21 The derivative as the rate of change of area under the curve.

make these intervals small enough (i.e., make N large enough), then within any interval $f(t) \approx \text{constant} = f(t_i)$. Now, the change in the function F across each small interval is $\Delta F_i \approx f(t_i)\Delta t$. To calculate the value of F at $t = t_1$ we just add up all these small increments:

$$F(t) = F(t_0) + \sum_{i=1}^{N} \Delta F_i = F(t_0) + \sum_{i=1}^{N} f(t_i)\Delta t.$$

Now, take the limit as $N \to \infty$, i.e., $\Delta t \to 0$, to give

$$F(t_1) = F(t_0) + \lim_{\Delta t \to 0} \sum_{i=1}^{N} f(t_i)\Delta t = F(t_0) + \int_{t_0}^{t_1} f(t)\,dt = F(t_0) + \int_{t_0}^{t_1} \frac{dF}{dt}\,dt, \qquad (2.38)$$

where we have introduced the integral sign[11] to denote the limit of the summation. This is interesting because it tells us that the integral of the derivative of $F(t)$ between t_0 and t_1 is simply $F(t_1) - F(t_0)$.

We can think of this in another way. Let us assume this time that we know the function $F(t)$ and we know the area (A) under the curve between $t = t_0$ and $t = t_1$ (Figure 2.21). Now, let us move a very small increment Δt to the right of t_1 and ask what is the corresponding change (ΔA) in A. If Δt is small enough, then we can approximate the new area as a rectangle with height $F(t_1)$ and width Δt. So $\Delta A \approx F(t_1)\Delta t$. Rearranging this equation and taking the limit as $\Delta t \to 0$ tells us that

$$F(t) = \lim_{\Delta t \to 0} \frac{\Delta A}{\Delta t} = \frac{dA}{dt}.$$

So, the function $F(t)$ is the derivative of the area under the curve, and using Equation (2.38) we find that

[11] This symbol was introduced by Gottfried Wilhelm Leibniz (1646–1716), who, independently from Isaac Newton, also invented differential and integral calculus. Newton's approach to calculus was hard to understand, and it is Leibniz's approach and notation that we use today. The integral sign is derived from the typographical long-s and was used by Leibniz to denote an infinite sum.

$$\int_{t_0}^{t_1} F(t)\,dt = A(t_1) - A(t_0);$$

in other words, the integral is the area under the curve.

These results demonstrate, in a rather nonrigorous way, the validity of what is called *the fundamental theorem of calculus* , which relates the two processes of integration and differentiation and which allows us to compute integrals. It also provides a rationale for calling an integral an *antiderivative*. To compute an integral of a function we have to find another function whose derivative is the function we are integrating, in other words

$$\int_{x=a}^{x=b} \frac{dg(x)}{dx}\,dx = g(x)|_{x=a}^{x=b} = g(x=b) - g(x=a), \tag{2.39}$$

which also defines the symbol $|_a^b$.

Example 2.23 We can use the fundamental theorem of calculus to evaluate integrals. For example, let us evaluate the definite integral

$$\int_0^1 (2x^2 - x + 1)\,dx.$$

We know that the derivative of a power of x is given by

$$\frac{dax^n}{dx} = a(n-1)x^{n-1},$$

so we can calculate the antiderivative of each term in the integral by working backwards

$$2x^2 \to \frac{2}{3}x^3, \quad x \to \frac{1}{2}x^2, \quad 1 \to x,$$

but there is something missing. The derivative of a constant is zero, so we could have a constant term without realizing it. So, our integral is

$$\int_0^1 (2x^2 - x + 1)\,dx = \left(\frac{2}{3}x^3 - \frac{1}{2}x^2 + x + c\right)\Big|_0^1 = \left(\frac{2}{3} - \frac{1}{2} + 1 + c\right) - (0 - 0 + 0 + c) = \frac{7}{6}.$$

Notice that the constant c cancelled out. However, if we had had an indefinite integral (i.e., the limits were not specified), then the constant c would have remained as a *constant of integration*.

The fundamental theorem of calculus is a very important theorem and, as we have seen, gives us a method for evaluating integrals between two values. However, we will very rapidly run out of integrals that are easy to evaluate in this manner. For example, it is unclear what function we would have to differentiate to evaluate the following integral:

$$\int_0^1 \frac{3x^2 \sin(x)}{\sqrt{x^2 + 12}}.$$

Perhaps thinking of the integral as an area under the curve will allow us to evaluate more integrals. Let us do this to evaluate the integral

$$\int_0^a \beta x^2\, dx.$$

The first thing to do is subdivide the interval $0 \le x \le a$ into N intervals, each of width $\Delta x = a/N$. The sum of the area of all the rectangles is

$$\begin{aligned} A &= \frac{a}{N}\beta(0)^2 + \frac{a}{N}\beta\left(\frac{a}{N}\right)^2 + \frac{a}{N}\beta\left(\frac{2a}{N}\right)^2 + \cdots + \frac{a}{N}\beta\left(\frac{(N-1)a}{N}\right)^2 \\ &= \frac{a}{N}\beta\left(\frac{a}{N}\right)^2\left((0)^2 + (1)^2 + 2^2 + \cdots + (N-1)^2\right) \\ &= \frac{a}{N}\beta\left(\frac{a}{N}\right)^2\left(\frac{N(N-1)(2N-1)}{6}\right) = \beta\frac{a^3}{6}\left(2 - \frac{3}{N} + \frac{1}{N^2}\right). \end{aligned}$$

Now, let $N \to \infty$ (i.e., $\Delta x \to 0$), so that

$$A = \lim_{N\to\infty} \beta\frac{a^3}{6}\left(2 - \frac{3}{N} + \frac{1}{N^2}\right) = \beta\frac{a^3}{6}\lim_{N\to\infty}\left(2 - \frac{3}{N} + \frac{1}{N^2}\right) = \beta\frac{a^3}{3}.$$

We could have done this for a more general power and found that

$$\int_a^b x^n\, dx = \left.\frac{1}{n+1}x^{n+1}\right|_a^b = \frac{1}{n+1}\left(b^{n+1} - a^{n+1}\right).$$

This process can also rapidly become rather tedious. So, generally we make use of a few fundamental integrals (see Appendix B) and use techniques for evaluating more complicated cases from them. However, unlike differentiation, there are many integrals that we cannot evaluate in terms of what are called *elementary functions* (i.e., powers, exponentials, logarithms, trigonometric functions, etc.). In those cases, we have to resort to numerical methods (see Section 2.16). First, we will look at some of the common techniques for evaluating integrals analytically, and then we will look at some numerical methods.

2.10.1 Properties of Integrals

Integrals can broadly be classified into *definite integrals* and *indefinite integrals*. A definite integral of a function has limits on the integral sign and evaluates to a number, the area under the curve represented by the function between the two limits. An indefinite integral does not have limits on the integral sign and evaluates to another function. For example, if we evaluate the definite integral

$$\int_0^1 x^2\, dx = \left.\frac{x^3}{3}\right|_0^1 = \left.\frac{x^3}{3}\right|_{x=1} - \left.\frac{x^3}{3}\right|_{x=0} = \frac{1}{3},$$

the result is a number. However, integrating the same function without specifying limits (an indefinite integral) gives

$$\int x^2\, dx = \frac{x^3}{3} + \text{constant},$$

where the constant arises because if we differentiate the right-hand side of this equation we get

$$\frac{d}{dx}\left(\frac{x^3}{3} + \text{constant}\right) = x^2 + 0,$$

so we can only evaluate an indefinite integral up to an unknown constant—we need to have more information to give a value to the constant.

The integral has some useful properties, some of which are obvious from the geometric interpretations we have been using. Firstly, integrals are *additive*, so if $f(x)$ and $g(x)$ are integrable functions in an interval $a \leq x \leq b$, then

$$\int_a^b (f(x) \pm g(x))\, dx = \int_a^b f(x)\, dx \pm \int_a^b g(x)\, dx. \tag{2.40}$$

If a function is multiplied by a constant, then so is the integral

$$\int_a^b \alpha f(x)\, dx = \alpha \int_a^b f(x)\, dx. \tag{2.41}$$

We can also add integrals over contiguous regions, so if $a \leq c \leq b$, we have

$$\int_a^b f(x)\, dx = \int_a^c f(x)\, dx + \int_c^b f(x)\, dx. \tag{2.42}$$

Also,

$$\int_a^a f(x)\, dx = 0 \tag{2.43}$$

and

$$\int_a^b f(x)\, dx = -\int_b^a f(x)\, dx. \tag{2.44}$$

With a little thought, you can see that the next two properties are also true; if, within the interval $a \leq x \leq b$ we have that $m \leq f(x) \leq M$, where m and M are constants, then

$$m(b-a) \leq \int_a^b f(x)\, dx \leq M(b-a). \tag{2.45}$$

This makes intuitive sense: if $f(x)$ is always larger than m, then the area under the curve of $f(x)$ must be larger than the area of the rectangle of sides m and $(b-a)$; similarly for the other bound. Next, if in the interval $a \leq x \leq b$ we have that $f(x) \leq g(x)$, then

$$\int_a^b f(x)\, dx \leq \int_a^b g(x)\, dx. \tag{2.46}$$

Exercise 2.10.1 Look at each of the Equations (2.40)–(2.43) and, using the geometric interpretation of the integral, convince yourself that they are true.

A definite integral is just a number. This means that

$$I = \int_a^b f(x)\, dx = \int_a^b f(y)\, dy = \int_a^b f(\zeta)\, d\zeta,$$

where the function being integrated is the same in each integral. For example,

$$I = \int_0^1 x^2\,dx = \int_0^1 y^2\,dy = \frac{1}{3}.$$

In each case, the parameter x, y, or ζ being integrated over is a *dummy variable*—it looks like a variable of the equation, but because the definite integral evaluates to a number, it vanishes from the final answer. In the following equation, x is not a dummy variable even though it appears as a limit of the integral,

$$y(x) = \int_0^x x\mathrm{e}^t\,dt = x\int_0^x \mathrm{e}^t\,dt = x(\mathrm{e}^t)\big|_0^x = x(\mathrm{e}^x - 1),$$

where we have used Equation (2.41) and the integral evaluates to a function of x (because of the upper limit of the integral), not a function of t.

It is always useful to know something about the function you are integrating because properties of the function can sometimes help simplify the integral considerably. This is especially true of even and odd functions. A function $f(x)$ is *even* if $f(-x) = f(x)$, and it is *odd* if $f(-x) = -f(x)$. For example, $f(x) = x^2$ is an even function because $f(-x) = (-x)^2 = x^2 = f(x)$, whereas $f(x) = x^3$ is an odd function. In fact, all even powers are even functions and odd powers are odd functions.[12] What happens if we multiply even and odd functions together? An even function multiplied by an even function produces another even function, and an odd function multiplied by an odd function also produces an even function, and an even function multiplied by an odd function produces an odd function. For example, if $f(x) = x^2$, which is even, and $g(x) = x^3$, which is odd, then $f(x) * f(x) = x^4$ and $g(x) * g(x) = x^6$, both of which are even, while $f(x) * g(x) = x^5$, which is odd. Lastly, any function can be written as a sum of an even and an odd function—i.e., for any function $f(x)$, we can write

$$f(x) = f_{\text{even}} + f_{\text{odd}} = \frac{1}{2}(f(x) + f(-x)) + \frac{1}{2}(f(x) - f(-x)).$$

The first function is even, because

$$f_{\text{even}}(-x) = \frac{1}{2}(f(-x) + f(x)) = f_{\text{even}}(-x).$$

Exercise 2.10.2 Show that $f_{\text{odd}}(x) = (f(x) - f(-x))/2$ is an odd function.

Example 2.24 As an example, let us decompose $f(x) = \mathrm{e}^x$ into even and odd components. Using the definitions of hyperbolic functions (Appendix B) we can write

$$\mathrm{e}^x = \frac{1}{2}(\mathrm{e}^x + \mathrm{e}^{-x}) + \frac{1}{2}(\mathrm{e}^x - \mathrm{e}^{-x}) = \cosh(x) + \sinh(x),$$

so the even function is $\cosh(x)$ and the odd function is $\sinh(x)$.

---

[12] This is where the terminology comes from!

Why is it useful to know if a function is even or odd? One important reason is that integrals of even and odd functions over an interval $-L \leq x \leq +L$ simplify considerably. The integral of an odd function over this interval is zero, and the integral of an even function is twice the integral of the same function between $0 \leq x \leq +L$. To see this, consider the integral of the function $f(x)$ between the limits $x = -L$ and $x = L$. We can split the integral into two parts using Equation (2.42),

$$\int_{-L}^{L} f(x)\,dx = \int_{-L}^{0} f(x)\,dx + \int_{0}^{L} f(x)\,dx$$

and make the substitution $u = -x$ in the first integral (remember, x is a dummy variable),

$$\int_{-L}^{L} f(x)\,dx = -\int_{u=L}^{0} f(-u)\,du + \int_{0}^{L} f(x)\,dx = \int_{0}^{L} f(-x)\,dx + \int_{0}^{L} f(x)\,dx.$$

If $f(x)$ is an even function, then $f(-x) = f(x)$, and we find

$$\int_{-L}^{L} f_{\text{even}}(x)\,dx = 2\int_{0}^{L} f(x)\,dx, \tag{2.47}$$

but if $f(x)$ is odd, then $f(-x) = -f(x)$, and we find

$$\int_{-L}^{L} f_{\text{odd}}(x)\,dx = 0. \tag{2.48}$$

Exercise 2.10.3 Verify that Equations (2.47) and (2.48) are true for the functions $f(x) = x^2$, $f(x) = x^3$, $f(\theta) = \sin(\theta)$, and $f(\theta) = \cos(\theta)$.

2.11 Techniques of Integration

Evaluating integrals in terms of elementary functions is not always easy and can require cunning, skill, understanding, and above all, a lot of patience. However, it is often worth the attempt to try. For example, numerically evaluating an integral on a computer can involve many separate calculations that can be avoided if we analytically evaluate the integral instead. In the following sections, we will explore some of the more common and useful techniques for evaluating integrals. There are many more techniques than can be covered here, and to learn more you should consult the references given in Section 2.17.

2.11.1 Partial Fractions

The method of partial fractions is not so much a method of integration in itself, rather it is a means of decomposing a rational function (i.e., a ratio of two polynomials) into sums of simpler rational functions, ones that hopefully we can integrate. The basic idea is that if we have a rational function of the form

$$f(x) = \frac{P(x)}{Q(x)}$$

and we can factorize the denominator so that

$$f(x) = \frac{P(x)}{g(x)h(x)},$$

then we can decompose $f(x)$ into an expression of the form

$$f(x) = \frac{U(x)}{g(x)} + \frac{V(x)}{h(x)}. \tag{2.49}$$

It is important to note that for this technique to work, the degree of the polynomial in the numerator should be less than that in the denominator. How do we choose $U(x)$ and $V(x)$? There are several rules of thumb to guide us:

- If $g(x)$ or $h(x)$ is linear (i.e., of the form $ax+b$ with a and b constants), then the numerator ($U(x)$ or $V(x)$) is a constant.
- If either $g(x)$ or $h(x)$ has the form $(ax + b)^n$, then we need a sum of multiple terms, one for each power of the linear term. For example, for a factor $(a + bx)^4$ we would use
$$\frac{P(x)}{(ax+b)^4} = \frac{A}{ax+b} + \frac{B}{(ax+b)^2} + \frac{C}{(ax+b)^3} + \frac{D}{(ax+b)^4}.$$
- If either $g(x)$ or $h(x)$ is a quadratic expression $(ax^2 + bx + c)$ that we cannot factorize, then we need to use a term of the form
$$\frac{P(x)}{ax^2+bx+c} = \frac{Ax+B}{ax^2+bx+c}.$$
- Lastly, if we have a quadratic expression we cannot factorize and it is raised to a power, we need a sum of terms again. For example, if $h(x)$ or $g(x)$ has the form $(ax^2 + bx + c)^2$, then we use a term of the form
$$\frac{P(x)}{(ax^2+bx+c)^2} = \frac{Ax+B}{ax^2+bx+c} + \frac{Cx+D}{(ax^2+bx+c)^2}.$$

To find the values of the constants (A, B, etc.) we put the right-hand side of the equation over a common denominator and equate powers of x on both sides of the equals sign. The different forms listed in these rules of thumb are designed to make sure that there are sufficient powers of x on the right-hand side of the equation to do this.

Example 2.25 To see how partial fractions work, let us evaluate the integral

$$\int \frac{2x+1}{x^2+2x-8}\,dx.$$

The first thing is to realize that we can factorize the denominator into two linear factors,

$$x^2 + 2x - 8 = (x-2)(x+4),$$

so we can write the integrand as

$$\frac{2x+1}{(x-2)(x+4)} = \frac{A}{(x-2)} + \frac{B}{(x+4)}.$$

Now, our job is to find A and B. To do that we rearrange the right-hand side of the equation:

$$\frac{2x+1}{(x-2)(x+4)} = \frac{A(x+4)+B(x-2)}{(x-2)(x+4)} = \frac{x(A+B)+4A-2B}{(x-2)(x+4)}.$$

If these expressions are to be equal, then we must have

$$A + B = 2 \qquad \text{and} \qquad 4A - 2B = 1,$$

or $A = 5/6$, $B = 7/6$. So now, we can do our integral:

$$\int \frac{2x+1}{x^2+2x-8}\,dx = \int \frac{5}{6(x-2)}\,dx + \int \frac{7}{6(x+4)} = \frac{5}{6}\ln|x-2| + \frac{7}{6}\ln|x+4| + c.$$

Exercise 2.11.1 We can have problems if we choose the wrong form for the partial fractions expansion. Evaluate the constants A, B, C, and D in the partial fractions expansion

$$\frac{2x+1}{(x-2)(x+4)} = \frac{Ax+B}{(x-2)} + \frac{Cx+D}{(x+4)}.$$

Exercise 2.11.2 Use partial fractions to evaluate the following integrals:

$$\text{a.} \int \frac{2x-1}{(x-1)(x+2)}\,dx, \qquad \text{b.} \int \frac{2x-1}{(x-1)(x^2+1)}\,dx.$$

2.11.2 Substitution of Variables

Many integrals that seem impossible to compute can be turned into ones we can evaluate by using a suitable substitution of variables. There are many substitutions that can be used, and the trick to choosing the right one is to know what integrals you *can* evaluate (see those in Appendix B for example) and choose substitutions that turn your integral into one of those—doing so may require more than one substitution.

Example 2.26 To see how this method works, let us use a substitution to evaluate

$$\int_0^1 (3x+5)^4\,dx.$$

A good substitution is one that will transform the integrand into something we know how to integrate. We know how to integrate u^4:

$$\int_a^b u^4\,du = \left.\frac{u^5}{5}\right|_a^b = \left.\frac{u^5}{5}\right|_{u=b} - \left.\frac{u^5}{5}\right|_{u=a} = \frac{1}{5}\left(b^5 - a^5\right),$$

so we should look for a substitution that turns our integral into this one; i.e., we want $u(x) = 3x + 5$. By making this substitution we also change the differential dx, but it does not change simply to du. Instead we have

$$du = \frac{du}{dx}\,dx = 3dx \qquad \text{so} \qquad dx = \frac{du}{3}.$$

The limits of the original integral are in terms of x, so we need to change them to the corresponding values of u; $u(x = 0) = 5$ and $u(x = 1) = 8$. The integral now becomes

$$\int_{x=0}^{x=1} (3x+5)^4\,dx = \int_{u=5}^{u=8} u^4 \frac{du}{3} = \frac{1}{3}\left(\frac{u^5}{5}\right)\Big|_5^8 \approx 1976.2.$$

Example 2.27 A slightly more involved example is given by the integral

$$\int \sin^n(\theta)\cos(\theta)\,d\theta,$$

where n is a constant. Here we have a mixture of sines and cosines, and we can take advantage of the fact that the derivative of a sine is a cosine (see Appendix B). If we make the substitution $x(\theta) = \sin(\theta)$, then

$$dx = \cos(\theta)\,d\theta,$$

which is actually part of the original integral. So,

$$\int \sin^n(\theta)\cos(\theta)\,d\theta = \int x^n\,dx = \frac{x^{n+1}}{n+1} + C = \frac{\sin^{(n+1)}(\theta)}{n+1} + C,$$

where C is the constant of integration that we get whenever we have an indefinite integral.

Example 2.28 As a last example, let us evaluate the integral

$$\int \frac{\sqrt{4-x^2}}{x^2}\,dx.$$

Integrals involving expressions like $\sqrt{a^2+x^2}$, $\sqrt{a^2-x^2}$, and $\sqrt{x^2-a^2}$ can often be evaluated by recalling a little bit of trigonometry (see Appendix B). If we want our answer to be a real number, then $x \leq \pm 2$, so we can make the substitution $x = 2\sin(\theta)$, and the integrand becomes

$$\frac{\sqrt{4-x^2}}{x^2}\,dx = \frac{\sqrt{4-4\sin^2(\theta)}}{4\sin^2(\theta)}2\cos(\theta)\,d\theta = \frac{\cos^2(\theta)}{\sin^2(\theta)}\,d\theta = \cot^2(\theta)\,d\theta.$$

We can use the formulae in Appendix B to evaluate this, giving

$$\int \cot^2(\theta)\,d\theta = \int (\csc^2(\theta) - 1)\,d\theta = -\cot(\theta) - \theta + C.$$

Now, we have to substitute back to get an equation in the original variables:

$$\int \frac{\sqrt{4-x^2}}{x^2}\,dx = -\cot\left(\arcsin\left(\frac{x}{2}\right)\right) - \arcsin\left(\frac{x}{2}\right) + C.$$

Exercise 2.11.3 Use substitution to evaluate the following integrals (the first two contain hints for the substitutions to use):

a. $\int \cos(4x-5)\,dx \quad [u = 4x-5]$, b. $\int \frac{\sin(\theta^{1/2})}{\theta^{1/2}}\,d\theta \quad [u = \theta^{1/2}]$,

c. $\int \frac{\sin(\theta)}{\cos^7(\theta)}\,d\theta$, d. $\int \frac{u^2}{\sqrt{u^3+6}}\,du$

2.11.3 Integration by Parts

In Section 2.2.1 we learned how to take the derivative of a product of functions. We can use that sometimes to evaluate integrals of products of functions using a technique called *integration by parts*. If we have a function $f(x) = u(x)v(x)$, then we can differentiate $f(x)$ using Equation (2.10) to get

$$\frac{d(u(x)v(x))}{dx} = u(x)\frac{dv(x)}{dx} + v(x)\frac{du(x)}{dx}.$$

We can integrate both sides of this equation to give

$$\int \frac{d(u(x)v(x))}{dx}\,dx = \int u(x)\frac{dv(x)}{dx}\,dx + \int v(x)\frac{du(x)}{dx}\,dx.$$

We can use Equation (2.39) to evaluate the integral on the left-hand side of the equation, and after rearranging we get

$$\int u(x)\frac{dv(x)}{dx}\,dx = u(x)v(x) + c - \int v(x)\frac{du(x)}{dx}\,dx. \tag{2.50}$$

We can now evaluate the integral of $g(x)h(x)$ by choosing which function we pick as $u(x)$ and $v'(x)$ and using Equation (2.50).

Example 2.29 We have to be judicious in how we choose our functions. For example, let us evaluate

$$\int xe^{2x}\,dx.$$

We know how to integrate both x and $\exp(2x)$ on their own, but we are apparently stuck when they are multiplied together. However, we can use integration by parts to simplify things. The first thing to do is to choose whether x or $\exp(2x)$ will be $u(x)$ or $v(x)$. If we look at Equation (2.50), we see that the function we choose as $u(x)$ gets differentiated in the integral on the right-hand side, and the function we choose as $v'(x)$ gets integrated. If we were to choose $u(x) = \exp(2x)$, then repeatedly differentiating $u(x)$ would keep giving us an exponential, but integrating x to get $v(x)$ would give us a quadratic term. This would make the integral on the right-hand side of Equation (2.50) more complicated than the original one! However, if we choose $u(x) = x$, then when we differentiate it we get a constant and the integral on the right-hand side of the equation becomes easier. So, let us pick $u(x) = x$, and $dv/dx = \exp(2x)$, then

$$\frac{du}{dx} = 1, \qquad v(x) = \int \exp(2x)\,dx = \frac{1}{2}e^{2x}.$$

Notice that we did not add any constants of integration at this stage because they can all be lumped into a single constant at the end of the calculation. Our integral now becomes

$$\int xe^{2x}\,dx = x\frac{1}{2}e^{2x} - \int \frac{1}{2}e^{2x}\,dx = \frac{x}{2}e^{2x} - \frac{1}{4}e^{2x} + c = \frac{1}{4}\,(2x-1)\,e^{2x} + c.$$

Example 2.30 Sometimes we have to use integration by parts more than once to evaluate an integral. This can often happen with integrals involving sines and cosines. For example, let us evaluate

$$\int e^x \cos(x)\, dx.$$

Choosing $u(x) = \exp(x)$ and $v(x) = \cos(x)$ we get

$$\int e^x \cos(x)\, dx = e^x \sin(x) - \int e^x \sin(x)\, dx,$$

which looks as if we have gotten nowhere. However, if we apply the integration by parts technique to the integral on the right-hand side of the equation, we get

$$\int e^x \cos(x)\, dx = e^x \sin(x) - \int e^x \sin(x)\, dx = e^x \sin(x) + e^x \cos(x) - \int e^x \cos(x)\, dx.$$

We can see what has happened by applying the technique a second time—because successive differentiations of $\sin(x)$ and $\cos(x)$ cycle through $\pm\sin(x)$ and $\pm\cos(x)$, we get back the integral we started with, but with a different sign. So, we can rearrange our equation to get

$$\int e^x \cos(x)\, dx = \frac{1}{2}\left(e^x \sin(x) + e^x \cos(x)\right) + C.$$

Exercise 2.11.4 Evaluate the following integrals using integration by parts:

$$\text{a. } \int u^3 e^u\, du, \quad \text{b. } \int x \ln(x)\, dx, \quad \text{c. } \int x \cos(x) \sin(x)\, dx.$$

2.11.4 Differentiation

Some techniques for evaluating integrals are not used very often, but can be very useful. One of these is differentiation with respect to a parameter, sometimes called differentiation under the integral sign. Our ability to do this arises from something called *Leibniz's rule*,[13] which, in its general form, states

$$\frac{d}{dt}\int_{a(t)}^{b(t)} g(x,t)\, dx = \int_{a(t)}^{b(t)} \frac{\partial g(x,t)}{\partial t}\, dx + g(b(t),t)\frac{db}{dt} - g(a(t),t)\frac{da}{dt}. \tag{2.51}$$

Let us see first what this implies if the limits on the integral $a(t)$ and $b(t)$ are not functions of t but constants. Then the last two terms on the right-hand side of the equation are zero, and we have

$$\frac{d}{dt}\int_a^b g(x,t)\, dx = \int_a^b \frac{\partial f(x,t)}{\partial t}\, dx.$$

This equation tells us that we can change the order in which we do the integration or differentiation as long as we are differentiating a variable that it not being integrated over.

[13] Named after Gottfried Wilhelm Leibniz.

We can make things a little more complicated by seeing what happens when $g(x,t)$ is just a function of x. Then, the first term on the right-hand side of Equation (2.51) is zero, and

$$\frac{d}{dt}\int_{a(t)}^{b(t)} g(x)\,dx = g(b(t))\frac{db}{dt} - g(a(t))\frac{da}{dt}.$$

Example 2.31 Leibniz's rule can be useful in evaluating integrals that might otherwise seem hopeless. For example, the integral

$$\mathcal{I}(\alpha) = \int_0^1 \frac{x^\alpha - 1}{\log x}\,dx$$

involves an unknown parameter α, which makes the integral hard to evaluate. However, we can use Leibniz's rule to differentiate with respect to α:

$$\frac{d}{d\alpha}\int_0^1 \frac{x^\alpha - 1}{\log x}\,dx = \int_0^1 \frac{\partial}{\partial\alpha}\left(\frac{x^\alpha - 1}{\log x}\right)dx = \int_0^1 \frac{x^\alpha \log x}{\log x}\,dx = \int_0^1 x^\alpha\,dx = \frac{1}{1+\alpha}.$$

Now, we can integrate both sides with respect to α to find

$$\mathcal{I}(\alpha) = \int_0^1 \frac{x^\alpha - 1}{\log x}\,dx = \log(1+\alpha) + c.$$

When $\alpha = 0$, $\mathcal{I}(0) = 0$ (because $x^0 = 1$), so the integrand is zero no matter what the value of x, thus $\mathcal{I}(0) = 0 = \log(1+0) + c = 0$, so $c = 0$, and we end with

$$\mathcal{I}(\alpha) = \int_0^1 \frac{x^\alpha - 1}{\log x}\,dx = \log(1+\alpha).$$

Example 2.32 Leibniz's rule can also help us take derivatives of integrals. For example, if

$$y(x) = \int_x^{x^2} xu^2\,du,$$

we can use Leibniz's rule to calculate dy/dx as

$$\frac{dy}{dx} = (x)(x^4)(2x) - (x)(x^2)(1) + \int_x^{x^2} u^2\,du = \frac{7x^6}{3} - \frac{4x^3}{3}.$$

2.11.5 Other Methods

Evaluating integrals on paper using techniques like the ones we have just explored can be tricky and time consuming. There are alternatives, but using these alternatives often requires you to be able to use some or all of the techniques we have explored. The first alternative is to use tables of integrals. Perhaps the most comprehensive and famous collection is the one compiled by Gradshteyn and Ryzhik (1980). This is a hefty tome that lists the evaluation of thousands of definite and indefinite integrals, but to use such a set of tables effectively you have to be able to transform the stubborn integral you are working on into a form that is in the tables. This means being familiar with using substitutions, at the very least.

Another alternative method for evaluating integrals is to use the capabilities of a computer algebra system.[14] These programs make use of the variants of the Risch algorithm (Risch, 1969, 1970), which is an ingenious method of converting an integral into an algebraic problem. The algorithm either evaluates the integral if it can be evaluated in terms of elementary functions, or tells you that such an evaluation does not exist. However, like the use of tables of integrals, you need to know how to transform and manipulate integrals to make effective use of such a system.

2.12 Proper and Improper Integrals

So far we have been a little cavalier with the functions we integrate, but in reality we need to exercise some caution. Most of the integrals we have looked at so far are called *proper* integrals. A proper integral,

$$\int_a^b f(x)\,dx,$$

is one where

- both of the limits a and b are finite
- the function $f(x)$ is not infinite on the interval $a \leq x \leq b$
- the function $f(x)$ has only a finite number of discontinuities on the interval $a \leq x \leq b$.

An *improper* integral is one that violates any one of these conditions. Why is this important? Well, recall that our definition of an integral (Equation (2.38)) involved taking a limit as $\Delta x \to 0$. Each of the conditions for proper integrals ensure that this is a valid thing to do. If we have infinities in the limits or integrand, or discontinuities in the integrand, then we may not be able to legitimately take this limit. This means that we have to actually examine the functions we are integrating. We can evaluate proper integrals using the techniques we have been studying, but to evaluate improper integrals we must explicitly take limits, splitting the integral into two or more parts.

Example 2.33 To see an example of this, let us evaluate the integral

$$\int_0^\pi \sec^2(\theta)\,d\theta.$$

This integral is an improper integral because the function $\sec^2(\theta) = 1/\cos^2(\theta)$ has a discontinuity at $\theta = \pi/2$, where $\cos(\theta) = 0$ and changes sign. This is in the middle of the interval we are integrating over (Figure 2.22).[15] We can overcome this problem by taking the interval from 0 to π and splitting it at the discontinuity, so that the integral becomes

[14] Such as Mathematica™ or Maple™.

[15] This really highlights the need to understand what a function looks like *before* we start using it, which means being able to sketch curves (see Section 2.7.1).

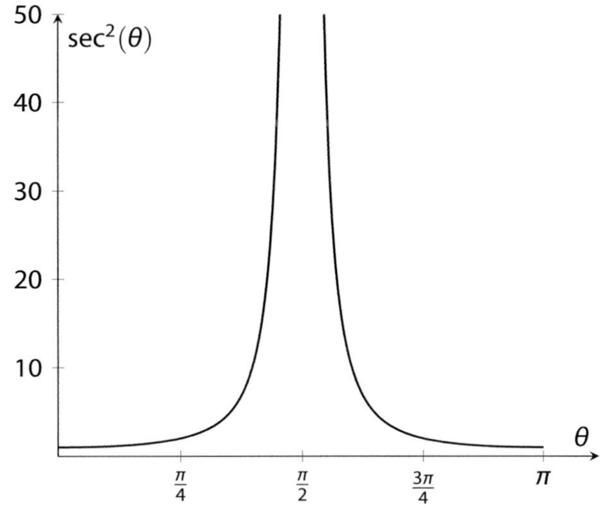

Figure 2.22 A plot of $\sec^2(\theta)$ showing the singularity at $\theta = \pi/2$ where the function becomes infinite.

$$\int_0^{\pi} \sec^2(\theta)\, d\theta = \int_0^{\pi/2} \sec^2(\theta)\, d\theta + \int_{\pi/2}^{\pi} \sec^2(\theta)\, d\theta.$$

Let us start by looking at the first integral on the right-hand side. We know the function $\sec^2(\theta)$ has a discontinuity at $\theta = \pi/2$, so we will take the upper limit of integration to be a parameter (η) and then take the limit as $\eta \to \pi/2$ from below,

$$\begin{aligned}\int_0^{\pi/2} \sec^2(\theta)\, d\theta &= \lim_{\eta\to\pi/2} \int_0^{\eta} \sec^2(\theta)\, d\theta = \lim_{\eta\to\pi/2} (\tan(\eta) - \tan(0)) \\ &= \lim_{\eta\to\pi/2} \tan(\eta) = \infty.\end{aligned}$$

The second integral has to be a positive number (because we are integrating a function squared), so we can only add to the value of the first integral. Therefore,

$$\int_0^{\pi} \sec^2(\theta)\, d\theta = \infty.$$

What would have happened if we had not recognized that $\sec^2(\theta)$ was discontinuous in the interval we were interested in? We would have written

$$\int_0^{\pi} \sec^2(\theta)\, d\theta = \tan(\theta)|_0^{\pi} = \tan(\pi) - \tan(0) = 0,$$

which would have been gloriously incorrect!

We might expect that improper integrals with infinite limits always evaluate to infinity. But this is not the case. Let us look at

$$\mathcal{I} = \int_1^{\infty} \frac{1}{x^p}\, dx, \quad p > 0. \tag{2.52}$$

We can evaluate this by replacing the infinite limit by a constant and then taking the limit as that constant tends to infinity,

$$\mathcal{I} = \lim_{a\to\infty} \int_1^a \frac{1}{x^p}\,dx = \lim_{a\to\infty} \left(\frac{x^{1-p}}{1-p} \right) \bigg|_1^a .$$

Now, we have to look at the sign of the exponent $(1-p)$. If $(1-p) > 0$, then $p < 1$ and

$$\mathcal{I} = \lim_{a\to\infty} \left(\frac{x^{1-p}}{1-p} \right) \bigg|_1^a = \lim_{a\to\infty} \left(\frac{a^{1-p}}{1-p} \right) - \frac{1}{1-p} = \infty,$$

and the integral *diverges*. If, on the other hand, $(1-p) < 0$, then $p > 1$ and

$$\mathcal{I} = \lim_{a\to\infty} \left(\frac{1}{(1-p)x^{p-1}} \right) \bigg|_1^a = \lim_{a\to\infty} \left(\frac{1}{(1-p)a^{p-1}} \right) - \frac{1}{1-p} = \frac{1}{p-1},$$

so the integral converges to a finite value.

Exercise 2.12.1 What is the value of $\mathcal{I}$ in Equation (2.52) if $p = 1$?

This is interesting. If $p = 0.9999999$, the integral diverges; but if $p = 1.00000001$, the integral converges. What is happening here? Why should the value of p make such a dramatic difference in the value of the integral? Figure 2.23 shows that the curve $y(x) = 1/x^p$ tends to zero at different rates depending on the value of p. Those functions that tend to zero sufficiently fast add increasingly smaller amounts to the area under the curve as x increases, so the total area remains finite. How can we tell if an improper integral with a limit of infinity is going to be convergent or divergent? One way is to make use of

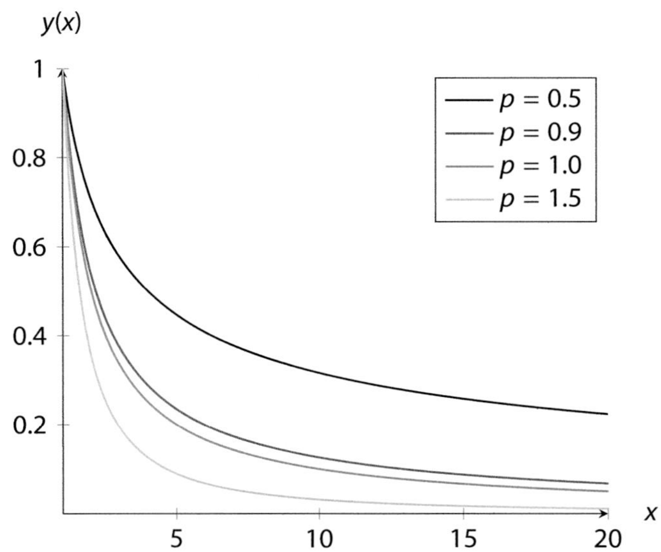

Figure 2.23 A plot of the function $y(x) = 1/x^p$ for different values of p, showing that the curves tend to zero at different rates.

Equation (2.46): if $f(x) < g(x)$ on an interval $a \leq x \leq b$, and if

$$\int_a^b g(x)\,dx \text{ is convergent, then } \int_a^b f(x)\,dx \text{ is also convergent,}$$

and if

$$\int_a^b f(x)\,dx \text{ is divergent, then } \int_a^b g(x)\,dx \text{ is also divergent.}$$

Being able to identify improper integrals is important because they can cause severe problems if we have to use numerical methods (see Section 2.16) to evaluate them (Acton, 1990, 1996), and computer algorithms generally misbehave badly when they encounter infinities.

Exercise 2.12.2 Evaluate the following improper integrals by taking the appropriate limits:

a. $\displaystyle\int_0^\infty e^{-x}\,dx$ (Replace the upper limit with a and take the limit as $a \to \infty$).

b. $\displaystyle\int_0^2 \frac{1}{\sqrt{2-x}}\,dx$

2.13 Mean Value Theorems

In Section 2.4 we discussed a mean value theorem for derivatives. There is also a mean value theorem for integrals. The theorem basically says that if $f(x)$ is a continuous function on the interval $a \leq x \leq b$, then there is a value of x, say $x = c$, lying between a and b such that

$$f(c) = \frac{1}{b-a}\int_a^b f(x)\,dx. \tag{2.53}$$

In other words, $f(c)$ is the *average* value of the function $f(x)$ over the interval $a \leq x \leq b$. This has a nice geometric interpretation (Figure 2.24). If we rewrite Equation (2.53) as

$$\int_a^b f(x)\,dx = f(c) \times (b-a),$$

we can see that the right-hand side is just the area of a rectangle of length $(b-a)$ and height $f(c)$. In other words, c is the value of x for which the area under the curve of $f(x)$ between $x = a$ and $x = b$ equals the area of a rectangle whose base is $(b-a)$ and height is $f(c)$.

Example 2.34 We can use the mean value theorem to calculate the mean value of $\sin(\theta)$ over the interval $0 \leq \theta \leq \pi$. Using Equation (2.53), the average value is given by

$$\frac{1}{\pi - 0}\int_0^\pi \sin(\theta)\,d\theta = \frac{1}{\pi}\left(-\cos(\theta)\right)\Big|_0^\pi = \frac{2}{\pi}.$$

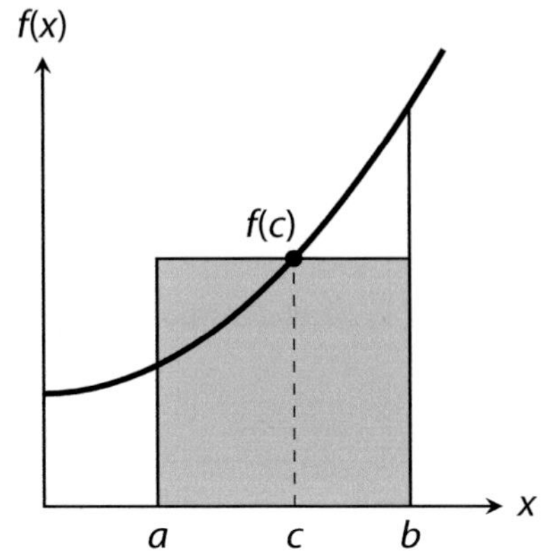

Figure 2.24 The meaning of the mean value theorem for integrals. The value of the function $f(x)$ at $x = c$ gives the height of a rectangle (the shaded area) that has the *same* area as the area under the curve $y = f(x)$ between the limits $x = a$ and $x = b$.

Example 2.35 If we know the mean value, then we can find the value of x at which it occurs. For example, we can calculate the mean value of the function $f(x) = x^2$ between $x = 0$ and $x = 2$ using Equation (2.53).

$$\frac{1}{2-0}\int_0^2 x^2\,dx = \frac{1}{2}\left(\frac{x^3}{3}\right)\bigg|_0^2 = \frac{4}{3}.$$

To find the value of x for which this value of $f(x)$ occurs, we have to solve $f(c) = c^2 = 4/3$, so that $c = 2/\sqrt{3}$; note that we could not choose the negative sign when taking the square root because $c = -2/\sqrt{3}$ lies outside of the range $0 \leq x \leq 2$.

Exercise 2.13.1 Calculate the mean value of $g(x) = \sin(x)$ and $f(x) = \sin^2(x)$ over the interval $0 \leq x \leq 2\pi$.

Exercise 2.13.2 Calculate the mean value ($\bar{y}$) of $y(x) = x^2$ on the interval $-2 \leq x \leq 2$, and find the values of x such that $y(x) = \bar{y}$.

2.14 Integrals, Areas, and Volumes

We have seen that we can interpret an integral as being the area under a curve. Let us return to the problem we posed at the beginning of Section 2.10. Integration can be useful for comparing quantities that, say, vary with depth at two different locations. Looking again at Figure 2.19, it is hard to know whether site A or site B has the greater production. However, we can integrate the curves (called *profiles*) over depth at the two locations and compare the *total* (i.e., integrated) production at each location. The profile at both locations is well approximated by a decaying exponential,

$$P(z) = P_0 \exp(-kz), \tag{2.54}$$

where $P(z)$ is the phytoplankton production at depth z (in meters), P_0 is the production at the surface, and k is a positive coefficient. For profile A, $k_A = 0.08$ m^{-1} and $P_0 =$ 5 mg C m^{-3} d^{-1}, whereas for profile B, $k_b = 0.4$ m^{-1} and $P_0 = 18$ mg C m^{-3} d^{-1}. The production is quite low for both profiles by the time we reach a depth of 50 m—$P(z = 50) = 0.09$ mg C m^{-3} d^{-1} for profile A and $P(z = 50) = 3.7 \times 10^{-8}$ mg C m^{-3} d^{-1} for profile B.

Exercise 2.14.1 Why does k have units of m^{-1}?

Integrating Equation (2.54) to a depth z gives us the total production (P_T) to that depth for a given profile,

$$P_T = P_0 \int_0^z e^{-k\tilde{z}}\, d\tilde{z} = -\frac{P_0}{k}\left(e^{-kz} - e^0\right) = \frac{P_0}{k}\left(1 - e^{-kz}\right),$$

where we have made use of the dummy variable $\tilde{z}$ in the integral. Using a value of $z = 50$ m, for profile A we have $P_T = 61.36$ mg C m^{-2} d^{-1} and for profile B, $P_T = 50.00$ mg C m^{-2} d^{-1}. So, the total production at site A is greater than at B, even though the surface production is much lower. This is because the rate of decrease of production with depth (given by the value k) in profile A is much smaller than in profile B. Integrating over the profile has given us a quantitative measure that we can use to compare locations with different spatial variability.

Exercise 2.14.2 Why are the units of P_T given as mg C m^{-2} d^{-1} whereas $P(z)$ has units of mg C m^{-3} d^{-1}?

This example has involved computing the area between a curve and an axis (in this case, the y axis), but we can also use integrals to calculate the area between two curves. However, we have to be careful to get the limits of integration right.

Example 2.36 To see what is involved, we can calculate the area between the straight line $y_a = 1$ and the curve $y_b(x) = x^2 - x - 1$ shown in Figure 2.25. The first thing we need to do is find out where the curve and line intersect. This occurs at the values of x when $y_a(x) = y_b(x)$, i.e., where $x^2 - x - 1 = 1$, which we can solve to give the two points $(x_1, y_1) = (-1, 1)$ and $(x_2, y_2) = (2, 1)$. We can still use the idea of an integral being the sum of lots of strips of area, but now the area is not between the x axis and the curve, but between the two curves. If the width (Δx) of the shaded area in Figure 2.25 is small enough, then we can approximate it as a rectangle with a width Δx and height $y_a(x) - y_b(x)$; notice that this is the height of the rectangle even though $y_b < 0$. So, our integral will be

$$\int_{-1}^{2} (y_a(x) - y_b(x))\, dx = \int_{-1}^{2} (1 - (x^2 - x - 1))\, dx = \int_{-1}^{2} (2 + x - x^2)\, dx = \frac{11}{2}.$$

Exercise 2.14.3 Calculate the area between the curves $y_1(x) = x^2 - 3x$ and $y_2(x) = 9 - x^2$.

We can also use integration to calculate the volumes of shapes that can be formed by rotating a curve around the x or y axis. We have to be careful to identify the area element we are considering and determine the limits of integration.

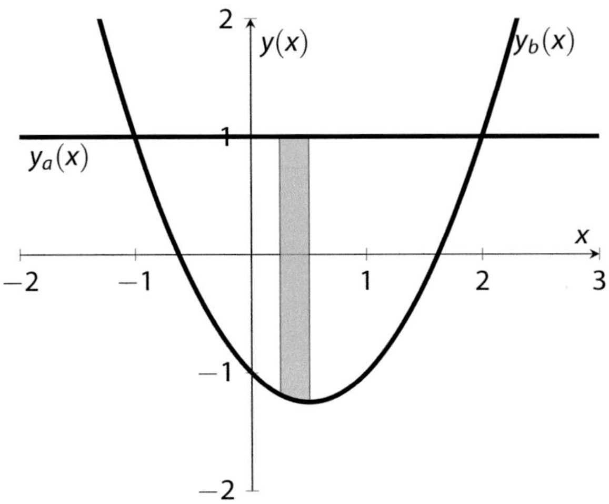

Figure 2.25 The integral between the straight line $y_a(x) = 1$ and the curve $y_b(x) = x^2 - x - 1$. The shaded area is an area element of width Δx and height $y_a(x) - y_b(x)$.

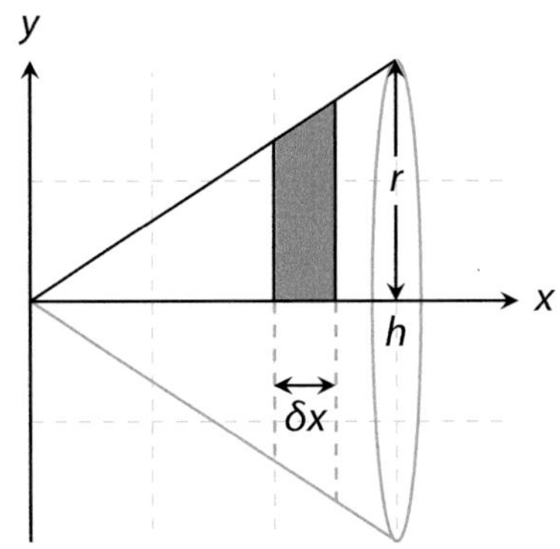

Figure 2.26 Generating a cone by rotating a straight line around the x axis. The straight line connecting the origin to the point $(x, y) = (h, r)$ is rotated through 2π about the x axis creating a cone. The shaded area of width δx becomes a slice through the cone parallel to the base.

Example 2.37 As an example, let us calculate the volume of a cone whose base has a radius r and whose height is h. One way to do this is to look at what happens if we rotate a line around the x axis (Figure 2.26). The shaded area in Figure 2.26 has a width δx. If we rotated that shaded area around the x axis, we would end up with a shape that looks like a disk, but whose edges are slanted; in doing this type of problem we have to remember that we will eventually take the limit as $\delta x \to 0$, so the difference in y values between the left and right sides of the shaded area will become zero. So, we ignore this difference in y values and approximate the volume of the disk to be $\delta V \approx \pi y^2 \delta x$. When we calculate the integral, δV will become a better and better approximation to the real volume as we take the limit of $\delta x \to 0$. Since δV is the volume of the disk, the total volume of the cone is obtained by summing the volumes of all the disks from $x = 0$ to $x = h$,

$$V = \int_0^h \delta V = \int_0^h \pi y^2 \, dx.$$

We can now use the equation of the straight line $y(x) = (r/h)x$ to write the integrand as a function of x, giving us

$$V = \pi \int_0^h y^2 \, dx = \frac{1}{3}\pi r^2 h.$$

We can simlarly rotate a curve about any line, not just a coordinate axis, to calculate a volume. As with all these problems, it is a very good idea to sketch the curves and lines (as we did in Figure 2.26), drawing in the elemental area and the limits of integration that are needed. This can be a great help in making sure we get the limits of the integrals correct.

Exercise 2.14.4 Calculate the volume obtained when you rotate the region enclosed by curve $y(x) = x^2 - 2x + 4$ and $y = 7$ about the line $y = 7$.

2.15 Integrating Multivariate Functions

So far the integrals we have looked at have all involved functions of a single variable. We have been able to do quite a lot with this, but we might wonder how we extend what we have learned to functions of more than one variable. Recalling our experience with partial derivatives, we might suspect that the number of possibilities we need to consider increases when we integrate functions of more than one variable. First, let us remind ourselves what such a function represents. A function of a single variable, $y = f(x)$, represents a curve in two dimensions (x and y). A function of two variables represents a surface $z = f(x, y)$ in three dimensions.

2.15.1 Line Integrals

In many cases, we are not interested in integrating with respect to a coordinate such as x or y, but rather we need to know the integral along a given path or trajectory. For example, we might want to know the integrated heat input to the surface waters of the Gulf Stream as they travel across the Atlantic. In this case, the water does not follow a line of constant latitude or longitude, but rather takes a complex path with changes in both latitude and longitude, and we need to be able to integrate along this path. Another example is that the function $g(x, y, z)$ might represent the oxygen concentration in a region of the ocean and we might want to know the integrated (or average) oxygen experienced by an organism as it swims through this region.

When we integrate a function of a single variable, $f(x)$ for example, we have little choice but to integrate along the x axis. We have more options if the function is of more than one variable. Consider a function of two variables, $g(x, y)$. We can integrate along the x axis, along the y axis, or along some path in the (x, y) plane. These integrals are called *line integrals* or *path integrals* because we are integrating along a path, whether that path is given by x = constant, y = constant, or some function of x and y. There are three corresponding types of integral that we have to consider:

$$\int_C g(x, y)\, dx \qquad \int_C g(x, y)\, dy \qquad \int_C g(x, y)\, ds,$$

where C represents the path we are integrating along, which we have to specify. The first two integrals are with respect to x and y. The third integral is an integral with respect to s, the *arc length*, which is a distance along a curve in the (x, y) plane. There are two immediate questions we need to answer: how do we evaluate such integrals, and how do we interpret them? Let us deal with evaluating them first.

The problem we face is that we have a function of more than one variable (e.g., $g(x, y)$), but we are integrating with respect to only one variable. So, what do we do with the other variable? For a line integral we have an additional piece of information, the curve C that defines the path we are integrating along. This means we have a relationship between x and y that allows us to write one variable in terms of the other or to combine both variables into a single variable, the arc length. If we are integrating with respect to x or y, then our goal should be to use the equation representing the curve along which we are integrating to write the integrand entirely in terms of x or y, whichever is required. Sometimes, it is also possible to write the equation of the curve in terms of another parameter, say t, and then we can evaluate the integral parametrically. What about integrating with respect to the arc length? Here we need to make use of Pythagoras' theorem to write

$$ds^2 = dx^2 + dy^2, \tag{2.55}$$

and this will allow us to write the integral in terms of x or y, whichever is more appropriate (Figure 2.27).

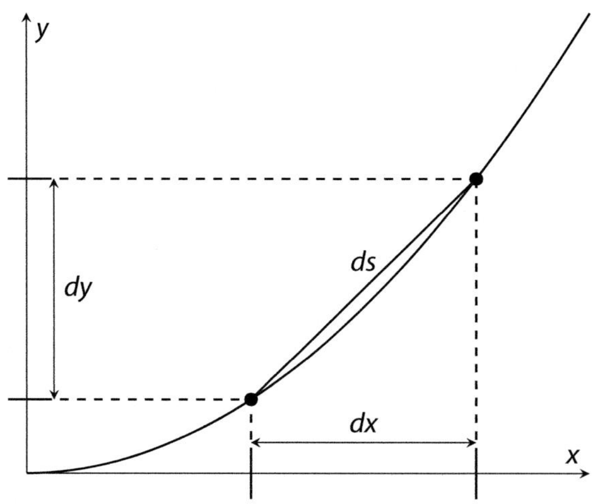

Figure 2.27 The elements dx, dy, and ds used in line integrals.

Example 2.38 As an example, we can evaluate the line integral of the function $f(x, y) = xy$ along a path (C) from $(x, y) = (0, 0)$ to $(1, 0)$ to $(1, 1)$, shown in Figure 2.28. To do this, we need to sum the small increments df as we move along the specified path, i.e.,

$$\int_C df = \int_C \left(\frac{\partial f}{\partial x}\, dx + \frac{\partial f}{\partial y}\, dy \right) = \int_C y\, dx + x\, dy.$$

To evaluate the integral, we split the path into two parts, as shown in Figure 2.28, giving us the two integrals

$$\int_{C_A} y\, dx + x\, dy + \int_{C_B} y\, dx + x\, dy.$$

Next, we can parameterize the two paths in terms of a parameter t. For the path C_A, we can write $x = t$, $y = 0$ and $0 \leq t \leq 1$, which tells us that $dx = dt$ and $dy = 0$ along that path. We can parameterize C_B in a similar way, with $y = t$, $x = 1$, $0 \leq t \leq 1$, so that $dx = 0$ and $dy = dt$. The integral along C_A is now

$$\int_{C_A} y\, dx + x\, dy = \int_0^1 (0 + 0)\, dt = 0,$$

and along C_B we have

$$\int_{C_B} y\, dx + x\, dy = \int_0^1 dt = [t]_0^1 = 1.$$

So, along the given path,

$$\int_C y\, dx + x\, dy = 1.$$

Example 2.39 The value of a line integral along the same path will be different depending on what variable we integrate with respect to. We can see this by evaluating the same function

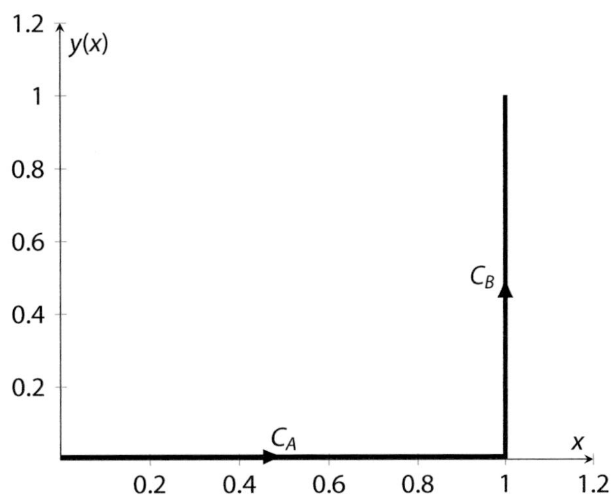

Figure 2.28 The curve C from $(x, y) = (0, 0)$ to $(1, 0)$ and from $(1, 0)$ to $(1, 1)$.

along the path C given by $y(x) = x^2$, $x = 0$ to $x = 1$ but integrate with respect to x, y, and arc length s,

$$I_1 = \int_C x(1+y^2)\,dx \qquad I_2 = \int_C x(1+y^2)\,dy \qquad I_3 = \int_C x(1+y^2)\,ds.$$

For the integral I_1, we can use the equation of the path to write the integrand entirely in terms of x (the variable we are integrating with respect to). So,

$$I_1 = \int_C x(1+y^2)\,dx = \int_{x=0}^{x=1} x(1+x^4)\,dx = \left(\frac{x^2}{2} + \frac{x^6}{6}\right)\Bigg|_0^1 = \frac{2}{3}.$$

For the integral I_2, we can again use the equation of the path to write the integrand entirely in terms of y, but we also need to remember to determine the limits of integration in terms of y, instead of x. Using the equation of the path we see that $y = 0$ when $x = 0$ and $y = 1$ when $x = 1$, so

$$I_2 = \int_C x(1+y^2)\,dy = \int_{y=0}^{y=1} (y^{1/2} + y^{5/2})\,dy = \frac{20}{21}.$$

The value of the integral differs from I_1 — we will see why shortly. To evaluate the integral I_3 we need to write the integrand and limits in terms of either x or y. For this example, we will choose to use x, so making use of Equation (2.55) and the equation of the path, and we can write the arc length as

$$ds = \sqrt{1 + \left(\frac{dy}{dx}\right)^2}\,dx = \sqrt{1+4x^2}\,dx$$

so that the integral becomes

$$I_3 = \int_C x(1+y^2)\,ds = \int_{x=0}^{x=1} x(1+x^4)\sqrt{1+4x^2}\,dx.$$

We can evaluate this integral using a substitution. Let $u = 1 + 4x^2$ so that $du = 8x\,dx$. Notice how the right-hand side of this includes the factor $x\,dx$, which also appears in the integral. Using this substitution the integral becomes

$$\begin{aligned} I_3 &= \int_{x=0}^{x=1} x(1+x^4)\sqrt{1+4x^2}\,dx = \frac{1}{8}\int_{u=1}^{u=5} \left(1 + \frac{(u-1)^2}{16}\right) u^{1/2}\,du \\ &= \frac{1}{128}\int_{u=1}^{u=5} \left(u^{5/2} + 17u^{1/2} - 2u^{3/2}\right)\,du \\ &= \frac{1}{128}\left[\frac{2}{7}u^{7/2} + \frac{34}{3}u^{3/2} - \frac{4}{5}u^{5/2}\right]_{u=1}^{u=5} \approx 18.8787. \end{aligned}$$

Even though we are integrating the same function, each of these integrals evaluates to a different value, telling us that the areas are different.

Exercise 2.15.1 Evaluate the following line integrals:

a) $I_1 = \int_C (x+y)\,ds$, where C is the straight line from $(x, y) = (1, 3)$ to $(x, y) = (5, -2)$,

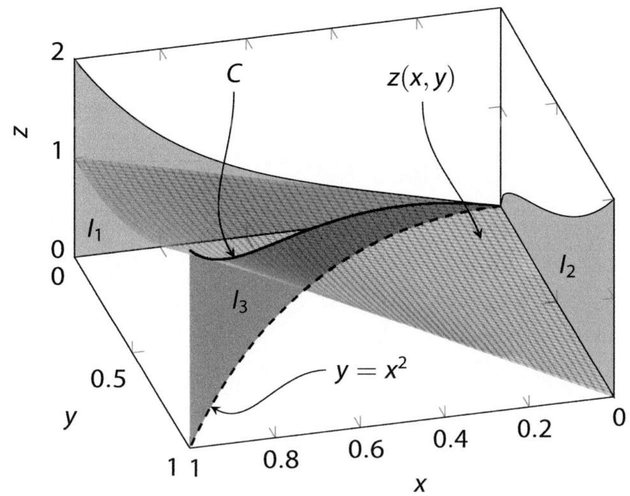

Figure 2.29 The relationship between the three integrals in Example 2.39. The gridded surface is the surface $z = x(1 + y^2)$, and the thick line is the path C that lies within the surface. The dashed line is the curve $y = x^2$ and lies in the (x, y) plane.

$$b)\ I_2 = \int_C x^3 y\, ds, \quad \text{where C is the curve } y = x^2 \text{ from } x = 0 \text{ to } x = 2,$$

$$c)\ I_3 = \int_C xy\, dx, \quad \text{where C is the curve } y = x^2 \text{ from } x = 0 \text{ to } x = 2.$$

Now we know *how* to do line integrals, we need to know what they mean and understand why the same integral has different values depending on which variable we use to integrate with. A function of two variables, such as $z = g(x, y)$ represents a surface, with the value z being the height of the surface above (or below) the (x, y) plane. The areas calculated by the different line integrals in Example 2.39 are shown in Figure 2.29. The integral I_3 with respect to the arc length is the area that lies between the path C in the surface and the curve $y = x^2$ in the plane $z = 0$. The integrals I_1 and I_2 are the areas under the projection[16] of the path C in the (x, z) plane and the (y, z) plane, respectively.

Let us consider another example and evaluate the line integral of the function $g(x, y) = z_0 = \text{constant}$ around the semicircle $x^2 + y^2 = r^2$, with $-r \leq x \leq r$ and $y > 0$ and $r > 0$.

Exercise 2.15.2 Sketch the function $g(x, y)$ and the path $x^2 + y^2 = r^2$, with $-r \leq x \leq r$ and $y > 0$ and $r > 0$.

We will first calculate the line integral with respect to the path length s. On the path we know that $x^2 + y^2 = r^2$, so $2xdx + 2ydy = 0$. We will choose to evaluate the integral in terms of x,

$$\int_C g(x, y)\, ds = \int_C z_0\, ds = \int_C z_0 \sqrt{1 - \left(\frac{dy}{dx}\right)^2}\, dx.$$

[16] You can think of this as the “shadow” the area under C would cast in either the (x, z) or (y, z) planes if a light shone perpendicular to that plane.

We do not have to do any work with the function because it is a constant. We can evaluate dy/dx along the curve, i.e.,

$$\frac{dy}{dx} = -\frac{x}{y} = -\frac{x}{\sqrt{r^2 - x^2}},$$

so that our integral becomes

$$\int_C z_0 \, ds = \int_{-r}^{r} z_0 \left[1 + \left(\frac{x^2}{r^2 - x^2}\right)\right]^{1/2} dx = z_0 \int_{-r}^{r} \left(\frac{r^2}{r^2 - x^2}\right)^{1/2} dx = z_0 \int_{-r}^{r} \frac{r}{\sqrt{r^2 - x^2}} \, dx$$
$$= z_0 r \left[\arcsin\left(\frac{x}{r}\right)\right]_{-r}^{r} = z_0 r \left(\frac{\pi}{2} - \left(-\frac{\pi}{2}\right)\right) = z_0 r \pi.$$

This is precisely the area of a half-cylinder, which is what we would expect. Now, let us do the integral with respect to x. In this case, we have

$$\int_C z_0 \, dx = z_0 \int_{-r}^{r} dx = 2 z_0 r,$$

which is the area of a rectangle of height z_0 and base $2r$. This is what we would expect to see from the projection of the half-cylinder onto the (x, z) plane.

Exercise 2.15.3 Evaluate the y integral and show that its value is $z_0 r$. Is this what you would have expected? (Hint: make sure you get the y limits correct!)

Line integrals along closed paths tend to occur quite often, and they can imply some very interesting results about the world we live in. We can evaluate a line integral along a closed path using these same techniques, but again we have to take care to get the limits correct because at some point the path must have a reversal in the direction of one of the coordinates.

Example 2.40 Let us evaluate the following integral,

$$\int_C x(1 + y^2) \, dx - 2xy \, dy,$$

along the path given by the curve $y(x) = x^2$ from $x = 0$ to $x = 2$, followed by the straight line $y = 2x$ from $x = 2$ to $x = 0$ (Figure 2.30). As before, the technique we use is to split the path into two sections (the quadratic curve and the straight line), evaluate the line integral separately along these paths, and then combine them to get our final answer. We shall parameterize both paths in terms of the parameter t. For the quadratic curve we have $x = t$ and $y = t^2$ with $0 \leq t \leq 2$, so $dx = dt$ and $dy = 2t dt$. The integral along this part of the closed path becomes

$$\int_{t=0}^{t=2} t(1 + t^4) \, dt - \int_{t=0}^{t=2} 4t^4 \, dt = \frac{78}{6} - \frac{128}{5} = -\frac{378}{30}.$$

Moving along the straight line we notice that the direction of the path is reversed; the x values are decreasing. Our parameterization is now $x = t$, $y = 2t$, $2 \geq t \geq 0$. Notice that t is decreasing from 2 to 0, indicating the reversal in direction. The integral along this part of the path becomes

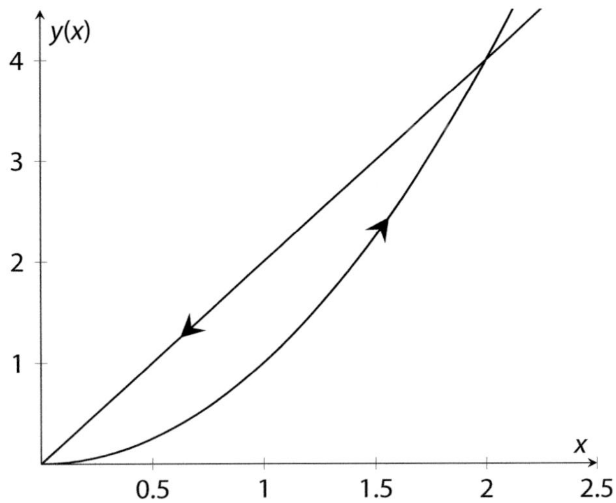

Figure 2.30 The closed path in the (x, y) plane used for evaluating the line integral in Example 2.40.

$$\int_{t=2}^{t=0} t(1+4t^2)\,dt - \int_{t=2}^{t=0} 6t^3\,dt = -2.$$

The value of the integral along the whole path is the sum of these two separate integrals, i.e., ≈ -14.6.

We can also evaluate line integrals in three dimensions. Many small crustaceans in the ocean swim in helical trajectories (Kiørboe, 2008; Heuschele and Selander, 2014), especially in response to chemical cues such as those from food or pheromones. Let us assume that the concentration of the chemical in the water is given by the function $f(x, y, z) = x^2 z$. We would like to know what is the average concentration of this chemical that the crustacean encounters as it swims a distance 8π. The helical path can be represented in terms of the parameter t as

$$x = \cos(t) \quad y = \sin(t) \quad z = 2t \qquad 0 \leq t \leq 4\pi. \tag{2.56}$$

Exercise 2.15.4 Sketch the helical curve given by Equation (2.56).

To calculate the average concentration we will need to calculate the integrated amount of chemical that the crustacean encounters as it swims along this path, and the total length of the path it swims. The total amount of chemical encountered is given by the line integral with respect to the arc length along the trajectory, which we can evaluate using integration by parts:

$$\begin{aligned}\int_C x^2 z\,ds &= \int_0^{4\pi} 2t\cos^2(t)\sqrt{\sin^2(t)+\cos^2(t)+4}\,dt = 2\sqrt{5}\int_0^{4\pi} t\cos^2(t)\,dt \\ &= 2\sqrt{5}\left[t\left(\frac{1}{4}\sin(2t)+\frac{t}{2}\right)\Bigg|_{t=0}^{t=4\pi} - \int_0^{4\pi}\left(\frac{1}{4}\sin(2t)+\frac{t}{2}\right)dt\right] \\ &= 2\sqrt{5}\left\{4\pi\left(\frac{1}{4}\sin(8\pi)+2\pi\right) - \left(-\frac{1}{8}\cos(2t)+\frac{t^2}{4}\right)\Bigg|_0^{4\pi}\right\} = 8\sqrt{5}\pi^2.\end{aligned}$$

Now we need to calculate the total distance travelled along the helical path. This is just the integral of the arc length:

$$\int_C ds = \int_0^{4\pi} \sqrt{\sin^2(t) + \cos^2(t) + 4}\, dt = \sqrt{5} \int_0^{4\pi} dt = 4\pi\sqrt{5}.$$

The average concentration along the path is then

$$\frac{8\sqrt{5}\pi^2}{4\sqrt{5}\pi} = 2\pi.$$

2.15.2 Multiple Integrals

So far we have been integrating a function with respect to a single variable, for example x, y, or arc length. This means we have been integrating along a one-dimensional path. The question we want to address now is, can we integrate with respect to more than one variable? In other words, can we integrate over a surface, or a volume? There are many situations where we might want to do this. For example, we can calculate the total flux of heat from the surface of the Earth if we have an equation that tells us how the temperature varies with latitude. Such calculations are common in simplified climate models.

To make our thoughts a little more concrete, we will consider a function $f(x, y) \geq 0$ that we want to integrate over the region $a \leq x \leq b$ and $c \leq y \leq d$. The integral of a function $y = g(x)$ of a single variable with respect to that variable is the area between the curve $y = g(x)$ and the x axis. By analogy (Figure 2.31), we expect that the integral of

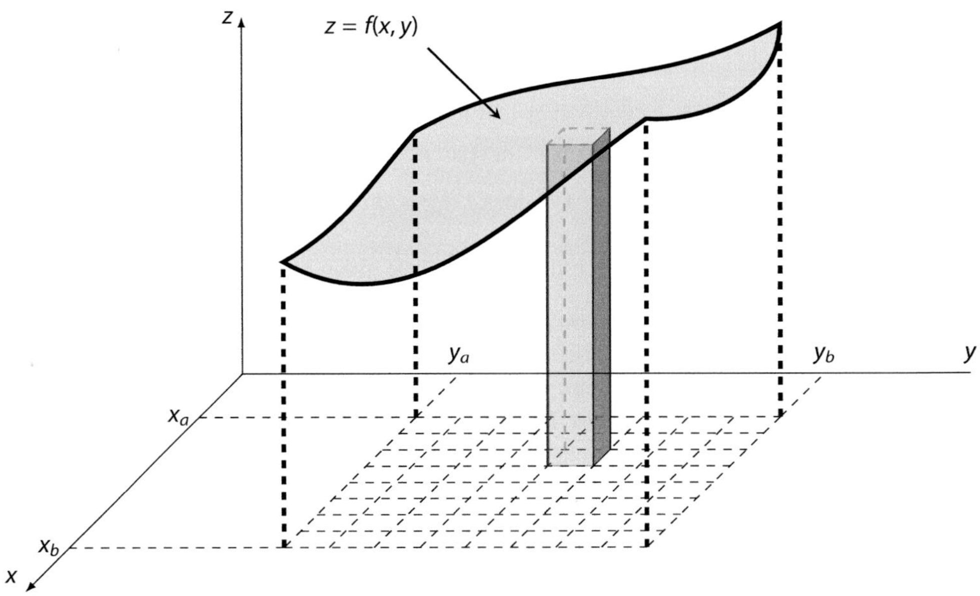

Figure 2.31 A two-dimensional integral consists of summing the rectangular volumes between the (x, y) plane and the surface $z = f(x, y)$. The result is the volume between the (x, y) plane and the surface $z = f(x, y)$.

$z = f(x, y)$ with respect to both x and y will be the volume between the surface $z = f(x, y)$ and the (x, y) plane (i.e., the surface $z = 0$). To evaluate the integral of $y = f(x)$, we divide up the area under the curve into rectangles and then take the limit as the width of the rectangles tends to zero. We can do the analogous thing here. To integrate between the limits x_a and x_b in the x direction and y_a and y_b in the y direction, we subdivide these ranges into N intervals of size Δx in the x direction and M intervals of size Δy in the y direction. Now, instead of the rectangles we had in the one-dimensional case, we have small volumes that are Δx in length, Δy in width, and have a height $f(x, y)$ (Figure 2.31). The volume of one of these shapes centered on the point (x_i, y_j) is $f(x_i, y_j)\Delta x \Delta y$. We can add up all these volumes to get an approximation to the volume under the surface,

$$V \approx \sum_{i=1}^{N}\sum_{j=1}^{M} f(x_i, y_j)\Delta x \Delta y.$$

Just as with the one-dimensional case, we take the limits as Δx and Δy tend to zero to get

$$\int_{y_a}^{y_b}\int_{x_a}^{x_b} f(x, y)\,dx\,dy = \lim_{\Delta x\to 0}\lim_{\Delta y\to\infty} f(x_i, y_j)\Delta x \Delta y. \tag{2.57}$$

We can think of evaluating this expression in two stages:

$$\int_{y_a}^{y_b}\int_{x_a}^{x_b} f(x, y)\,dx\,dy = \int_{y_a}^{y_b}\left(\int_{x_a}^{x_b} f(x, y)\,dx\right)dy = \lim_{\Delta y\to\infty}\left(\lim_{\Delta x\to 0} f(x_i, y_j)\Delta x\right)\Delta y.$$

We first evaluate the inner integral (in the above case, that is the x integral), treating the other variable (y) as if it is a constant. We then evaluate the outer integral. If the integrals are not definite integrals, then we treat x as a constant when we evaluate the y integral. This procedure works because of *Fubini's theorem*.[17]

Theorem 2.6 (Fubini's Theorem) *If the function $z = f(x, y)$ is continuous over a region A defined by $a \le x \le b$ and $c \le y \le d$, then*

$$\iint_A f(x, y)\,dx\,dy = \int_a^b\left(\int_c^d f(x, y)\,dy\right)dx = \int_c^d\left(\int_a^b f(x, y)\,dx\right)dy.$$

Fubini's theorem tells us that for a rectangular region we can evaluate the double integral as two separate integrals over single variables, and we can also change the order in which we perform these integrals.

Example 2.41 To see how this works for a definite integral, let us evaluate the integral

$$I = \int_{y=0}^{y=1}\int_{x=0}^{x=2}(1 + x^2 y + 3x - 2y)\,dx\,dy.$$

We first evaluate the inner integral (the x integral in this case), treating y as a constant, and then evaluate the outer integral, giving

[17] Named after the Italian mathematician Guido Fubini (1879–1943).

$$I = \int_{y=0}^{y=2}\left(\int_{x=0}^{x=1}(1+x^2y+3x-2y)\,dx\right)dy = \int_{y=0}^{y=2}\left(\left[x+\frac{1}{3}x^3y+\frac{3}{2}x^2-2yx\right]_{x=0}^{x=1}\right)dy$$

$$= \int_{y=0}^{y=2}\left(\frac{1}{3}y+\frac{5}{2}-2y\right)dy = \left[\frac{1}{6}y^2+\frac{5}{2}y-y^2\right]_{y=0}^{y=2} = \frac{5}{3}.$$

Example 2.42 Evaluating an indefinite integral will produce another function, but instead of integration constants we will get functions. Let us integrate

$$I = \iint(1+x+y^2)\,dx\,dy.$$

When we integrate with respect to x, we will get an unknown function of y because this would be zero if we were to differentiate it with respect to x. This function will then have to be integrated with respect to y to give

$$\iint(1+x+y^2)\,dx\,dy = \int\left(x+\frac{x^2}{2}+xy^2+a(y)\right)dy = xy+\frac{x^2y}{2}+\frac{xy^3}{3}+\int a(y)\,dy+b,$$

where b is a constant.

Exercise 2.15.5 Evaluate the integral in Example 2.41 by doing the two integrals in the opposite order and compare your answer with that derived in Example 2.41.

This is all well and good if we want to always integrate over rectangular regions whose sides correspond with the x and y axes. But what if we want to integrate over regions with different shapes, such as circles or spheres? Integrating over the volume or surface of a sphere is very common in the geosciences. For example, we might need to calculate the integrated heat production via radioactive decay throughout the interior of the Earth, or the total amount of water loss from an evaporating raindrop as it falls through the air. We can try using rectangular coordinates (x, y, z), but it is often far more convenient to choose a coordinate system that is suited to the problem at hand, and this will involve a change of coordinates. For example, consider calculating the area of a disk (Figure 2.32a). Using (x, y) coordinates we know that the equation of a circle with radius r is $x^2 + y^2 = r^2$. Therefore, we can write the integral with limits such that y takes values between $y = -r$ and $y = +r$, and then x will take values from $x = -\sqrt{r^2 - y^2}$ to $x = +\sqrt{r^2 - y^2}$, so that the integral becomes

$$\int_{y=-r}^{y=r}\int_{x=-\sqrt{r^2-y^2}}^{x=\sqrt{r^2-y^2}} dx\,dy.$$

Evaluating the x integral first leaves

$$2\int_{y=-r}^{y=r}\sqrt{r^2-y^2}\,dy.$$

We can evaluate this using a trigonometric substitution, $y = r\sin(\theta)$, $dy = r\cos(\theta)d\theta$.

$$2\int_{y=-r}^{y=r}\sqrt{r^2-y^2}\,dy = 2\int_{\theta=-\frac{\pi}{2}}^{\theta=\frac{\pi}{2}}\sqrt{r^2-r^2\sin^2(\theta)}\,r\cos(\theta)\,d\theta = 2\int_{\theta=-\frac{\pi}{2}}^{\theta=\frac{\pi}{2}} r^2\cos(\theta)\,d\theta$$

$$= r^2\int_{\theta=-\frac{\pi}{2}}^{\theta=\frac{\pi}{2}}(1+\cos(2\theta))\,d\theta = r^2\left(\theta+\frac{1}{2}\sin(2\theta)\right)\Bigg|_{-\frac{\pi}{2}}^{\frac{\pi}{2}} = r^2\pi.$$

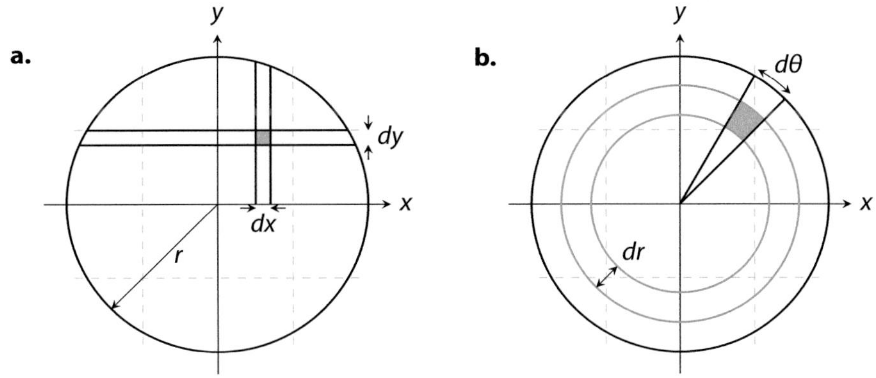

Figure 2.32 Using double integration to find the area of a disk. **a.** We consider a strip of width *dy* and a strip of width *dx*. These intersect and create the shaded region. We then allow *x* and *y* to vary, constrained by the equation of the circle, so that the the whole circle is covered. **b.** We use polar coordinates and consider an area between *r* and $r + dr$ and θ and $\theta + d\theta$ and integrate over *r* and θ.

Alternatively, instead of taking an area element $dA = dxdy$, we could have worked from the start in polar coordinates (r, θ), where an area element is $dA = rdrd\theta$. Then, our integral becomes

$$\int_{\tilde{r}=0}^{\tilde{r}=r} \int_{\theta=0}^{\theta=2\pi} r\,dr\,d\theta = \frac{r^2}{2} \int_{\theta=0}^{\theta=2\pi} = \pi r^2,$$

which is far easier that using rectangular coordinates. In fact, we can frequently make our mathematical lives easier by choosing to work in coordinate systems that reflect the symmetry of the problem we are working on. For example, if our problem involves integrals over spheres, then working in spherical coordinates will make the calculations simpler.

Exercise 2.15.6 Show that in spherical coordinates (r, θ, ϕ) the area element is $dA = r^2 \sin(\theta) d\theta d\phi$ and that the volume element is $dV = r^2 \sin(\theta) d\theta d\phi dr$.

Exercise 2.15.7 Use multiple integration to derive formulae for the volume and surface area of a sphere of radius r.

2.15.3 Green's Theorem

There is an important theorem that links line integrals along closed paths to double integrals over a surface. This is called *Green's theorem*[18] and basically states that if we have two functions, $P(x, y)$ and $Q(x, y)$, then

$$\oint_C P(x, y)\,dx + Q(x, y)\,dy = \iint_D \left(\frac{\partial Q}{\partial x} - \frac{\partial P}{\partial y} \right) dx\,dy, \qquad (2.58)$$

where the symbol $\oint$ signifies a line integral along a *closed* path and $\iint_D$ is the double integral over the region D enclosed by the closed curve C. This is definitely something

[18] The theorem is named after the British theoretical physicist George Green (1793–1841).

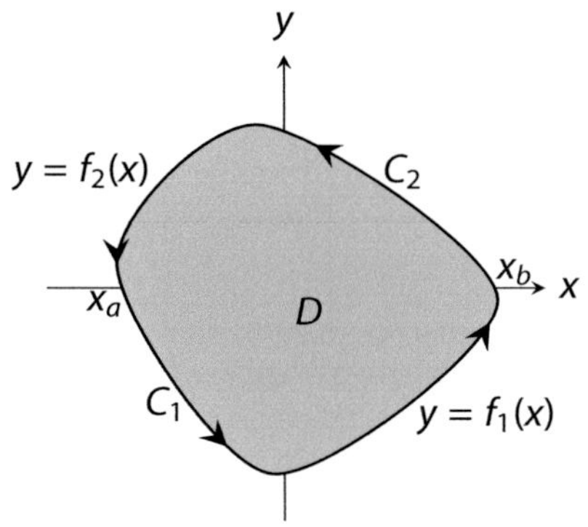

Figure 2.33 A closed curve C is divided into two parts: C_1, which lies below the x axis and is described by the function $y = f_1(x)$; and C_2, which lies above the x axis and is described by the function $y = f_2(x)$. The curve crosses the x axis at the points x_a and x_b. The shaded area inside the curve is the area D in Equation (2.58).

that is not obvious at first sight. We can provide a simple proof in the special case of a simple closed path such as that shown in Figure 2.33. To do this, we start by looking at the second term on the right-hand side of Equation (2.58),

$$\begin{aligned}\iint_D \frac{\partial P}{\partial y}\,dx\,dy &= \iint_D \frac{\partial P}{\partial y}\,dy\,dx = \int_{x=x_a}^{x=x_b}\left\{\int_{y=f_1(x)}^{y=f_2(x)} \frac{\partial P}{\partial y}\,dy\right\}dx \\ &= \int_{x=x_a}^{x=x_b} P(x, f_2(x))\,dx - \int_{x=x_a}^{x=x_b} P(x, f_1(x))\,dx \\ &= \int_{-C_2} P(x,y)\,dx - \int_{C_1} P(x,y)\,dx \\ &= -\left(\int_{C_2} P(x,y)\,dx + \int_{C_1} P(x,y)\,dx\right) = -\oint_C P(x,y)\,dx. \qquad (2.59)\end{aligned}$$

There are a couple of things to notice in this calculation. On the first line, we have changed the order of the integration. The reason for doing this is that the partial derivative is with respect to the variable y, but the inner integral is with respect to x. So, by swapping the order of the integration both the derivative and the integral are with respect to the same variable (y), and we can use the fundamental theorem of calculus to evaluate the integral. In the second line of the calculation, we have realized that the lower limit of the integral is x_a and the upper limit is x_b. But the direction of the path C_2 in Figure 2.33 is in the opposite sense, from positive to negative values of x. We write this as $-C_2$ to remind ourselves of this difference in direction. On the last line of the calculation, we switch the direction of the integral back again so that we traverse the path C_2 in the direction shown in Figure 2.33, which, using the properties of the integral (Equation (2.44)), introduces a minus sign in front of the integral.

Exercise 2.15.8 Use similar arguments to those used in deriving Equation (2.59) to show that

$$\oint_C Q(x,y)\,dy = \iint_D \frac{\partial Q}{\partial x}\,dx\,dy. \qquad (2.60)$$

Combining Equation (2.59) and Equation (2.60) gives us Equation (2.58). To prove this rigorously we would need to take into account the possibility that D might have a complex shape with holes in it, and such proofs can be found in the references given in Section 2.17. One reason why this is a useful theorem is that it can often be much easier to evaluate line integrals using Green's theorem, rather than by evaluating the line integral directly.

Example 2.43 As an example of the advantages of using Green's theorem, let us evaluate the line integral

$$\oint 2y\,dx - x\,dy$$

along the closed path given by a half-circle of radius 2 for $y \geq 0$, and the x axis (Figure 2.34), both using and without using Green's theorem. Let us first evaluate the integral without using Green's theorem. We can parameterize the path such that along the x axis (C_1) we have $x = t$, $y = 0$ with $-2 \leq t \leq 2$. We can parameterize the half-circle (C_2) using the equation of a circle, but restricting the angle so we only cover the upper half-circle, i.e., $x = 2\cos(t)$, $y = 2\sin(t)$ with $0 \leq t \leq \pi$. Writing the integrals in terms of the parameter t we get

$$\begin{aligned}\oint 2y\,dx - x\,dy &= \int_{C_1} 2y\,dx - x\,dy + \int_{C_2} 2y\,dx - x\,dy \\ &= \int_{t=-2}^{t=2} 2 \times 0\,dt - \int_{t=-2}^{t=2} t \times 0\,dt - 6\int_{t=0}^{t=\pi} \sin^2(t)\,dt - 4\int_{t=0}^{t=\pi} \cos^2(t)\,dt \\ &= -6\pi.\end{aligned}$$

To use Green's theorem we need to set $P(x, y) = 2y$ and $Q(x, y) = -x$, so that, using polar coordinates, we have

$$\begin{aligned}\oint 2y\,dx - x\,dy &= \iint \left(\frac{\partial Q}{\partial x} - \frac{\partial P}{\partial y}\right) dx\,dy = \iint (-1 - 2)\,dx\,dy \\ &= -3\int_{\tilde{r}=0}^{\tilde{r}=2} \int_{\theta=0}^{\theta=\pi} \tilde{r}\,d\tilde{r}\,d\theta = -6\pi.\end{aligned}$$

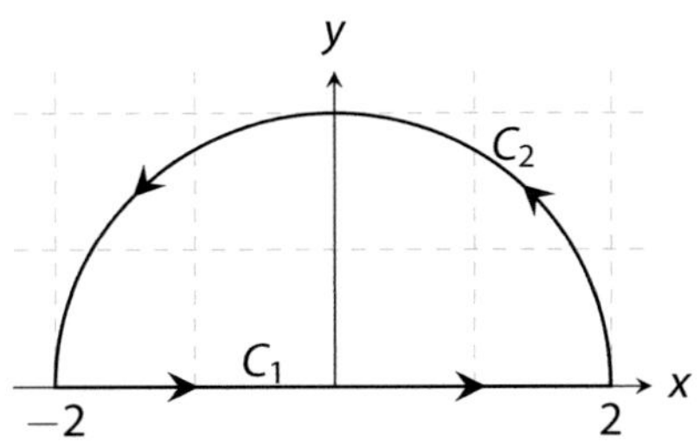

Figure 2.34 The closed path used to evaluate the integral in Example 2.43.

2.16 Numerical Evaluation of Integrals

It is not always possible to evaluate an integral in terms of elementary functions.[19] In fact, most integrals cannot be evaluated in this way, so we often have to resort to numerical methods. Fortunately, there are many well-developed algorithms for computing the values of integrals, and many of these are available in multiple computer languages such as Python, Fortran, C, and others. It is not a great idea, however, to rely solely on numerical methods and forego the techniques we have been exploring. There are several reasons for this. Firstly, by their very nature numerical algorithms provide us with *approximate* solutions—these are often very good approximations, but approximations nonetheless. It is up to us to determine if a particular approximation is good enough for what we need. Secondly, studying an integral on paper can give us insight into how best to evaluate it numerically. For example, the function we are integrating might have a discontinuity that would cause problems for a numerical algorithm. Lastly, many large computer models (e.g., global atmospheric aerosol models, or ocean biogeochemical models) may require the evaluation of an integral many, many times, and using an analytical rather than numerical solution may help reduce the time the simulation needs to run. There are also times when we need to integrate over a set of data points rather than a mathematical function. For example, we may have measurements of ozone concentrations at different heights in the atmosphere and need to know the vertically integrated amount of ozone. In such cases, we can evaluate the integral numerically.

In this section, we will explore some of the simpler numerical algorithms for evaluating integrals. We should know something of how these methods work, rather than treating them as black boxes that magically give us answers. This is so we can choose a numerical method that works well for the problem at hand and understand the factors that determine its accuracy.

2.16.1 Rectangle Rules

The simplest way to think about numerically evaluating an integral is to go back to our original discussion of what an integral is (Section 2.10). To integrate a function $f(x)$ between the limits $x = a$ and $x = b$ we divided up the area under the curve between these limits into small rectangles, each having a width Δx. We then looked at the limit of the sum of these areas as Δx tended to zero.

We can do something similar on a computer, except that we cannot let $\Delta x \to 0$, but we can make it very small. If we specify a value N for how many rectangles we want, and also the limits (x_a and x_b) of the integral, then the computer can calculate the corresponding value of Δx and the sum of the areas of all the rectangles between x_a and x_b, giving us a value for the integral. We can see intuitively that the smaller we can make the value of Δx, the more accurate our answer will be. However, the smaller we make Δx, the more

[19] Elementary functions include polynomials, trigonometric functions, rational functions, and logarithmic and exponential functions.

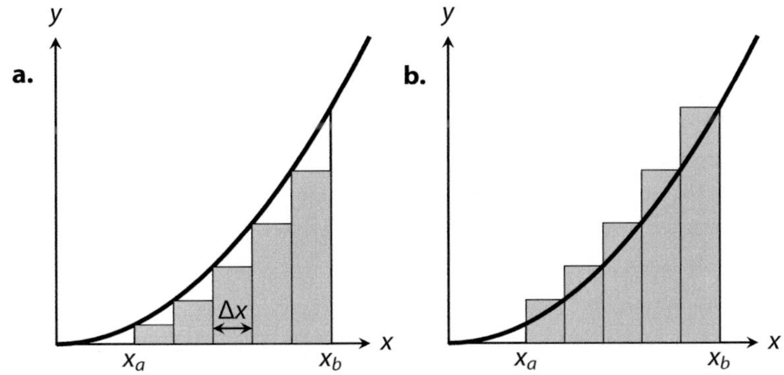

Figure 2.35 The rectangle rule approximates the area under the curve as a sum of rectangles that have one upper corner lying on the curve. The left-point rectangle rule **a.** has the upper left-hand corner of each rectangle on the curve. We can also have a right-point rectangle rule **b.** where the upper right-hand corner of each rectangle lies on the curve.

rectangles we have and the more calculations are needed, making the computer program take longer to complete. So, using a numerical method often involves a compromise between the accuracy of the answer we get and how long we are willing to wait for that answer.

To integrate a function $f(x)$ between the limits $x = a$ and $x = b$ we start with the value of the function at the lower limit of the integral ($f(x = a)$) and use that as the upper left-hand corner of our first rectangle (Figure 2.35a). We then draw a rectangle of width Δx and use the next point $f(x + \Delta x)$ as the upper left-hand corner of our next rectangle, and so on. We choose Δx so that we will have N rectangles between $x = a$ and $x = b$,

$$\Delta x = \frac{b-a}{N}.$$

We can now add up the areas of all the rectangles to give an approximation of the total area (A) under the curve:

$$A \approx \sum_{i=1}^{N} f(x_i)\Delta x, \qquad \text{where} \qquad x_i = x_a + (i-1)\Delta x, \qquad i = 1, \ldots, N. \tag{2.61}$$

This approximation is called the *left-point rectangle rule*. You may see that we have a problem here. If our function is continually increasing, then $f(x_i) < f(x_i + \Delta x)$ and the area of the rectangle will always be an underestimate of the actual area under the curve between x and $x + \Delta x$ (Figure 2.35a). We can change this by choosing the rectangles such that the curve intersects the upper right-hand corner of each rectangle (Figure 2.35b). This is called the *right-point rectangle rule*, but now we have the opposite problem: if $f(x_i) < f(x_i + \Delta x)$, our approximation will consistently overestimate the value of the integral.

The left- and right-point rectangle rules are simple and easy to apply, but they are not very accurate. One way to improve the accuracy is to try and balance out the overestimated values and underestimated values by having the curve intersect with the mid-point of the upper edge of the rectangle rather than either of the two upper corners (Figure 2.36).

Table 2.1 Comparison of the accuracy of the left- and right-point rectangle rules and the midpoint rule in computing the value of the integral of x^2

		Left-point rule		Right-point rule		Midpoint rule	
N	Δx	Value	Error	Value	Error	Value	Error
10	0.4	18.240	3.0933	24.640	3.3067	21.280	5.3333×10^{-2}
50	0.08	20.698	0.6357	21.978	0.6644	21.331	2.1333×10^{-3}
100	0.04	21.014	0.3189	21.654	0.3211	21.333	5.3333×10^{-4}
500	0.008	21.269	0.0639	21.397	0.0640	21.333	2.1333×10^{-5}
1000	0.004	21.301	0.0319	21.365	0.0320	21.333	5.3333×10^{-6}
10000	0.0004	21.330	0.0032	21.337	0.0032	21.333	5.3333×10^{-8}

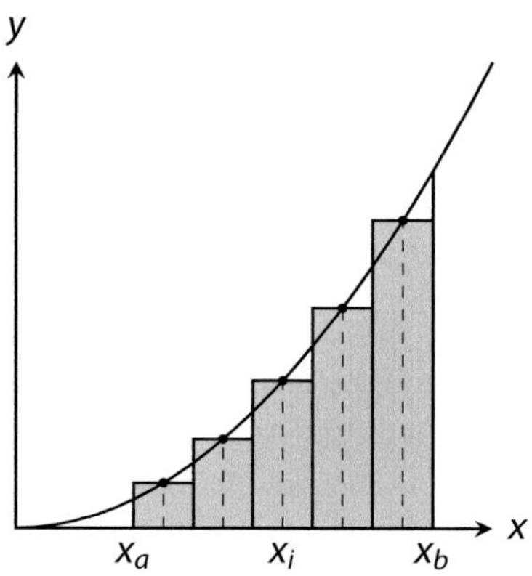

Figure 2.36 The midpoint rule still uses rectangles to approximate the area under the curve, but now the height of each rectangle is given by the midpoint of the upper edge of the rectangle intersecting with the curve.

We still evaluate the integral as the sum of the areas of all the rectangles, except that now we have

$$A \approx \sum_{i=1}^{N} f(x_i)\Delta x, \qquad \text{where} \qquad x_i = x_a + \frac{(2i-1)\Delta x}{2}, \quad i = 1, \ldots, N. \tag{2.62}$$

This rule is called the *midpoint rule*, for obvious reasons.

We can test the accuracy of these three approximations using an integral we know how to evaluate analytically. For example, Table 2.1 shows the results of left-point, right-point, and midpoint rules when calculating the value of

$$I = \int_0^4 x^2\, dx = \frac{64}{3} \approx 21.3333. \tag{2.63}$$

As we determined earlier, the left-point rule produces values that underestimate the real value of the integral, and the right-point rule gives values that overestimate it; the value given by the midpoint rule is in close agreement with the actual value. The absolute error (i.e., the absolute value of the difference between the numerical and exact values) decreases as Δx decreases (i.e., as N increases), again as we expected. The absolute error for the midpoint rule is far smaller than that of either the left- or right-point rules, and decreases far more rapidly as Δx decreases. So, we can conclude that to compute the integral in Equation (2.63), the midpoint method gives us the most accurate answer with the least amount of computation. But, will this always be the case for all functions?

To answer this question, we need to look at the approximations underlying these methods and how they affect the computed value of the integral. To do this, we are going to calculate the worst error each method can give—this is what is called an *upper bound*. If A_0 is the true area under the curve and A is the value we get from a numerical computation, then the error is the difference between these values, $\epsilon = A - A_0$. We want to derive a general formula for ϵ without knowing the function $f(x)$ that we are integrating or the value of A_0. This may seem a tall order, but let us start from the definition of the integral that gives the true area,

$$A_0 = \int_{x=a}^{x=b} f(x)\, dx = F(b) - F(a).$$

Now, think about how we can estimate that area using rectangles. The worst approximation we can make for A_0 comes from using a single rectangle of width $\Delta x = b - a$. Recall that any analytic function can be represented exactly as an infinite Taylor series. So, if we concentrate on the left-point rule for the moment, we can expand $F(b)$ as a Taylor series about $x = a$ and let $(b - a) = \Delta x$, so that

$$F(b) = F(a) + \Delta x \left.\frac{dF}{dx}\right|_{x=a} + \frac{\Delta x^2}{2!}\left.\frac{d^2F}{dx^2}\right|_{x=a} + \frac{\Delta x^3}{3!}\left.\frac{d^3F}{dx^3}\right|_{x=a} + \cdots,$$

and using the fundamental theorem of calculus we can write

$$\begin{aligned} A_0 = F(b) - F(a) &= \Delta x \left.\frac{dF}{dx}\right|_{x=a} + \frac{\Delta x^2}{2!}\left.\frac{d^2F}{dx^2}\right|_{x=a} + \frac{\Delta x^3}{3!}\left.\frac{d^3F}{dx^3}\right|_{x=a} + \cdots \\ &= \Delta x f(a) + \frac{\Delta x^2}{2!}\left.\frac{df}{dx}\right|_{x=a} + \frac{\Delta x^3}{3!}\left.\frac{d^2f}{dx^2}\right|_{x=a} + \cdots. \end{aligned}$$

Now, the numerical approximation to A_0 that we get by using just a single rectangle is $A = f(a)\Delta x$, so for the left-point method we have an error

$$\epsilon_{\text{left}} = A - A_0 = -\frac{\Delta x^2}{2!}\left.\frac{df}{dx}\right|_{x=a} - \frac{\Delta x^3}{3!}\left.\frac{d^2f}{dx^2}\right|_{x=a} - \cdots, \tag{2.64}$$

If we decrease the distance between x_a and x_b so that Δx becomes smaller and smaller, eventually the Δx^2 term will dominate all the higher-order terms, so it is usual to neglect them and write

$$\epsilon_{\text{left}} \approx -\frac{\Delta x^2}{2!}\left.\frac{df}{dx}\right|_{x=a}.$$

Exercise 2.16.1 Repeat the above analysis for the right-point rectangle rule and show that the error is

$$\epsilon_{\text{right}} \approx \frac{\Delta x^2}{2!} \left.\frac{df}{dx}\right|_{x=b} + \frac{\Delta x^3}{3!} \left.\frac{d^2 f}{dx^2}\right|_{x=b} + \text{higher-order terms.} \tag{2.65}$$

Exercise 2.16.2 What is the meaning of the positive and negative signs in Equations (2.64) and (2.65)?

Now, what happens in the case of the midpoint rule? If we compare Figure 2.35 with Figure 2.36, we see that we can think of the midpoint rule as applying the right-point rule followed by the left-point rule on two successive rectangles of width $\Delta x/2$. So, to get an equation for the error of the midpoint rule we can simply add Equation (2.64) and Equation (2.65), remembering that each has a rectangle of width $\Delta x/2$. The even powers of Δx cancel out, because they have opposite signs, leaving the dominant term as

$$|\epsilon_{\text{mid}}| \lessapprox 2\left(\frac{1}{3!}\left(\frac{\Delta x^3}{8}\right)\frac{d^2 f}{dx^2} + \cdots\right). \tag{2.66}$$

Now we can see the advantage that the midpoint method has over the right-point and left-point methods. The error term is much smaller, and it decreases faster as Δx decreases. But, can we do better?

2.16.2 Trapezium Rule

The rectangle and midpoint rules approximate the function $f(x)$ using a series of rectangles of width Δx. Within each step, x to $x + \Delta x$, the function is approximated as a constant. The difference between the rules depends on where each step starts in relationship to the curve. Another approach is to actually connect the points on the curve. The simplest thing we can do is to select points at regular intervals of Δx along the curve and join successive points with a straight line, giving a series of trapeziums instead of rectangles (Figure 2.37).

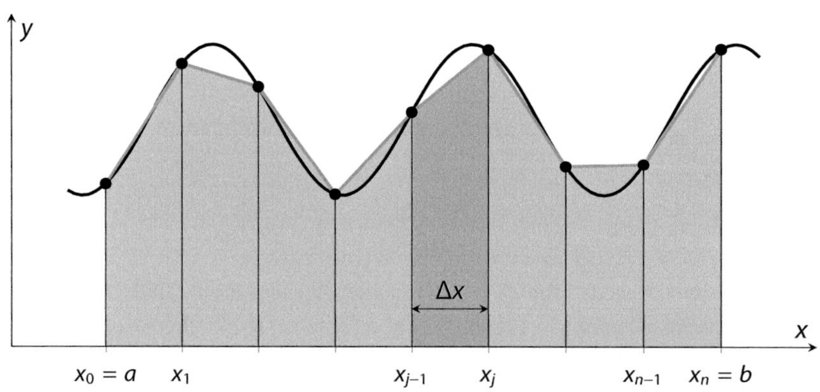

Figure 2.37 The trapezium rule joins points on the curve with sloping straight lines, rather than horizontal ones, thereby forming trapeziums rather than rectangles.

To integrate the function $f(x)$ over the interval $a \leq x \leq b$, we first divide the interval into N equally spaced intervals such that their width is

$$\Delta x = \frac{b-a}{N}$$

giving us $N + 1$ points

$$x_0 = a,\ x_1 = a + \Delta x,\ x_2 = a + 2\Delta x, \dots x_n = b = a + N\Delta x.$$

We then calculate the values of $f(x)$ at these points. The area of each trapezium is given by

$$A_i = f(x_i)\Delta x + \frac{1}{2}[f(x_{i+1}) - f(x_i)]\Delta x = \frac{1}{2}[f(x_{i+1}) + f(x_i)]\Delta x.$$

As before, we approximate the integral by summing the areas of the individual trapeziums,

$$\int_a^b f(x)\,dx \approx \frac{1}{2}(f(x_0) + f(x_1))\Delta x + \frac{1}{2}(f(x_1) + f(x_2))\Delta x + \cdots + \frac{1}{2}(f(x_{n-1}) + f(x_n))$$

$$\approx \frac{\Delta x}{2}(f(x_0) + 2f(x_1) + 2f(x_2) + \cdots + 2f(x_{n-1}) + f(x_n)). \tag{2.67}$$

Equation (2.67) is the *trapezium rule* for evaluating an integral. The numbers multiplying each $f(x_i)$ term are called *weights*, and as we shall see shortly, different numerical schemes have different sets of weights. Table 2.2 compares the results of using the midpoint and trapezium rules to evaluate the integral in Equation (2.63). We may be surprised to see that the trapezium rule has a larger error than the midpoint rule. This does not seem right, because it would appear at first glance that trapeziums should approximate the area under the curve better than rectangles. This is one place where our intuition can deceive us, and it is worth digging into to find out why. Following a similar argument to those we used to derive Equation (2.66) (the midpoint rule), the error bounds for the midpoint rule and trapezium rule can be written as

$$|\epsilon_{\text{mid}}| \leq \xi\frac{(b-a)^3}{24N^2}, \qquad |\epsilon_{\text{trap}}| \leq \xi\frac{(b-a)^3}{12N^2}, \tag{2.68}$$

where ξ is the maximum value of the second derivative of $f(x)$ in the interval $a \leq x \leq b$. This shows that, surprisingly, the error bound for the trapezium rule is larger than for

Table 2.2 Comparison of the accuracy of the midpoint rule and trapezium rule in computing the value of the integral of x^2

		Midpoint rule		Trapezium rule	
N	Δx	Value	Error	Value	Error
10	0.4	21.280	5.3333×10^{-2}	21.440	1.0667×10^{-1}
50	0.08	21.331	2.1333×10^{-3}	21.338	4.2667×10^{-3}
100	0.04	21.333	5.3333×10^{-4}	21.334	1.0667×10^{-3}
500	0.008	21.333	2.1333×10^{-5}	21.333	4.2667×10^{-5}
1000	0.004	21.333	5.3333×10^{-6}	21.333	1.0667×10^{-5}
10000	0.0004	21.333	5.3333×10^{-8}	21.333	1.0667×10^{-7}

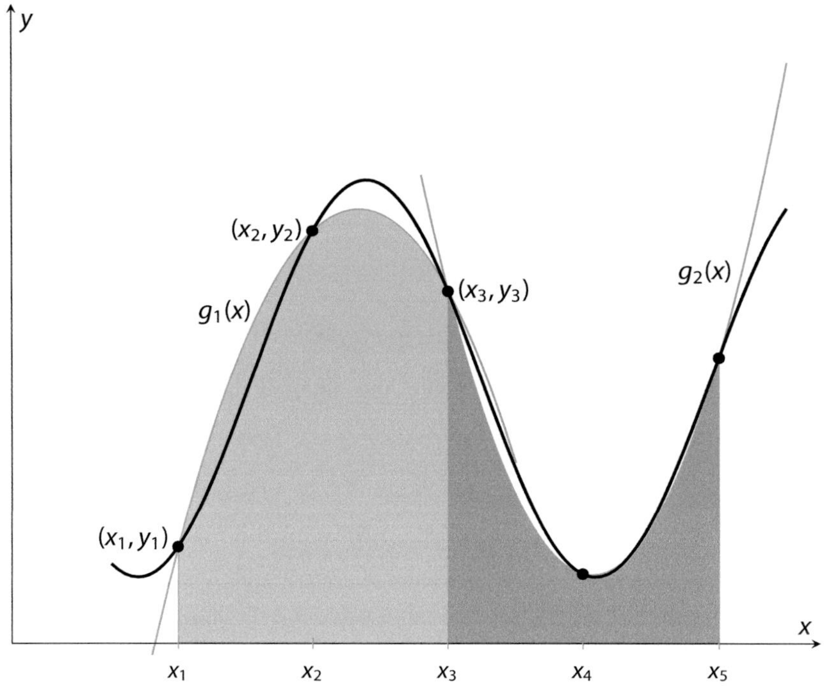

Figure 2.38 Simpson's rule uses quadratic functions instead of straight lines to approximate the curve. The quadratic function ($g_1(x)$) is constructed such that it passes through the points $x_1 \ldots x_3$ on the curve, the second quadratic ($g_2(x)$) passes through the points $x_3 \ldots x_5$.

the midpoint rule, though both perform better than the left- and right-point rectangle rules.

2.16.3 Simpson's Rule

For the rectangle, midpoint, and trapezium rules, we used various straight lines to approximate the function $f(x)$ in an interval Δx. Perhaps we can obtain better approximations to the integral by using curves to approximate $f(x)$ instead. The simplest curve we can use is a quadratic curve. But just as we need two points to define a straight line, we need three points to uniquely determine a quadratic. The rule we are going derive is called *Simpson's rule*.[20]

To derive this rule, consider the three points x_1, x_2, and x_3 in Figure 2.38. The distance between these points is the same, so $x_2 = x_1 + h$, $x_3 = x_2 + h$. We will think of $x_2 = x$ as our "zero point" so that $x_1 = x - h$ and $x_3 = x + h$. This is just for convenience and to simplify the notation. We can evaluate the integral of a quadratic function analytically,

$$A_S = \int_{x-h}^{x+h} (ax^2 + bx + c)\, dx = \frac{h}{3}(6ax^2 + 2ah^2 + 6bx + 6c). \tag{2.69}$$

[20] This technique is named after Thomas Simpson (1710–1761) even though it was used by Johannes Kepler a century earlier.

Exercise 2.16.3 Show that Equation (2.69) is correct.

and if we repeat this procedure with the points (x_3, y_3), (x_4, y_4), (x_5, y_5), and so on, we can sum these areas to get the area under the whole curve. To do this we first need to find the quadratic equation $g(x) = ax^2 + bx + c$ that passes through the three points (x_1, y_1), (x_2, y_2), and (x_3, y_3), in other words we need to find the values of a, b, and c such that $g(x)$ passes through these three points. This gives us three equations,

$$y_1 = a(x-h)^2 + b(x-h) + c, \quad y_2 = ax^2 + bx + c, \quad \text{and} \quad y_3 = a(x+h)^2 + b(x+h) + c,$$

which we can solve for a, b, and c. However, it turns out that we do not have to actually solve these equations.

Exercise 2.16.4 Show that $y_1 + y_3 = 2ax^2 + 2ah^2 + 2bx + 2c$.

We can now write Equation (2.69) in terms of the points y_1, y_2, and y_3 on the curve:

$$A_S = \frac{h}{3}((2ax^2 + 2ah^2 + 2bx + 2c) + 4ax^2 + 4bx + 4c) = \frac{h}{3}((y_1 + y_3) + 4y_2). \quad (2.70)$$

Notice that we now need three points to specify our approximation to $f(x)$, so we need to have two x intervals—x_2 to x_1 and x_3 to x_2. So, if we are integrating from $x = x_a$ to $x = x_b$, we first divide the interval into an even number ($N = 2m$) of intervals (or equivalently an odd number of points) $x_0, x_1, x_2, x_3, x_4 \ldots x_{N-2}, x_{N-1}, x_N$, where $x_0 = x_a$ and $n_N = x_b$. We then apply Equation (2.70) to each interval in turn:

$$A_S = \frac{h}{3}\left[(f(x_0) + f(x_n)) + 4(f(x_1) + f(x_3) + \cdots + f(x_{n-1})) + 2(f(x_2) + f(x_2) + \cdots + f(x_{n-2}))\right]. \quad (2.71)$$

Equation (2.71) is Simpson's rule. We can also see that the weights (the factors of 1, 4, and 2) differ from those in the previous rules. Table 2.3 compares the results of using the midpoint, trapezium, and Simpson's rules to compute the value of the integral

$$\int_0^4 \mathrm{e}^x\, dx = \mathrm{e}^4 - \mathrm{e}^0 \approx 53.598,$$

Table 2.3 Comparison of the accuracy of the midpoint, trapezium and Simpson's rules in computing the value of the integral of x^2

		Midpoint rule		Trapezium rule		Simpson's rule	
N	Δx	Value	Error	Value	Error	Value	Error
10	0.4	53.242	3.6×10^{-1}	54.311	7.1×10^{-1}	53.606	7.5×10^{-3}
50	0.08	53.584	1.4×10^{-2}	53.627	2.9×10^{-2}	53.598	1.2×10^{-5}
100	0.04	53.595	3.6×10^{-3}	53.605	7.1×10^{-3}	53.598	7.6×10^{-7}
500	0.008	53.598	1.4×10^{-4}	53.598	2.9×10^{-4}	53.598	1.2×10^{-9}
1000	0.004	53.598	3.6×10^{-5}	53.598	7.1×10^{-5}	53.598	7.6×10^{-11}
10000	0.0004	53.598	3.6×10^{-7}	53.598	7.1×10^{-7}	53.598	5.0×10^{-14}

and we can see that Simpson's rule provides the more accurate numerical solution, but what is more, the error decreases more rapidly as N increases. The error bound for Simpson's rule can be shown to be

$$|\epsilon_{\text{mid}}| \leq \eta \frac{(b-a)^5}{2880N^4}, \tag{2.72}$$

where η is an estimate of the maximum value of the fourth derivative of the function $f(x)$. The fifth power in the error term shows why the error term decreases rapidly as the interval decreases. Simpson's rule is a good algorithm that can provide accurate numerical estimates for a wide range of integrals, but it is not a universal tool and can fail spectacularly. In general, it will work well with smooth functions.

The methods for numerically evaluating integrals that we have looked are often called *quadrature rules*. There are other quadrature rules that can be derived using a similar framework with different functions to provide highly accurate approximations for many integrals. These techniques are very successful, but may break down if the function we are integrating is not well approximated by a polynomial. There are ways to overcome these issues (some are described in the suggested further reading for this chapter), but it is worth sketching the function you want to integrate, then looking at its behavior to see if there are any places that may trip up a simple quadrature rule.

2.17 Further Reading

We have taken a rather practical approach to calculus in this chapter. However, calculus is a very fascinating and rich topic that can be invaluable. There are many wonderful books to learn this topic from, most of them titled *Calculus*. One of the author's favorites is the book *Calculus* by Michael Spivak (2008), which covers the foundations of calculus with a more mathematically rigorous approach than ours. Another is the book *Calculus* by Gilbert Strang (2017), which is rigorous but aims to develop a practical intuition in the reader as well. An older but excellent book that moves at a sedate pace and contains many practical examples is *Calculus* by Morris Kline (1977).

Although most integrals cannot be evaluated analytically, many can, and having access to a book of integral tables is a very good idea. One of the most venerable is the huge (over 1000 pages) *Table of Integrals, Series, and Products* by Gradshteyn and Ryzhik (1980); this is arguably the gold standard, but there are also many others that are less intimidating. There are some good collections of analytical and numerical techniques that can be used to evaluate integrals. *Handbook of Integration* by Daniel Zwillinger (1992) is quite comprehensive; more detailed and involved techniques are demonstrated in *Inside Interesting Integrals* by Paul Nahin (2014). Computer software packages such as Mathematica and Maple can perform many integrals and generally make our lives easier. However, depending on the integral, varying degrees of user input are required: you may need to tell the program what substitutions to use for example, or make sure that infinities and discontinuities are dealt with properly. This means that you need to know and understand different techniques of integration to be able to use these programs effectively.

Numerical methods for calculating derivatives and integrals are very common, and most scientific software packages will come with a selection of routines to use. This is a good thing because it allows us to relatively easily compute integrals that cannot be solved analytically. But it is also a dangerous thing, because we can forget to examine the function to make sure the assumptions behind the method we choose are satisfied. Some good resources for learning more about numerical methods, their advantages, and pitfalls to look out for, include the two books *Numerical Methods That (Usually) Work* and *Real Computing Made Real*, both by Forman Acton 1990, 1996. The *Numerical Recipes* series of books is also a wonderful resource; each volume contains essentially the same text but differs in the computer language used for the computer codes (Press et al., 1992).

Problems

2.1 Calculate the following derivatives:

1. $\dfrac{d}{d\theta}\sin(2\theta)\tan(\theta+\pi)$
2. $\dfrac{d}{dx}\ln(2x^2+3)$
3. $\dfrac{d}{dy}\left(\sqrt{y}+\dfrac{1}{\sqrt{y}}\right)$
4. $\dfrac{d}{dx}\dfrac{2x^2+3x+2}{(x^2-1)^2}$
5. $\dfrac{d}{dx}\dfrac{e^x}{e^{-x}+1}$
6. $\dfrac{d}{dx}x^2 e^{\ln(x^3)}$
7. $\dfrac{d}{dx}e^{\sin(2x+3)}$
8. $\dfrac{d}{dx}\left(\dfrac{\cos(x+2)}{1+\sin(x-2)}\right)^2$
9. $\dfrac{\partial}{\partial x}2xy^2+3x^2y+2y$
10. $\dfrac{\partial}{\partial y}e^{x^2+3xy+2y^2}$
11. $\dfrac{\partial^2}{\partial x^2}e^{-xt}\sin(3x-2xt)$
12. $\dfrac{\partial^2}{\partial x\partial y}\dfrac{2x^2+3xy}{4xy^2}$

2.2 Calculate the following integrals:

1. $\displaystyle\int\frac{\sin(\theta)}{\cos(\theta)}\,d\theta$
2. $\displaystyle\int 3\sin^2(x)\cos(x)\,dx$
3. $\displaystyle\int_1^2 x^2\sqrt{3-x}\,dx$
4. $\displaystyle\int x\ln(|x|)\,dx$
5. $\displaystyle\int_0^{\pi/4} x^2\cos(2x)\,dx$
6. $\displaystyle\int_1^2\frac{(\ln(x))^2}{x^3}\,dx$

2.3 Evaluate the following line integrals:

1. $\displaystyle\int_C 5x\,dy$, where C is the curve $y=x^2$ from $x=0$ to $x=4$.
2. $\displaystyle\int_C (e^{-2x}\sin(x)+7y)\,dx+(3x-y\sqrt{y-1})\,dy$, where C is the circle $x^2+y^2=4$.

2.4 Calculate the following areas or volumes of revolution:

1. The area bounded by the curves $y(x) = x^2 - 1$ and $y(x) = 2x + 8$.
2. The area bounded by the curves $y(x) = x^2$ and $y = 3x$.
3. The volume you would get by rotating the area bounded by $y(x) = x^2$ and $y = 3x$ about the y-axis.

2.5 Calculate the following multiple integrals: (Hint: some of these integrals are best done by converting to polar or spherical coordinates first.)

1. $\int_0^1 \int_y^{y^2} xy^2 \, dx \, dy$
2. $\iiint (1 + x^2 + y^2 + z^2) \, dx \, dy \, dz$ over the sphere $x^2 + y^2 + z^2 = 1$.

2.6 The number of hours of sunlight received on a (cloudless) day depends on where you are on the Earth and the time of year. Knowing how this varies is important for understanding plant growth. Assume that the number of hours of sunlight per day at a given location varies with time according to

$$s(t) = 12 + 2 \sin\left(\frac{2\pi(t - t_{eq})}{365}\right)$$

where t is the time in days (from 0 to 365) and t_{eq} is the day of the year of the Spring Equinox (when $s(t) = 12$ hours). Calculate the average number of hours of sunlight per day from the Spring Equinox (day 79) to the Vernal Equinox (day 265).

2.7 Simple climate models represent the albedo[21] (α) of the Earth as a function of temperature (T)

$$\alpha(T) = A - B \tanh\left(\frac{T - T_0}{C}\right)$$

where A, B, C, and T_0 are constants. Sketch the curve and determine how each parameter changes the shape of the curve.

2.8 Sometimes we need to develop a function that has certain characteristics and stationary points in order to represent some known phenomenon (or to derive equations for problems to solve in a textbook). Let us consider Example 2.19. Start with an equation of the form $g(x, y) = ax^3 + bxy^2 + cy^3 + dx$ and take first- and second-order partial derivatives and use the conditions for stationary points and their nature to derive conditions for the constants a through d.

2.9 The total daily solar radiation striking a horizontal surface at the top of the atmosphere (the *extraterrestrial radiation*) is

$$\frac{24IE}{\pi} \int_0^{\omega_s} (\sin(\delta)\sin(\phi) + \cos(\delta)\cos(\phi)\cos(\omega)) \, d\omega$$

[21] The fraction of energy from the Sun that the Earth reflects back into space.

where I is the solar constant, E the distance of the Earth from the Sun, δ the *solar declination*, ϕ the latitude of the surface, and ω is the *hour angle* of the Sun. Calculate this integral.

2.10 Gravity-driven current flows include dust storms, the water flow out of estuaries, deep ocean turbidity currents, and volcanic pyroclastic flows. The length (L) of such a current flow can be given by

$$\frac{2}{3}L^{3/2} = \int \mathrm{Fr}(\bar{g}h_c)^{1/2}\,dt$$

where Fr is the (dimensionless) Froude Number which gives the ratio of inertial forces to buoyancy forces, $\bar{g}$ is the *reduced gravity* and can be taken as a constant, and h_c is the time-dependent height of the current which is given by

$$h_c = Qt^{\lambda}$$

where Q and λ are constants. Show that

$$L = \left(\frac{6\mathrm{Fr}}{2(\lambda+2)}\right)^{2/3}(\bar{g}Q)^{1/3}t^{(\lambda+2)/3}$$

2.11 Evaluate the integral

$$\int e^{\kappa t}\sin(\omega t)\,dt,$$

where ω and κ are constants.

2.12 Blooms of phytoplankton occur throughout the oceans. One model for their occurrence looks for a balance between production and respiration (Figure 2.39). Assume that respiration is constant (R_0) with depth (z) and production varies according to $P(z) = P_0 e^{-kz}$ where P_0 is the(constant) production at the surface ($z = 0$) and k is the constant attenuation factor. Derive an equation for the the depth, z_c, at which the integrated production equals the integrated respiration. If $P_0 = 10$ mgC m^{-3} d^{-1}, $R_0 = 5$ mgC m^{-3} d^{-1}, and $k = 0.8$ m^{-1}, use Newton's method to calculate the value of z_c.

2.13 The rate of photosynthesis (P) can be related to the irradiance (I) received by the plant by

$$P = P_{\max}\left(1 - e^{-\alpha I/P_{\max}}\right)e^{-\beta I/P_{\max}}$$

where α determines the response of the plant at low irradiance, and β determines the effects of photoinhibition at high irradiance.

1. Determine the value of I for which P is a maximum.
2. What is the value of P at this value of I?

2.14 Determining the impact of a fishery on fish stocks is crucial for successful conservation. Since such impacts may take years or decades to be apparent, population models have to be used to make quantitative predictions of the sustainable stock. One such (simple) model is

$$Y = qP_{\mathrm{nat}}X - qkX^2$$

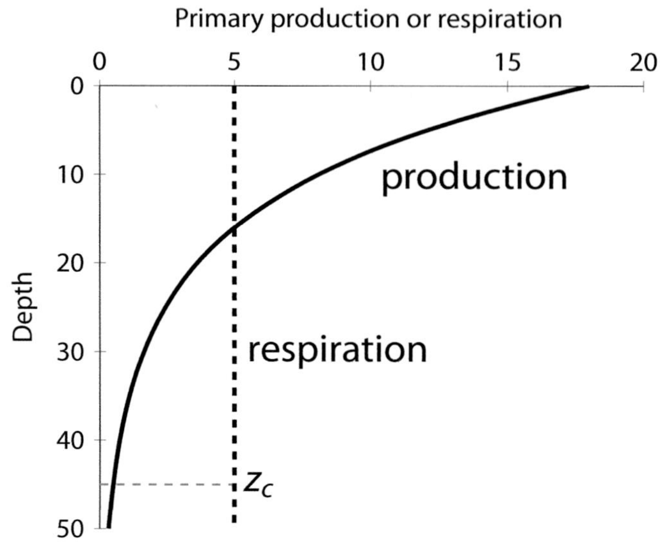

Figure 2.39 A model for a phytoplankton bloom in the oceans. The solid line represents the rate of production of phytoplankton with depth in the upper ocean and the dashed line the rate of respiration of phytoplankton. The depth z_C is the depth at which the depth integrated production equals the depth integrated production.

where Y is the yield per unit fishing effort X, q is the fraction of stock removed by a unit effort, k is a constant, and P_{nat} is an estimate of the population if there were no fishing.

1. Determine fishing effort X that gives the maximum yield Y.
2. What is the value of the maximum yield?

2.15 An integral that appears often is what is called a *Gaussian Integral*,

$$I = \int_{-\infty}^{\infty} e^{-\lambda x^2}\, dx \tag{2.73}$$

where λ is a constant. Evaluate

$$I^2 = \left(\int_{-\infty}^{\infty} e^{-\lambda x^2}\, dx\right)\left(\int_{-\infty}^{\infty} e^{-\lambda y^2}\, dy\right)$$

by rearranging the integrals and transforming to polar coordinates $x = r\cos(\theta)$, $y = r\sin(\theta)$ and show that

$$I = \int_{-\infty}^{\infty} e^{-\lambda x^2}\, dx = \sqrt{\frac{\pi}{\lambda}}.$$

By differentiating Equation (2.73) with respect to λ, show that

$$\int_{-\infty}^{\infty} x^2 e^{-\lambda x^2}\, dx = \frac{1}{2}\sqrt{\frac{\pi}{\lambda^3}}.$$

2.16 Turbulent mixing occurs in many fluids in the environment. Imagine that a substance is injected into the flow at a certain location. This substance (a pollutant, a gas, etc.)

moves with the fluid and does not react with other components of the fluid (it behaves as a so-called *conserved passive tracer*). As the substance is moved about by the fluid, its concentration (C) evolves over time and with space (x) according to

$$C(x,t) = \frac{M}{\sqrt{4\pi Kt}} e^{-x^2/(4Kt)}$$

where K is a measure of the turbulence and is called the *eddy diffusivity*. The size of the patch is given by the variance of $C(x,t)$

$$\sigma^2 = \frac{\int_{-\infty}^{\infty} x^2 C(x,t)\,dx}{\int_{-\infty}^{\infty} C(x,t)\,dx}.$$

Show that $\sigma = \sqrt{2Kt}$.

2.17 The particle size spectrum ($n(r)$) tells us how the concentration of particles varies with particle radius (r), and is used in characterizing aerosols in the atmosphere. The size spectrum has dimensions of $[L]^{-4}$, in other words a number of particles per unit volume per unit size of particles. The total number of aerosol particles with radius between r_1 and r_2 per unit volume of atmosphere is

$$N = \int_{r_1}^{r_2} n(r)\,dr$$

and average aerosol radius and total is

$$\bar{r} = \frac{1}{N}\int_{r_1}^{r_2} rn(r)\,dr.$$

1. If $n(r) = 10^{-10} r^{-4}$, calculate the total number of aerosol particles per cubic meter and the average radius for particles with radius between 0.1 μm and 100 μm.
2. If the surface area of a single aerosol particle is $A(r)$ and the volume of a single aerosol particle is $V(r)$, then the total surface area and volume of all particles is given by

$$A_T = \int_{r_1}^{r_2} A(r)n(r)\,dr \quad \text{and} \quad V_T = \int_{r_1}^{r_2} V(r)n(r)\,dr.$$

   Calculate the total surface area and volume of the spherical aerosol particles for the distribution described above.

2.18 Show by integration that half of the surface area of the Earth (assumed to be a perfect sphere) lies between the latitudes of 30°N and 30°S.

2.19 Consider a sphere of radius R that is sliced by two planes separated by a height h (Figure 2.40). Show that the surface area of the sphere between these two planes is $2\pi Rh$, which is independent of where the slices are made on the sphere.

2.20 In Chapter 1 we estimated the mass of the Earth's atmosphere using the surface atmospheric pressure, the surface area of the Earth, and the acceleration due to gravity. A slightly better calculation takes into account the changing density of the

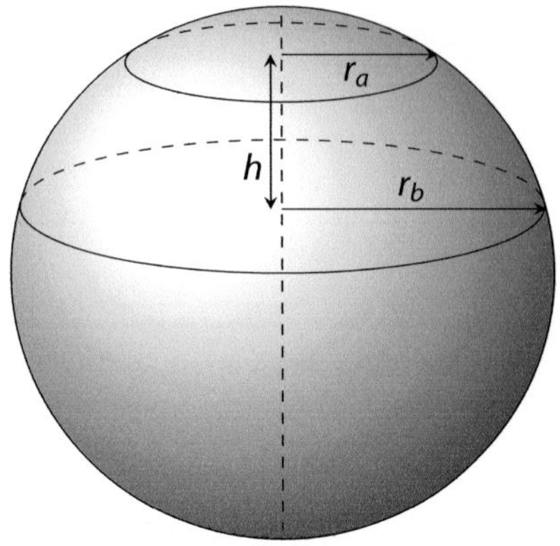

Figure 2.40 A sphere of radius R is cut by two planes forming a slice whose upper surface is a circle of radius r_a and lower surface is a circle of radius r_b.

atmosphere with height. Assume that the Earth's atmosphere can be described as an *isothermal* atmosphere so that

$$\rho(z) = \frac{p_s}{gH} e^{-z/H}$$

where $\rho(z)$ is the atmospheric density as a function of height above the surface of the Earth (z), p_s is the atmospheric pressure at the surface of the Earth, g is the acceleration due to gravity, and H is a constant called the *scale height*.

1. Write down an expression for the total mass of the atmosphere using a triple integral in spherical coordinates.
2. Use integration by parts to evaluate the integral and show that the assumption of an isothermal atmosphere gives a higher estimate of the mass of the atmosphere than the one used in Chapter 1.

2.21 A geological fault is where a fracture in the rock leads to one part of the rock moving with respect to the other. Subsurface fractures are often not vertical or horizontal, but frequently occur at an angle of approximately 60° to the horizontal. When motion occurs along such a fault it increases the total length (L) of crust by an amount ΔL (Figure 2.41). On theory for why these faults occur at an angle of 60°is that this angle minimizes the amount of work that the rock has to do to increase the length of crust. The amount of work (W) done is given by

$$W = \frac{\alpha T \Delta L}{\cos(\theta)\sin(\theta)}$$

where T is the thickness of the crust, θ is the angle of the fault to the horizontal, and α is a constant. Show that W is a minimum value when $\theta = 45°$, thereby showing that this theory does not agree with observations.

2.22 The amount of solar radiation intercepted by an object varies according to the angular distribution of light over the sky, and this affect the amount of radiation received by

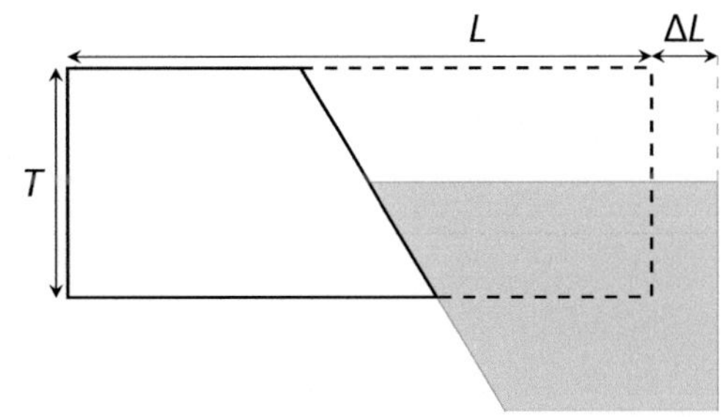

Figure 2.41 A schematic of the change in crust length along a fault line. The dashed area represents the section of crust before movement has occurred, and shaded area shows it after movement has occurred. The increase in the length of surface is ΔL.

a plant canopy, the input of heat to the land and ocean, and the amount of radiation measured by a detector. Two important quantities are the *plane downward irradiance* (E_d) and the *scalar downward irradiance* (E_{od}) (Mobley, 1994)

$$E_d = \int_0^{2\pi}\int_0^{\pi/2} L(\theta,\phi)\cos(\theta)\sin(\theta)\,d\theta, d\phi, \quad E_{od} = \int_0^{2\pi}\int_0^{\pi/2} L(\theta,\phi)\sin(\theta)\,d\theta, d\phi.$$

1. Calculate E_d and E_{od} if $L(\theta,\phi,t) = L_0 =$ constant—this is an *isotropic* distribution, i.e., it does not depend on angle.
2. In the deep ocean, $L \approx L_0/(1-\epsilon\cos(\theta))$, where ϵ is a constant. Calculate E_d and E_{od} for such a distribution.

2.23 You want to numerically evaluate the integral

$$I = \int_0^1 \frac{1}{1+x}\,dx.$$

Use the equations for the maximum error bounds to estimate the number of times you have to evaluate the function $1/(1+x)$ to obtain an answer that has an error smaller than 10^{-3}, 10^{-6}, 10^{-9} using the Midpoint Rule and Simpson's Rule.

2.24 Numerical integration is frequently used to integrate observed data. A colleague collects the following data of microbial production with depth in the Arctic tundra (Table 2.4). Use the Midpoint Rule, Trapezium Rule, and Simpson's Rule to numerically integrate the data with depth and compare the results of the three methods.

2.25 The velocity and depth h of flow in a river have to adjust to the height of the river bed z. If x is the distance along a river, then the relationship between the depth of the river and the shape of the river bed is given by

$$z(x) = \frac{U^2}{2g} + H - h(x) - \frac{(UH)^2}{2gh(x)^2}$$

where U and H are the constant upstream water velocity and depth, $h(x)$ is the water depth as a function of x as water passes over changes in the height of the river

Table 2.4 Data from a hypothetical set of observations of microbial production rates in soils

Depth [cm]	Production [μg C L^{-1} h^{-1}
0.5	1.632
1.5	1.271
2.5	0.928
3.5	0.763
4.5	0.628
5.5	0.495
6.5	0.198
7.5	0.218
8.5	0.0347
9.5	0.0043
10.5	0.085

bed ($z(x)$), and g is the acceleration due to gravity. Show that the minimum of z considered as a function of h occurs when

$$h = h_0 = \left(\frac{(UH)^2}{g}\right)^{1/3}$$

and that at this value

$$z_0 = H\left(1 + \frac{F^2}{2} - \frac{3F^{2/3}}{2}\right),$$

where $F = U/\sqrt{gH}$.

2.26 Find the average value of the function $g(\theta, \phi) = \sin(\theta + \phi)$ over $0 \leq \theta \leq \pi$, $0 \leq \phi \leq \pi/2$.

3 Series and Summations

3.1 Sequences and Series

Many of the problems we encounter in the Earth and environmental sciences involve complicated mathematical expressions that can be hard to work with. Fortunately, we can often make these expressions easier to work with using series, very useful tools that can help us find solutions to seemingly intractable problems.

Series can also be useful for describing phenomena that involve many repetitions of the same process. For example, reflection of solar radiation from clouds and aerosols in the atmosphere is an important process in atmospheric and climate studies. Not all the solar radiation that arrives at the top of the Earth's atmosphere reaches the ground; some of it is absorbed as it passes through the atmosphere, and some is reflected back into space. The proportion of incident radiation reflected back into space is called the *albedo*, and this is an important factor for understanding the heating of the planet by the Sun. Two major components of the atmosphere that reflect incident solar radiation are clouds and aerosols, and we can make a simple model of the atmosphere by assuming that the aerosols are contained in a layer at high altitude in the atmosphere, with clouds being contained in a lower layer (Figure 3.1).[1] Some solar radiation is reflected by the higher aerosol layer, but some passes through to the lower cloud layer, where it is either reflected or transmitted. The radiation reflected from the upper surface of the cloud layer either passes through the aerosol layer or is reflected back downward to the cloud layer and so on. To calculate the total albedo we have to sum all the contributions from the multiple reflections to calculate the total radiation heading back into space. But, with each multiple reflection and transmission, the amount of radiation passing through the aerosol layer into space decreases. We can use series to take this into account and calculate the total albedo, as we shall see a bit later. So, a knowledge of series can help us solve problems we are interested in.

In order to define what we mean by a series, we will start by defining a related concept, a *sequence*. This is basically a sequence of mathematical terms that exhibit a pattern allowing us to calculate the next member of the sequence. For example, the following numbers,

$$1, \frac{1}{2}, \frac{1}{3}, \frac{1}{4}, \dots,$$

[1] This is an oversimplification, but a good tactic for tackling a new problem is to always start simple.

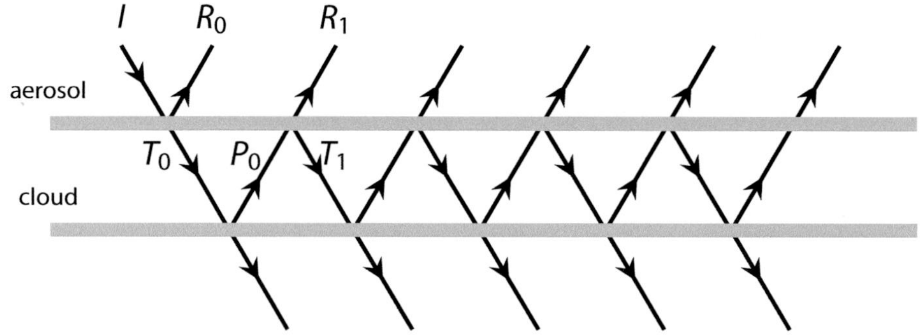

Figure 3.1 A simple two-layer model of the atmosphere with a layer of aerosols high in the atmosphere, and a cloud layer lower in the atmosphere. Some of the solar radiation (I) incident on the aerosol layer is reflected back into space (R_0), and some is transmitted (T_0) to the cloud layer below. Some of the solar radiation incident on the cloud layer passes through to the Earth below, and some is reflected back to the aerosol layer (P_0). Some of this reflected radiation passes through the aerosol layer, and some is reflected back to the cloud layer. These multiple reflections and transmissions affect the total albedo of the planet.

form a sequence in which the nth term in the sequence is $1/n$. A *series* is mathematically defined as the sum of the terms in a sequence, i.e.,

$$S = 1 + \frac{1}{2} + \frac{1}{3} + \frac{1}{4} + \ldots$$

3.2 Arithmetic and Geometric Series

Two of the simplest kinds of series are arithmetic and geometric series. An *arithmetic series* is defined such that each term is obtained from the previous one by adding a constant to it. So, we can write a general arithmetic series as

$$A = a + (a + \delta) + (a + 2\delta) + (a + 3\delta) + \cdots, \tag{3.1}$$

where a is a constant, and the constant δ, which can be positive or negative, is added to a term to obtain the next term in the series. Let us look at how each term is constructed. To obtain the second term, we add δ to the starting value (a), to get the third term we add 2δ, for the fourth we add 3δ, and so on. We can see from this that the nth term in the series will be $(a + (n - 1)\delta)$, and this is the form of the general term in the series. As an example, imagine the accumulation of sediments on the seafloor. If the rate at which new sediment is added (the sedimentation rate) is constant, then each layer of sediment will be of constant thickness (δ) and we can determine the age of any layer in the sediment using an arithmetic series.

Now we know what the general form of an arithmetic series is, we can ask what its value is, i.e., what the sum of all the terms in the series is, which will depend on how many terms there are. The sum (S_N) of a finite arithmetic series containing N terms is

$$S_N = a + (a + \delta) + (a + 2\delta) + (a + 3\delta) + \cdots + (a + (N - 1)\delta). \tag{3.2}$$

To evaluate this sum, let us write it out twice, but the second time we will reverse the ordering of the terms. The reason for doing this is that we want to eliminate as many terms as we can from Equation (3.2) to leave us with a formula that is easy to use:

$$\begin{aligned} S_N &= a + (a+\delta) + (a+2\delta) + \cdots + (a+(N-1)\delta) \\ S_N &= (a+(N-1)\delta) + (a+(N-2)\delta) + (a+(N-3)\delta) + \cdots + a \quad . \end{aligned}$$

If we add these two equations together, a lot of the terms cancel out to give

$$2S_N = (2a+(N-1)\delta) + (2a+(N-1)\delta) + \cdots + (2a+(N-1)\delta) = N(2a+(N-1)\delta),$$

so that

$$S_N = \frac{N}{2}(2a+(N-1)\delta). \tag{3.3}$$

Knowing the values of N, a, and δ we can use Equation (3.3) to calculate the value of the series.

In a *geometric series*, each term is obtained from the previous one by multiplying it by a constant factor, so that for a finite series with N terms

$$S_N = a + a\delta + a\delta^2 + \cdots + a\delta^{N-1}. \tag{3.4}$$

Exercise 3.2.1 Using a similar argument to the one used for the arithmetic series, show that the Nth term of the geometric series in Equation (3.4) is $a\delta^{N-1}$.

The standard example of a geometric series is the growth of a population of microbial cells, where each cell divides into two (i.e., $\delta = 2$) and N is the number of divisions that have occurred. Geometric and arithmetic series also arise whenever we divide a range of values into linear or logarithmic subintervals.

Exercise 3.2.2 Consider the general term of a geometric series $a\delta^{n-1}$. Show that by taking logarithms to base δ that you can convert this into the general term of an arithmetic series and identify the additive constant.

To find the value of the sum to N terms of a geometric series we employ a slightly different strategy to the one we used for arithmetic series, but for a similar reason. First, we multiply Equation (3.4) by δ:

$$\delta S_N = a\delta + a\delta^2 + a\delta^3 + \cdots + a\delta^N. \tag{3.5}$$

Subtracting Equation (3.5) from Equation (3.4) gives

$$(1-\delta)S_N = a - a\delta^N,$$

so that

$$S_N = \frac{a(1-\delta^N)}{1-\delta}. \tag{3.6}$$

Exercise 3.2.3 How many terms are there in the arithmetic series $2.1 + 3.6 + \cdots + 20.1$?

Exercise 3.2.4 If a population of microbial cells starts with two cells and each cell divides in two twice a day, how many days will it take for the population size to reach
a. 10^6 cells, b. 10^{10} cells, and c. 10^{20} cells?

Exercise 3.2.5 If L is the last term and A is the first term in a finite, arithmetic series, show that $S_N = (1/2)N(A + L)$.

Equations (3.3) and (3.6) are useful for series with a finite number of terms, but what happens if $N \to \infty$? Let us consider a geometric series as an example. The value of the series will depend on the value of δ. If $-1 < \delta < 1$, then each successive term in the series gets smaller and smaller and $\delta^N \to 0$ as $N \to \infty$. In this case, Equation (3.6) becomes

$$\lim_{N\to\infty} S_N = \frac{a}{1-\delta}, \tag{3.7}$$

which is indeed finite; we say that the series *converges* to this finite value (Figure 3.2). If, on the other hand, $\delta > +1$, then $\delta^N \to \infty$ as $N \to \infty$ and $S_N \to \infty$, and the series *diverges* as $N \to \infty$.

What happens if $\delta \leq -1$? In this case as N becomes very large, δ^N is either a very, very large positive or negative number depending on whether N is even or odd (Figure 3.3). Therefore, S_N alternates between increasingly large positive and negative numbers and the series does not converge to a single value. Such a series is an *oscillating series* or *alternating series*. Let us examine the value of an alternating series in a bit more detail by looking at an example. Consider the geometric series that we get by starting with the number 1 and multiplying it by -1; i.e. $a = 1$, $\delta = -1$ in Equation (3.4). The sum for a finite number of terms is easy to calculate by rearranging terms, for example:

$$\begin{aligned}\sum_{i=1}^{10}(-1)^{i-1} &= 1-1+1-1+1-1+1-1+1-1\\ &= (1-1)+(1-1)+(1-1)+(1-1)+(1-1) = 0,\end{aligned}$$

and similarly $\sum_{i=1}^{11}(-1)^{i-1} = 1$, so the series oscillates between the values 0 and 1 as we add successive terms. However, if we try the same method with an infinite geometric series,

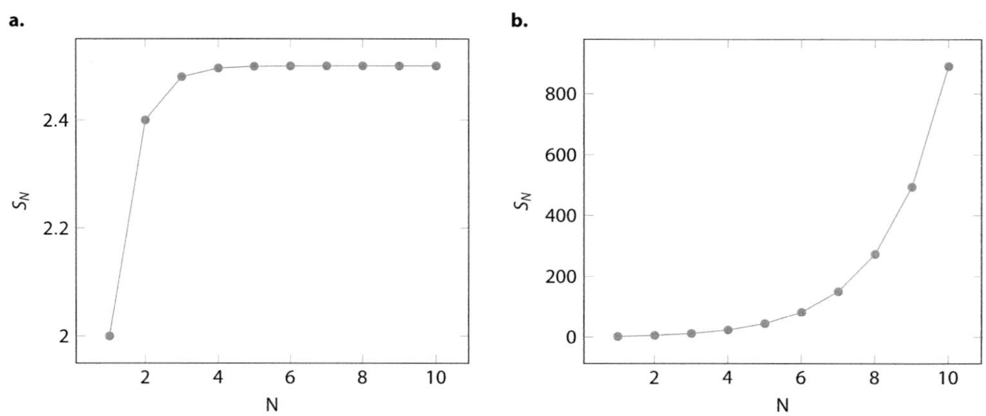

Figure 3.2 Examples of a convergent geometric series with $a = 2$, $\delta = 0.2$ (a.) and a divergent geometric series with $a = 2$, $\delta = 1.8$ (b.).

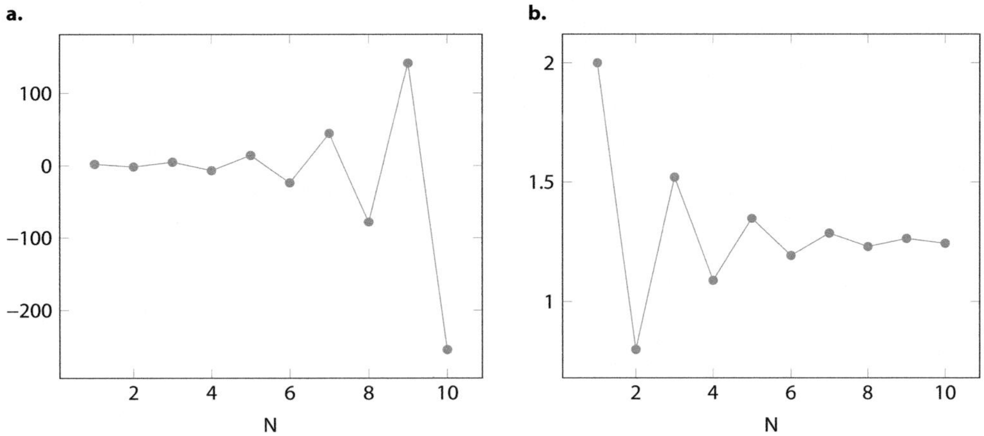

Figure 3.3 Examples of a divergent alternating geometric series with $a = 2, \delta = -1.8$ (a.) and a convergent oscillating geometric series with $a = 2, \delta = -0.6$ (b.).

then things become a little more complicated because we can group the terms in the series in different ways, which gives different answers. For example, we might try

$$\sum_{i=1}^{\infty}(-1)^{i-1} = (1-1) + (1-1) + (1-1) + \ldots = 0$$

and

$$\sum_{i=1}^{\infty}(-1)^{i-1} = 1 - (1-1) - (1-1) - (1-1) - \ldots = 1.$$

Having two different answers for the same calculation is not a good situation to be in, and it highlights the fact that we must be careful when dealing with infinities. For a finite sum, it does not matter how we arrange the terms; we will always get the same answer. For the infinite sum, the problem has arisen because we have not been given a rule for *how* to take the sum of an infinite number of terms, and our naive approach does not work.

We can define a consistent procedure for calculating the value of an infinite series by using sequences. We know that we can calculate the sum of a series with a finite number of terms. So, one way to define the sum of an infinite series is to first calculate the sum of the series truncated at a finite number (N) of terms, call this sum S_N. Then we add another term to the series and calculate S_{N+1}. We then add another term to get S_{N+2}, and so on. In this way, we create a *sequence* of values, each being the sum of a series of increasing length. The individual sums are called *partial sums*, so that if we have an infinite series $\sum u_i = u_0 + u_1 + u_2 + \ldots$, then the Nth partial sum is the sum to N terms of the series

$$S_N = \sum_{i=1}^{N} u_i. \tag{3.8}$$

If the limit of the sequence of partial sums as $N \to \infty$ is a single, specific number (s), then we say that the series is convergent and converges to the value s. If the limit of the sequence of partial sums does not have a single, unique value, then we say the series diverges.

Example 3.1 Let us use this method to see if the series $\frac{1}{2} + \frac{1}{4} + \frac{1}{8} + \ldots$ is convergent or divergent. The partial sums of the series are:

$$S_1 = \frac{1}{2}, \quad S_2 = \frac{1}{2} + \frac{1}{4} = \frac{3}{4}, \quad S_3 = \frac{1}{2} + \frac{1}{4} + \frac{1}{8} = \frac{7}{8}, \quad S_4 = \frac{1}{2} + \frac{1}{4} + \frac{1}{8} + \frac{1}{16} = \frac{15}{16} \cdots ,$$

which forms a sequence of terms in which the Nth term is $(2^N - 1)/2^N$, and

$$\lim_{N\to\infty} \frac{2^N - 1}{2^N} = \lim_{N\to\infty} 1 - \frac{1}{2^N} = 1,$$

showing that the original series converges to the value 1.

We can see that for a series to converge as $N \to \infty$, each successive term added to the sum must get smaller and smaller so as to make smaller and smaller contributions to the total sum. Now that we have some simple examples of series, we can look at some more complicated examples. But first, let us familiarize ourselves with an extremely useful theorem.

3.3 Binomial Theorem and Binomial Series

Let us start with the *binomial theorem*. It is quite easy to explicitly expand expressions such as $(1 + x)^3$ and $(1 + x)^5$. So, for example,

$$\begin{aligned}
(1 + x)^2 &= x^2 + 2x + 1 \\
(1 + x)^3 &= x^3 + 3x^2 + 3x + 1 \\
(1 + x)^4 &= x^4 + 4x^3 + 6x^2 + 4x + 1 \\
(1 + x)^5 &= x^5 + 5x^4 + 10x^3 + 10x^2 + 5x + 1.
\end{aligned}$$

In principle we can also expand expressions like $(1 + x)^{24}$, though it is a rather tedious calculation. However, if we look carefully at the expansions above, we can discern a pattern that can make our lives a lot easier, which is a good thing. Let us look at the expansion of $(1 + x)^3$ in detail to see how the final coefficients arise. To do this, we will label each x with a subscript depending on which factor of $(1 + x)$ it comes from:

$$\begin{aligned}
(1 + x)^3 &= (1 + x_1) \times (1 + x_2) \times (1 + x_3) \\
&= (1 + x_1)(1 + x_3 + x_2 + x_2x_3) \\
&= (1 + x_3 + x_2 + x_2x_3) + (x_1 + x_1x_3 + x_1x_2 + x_1x_2x_3) \\
&= 1 + (x_1 + x_2 + x_3) + (x_1x_2 + x_2x_3 + x_1x_3) + x_1x_2x_3.
\end{aligned}$$

This is just the equation $(1+x)^3 = 1+3x+3x^2+x^3$ because $x_1 = x_2 = x_3 = x$. We can see from this that the three x^2 terms (i.e., the terms x_1x_2, x_1x_3, and x_2x_3) arise from picking two of the possible x in the original expression and multiplying them together. There are three ways to pick two objects (i.e., the x_i) out of three possibilities, and this gives us the coefficient of the x^2 term in the final expression. This kind of argument will apply for each term in the expansion, so we can write that for a positive integer n:

$$(1+x)^n = 1 + \binom{n}{1}x + \binom{n}{2}x^2 + \binom{n}{3}x^3 + \cdots + \binom{n}{r}x^r + \cdots + x^n, \qquad (3.9)$$

which is called the *binomial expansion*, and where

$$\binom{n}{r} = {}^nC_r = \frac{n!}{(n-r)!\,r!} \qquad (3.10)$$

is the *binomial coefficient*, which tells us how many ways we can choose r objects from a collection of n. At first glance the binomial theorem might seem of limited use, but we can do quite a lot with it. For example, many natural phenomena that we come across can be represented mathematically as power laws, so we can use Equation (3.9) to quickly expand $(a+x)^n$, where a and n are constants:

$$\begin{aligned}(a+x)^n &= a^n\left(1+\frac{x}{a}\right)^n = a^n\left\{1 + {}^nC_1\frac{x}{a} + {}^nC_2\left(\frac{x}{a}\right)^2 + \cdots + \left(\frac{x}{a}\right)^n\right\}\\ &= a^n + {}^nC_1a^{n-1}x + {}^nC_2a^{n-2}x^2 + \cdots + x^n = \sum_{i=0}^{n}{}^nC_ia^{n-i}x^i. \qquad (3.11)\end{aligned}$$

This may seem quite abstract, but we can make use of the binomial theorem to approximate functions and numbers to large powers, something that can be useful for numerically evaluating expressions.

Example 3.2 Binomial expansions are useful for finding approximate values of expressions such as $(1-x)^{10}$. For example, let us calculate the first five terms in the expansion of $(1-x)^{10}$ and then use this to calculate $(0.998)^{10}$ to four decimal places. First, we use the binomial theorem to write

$$\begin{aligned}(1-x)^{10} &= 1 - \binom{10}{1}x + \binom{10}{2}x^2 - \binom{10}{3}x^3 + \binom{10}{4}x^4 + \ldots\\ &= 1 - \frac{10!}{9!\,1!}x + \frac{10!}{8!\,2!}x^2 - \frac{10!}{7!\,3!}x^3 + \frac{10!}{6!\,4!}x^4 + \ldots\\ &= 1 - 10x + \frac{90}{2}x^2 - \frac{720}{6}x^3 + \frac{5040}{24}x^4 + \ldots\\ &= 1 - 10x + 45x^2 - 120x^3 + 210x^4 + \ldots\end{aligned}$$

To calculate $(0.998)^{10}$, we put $x = 0.002$ in our expression to get

$$\begin{aligned}(0.998)^{10} &= 1 - 10(2\times10^{-3}) + 45(2\times10^{-3})^2 - 120(2\times10^{-3})^3 + 210(2\times10^{-3})^4 + \ldots\\ &= 1 - 2\times10^{-2} + 18\times10^{-5} - 960\times10^{-9} + 3360\times10^{-12} + \ldots\\ &= 1 - 0.02 + 0.00018 - 0.00000096 + \ldots\\ &= 0.9802 \quad \text{to four decimal places.}\end{aligned}$$

The next term (the x^5 term) in the series has a value of 8.064×10^{-12}, which will not affect the fourth decimal place of the answer.

Exercise 3.3.1 What is the coefficient of the y^7th term in the binomial expansion of

$$\left(y - \frac{2}{y}\right)^9?$$

Hint: Let $a = y$ and $x = 2/y$ in Equation (3.11).

Exercise 3.3.2 Use the binomial theorem to expand $(2a - 3c)^4$.

Exercise 3.3.3 Use the binomial theorem to calculate $(103)^5$.

We can now return to the example shown in Figure 3.1 and use series to calculate the total albedo of a planet. We let the solar radiation incident on the top of the atmosphere be I, assume that the albedos of each layer in the atmosphere are constant, and have values $0 < \alpha_a < 1$ for the aerosol layer and $0 < \alpha_c < 1$ for the cloud layer. We will also assume that there is no absorption of radiation as it propagates through the atmosphere, only reflection and transmission. After the very first reflection, the radiation reflected back into space from the aerosol layer is $R = \alpha_a I$, and the radiation that passes through the aerosol layer is $T_0 = (1 - \alpha_a)I$. This transmitted radiation propagates to the top of the cloud layer, where an amount $P_0 = \alpha_c T_0$ is reflected back up to the underside of the aerosol layer. Then, $R_1 = (1 - \alpha_a)P_0$ is transmitted through the aerosol layer into outer space, and T_1 is reflected back to the cloud layer. From Figure 3.1 we can see that for the nth set of reflections

$$P_n = \alpha_c T_n, \quad T_{n+1} = \alpha_a P_n = \alpha_a \alpha_c T_n, \text{ and } \quad R_{n+1} = (1 - \alpha_a)P_n.$$

The albedo is the fraction of incident radiation reflected back into space (i.e., $I/\sum_n R_n$), so we want to be able to write the equation for R_{n+1} in terms of I. However, the equation for R_{n+1} involves P_n, which in turn depends on T_n; this does not appear to be a fruitful way to find R_{n+1} as a function of I. However, the equation for T_{n+1} contains only constants and T_n, so we should be able to write T_{n+1} in terms of I because $T_0 = (1 - \alpha_a)I$. Let us look at an explicit case, say T_4, to see if we can figure out the general solution:

$$T_4 = (\alpha_a\alpha_c)T_3 = (\alpha_a\alpha_c)^2 T_2 = (\alpha_a\alpha_c)^3 T_1 = (\alpha_a\alpha_c)^4 T_0,$$

which tells us that we can write $T_n = (\alpha_a\alpha_c)^n T_0$. We have already seen that we can write R_{n+1} in terms of T_n, so

$$R_{n+1} = (1 - \alpha_a)P_n = (1 - \alpha_a)\alpha_c T_n = \alpha_c(1 - \alpha_a)(\alpha_a\alpha_c)^n T_0 = \alpha_c(1 - \alpha_a)^2(\alpha_a\alpha_c)^n I.$$

The total reflected light from the top of the atmosphere is then the sum of *all* the R_n values,

$$R = R_0 + \sum_{n=0}^{\infty} R_{n+1} = R_0 + \alpha_c(1 - \alpha_a)^2 I \sum_{n=0}^{\infty} (\alpha_a\alpha_c)^n. \tag{3.12}$$

The term $(\alpha_a\alpha_c)$ is a constant, so this is a geometric series with $a = 1$ (because $(\alpha_a\alpha_c)^0 = 1$) and $\delta = (\alpha_a\alpha_c)$, which lies between 0 and 1. We have taken the upper end of the summation as being infinity, which may see a bit presumptuous; surely there will not be light that undergoes an infinite number of reflections between the cloud and aerosol layers. However, we might expect that the amount of light undergoing reflection decreases as n increases, and will be negligible for very large values of n. This means that extending the summation to infinity incurs very little error. But why should we do it in the first place? If we took the sum to N_{large}, a large but finite value of N, then Equation (3.6) tells us that R would depend on N_{large}, a number that we cannot really know. However, if we take the sum to infinity, then by using Equation (3.6) our value of R will be independent of the number of reflections and transmissions. Using Equation (3.7) for the sum of an infinite geometric series, Equation (3.12) becomes

$$R_0 + \frac{\alpha_c(1-\alpha_a)^2}{1-\alpha_a\alpha_c}I = \left(\alpha_a + \frac{\alpha_c(1-\alpha_a)^2}{1-\alpha_a\alpha_c}\right)I.$$

This equation tells us the fraction of incident solar radiation that is reflected back into outer space, so the factor in parentheses is the total albedo, α. Now, we can use the binomial theorem to expand α to first order in α_a:

$$\begin{aligned}\alpha = \alpha_a + \frac{\alpha_c(1-\alpha_a)^2}{1-\alpha_a\alpha_c} &= \alpha_a + \alpha_c(1-\alpha_a)^2(1-\alpha_a\alpha_c)^{-1} \\ &\approx \alpha_a + \alpha_c(1-2\alpha_a)(1+\alpha_a\alpha_c) = \alpha_c + \alpha_a(1-\alpha_c)^2 \neq \alpha_a + \alpha_c.\end{aligned}$$

This shows us that we cannot simply sum the individual albedos of the two layers to get the total albedo, but the multiple reflections make the total albedo a more complicated function of α_a and α_c.

The binomial theorem, Equation (3.9), only works if n is a positive integer, because at some point the series of values will end when the binomial coefficient has a value of 1. We might wonder if there is an equivalent expansion for negative values of n, or even for cases when n is not an integer. It turns out there is, but we have to be comfortable with an added layer of complexity: the expansion has an infinite number of terms. For a general value of n, the binomial expansion is

$$\begin{aligned}(1+x)^n &= 1 + nx + \frac{n(n-1)}{2!}x^2 + \frac{n(n-1)(n-2)}{3!}x^3 + \cdots + \frac{n(n-1)\cdots(n-r+1)}{r!}x^r + \cdots \\ &= \sum_{n=k}^{\infty}\binom{n}{k}x^k.\end{aligned} \tag{3.13}$$

We have already seen that if n is a positive integer, the series will have a finite number of terms. If n is negative or a noninteger, then none of the terms in the expansion will be zero, and the series will have an infinite number of terms. This means we again face the problem of determining whether or not the series converges or diverges. We will learn how to do this shortly, but for the time being we will take it as a given that Equation (3.13) will converge if $|x| < 1$. This means that the binomial expansion, (Equation (3.13)), can be very useful in situations where x is small (i.e., less than 1).

Example 3.3 We can use Equation (3.13) to expand expressions like $(1+x)^{-1/2}$ as an infinite series. By comparison with Equation (3.13) we can see that $n = -1/2$, so we get

$$\begin{aligned}\frac{1}{(1+x)^{1/2}} &= 1 - \frac{1}{2}x + \frac{1}{2!}\left(-\frac{1}{2}\right)\left(-\frac{3}{2}\right)x^2 + \frac{1}{3!}\left(-\frac{1}{2}\right)\left(-\frac{3}{2}\right)\left(-\frac{5}{2}\right)x^3 + \cdots \\ &= 1 - \frac{1}{2}x + \frac{3}{8}x^2 - \frac{5}{16}x^3 + \cdots\end{aligned}$$

If $x = 0.1$ (i.e., $|x| < 1$, so we expect the series to converge), then $(1+x)^{-1/2} \approx 0.95346$, and the series expansion to the x^3 term gives $(1+x)^{-1/2} \approx 0.95313$, which is quite close.

Exercise 3.3.4 Continue the expansion in Example 3.3 up to and including the x^5 term. Compare and comment on the values of these two expansions with the value of $(1+x)^{-1/2}$ calculated by a computer for the three values $x = 0.1$, $x = 0.9$, and $x = 1.1$.

Binomial expansions can be very useful in solving problems that involve a parameter with has a small value. As an example, let us consider the effect that the gravity of the Moon has on the Earth. The gravitational field of a body can be described in terms of a quantity called the *gravitational potential*, and the gravitational potential at a point a distance r from a sphere of uniform density is

$$U = -\frac{GM}{r},$$

where M is the mass of the sphere and $G = 6.673 \times 10^{-11}\ \mathrm{N\ m^2\ kg^{-2}}$, which is Newton's gravitational constant. The gravitational field of the Moon is the primary cause of the tides on Earth. However, the Earth as seen from the Moon is not a point but an extended body, so the gravitational effects of the Moon will be different at different locations on the surface of the Earth (Figure 3.4). Let us consider a point P on the surface of the Earth. The distance from P to the center of the Moon is b, which is slightly greater than r, and the gravitational potential at P is

$$U = -\frac{G}{b}M.$$

We can use the *law of cosines* (see Appendix B) to write b in terms of the distance r between the center of the Earth and the center of the Moon and the radius a of the Earth:

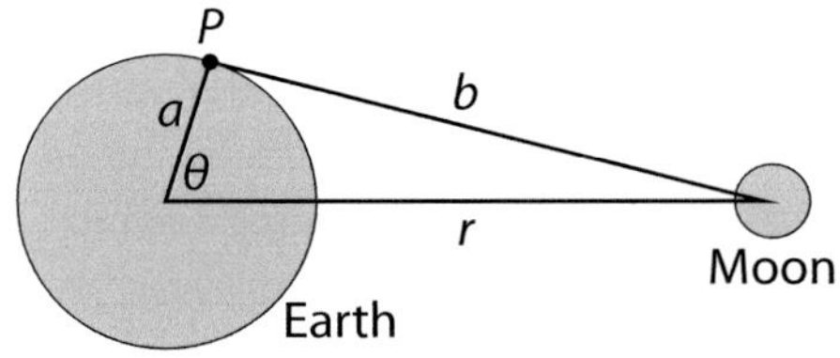

Figure 3.4 A point P on the surface of the Earth (radius a) is a distance b from the center of the Moon, which in turn is a distance r from the center of the Earth.

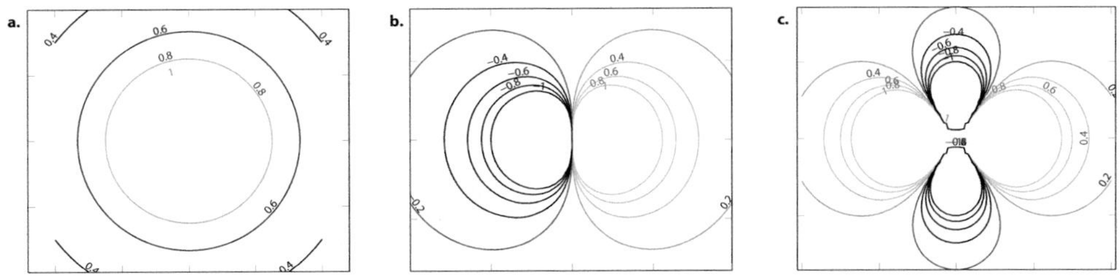

Figure 3.5 The shapes of the first three terms in a multipole expansion: the monopole (**a.**), dipole (**b.**), and quadrupole (**c.**)

$b^2 = r^2 + a^2 - 2ra\cos(\theta)$. Now, if we assume that $r \gg a$, then $a/r \ll 1$, and we can write the potential as

$$U = -\frac{G}{r\left(1 + \left(\frac{a}{r}\right)^2 - 2\frac{a}{r}\cos(\theta)\right)^{1/2}} M. \tag{3.14}$$

If we define $x = (a/r)^2 - 2(a/r)\cos(\theta)$, then we can apply the result from Example 3.3, keeping only those terms of the expansion up to $(a/r)^2$ (higher-order terms will have increasingly smaller values, so we can safely ignore them for this calculation), to get:

$$\begin{aligned}\left[1 + \left(\frac{a}{r}\right)^2 - 2\frac{a}{r}\cos(\theta)\right]^{1/2} &= 1 - \frac{1}{2}\left[\left(\frac{a}{r}\right)^2 - 2\frac{a}{r}\cos(\theta)\right] + \frac{3}{8}\left[\left(\frac{a}{r}\right)^2 - 2\frac{a}{r}\cos(\theta)\right]^2 + \cdots \\ &= 1 - \frac{1}{2}\left(\frac{a}{r}\right)^2 + \frac{a}{r}\cos(\theta) + \frac{3}{2}\left(\frac{a}{r}\right)^2\cos^2(\theta) + \cdots \\ &= 1 + \left(\frac{a}{r}\right)\cos(\theta) + \frac{1}{2}\left(\frac{a}{r}\right)^2(3\cos^2(\theta) - 1) + \cdots,\end{aligned}$$

so

$$U \approx -\frac{GM}{r}\left[1 + \left(\frac{a}{r}\right)\cos(\theta) + \frac{1}{2}\left(\frac{a}{r}\right)^2(3\cos^2(\theta) - 1)\right]. \tag{3.15}$$

This shows us that the gravitational potential of the Moon is not uniform across the surface of the Earth, and the binomial theorem has allowed us to write the potential as a sum of terms of increasing powers of (a/r) and the location on the Earth's surface (using the angle θ).[2]

Equation (3.15) is called a *multipole expansion*. The *monopole* term (GM/r) describes the field of a perfect sphere and depends only on the distance you are from the center of the sphere (Figure 3.5a). The *dipole* term $((GMa\cos(\theta))/r)$ has an angular dependence and is reminiscent of the magnetic field lines of a simple bar magnet with a north and south pole (Figure 3.5b). The last term in Equation (3.15) is called the *quadrupole* and has a yet more complicated spatial pattern (Figure 3.5c).[3] Each of these patterns has a magnitude that is

[2] This description of the forces generating tides is called the *equilibrium theory* of tides. This theory is unable to explain all the features of the Earth's tides because it does not include the time it takes for the oceans to respond to changes in the gravitational force of the Moon as it orbits the Earth. These effects are taken into account in the *dynamical theory* of ocean tides (Butikov, 2002), and the history of this problem is nicely described in Darrigol (2005).

[3] The names given to each of these terms come from the field of electromagnetism and describe the fields that result from different configurations of positive and negative charges, or magnetic poles.

smaller than the previous one because $(a/r) \ll 1$, so each term is a *perturbation* on the monopole term.

3.4 Power Series

The geometric series and Taylor series are all examples of *power series* because they involve terms with an argument (e.g., x) raised to different powers. These are among the most common and most useful types of series that we will come across. As we have seen with Equation (3.15), they are frequently used to approximate particularly unpleasant functions; what is more, it is often easier to differentiate and integrate a power series expansion. Power series expansions can also help us gain a greater understanding of a problem. In our gravitational potential example in the previous section, we saw that the expansion had terms that were successively more complicated in their spatial pattern but decreasing in magnitude, showing the different terms that make up the potential.

A general power series will have the form

$$S = a_0 + a_1(x - x_0) + a_2(x - x_0)^2 + \cdots + a_N(x - x_0)^N = \sum_{k=1}^{N} a_k(x - x_0)^k, \tag{3.16}$$

where a_k are called the *coefficients*, x_0 is the *center*, and N can be finite or infinite. If N is finite, then the power series is a polynomial of order N, whereas if N is infinite, we can think of the power series as being a polynomial of infinite order. For an infinite power series we again have to confront the issue of whether or not it converges. This is especially important if we are using the power series as an approximation to a function (such as with a Taylor series), because the series may converge for only a specific range of values of x, restricting our use of the approximation. For example, the Taylor series for $(1 + x)^{-1}$ is

$$f(x) = \frac{1}{1 + x} \approx 1 - x + x^2 - x^3 + x^4 - x^5 + \cdots \tag{3.17}$$

and converges only if $|x| < 1$. What is more, as x gets closer to ± 1, we have to use more and more terms of the series to obtain a good approximation to the function (Table 3.1). We could have guessed that there would be problems at these values by looking at what happens to both the function and the expansion at $x = 1$. The value of the function is 1/2, but the series expansion alternates between the values 1 and 0 as additional terms are added. This idea of a limited range of x where the expansion converges can be formalized in terms of the *radius of convergence* of a series (Figure 3.6). If we expand a function $f(x)$ about a point $x = x_0$, and if the series converges for $|x - x_0| < R$, R is called the radius of convergence. It is basically how far we can move away from the point $x = x_0$ in any direction and still have the series converge to a finite value. As an example, let us look at the convergence of the series shown in Equation (3.17) (Table 3.1). For values of x close to $x = 0$, the series converges quite quickly; for $x = \pm 0.1$ the series expansion agrees with the actual value of $f(x)$ to four decimal places using only five terms of the expansion. However, as we approach $x = \pm 1$ from below, we need more and more terms

Table 3.1 The value of the power series Equation (3.17) for different values of x and number of terms used in the expansion (N)

	$f(x)$	$N = 2$	$N = 5$	$N = 10$	$N = 15$
$x = -1.5$	−2.0	2.5	13.1875	113.33	873.79
$x = -0.8$	5.0	1.8	3.3616	4.4631	4.8241
$x = -0.5$	2.0	1.5000	1.9375	1.9980	1.9999
$x = -0.2$	1.25	1.2000	1.2496	1.2500	1.2500
$x = -0.1$	1.1111	1.1000	1.1111	1.1111	1.1111
$x = 0.1$	0.9091	0.9000	0.9091	0.9091	0.9091
$x = 0.2$	0.8333	0.8000	0.8336	0.8333	0.8333
$x = 0.5$	0.6667	0.5000	0.6875	0.6660	0.6667
$x = 0.8$	0.5556	0.2000	0.7376	0.4959	0.5751
$x = 1.5$	0.4	−0.5000	3.4375	−22.6660	175.5576

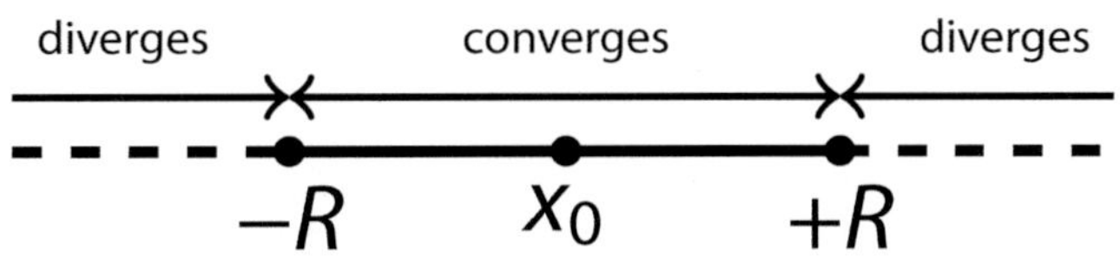

Figure 3.6 A power series expansion of a function $f(x)$ about a value $x = x_0$ may converge for only a small range of values of x, $x = x_0 - R$ to $x = x_0 + R$. R is the radius of convergence.

of the series to get an accurate answer (e.g., for $x = 0.8$ we need more than fifteen terms in the expansion to get a value that agrees with $f(x)$ to four decimal places. For values of $x > \pm 1$, the series rapidly gives values that are very different from the value of the function, and what is more, the difference increases alarmingly as we add more terms to the series. Table 3.1 shows that for $|x| > 1$, the power series in Equation (3.17) does not converge, and we would be in error to use it as an approximation to the function in such cases. For $|x| < 1$, the Taylor series converges, but we have to use more terms in the series to get an accurate approximation to $f(x)$ as x gets closer to ± 1. This demonstrates that the *rate of convergence* (i.e., how many terms we need in the expansion to obtain an accurate approximation) changes as the value of x changes.

Power series have some very useful properties that allow us to easily manipulate them. For example, if

$$f(x) = \sum a_k (x - x_0)^k \quad \text{and} \quad g(x) = \sum b_k (x - x_0)^k$$

are two power series, then we can add, subtract, and multiply the two series term-by-term so that

$$\begin{aligned} f(x) \pm g(x) &= (a_0 + a_1(x - x_0) + a_2(x - x_0) + \cdots) \pm (b_0 + b_1(x - x_0) + b_2(x - x_0) + \cdots) \\ &= (a_0 \pm b_0) + (a_1 \pm b_1)(x - x_0) + (a_2 \pm b_2)(x - x_0)^2 + \cdots \\ &= \sum (a_k \pm b_k)(x - x_0)^k. \end{aligned} \tag{3.18}$$

In addition, if the radius of convergence of the series $f(x)$ is R and $|x - x_0| < R$, then we can differentiate and integrate the power series term-by-term and, what is more, the new series will represent the derivative (or integral) of the original function and have the same radius of convergence (R). So, for example,

$$\frac{df(x)}{dx} = \frac{d}{dx}\sum a_k(x - x_0)^k = \frac{d}{dx}(a_0 + a_1(x - x_0) + a_2(x - x_0)^2 + a_3(x - x_0)^3 + \cdots$$
$$= a_1 + 2a_2(x - x_0) + 3a_3(x - x_0)^2 + \cdots = \sum a_k k(x - x_0)^{k-1}. \tag{3.19}$$

As another example, the Taylor series for $\sin(x)$ and $\cos(x)$ are (see Appendix B)

$$\sin(x) = x - \frac{x^3}{3!} + \frac{x^5}{5!} - \frac{x^7}{7!} + \cdots = \sum_{k=0}^{\infty}(-1)^k \frac{x^{2k+1}}{(2k+1)!}$$
$$\cos(x) = 1 - \frac{x^2}{2!} + \frac{x^4}{4!} - \frac{x^6}{6!} + \cdots = \sum_{k=0}^{\infty}(-1)^k \frac{x^{2k}}{(2k)!}$$

Taking the derivative with respect to x of the series for $\sin(x)$ gives

$$1 - \frac{3x^2}{3!} + \frac{5x^4}{5!} - \frac{7x^6}{7!} + \cdots = \sum_{k=0}^{\infty}(-1)^k \frac{(2k+1)x^{2k}}{(2k+1)!} = \sum_{k=0}^{\infty}(-1)^k \frac{x^{2k}}{(2k)!},$$

which is the series expansion for $\cos(x)$, the derivative of $\sin(x)$. Power series are very useful because the ease with which we can manipulate them makes them more friendly to work with than complicated mathematical expressions.

Exercise 3.4.1 For the two power series $f(x)$ and $g(x)$ given in this section, calculate an expression for $f(x)g(x)$.

Exercise 3.4.2 For a power series $f(x)$, calculate an expression for the integral of $f(x)$.

Exercise 3.4.3 Show that the derivative of the Taylor series for $\cos(x)$ is $-\sin(x)$.

Exercise 3.4.4 The gravitational force can be calculated from the gravitation potential by differentiation. Take the derivative of U given by Equation (3.15) with respect to r, and then take the derivative of U given by Equation (3.14) with respect to r.

3.5 Convergence Criteria

We have talked a great deal about the convergence of infinite series and its importance, especially if we use a series expansion to approximate a function. Now it is time to examine some methods we can use to discover under what conditions an infinite series will converge or not. Let us first think a little about what it means for an infinite series to converge. If we have an infinite series

$$S = \sum_{i=1}^{\infty} u_n,$$

then, intuitively, for the series to converge we require that the magnitude of each new term gets smaller and smaller, i.e., $u_{n+1} < u_n$. However, there are times when our intuition can fail us, and this can be one of those times. For example, let us consider the *harmonic series*

$$S = \sum_{n=1}^{\infty} \frac{1}{n} = 1 + \frac{1}{2} + \frac{1}{3} + \frac{1}{4} + \cdots$$

Even though each term of the series is smaller than the previous one, this series diverges. We can show this by rearranging the terms:

$$S = 1 + \frac{1}{2} + \left(\frac{1}{3} + \frac{1}{4}\right) + \left(\frac{1}{5} + \frac{1}{6} + \frac{1}{7} + \frac{1}{8}\right) + \cdots$$

Each of the terms in parentheses is greater than or equal to 1/2, so

$$S > S_a = 1 + \frac{1}{2} + \frac{1}{2} + \frac{1}{2} + \cdots$$

But the series S_a diverges (we are adding an infinite number of constant values), so S must also diverge. This shows that we have to be careful when examining the convergence of an infinite series; the condition that $u_{n+1} < u_n$ has to hold if the series converges, but it does not *guarantee* that it will converge. The problem with the harmonic series is that successive terms in the series do not become small fast enough for the series to converge. To motivate a stronger condition for convergence, let us look at some other, similar series. First, consider the series

$$S_a = \sum_{n=1}^{\infty} \frac{1}{n^2}, \tag{3.20}$$

which seems to satisfy our intuitive condition for convergence, and indeed, each successive term in the series gets smaller and smaller fast enough that the series converges to a finite value (Figure 3.7). Now consider the alternating series

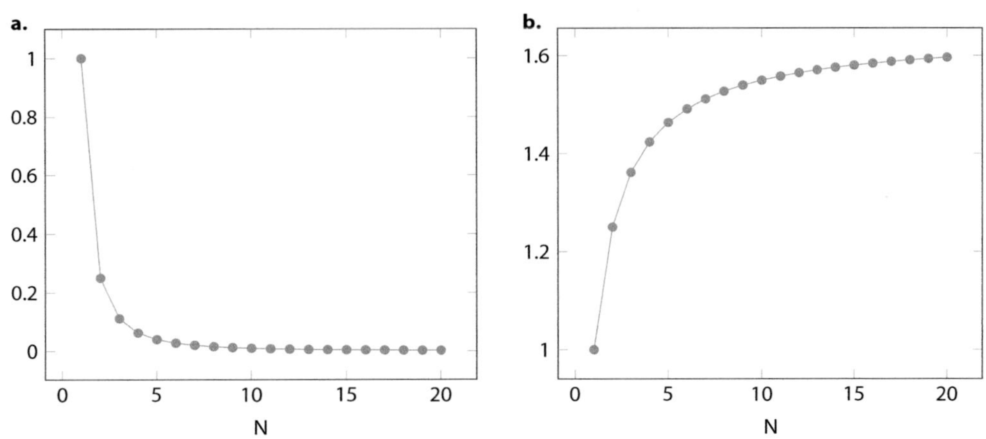

Figure 3.7 The individual terms (a.) and partial sums (b.) of the power series Equation (3.20).

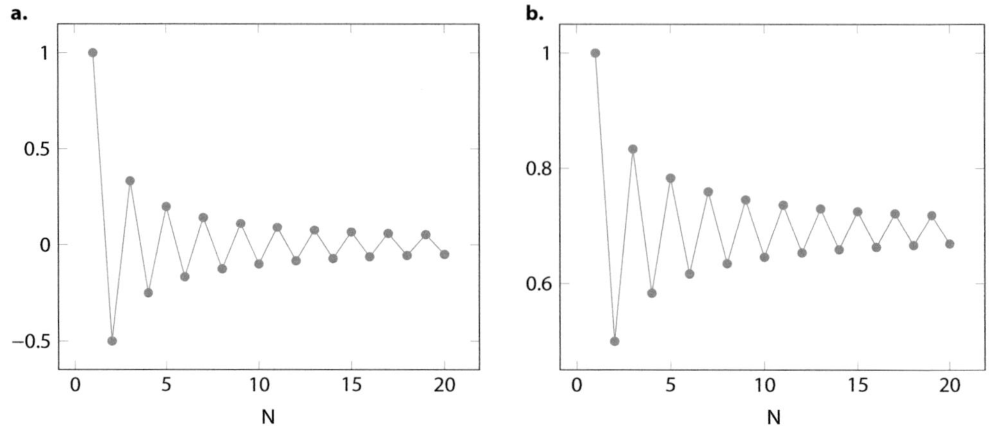

Figure 3.8 The individual terms (a.) and partial sums (b.) of the power series Equation (3.21).

$$S_b = \sum_{n=1}^{\infty} (-1)^{n-1} \frac{1}{n}, \tag{3.21}$$

which is similar to the harmonic series, which we know diverges. However, the even terms in this series are negative and just large enough to cancel the divergence from the positive terms, resulting in a series in which successive terms alternate in sign, but overall the series converges (Figure 3.8). This suggests that we can make the condition for convergence stronger by taking the absolute value of each term; in other words, if the series $\sum |a_n|$ converges, then so does the series $\sum a_n$. If this is true, then we say that the series $\sum a_n$ is *absolutely convergent*, and if $\sum |a_n|$ diverges but $\sum a_n$ converges, then we say that $\sum a_n$ is *conditionally convergent*.

A simple test for convergence of a series is the *ratio test*. An infinite series $\sum_n a_n$ converges if

$$r = \lim_{n\to\infty} \left| \frac{a_{n+1}}{a_n} \right| < 1 \tag{3.22}$$

and the number r is independent of n, otherwise it diverges. There is a special case that occurs when $r = 1$, and in this situation we will need to resort to a more sensitive test.

Example 3.4 Let us use the ratio test to see if the following series converge or diverge:

$$\text{a. } \sum_{n=1}^{\infty} \frac{2^n}{n!} \quad \text{b. } \sum_{n=1}^{\infty} \frac{1}{n}.$$

a. To apply the ratio test we have to calculate the limit

$$\lim_{n\to\infty} \left| \frac{a_{n+1}}{a_n} \right| = \lim_{n\to\infty} \left| \frac{2}{n+1} \right| = 0,$$

so the series converges.

b. For this series we have

$$\lim_{n\to\infty}\left|\frac{a_{n+1}}{a_n}\right| = \lim_{n\to\infty}\left|\frac{n}{n+1}\right| = 1,$$

and the ratio test cannot tell us if the series converges or diverges. In such cases, we have to resort to a more sensitive convergence test.

Questions of convergence become a little more complicated for power series such as Equation (3.16) because the value of the series depends on the value of x. Consequently, the conditions for a given power series to converge will also depend on x, and this brings us back to the concept of a radius of convergence. Each term in the series now depends on the value of x, so in order for the series to converge we will have to require the convergence condition $|a_{k+1}(x-x_0)^{k+1}| < |a_k(x-x_0)^k|$ instead of $|a_{k+1}| < |a_k|$.

Exercise 3.5.1 Why did we require the absolute values in the condition $| a_{k+1}(x-x_0)^{k+1} | < | a_k(x-x_0)^k |$?

Example 3.5 Use the ratio test to find the radius of convergence of the series

$$\sum_{k=0}^{k=\infty} \frac{x^k}{k!}.$$

In other words, we want to find the values of x for which the series converges. Using the ratio test we have

$$\lim_{k\to\infty}\left|\frac{x^{k+1}}{(k+1)!}\frac{k!}{x^k}\right| = \lim_{k\to\infty}\left|\frac{x}{k+1}\right| = 0,$$

so this series converges for all values of x.

Example 3.6 Let us look at a slightly more involved example and find the values of x for which the series

$$\sum_{n=1}^{\infty} \frac{nx^n}{n^2+1}$$

converges. Using the ratio test we have

$$\lim_{n\to\infty}\left|\frac{a_{n+1}}{a_n}\right| = \left|\frac{n^2+1}{(n+1)^2+1}\frac{(n+1)x^{n+1}}{nx^n}\right| = \left|\frac{(n+1)}{n}\frac{n^2+1}{n^2+2n+2}x\right|.$$

Now, we use the properties of limits that we learned in Chapter 2 and calculate the limits of the two factors separately:

$$\lim_{n\to\infty}\frac{(n+1)}{n} = 1, \text{ and } \lim_{n\to\infty}\frac{n^2+1}{n^2+2n+2} = 1.$$

This leaves us with

$$\lim_{n\to\infty}\left|\frac{a_{n+1}}{a_n}\right| = |x|.$$

So, this series converges if $|x| < 1$ and the radius of convergence of the series is $R = 1$.

Exercise 3.5.2 Use the ratio test to determine if the following series converge or diverge:

$$\text{a. } \sum_{n=1}^{\infty} \frac{1}{n}, \quad \text{b. } \sum_{n=1}^{\infty} \frac{(n-1)!}{(n+1)^2}.$$

Exercise 3.5.3 Use the ratio test to determine if the following power series converge, and if they do, find their radius of convergence:

$$\text{a. } \sum_{n=2}^{\infty} (-1)^n \frac{x^n}{10^n \ln(n)}, \quad \text{b. } \sum_{n=1}^{\infty} \frac{(5x-3)^n}{n5^n}.$$

3.5.1 Root Test

The root test is a test for convergence that seems rather mysterious at first sight. It states that if we have an infinite series $\sum_n^{\infty} a_n$ and if

$$\lim_{n\to\infty} (a_n)^{1/n} < 1,$$

then the series is convergent. If the limit is greater than 1, then the series diverges, and if it equals 1, then the root test is inconclusive. Why should this be so? To help us see why this test works and is useful, let us look at the convergence of the series

$$\sum_{n=1}^{\infty} \frac{n^5}{2^n}.$$

At first sight we might suspect that as n gets larger the series diverges because the factor n^5 will get very large very quickly. However, one should not underestimate the power of the exponential! In fact, $n^5 > 2^n$ for $n < 23$, but for $n \geq 23$ the inequality is reversed, and by the time $n = 50$, $2^n = 1.1259 \times 10^{15}$, whereas $n^5 = 3.125 \times 10^8$ (Figure 3.9). In other words, after some value of n, the factor of 2^n will dominate and each term in the series will

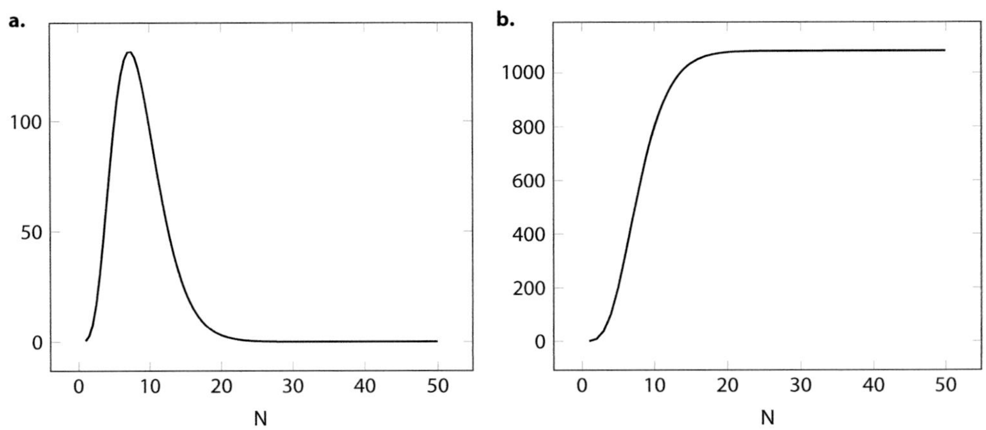

Figure 3.9 The convergence of the series $\sum_N (n^5)/2^n$. The value of each term initially increases and then decreases (a.) while the partial sums (b.) show an initial increase before converging to a finite value.

get smaller and smaller because 2^n grows so rapidly, and the series will converge. By taking the nth root of the individual terms, the root test highlights the effect of the exponential factor. As another example, in a geometric series, $a_n = \alpha r^n$, so $(a_n)^{1/n} = (\alpha)^{1/n} r$ and as $n \to \infty$, $(\alpha)^{1/n} \to 1$, so

$$\lim_{n\to\infty} (a_n)^{1/n} = r,$$

and we have the result that a geometric series converges if $r < 1$.

Example 3.7 Let us use the root test to determine whether the following series converge:

$$\text{a. } \sum_{n=0}^{\infty} \frac{5^n}{2^{n+1}}, \quad \text{b. } \sum_{n=1}^{\infty} (\arctan(n))^n, \quad \text{c. } \sum_{n=0}^{\infty} \frac{3^n}{4^n(n^7+3)}.$$

a. For this series we need to look at

$$\lim_{n\to\infty} \left(\frac{5^n}{2^{n+1}}\right)^{\frac{1}{n}} = \lim_{n\to\infty} \frac{5}{2^{(n+1)/n}} = \lim_{n\to\infty} \frac{5}{2^{1+1/n}} = \frac{5}{2} > 1,$$

so this series diverges.

b. Applying the ratio test we have

$$\lim_{n\to\infty} [(\arctan(n))^n]^{1/n} = \lim_{n\to\infty} \arctan(n) = \frac{\pi}{2} > 1,$$

so the series diverges.

c. We have in this case

$$\lim_{n\to\infty} \left(\frac{3^n}{4^n(n^7+3)}\right)^{1/n} = \lim_{n\to\infty} \frac{(3^n)^{1/n}}{(4^n)^{1/n}(n^7+3)^{1/n}} = \frac{3}{4\times 1} < 1,$$

so this series also converges.

Exercise 3.5.4 Use the ratio test to determine if the following series converge or diverge:

$$\text{a. } \sum_{n=1}^{\infty} \frac{2^n}{n5^{n+1}}, \quad \text{b. } \sum_{n=1}^{\infty} \frac{15^n}{n!\, 2^n}, \quad \text{c. } \sum_{n=1}^{\infty} \left(\frac{n}{2n+1}\right)^n.$$

3.5.2 Integral Test

There are many times when we are not interested in knowing the value that a series converges to, but we would still like to know *if* it converges. In such cases, we can use convergence tests that compare the series with something that we know converges (or is always finite) or diverges. The integral test is such a test. It states that if we have an infinite series

$$S = \sum_{n=1}^{\infty} a_n$$

and we choose a continuous function $f(x)$ that is monotonically decreasing and such that $f(n) = a_n$, then the series S converges if the integral

$$I = \int_{n=1}^{\infty} f(x)\, dx$$

is finite, and the series diverges if the integral is infinite.

Let us do an example before we look at why this works. We will use the integral test to see if the series

$$\sum_{n=2}^{\infty} a_n = \sum_{n=2}^{\infty} \frac{1}{n \ln(n)}$$

converges or not. Let us first look at the behavior of the individual terms. As n increases, a_n gets progressively smaller, which is promising, but as we have seen, it is not a guarantee that the series does in fact converge. For the integral test, we need to find a function $f(n) = a_n$, which instead of taking discrete values is a continuous function that we can integrate. For our series, we could choose

$$f = \frac{1}{x \ln(x)},$$

where we have simply replaced the discrete variable n with the variable x. If x takes only integer values (n), then this equation becomes $f(x) = a_n$. However, we allow x to take any value, not just integer values, so that it is a continuous rather than discrete variable. The integral test asks us to determine whether or not the integral

$$\int_2^{\infty} \frac{1}{x \ln(x)}\, dx$$

converges. We can evaluate the integral using the fact that (Appendix B)

$$\frac{d}{dx} \ln(x) = \frac{1}{x},$$

giving

$$\int_2^{\infty} \frac{1}{x \ln(x)}\, dx = \int_2^{\infty} \frac{1}{\ln(x)}\, d(\ln(x)) = \ln(\ln(x))|_2^{\infty} = \infty.$$

The integral is infinite, so the series diverges.

To see why the integral works let us go back to the definition of the Riemann integral (Section 2.10). We can see from Figure 3.10 that the integral of $f(x)$ between the limits $x = n$ and $x = n + 1$ is always less than the area of the rectangle of width $\Delta x = 1$ and height a_n. The nth partial sum, S_n, is just the sum of the areas of the first $n + 1$ rectangles, and this is always greater than the integral $f(x)$ up to the limit $x = n$. That is

$$s_n \geq \int_1^{n+1} f(x)\, dx,$$

so if the integral diverges as $n \to \infty$, then so must the series.

Exercise 3.5.5 How would you modify the argument for why the integral test can show if a series diverges to show that a series converges instead?

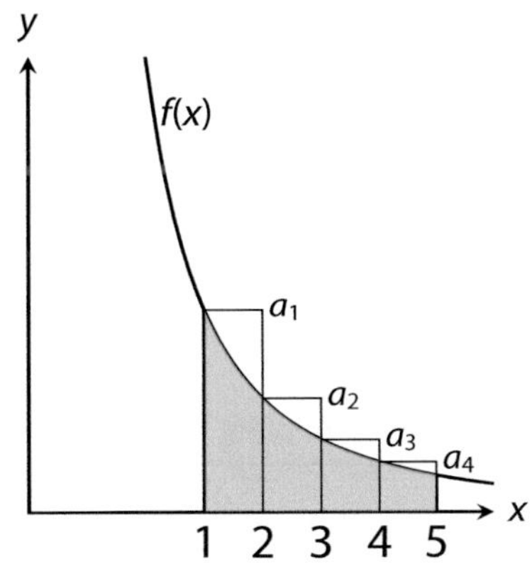

Figure 3.10 The geometry of the integral test. The partial sums of the series are given by the sum of the areas of the rectangles. The integral of the function is given by the shaded area.

Exercise 3.5.6 Use the integral test to determine if the following infinite series converge or diverge:

$$\text{a. } \sum_{n=2}^{\infty} \frac{1}{n(\ln(n))^2}, \quad \text{b. } \sum_{n=0}^{\infty} ne^{-n^2}, \quad \text{c. } \sum_{n=1}^{\infty} \frac{1}{1+n^2}.$$

Exercise 3.5.7 Use the integral test to show that if $m > 0$, then the so-called p-series

$$\sum_{n=m}^{\infty} \frac{1}{n^p}$$

converges if $p > 1$ and diverges if $p \leq 1$.

3.5.3 Comparison Test

The comparison test takes the logic of the integral test and applies it more generally. We can often compare a new series to one whose convergence properties we already know. If we know that a series $\sum_n a_n$ converges, then any other series, $\sum_n u_n$ whose individual terms (u_n) are always smaller than or equal to a_n (i.e., $u_n \leq a_n$) must also converge. This is because, as n gets larger, the terms u_n get smaller faster than, or at the same rate as the terms a_n, and since we know $\sum a_n$ converges, then so must $\sum u_n$. We can state this idea more formally as follows: consider two series $\sum u_n$ and $\sum a_n$, where $0 \leq u_n \leq a_n$ for all terms in both series and where $\sum a_n$ converges, then $\sum u_n$ also converges. The opposite is also true: if $0 \leq a_n \leq u_n$ and a_n is divergent, then u_n is also divergent.

Example 3.8 In order to use the comparison test we have to have a suitable series that we know converges or diverges. An an example, let us see if the series

$$\sum_{n=1}^{n=\infty} \frac{1}{n^n} = 1 + \sum_{n=2}^{\infty} \frac{1}{n^n}$$

converges or diverges. We can compare this series with the series

$$\sum_{n=0}^{\infty} \frac{1}{2^n},$$

which is a geometric series with a factor smaller than 1, so we know it converges and the sum of the series is 2. What is more, for each term of the series

$$\frac{1}{n^n} \leq \frac{1}{2^n},$$

so the series converges to a value that is less than 3.

Exercise 3.5.8 Use the comparison test to determine whether the following series converge or diverge:

$$\text{a. } \sum_{n=1}^{\infty} \frac{1}{n^2 2^n}, \qquad \text{b. } \sum_{n=2}^{\infty} \frac{1}{\ln(n)}.$$

3.5.4 Alternating Series

All the tests in Sections 3.5.1–3.5.3 are restricted to the case when $a_n > 0$. But what if we have an alternating series? We would expect that the influence of the negative terms would provide at least a partial cancellation of the positive terms, so convergence would be faster and hence easier to recognize. For this kind of series we can use the *Leibniz criterion*[4] to test for convergence. This applies to series of the form

$$\sum_{n=1}^{\infty} (-1)^{n+1} a_n, \qquad a_n > 0,$$

and says that if a_n is a monotonically decreasing function for sufficiently large n and that $\lim_{n\to\infty} a_n = 0$, then the series converges.

Exercise 3.5.9 Determine if the following alternating series converge or diverge:

$$\text{a. } \sum_{n=2}^{\infty} (-1)^n \frac{\ln(n)}{n}, \qquad \text{b. } \sum_{n=2}^{\infty} \frac{\cos(n\pi)}{\sqrt{n}}.$$

3.6 Double Series

All of the series that we have looked at so far use a summation over a single index. But we are not restricted to this, and we can have summations over multiple indices. For example,

$$S = \sum_{i=0}^{N} \sum_{j=0}^{M} a_{ij} = \sum_{i=0}^{N} \left(\sum_{j=0}^{M} a_{ij} \right).$$

[4] This was first used by Gottfried Wilhelm Leibniz (1646–1716).

The way to think about such an expression is that the index of the second summation cycles more rapidly than that of the first summation. So, for example,

$$\sum_{i=0}^{2}\sum_{j=0}^{3} a_{ij} = (a_{00} + a_{01} + a_{02} + a_{03}) + (a_{10} + a_{11} + a_{12} + a_{13}) + (a_{20} + a_{21} + a_{22} + a_{23}).$$

Notice that the j index cycles three times for every value of the i index. For a finite summation, we can easily swap the order in which we do the summations without affecting the value of the sum. For example, swapping the order of the summations in the previous double sum gives

$$\sum_{j=0}^{3}\sum_{i=0}^{2} a_{ij} = (a_{00} + a_{10} + a_{20}) + (a_{01} + a_{11} + a_{21}) + (a_{02} + a_{12} + a_{22}) + (a_{03} + a_{13} + a_{23}),$$

and you can see that the same terms are summed, only in a different order.

Example 3.9 To see how to evaluate a double summation, let us find the value of the series

$$S = \sum_{i=1}^{5}\sum_{j=1}^{2}(3i - 2j).$$

The method here is to deal with each index individually:

$$S = \sum_{i=1}^{5}\sum_{j=1}^{2}(3i - 2j) = \sum_{i=1}^{5}\left(\sum_{j=1}^{2} 3i - \sum_{j=1}^{2} 2j\right) = \sum_{i=1}^{5}\left(3i\sum_{j=1}^{2}(1) - 2\sum_{j=1}^{2} j\right)$$
$$= \sum_{i=1}^{5}(3i \times 2 - 2 \times 3) = \sum_{i=1}^{5}(6i - 6) = 6\sum_{i=1}^{5} i - 6\sum_{i=1}^{5}(1) = 60.$$

Exercise 3.6.1 Evaluate the sums

$$\text{a. } \sum_{i=1}^{2}\sum_{j=1}^{2} 4(2i - j), \quad \text{b. } \sum_{j=1}^{4}\sum_{i=1}^{2} 2(2i - j).$$

If the series is an infinite series, then we can algebraically manipulate the indices (sometimes referred to as index gymnastics) to simplify the double summations. However, we have to be a little careful, because we can only do this for series that are absolutely convergent. To see how this works, let us look at the order of summing terms in an infinite double summation

$$S = \sum_{m=0}^{\infty}\sum_{n=0}^{\infty} a_{nm}. \tag{3.23}$$

If we lay out the values of n and m in a grid, where the n values run along the horizontal axis and the m values along the vertical axis, then the a_{nm} values sit at the intersections of the n and m constant lines. The summation proceeds along the rows, starting at the bottom

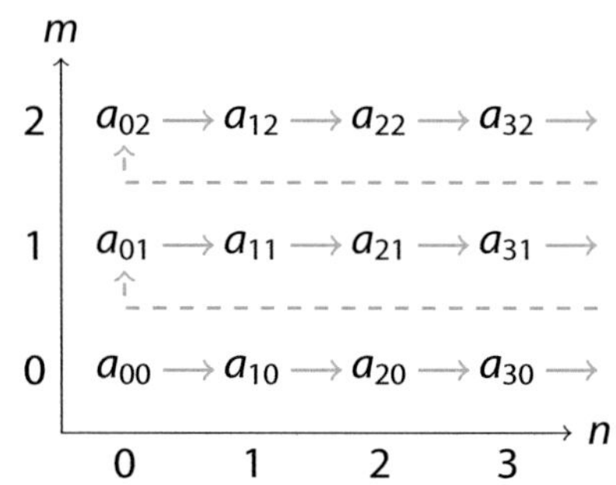

Figure 3.11 The order of terms in the infinite double sum Equation (3.23). We move along a row incrementing the n index until $n = \infty$, and then we move to the next row.

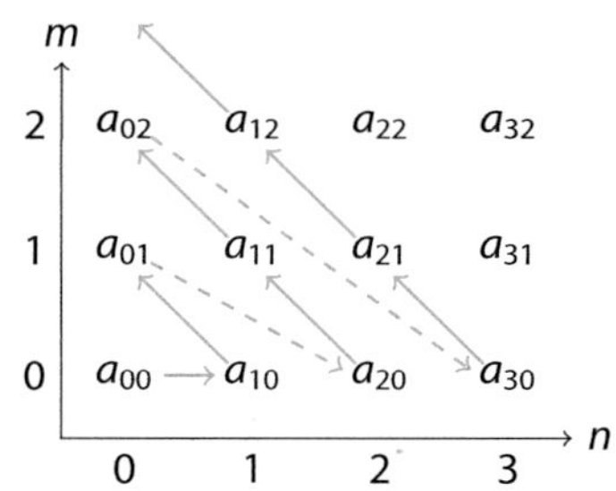

Figure 3.12 The ordering of summing the terms in the summation Equation (3.24).

left-hand corner; recall that the second index of the summation cycles faster. Once we get to $n = \infty$ on the first row, the m index increases by 1 and we move to the next row. This process covers the whole grid (Figure 3.11). If the series is absolutely convergent, then we can try to rearrange the order in which we sum the terms. Doing this can sometimes make the overall summation easier. For example, consider the double series

$$S = \sum_{j=0}^{\infty} \sum_{i=0}^{\infty} a_{ij}.$$

We can make the substitution $j = q$, $i = p - q$, but now have to think a bit about the limits on the summations so that we do not leave any a_{ij} terms out of the summation. The indices i and j cover the upper right-hand quadrant of the plane (see, e.g., Figure 3.11). In order to cover the same range, we can let p range between 0 and ∞, but then q will have to range from 0 to p. So, the summation becomes

$$S = \sum_{p=0}^{\infty} \sum_{q=0}^{p} a_{p-q,q}. \tag{3.24}$$

This means that the a_{ij} values will be added in a different order (Figure 3.12), but the upper right-hand quadrant in the (i, j) plane will still be entirely covered.

3.7 Further Reading

The topic of series and sequences is covered in mathematical books on analysis and calculus as well as books with titles like *Mathematical Methods in Physics* or *Mathematical Methods in Engineering*. Of these, the book by Mary Boas (2006) is particularly accessible. More information on multipole expansions and their uses can be found in standard books on electromagnetism or geophysics.

Problems

3.1 Show that

$$\sum_{i=1}^{n}\sum_{j=1}^{i} b_j = \sum_{i=1}^{n} b_i(n-i+1).$$

3.2 Use a Taylor series expansion of $\tanh(x)$ to evaluate to two decimal places

$$\int_0^{\frac{\pi}{4}} \tanh(\sin(x))\,dx.$$

Can the following integral also be evaluated in the same way, and if not, why?

$$\int_0^{\pi} \tanh(A\sin(x))\,dx, \quad A > 1.$$

3.3 Use suitable tests to determine whether the following series are convergent or divergent. (Note that different tests can be used in each example, but some tests will be easier to apply than others.)

1. $\displaystyle\sum_{n=0}^{\infty} 2e^{-4n}$
2. $\displaystyle\sum_{n=1}^{\infty} \frac{n^3}{n^4 - \sin^{10}(n)}$
3. $\displaystyle\sum_{n=1}^{\infty} \frac{1}{n!}$
4. $\displaystyle\sum_{n=1}^{\infty} \frac{n^n}{n!}$
5. $\displaystyle\sum_{n=0}^{\infty} \frac{x^n}{n!}$
6. $\displaystyle\sum_{n=1}^{\infty} \frac{8}{1+n}$
7. $\displaystyle\sum_{n=1}^{\infty} \frac{8}{1+n^2}$
8. $\displaystyle\sum_{n=0}^{\infty} x^n n!$
9. $\displaystyle\sum_{n=1}^{\infty} \frac{x^n}{n4^n}$

3.4 Calculate the power series expansion and radius of convergence for the functions

1. $f(x) = \left(\dfrac{x}{1-x}\right)^3$
2. $f(x) = \ln(1+x^2)$

3.5 Show that

$$\left(\sum_{i=1}^{n} a_i\right)\left(\sum_{j=1}^{n} a_j\right) = \sum_{i=1}^{n} a_i^2 + 2\sum_{j=1}^{n}\sum_{i<j} a_i a_j.$$

3.6 The velocity (v) of a water wave with wave length λ propagating on an ocean with depth h is

$$v = \sqrt{\frac{g\lambda}{2\pi}}\left(\tanh\left(\frac{2\pi h}{\lambda}\right)\right)^{1/2}.$$

Show that, for shallow waves, $v = \sqrt{gh}$, and find a relationship between h and λ that needs to hold if this approximation is to be valid.

3.7 Blackbody radiation is given by Planck's law,

$$I(\lambda) = 8\pi hc\frac{\lambda^{-5}}{e^{hc/(\lambda kT)} - 1},$$

where λ is the wavelength of energy, T is the temperature of the body, h is a constant (Planck's constant), c is the speed of light, and k is Boltzmann's constant. Show that for large wavelengths

$$I(\lambda) \approx \frac{8\pi kT}{\lambda^4},$$

which is called the Rayleigh–Jeans law.

3.8 Use a suitable convergence test to determine whether the following series converge or diverge.

1. $\displaystyle\sum_{n=1}^{\infty}\frac{n}{2^n}$
2. $\displaystyle\sum_{n=1}^{\infty}\frac{n!\,(n+1)!}{2n}$
3. $\displaystyle\sum_{n=1}^{\infty}(2+\sin(n))e^{-n}$

3.9 Determine the radius of convergence of the following series.

1. $\displaystyle\sum_{n=1}^{\infty}\frac{2^n x^n}{n^3}$
2. $\displaystyle\sum_{n=0}^{\infty}(-1)^n\frac{(x-5)^n}{5n+3}$
3. $\displaystyle\sum_{n=0}^{\infty}3^n x^n$

3.10 Consider the series

$$S = \sum_{n=1}^{\infty} nx^n.$$

1. Show that this series converges for $|x| < 1$.
2. Use the fact that

$$\sum_{n=0}^{\infty} x^n = \frac{1}{1-x}$$

to show that

$$\sum_{n=1}^{\infty} nx^n = \frac{x}{(1-x)^2}.$$

3.11 Geometric growth can be deceptively rapid. Consider the geometric growth of a bacterial population starting from a single cell of mass 500×10^{-15} g of dry mass (i.e., the mass of the cell with all the water removed) that doubles once per day. How long would it take for the dry mass of bacteria to equal the mass of the Earth? If the cell has a diameter of 1 μm, estimate how long it will take for a sphere containing all these bacteria to be expanding at the speed of light (approximately 3×10^8 m s^{-1}).

3.12 The acceleration due to gravity, g, changes as your elevation increases according to

$$g(h) = \frac{GM}{(R+h)^2},$$

where G is Newton's constant, h is your elevation, and R is the radius of the Earth. If $h \ll R$, show that

$$g(h) \approx \frac{GM}{R^2}\left(1 - 2\frac{h}{R}\right).$$

3.13 The number of atoms (N) of a radioactive element in a sample decreases with time (t) according to $N(t) = N_0 e^{-\lambda t}$, where N_0 is the number of atoms in the sample at $t = 0$ and λ is the radioactive decay constant.

1. If $t_{1/2}$ is the time it takes for the number of radioactive atoms to decay to half its initial value, show that $\lambda = \ln(2)/t_{1/2}$ ($t_{1/2}$ is the half-life of the element).
2. Show that the number of radioactive atoms remaining after n half-lives is given by the sequence $N(n) = N_0(1/2)^n$.
3. The isotope carbon-14 (^{14}C) is formed naturally in the atmosphere at a constant rate, and photosynthesizing organisms incorporate some of this material while they are alive. After they die, the ^{14}C decays, and by measuring the remaining ^{14}C in a sample, we can estimate its age. If a sample of organic material is determined to have only 30% of its original ^{14}C remaining, what is the age of the sample, given that the half-life of ^{14}C is 5730 years?

3.14 The transport of sediment is strongly affected by the size of the sediment particles. A commonly used scale for particle size is the Wentworth scale (Wentworth, 1922), which puts particles into classes varying from dust to boulders, with the boundaries between classes being powers of 2; for example, very coarse sand is 1–2 mm in diameter, granules are 2–4 mm in diameter, etc. Show that by taking logarithms to a suitable base we can define a new scale, the Krumbein scale (Krumbein and Aberdeen, 1937), which is an arithmetic sequence.

3.15 Find the approximate value of the integral

$$\int_0^a \frac{e^{-x}}{(1-(x/b))^2}\,dx$$

if a and b are constants, with $b \gg x$.

4 Scalars, Vectors, and Matrices

In this chapter we will look at vectors and matrices and how they can be used. The study of vectors and matrices is part of *linear algebra*, an area of mathematics that is so useful that it has become a standard tool for most scientists. As we shall see, vectors can be used to represent many of the physical quantities that are of interest to us. Manipulating vectors is more complicated than manipulating simple numbers, but with this additional complexity comes a greater scope for representing, understanding, and predicting the complicated phenomena we observe in the natural world. This makes an understanding of vectors very valuable indeed.

Matrices are closely related to vectors and also appear in many areas of modern science. For example, modern environmental data sets often contain many variables and are analyzed using *multivariate data analysis* techniques in which the data are arranged as a matrix. Analyzing the structure of that matrix can reveal interesting patterns and relationships in the data that would have been hard, or impossible, to see otherwise. When analyzing large data sets or numerically solving equations, we often find ourselves needing to solve large systems of coupled, linear equations. Representing these equations using matrices helps us to find solutions to the equations, or to discover that no solution exists. In this chapter we will study vectors first before moving on to matrices. Along the journey, we will also look at coordinate transformations.

4.1 Scalars and Vectors

Many quantities that we measure require only a single number, the magnitude of the quantity, to completely specify them; for example, air temperature, salinity in the oceans, or the concentration of a pollutant. Such quantities are called *scalar* quantities, and they can be represented by a single, normal number. What is more, the value of a scalar quantity does not change if we change the way we specify the location of our measurement. For example, say we use conventional latitude and longitude to specify location on the surface of the Earth, and we measure the air temperature at a given location over time. If we change the way we specify location by redefining the zero of longitude to pass through the middle of the Atlantic Ocean instead of through Greenwich, England, then the value of the temperature we have measured does not change; it was −89.2°C on July 21, 1983, at Vostok Station in Antarctica irrespective of where the line of zero longitude is.

However, there are many quantities that we measure that require more than a single number to specify them. For example, to specify a velocity we need to give a magnitude (the speed) and a direction (e.g., due east). Quantities that are specified by a magnitude (which is a scalar) and a direction are called *vectors*. We require some new mathematical tools if we are to manipulate and work with vectors because vectors will change if we change how we specify location. The reason for this is because in order to specify a direction we need to have some agreed upon set of coordinates with which to specify that direction.[1] The magnitude of the vector will not change, but the direction may change.

Scalar or vector quantities can be defined at a single point, but it is often useful to be able to have them defined everywhere throughout a region. For example, we can measure air temperature and wind velocity at a single location (e.g., Amundsen South Pole Station in Antarctica), but it can also be useful to specify them at every point across the whole globe, giving us a *scalar field* and *vector field*.

4.2 Vector Algebra

To specify a vector we need both its magnitude and its direction. We can represent this pictorially as an arrow, where the length of the arrow represents the magnitude of the vector and the direction the arrow points in is the direction of the vector (Figure 4.1a). Some examples of variables that are vectors include:

- Displacement: an object moves a certain distance (the magnitude of the vector) in a given direction.
- Velocity: an object moves at a certain speed (the magnitude of the vector) in a given direction.
- Force: a force of a given magnitude in a given direction is applied to an object.

To distinguish vectors from simple numbers, we will write a vector using a bold letter. For example, we write the velocity of an ocean current as $\mathbf{u}$, and its speed as u; $\mathbf{u}$ is the vector that represents both the speed and direction of the velocity, whereas u, its speed, is just the magnitude of the velocity (i.e., the length of the vector).

We can perform simple arithmetic operations on vectors. For example, multiplication of a vector by a positive scalar (i.e., by just a positive number) changes the length of the vector but preserves its direction (Figure 4.1b); if the scalar is greater than 1, then the new vector is longer than the original, whereas if the scalar is less than 1, the new vector has a smaller magnitude than the original one. This makes sense because if we double our speed we double the magnitude of our velocity but not the direction we are traveling in. Multiplying a vector by a negative number produces a vector in the opposite direction of the original one and with a magnitude scaled by the size of the number (Figure 4.1b).

[1] An exercise that is often used to help develop teamwork in Antarctica is to find a colleague who is lost outside in whiteout blizzard conditions. To simulate these conditions, members of the search party are all roped together in a line and wear white buckets over their heads so they cannot see anything except their own feet. If the team leader tells everyone to turn right, you can turn to your right, but your colleagues in the search party might all be facing in different directions. As a result, everyone ends up facing in a different direction because there is no common set of coordinates or reference points that everyone can use.

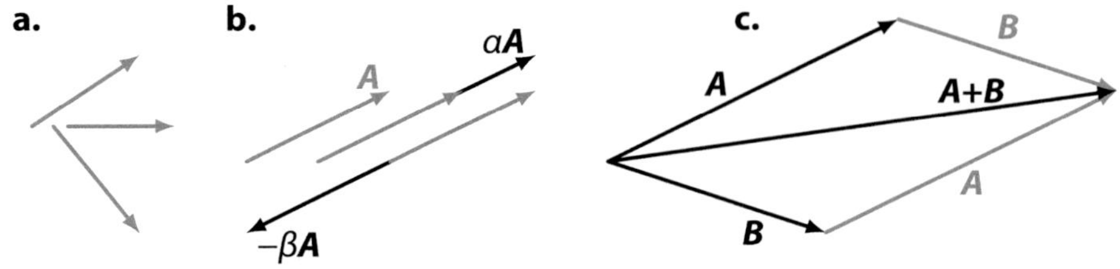

Figure 4.1 A vector can be represented as an arrow pointing in the direction of the vector with the length of the arrow representing the magnitude of the vector (a.). Multiplying the vector by a positive number changes the magnitude of the vector, whereas multiplication by a negative number (e.g., −1) results in a vector pointing in the opposite direction with a magnitude scaled by the size of the number (b.)—the gray arrow in (b.) represents the same vector (**A**). To add two vectors **A** and **B**, we replicate the two vectors to form a parallelogram (the gray arrows) and draw a new vector along the diagonal of the parallelogram. This new vector is the resultant of adding **A** and **B** (c.).

We can represent vector addition using the *parallelogram method*[2] that you likely learned in high school (Figure 4.1c). This method makes intuitive sense if we think of the vector in terms of a displacement: if we walk in the direction **A** for a distance given by the length of that vector, and then we walk in the direction given by the vector **B** for a distance given by the length of that vector, our resultant motion will have been in the direction given by the vector **A** + **B** for a distance given by its length. This method can be used to add any vector, not just displacements.

Example 4.1 Let us look at a simple example. A wind blows in the direction 30° north of due east at 3 m s^{-1} and a sail boat is moved by a surface current in the ocean with a speed of 2 m s^{-1} moving in a direction 30° east of due north. We want to use vector addition to find the resultant velocity (its magnitude and direction) of the boat. To solve this problem using the parallelogram method, we need to draw a figure and use our knowledge of basic trigonometry. From Figure 4.2 we see that we need to find the length of the vector **OC** and the angle it makes with either axis. The lengths of two of the sides of the triangle OBC are the magnitudes of the current and the wind; **OB** has length 2 and **BC** has length 3. Similarly, we can find the angles between **OB** and **BC** (150°) and the angle between **OB** and **OC** ($\alpha - 30°$). We can use the cosine rule (Appendix B) to find the length of **OC**:

$$(\mathrm{OC})^2 = (\mathrm{OB})^2 + (\mathrm{BC})^2 - 2(\mathrm{OB})(\mathrm{BC})\sin(150°),$$

giving (OC) = 4.8366, which is the resultant velocity of the boat in units of m s^{-1}. To find the angle between **OB** and **OC** we can make use of the sine rule (Appendix B):

$$\frac{\sin(150°)}{\mathrm{OC}} = \frac{\sin(\alpha - 30°)}{\mathrm{BC}},$$

which gives $\alpha = 48.0675°$. So, the resultant motion of the boat is in the direction 48.0675° east of due north.

[2] Although the mathematical concept of a vector was not developed until the nineteenth century, Newton showed that forces could be added using the parallelogram method (Crowe, 1994).

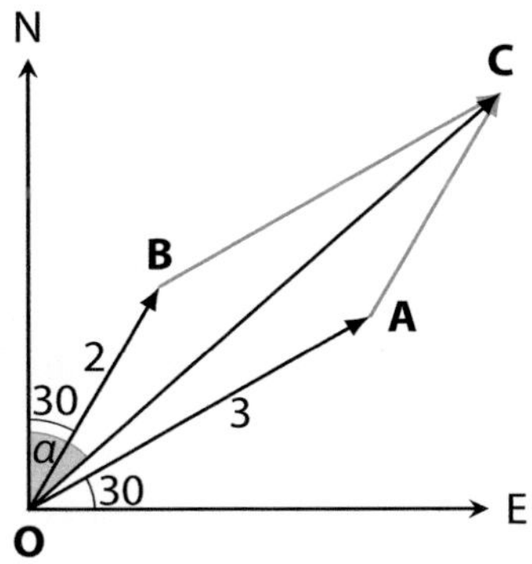

Figure 4.2 The vector **OB** represents the velocity of the ocean current, and the vector **OA** represents the velocity of the wind. The parallelogram rule for vector addition tells us that the vector **OC** is the resultant of these two velocities.

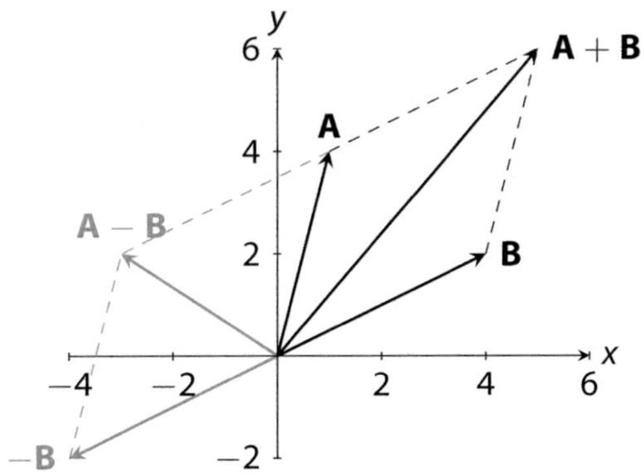

Figure 4.3 Addition and subtraction of the vector **A** and **B** using the parallelogram rule. The addition **A** + **B** of the two vectors is shown in black, and the subtraction **A** − **B** is shown in gray.

Example 4.2 Consider an example of a vector **A** that represents a displacement of 1 unit in the x direction and 4 units in the y direction (Figure 4.3) and a vector **B** that represents a displacement of 4 units in the x-direction and 2 in the y-direction. It is important to note that although the base of both vectors is pictured as being at the origin of coordinates, this need not be the case. So, for example, we can envisage walking from the origin to a point $x = 1$, $y = 4$ represented by the tip of **A**, and from there walking a further 4 units in the x direction and 2 units in the y direction. This is the same as shifting the base of the vector **B** from the origin to the tip of **A**, from which we can form a parallelogram and calculate the net displacement of $\mathbf{A} + \mathbf{B}$. To subtract **B** from **A** we basically add the vector $-\mathbf{B}$ to **A**, where $-\mathbf{B}$ is a vector of the same length as **B** but pointing in the opposite direction.

Vector addition, and multiplication of a vector by a scalar, obey certain rules. These are intuitively obvious if we think of them in terms of displacements, but like the parallelogram method, they hold for all vectors. These rules can be summarized as:

$$\mathbf{A} + \mathbf{B} = \mathbf{B} + \mathbf{A} \qquad \text{commutative law,} \tag{4.1a}$$

$$\mathbf{A} + (\mathbf{B} + \mathbf{C}) = (\mathbf{A} + \mathbf{B}) + \mathbf{C} \qquad \text{associative law,} \tag{4.1b}$$

$$(\alpha + \beta)\mathbf{A} = \alpha\mathbf{A} + \beta\mathbf{A} \qquad \text{distributive law,} \tag{4.1c}$$

$$\alpha(\mathbf{A} + \mathbf{B}) = \alpha\mathbf{A} + \alpha\mathbf{B} \qquad \text{distributive law.} \tag{4.1d}$$

We can see from Figure 4.1c that the commutative law for addition is true: it does not matter if the displacement **A** happens before the displacement **B** or vice versa, both situations give the same resultant displacement.

Exercise 4.2.1 Use the geometrical picture of a vector as an arrow to convince yourself of the validity of the other rules of vector addition and multiplication by a scalar.

The process of vector addition can be extended to include subtraction by reversing the direction of the appropriate vector. Note that this will change the direction of the arrow but not the magnitude of the vector.

If we divide a vector **A** by its length we will end up with a vector **a** that points in the same direction as **A**, but has a length of 1. Such a vector is called a *unit vector*. These vectors are very useful because, by multiplying them by a scalar we can create a vector of any desired length pointing in the direction given by **a**. This is useful in its own right, but it becomes especially useful if we have directions defining a coordinate system. For example, if we set up standard rectangular (x, y) coordinates on a plane, then we can draw a unit vector $\hat{\mathbf{i}}$ along the direction of the x axis and another, $\hat{\mathbf{j}}$, along the direction of the y axis. These vectors define the directions of the x and y coordinates and are called *basis vectors*, and we will have more to say about them shortly. For any vector **A** we can draw perpendicular lines from the end of the vector to the x- and y-axes (Figure 4.4), and these numbers (x_A and y_A) are called the x- and *y-components* of the vector **A**. The components x_A and y_A tell us how far along the $\hat{\mathbf{i}}$ and $\hat{\mathbf{j}}$ directions we have to move to get to the tip of the vector **A**. Recalling vector addition, we can see that multiplying the components of **A** by their corresponding basis vector gives us two vectors that, when added together, give us the vector **A** (Figure 4.4):

$$\mathbf{A} = x_A\hat{\mathbf{i}} + y_A\hat{\mathbf{j}}, \tag{4.2}$$

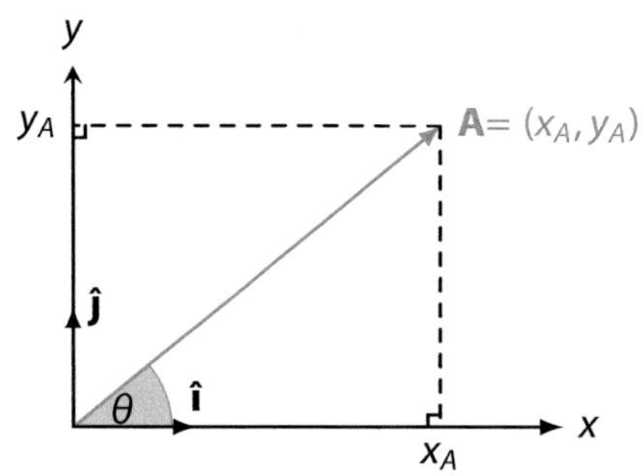

Figure 4.4 Unit vectors $\hat{\mathbf{i}}$ and $\hat{\mathbf{j}}$ can be defined along the directions of the x and y axes, and these allow us to define the x- and y-components, x_A and y_A, of a vector **A**.

which is also often written $\mathbf{A} = (x_A, y_A)$ if the basis vectors have been already been specified. The vector $x_A\hat{\mathbf{i}}$ is a vector of length x_A along the x axis (i.e., in the direction of $\hat{\mathbf{i}}$) and is called the *projection* of the vector $\mathbf{A}$ along $\hat{\mathbf{i}}$. Similarly, the vector $y_A\hat{\mathbf{j}}$ is the projection of $\mathbf{A}$ along $\hat{\mathbf{j}}$. This idea can easily be extended to three dimensions, where the unit vector along the z axis is denoted by $\hat{\mathbf{k}}$, so that $\mathbf{A} = x_A\hat{\mathbf{i}} + y_A\hat{\mathbf{j}} + z_A\hat{\mathbf{k}} = (x_A, y_A, z_A)$.

Now we can look at vector arithmetic in terms of the components of the vectors. If we have vectors $\mathbf{A} = (x_A, y_A, z_A)$ and $\mathbf{B} = (x_B, y_B, z_B)$ in three dimensions, then

$$\mathbf{A} + \mathbf{B} = (x_A + x_B)\hat{\mathbf{i}} + (y_A + y_B)\hat{\mathbf{j}} + (z_A + z_B)\hat{\mathbf{k}} \tag{4.3a}$$

$$\alpha\mathbf{A} = \alpha x_A\hat{\mathbf{i}} + \alpha y_A\hat{\mathbf{j}} + \alpha z_A\hat{\mathbf{k}}. \tag{4.3b}$$

That is, we add (or subtract) vectors by adding (or subtracting) their components, and to multiply a vector by a scalar, we multiply all of its components by the same scalar.

Exercise 4.2.2 Use a diagram similar to that in Figure 4.3 to show that Equations (4.3a) and (4.3b) are correct in two dimensions (x, y).

We can also use Figure 4.4 to see how we calculate the length of a vector using its components. Lines from the tip of the vector perpendicular to the coordinate axes form two right-angled triangles, so we can use Pythagoras' theorem to find the length of the vector $\mathbf{A}$ in terms of its components (x_A, y_A):

$$||\mathbf{A}|| = \text{Length of } \mathbf{A} = \sqrt{x_A^2 + y_A^2}, \tag{4.4}$$

where the notation $||\cdots||$ indicates the magnitude (i.e., length) of the vector (this is also sometimes written with single vertical lines $|\cdots|$). This generalizes to three dimensions:

$$||\mathbf{A}|| = \text{Length of } \mathbf{A} = \sqrt{x_A^2 + y_A^2 + z_A^2}.$$

Exercise 4.2.3 Consider a vector $\mathbf{A} = x\hat{\mathbf{i}} + y\hat{\mathbf{j}}$. If the vector is multiplied by a scalar quantity β, show that the length of the new vector is β multiplied by the length of $\mathbf{A}$.

We can now define a *unit vector* in the direction of $\mathbf{A}$ by

$$\mathbf{a} = \frac{\mathbf{A}}{||\mathbf{A}||}, \tag{4.5}$$

so that $\mathbf{a}$ is a vector that has a length of 1 unit in the direction of $\mathbf{A}$.

We have mentioned that a useful way to think of the components of a vector $\mathbf{A}$ is as the *projection* of the vector onto the coordinate axes. What does this mean? If we think of the vector $\mathbf{A}$ as being a stick, then the projection of $\mathbf{A}$ onto the x axis is simply the shadow cast by the stick if it is illuminated from directly above. A similar idea holds for the y axis, but the vector has to be illuminated from the side. If we look at Figure 4.4 and use some trigonometry, we can see that the components of $\mathbf{A}$ are given by the equations

$$x_A = ||\mathbf{A}|| \cos(\theta), \qquad y_A = ||\mathbf{A}|| \sin(\theta), \tag{4.6}$$

where θ is the angle that $\mathbf{A}$ makes with the x axis.

Example 4.3 Floating drifters are used in oceanography to track the movement of surface water. Unfortunately, the trajectory of a badly designed drifter can be influenced by the movement of both the water and the winds, and the resultant motion of the drifter will be a combination of the vectors representing these motions. Imagine we are tracking the motion of a surface drifter that is moved by a surface current with a speed of 2 m s^{-1} moving in a direction 30° east of due north. A wind blows in the direction 30° north of due east at 3 m s^{-1}. We can use vectors to find the components in the north (y) and east (x) directions of both vectors and calculate the sum of the two vectors, then compare this with the motion of the surface current. From the geometry of Figure 4.4 we can see that the x- and y-components of a general vector $\mathbf{A}$ in two dimensions are given by

$$x_A = ||\mathbf{A}|| \cos(\theta), \qquad y_A = ||\mathbf{A}|| \sin(\theta).$$

For the ocean current vector ($\mathbf{A}$) we have $||\mathbf{A}|| = 2$ and $\theta = 60°$, whereas for the wind velocity vector ($\mathbf{B}$) we have $||\mathbf{B}|| = 3$ and $\theta = 30°$. Therefore, the components of the two vectors are

$$(x_A, y_A) = (2\cos(60), 2\sin(60)) = (1.0, \sqrt{3})$$

$$(x_B, y_B) = (3\cos(30), 3\sin(30)) = \left(\frac{3\sqrt{3}}{2}, \frac{3}{2}\right).$$

The resultant vector has components

$$\mathbf{R} = \left(1 + \frac{3\sqrt{3}}{2}, \sqrt{3} + \frac{3}{2}\right) = (3.598, 3.232).$$

This is a vector that makes an angle of $\phi = \tan^{-1}(3.232/3.598) \approx 42°$ with the x axis, and has a length ≈ 4.8. So, the drifter is moving much faster than the current and at an angle of approximately 18° east of the current; these values are the same as in Example 4.1.

The cosines of the directions that a vector $\mathbf{A} = x_A\mathbf{i} + y_A\mathbf{j} + z_A\mathbf{k}$ makes with the x, y, and z axes are called *direction cosines*:

$$\cos\theta = \frac{x_A}{||\mathbf{A}||}, \qquad \cos\phi = \frac{y_A}{||\mathbf{A}||}, \qquad \cos\psi = \frac{z_A}{||\mathbf{A}||}. \tag{4.7}$$

Direction cosines are useful precisely because they specify the direction of a vector, and they have some nice properties. For example, if $\mathbf{a}$ is a unit vector, then the direction cosines are the components of $\mathbf{a}$ along the coordinate axes, and Pythagoras' theorem leads us to

$$\cos^2(\theta) + \cos^2(\phi) + \cos^2(\psi) = 1.$$

Direction cosines are useful for determining the directions in geological formations (Pollard and Fletcher, 2005), and in determining leaf orientations for calculating the reflection of light from plant canopies in remote sensing applications and flux calculations (Monson and Baldocchi, 2014).

Another common way of writing a vector $\mathbf{A} = x_A\hat{\mathbf{i}} + y_A\hat{\mathbf{j}} + z_A\hat{\mathbf{k}}$ is as a *column vector*,

$$\mathbf{A} = \begin{pmatrix} x_A \\ y_A \\ z_A \end{pmatrix},$$

where it is implicit that x_A is the component of $\mathbf{A}$ along the x axis and so on. Using this notation we have

$$\mathbf{A} + \mathbf{B} = \begin{pmatrix} x_A \\ y_A \\ z_A \end{pmatrix} + \begin{pmatrix} x_B \\ y_B \\ z_B \end{pmatrix} = \begin{pmatrix} x_A + x_B \\ y_A + y_B \\ z_A + z_B \end{pmatrix}, \qquad \alpha\mathbf{A} = \alpha \begin{pmatrix} x_A \\ y_A \\ z_A \end{pmatrix} = \begin{pmatrix} \alpha x_A \\ \alpha y_A \\ \alpha z_A \end{pmatrix}.$$

We now have several equivalent ways of writing a vector, and each has its advantages and disadvantages. If the specific coordinate system we are using is not important, then we can write a vector as a bold letter (e.g., $\mathbf{A}$). For example, the relationship $\mathbf{A} + \mathbf{B} = \mathbf{B} + \mathbf{A}$ holds irrespective of what coordinates we use. In other situations we might be interested in knowing the component of a vector in a specific direction, so knowing the components of a vector with respect to a given coordinate system can make this calculation easier.

Exercise 4.2.4 Calculate $\mathbf{A} - \mathbf{B}$ using components when $\mathbf{A} = 3\hat{\mathbf{i}} - 2\hat{\mathbf{j}}$ and $\mathbf{B} = 5\hat{\mathbf{i}} + 2\hat{\mathbf{j}} - \hat{\mathbf{k}}$.

Exercise 4.2.5 Draw the following vectors: a. $\mathbf{C} = -2\hat{\mathbf{i}} - 4\hat{\mathbf{j}}$; b. $\mathbf{D} = 0.5\mathbf{C}$; c. $-\mathbf{C}$.

Exercise 4.2.6 Calculate the direction cosines for the following vectors: a. $\mathbf{A} = -3\hat{\mathbf{i}} + 3\hat{\mathbf{j}}$; b. $\mathbf{B} = \hat{\mathbf{i}} + \hat{\mathbf{j}} + 2\hat{\mathbf{k}}$; c. $\mathbf{A} = -\hat{\mathbf{k}}$.

4.2.1 Linear Independence and Basis Vectors

The unit vectors $\hat{\mathbf{i}}$, $\hat{\mathbf{j}}$, and $\hat{\mathbf{k}}$ form a set of basis vectors for vectors in a three-dimensional, rectangular coordinate system. The concept of a basis is an important one and will appear many times in different guises throughout this book, so it is worth spending a little time thinking about it here. To do this, we first need to introduce another important concept, *linear independence*.

A set of vectors $\mathbf{a}$, $\mathbf{b} \dots \mathbf{n}$, is *linearly independent* if the only solution to the equation

$$\lambda\mathbf{a} + \mu\mathbf{b} + \dots + \nu\mathbf{n} = 0 \tag{4.8}$$

is $\lambda = \mu = \nu = \cdots = 0$, where λ, μ, $\cdots$ are scalars. If this is the case, then we cannot pick any one of these vectors (say $\mathbf{b}$) and write it by adding and subtracting the remaining vectors. We can see this intuitively by looking at the basis vectors $\hat{\mathbf{i}}$ and $\hat{\mathbf{j}}$ in two dimensions (Figure 4.4). We can write $\hat{\mathbf{j}}$ in terms of $\hat{\mathbf{i}}$ only if there is some component of $\hat{\mathbf{j}}$ in the $\hat{\mathbf{i}}$ direction. However, this is not the case, and in fact the x-component of $\hat{\mathbf{j}}$ is $\|\hat{\mathbf{j}}\| \cos(\pi/2) = 0$. The vectors are at right angles to each other and have a length of 1 and form an *orthonormal basis* (i.e., an orthogonal, normalized basis). Similarly for the vectors $\hat{\mathbf{i}}$, $\hat{\mathbf{j}}$, and $\hat{\mathbf{k}}$ in three dimensions. In fact, we can define the dimension (N) of a space as being the largest number of linearly independent vectors we can define for that space. A set of N linearly independent vectors in an N-dimensional space forms a *complete basis* for that space. This means that we can write a vector $\mathbf{A}$ as a linear combination of these basis vectors, with the coefficients being the components of the vector in that basis.

For example, an arbitrary vector in two dimensions can be written $\mathbf{A} = x_A\hat{\mathbf{\imath}}+y_A\hat{\mathbf{\jmath}}$. Similarly, in three dimensions, $\mathbf{V} = x_V\hat{\mathbf{\imath}} + y_V\hat{\mathbf{\jmath}} + z_V\hat{\mathbf{k}}$. This is a useful thing to do because once we have identified a set of basis vectors for our problem, we can immediately start working with the components of the vectors, and this makes calculations a lot easier.

Example 4.4 We have claimed several times that the vectors $\hat{\mathbf{\imath}}$ and $\hat{\mathbf{\jmath}}$ form a basis in two dimensions. This means that these vectors must be linearly independent. We can check this using Equation (4.8),

$$\lambda\hat{\mathbf{\imath}} + \mu\hat{\mathbf{\jmath}} = 0,$$

which is just vector addition and gives us $\lambda = \mu = 0$, so these vectors are linearly independent. We can try another two vectors, $\mathbf{u} = \hat{\mathbf{\imath}}+2\hat{\mathbf{\jmath}}$ and $\mathbf{v} = 2\hat{\mathbf{\imath}}+4\hat{\mathbf{\jmath}}$. Using Equation (4.8) we get $(\lambda + 2\mu)\hat{\mathbf{\imath}} + (2\lambda + 4\mu)\hat{\mathbf{\jmath}} = 0$. Both components must be zero to satisfy this equation, but we notice that the y component is twice the x component, so we have only one equation to satisfy, $\lambda + 2\mu = 0$, or $\lambda = -2\mu$. So, these vectors are not linearly independent.

The set of vectors $\hat{\mathbf{\imath}}$, $\hat{\mathbf{\jmath}}$, and $\hat{\mathbf{k}}$ is not the only basis there is for three-dimensional space. Consider the vectors $\mathbf{e}_1 = \hat{\mathbf{\imath}} + 2\hat{\mathbf{\jmath}} - \hat{\mathbf{k}}$, $\mathbf{e}_2 = \hat{\mathbf{\imath}} + 3\hat{\mathbf{\jmath}} + \hat{\mathbf{k}}$, and $\mathbf{e}_3 = 6\hat{\mathbf{\imath}}$. To show that these three vectors form a basis for three-dimensional space, we need to show that the only solution to the equation

$$\lambda\hat{\mathbf{e}}_1 + \mu\hat{\mathbf{e}}_2 + \nu\hat{\mathbf{e}}_3 = 0$$

is $\lambda = \mu = \nu = 0$ where λ, μ, and ν are constants. Substituting the values for the three vectors we get the equation

$$\lambda(\hat{\mathbf{\imath}} + 2\hat{\mathbf{\jmath}} - \hat{\mathbf{k}}) + \mu(\hat{\mathbf{\imath}} + 3\hat{\mathbf{\jmath}} + \hat{\mathbf{k}}) + \nu(6\hat{\mathbf{\imath}}) = (\lambda + \mu + 6\nu)\hat{\mathbf{\imath}} + (2\lambda + 3\mu)\hat{\mathbf{\jmath}} + (\mu - \lambda)\hat{\mathbf{k}} = 0.$$

We know that the three basis vectors $\hat{\mathbf{\imath}}$, $\hat{\mathbf{\jmath}}$, and $\hat{\mathbf{k}}$ are linearly independent, so this equation can only be true if the coefficients of these vectors are also zero. In other words,

$$\lambda + \mu + 6\nu = 0, \qquad 2\lambda + 3\mu = 0, \qquad -\lambda + \mu = 0,$$

and the solution to these equations is $\lambda = \mu = \nu = 0$. Therefore, the vectors $\mathbf{e}_1$, $\mathbf{e}_2$, and $\mathbf{e}_3$ are linearly independent. We have three linearly independent vectors in three dimensions, so they form a complete basis. However, $\hat{\mathbf{e}}_1$, $\hat{\mathbf{e}}_2$, and $\hat{\mathbf{e}}_3$ are not unit vectors because they have lengths $\sqrt{6}$, $\sqrt{11}$, and 6 respectively. We can make them into unit vectors by simply dividing each vector by its length:

$$\hat{\mathbf{e}}_1 = \frac{1}{\sqrt{6}}(\hat{\mathbf{\imath}} + 2\hat{\mathbf{\jmath}} - \hat{\mathbf{k}}), \qquad \hat{\mathbf{e}}_2 = \frac{1}{\sqrt{11}}(\hat{\mathbf{\imath}} + 3\hat{\mathbf{\jmath}} + \hat{\mathbf{k}}), \qquad \hat{\mathbf{e}}_3 = \hat{\mathbf{k}}.$$

Why do we not use these vectors as a basis for vectors in three dimensions? We could do, but the set of vectors $\hat{\mathbf{\imath}}$, $\hat{\mathbf{\jmath}}$, and $\hat{\mathbf{k}}$ has an additional property that makes calculations using them a lot easier; they are all at right angles to each other, whereas the angle between $\hat{\mathbf{e}}_1$ and $\hat{\mathbf{e}}_2$ is approximately 42°. As we will see soon, using an orthonormal basis of vectors simplifies our lives a lot.

Exercise 4.2.7 Are the vectors $\mathbf{e}_1 = 3\hat{\imath} + 2\hat{\jmath}$ and $\mathbf{e}_2 = \hat{\imath} - \hat{\jmath}$ linearly independent? If so, do they form a basis for three-dimensional space?

4.2.2 Transformations of Vectors

We have seen that the components of a vector are intimately connected with the basis vectors and the system of coordinates they describe. What happens to a vector if we change the coordinates and the basis vectors? Let us first look at a simpler case and try to understand what happens to a scalar when we transform coordinates. To do this, we will examine a simple thought experiment. Consider a map of the Earth overlain with a map of surface temperatures. Each location on the map is specified by giving its latitude and longitude. So, for example, the temperature at a given location in Cambridge, England (52.1882N, 0.132E), might be 18°C. Now, let us change our definition of latitude and longitude such that Cambridge is now at the new North Pole (i.e., at 90N, 0E). The temperature will not have changed — we have not moved the city of Cambridge, only rotated the lines of latitude such that Cambridge is now at 90N—temperature is said to be *invariant* under the coordinate transformation. So, the value of a scalar does not depend on the coordinate system we use.

Now, consider what happens to a vector. On top of the map of temperature we also overlay a map of wind speed and direction; each location has an arrow with the length of the arrow representing the wind speed and the direction of the arrow being the wind direction. In our conventional latitude and longitude coordinates, the wind at Cambridge is 1 m s^{-1} blowing directly due north. In our new coordinates, however, the wind will be blowing due south! However, neither the physical direction of the wind nor its speed has changed, but the components of the wind vector in the different coordinates have changed. So, the vector itself is invariant under the change in coordinates, but the components of the vector are not.

To make this discussion more concrete, we will look at a particular type of transformation that is very useful, a rotation. Let us start by looking at what happens if we keep the coordinates fixed and rotate a vector through an angle ψ (Figure 4.5). This is going to be related in some way to holding the vector constant and rotating the coordinates. Rotating a vector through an angle ψ about the origin will not change the magnitude of the vector, only its direction. Recalling some trigonometry (Appendix B), we can calculate the components of the new vector (**B**) in terms of the components of the original vector (**A**),

$$\begin{aligned} x_B &= ||\mathbf{A}||\cos(\psi+\theta) = ||\mathbf{A}||\cos(\psi)\cos(\theta) - ||\mathbf{A}||\sin(\psi)\sin(\theta) \\ &= x_A\cos(\psi) - y_A\sin(\psi), \qquad (4.9a) \\ y_B &= ||\mathbf{A}||\sin(\psi+\theta) = ||\mathbf{A}||\sin(\psi)\cos(\theta) + ||\mathbf{A}||\cos(\psi)\sin(\theta) \\ &= x_A\sin(\psi) + y_A\cos(\psi), \qquad (4.9b) \end{aligned}$$

where we have used that $x_A = ||\mathbf{A}||\cos(\theta)$ and $y_A = ||\mathbf{A}||\sin(\theta)$. These equations tell us how the individual components of the vector change under this transformation.

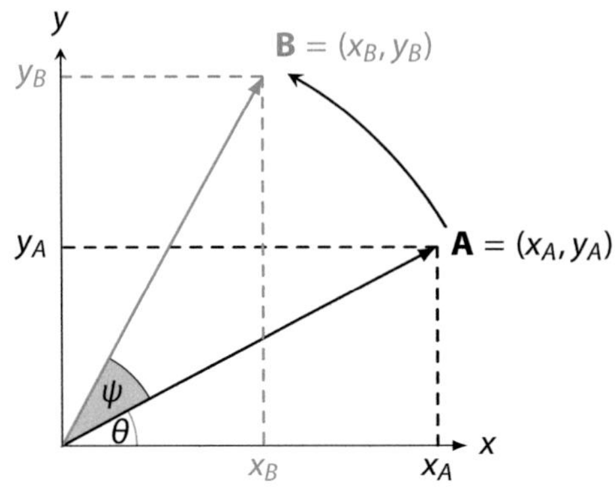

Figure 4.5 A vector $\mathbf{A} = (x_A, y_A)$, making an angle θ with the x axis, is rotated through an angle ψ to obtain a new vector $\mathbf{B} = (x_B, y_B)$.

Exercise 4.2.8 Show that if we take a vector $\mathbf{A}$ and displace it by $\mathbf{D}$, then the new vector is $\mathbf{B} = \mathbf{A} + \mathbf{D}$.

Exercise 4.2.9 Use Equations (4.9) to show that $x_A^2 + y_A^2 = x_B^2 + y_B^2$ and so conclude that the length of the vector $\mathbf{A}$ does not change when it is rotated about the origin.

Rotations and scalings are examples of special kind of transformations called *linear transformations*. Linear transformations also have some nice properties. First, if we add two vectors (say, $\mathbf{u}$ and $\mathbf{v}$) and then rotate the resulting vector, this will be the same as if we first rotate $\mathbf{u}$, then rotate $\mathbf{v}$, and add the resulting two vectors. Second, if we first scale $\mathbf{u}$ by multiplying it by a constant factor α and then transform the result, this will be same as if we first transformed $\mathbf{u}$ and then scaled the answer. These properties can be written in a nice, compact form if we represent the act of making a transformation of a vector $\mathbf{u}$ by $T(\mathbf{u})$; for example, if the transformation we are concerned with is a rotation about the origin through an angle θ, then $T(\mathbf{u})$ is shorthand for rotating $\mathbf{u}$ through an angle θ (i.e., applying Equations 4.9), and $T(\mathbf{u} + \mathbf{v})$ is shorthand for first adding the vectors $\mathbf{u}$ and $\mathbf{v}$ and then rotating the resultant vector through an angle θ. We can write the properties we have just described in a compact form using this notation,

$$T(\mathbf{u} + \mathbf{v}) = T(\mathbf{u}) + T(\mathbf{v}), \qquad T(\alpha\mathbf{u}) = \alpha T(\mathbf{u}). \tag{4.10}$$

Example 4.5 If $\mathbf{X}$ is a vector with components (x, y), let us show that the transformation $x' = x$, $y' = -y$ is a linear transformation. To do this, we consider the vectors

$$\mathbf{X} = x\hat{\mathbf{i}} + y\hat{\mathbf{j}} = \begin{pmatrix} x \\ y \end{pmatrix} \qquad \mathbf{U} = u\hat{\mathbf{i}} + v\hat{\mathbf{j}} = \begin{pmatrix} u \\ v \end{pmatrix}.$$

The transformation takes the components of $\mathbf{X}$ and multiplies the y component by -1, so that

$$T(\mathbf{X}) = T\begin{pmatrix} x \\ y \end{pmatrix} = \begin{pmatrix} x \\ -y. \end{pmatrix}, \quad \text{i.e.,} \quad T(x\hat{\mathbf{i}} + y\hat{\mathbf{j}}) = x\hat{\mathbf{i}} - y\hat{\mathbf{j}}.$$

To show that this is a linear transformation, we have to show that it obeys the properties in Equations (4.10). Consider the first condition:

$$T(\mathbf{X}+\mathbf{U}) = T\left(\begin{pmatrix} x \\ y \end{pmatrix} + \begin{pmatrix} u \\ v \end{pmatrix}\right) = T\begin{pmatrix} x+u \\ y+v \end{pmatrix} = \begin{pmatrix} x+u \\ -y-v \end{pmatrix} = \begin{pmatrix} x \\ -y \end{pmatrix} + \begin{pmatrix} u \\ -v \end{pmatrix} = T(\mathbf{X})+T(\mathbf{U}).$$

We can also show that the second property is obeyed:

$$T(\alpha\mathbf{X}) = T\begin{pmatrix} \alpha x \\ \alpha y \end{pmatrix} = \begin{pmatrix} \alpha x \\ -\alpha y \end{pmatrix} = \alpha\begin{pmatrix} x \\ -y \end{pmatrix} = \alpha T(\mathbf{X}).$$

So, the transformation $x' = x$, $y' = -y$ is a linear transformation.

Linear transformations have other useful properties. For example, $T(-\mathbf{A}) = -T(\mathbf{A})$. We can show that this must be true using Equations (4.10) because

$$T(-\mathbf{A}) = T((-1)\mathbf{A}) = (-1)T(\mathbf{A}) = -T(\mathbf{A}).$$

Another consequence of Equations (4.10) is that $T(\mathbf{0}) = 0$, where $\mathbf{0}$ is the *zero vector*; that is, $\mathbf{0}$ is a vector in which all the components are zero.

Exercise 4.2.10 Show that rotating a vector through an angle θ about the origin is a linear transformation.

Exercise 4.2.11 Show that for a linear transformation, $T(\mathbf{0}) = 0$.

Exercise 4.2.12 If $T(\cdot)$ is a linear transformation, show that $T(\mathbf{u} - \mathbf{v}) = T(\mathbf{u}) - T(\mathbf{v})$.

Exercise 4.2.13 Show that the transformation resulting from translating a vector by a constant vector $\mathbf{A} = (a, b)$ (i.e., $\mathbf{X}$ becomes $\mathbf{X} + \mathbf{A}$) is not a linear transformation.

So far we have transformed the vector but kept the coordinates and their basis vectors unchanged. We can also look at transformations in another way. Instead of changing the vector we can transform the coordinates and keep the vector unchanged. We can then look at a transformation in terms of how the basis vectors transform. For example, assume that we have a set of basis vector $\mathbf{B} = \{\mathbf{e}_1, \dots, \mathbf{e}_n\}$. We know that, because this is a basis, we can write any vector $\mathbf{x}$ using these basis vectors as

$$\mathbf{x} = x_1\mathbf{e}_1 + x_2\mathbf{e}_2 + \cdots + x_n\mathbf{e}_n,$$

where $(x_1, \dots, x_n)$ are the components of the vector $\mathbf{x}$ with respect to that basis. Let us now transform the coordinates giving a new set of basis vectors $\tilde{\mathbf{B}} = \{\tilde{\mathbf{e}}_1, \dots, \tilde{\mathbf{e}}_n\}$. For example, we could get the new basis by rotating all the original basis vectors through the same angle. As we have seen, this means that although the vector itself does not change, its components in the new basis will be different from the components in the old one. In other words

$$\mathbf{x} = x_1\mathbf{e}_1 + x_2\mathbf{e}_2 + \cdots + x_n\mathbf{e}_n = \tilde{\mathbf{x}} = \tilde{x}_1\tilde{\mathbf{e}}_1 + \tilde{x}_2\tilde{\mathbf{e}}_2 + \cdots + \tilde{x}_n\tilde{\mathbf{e}}_n. \quad (4.11)$$

We would like to know what the relationship is between the components of the vector in the basis $\mathbf{B}$ and its components in the basis $\tilde{\mathbf{B}}$. The basis vectors are themselves vectors, so we should be able to represent the old basis vectors in terms of the new ones, i.e.,

$$\begin{aligned}
\mathbf{e}_1 &= a_{11}\tilde{\mathbf{e}}_1 + a_{21}\tilde{\mathbf{e}}_2 + \cdots + a_{n1}\tilde{\mathbf{e}}_n, \\
\mathbf{e}_2 &= a_{12}\tilde{\mathbf{e}}_1 + a_{22}\tilde{\mathbf{e}}_2 + \cdots + a_{n2}\tilde{\mathbf{e}}_n, \\
\vdots &= \qquad \vdots \\
\mathbf{e}_n &= a_{1n}\tilde{\mathbf{e}}_1 + a_{2n}\tilde{\mathbf{e}}_2 + \cdots + a_{nn}\tilde{\mathbf{e}}_n,
\end{aligned}$$

which we can substitute into Equation (4.11) for $\mathbf{x}$ to give

$$\begin{aligned}
\mathbf{x} = x_1\left(a_{11}\tilde{\mathbf{e}}_1 + a_{21}\tilde{\mathbf{e}}_2 + \cdots + a_{n1}\tilde{\mathbf{e}}_n\right) + x_2\left(a_{12}\tilde{\mathbf{e}}_1 + a_{22}\tilde{\mathbf{e}}_2 + \cdots + a_{n2}\tilde{\mathbf{e}}_n\right) \\
+ \cdots x_n\left(a_{1n}\tilde{\mathbf{e}}_1 + a_{2n}\tilde{\mathbf{e}}_2 + \cdots + a_{nn}\tilde{\mathbf{e}}_n\right).
\end{aligned}$$

We can make this equation look more like the one for $\mathbf{x}$ in the other basis by collecting all the terms containing the different basis vectors:

$$\mathbf{x} = (x_1a_{11} + \cdots + x_na_{1n})\tilde{\mathbf{e}}_1 + (x_2a_{21} + \cdots + x_na_{2n})\tilde{\mathbf{e}}_2 + \cdots (x_na_{n1} + \cdots + x_na_{nn})\tilde{\mathbf{e}}_n. \quad (4.12)$$

We can now compare Equation (4.12) with Equation (4.11). Both are equations for $\mathbf{x}$ in terms of the basis vectors $\{\tilde{\mathbf{e}}_1, \ldots, \tilde{\mathbf{e}}_n\}$, and the only way both equations can be true is if the coefficients multiplying each basis vector are the same in both equations. In other words

$$\left.\begin{aligned}
\tilde{x}_1 &= x_1a_{11} + \cdots + x_na_{1n} \\
\tilde{x}_2 &= x_1a_{21} + \cdots + x_na_{2n} \\
\vdots &= \qquad \vdots \\
\tilde{x}_n &= x_1a_{n1} + \cdots + x_na_{nn}
\end{aligned}\right\} \qquad (4.13)$$

Example 4.6 Let us calculate the transformation of the components of a vector $\mathbf{X}$ in two dimensions if the unit basis vectors $\mathbf{e}_x$ and $\mathbf{e}_y$ are rotated about the origin though an angle θ to produce new basis vectors $\tilde{\mathbf{e}}_x$ and $\tilde{\mathbf{e}}_y$. It is a good idea to first draw a diagram (Figure 4.6). Using basic trigonometry and the fact that unit basis vectors have a length of 1, we have

$$\mathbf{e}_x = \cos(\theta)\tilde{\mathbf{e}}_x - \sin(\theta)\tilde{\mathbf{e}}_y, \qquad \mathbf{e}_y = \sin(\theta)\tilde{\mathbf{e}}_x + \cos(\theta)\tilde{\mathbf{e}}_y. \qquad (4.14)$$

Then, using Equation (4.13), we find that for a vector $\mathbf{X} = x\mathbf{e}_x + y\mathbf{e}_y$ the components in the rotated coordinates will be

$$\begin{aligned}
\tilde{x} &= x\cos(\theta) + y\sin(\theta), \\
\tilde{y} &= -x\sin(\theta) + y\cos(\theta).
\end{aligned}$$

Notice that the location of the minus sign differs from Equation (4.9). Both equations are correct, but they are looking at the same problem from different points of view: rotating the vector while keeping the basis fixed, or rotating the basis while keeping the vector fixed. So, we need to keep our wits about us when making calculations using rotations!

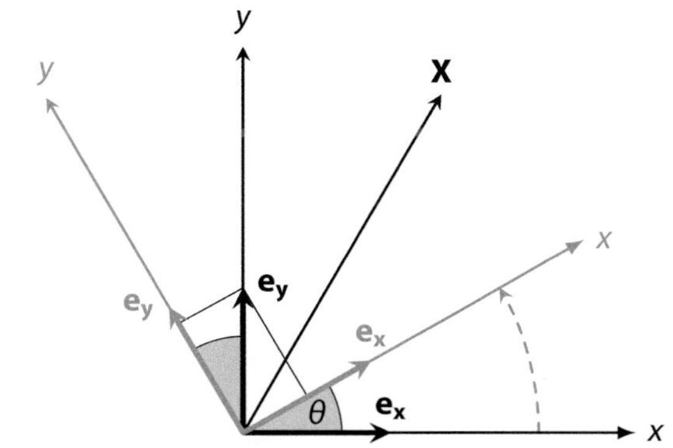

Figure 4.6 Rotating the x and y coordinate axes and corresponding basis vectors $\mathbf{e}_x$ and $\mathbf{e}_y$ (in black) through an angle θ produces a new set of coordinate axes and corresponding basis vectors (in gray). The components of the old basis vectors can be determined in terms of the new basis vectors by trigonometry.

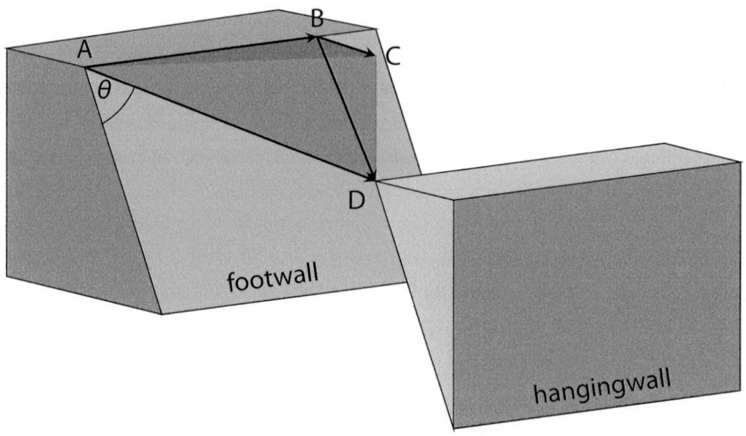

Figure 4.7 An oblique-slip fault in which a fracture has resulted in part of the rock moving laterally and vertically with respect to the other. The relative motion of the hangingwall and footwall can be described by a set of planes and lines. The horizontal motion occurs in a plane containing the vectors connecting A and B (the *strike-slip*), and B and C (the *heave*). The vertical motion occurs in a plane containing the heave and the vector connecting B and D (the *dip-slip*). The resultant motion is at an angle θ (the *rake*), and the length of the vector connecting A and D is called the *slip*.

4.2.3 Describing Lines and Curves Using Vectors

We can use vectors to describe geometric objects such as lines, curves, and planes, which can be useful for understanding the geometry of phenomena we see. For example, a geologist might want to describe the configuration of an *oblique-slip fault* in a rock (Figure 4.7), where a fracture has resulted in part of the rock, the hangingwall, moving laterally and vertically with respect to the other, the footwall (Pollard and Fletcher, 2005; Allmendinger et al., 2012).

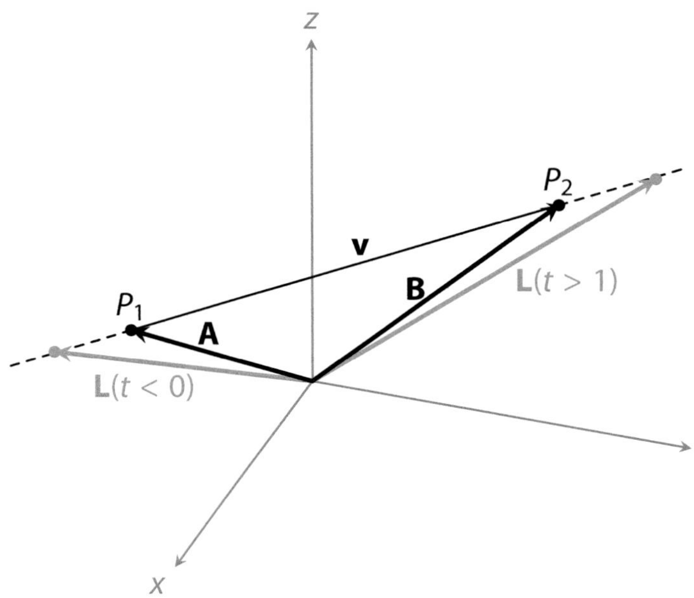

Figure 4.8 The points P_1 and P_2 lie on a straight line, and the vectors **A** and **B** connect the origin to P_1 and P_2, respectively. The vector **v** lies along the straight line that connects P_1 and P_2. The vectors $\mathbf{L}(t < 0)$ and $\mathbf{L}(t > 1)$ connect the origin to points on the line that lie outside the interval connecting P_1 and P_2 (see Equation (4.15)).

To describe a straight line using a vector (Figure 4.8) we can consider two points (P_1 and P_2) that lie on a straight line in three dimensions. We can write down a vector from the origin to the point P_1 and call it **A**; this is called a *position vector*. Using the rules for addition of vectors, we can write the position vector **B** from the origin to the point P_2 as

$$\mathbf{B} = \mathbf{A} + \mathbf{v}, \tag{4.15}$$

where **v** is the vector connecting P_1 and P_2. The length of **v** varies as P_1 and P_2 change. For example, if P_1 and P_2 are coincident, then **v** has zero length, and as P_2 moves away from P_1, the length of **v** increases. The vector **v** lies along the straight line, so the position vector **L** of any point on the line between P_1 and P_2 can now be written in terms of the vectors **A** and **v**:

$$\mathbf{L} = \mathbf{A} + t\mathbf{v}, \tag{4.16}$$

where the parameter t varies between $t = 0$ and $t = 1$. When $t = 1$, **L** connects the origin to P_2 (and so it is identical to the vector **B**); and when $t = 0$, it connects the origin to P_1 and is identical to the vector **A**. So, as t varies, the tip of **L** moves along the straight line connecting P_1 and P_2. What happens if $t > 1$? The vector **v** points along the direction of the straight line connecting P_1 and P_2 (we constructed it to be that way), so if $t > 1$, **L** connects the origin to a point on the straight line that is further from P_1 than P_2.

Example 4.7 We can use Equation (4.16) to find the vector representation of the straight line that passes through the points $P_1 = (0, 1, 4)$ and $P_2 = (4, 2, 1)$. If $\mathbf{A} = \hat{\jmath} + 4\hat{\mathbf{k}}$ connects the origin to P_1 and $\mathbf{B} = 4\hat{\imath} + 2\hat{\jmath} + \hat{\mathbf{k}}$ connects the origin to P_2, then the vector $\mathbf{v}$ connecting P_1 and P_2 is given by

$$\mathbf{v} = \mathbf{B} - \mathbf{A} = (4 - 0)\hat{\imath} + (2 - 1)\hat{\jmath} + (1 - 4)\hat{\mathbf{k}} = 4\hat{\imath} + \hat{\jmath} - 3\hat{\mathbf{k}},$$

so the vector representing the line connecting P_1 and P_2 is

$$\mathbf{L} = (0\hat{\imath} + 1\hat{\jmath} + 4\hat{\mathbf{k}}) + t(4\hat{\imath} + \hat{\jmath} - 3\hat{\mathbf{k}}) = 4t\hat{\imath} + (1 + t)\hat{\jmath} + (4 - 3t)\hat{\mathbf{k}}).$$

As a check, substituting $t = 0$ gives that $\mathbf{L} = \hat{\jmath} + 4\hat{\mathbf{k}}$, which is the position vector pointing to P_1, and substituting $t = 1$ gives a vector $4\hat{\imath} + 2\hat{\jmath} + \hat{\mathbf{k}}$, which is the position vector pointing to P_2.

Exercise 4.2.14 Find the vector representation of the straight line that passes through the points $P_1 = (1, 2, -1)$ and $P_2 = (3, 1, 2)$. What are the coordinates of the midpoint between P_1 and P_2?

Each component of the vector $\mathbf{L}$ we obtained in Example 4.7 is linear in the parameter t, and it is this feature that gives us the straight line. But if the factors multiplying the basis vectors were nonlinear functions of t, then the tip of $\mathbf{L}$ would trace out a curve. For example, the vector

$$\mathbf{L} = \cos(t)\hat{\imath} + \sin(t)\hat{\imath} + 0\hat{\mathbf{k}} \tag{4.17}$$

traces out a circle in the (x, y) plane. We can see this by looking at the components as t varies. First, the z component of the $\mathbf{L}$ is always zero, so the vector stays in the (x, y) plane. Now, when $t = 0$, the x component of $\mathbf{L}$ is 1, the y component is 0, and the z component is 0, and $\mathbf{L} = \hat{\imath}$. When $t = \pi/2$, $\mathbf{L} = \hat{\jmath}$ and points along the y axis, and so on. As t increases, the tip of the vector moves in a counterclockwise direction in a circle of radius 1, with the center being the origin of the coordinates.

Exercise 4.2.15 Describe the curve that is traced by the vector $\mathbf{L} = \cos(t)\hat{\imath} + \sin(t)\hat{\jmath} + 5t\hat{\mathbf{k}}$.

Exercise 4.2.16 Describe the curve that is traced by the vector $\mathbf{L} = \cos(t)\hat{\imath}+2\sin(t)\hat{\jmath}+\cos(t)\hat{\mathbf{k}}$.

Equations such as (4.17), where the components of the vector are functions of a parameter, are called *vector functions*. In three-dimensional rectangular coordinates, we can write a general vector function as $\mathbf{L} = u(t)\hat{\imath} + v(t)\hat{\jmath} + w(t)\hat{\mathbf{k}}$ or as components $\mathbf{L} = (u(t), v(t), w(t))$. In other words, we can think of the vector function as really three functions that give the components of the vector as the parameter changes. You can imagine that the tip of the vector traces out the trajectory of an object in space (such as a parcel of air, or a drifter in the ocean) as time (the parameter) changes, so vector functions are useful in visualizing and analyzing curves and trajectories in space. We will see later (Chapter 7) that we can differentiate and integrate these functions, allowing us to calculate quantities such as the velocity and acceleration of objects with complex trajectories. We can also use vectors to define a plane, but before we do that we have to understand how we multiply vectors.

4.3 Multiplying Vectors Together

We have already looked at addition and subtraction of vectors, and now we need to examine how we multiply two vectors together. Multiplying two scalars is simply a matter of multiplying two numbers, and we know how to interpret that. For example, multiplying a length of 2 m by a length of 3 m produces an area of 6 m^2. Multiplying two vectors together is more complicated because each vector is described by more than a single number; e.g., a magnitude and a direction, or the components of the vector with respect to a coordinate system. So, which numbers do we multiply together, and how do we interpret the result? As we shall see, there are different ways in which we can multiplying vectors together.

4.3.1 Scalar Product

We have seen that the components of a vector **A** can be thought of as the projections of the vector **A** along the directions given by the basis vectors describing the coordinate system. We can generalize this and ask what the projection of **A** is along another nonbasis vector **B**. In other words, we are asking what the projection of **A** is in the *direction* of the vector **B**. Since we are only interested in the direction of **B**, we can define a unit vector in that direction by dividing **B** by its magnitude as we did in Equation (4.5), so that the projection of **A** onto the direction of **B** is $\|\mathbf{A}\| \cos(\theta)$, where θ is the angle between **A** and **B**. This is telling us how much of the length of vector **A** is along the direction of vector **B** (Figure 4.9). We then define the *scalar product* of **A** with the direction of **B** by the equation

$$\mathbf{A} \cdot \left(\frac{\mathbf{B}}{\|\mathbf{B}\|} \right) = \|\mathbf{A}\| \cos(\theta),$$

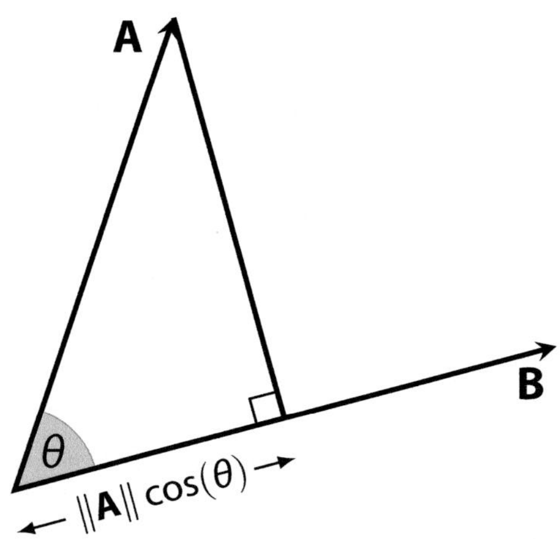

Figure 4.9 The geometry of taking the scalar product. The scalar product of **A** with the unit vector in the direction of vector **B** is $\|\mathbf{A}\| \cos(\theta)$, where θ is the angle between **A** and **B**.

and multiplying this equation through by $||\mathbf{B}||$, which is a scalar, gives the equation for the scalar product of **A** and **B**:

$$\mathbf{A} \cdot \mathbf{B} = ||\mathbf{A}||\, ||\mathbf{B}|| \cos(\theta). \tag{4.18}$$

Notice that the scalar product is represented using a dot centered between the two vectors, $\mathbf{A} \cdot \mathbf{B}$, and is sometimes called the *dot product*. The scalar product is a scalar, so like any other scalar it is invariant under coordinate transformations. It is a useful quantity because if we know **A** and **B**, we can use the scalar product to calculate the angle between the two vectors. If, **A** and **B** are orthogonal (i.e., at right angles) to each other, then $\cos(\theta) = \cos \pi/2 = 0$ (because $\cos(90°) = 0$, which implies that if **A** and **B** are at right angles to each other, then there is no amount of **A** in the direction of **B** and vice versa.[3]

We can apply the scalar product to the set of basis vectors for rectangular coordinates, and because they have unit length and are orthogonal to each other, we find that

$$\hat{\mathbf{\imath}} \cdot \hat{\mathbf{\imath}} = \hat{\mathbf{\jmath}} \cdot \hat{\mathbf{\jmath}} = \hat{\mathbf{k}} \cdot \hat{\mathbf{k}} = 1, \qquad \hat{\mathbf{\imath}} \cdot \hat{\mathbf{\jmath}} = \hat{\mathbf{\imath}} \cdot \hat{\mathbf{k}} = \hat{\mathbf{\jmath}} \cdot \hat{\mathbf{k}} = 0. \tag{4.19}$$

A basis for which all the basis vectors are orthogonal to each other ($\hat{\mathbf{\imath}} \cdot \hat{\mathbf{\jmath}} = 0$, etc.) is an orthogonal basis and if the basis vectors are all of length 1 ($\hat{\mathbf{\imath}} \cdot \hat{\mathbf{\imath}} = 1$, etc.), the basis is called an orthonormal basis.

Example 4.8 The vectors $\hat{\mathbf{e}}_1 = \hat{\mathbf{\imath}} + 2\hat{\mathbf{\jmath}} - \hat{\mathbf{k}}$, $\hat{\mathbf{e}}_2 = \hat{\mathbf{\imath}} + 3\hat{\mathbf{\jmath}} + \hat{\mathbf{k}}$, and $\hat{\mathbf{e}}_3 = 6\hat{\mathbf{\imath}}$ form a basis for three-dimensional Cartesian space. Is this an orthogonal basis? The first thing we have to do is to show that the three vectors are linearly independent, which we can do by showing that the only solution to Equation (4.8) is $\lambda = \mu = \nu = 0$. This gives a set of three equations: $\lambda + \mu = 0$, $2\lambda + 3\mu = 0$, and $-\lambda + \mu + \nu = 0$, which imply $\lambda = \mu = \nu = 0$, so the vectors are linearly independent. To show that they are orthogonal we need to show that $\hat{\mathbf{e}}_i \cdot \hat{\mathbf{e}}_j = 0$ if $i \neq j$,

$$\begin{aligned}
\hat{\mathbf{e}}_1 \cdot \hat{\mathbf{e}}_2 &= (\hat{\mathbf{\imath}} + 2\hat{\mathbf{\jmath}} - \hat{\mathbf{k}}) \cdot (\hat{\mathbf{\imath}} + 3\hat{\mathbf{\jmath}} + \hat{\mathbf{k}}) = 1 \\
\hat{\mathbf{e}}_1 \cdot \hat{\mathbf{e}}_3 &= (\hat{\mathbf{\imath}} + 2\hat{\mathbf{\jmath}} - \hat{\mathbf{k}}) \cdot (6\hat{\mathbf{\imath}}) = 6 \\
\hat{\mathbf{e}}_2 \cdot \hat{\mathbf{e}}_3 &= (\hat{\mathbf{\imath}} + 3\hat{\mathbf{\jmath}} + \hat{\mathbf{k}}) \cdot (6\hat{\mathbf{\imath}}) = 6.
\end{aligned}$$

We can see that this particular basis is not orthogonal, because $\hat{\mathbf{e}}_i \cdot \hat{\mathbf{e}}_j \neq 0$ when $i \neq j$.

Using equations (4.19), we can see that the components of a vector $\mathbf{A} = a_x\hat{\mathbf{\imath}} + a_y\hat{\mathbf{\jmath}} + a_z\hat{\mathbf{k}}$ are simply the dot products of **A** with each of the unit vectors in turn; for example,

$$\mathbf{A} \cdot \hat{\mathbf{\imath}} = (a_x\hat{\mathbf{\imath}} + a_y\hat{\mathbf{\jmath}} + a_z\hat{\mathbf{k}}) \cdot \hat{\mathbf{\imath}} = a_x\hat{\mathbf{\imath}} \cdot \hat{\mathbf{\imath}} = a_x. \tag{4.20}$$

Exercise 4.3.1 Prove Equations 4.19.
Exercise 4.3.2 Is the basis in Exercise 4.2.7 an orthogonal basis?

These results allow us to represent the scalar product of two vectors in terms of the components of those vectors:

[3] If we think of the projection of **A** onto **B** as being the shadow cast when we shine a light perpendicular to **B**, then this says that **A** does not cast a shadow on **B**.

$$\mathbf{A}\cdot\mathbf{B} = (a_x\hat{\mathbf{i}} + a_y\hat{\mathbf{j}} + a_z\hat{\mathbf{k}})\cdot(b_x\hat{\mathbf{i}} + b_y\hat{\mathbf{j}} + b_z\hat{\mathbf{k}}) = a_xb_x\hat{\mathbf{i}}\cdot\hat{\mathbf{i}} + a_yb_y\hat{\mathbf{j}}\cdot\hat{\mathbf{j}} + a_zb_z\hat{\mathbf{k}}\cdot\hat{\mathbf{k}}$$
$$= a_xb_x + a_yb_y + a_zb_z. \tag{4.21}$$

If instead of using the basis $(\hat{\mathbf{i}}, \hat{\mathbf{j}}, \hat{\mathbf{k}})$ we had used a different orthonormal basis, $\hat{\mathbf{e}}_1, \hat{\mathbf{e}}_2, \hat{\mathbf{e}}_3$, so that $\mathbf{A} = a_1\hat{\mathbf{e}}_1 + a_2\hat{\mathbf{e}}_2 + a_3\hat{\mathbf{e}}_3$, then Equation (4.21) would be

$$\mathbf{A}\cdot\mathbf{B} = a_1b_1 + a_2b_2 + a_3b_3 = \sum_{i=1}^{3} a_ib_i. \tag{4.22}$$

Notice that if the basis is not orthonormal, then Equation (4.22) does not hold because, for example, $\hat{\mathbf{e}}_i\cdot\hat{\mathbf{e}}_j$ is not necessarily zero for $i \neq j$.

Exercise 4.3.3 Calculate the scalar product of the vectors $\mathbf{A} = 3\hat{\mathbf{i}} - 2\hat{\mathbf{j}} + 5\hat{\mathbf{k}}$ and $\mathbf{B} = -2\hat{\mathbf{i}} + 3\hat{\mathbf{j}}$.

Exercise 4.3.4 Show that the vectors $3\hat{\mathbf{i}} + 2\hat{\mathbf{j}} + \hat{\mathbf{k}}$ and $\hat{\mathbf{i}} - 4\hat{\mathbf{j}} + 5\hat{\mathbf{k}}$ are orthogonal to each other.

The scalar product also has some nice properties. First, it is *commutative*, that is

$$\mathbf{A}\cdot\mathbf{B} = ||\mathbf{A}||\,||\mathbf{B}||\cos(\theta) = ||\mathbf{B}||\,||\mathbf{A}||\cos(\theta) = \mathbf{B}\cdot\mathbf{A}, \tag{4.23}$$

which makes sense because $||\mathbf{A}||$, $||\mathbf{B}||$, and $\cos\theta$ are all just numbers, and we know that multiplication of numbers is commutative. Second, it is *distributive over addition*, i.e.,

$$\mathbf{A}\cdot(\mathbf{B}+\mathbf{C}) = \mathbf{A}\cdot\mathbf{B} + \mathbf{A}\cdot\mathbf{C}. \tag{4.24}$$

There are many of these types of vector equations, and they are useful for proving general relationships between vectors. In general, they are often easiest to prove by writing the vectors in terms of components. This is because a vector equation such as Equation (4.24) must hold in any coordinate system. Therefore, if we can show it is true in one coordinate system, such as rectangular coordinates with an orthonormal basis, then it must be true generally.

Example 4.9 Let us prove that the scalar product is distributive over addition. If we let $\mathbf{A} = a_x\hat{\mathbf{i}} + a_y\hat{\mathbf{j}} + a_z\hat{\mathbf{k}}$, $\mathbf{B} = b_x\hat{\mathbf{i}} + b_y\hat{\mathbf{j}} + b_z\hat{\mathbf{k}}$, and $\mathbf{C} = c_x\hat{\mathbf{i}} + c_y\hat{\mathbf{j}} + c_z\hat{\mathbf{k}}$, then

$$\mathbf{A}\cdot(\mathbf{B}+\mathbf{C}) = (a_x\hat{\mathbf{i}} + a_y\hat{\mathbf{j}} + a_z\hat{\mathbf{k}})\cdot(b_x\hat{\mathbf{i}} + b_y\hat{\mathbf{j}} + b_z\hat{\mathbf{k}} + c_x\hat{\mathbf{i}} + c_y\hat{\mathbf{j}} + c_z\hat{\mathbf{k}})$$
$$= a_x(b_x + c_x) + a_y(b_y + c_y) + a_z(b_z + c_z)$$
$$= (a_xb_x + a_yb_y + a_zb_z) + (a_xc_x + a_yc_y + a_zc_z). \tag{4.25}$$

Also,

$$\mathbf{A}\cdot\mathbf{B} + \mathbf{A}\cdot\mathbf{C} = (a_x\hat{\mathbf{i}} + a_y\hat{\mathbf{j}} + a_z\hat{\mathbf{k}})\cdot(b_x\hat{\mathbf{i}} + b_y\hat{\mathbf{j}} + b_z\hat{\mathbf{k}}) + (a_x\hat{\mathbf{i}} + a_y\hat{\mathbf{j}} + a_z\hat{\mathbf{k}})\cdot(c_x\hat{\mathbf{i}} + c_y\hat{\mathbf{j}} + c_z\hat{\mathbf{k}})$$
$$= (a_xb_x + a_yb_y + a_zb_z) + (a_xc_x + a_yc_y + a_zc_z). \tag{4.26}$$

Equations (4.25) and (4.26) are the same, so the scalar product is distributive over addition.

4.3.1.1 Applications of the Scalar Product

We have already seen that the scalar product can be used to find the angle θ between two vectors $\mathbf{A}$ and $\mathbf{B}$:

$$\cos(\theta) = \frac{\mathbf{A}\cdot\mathbf{B}}{||\mathbf{A}||\ ||\mathbf{B}||}. \tag{4.27}$$

It also has many other uses. For example, we can use it to easily find the length of a vector by taking the scalar product of the vector with itself:

$$\mathbf{A}\cdot\mathbf{A} = ||\mathbf{A}||\ ||\mathbf{A}||\cos(0) = (||\mathbf{A}||)^2\,. \tag{4.28}$$

Example 4.10 Let us use the scalar product to find the angle between the vectors $\mathbf{A} = 4\hat{\imath} + 6\hat{\jmath} - 2\hat{\mathbf{k}}$ and $\mathbf{B} = -2\hat{\imath} + 4\hat{\jmath} + \hat{\mathbf{k}}$. To use Equation (4.27) we need to calculate the scalar product of the two vectors as well as the two scalar products of the vectors with themselves. These are $\mathbf{A}\cdot\mathbf{B} = 14$, $||\mathbf{A}|| = \sqrt{\mathbf{A}\cdot\mathbf{A}} = \sqrt{54}$, and $||\mathbf{B}|| = \sqrt{21}$. So, using Equation (4.27) gives

$$\cos(\theta) = \frac{14}{\sqrt{54}\sqrt{21}}, \tag{4.29}$$

and the angle between the two vectors is $\theta \approx 65.43°$.

Exercise 4.3.5 Consider three vectors $\mathbf{A}$, $\mathbf{B}$, and $\mathbf{C}$ such that $\mathbf{C} = \mathbf{A} - \mathbf{B}$. Show that $||\mathbf{C}||^2 = ||\mathbf{A}||^2 + ||\mathbf{B}||^2 - 2||\mathbf{A}||\,||\mathbf{B}||\cos(\theta)$, where θ is the angle between the vectors $\mathbf{A}$ and $\mathbf{B}$. This is simply the law of cosines (Appendix B).

We can also use the scalar product to find the magnitude of the projection of one vector ($\mathbf{A}$) in the direction of another ($\mathbf{B}$), where θ is the angle between them (Figure 4.10a):

$$l = ||\mathbf{A}||\cos(\theta) = ||\mathbf{A}||\frac{\mathbf{A}\cdot\mathbf{B}}{||\mathbf{A}||\ ||\mathbf{B}||} = \frac{\mathbf{A}\cdot\mathbf{B}}{||\mathbf{B}||}. \tag{4.30}$$

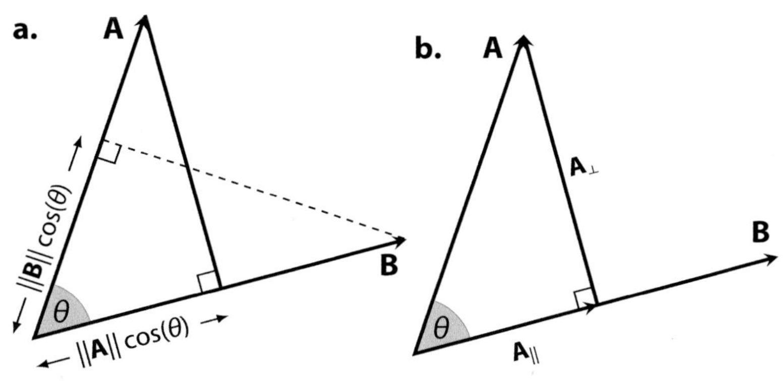

Figure 4.10 The projection of vector **A** onto vector **B** is $||\mathbf{A}||\cos(\theta)$, and the projection of **B** onto **A** is $||\mathbf{B}||\cos(\theta)$ (a.). We can use this to decompose the vector **A** into a part that is parallel to the vector **B** and a part that is perpendicular to **B** (b.).

Let us explore the idea of taking the projection of **A** onto **B** in a little more depth. Equation (4.30) tells us the magnitude of the projection of **A** onto **B**. However, we know that this projection has a direction that lies along the direction of **B**. So, let us define a new vector

$$\mathbf{A}_{\parallel} = \frac{\mathbf{A}\cdot\mathbf{B}}{\mathbf{B}\cdot\mathbf{B}}\mathbf{B},$$

which has the magnitude of the projection of **A** onto **B** and which lies in the direction of **B**. We can define another vector (Figure 4.10b)

$$\mathbf{A}_{\perp} = \mathbf{A} - \mathbf{A}_{\parallel}.$$

We can calculate the direction of this new vector with respect to the vector **B** by taking the dot product:

$$\mathbf{A}_{\perp}\cdot\mathbf{B} = (\mathbf{A} - \mathbf{A}_{\parallel})\cdot\mathbf{B} = \mathbf{A}\cdot\mathbf{B} - \frac{\mathbf{A}\cdot\mathbf{B}}{\mathbf{B}\cdot\mathbf{B}}\mathbf{B}\cdot\mathbf{B} = \mathbf{A}\cdot\mathbf{B} - \mathbf{A}\cdot\mathbf{B} = 0,$$

showing that $\mathbf{A}_{\perp}$ is orthogonal to **B**. What we have done here is to define a *decomposition* of **A** into two vectors: $\mathbf{A}_{\parallel}$, which is parallel to **B**, and $\mathbf{A}_{\perp}$, which is at right angles to **B**; such that

$$\mathbf{A} = \mathbf{A}_{\parallel} + \mathbf{A}_{\perp}. \tag{4.31}$$

There are many situations in which we want to decompose a vector in such a manner. For example, ocean currents close to a shoreline can be decomposed into an alongshore component moving parallel to the coast and an across-shore component moving away or toward the shore. Being able to do this helps us to understand coastal erosion and the transport of sediments and sand from beach to beach.

The decomposition in Equation (4.31) is unique. We will show this using a standard mathematical strategy for proving statetments. We start by assuming that the decomposition is *not* unique. In other words, we assume that we can find *different* vectors, $\mathbf{a}_{\parallel}$ and $\mathbf{a}_{\perp}$, such that $\mathbf{A} = \mathbf{a}_{\parallel} + \mathbf{a}_{\perp}$, where $\mathbf{a}_{\parallel}$ lies along **B** (i.e., $\mathbf{a}_{\parallel} = \mu\mathbf{B}$), and $\mathbf{a}_{\perp}$ is orthogonal to **B**. Then, $\mathbf{a}_{\parallel}\cdot\mathbf{B} = \mu\mathbf{B}\cdot\mathbf{B}$ and

$$\mathbf{a}_{\parallel} = \mu\mathbf{B} = \frac{\mathbf{a}_{\parallel}\cdot\mathbf{B}}{\mathbf{B}\cdot\mathbf{B}}\mathbf{B} = \frac{\mathbf{A}\cdot\mathbf{B}}{\mathbf{B}\cdot\mathbf{B}}\mathbf{B} = \mathbf{A}_{\parallel},$$

where we have used the fact that $\mathbf{a}_{\parallel}\cdot\mathbf{B} = (\mathbf{A} - \mathbf{a}_{\perp})\cdot\mathbf{B} = \mathbf{A}\cdot\mathbf{B}$. Using this result, we also see that $\mathbf{a}_{\perp} = \mathbf{A} - \mathbf{a}_{\parallel} = \mathbf{A} - \mathbf{A}_{\parallel} = \mathbf{A}_{\perp}$. Thus, we have shown that this decomposition of **A** is unique. The vector $\mathbf{A}_{\parallel}$ is the projection of **A** onto **B** and is sometimes written as $\text{Proj}_{\mathbf{B}}\mathbf{A}$.

Example 4.11 Let us calculate $\text{Proj}_{\hat{\mathbf{e}}_2}\hat{\mathbf{e}}_1$ using the vectors from Example 4.8, and show that $\hat{\mathbf{e}}_1 - \text{Proj}_{\hat{\mathbf{e}}_2}\hat{\mathbf{e}}_1$ is orthogonal to $\hat{\mathbf{e}}_2$. The projection of $\hat{\mathbf{e}}_1$ onto $\hat{\mathbf{e}}_2$ is

$$\text{Proj}_{\hat{\mathbf{e}}_2}\hat{\mathbf{e}}_1 = \frac{\hat{\mathbf{e}}_2\cdot\hat{\mathbf{e}}_2}{\hat{\mathbf{e}}_2\cdot\hat{\mathbf{e}}_2}\hat{\mathbf{e}}_2 = \frac{6}{11}(\hat{\mathbf{i}} + 3\hat{\mathbf{j}} + \hat{\mathbf{k}}).$$

So,

$$\hat{\mathbf{e}}_1 - \text{Proj}_{\hat{\mathbf{e}}_2}\hat{\mathbf{e}}_1 = (\hat{\mathbf{i}} + 2\hat{\mathbf{j}} - \hat{\mathbf{k}}) - (\frac{6}{11}(\hat{\mathbf{i}} + 3\hat{\mathbf{j}} + \hat{\mathbf{k}}) = \frac{1}{11}(5\hat{\mathbf{i}} + 4\hat{\mathbf{j}} - 17\hat{\mathbf{k}}).$$

To show that this vector is orthogonal to $\hat{\mathbf{e}}_2$, we take the scalar product of the two vectors

$$\hat{\mathbf{e}}_2 \cdot (\hat{\mathbf{e}}_1 - \text{Proj}_{\hat{\mathbf{e}}_2}\hat{\mathbf{e}}_1) = (\hat{\mathbf{\imath}} + 3\hat{\mathbf{\jmath}} + \hat{\mathbf{k}}) \cdot \frac{1}{11}(5\hat{\mathbf{\imath}} + 4\hat{\mathbf{\jmath}} - 17\hat{\mathbf{k}}) = 0,$$

showing that these vectors are indeed orthogonal.

We can see geometrically what we have done by looking at Figure 4.10. The projection of **A** onto **B** is the component of **A** that is parallel to **B**. If we subtract $\text{Proj}_{\mathbf{B}}\mathbf{A}$ from **A**, we are left with the component of **A** that is orthogonal to **B**. If we have more than two vectors, then we can continue this process and thereby construct a set of vectors that are mutually orthogonal, and if the original vectors form a nonorthogonal basis, then these new vectors will form an orthogonal basis.

To be more concrete, assume that we have a nonorthogonal basis **u**, **v**, and **w**, for a three-dimensional Cartesian coordinate system, and we wish to construct an orthogonal basis, $\mathbf{e}_1$, $\mathbf{e}_2$, and $\mathbf{e}_3$. We choose one of the original basis vectors and take projections of the others onto it, forming the following three vectors:

$$\left.\begin{aligned}
\mathbf{e}_1 &= \mathbf{u} \\
\mathbf{e}_2 &= \mathbf{v} - \text{Proj}_{\mathbf{e}_1}\mathbf{v} = \mathbf{v} - \frac{\mathbf{e}_1 \cdot \mathbf{v}}{\mathbf{e}_1 \cdot \mathbf{e}_1}\mathbf{e}_1 \\
\mathbf{e}_3 &= \mathbf{w} - \text{Proj}_{\mathbf{e}_1}\mathbf{w} - \text{Proj}_{\mathbf{e}_2}\mathbf{w} = \mathbf{w} - \frac{\mathbf{e}_1 \cdot \mathbf{w}}{\mathbf{e}_1 \cdot \mathbf{e}_1}\mathbf{e}_1 - \frac{\mathbf{e}_2 \cdot \mathbf{w}}{\mathbf{e}_2 \cdot \mathbf{e}_2}\mathbf{e}_2
\end{aligned}\right\} \tag{4.32}$$

These new vectors are mutually orthogonal (because we have constructed them that way) and so form an orthogonal basis for the system. This is called the *Gram–Schmidt procedure*.[4] Once we have an orthogonal basis, we can easily convert it to an orthonormal basis by dividing each basis vector by its length.

Why is it useful to have an orthogonal basis? If we have a nonorthogonal basis (**u**, **v**, **w**), we know from the property of linear independence that we can write any other vector in the system as a linear combination of these three: i.e., for any vector **X** in the system, $\mathbf{X} = \lambda_1\mathbf{u} + \lambda_2\mathbf{v} + \lambda_3\mathbf{w}$. However, we do not have an easy way to find the numbers λ_1, λ_2, and λ_3. If the basis is an orthogonal basis, then we can take the scalar product of **X** with each of the vectors in the orthogonal basis to find λ_1 etc. This is the advantage of a system of basis vectors like $(\hat{\mathbf{\imath}}, \hat{\mathbf{\jmath}}, \hat{\mathbf{k}})$ for three-dimensional Cartesian space.

Example 4.12 Starting with the nonorthogonal basis $\mathbf{u} = \hat{\mathbf{\imath}} + 2\hat{\mathbf{\jmath}} - \hat{\mathbf{k}}$, $\mathbf{v} = \hat{\mathbf{\imath}} + 3\hat{\mathbf{\jmath}} + \hat{\mathbf{k}}$, and $\mathbf{w} = 6\hat{\mathbf{\imath}}$, we can use the Gram–Schmidt procedure to derive an orthogonal basis and demonstrate that the new vectors are indeed orthogonal. To start with we see that **w** only contains a single component, so we can make our calculations easier by choosing $\mathbf{e}_1 = \mathbf{w}$. Then

$$\mathbf{e}_2 = \mathbf{u} - \text{Proj}_{\mathbf{e}_1}\mathbf{u} = \mathbf{u} - \frac{\mathbf{e}_1 \cdot \mathbf{u}}{\mathbf{e}_1 \cdot \mathbf{e}_1}\mathbf{e}_1 = (\hat{\mathbf{\imath}} + 2\hat{\mathbf{\jmath}} - \hat{\mathbf{k}}) - \hat{\mathbf{\imath}} = 2\hat{\mathbf{\jmath}} - \hat{\mathbf{k}},$$

and the third vector is

$$\mathbf{e}_3 = \mathbf{v} - \text{Proj}_{\mathbf{e}_1}\mathbf{v} - \text{Proj}_{\mathbf{e}_2}\mathbf{v} = (\hat{\mathbf{\imath}} + 3\hat{\mathbf{\jmath}} + \hat{\mathbf{k}}) - \hat{\mathbf{\imath}} - (2\hat{\mathbf{\jmath}} - \hat{\mathbf{k}}) = \hat{\mathbf{\jmath}} + 2\hat{\mathbf{k}}.$$

[4] Named after Danish mathematician Jørgen Pedersen Gram (1850–1916) and Baltic German mathematician Erhard Schmidt (1876–1959).

Because $\mathbf{e}_2$ and $\mathbf{e}_3$ do not contain an $\hat{\mathbf{\imath}}$ term, we see that $\mathbf{e}_1 \cdot \mathbf{e}_2 = \mathbf{e}_1 \cdot \mathbf{e}_3 = 0$. We then have to calculate $\mathbf{e}_2 \cdot \mathbf{e}_3 = (2\hat{\mathbf{\jmath}} - \hat{\mathbf{k}}) \cdot (\hat{\mathbf{\jmath}} + 2\hat{\mathbf{k}}) = 2 - 2 = 0$. So, all the new vectors are indeed orthogonal. The orthonormal basis is then

$$\mathbf{e}_1 = \hat{\mathbf{\imath}}, \quad \mathbf{e}_2 = \frac{1}{\sqrt{5}}(2\hat{\mathbf{\jmath}} - \hat{\mathbf{k}}), \quad \mathbf{e}_3 = \frac{1}{\sqrt{5}}(\hat{\mathbf{\jmath}} + 2\hat{\mathbf{k}}).$$

Recall that the definition of the scalar product of two vectors is $\mathbf{A} \cdot \mathbf{B} = ||\mathbf{A}||\,||\mathbf{B}|| \cos\theta$. Consequently, $||\mathbf{A} \cdot \mathbf{B}|| = ||\mathbf{A}||\,||\mathbf{B}||\,||\cos\theta||$, and since $||\cos(\theta)|| \le 1$, we have

$$||\mathbf{A} \cdot \mathbf{B}|| \le ||\mathbf{A}||\,||\mathbf{B}||; \tag{4.33}$$

that is, the absolute value of $\mathbf{A} \cdot \mathbf{B}$ is less than or equal to the product of the magnitudes of the individual vectors. Notice that $\mathbf{A} \cdot \mathbf{B}$ is just a number (a scalar) and the left-hand side of this equation refers to the absolute value of this number and so uses $|\cdot|$. This is called the *Cauchy–Schwarz inequality*,[5] and it is true.[6]

Exercise 4.3.6 Use equations (4.22) and (4.33) to show that

$$\left(\sum_{i=1}^{3} a_i b_i\right)^2 \le \left(\sum_{i=1}^{3} a_i\right)^2 \left(\sum_{i=1}^{3} b_i\right)^2.$$

Let us see what happens if we use the Cauchy–Schwarz inequality on the magnitude of the sum of two vectors $\mathbf{A}$ and $\mathbf{B}$,

$$\begin{aligned} ||\mathbf{A} + \mathbf{B}|| &= (\mathbf{A} + \mathbf{B}) \cdot (\mathbf{A} + \mathbf{B}) = \mathbf{A} \cdot \mathbf{A} + 2\mathbf{A} \cdot \mathbf{B} + \mathbf{B} \cdot \mathbf{B} = ||\mathbf{A}|| + 2\mathbf{A} \cdot \mathbf{B} + ||\mathbf{B}|| \\ &\le ||\mathbf{A}|| + 2||\mathbf{A}||\,||\mathbf{B}|| + ||\mathbf{B}|| = (||\mathbf{A}|| + ||\mathbf{B}||)^2, \end{aligned} \tag{4.34}$$

where in the second line we have used the fact that the scalar product is commutative (Equation (4.23)). To get the last line, we have made use of the Cauchy–Schwarz inequality and recognized that if $||\mathbf{A} \cdot \mathbf{B}|| \le ||\mathbf{A}||\,||\mathbf{B}||$, then $\mathbf{A} \cdot \mathbf{B} \le ||\mathbf{A}||\,||\mathbf{B}||$. So, we end up with

$$||\mathbf{A} + \mathbf{B}|| \le (||\mathbf{A}|| + ||\mathbf{B}||), \tag{4.35}$$

which is called the *triangle inequality*.

Exercise 4.3.7 Show that $||\mathbf{A} + \mathbf{B}||^2 + ||\mathbf{A} - \mathbf{B}||^2 = 2(||\mathbf{A}||^2 + ||\mathbf{B}||^2)$.

Exercise 4.3.8 Show that if $\mathbf{A}$ and $\mathbf{B}$ are orthogonal, then $||\mathbf{A} + \mathbf{B}||^2 = ||\mathbf{A}||^2 + ||\mathbf{B}||^2$.

4.3.2 Vector Product

With the dot product we were able to combine vectors in such a way so as to create a scalar. You will probably not be too surprised to learn that we can take a product of vectors to also produce a new vector. This is called the *vector product*, or cross product, and we can define it by the equation

[5] French mathematician Augustin-Louis Cauchy (1789–1857) was the first to publish a form of this inequality for sums, and German mathematician Karl Hermann Amandus Schwarz (1843–1921) developed a form of the inequality for integrals.

[6] Although the Cauchy–Schwarz inequality is true in general, the demonstration we have used here does not work if the dimension of the space is greater than three.

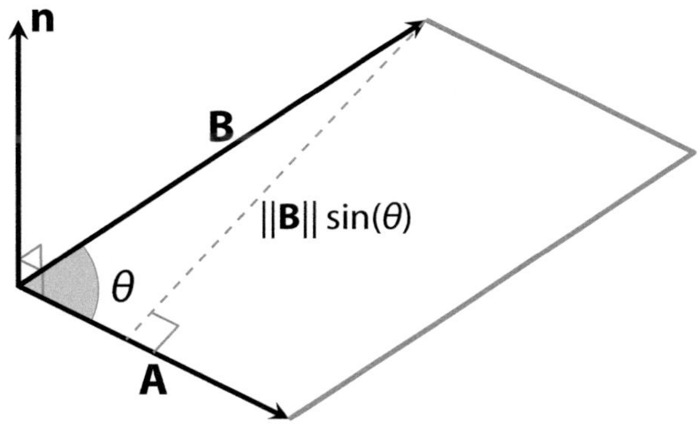

Figure 4.11 The vector product of vectors **A** and **B** is $\|\mathbf{A}\|\,\|\mathbf{B}\|\,\sin(\theta)\mathbf{n}$. The vector **n** is at right angles to both **A** and **B** and in the direction given by the right-hand rule. The perpendicular height of the parallelogram is $\|\mathbf{B}\|\,\sin(\theta)$.

$$\mathbf{A}\times\mathbf{B} = \|\mathbf{A}\|\,\|\mathbf{B}\|\,\sin(\theta)\mathbf{n}, \tag{4.36}$$

where **n** is a unit vector that is mutually perpendicular to both **A** and **B** (Figure 4.11). This definition is very similar to the scalar product, except that we have replaced $\cos(\theta)$ with $\sin(\theta)$ and we have a new vector, $\hat{\mathbf{n}}$. How do we determine the direction of $\hat{\mathbf{n}}$? By definition, the direction is chosen to be at right angles to the two vectors **A** and **B**, which also means that it is at right angles to the plane defined by those two vectors. This gives us a choice of two possible directions, and by convention we choose the direction given by the right-hand rule.[7]

We can gain a geometric intuition for the meaning of the vector product by looking at Figure 4.11. The area of the parallelogram formed by the vectors **A** and **B** is simply $\|\mathbf{A}\|\,\|\mathbf{B}\|\,\sin(\theta)$.[8] So the *magnitude* of the vector product $\mathbf{A}\times\mathbf{B}$ is just the area of the parallelogram formed by the two vectors, and the direction of the vector $\mathbf{A}\times\mathbf{B}$ is given by the right-hand rule.

This is interesting because it shows us that we can represent an area as a vector, which is something we might not have expected. The magnitude of the vector is the same as the size of the area, and the direction of the vector is determined (by convention) using the right-hand rule. It may seem strange to think of an area as having a direction, but it is a very useful concept, particularly when dealing with the transport or flow of energy or substances. For example, we might want to know the amount of radiant energy from the Sun falling on a unit area of the Earth's surface (Figure 4.12) and how this varies with the angle (θ) of the Sun from the vertical (this is called the *zenith angle*). This is needed for calculating photosynthesis rates of plants, or the heat being absorbed by the Earth. In Figure 4.12, the direction of the vector **S** tells us the direction from the Sun to the location

[7] There are several ways to think of this. Some find it useful to think of holding a screwdriver in the right hand and turning it from **A** to **B**, the direction of the vector is then the direction that the screwdriver is pointing in. Another way to look at this is to curl your fingers of your right hand in the direction from **A** to **B**, your thumb will then point in the direction of the vector. A third way is to point the first finger of your right hand along **A**, the middle finger of your right hand along **B**, then your thumb will point in the direction of the vector.

[8] Remember that the area of a parallelogram is the length of the base multiplied by the perpendicular height.

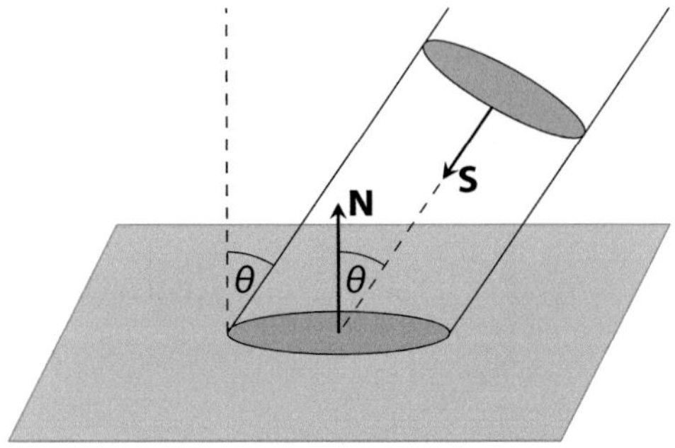

Figure 4.12 The dark gray area shown has a vector **N** which points upward and is at right angles to the flat surface. A flow of material enters this area from above at a certain angle and is represented by the vector **S**.

on the Earth we are interested in, and its length tells us the amount of radiant energy passing through a disk of unit area (the light gray disk) that is perpendicular to the direction of the Sun's rays. The vector **N** is a unit vector that is perpendicular to the surface of the Earth. The amount of energy falling on a unit area of the Earth is then $\mathbf{S} \cdot \mathbf{N} = ||\mathbf{S}|| \cos(\theta)$. Since $\cos(\theta) \leq 1$, this is always less than or equal to $||\mathbf{S}||$, and is only equal to $||\mathbf{S}||$ when the Sun is directly overhead ($\theta = 0$). The reason for this is that as θ increases, the area of the Sun's disk projected on the surface of the Earth increases so that the same amount of radiant energy is spread over a larger area.

The vector product has some interesting properties. The first is that it *does not commute*; in fact

$$\mathbf{A} \times \mathbf{B} = -\mathbf{B} \times \mathbf{A}. \tag{4.37}$$

This makes sense from our definition of the vector product, remembering that we defined the direction of the normal $\mathbf{n} = \mathbf{A} \times \mathbf{B}$ according to the right-hand rule. If we reverse the ordering of the two vectors **A** and **B**, then according to our right-hand rule, we also reverse the direction of **n**. Another useful property of the vector product is the fact that it is distributive, i.e., $\mathbf{A} \times (\mathbf{B} + \mathbf{C}) = \mathbf{A} \times \mathbf{B} + \mathbf{A} \times \mathbf{C}$. Also, if we multiply one of the vectors in the vector product by a scalar (λ), then $(\lambda\mathbf{A}) \times \mathbf{B} = \mathbf{A} \times (\lambda\mathbf{B}) = \lambda(\mathbf{A} \times \mathbf{B})$. Lastly, from the definition of the vector product (Equation (4.36)), we can see that if $\mathbf{A} \times \mathbf{B} = 0$, then **A** and **B** are parallel if neither **A** nor **B** are zero.

We know that the scalar product of the basis vectors $\hat{\mathbf{\imath}}$, $\hat{\mathbf{\jmath}}$, and $\hat{\mathbf{k}}$ produces some useful relationships, so let us see what happens with the vector product. First, the vector product of any of these vectors with itself is zero (because $\sin(0) = 0$), but what about products such as $\hat{\mathbf{\imath}} \times \hat{\mathbf{\jmath}}$. These are both unit vectors and the angle between them is 90°, so by Equation (4.36), the vector $\hat{\mathbf{\imath}} \times \hat{\mathbf{\jmath}}$ is also a unit vector, and it is mutually orthogonal to both $\hat{\mathbf{\imath}}$ and $\hat{\mathbf{\jmath}}$. Using the right-hand rule to get the orientation correct, we find that $\hat{\mathbf{\imath}} \times \hat{\mathbf{\jmath}} = \hat{\mathbf{k}}$. Similar arguments allow us to write the results of the other vector products so that

$$\left.\begin{array}{c} \hat{\mathbf{\imath}} \times \hat{\mathbf{\imath}} = \hat{\mathbf{\jmath}} \times \hat{\mathbf{\jmath}} = \hat{\mathbf{k}} \times \hat{\mathbf{k}} = 0 \\ \hat{\mathbf{\imath}} \times \hat{\mathbf{\jmath}} = \hat{\mathbf{k}}, \qquad \hat{\mathbf{\jmath}} \times \hat{\mathbf{k}} = \hat{\mathbf{\imath}}, \qquad \hat{\mathbf{k}} \times \hat{\mathbf{\imath}} = \hat{\mathbf{\jmath}} \end{array}\right\} \tag{4.38}$$

Notice that if we order the unit vectors as $\hat{\mathbf{\imath}}$, $\hat{\mathbf{\jmath}}$, $\hat{\mathbf{k}}$, then the ordering of the mixed vector products follows the same cycles because of the right-hand rule. If we want to calculate $\hat{\mathbf{k}} \times \hat{\mathbf{\jmath}}$, then we have flipped the order of unit vectors, so we introduce a negative sign (this is really just using the fact that we set up the coordinates using the right-hand rule, and taking into account the anticommutative nature of the vector product).

Now that we know how to take the vector product of the basis vectors $\hat{\mathbf{\imath}}$, $\hat{\mathbf{\jmath}}$, $\hat{\mathbf{k}}$, we can calculate the components of the vector product of two vectors:

$$\begin{aligned}
\mathbf{A} \times \mathbf{B} &= (a_x\hat{\mathbf{\imath}} + a_y\hat{\mathbf{\jmath}} + a_z\hat{\mathbf{k}}) \times (b_x\hat{\mathbf{\imath}} + b_y\hat{\mathbf{\jmath}} + b_z\hat{\mathbf{k}}) \\
&= a_x b_x \hat{\mathbf{\imath}} \times \hat{\mathbf{\imath}} + a_y b_y \hat{\mathbf{\jmath}} \times \hat{\mathbf{\jmath}} + a_z b_z \hat{\mathbf{k}} \times \hat{\mathbf{k}} \\
&+ a_x b_y \hat{\mathbf{\imath}} \times \hat{\mathbf{\jmath}} + a_x b_z \hat{\mathbf{\imath}} \times \hat{\mathbf{k}} \\
&+ a_y b_x \hat{\mathbf{\jmath}} \times \hat{\mathbf{\imath}} + a_y b_z \hat{\mathbf{\jmath}} \times \hat{\mathbf{k}} \\
&+ a_z b_x \hat{\mathbf{k}} \times \hat{\mathbf{\imath}} + a_z b_y \hat{\mathbf{k}} \times \hat{\mathbf{\jmath}} \\
&= (a_y b_z - a_z b_y)\hat{\mathbf{\imath}} + (a_z b_x - a_x b_z)\hat{\mathbf{\jmath}} + (a_x b_y - a_y b_x)\hat{\mathbf{k}}.
\end{aligned} \tag{4.39}$$

By comparing this equation with the one for the scalar product in component form (Equation (4.21)) we can see that the vector product mixes up the different components of the two vectors.

Example 4.13 As an example, let us calculate the vector product of the two vectors $\mathbf{A} = 2\hat{\mathbf{\imath}} + 3\hat{\mathbf{\jmath}} - \hat{\mathbf{k}}$ and $\mathbf{B} = \hat{\mathbf{\imath}} - \hat{\mathbf{\jmath}} + 2\hat{\mathbf{k}}$ and show that it is orthogonal to both $\mathbf{A}$ and $\mathbf{B}$. The vector product is

$$\begin{aligned}
\mathbf{C} = \mathbf{A} \times \mathbf{B} &= (2\hat{\mathbf{\imath}} + 3\hat{\mathbf{\jmath}} - \hat{\mathbf{k}}) \times (\hat{\mathbf{\imath}} - \hat{\mathbf{\jmath}} + 2\hat{\mathbf{k}}) \\
&= (2 \times 3 - (-1) \times (-1))\hat{\mathbf{\imath}} + ((-1) \times (1) - 2 \times 2)\hat{\mathbf{\jmath}} + (2 \times (-1) - 3 \times 1)\hat{\mathbf{k}} \\
&= 5\hat{\mathbf{\imath}} - 5\hat{\mathbf{\jmath}} - 5\hat{\mathbf{k}}.
\end{aligned}$$

We can use the scalar product to show that this is orthogonal to both $\mathbf{A}$ and $\mathbf{B}$:

$$\begin{aligned}
\mathbf{C} \cdot \mathbf{A} &= (5\hat{\mathbf{\imath}} - 5\hat{\mathbf{\jmath}} - 5\hat{\mathbf{k}}) \cdot (2\hat{\mathbf{\imath}} + 3\hat{\mathbf{\jmath}} - \hat{\mathbf{k}}) = 10 - 15 + 5 = 0, \\
\mathbf{C} \cdot \mathbf{B} &= (5\hat{\mathbf{\imath}} - 5\hat{\mathbf{\jmath}} - 5\hat{\mathbf{k}}) \cdot (\hat{\mathbf{\imath}} - \hat{\mathbf{\jmath}} + 2\hat{\mathbf{k}}) = 5 + 5 - 10 = 0,
\end{aligned}$$

so $\mathbf{A} \times \mathbf{B}$ is orthogonal to both $\mathbf{A}$ and $\mathbf{B}$.

Equation (4.39) is something of a mess and a little difficult to memorize, but there is a convenient way we can remember the ordering of the terms. Let us write the vector product of the two vectors $\mathbf{A} = a_x\hat{\mathbf{\imath}} + a_y\hat{\mathbf{\jmath}} + a_z\hat{\mathbf{k}}$ and $\mathbf{B} = b_x\hat{\mathbf{\imath}} + b_y\hat{\mathbf{\jmath}} + b_z\hat{\mathbf{k}}$ in the following way. First we write down the unit vectors $\hat{\mathbf{\imath}}$, $\hat{\mathbf{\jmath}}$, and $\hat{\mathbf{k}}$ in a row. Underneath each unit vector we write the corresponding components of the two vectors, one vector on each row:

$$\mathbf{A} \times \mathbf{B} = \begin{vmatrix} \hat{\mathbf{\imath}} & \hat{\mathbf{\jmath}} & \hat{\mathbf{k}} \\ a_x & a_y & a_z \\ b_x & b_y & b_z \end{vmatrix}. \tag{4.40}$$

To evaluate this we work along the top row starting with the first column ($\hat{\mathbf{i}}$) and disregarding the rest of the first row and the first column leaving a block of four elements (a_y, a_z, b_y, and b_z) as seen in Equation (4.41). We multiply the two diagonal elements ($a_y b_z$) and subtract the product of the two remaining elements ($a_z b_y$).

$$\begin{vmatrix} \hat{\mathbf{i}} & \hat{\mathbf{j}} & \hat{\mathbf{k}} \\ a_x & a_y & a_z \\ b_x & b_y & b_z \end{vmatrix}, \quad \begin{vmatrix} \hat{\mathbf{i}} & \hat{\mathbf{j}} & \hat{\mathbf{k}} \\ a_x & a_y & a_z \\ b_x & b_y & b_z \end{vmatrix}, \quad \begin{vmatrix} \hat{\mathbf{i}} & \hat{\mathbf{j}} & \hat{\mathbf{k}} \\ a_x & a_y & a_z \\ b_x & b_y & b_z \end{vmatrix}. \tag{4.41}$$

This gives the factor for the $\hat{\mathbf{i}}$ term. We then move to the second element of the top row ($\hat{\mathbf{j}}$) and follow a similar procedure but multiply the resultant term by -1. Lastly, we add on the result of following the same process for the last element in the top row ($\hat{\mathbf{k}}$). The whole process looks like this

$$\begin{aligned} \begin{vmatrix} \hat{\mathbf{i}} & \hat{\mathbf{j}} & \hat{\mathbf{k}} \\ a_x & a_y & a_z \\ b_x & b_y & b_z \end{vmatrix} &= \hat{\mathbf{i}} \begin{vmatrix} a_y & a_z \\ b_y & b_z \end{vmatrix} - \hat{\mathbf{j}} \begin{vmatrix} a_x & a_z \\ b_x & b_z \end{vmatrix} + \hat{\mathbf{k}} \begin{vmatrix} a_x & a_y \\ b_x & b_y \end{vmatrix} \\ &= \hat{\mathbf{i}}(a_y b_z - a_z b_y) - \hat{\mathbf{j}}(a_x b_z - a_z b_x) + \hat{\mathbf{k}}(a_x b_y - a_y b_x). \end{aligned} \tag{4.42}$$

The object in Equation (4.42) is called a *determinant*, and we will have more to say about it in Section 4.5.1.

Exercise 4.3.9 Calculate the vector products $\mathbf{A} \times \mathbf{B}$ and $\mathbf{B} \times \mathbf{A}$ where $\mathbf{A} = 3\hat{\mathbf{i}} - \hat{\mathbf{j}} + 2\hat{\mathbf{k}}$ and $\mathbf{B} = \hat{\mathbf{i}} + 4\hat{\mathbf{j}} - \hat{\mathbf{k}}$.

Exercise 4.3.10 Show that $|\mathbf{A} \times \mathbf{B}|^2 + (\mathbf{A} \cdot \mathbf{B})^2 = |\mathbf{A}|^2 |\mathbf{B}|^2$.

4.3.2.1 Applications of the Vector Product

Many problems in the Earth and environmental sciences require us to work with complicated geometries. For example, geological formations are the result of multiple processes that move and deform rocks. As we saw in Figure 4.7, by examining the geometry of these formations using vectors, we can understand how these processes lead to the structures we observe today (Pollard and Fletcher, 2005). In particular, we may need to use vectors to define a plane in three dimensions.

We can use the definition of the vector product to define a two-dimensional plane in three dimensions. To do this, we need two vectors that lie within the plane, which means that we need three points (e.g., P, Q, and R in Figure 4.13) lying in the plane, the two vectors being the vectors connecting pairs of the points (i.e., the vectors connecting the points P and Q, and P and R). The vector product of these two vectors is a vector that is at right angles to both vectors in the plane, so it is the normal vector ($\mathbf{n}$) to the plane. To get the equation of the plane, we can consider an arbitrary point in the plane (e.g., $A = (x, y, z)$). The scalar product of the vector $\mathbf{A} = x\hat{\mathbf{i}} + y\hat{\mathbf{j}} + z\hat{\mathbf{k}}$ with the normal vector will give us the desired equation, because $\mathbf{A}$ must be normal to $\mathbf{n}$.

One thing we need to think about is what determines the direction of the vector representing the area? This is determined by convention, which says that the direction

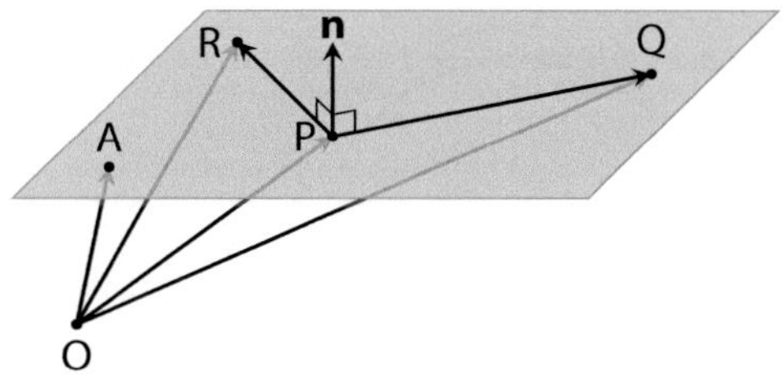

Figure 4.13 Three points P, Q, and R lie in a plane. The vectors connecting point P to points Q and R also lie in the plane, which has a normal **n**. The point A is an arbitrary point lying in the plane.

of the vector is determined by the direction you traverse the rim of the area and use of the right-hand rule.

So, if P_0 is the point at which we define the normal, and P is any arbitrary point in the plane, with $\mathbf{r_0}$ and $\mathbf{r}$ being position vectors of these two points (Figure 4.13), then the normal $\mathbf{n}$ is given by the equation

$$\mathbf{n} \cdot (\mathbf{r} - \mathbf{r_0}) = 0, \tag{4.43}$$

or

$$a(x - x_0) + b(y - y_0) + c(z - z_0) = 0.$$

Example 4.14 We can use Equation (4.43) to calculate the normal vector to the plane defined by the points $P = (0, 1, -1)$, $Q = (2, 2, 1)$, $R = (1, 1/2, -1/2)$. We first choose one of the points and calculate the vectors between that point and the other two. If we choose P to be the point, then the vectors are

$$\mathbf{PQ} = (2 - 0)\hat{\mathbf{i}} + (2 - 1)\hat{\mathbf{j}} + (1 + 1)\hat{\mathbf{k}} = 2\hat{\mathbf{i}} + \hat{\mathbf{j}} + 2\hat{\mathbf{k}}$$

and

$$\mathbf{PR} = (1 - 0)\hat{\mathbf{i}} + \left(\frac{1}{2} - 1\right)\hat{\mathbf{j}} + \left(-\frac{1}{2} + 1\right)\hat{\mathbf{k}} = \hat{\mathbf{i}} - \frac{1}{2}\hat{\mathbf{j}} + \frac{1}{2}\hat{\mathbf{k}}.$$

These two vectors must lie in the plane we want to find, and the normal to that plane is given by

$$\mathbf{n} = \mathbf{PQ} \times \mathbf{PR} = \begin{vmatrix} \hat{\mathbf{i}} & \hat{\mathbf{j}} & \hat{\mathbf{k}} \\ 2 & 1 & 2 \\ 1 & -1/2 & 1/2 \end{vmatrix} = \frac{3}{2}\hat{\mathbf{i}} + \hat{\mathbf{j}} - 2\hat{\mathbf{k}}.$$

If $A = (x, y, z)$ is an arbitrary point in the plane, then the vector $\mathbf{PA}$ must be orthogonal to the normal, so

$$\mathbf{PA} \cdot \mathbf{n} = \frac{3}{2}(x - 0) + 1(y - 1) - 2(z + 1) = 0$$

or, simplifying,

$$3x + 2y - 4z = 6,$$

which is the equation of the plane containing the points P, Q, and R in Figure 4.13.

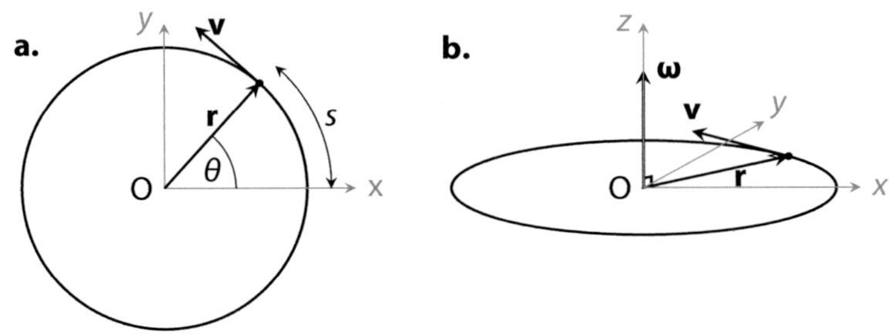

Figure 4.14 An object moves in a circle about a point O. Viewed from above (a.), the vector **r** from O to the particle has a constant length (the radius of the circle), but the arclength *s* and the angle θ that **r** makes with the *x* axis change over time. The vector $\boldsymbol{\omega}$ is perpendicular to the plane containing **r** and **v**, and because of the right-hand rule lies in the positive *z* direction (b.).

The vector product provides a useful way to describe rotational motion. For simplicity, let us consider the motion of an object moving in a circle. The velocity of the object as it moves around the circle is the rate of change of *arclength* s (Figure 4.14a). But the arclength is, by definition of the angle, the ratio of the arclength to the radius r of the circle, and because r is constant for circular motion, we have

$$v = \frac{ds}{dt} = r\frac{d\theta}{dt} = r\omega,$$

where $\omega = \dot{\theta}$ is the rate of change of the angle, or angular velocity. Let us look at this from a slightly different perspective (Figure 4.14b) where we consider the circular motion to be occurring in three dimensions. For example, we could be thinking of the motion of a body on the surface at the equator of the Earth as the planet spins on its axis, in which case the origin of the coordinates would be the center of the Earth and the x and y coordinates would define the position of the body on the equator. The vector giving the position of the object at an instant in time is, in polar coordinates, $\mathbf{r} = (r\cos(\theta(t)), r\sin(\theta(t)), 0)$ where we have written $\|\mathbf{r}\| = r$. The x- and y-components of the vector change with time as the body moves in the circle, so we can differentiate them:

$$\dot{x} = -r\dot{\theta}\sin(\theta), \quad \dot{y} = r\dot{\theta}\cos(\theta), \quad \dot{z} = 0.$$

These make up the components of the velocity vector of the object, so we can write $\dot{\mathbf{r}} = (-r\dot{\theta}\sin(\theta))\hat{\mathbf{i}} + (r\dot{\theta}\cos(\theta))\hat{\mathbf{j}} + 0\hat{\mathbf{k}}$.

Now, let us define a vector $\boldsymbol{\omega} = \dot{\theta}\hat{\mathbf{k}}$ that represents the angular velocity of the body. This is a vector whose magnitude is the magnitude of the angular velocity and that points along the z axis in the positive direction. We can now calculate

$$\boldsymbol{\omega} \times \boldsymbol{r} = (-r\dot{\theta}\sin(\theta))\hat{\mathbf{i}} + (r\dot{\theta}\cos(\theta))\hat{\mathbf{j}} + 0\hat{\mathbf{k}}.$$

This is precisely the same expression that we obtained for $\dot{\mathbf{r}}$, so we can write

$$\dot{\mathbf{r}} = \boldsymbol{\omega} \times \boldsymbol{r}.$$

The direction in which the body moves around the circle is given by the direction of the vector $\boldsymbol{\omega}$. Recall that we use the right-hand rule to define the direction of the vector product, so if ω points in the positive z direction, then the object is moving counterclockwise around the circle, and if ω points in the negative z direction, then the object is moving in a clockwise direction.

4.3.3 Triple Product

As you might guess there are also two types of triple products (where we take the product of three vectors). One is the *scalar triple product* and the other is the *vector triple product*, so named because the result is a scalar in the first case and a vector in the second.

The scalar triple product of three vectors **A**, **B**, **C** is written $\mathbf{A} \cdot (\mathbf{B} \times \mathbf{C})$. Using Equation (4.40), we can write this in three-dimensional rectangular coordinates as

$$\begin{aligned} \mathbf{A} \cdot \mathbf{B} \times \mathbf{C} &= \mathbf{A} \cdot \begin{vmatrix} \hat{\mathbf{i}} & \hat{\mathbf{j}} & \hat{\mathbf{k}} \\ b_x & b_y & b_z \\ c_x & c_y & c_z \end{vmatrix} \\ &= (a_x\hat{\mathbf{i}} + a_y\hat{\mathbf{j}} + a_z\hat{\mathbf{k}}) \cdot \left(\hat{\mathbf{i}} \begin{vmatrix} b_y & b_z \\ c_y & c_z \end{vmatrix} - \hat{\mathbf{j}} \begin{vmatrix} b_x & b_z \\ c_x & c_z \end{vmatrix} + \hat{\mathbf{k}} \begin{vmatrix} b_x & b_y \\ c_x & c_y \end{vmatrix} \right) \\ &= \begin{vmatrix} a_x & a_y & a_z \\ b_x & b_y & b_z \\ c_x & c_y & c_z \end{vmatrix}, \end{aligned} \tag{4.44}$$

where we have used Equation (4.19) for the scalar product of orthonormal basis vectors. The scalar triple product has a nice cyclic symmetry, so that

$$\mathbf{A} \cdot (\mathbf{B} \times \mathbf{C}) = \mathbf{B} \cdot (\mathbf{C} \times \mathbf{A}) = \mathbf{C} \cdot (\mathbf{A} \times \mathbf{B}),$$

and if we swap the order of the vector in any of the cross products, then we introduce a minus sign into the expression, so that

$$\mathbf{A} \cdot (\mathbf{B} \times \mathbf{C}) = -\mathbf{A}(\cdot \mathbf{C} \times \mathbf{B}), \qquad \text{etc.}$$

Exercise 4.3.11 Show that $\mathbf{A} \cdot (\mathbf{B} \times \mathbf{C}) = \mathbf{B} \cdot (\mathbf{C} \times \mathbf{A}) = \mathbf{C} \cdot (\mathbf{A} \times \mathbf{B})$.
Exercise 4.3.12 Show that $\mathbf{A} \cdot (\mathbf{B} \times \mathbf{C}) = -\mathbf{A}(\cdot \mathbf{C} \times \mathbf{B})$.

The scalar triple product gives us the volume of a parallelepiped (Figure 4.15) that is defined by the three vectors **A**, **B**, and **C**. Recall that the volume of a parallelepiped is the product of the area of the base and the perpendicular height of the parallelepiped. The vector product of **B** and **C** is a vector whose magnitude ($||\mathbf{B} \times \mathbf{C}||$) is equal to the area of the base of the parallelepiped and that is orthogonal to both **B** and **C**. To get the perpendicular height of the volume, we take the scalar product of the vector **A** and the unit vector in the direction of $\mathbf{B} \times \mathbf{C}$, that is the projection of **A** onto $\mathbf{B} \times \mathbf{C}$:

$$\mathbf{A} \cdot \frac{\mathbf{B} \times \mathbf{C}}{||\mathbf{B} \times \mathbf{C}||}.$$

So, we end up with

$$\text{volume} = ||\mathbf{B} \times \mathbf{C}|| \left(\mathbf{A} \cdot \frac{\mathbf{B} \times \mathbf{C}}{||\mathbf{B} \times \mathbf{C}||} \right) = \mathbf{A} \cdot (\mathbf{B} \times \mathbf{C}).$$

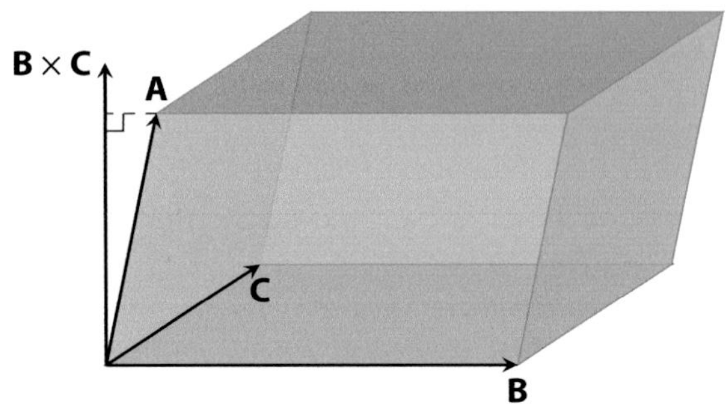

Figure 4.15 The scalar triple product tells us the volume of a parallelepiped with vectors **A**, **B**, and **C** along its edges.

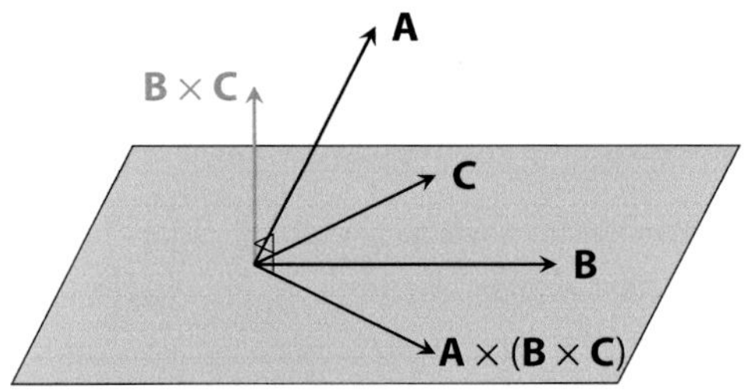

Figure 4.16 The vector triple product of three vectors **A**, **B**, and **C**. The vector **B** × **C** is perpendicular to the plane containing **B** and **C**, and the vector **A** × (**B** × **C**) is perpendicular to **B** × **C** and so lies in the plane containing **B** and **C**.

One of the consequences of this equation is that if all the vectors **A**, **B**, and **C** lie in the same plane (i.e., they are *coplanar*), then the volume of the parallelepiped is zero (because the perpendicular height of the parallelepiped is zero), so $\mathbf{A} \cdot (\mathbf{B} \times \mathbf{C}) = 0$, providing a nice test for coplanar vectors. The converse of this is that three vectors are linearly independent if their scalar triple product is nonzero.

Exercise 4.3.13 Show that if $\mathbf{u}$ is a linear combination of $\mathbf{v}$ and $\mathbf{w}$, then $\mathbf{u} \cdot (\mathbf{v} \times \mathbf{w}) = 0$.

Exercise 4.3.14 Show that the vectors $\mathbf{u} = \hat{\mathbf{\imath}} + \hat{\mathbf{\jmath}} - \hat{\mathbf{k}}$, $\mathbf{v} = 2\hat{\mathbf{\imath}} - \hat{\mathbf{\jmath}} + 3\hat{\mathbf{k}}$, and $\mathbf{w} = 3\hat{\mathbf{\imath}} + 2\hat{\mathbf{k}}$ are coplanar.

The last vector product we will talk about is the vector triple product. This is something of a complicated beast. If we have three vectors **A**, **B**, and **C**, then the vector triple product is $\mathbf{A} \times (\mathbf{B} \times \mathbf{C})$. The vector $\mathbf{U} = \mathbf{B} \times \mathbf{C}$ is a vector that is perpendicular to both **B** and **C**. So, the vector $\mathbf{A} \times (\mathbf{B} \times \mathbf{C}) = \mathbf{A} \times \mathbf{U}$ is a vector that is perpendicular to both **A** and $\mathbf{B} \times \mathbf{C}$, and so must lie in the same plane as **B** and **C** (Figure 4.16). The vector triple product can be expressed as the difference of two vectors,

$$\mathbf{A} \times (\mathbf{B} \times \mathbf{C}) = (\mathbf{A} \cdot \mathbf{C})\mathbf{B} - (\mathbf{A} \cdot \mathbf{B})\mathbf{C}. \tag{4.45}$$

Equation (4.45) may look a little strange at first, but recall that the scalar product of two vectors is just a number, so the vector triple product $\mathbf{A} \times (\mathbf{B} \times \mathbf{C})$ is a linear combination of the vectors **B** and **C**, which we can also see from Figure 4.16. As we can see from Equation (4.45), the placement of the parentheses is very important. In fact, because the vector cross

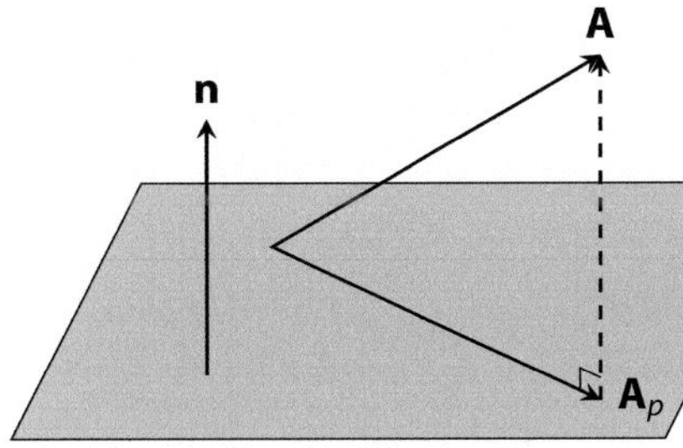

Figure 4.17 The projection of the vector **A** into a plane that has a unit normal **n** is $\mathbf{A}_p$.

product does not commute (i.e., $\mathbf{A} \times \mathbf{B} = -\mathbf{B} \times \mathbf{A}$), we can see that, for example,

$$\mathbf{A} \times (\mathbf{B} \times \mathbf{C}) = -(\mathbf{B} \times \mathbf{C}) \times \mathbf{A}.$$

We can use the vector triple product to calculate the projection ($\mathbf{A}_p$) of a vector **A** into a plane that has a unit normal vector **n** (Figure 4.17). The simplest way to calculate $\mathbf{A}_p$ is to realize that the component of **A** perpendicular to the plane is just $\mathbf{A} \cdot \mathbf{n}$, and then $\mathbf{A}_p = \mathbf{A} - (\mathbf{A} \cdot \mathbf{n})\mathbf{n}$. But we can also use the vector triple product, because using Equation (4.45) we find

$$\mathbf{n} \times (\mathbf{A} \times \mathbf{n}) = (\mathbf{n} \cdot \mathbf{n})\mathbf{A} - (\mathbf{A} \cdot \mathbf{n})\mathbf{n} = \mathbf{A} - (\mathbf{A} \cdot \mathbf{n})\mathbf{n},$$

which is the same formula we had before.

Exercise 4.3.15 Consider the vectors $\mathbf{A} = a_x\hat{\mathbf{i}} + a_y\hat{\mathbf{j}} + a_z\hat{\mathbf{k}}$, $\mathbf{B} = b_x\hat{\mathbf{i}} + b_y\hat{\mathbf{j}} + b_z\hat{\mathbf{k}}$, and $\mathbf{C} = c_x\hat{\mathbf{i}} + c_y\hat{\mathbf{j}} + c_z\hat{\mathbf{k}}$. Show that the x component of $\mathbf{A} \times (\mathbf{B} \times \mathbf{C})$ is $(\mathbf{A} \cdot \mathbf{C})b_x - (\mathbf{A} \cdot \mathbf{B})c_x$.

Exercise 4.3.16 Show that $\mathbf{A} \times (\mathbf{B} \times \mathbf{C}) + \mathbf{B} \times (\mathbf{C} \times \mathbf{A}) + \mathbf{C} \times (\mathbf{A} \times \mathbf{B}) = 0$.

Many of the examples we have used in this chapter have involved geometry or the use of vectors to understand and describe natural phenomena. However, it is worth bearing in mind that the vectors we have looked at belong to a broader mathematical structure called a *linear vector space* (see Box 4.1). Such abstractions can unify apparently disconnected mathematical objects; the mathematical machinery that we have developed here will also apply to other apparently unrelated objects. This is useful because while vectors are easy to visualize and allow us to hone our intuition, some vector spaces contain mathematical entities that are harder to visualize and think about. We will not delve into the formalism of linear vector spaces in this book (see Section 4.9), but we will mention them from time to time, and when we do, you may find it useful to think about vectors.

4.4 Matrices

Matrices are closely related to vectors and provide us with a useful structure for many types of calculations we want to do as scientists. The analysis of large data sets that include many different types of variables is becoming commonplace in the Earth and environmental

Box 4.1 **Vector Spaces**

The vectors that we have looked at so far have represented entities, such as a displacement or a velocity, that we can easily visualize as arrows having a certain length and direction. We have also defined a set of rules for manipulating these objects and required these rules to satisfy certain conditions: for example, addition and subtraction of vectors are commutative and associative. The set of such vectors and rules to manipulate them forms a mathematical structure that is called a *vector space*. This is a useful concept because it can be generalized beyond objects that we can represent as arrows to more abstract entities that are harder to visualize. However, the relationships (e.g., Equation (4.33), the Cauchy–Schwartz inequality) that we can prove in one vector space will hold in all vector spaces.

If you will allow a bit of formality for a minute, we can define a vector space as a set $\mathcal{V}$ of mathematical objects called vectors (**a**, **b**, **c**, etc.) that has the following properties:

1. We can define addition of vectors such that if **a** and **b** are members of the set $\mathcal{V}$, then $\mathbf{a} + \mathbf{b}$ is also a member of the set $\mathcal{V}$ (i.e., the set $\mathcal{V}$ is *closed* under addition), and addition is both commutative (i.e., $\mathbf{a} + \mathbf{b} = \mathbf{b} + \mathbf{a}$) and associative (i.e., $\mathbf{a} + (\mathbf{b} + \mathbf{c}) = (\mathbf{a} + \mathbf{b}) + \mathbf{c}$).
2. The set $\mathcal{V}$ contains a unique vector, the zero vector **0**, such that $\mathbf{a} + \mathbf{0} = \mathbf{a}$.
3. For each vector **a** in $\mathcal{V}$, there is a unique negative ($-\mathbf{a}$), which is also a vector in the set $\mathcal{V}$ and is defined such that $\mathbf{a} + (-\mathbf{a}) = 0$.
4. Scalar multiplication of vectors in $\mathcal{V}$ is defined such that if μ and ν are scalars, then

    1. $\mu(\nu\mathbf{a}) = (\mu\nu)\mathbf{a}$ (Associative property)
    2. $(\mu + \nu)\mathbf{a} = \mu\mathbf{a} + \nu\mathbf{a}$ (Distributive)
    3. $\mu(\mathbf{a} + \mathbf{b}) = \mu\mathbf{a} + \mu\mathbf{a}$ (Distributive)
    4. $1\mathbf{a} = \mathbf{a}$.

Notice that these requirements do not tell us *how* to define addition and subtraction, only that however we do define them, they must obey these conditions.

An example of a more abstract vector space is the set of all quadratic polynomials $P(x) = ax^2 + bx + c$. We can define addition such that we simply add the coefficients of the two polynomials, and we can define multiplication by a scalar as multiplying each coefficient by the scalar:

$$P_1(x) + P_2(x) = (a_1x^2 + b_1x + c_1) + (a_2x^2 + b_2x + c_2) = (a_1 + a_2)x^2 + (b_1 + b_2x + (c_1 + c_2),$$

$$\mu P(x) = \mu ax^2 + \mu bx + \mu c,$$

and these satisfy all of the conditions given. So, the set of all quadratic polynomials forms a vector space!

sciences, and many of the techniques that are used to analyze such data rely on matrices and matrix multiplication. Matrices are also used for numerically solving different types of equations, from large systems of simultaneous linear equations to partial differential equations (see Chapter 10).

A matrix is simply a rectangular array of numbers and is often written using a bold capital letter. Some examples are

$$\mathbf{A} = \begin{pmatrix} 1 & 2 & 3 \\ 4 & 5 & 6 \\ 7 & 8 & 9 \end{pmatrix} \qquad \mathbf{B} = \begin{pmatrix} 1 & 2 & 3 \end{pmatrix} \qquad \mathbf{C} = \begin{pmatrix} 1 \\ 2 \\ 3 \end{pmatrix} \qquad \mathbf{D} = \begin{pmatrix} 1 & 2 \\ 3 & 4 \\ 5 & 6 \end{pmatrix}. \qquad (4.46)$$

The *shape* of a matrix is described by its number of rows and columns and provides an important but simple method of classification. When we specify the shape of a matrix we follow the convention that we always put the number of rows first, then the number of columns, writing it as (number of rows × number of columns). Matrices that have only a single column or a single row are called *column vectors* and *row vectors*, respectively (we already met these in Section 4.1). A matrix with the same number of rows as columns (e.g., **A** in Equation (4.46)) is called a *square matrix*. Following these conventions, we can always unambiguously specify the shape of a matrix.

The entries within a matrix are called the *elements* of the matrix. These are often written using the nonbold, lowercase letter representing the matrix and with subscripts representing the position of the element within the matrix, e.g., a_{ij} represents the element in the ith row and jth column of matrix **A**. Using this notation is useful when we want to develop relationships between generic matrices and do not need to specify actual values for the elements. The ordering of the subscripts is again very important and follows the same row-first, column-second convention we used to specify the shape of the matrix to avoid ambiguity. For example, the element $a_{23} = 6$ in matrix **A** in Equation (4.46) whereas the element $a_{32} = 8$.

Exercise 4.4.1 Consider these matrices:

$$\mathbf{A} = \begin{pmatrix} 1 & 3 & 2 \\ 9 & -1 & 5 \end{pmatrix} \qquad \mathbf{B} = \begin{pmatrix} 1 \\ 9 \\ -1 \\ 5 \end{pmatrix} \qquad \mathbf{C} = \begin{pmatrix} 1 & -2 \\ 9 & 0 \\ -1 & -1 \\ 5 & 3 \end{pmatrix}$$

What are the shapes of the matrices **A**, **B**, and **C**? What are the values of the elements a_{22}, a_{23}, b_{31}, b_{13}, and c_{42}?

4.4.1 Matrix Arithmetic

We add and subtract matrices by adding and subtracting the corresponding elements of the matrices. For example

$$\begin{pmatrix} 0 & 1 \\ 4 & 3 \end{pmatrix} + \begin{pmatrix} 3 & -2 \\ 5 & 9 \end{pmatrix} = \begin{pmatrix} 0+3 & 1-2 \\ 4+5 & 3+9 \end{pmatrix} = \begin{pmatrix} 3 & -1 \\ 9 & 12 \end{pmatrix}.$$

This only makes sense if all the matrices in the equation have the same shape (i.e., the same number of rows and columns). It does not make any sense to add a matrix with size (3 × 2) to one with size (2 × 3). More generally, if **A**, **B**, and **C** are all $(m \times n)$ matrices, then $\mathbf{C} = \mathbf{A} \pm \mathbf{B}$ is the same as

$$\begin{pmatrix} c_{11} & c_{12} & \dots & c_{1n} \\ c_{21} & c_{22} & \dots & c_{2n} \\ \vdots & \vdots & & \vdots \\ c_{m1} & c_{m2} & \dots & c_{mn} \end{pmatrix} = \begin{pmatrix} a_{11} & a_{12} & \dots & a_{1n} \\ a_{21} & a_{22} & \dots & a_{2n} \\ \vdots & \vdots & & \vdots \\ a_{m1} & a_{m2} & \dots & a_{mn} \end{pmatrix} \pm \begin{pmatrix} b_{11} & b_{12} & \dots & b_{1n} \\ b_{21} & b_{22} & \dots & b_{2n} \\ \vdots & \vdots & & \vdots \\ b_{m1} & b_{m2} & \dots & b_{mn} \end{pmatrix}$$
$$= \begin{pmatrix} a_{11} \pm b_{11} & a_{12} \pm b_{12} & \dots & a_{1n} \pm b_{1n} \\ a_{21} \pm b_{21} & a_{22} \pm b_{22} & \dots & a_{2n} \pm b_{2n} \\ \vdots & \vdots & & \vdots \\ a_{m1} \pm b_{m1} & a_{m2} \pm b_{m2} & \dots & a_{mn} \pm b_{mn} \end{pmatrix}. \tag{4.47}$$

Because we add and subtract matrices using the same rules of normal algebra or arithmetic, addition (and subtraction) is both *commutative* ($\mathbf{A}+\mathbf{B} = \mathbf{B}+\mathbf{A}$) and *associative* ($(\mathbf{A}+\mathbf{B})+\mathbf{C} = \mathbf{A}+(\mathbf{B}+\mathbf{C})$).

We can define a *null matrix* as an $(m \times n)$ matrix that has all of its elements equal to zero,

$$\mathbf{0} = \begin{pmatrix} 0 & 0 & \dots & 0 \\ 0 & 0 & \dots & 0 \\ \vdots & \vdots & \vdots & \vdots \\ 0 & 0 & \dots & 0 \end{pmatrix}. \tag{4.48}$$

This matrix plays the same role as the number zero in the addition and subtraction of normal numbers. Adding the null matrix to any matrix $\mathbf{A}$ leaves $\mathbf{A}$ unchanged: $\mathbf{A} + \mathbf{0} = \mathbf{A}$. Another useful matrix is called the *unit matrix* and is written as $\mathbf{I}$. This is a square matrix in which all the elements have the value 0 except those along the main diagonal, which have the value 1. For example, the (2×2) and (3×3) unit matrices are respectively

$$\mathbf{I} = \begin{pmatrix} 1 & 0 \\ 0 & 1 \end{pmatrix} \quad \text{and} \quad \mathbf{I} = \begin{pmatrix} 1 & 0 & 0 \\ 0 & 1 & 0 \\ 0 & 0 & 1 \end{pmatrix}. \tag{4.49}$$

Exercise 4.4.2 What is the result of adding a (2×2) unit matrix $\mathbf{I}$ to a generic (2×2) matrix $\mathbf{A}$?

We say that two matrices $\mathbf{A}$ and $\mathbf{B}$ are equal if all the corresponding elements of the two matrices are equal; in other words, $a_{ij} = b_{ij}$ for all values of i and j. This means that two matrices can only be equal if they have the same shape, because if they are not, then one matrix contains rows or columns that do not exist in the other.

To multiply a matrix by a number (a scalar) we multiply each element of the matrix by that number. For example, if

$$\mathbf{B} = \begin{pmatrix} 1 & 2 & 3 \\ 4 & 5 & 6 \end{pmatrix},$$

then

$$2\mathbf{B} = 2\begin{pmatrix} 1 & 2 & 3 \\ 4 & 5 & 6 \end{pmatrix} = \begin{pmatrix} 2 & 4 & 6 \\ 8 & 10 & 12 \end{pmatrix}.$$

In other words, if $\mathbf{C} = \alpha\mathbf{B}$, then $c_{ij} = \alpha b_{ij}$. Multiplication of a matrix by a scalar is distributive, just like for vectors (see equations (4.1c) and (4.1d)).

Exercise 4.4.3 Consider the following matrices:

$$\mathbf{A} = \begin{pmatrix} -1 & 0 & 4 \\ 3 & 2 & 7 \\ 12 & -4 & 1 \end{pmatrix} \quad \mathbf{B} = \begin{pmatrix} 6 & 3 & -6 \\ 12 & 0 & 2 \\ 9 & -7 & 1 \end{pmatrix} \quad \mathbf{C} = \begin{pmatrix} 0 & 0 \\ 1 & -15 \\ 24 & -5 \end{pmatrix} \quad \mathbf{D} = \begin{pmatrix} 13 & 7 & 11 \\ -5 & 1 & 19 \end{pmatrix}$$

1. Evaluate $\mathbf{A} + \mathbf{B}$.
2. Evaluate $2\mathbf{D}$.
3. Is it possible to add matrices $\mathbf{C}$ and $\mathbf{D}$, and if so, what is the answer?

4.4.2 Linear Transformations and Matrix Multiplication

Although addition and subtraction of matrices works in a similar manner to addition and subtraction of numbers, *matrix multiplication* can seem very strange at first sight. To understand matrix multiplication, we need to understand how matrices are connected to vectors. Let us start by considering a point (x, y) in two dimensions (Figure 4.18). We can represent this point by a vector using unit vectors along the x and y axes:

$$\mathbf{A} = x\hat{\mathbf{i}} + y\hat{\mathbf{j}} = x\begin{pmatrix} 1 \\ 0 \end{pmatrix} + y\begin{pmatrix} 0 \\ 1 \end{pmatrix} = \begin{pmatrix} x \\ y \end{pmatrix}.$$

Now let us apply a transformation to this point and reflect it in the y axis so that the coordinates of the point become $(-x, y)$. What we want to do is find a mathematical object that will represent this transformation and allow us to calculate the new coordinates. We can see from Figure 4.18 that the unit vector along the x axis gets flipped so that it points in the negative x direction whereas the unit vector along the y axis remains unchanged. So, we can write the reflected point as

$$\mathbf{B} = -x\hat{\mathbf{i}} + y\hat{\mathbf{j}} = x\begin{pmatrix} -1 \\ 0 \end{pmatrix} + y\begin{pmatrix} 0 \\ 1 \end{pmatrix} = \begin{pmatrix} -x \\ y \end{pmatrix}. \tag{4.50}$$

This is an example of a *linear transformation*, and we can use matrices to write it in a compact way. To do this we construct a matrix whose columns are given by the transformed basis vectors,

$$\mathbf{M} = \left[\begin{pmatrix} -1 \\ 0 \end{pmatrix} \begin{pmatrix} 0 \\ 1 \end{pmatrix}\right] = \begin{pmatrix} -1 & 0 \\ 0 & 1 \end{pmatrix},$$

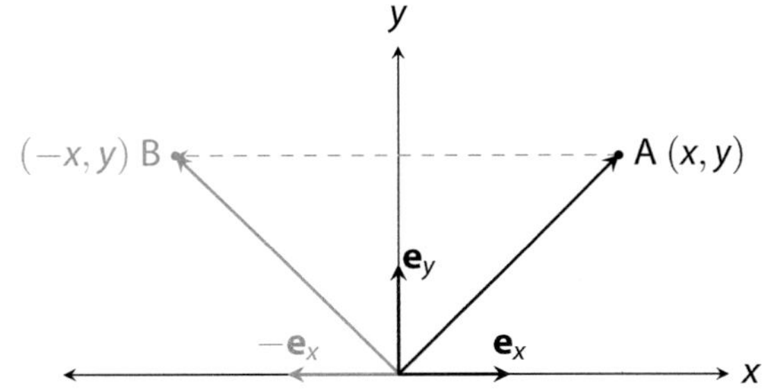

Figure 4.18 The reflection of point A (x, y) in the y axis is the point B $(-x, y)$. The unit vector along the x axis ($\mathbf{e}_x$) is transformed to $-\mathbf{e}_x$, whereas $\mathbf{e}_y$ remains unchanged.

and write our transformation as

$$\begin{pmatrix} -1 & 0 \\ 0 & 1 \end{pmatrix} \begin{pmatrix} x \\ y \end{pmatrix} = \begin{pmatrix} -x \\ y \end{pmatrix}, \tag{4.51}$$

which is just another way of writing Equation (4.50). However, whereas we know how to perform the addition and multiplications by scalars in Equation (4.50), we have not defined a mechanism for getting the right-hand side of Equation (4.51) from the left-hand side. So, we *define* matrix multiplication such that Equation (4.51) is true. Using the notation from the previous section we can write

$$\mathbf{MU} = \begin{pmatrix} m_{11} & m_{12} \\ m_{21} & m_{22} \end{pmatrix} \begin{pmatrix} u_1 \\ u_2 \end{pmatrix} = \begin{pmatrix} m_{11}u_1 + m_{12}u_2 \\ m_{21}u_1 + m_{22}u_2 \end{pmatrix}. \tag{4.52}$$

The rule for matrix multiplication that arises from this is that we take the first row of the matrix **M** and multiply the first element of that row (m_{11}) by the first element (u_1) in the column vector **U**. We then multiply the second element of the first row of **M** by the second element of **U** and add this to our previous calculation. We continue in this way until we have reached the end of the first row of **M**, and the result is the first element of the new vector. We repeat this procedure for the second row and so on.

Exercise 4.4.4 Show that Equation (4.51) is satisfied if we define matrix multiplication according to Equation (4.52).

What we have done is to define matrix multiplication such that a matrix multiplying a column vector is the same as applying a linear transformation to the vector. Recall that a linear transformation is a mathematical operation that acts on vectors and that satisfies the following constraints: if $L(\mathbf{v})$ is a linear transformation acting on a vector **v**, then

1. $L(\mathbf{v}_1 + \mathbf{v}_2) = L(\mathbf{v}_1) + L(\mathbf{v}_2)$,
2. $L(b\mathbf{v}) = bL(\mathbf{v})$.

Let us show that the reflection in the y axis satisfies these conditions. Consider two general vectors

$$\mathbf{v}_1 = \begin{pmatrix} x_1 \\ y_1 \end{pmatrix}, \qquad \mathbf{v}_2 = \begin{pmatrix} x_2 \\ y_2 \end{pmatrix},$$

then

$$L(\mathbf{v}_1 + \mathbf{v}_2) = \mathbf{M}(\mathbf{v}_1 + \mathbf{v}_2) = \begin{pmatrix} -1 & 0 \\ 0 & 1 \end{pmatrix} \begin{pmatrix} x_1 + x_2 \\ y_1 + y_2 \end{pmatrix} = \begin{pmatrix} -x_1 - x_2 \\ y_1 + y_2 \end{pmatrix}.$$

Also

$$L(\mathbf{v}_1) = \begin{pmatrix} -1 & 0 \\ 0 & 1 \end{pmatrix} \begin{pmatrix} x_1 \\ y_1 \end{pmatrix} = \begin{pmatrix} -x_1 \\ y_1 \end{pmatrix},$$

and similarly for $L(\mathbf{v}_2)$, so

$$L(\mathbf{v}_1) + L(\mathbf{v}_2) = \begin{pmatrix} -x_1 \\ y_1 \end{pmatrix} + \begin{pmatrix} -x_2 \\ y_2 \end{pmatrix} = \begin{pmatrix} -x_1 - x_2 \\ y_1 + y_2 \end{pmatrix},$$

and so the first condition, $L(\mathbf{v}_1 + \mathbf{v}_2) = L(\mathbf{v}_1) + L(\mathbf{v}_2)$, is satisfied.

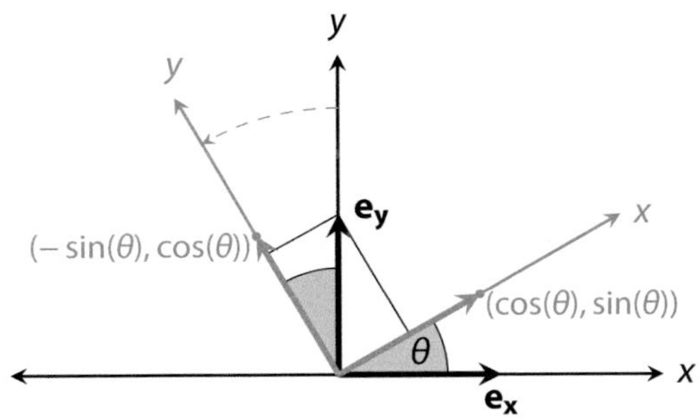

Figure 4.19 Under a counterclockwise rotation of the coordinates through an angle θ, the tip of the unit vector along the x axis is transformed to a point $x = \cos(\theta), y = \sin(\theta)$, and the tip of the unit vector along the y axis is transformed to a point $x = -\sin(\theta), y = \cos(\theta)$.

Exercise 4.4.5 Show that reflections in the y axis also satisfy $L(b\mathbf{v}) = bL(\mathbf{v})$.

Let us look at a slightly more complicated example: rotations about the origin in two dimensions. We know that if we rotate a vector through an angle θ, then the x and y components are transformed to new coordinates x' and y' such that

$$x' = x\cos(\theta) - y\sin(\theta), \qquad y' = x\sin(\theta) + y\cos(\theta).$$

When we looked at the reflection in the y axis, we constructed the matrix representing the linear transformation using the transformed vectors. Let us do the same again here. From (Figure 4.19) we can construct the matrix representing the transformation,

$$\mathbf{M} = \begin{pmatrix} \cos(\theta) & -\sin(\theta) \\ \sin(\theta) & \cos(\theta) \end{pmatrix}, \tag{4.53}$$

so an arbitrary point (x, y) will be transformed to the point

$$\begin{pmatrix} \cos(\theta) & -\sin(\theta) \\ \sin(\theta) & \cos(\theta) \end{pmatrix} \begin{pmatrix} x \\ y \end{pmatrix} = \begin{pmatrix} x\cos(\theta) - y\sin(\theta) \\ x\sin(\theta) + y\cos(\theta) \end{pmatrix},$$

which is what we had before (cf. Equation (4.14)).

Exercise 4.4.6 Show that a rotation in two dimensions satisfies the conditions to be a linear transformation.

Exercise 4.4.7 Construct a matrix representing a linear transformation that corresponds to a reflection about the line $y = x$.

Exercise 4.4.8 Show that the origin always remains unchanged under a linear transformation.

These examples suggest that we can think of a matrix as a machine that performs a linear transformation on vectors (or points in space). We can use matrices to represent other useful linear transformations such as scaling and shear (Figure 4.20). If we multiply the x or y coordinates, or both, by a constant factor k, the result will be an *expansion* (for $k > 1$) or *contraction* ($k < 1$) in that direction (Figure 4.20a). For example, if we were to scale only

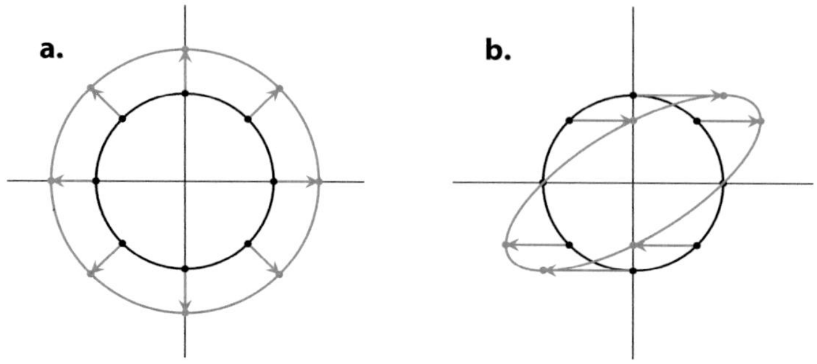

Figure 4.20 The linear transformations of expansion and contraction (a.), and shear (b.).

the x coordinates by a factor of 2, then the matrix that represents this linear transformation is (in two dimensions)

$$\mathbf{M} = \begin{pmatrix} 2 & 0 \\ 0 & 1 \end{pmatrix} \quad \text{such that} \quad \begin{pmatrix} 2 & 0 \\ 0 & 1 \end{pmatrix}\begin{pmatrix} x \\ y \end{pmatrix} = \begin{pmatrix} 2x \\ y \end{pmatrix}.$$

Another important transformation is a *shear* (Figure 4.20b). The transformations we have looked at so far preserve shapes, but a shear distorts shapes. In general, if a and b are constants, then the matrix representing a shear in two dimensions can be written

$$\mathbf{M} = \begin{pmatrix} 1 & a \\ b & 1 \end{pmatrix} \quad \text{such that} \quad \mathbf{M} = \begin{pmatrix} 1 & a \\ b & 1 \end{pmatrix}\begin{pmatrix} x \\ y \end{pmatrix} = \begin{pmatrix} x + ay \\ y + bx \end{pmatrix}.$$

Shear is an important aspect of the mechanics of materials that can be deformed when the forces acting on them vary spatially. Fluids such as air and water can be easily deformed, and their motions are strongly affected by shear. However, even material such as rock can undergo shear, and this can be particularly important near the boundaries of tectonic plates.

These transformations have shown us how to multiply a matrix and a vector, but now let us examine matrix multiplication a little more generally. If we have an $(m \times n)$ matrix $\mathbf{A}$ and an $(n \times p)$ matrix $\mathbf{B}$, then their product $\mathbf{C} = \mathbf{AB}$ is an $(m \times p)$ matrix defined by

$$\mathbf{C} = c_{ij} = \mathbf{AB} = \sum_{k=1}^{n} a_{ik} b_{kj}. \tag{4.54}$$

To see what this equation is telling us, let us look at that last term in more detail. To calculate the (i, j) element of matrix $\mathbf{C}$ we have to take a sum over the elements of the ith row of $\mathbf{A}$ with each element in that row being multiplied by the corresponding element in the jth column of $\mathbf{B}$. In other words,

$$c_{ij} = a_{i1}b_{1j} + a_{i2}b_{2j} + a_{i3}b_{3j} + \ldots a_{im}b_{mj}.$$

We know that when we refer to an element of a matrix we specify the row first, then the column. This same ordering of indices (row, then column) appears with matrix multiplication; we are moving along a row of matrix $\mathbf{A}$, taking successive elements and multiplying them by the corresponding element in a column of $\mathbf{B}$. For example, if $\mathbf{A}$ and

B are both (3 × 3) matrices and $\mathbf{C} = \mathbf{AB}$, then we calculate c_{23} (the element in row 2 and column 3 of **C**) as $c_{23} = a_{21}b_{13} + a_{22}b_{23} + a_{23}b_{33}$,

$$\begin{pmatrix} c_{11} & c_{12} & c_{13} \\ c_{21} & c_{22} & \mathbf{c_{23}} \\ c_{31} & c_{32} & c_{33} \end{pmatrix} = \begin{pmatrix} a_{11} & a_{12} & a_{13} \\ \mathbf{a_{21}} & \mathbf{a_{22}} & \mathbf{a_{23}} \\ a_{31} & a_{32} & a_{33} \end{pmatrix} \begin{pmatrix} b_{11} & b_{12} & \mathbf{b_{13}} \\ b_{21} & b_{22} & \mathbf{b_{23}} \\ b_{31} & b_{32} & \mathbf{b_{33}} \end{pmatrix}$$

$$= \begin{pmatrix} c_{11} & c_{12} & c_{13} \\ c_{21} & c_{22} & \mathbf{a_{21}b_{13}} + \mathbf{a_{22}b_{23}} + \mathbf{a_{23}b_{33}} \\ c_{31} & c_{32} & c_{33} \end{pmatrix}.$$

One way to remember this is that we always multiply the elements of a **row** by a **column**. The indices help us with the bookkeeping to make sure we are doing everything correctly.

There are some important consequences to the way we have defined matrix multiplication. The first, and one of the most important, is that you cannot always multiply any two matrices together. We can see the reason for this from Equation (4.54). That equation tells us that the number of columns in matrix **A** must be equal to the number of rows in matrix **B**. If they were not, say there were more columns in **A**, then in the summation we would run out of rows in **B**. Matrices that have compatible shapes that allow them to be multiplied are called *conformable for multiplication*.

Example 4.15 Consider the matrices

$$\mathbf{A} = \begin{pmatrix} 1 & 2 & 0 \\ 3 & 1 & 4 \\ -1 & 2 & 1 \end{pmatrix}, \qquad \mathbf{B} = \begin{pmatrix} 3 & 1 \\ 1 & 2 \\ 2 & 1 \end{pmatrix}.$$

If we want to multiply them, we need to ask if the matrices are conformable for multiplication before we can evaluate the products **AB** and **BA**. The number of columns in **A** is the same as the number of rows in **B**, so we can evaluate the product **AB**:

$$\mathbf{AB} = \begin{pmatrix} 1 & 2 & 0 \\ 3 & 1 & 4 \\ -1 & 2 & 1 \end{pmatrix} \begin{pmatrix} 3 & 1 \\ 1 & 2 \\ 2 & 1 \end{pmatrix} = \begin{pmatrix} 1\times 3 + 2\times 1 + 0\times 2 & 1\times 1 + 2\times 2 + 0\times 1 \\ 3\times 2 + 1\times 1 + 4\times 2 & 3\times 1 + 1\times 2 + 4\times 1 \\ -1\times 3 + 2\times 1 + 1\times 2 & -1\times 1 + 2\times 2 + 4\times 1 \end{pmatrix} = \begin{pmatrix} 5 & 5 \\ 18 & 9 \\ 1 & 4 \end{pmatrix}.$$

However, the number of columns in **B** is less than the number of rows in **A**, so we cannot evaluate the product **BA**.

Exercise 4.4.9 Consider the following matrices:

$$\mathbf{A} = \begin{pmatrix} 1 & -3 & 2 & 4 \\ 12 & 0 & 9 & -1 \\ -36 & 24 & 3 & 1 \\ -1 & 2 & 4 & 8 \end{pmatrix}, \quad \mathbf{B} = \begin{pmatrix} 1 & 4 & 6 & 2 \\ 4 & 9 & 8 & 2 \end{pmatrix}, \quad \mathbf{C} = \begin{pmatrix} 4 \\ 8 \\ 3 \\ 2 \end{pmatrix}, \quad \mathbf{D} = \begin{pmatrix} 1 & 0 & 0 \\ 0 & 1 & 0 \\ 0 & 0 & 1 \\ 0 & 0 & 0 \end{pmatrix}.$$

Determine which combinations of the matrices are conformable, and if they are conformable, evaluate their product.

Example 4.15 shows something very interesting, namely that even if we can form the product **AB**, this does not mean that we can form the product **BA**. This also implies that matrix multiplication does not necessarily *commute*. In fact, we can see from the definition of matrix multiplication that a *necessary* condition for **AB** = **BA** is that both matrices must have the same shape and have the same number of rows and columns; i.e., they are square matrices. This is a necessary but not sufficient condition (i.e., there are square matrices that do not commute) for two matrices to commute, but in general, matrix multiplication is not commutative, so multiplying **A** on the right of **B** (**AB**) will give a different result from multiplying **A** on the left of **B** (**BA**).

Example 4.16 Let us compare the products **AB** and **BA** where

$$\mathbf{A} = \begin{pmatrix} 1 & 3 \\ -1 & 2 \end{pmatrix}, \qquad \mathbf{B} = \begin{pmatrix} 2 & 1 \\ 2 & 5 \end{pmatrix}.$$

We can form the product **AB** (i.e., multiplying **A** on the left of **B**) and **BA** (multiplying **A** on the right of **B**) to get

$$\mathbf{AB} = \begin{pmatrix} 8 & 16 \\ 2 & 9 \end{pmatrix}, \qquad \mathbf{BA} = \begin{pmatrix} 1 & 8 \\ -3 & 16 \end{pmatrix}.$$

Thus, the two matrices do not commute, and **AB** ≠ **BA**.

Example 4.17 As an example of two matrices that do commute, consider

$$\mathbf{A} = \begin{pmatrix} 1 & 3 \\ -1 & 2 \end{pmatrix}, \qquad \mathbf{I} = \begin{pmatrix} 1 & 0 \\ 0 & 1 \end{pmatrix}.$$

Multiplying the matrices we find

$$\mathbf{AI} = \begin{pmatrix} 1 & 3 \\ -1 & 2 \end{pmatrix}\begin{pmatrix} 1 & 0 \\ 0 & 1 \end{pmatrix} = \begin{pmatrix} 1 & 3 \\ -1 & 2 \end{pmatrix}, \qquad \mathbf{IA} = \begin{pmatrix} 1 & 0 \\ 0 & 1 \end{pmatrix}\begin{pmatrix} 1 & 3 \\ -1 & 2 \end{pmatrix} = \begin{pmatrix} 1 & 3 \\ -1 & 2 \end{pmatrix}.$$

So, these matrices commute.

There are some further properties of matrix multiplication that stem from equation (4.54). For example, the matrix equation **AB** = 0 does not necessarily imply that either **A** = 0 or **B** = 0. To see why, let us look at the general result of multiplying two (2 × 2) matrices,

$$\mathbf{AB} = \begin{pmatrix} a_{11} & a_{12} \\ a_{21} & a_{22} \end{pmatrix}\begin{pmatrix} b_{11} & b_{12} \\ b_{21} & b_{22} \end{pmatrix} = \begin{pmatrix} a_{11}b_{11} + a_{12}b_{21} & a_{11}b_{12} + a_{12}b_{22} \\ a_{21}b_{11} + a_{22}b_{21} & a_{21}b_{12} + a_{22}b_{22} \end{pmatrix}.$$

The equation **AB** = 0 only says that all the elements in the matrix on the right-hand side must be zero, i.e.,

$$a_{11}b_{11} + a_{12}b_{21} = 0, \qquad a_{11}b_{12} + a_{12}b_{22} = 0,$$
$$a_{21}b_{11} + a_{22}b_{21} = 0, \qquad a_{21}b_{12} + a_{22}b_{22} = 0,$$

and this does not necessarily mean that $a_{ij} = 0$ or $b_{ij} = 0$ for all values of i and j. So, when we are dealing with matrices, we need to be careful because our intuition about the arithmetic of numbers does not always carry over to matrices. Multiplying large matrices together is a job best left to a computer, but again we need to exercise caution to make sure that the matrices are conformable for multiplication.

4.4.3 Inverse Matrix

There is one arithmetic operation we have not mentioned yet, and that is division. There is no operation of matrix division, instead there is multiplication by a matrix called the *inverse*, which plays a similar role. We can understand this by again thinking of matrices as representing linear transformations. For example, if we have a matrix $\mathbf{A}$ that represents a shear deformation, then multiplying a position vector $\mathbf{r}$ by $\mathbf{A}$ will move points in space and distort shapes. The inverse matrix of $\mathbf{A}$, written as $\mathbf{A}^{-1}$, applied to the transformed points will undo the distortion from the shear, returning the points back to their original locations. So, multiplying a vector by $\mathbf{A}$ and then multiplying the result by $\mathbf{A}^{-1}$ is the same as doing nothing, in other words, multiplying the vector by $\mathbf{I}$.

Formally, for a matrix $\mathbf{A}$, the inverse $\mathbf{A}^{-1}$ is defined such that $\mathbf{A}\mathbf{A}^{-1} = \mathbf{I}$, where $\mathbf{I}$ is the unit matrix that we introduced in Equation (4.49). Finding the inverse of a matrix is a tedious operation, especially for large matrices, but it is a task that is easily accomplished by a computer program. However, it is instructive to work through a simple case by hand. Consider the matrix

$$\mathbf{A} = \begin{pmatrix} a & b \\ c & d \end{pmatrix}. \tag{4.55}$$

In order to find the inverse of this matrix, we will assume that the inverse of the matrix $\mathbf{A}$ is the matrix

$$\mathbf{Z} = \begin{pmatrix} w & x \\ y & z \end{pmatrix},$$

where we do not yet know what w, x, y, and z are. Our job is to find what these are in terms of the things we do know, i.e., a, b, c, and d. The only other thing we know is that $\mathbf{AZ} = \mathbf{I}$. So, let us write that out in full:

$$\begin{pmatrix} w & x \\ y & z \end{pmatrix}\begin{pmatrix} a & b \\ c & d \end{pmatrix} = \begin{pmatrix} wa + xc & wb + xd \\ ya + zc & yb + zd \end{pmatrix} = \begin{pmatrix} 1 & 0 \\ 0 & 1 \end{pmatrix}. \tag{4.56}$$

For the last equality to be true, the corresponding elements of the two matrices must be equal. So, this gives us a series of four equations:

$$aw + cx = 1, \qquad bw + dx = 0, \qquad ay + cz = 0, \qquad by + dz = 1.$$

We can solve these equations to give

$$w = \frac{d}{ad - cb}, \qquad x = \frac{-b}{ad - bc}, \qquad y = \frac{-c}{ad - bc}, \qquad \frac{a}{ad - bc},$$

so that

$$\mathbf{Z} = \mathbf{A}^{-1} = \frac{1}{ad - bc}\begin{pmatrix} d & -b \\ -c & a \end{pmatrix}. \tag{4.57}$$

If $(ad - bc) = 0$, then all the elements of $\mathbf{A}^{-1}$ are infinite and the matrix $\mathbf{A}$ has no inverse. So, not all matrices have inverses!

Exercise 4.4.10 Find the inverse of the two-dimensional rotation matrix, Equation (4.53).

Exercise 4.4.11 Find the inverse of the two-dimensional unit matrix.

We have seen the quantity $(ad - bc)$ in another guise; it is the *determinant* that we met in Equation (4.42) and Equation (4.44), although in those examples we looked at (3×3) determinants. We will return to determinants shortly, but we need to note here that a matrix with a zero determinant has no inverse and is called a *singular matrix*. Before we return to determinants in Section 4.5.1, we will look at some special types of matrix, many of which occur frequently in scientific calculations.

4.4.4 Special Matrices

Some types of matrices have important properties that make calculations simpler or reflect specific aspects of a phenomenon. Many of these special matrices are defined by the distribution of zero and nonzero elements in the matrix. The first such matrix we will consider is called a *diagonal matrix*. This is a square matrix with nonzero elements only along the main diagonal. We have already met one type of diagonal matrix, the unit matrix (Equation (4.49)), where all the diagonal elements have the value 1. In general, each element on the diagonal can have a different value, so for example, a diagonal (3×3) matrix would look like

$$\begin{pmatrix} a & 0 & 0 \\ 0 & b & 0 \\ 0 & 0 & c \end{pmatrix}. \tag{4.58}$$

Exercise 4.4.12 Consider the matrices

$$\mathbf{A} = \begin{pmatrix} 1 & 2 & 3 \\ 4 & 5 & 6 \\ 7 & 8 & 9 \end{pmatrix}, \qquad \mathbf{D} = \begin{pmatrix} 1 & 0 & 0 \\ 0 & 2 & 0 \\ 0 & 0 & 3 \end{pmatrix},$$

and calculate the following, commenting on how the result relates to the matrix $\mathbf{A}$: a. $\mathbf{A} + \mathbf{D}$, b. $\mathbf{AD}$, c. $\mathbf{DA}$.

Exercise 4.4.13 What type of linear transformation is represented by a diagonal matrix?

A useful operation to perform on a matrix $\mathbf{A}$ is to calculate its *transpose* ($\mathbf{A}^{\mathrm{T}}$). This can be used to make a matrix conformable for multiplication, for example. The transpose of a matrix is obtained by swapping the rows and columns of the original matrix. For example, if

$$\mathbf{A} = \begin{pmatrix} 3 & -1 & 2 \\ 9 & -6 & 5 \\ -12 & 4 & 7 \end{pmatrix}, \quad \text{then} \quad \mathbf{A}^{\mathrm{T}} = \begin{pmatrix} 3 & 9 & -12 \\ -1 & -6 & 4 \\ 2 & 5 & 7 \end{pmatrix}.$$

If we write the elements of a matrix **A** as a_{ij}, then the elements of $\mathbf{A}^{\mathrm{T}}$ are a_{ji}. Notice that the elements along the diagonal do not move (if we swap the order of the subscripts on the element a_{ii}, the element is still in the ith row and ith column). Remember that we can write the components of a vector as a matrix with a single column (a column vector). Let us see what happens if we take the transpose of such a column vector. If a vector **v** has components (x_v, y_v, z_v), then

$$\mathbf{v} = \begin{pmatrix} x_v \\ y_v \\ z_v \end{pmatrix} \quad \text{and} \quad \mathbf{v}^{\mathrm{T}} = \begin{pmatrix} x_v & y_v & z_v \end{pmatrix},$$

so taking the transpose converts a column vector into a row vector. If we now multiply the transpose by the original vector, we find

$$\mathbf{v}^{\mathrm{T}}\mathbf{v} = \begin{pmatrix} x_v & y_v & z_v \end{pmatrix} \begin{pmatrix} x_v \\ y_v \\ z_v \end{pmatrix} = x_v^2 + y_v^2 + z_v^2 = \mathbf{v} \cdot \mathbf{v} = \|\mathbf{v}\|^2.$$

So, multiplying the transpose of a vector by itself is the same as taking the scalar product of the vector with itself. Given this, you will not be surprised to find that if the vector **u** has components (x_u, y_u, z_u), then

$$\mathbf{u}^{\mathrm{T}}\mathbf{v} = \begin{pmatrix} x_u & y_u & z_u \end{pmatrix} \begin{pmatrix} x_v \\ y_v \\ z_v \end{pmatrix} = u_x v_x + u_y v_y + u_z v_z = \mathbf{u} \cdot \mathbf{v}; \tag{4.59}$$

in other words, $\mathbf{u}^{\mathrm{T}}\mathbf{v}$ is the same as calculating the scalar product of **u** and **v**.

If we write the elements of the matrix **A** as a_{ij}, then we can see that

$$(\mathbf{A}^{\mathrm{T}})^{\mathrm{T}} = (a_{ij}^{\mathrm{T}})^{\mathrm{T}} = (a_{ji})^{\mathrm{T}} = a_{ij},$$

so that taking the transpose of the transpose of a matrix returns us to the original matrix.

Exercise 4.4.14 Show that $(\mathbf{A} + \mathbf{B})^{\mathrm{T}} = \mathbf{A}^{\mathrm{T}} + \mathbf{A}^{\mathrm{T}}$.
Exercise 4.4.15 Show that $(\gamma\mathbf{A})^{\mathrm{T}} = \gamma\mathbf{A}^{\mathrm{T}}$.
Exercise 4.4.16 Show that $(\mathbf{AB})^{\mathrm{T}} = \mathbf{B}^{\mathrm{T}}\mathbf{A}^{\mathrm{T}}$.
Exercise 4.4.17 Show that for a matrix **A** and vectors **u** and **v** that $\mathbf{u} \cdot \mathbf{A}\mathbf{v} = \mathbf{A}^{\mathrm{T}}\mathbf{u} \cdot \mathbf{v}$.

A *symmetric matrix* is a square matrix **A** that is equal to its own transpose, $\mathbf{A} = \mathbf{A}^{\mathrm{T}}$, which implies $a_{ij} = a_{ji}$. For example, consider the matrix

$$\begin{pmatrix} 3 & -4 & 2 \\ -4 & 12 & 21 \\ 2 & 21 & 0 \end{pmatrix}. \tag{4.60}$$

An *antisymmetric* (sometimes also called *skew symmetric*) *matrix* has $a_{ij} = -a_{ji}$ or $\mathbf{A} = -\mathbf{A}^{\mathrm{T}}$.

Now, let us take a square matrix **A** and calculate

$$(\mathbf{A} + \mathbf{A}^{\mathrm{T}})^{\mathrm{T}} = (\mathbf{A})^{\mathrm{T}} + (\mathbf{A}^{\mathrm{T}})^{\mathrm{T}} = \mathbf{A}^{\mathrm{T}} + \mathbf{A};$$

in other words, $(\mathbf{A} + \mathbf{A}^{\mathrm{T}})$ is a symmetric matrix. Similarly $(\mathbf{A} - \mathbf{A}^{\mathrm{T}})$ is antisymmetric. This means that we can write any square matrix as a sum of a purely symmetric matrix and a purely antisymmetric matrix:

$$\mathbf{A} = \frac{1}{2}(\mathbf{A} + \mathbf{A}^{\mathrm{T}}) + \frac{1}{2}(\mathbf{A} - \mathbf{A}^{\mathrm{T}}). \tag{4.61}$$

Exercise 4.4.18 Decompose the matrix

$$\mathbf{A} = \begin{pmatrix} 1 & 2 & 3 \\ 4 & 5 & 6 \\ 7 & 8 & 9 \end{pmatrix}$$

into a symmetric and antisymmetric matrix.

Exercise 4.4.19 If b_1, b_2, and b_3 are arbitrary numbers, write down the most general (3×3) antisymmetric matrix in terms of b_1, b_2, and b_3.

A square matrix $\mathbf{A}$ that satisfies the condition $\mathbf{A}\mathbf{A}^{\mathrm{T}} = \mathbf{I}$ is called an *orthogonal matrix*. Notice that if we multiply this equation on the left by $\mathbf{A}^{-1}$ we get

$$\begin{aligned} \mathbf{A}^{-1}\mathbf{A}\mathbf{A}^{\mathrm{T}} &= \mathbf{I}\mathbf{A}^{\mathrm{T}} = \mathbf{A}^{\mathrm{T}} \\ &= \mathbf{A}^{-1}\mathbf{I} = \mathbf{A}^{-1}, \end{aligned}$$

so that for an orthogonal matrix $\mathbf{A}^{\mathrm{T}} = \mathbf{A}^{-1}$. Orthogonal matrices have interesting properties and are useful because some linear transformations are represented by orthogonal matrices.

Exercise 4.4.20 Show that the rotation matrix in two dimensions is an orthogonal matrix.

Exercise 4.4.21 Show that the matrix representing a reflection about the y axis is an orthogonal matrix.

Exercise 4.4.22 Consider a point (x, y) and rotate the coordinate system through an angle θ so that the coordinates of the point in the new coordinates are (x', y') with

$$\begin{pmatrix} x' \\ y' \end{pmatrix} = \begin{pmatrix} \cos(\theta) & \sin(\theta) \\ -\sin(\theta) & \cos(\theta) \end{pmatrix} \begin{pmatrix} x \\ y \end{pmatrix}.$$

Show that the inverse transformation is given by

$$\begin{pmatrix} x \\ y \end{pmatrix} = \begin{pmatrix} \cos(\theta) & -\sin(\theta) \\ \sin(\theta) & \cos(\theta) \end{pmatrix} \begin{pmatrix} x' \\ y' \end{pmatrix}.$$

Note that, if we know a matrix is orthogonal, it is far easier to calculate its inverse by calculating its transpose than it is by calculating it using the methods in Section 4.4.3, especially for very large matrices with many rows and columns.

Orthogonal matrices have some important geometric properties. Let us consider an orthogonal matrix $\mathbf{A}$ acting on a vector $\mathbf{v}$. Recalling Equation (4.59), we can write that the length of the vector $\mathbf{A}\mathbf{v}$ is

$$||(\mathbf{A}\mathbf{v})||^2 = (\mathbf{A}\mathbf{v}) \cdot (\mathbf{A}\mathbf{v}) = \mathbf{A}^{\mathrm{T}}(\mathbf{A}\mathbf{v}) \cdot \mathbf{v} = \mathbf{v} \cdot \mathbf{v} = ||\mathbf{v}||^2,$$

where we have used results from Exercise 4.4.22. This equation tells us that the length of a vector is preserved when we multiply the vector by an orthogonal matrix. However, there is more. If **u** is another vector, then we know that the angle θ between **u** and **v** is given by

$$\cos(\theta) = \frac{\mathbf{u} \cdot \mathbf{v}}{||\mathbf{u}||\,||\mathbf{v}||}.$$

If we apply the orthogonal matrix **A** to both vectors, then we can calculate the angle between the transformed vectors, i.e., between **Au** and **Av**. Let us call this angle ϕ, then using our previous result,

$$\cos(\phi) = \frac{(\mathbf{Au}) \cdot (\mathbf{Av})}{||\mathbf{Au}||\,||\mathbf{Av}||} = \frac{\mathbf{A}^{\mathrm{T}}(\mathbf{Au}) \cdot \mathbf{v}}{||\mathbf{u}||\,||\mathbf{v}||} = \frac{\mathbf{u}) \cdot \mathbf{v}}{||\mathbf{u}||\,||\mathbf{v}||} = \cos(\theta).$$

So, an orthogonal matrix preserves not only the lengths of vectors, but also the angles between them; in other words, it preserves shapes. Matrices representing rotations and reflections are orthogonal matrices, but matrices representing expansion, contraction, or pure shear are not (they either do not preserve lengths or angles).

Why are orthogonal matrices called *orthogonal* matrices? What is at right angles to what? Recall that when we introduced linear transformations (Section 4.4.2) we constructed the matrix for the transformation by using column vectors representing how basis vectors changed under the transformation. Let us represent the columns of the square $(n \times n)$ orthogonal matrix **A** by the vectors $\mathbf{a}_1, \mathbf{a}_2, \ldots \mathbf{a}_n$, where each column vector has n elements. When we take the transpose of **A**, each of these column vectors becomes a row vector, so we can write

$$\mathbf{A}^{\mathrm{T}}\mathbf{A} = \begin{pmatrix} \mathbf{a}_1^{\mathrm{T}} \\ \mathbf{a}_2^{\mathrm{T}} \\ \vdots \\ \mathbf{a}_n^{\mathrm{T}} \end{pmatrix} \begin{pmatrix} \mathbf{a}_1 & \mathbf{a}_1 & \cdots & \mathbf{a}_1 \end{pmatrix} = \begin{pmatrix} \mathbf{a}_1^{\mathrm{T}}\mathbf{a}_1 & \mathbf{a}_1^{\mathrm{T}}\mathbf{a}_2 & \cdots & \mathbf{a}_1^{\mathrm{T}}\mathbf{a}_n \\ \mathbf{a}_2^{\mathrm{T}}\mathbf{a}_1 & \mathbf{a}_2^{\mathrm{T}}\mathbf{a}_2 & \cdots & \mathbf{a}_2^{\mathrm{T}}\mathbf{a}_n \\ \vdots & \vdots & \vdots & \vdots \\ \mathbf{a}_n^{\mathrm{T}}\mathbf{a}_1 & \mathbf{a}_n^{\mathrm{T}}\mathbf{a}_2 & \cdots & \mathbf{a}_n^{\mathrm{T}}\mathbf{a}_n \end{pmatrix} = \begin{pmatrix} 1 & 0 & \cdots & 0 \\ 0 & 1 & \cdots & 0 \\ \vdots & \vdots & \vdots & \vdots \\ 0 & 0 & \cdots & 1 \end{pmatrix}.$$

This equation is telling us that if **A** is orthogonal, then

$$\mathbf{a}_i^{\mathrm{T}}\mathbf{a}_j = \mathbf{a}_i \cdot \mathbf{a}_j = \begin{cases} 1 & i = j \\ 0 & i \neq j \end{cases};$$

in other words, the columns of **A** are orthonormal vectors and form a basis—the same is also true of the rows of an orthogonal matrix.

Exercise 4.4.23 Confirm that the columns of a two-dimensional rotation matrix are orthogonal.

Many of the transformation matrix examples we have looked at have been in two dimensions. But the Earth is a three-dimensional object. However, the general properties of these matrices hold in higher dimensions. For example, let us look at rotations in three dimensions. The familiar two-dimensional counterclockwise rotation in the (x, y) plane by an angle θ_z about the z axis can be written as

$$\begin{pmatrix} x' \\ y' \\ z' \end{pmatrix} = \begin{pmatrix} \cos(\theta_z) & -\sin(\theta_z) & 0 \\ \sin(\theta_z) & \cos(\theta_z) & 0 \\ 0 & 0 & 1 \end{pmatrix} = \mathbf{R}_z(\theta_z)\mathbf{x} \tag{4.62}$$

because the z coordinates of points will not change under such a rotation. The matrix $\mathbf{R}_z(\theta_z)$ is an orthogonal matrix and so has all the properties that we have discussed. Similarly, rotation matrices representing counterclockwise rotations about the x and y axes are

$$\mathbf{R}_x(\theta_x) = \begin{pmatrix} 1 & 0 & 0 \\ 0 & \cos(\theta_x) & -\sin(\theta_x) \\ 0 & \sin(\theta_x) & \cos(\theta_x) \end{pmatrix}, \qquad \mathbf{R}_y(\theta_y) = \begin{pmatrix} \cos(\theta_y) & 0 & \sin(\theta_y) \\ 0 & 1 & \\ -\sin(\theta_y) & 0 & \cos(\theta_y) \end{pmatrix}. \tag{4.63}$$

We can use these matrices successively to describe complicated rotations in three-dimensional space, but we need to take care that we perform the matrix multiplication in the correct order.

Exercise 4.4.24 Show that the rotation matrices $\mathbf{R}_x(\theta_x)$, $\mathbf{R}_y(\theta_y)$, and $\mathbf{R}_z(\theta_z)$ are orthogonal.

Rotation matrices are useful for visualizing and parameterizing motions on the surface of the Earth, and this is because of a theorem called *Euler's theorem*.[9] The gist of the theorem is that any motion of a rigid body (such as a plate of the Earth's crust) on the surface of a sphere can be described as a rotation about some axis that passes through the center of the sphere, though we will have to wait a little before we can prove this theorem.

4.5 Solving Linear Equations with Matrices

We know how to solve linear systems of algebraic equations such as

$$3x + 2y = 16, \qquad x + y = 6, \tag{4.64}$$

and solving such systems is a common task in science. For example, stoichiometric equations arise when balancing chemical equations or when determining the composition of phytoplankton communities from their pigment abundances (Mackey et al., 1996). As we will see in Chapter 10, the numerical solution the equations used to describe the motions of fluids in the environment also involves solving very large systems of linear equations. In all these cases, matrices provide a very powerful tool for helping us analyze and solve these systems.

To see how this works, let us return to Equation (4.64). We can write this as a matrix equation:

$$\begin{pmatrix} 3 & 2 \\ 1 & 1 \end{pmatrix} \begin{pmatrix} x \\ y \end{pmatrix} = \begin{pmatrix} 16 \\ 6 \end{pmatrix}$$

or

$$\mathbf{AX} = \mathbf{B}, \quad \text{where} \quad \mathbf{A} = \begin{pmatrix} 3 & 2 \\ 1 & 1 \end{pmatrix}, \quad \mathbf{X} = \begin{pmatrix} x \\ y \end{pmatrix}, \quad \mathbf{B} = \begin{pmatrix} 16 \\ 6 \end{pmatrix}.$$

[9] This is sometimes referred to as Euler's rotation theorem and is named after Leonhard Euler (1707–1783), one of the greatest mathematicians of his time.

If we multiply both sides of the equation on the left by the inverse matrix $\mathbf{A}^{-1}$, we get

$$\begin{aligned}\mathbf{A}^{-1}\mathbf{A}\mathbf{X} &= \mathbf{A}^{-1}\mathbf{B}\\ \mathbf{I}\mathbf{X} &= \mathbf{A}^{-1}\mathbf{B}\\ \mathbf{X} &= \mathbf{A}^{-1}\mathbf{B}.\end{aligned}$$

So, to solve the system of equations we only need to calculate the inverse of **A** and multiply it by **B**.

Exercise 4.5.1 Show that

$$\mathbf{A}^{-1} = \begin{pmatrix} 1 & -2 \\ -1 & 3 \end{pmatrix}$$

and hence that $x = 4$, $y = 2$ is a solution to Equation (4.64).

We can extend this to any number of linear equations. For n equations in n variables, we will have to find the inverse of an $(n \times n)$ matrix, and as we have seen (Section 4.4.3), this involves calculating the determinant. For a (2×2) or (3×3) matrix, calculating the determinant is not too much bother, but as the size of the matrix increases it becomes more and more tedious and the likelihood of our making a mistake increases. So, let us look at a very practical alternative method for solving systems of linear equations.

As an example, we will consider the following three-dimensional system of linear equations

$$x + y + z = 6, \tag{4.65a}$$

$$2x + 3y - z = 5, \tag{4.65b}$$

$$5x - y + 2z = 9, \tag{4.65c}$$

which we can also write as a matrix equation $\mathbf{Ax} = \mathbf{b}$, where

$$\mathbf{A} = \begin{pmatrix} 1 & 1 & 1 \\ 2 & 3 & -1 \\ 5 & -1 & 2 \end{pmatrix} \quad \mathbf{x} = \begin{pmatrix} x \\ y \\ z \end{pmatrix} \quad \mathbf{b} = \begin{pmatrix} 6 \\ 5 \\ 9 \end{pmatrix}. \tag{4.66}$$

We will solve these equations by rearranging them such that the first equation has all three variables (x, y, and z), the second has only two (y and z), and the last equation has only one (z). We will then automatically have a solution for z, and we can use *back substitute* to find the values of x, y, and z. First, we make sure that the equations are arranged such that the x term is first, the y term second, and the z term third in each equation (the order does not matter, so long as each equation is ordered in the same way). Next, we eliminate the x term from the second and third equations by multiplying Equation (4.65a) by 2 and subtracting it from Equation (4.65b), and then subtracting 3 multiplied by Equation (4.65a) from Equation (4.65c), giving

$$x + y + z = 6, \tag{4.67a}$$

$$y - 3z = -7, \tag{4.67b}$$

$$-6y - 3z = -21. \tag{4.67c}$$

Next, we use Equation (4.67b) to eliminate the y term from Equation (4.67c), giving

$$x + y + z = 6, \tag{4.68a}$$

$$y - 3z = -7, \tag{4.68b}$$

$$-21z = -63. \tag{4.68c}$$

The equations are now in the correct form, and we can immediately see that $z = 3$, and by backsubstituting this value into Equation (4.68b) we find that $y = 2$, and finally by substituting for z and y in Equation (4.68b) we find that $x = 1$. This algorithm that we have just followed is called *Gaussian elimination*,[10] and it provides a very convenient and easy way to numerically solve systems of linear equations.

What we have done is use a series of multiplications, additions, and subtractions acting on equations to change the form of the equations without changing the solution. If we think of this in terms of the matrices in Equation (4.66), we have manipulated the rows of the matrices to change $\mathbf{A}$ and $\mathbf{b}$ into the matrices $\mathbf{A}'\mathbf{x} = \mathbf{b}'$, where

$$\mathbf{A}' = \begin{pmatrix} 1 & 1 & 1 \\ 0 & 1 & -3 \\ 0 & 0 & -21 \end{pmatrix} \qquad \mathbf{x} = \begin{pmatrix} x \\ y \\ z \end{pmatrix} \qquad \mathbf{b}' = \begin{pmatrix} 6 \\ -7 \\ -63 \end{pmatrix}. \tag{4.69}$$

The matrix $\mathbf{A}'$ is said to be in *upper triangle form* — all entries below the main diagonal of the matrix are zero. So, the aim of Gaussian elimination is to manipulate the rows of $\mathbf{A}$ with the aim of writing it in upper triangle form because from that we can use back substitution to find the solution to the equations.

We can make this procedure more algorithmic, and so more suitable for use with a computer, by working with the *augmented matrix*. This is a matrix formed by simply adding $\mathbf{b}$ as an additional column to the matrix $\mathbf{A}$; but to denote that it is not really part of the original matrix, we use a vertical line to separate $\mathbf{b}$ from $\mathbf{A}$:

$$\tilde{\mathbf{A}} = [\mathbf{A} \mid \mathbf{b}] = \left(\begin{array}{ccc|c} 1 & 1 & 1 & 6 \\ 2 & 3 & -1 & 5 \\ 5 & -1 & 2 & 9 \end{array}\right). \tag{4.70}$$

Then, performing precisely the same row operations as we did with the equations, we end up with an augmented matrix that looks like

$$\left(\begin{array}{ccc|c} 1 & 1 & 1 & 6 \\ 0 & 1 & -3 & -7 \\ 0 & 0 & -21 & -63 \end{array}\right). \tag{4.71}$$

It is important to realize that whatever we do to the elements of the (3×3) matrix, we do the same thing to the added column in the augmented matrix. We now need to translate the operations we used to get Equation (4.68) from Equation (4.65) into a set of rules for manipulating the matrix $\tilde{\mathbf{A}}$. These rules are called *row operations* and, like the operations we used above, they do not change the solution of the equations:

[10] This technique is named after the German mathematician Johann Carl Friedrich Gauss (1777–1855) who worked in many areas of mathematics and is one of the most influential mathematicians of all time. His name will appear several times in this book.

1. We can swap whole rows in $\tilde{\mathbf{A}}$ (including the part that comes from **b**). This just changes the order of the equations, so will not affect the solution.
2. We can multiply a row by a constant. This is the same as multiplying an equation, e.g., $2x + 3y = 4$, by a constant such as 6, giving $12x + 18y = 24$; it does not affect the solution of the equation.
3. We can add a multiple of one row to another row. The new row is now a linear combination of the two rows, so also does not affect the solution of the system of equations.

Our aim now is to systematically use these operations to get the **A** part of the augmented matrix into a form that corresponds to Equation (4.68). In other words, if we started with an augmented matrix formed from **A** and **b** from Equation (4.66), we want to end up with an augmented matrix formed from **A′** and **b′** in Equation (4.69). Let us see how we get there. From the augmented matrix in Equation (4.70) we first want to use the row operations to make all the elements below the element $\tilde{a}_{11}$ equal to zero—this element is called the *pivot*. The first element in the second row can be made zero by subtracting twice the first row from it (we will write such an operation as $R^{(2)} - 2R^{(1)}$, where $R^{(k)}$ represents the kth row), giving

$$\left(\begin{array}{ccc|c} 1 & 1 & 1 & 6 \\ 0 & 1 & -3 & -7 \\ 5 & -1 & 2 & 9 \end{array}\right), \qquad (R^{(2)} - 2R^{(1)})$$

and similarly using the row operation $R^{(3)} - 5R^{(1)}$ gives

$$\left(\begin{array}{ccc|c} 1 & 1 & 1 & 6 \\ 0 & 1 & -3 & -7 \\ 0 & -6 & -3 & -21 \end{array}\right). \qquad (R^{(3)} - 5R^{(1)})$$

Now, the element $\tilde{a}_{22}$ becomes the pivot, and we want to make all the elements that lie below $\tilde{a}_{22}$ in that column equal to zero. This can be done with the operation $R^{(3)} + 6R^{(2)}$ to give

$$\left(\begin{array}{ccc|c} 1 & 1 & 1 & 6 \\ 0 & 1 & -3 & -7 \\ 0 & 0 & -21 & -63 \end{array}\right), \qquad (R^{(3)} + 6R^{(2)}) \tag{4.72}$$

from which we can use backsubstitution as before to solve the system of equations. This final form of the augmented matrix is called *row echelon form*. What happens if a pivot is zero before we apply a row operation? Let us look at this by modifying our linear system of equations to

$$y + z = 6, \qquad 2x + 3y - z = 5, \qquad 5x - y + 2z = 9,$$

so that our initial augmented matrix becomes

$$\left(\begin{array}{ccc|c} 0 & 1 & 1 & 6 \\ 2 & 3 & -1 & 5 \\ 5 & -1 & 2 & 9 \end{array}\right).$$

If we were to proceed as before you can see we quickly run into problems; we can eliminate either the 2 or the 5 in the first column, but not both, so we cannot get the augmented matrix into row echelon form. The way around this is simple, we swap two rows such that the top row has a nonzero element in the first column. For example, we could swap the first and third rows to get the augmented matrix

$$\left(\begin{array}{ccc|c} 5 & -1 & 2 & 9 \\ 2 & 3 & -1 & 5 \\ 0 & 1 & 1 & 6 \end{array}\right),$$

and use row operations to put this in row echelon form.

What happens if we arrive at an augmented matrix in row echelon form only to find that one of the rows produces an inconsistency? For example, if we ended up with a row echelon equation in the form

$$\left(\begin{array}{ccc|c} 2 & -1 & 1 & 3 \\ 0 & 1 & -1 & 5 \\ 0 & 0 & 0 & 6 \end{array}\right),$$

and then tried to use back substitution, we would find $0 = 6$, which is a contradiction. If this happens, then Gaussian elimination is telling us that there is no solution to the equations. Another possibility is that we end up with a matrix of the form

$$\left(\begin{array}{ccc|c} 2 & -1 & 1 & 3 \\ 0 & 1 & -1 & 5 \\ 0 & 0 & 0 & 0 \end{array}\right),$$

which would tell us that $0 = 0$. This is not a contradiction, but in this case Gaussian elimination is telling us that this system of equations does not give a value for the third variable (e.g., z). However, we can use the other two equations to write the first two variables (e.g., x and y) in terms of the third, unspecified variable; i.e., we have a parameterized solution. There are an infinite number of possible solutions, one for each value of the third variable.

Exercise 4.5.2 Use Gaussian elimination to solve the following equations or show that a solution does not exist.

1. $y - z = 3, \quad 2x - 3y + 4z = 1, \quad x + 2y - z = 0.$
2. $2x + 3y - z = 5, \quad 4x - y + 2z = 8, \quad 6x + 2y + z = 15.$
3. $2x - 4z = 2, \quad y + 3z = 2, \quad 2x + y - z = 4.$

Whether or not a system of equations has a unique solution, an infinite number of solutions, or no solution can be determined by comparing the *rank* of the matrix representing the coefficients in the equations ($\mathbf{A}$) with the rank of the augmented matrix ($\mathbf{A}|\mathbf{b}$). To calculate the rank of an $(m \times n)$ matrix we first put it into row echelon form. The rank is then the number of rows that have at least one element that is not zero. If we are solving m equations in n unknowns, then the three possibilities are:

1. If the rank of $\mathbf{A}$ equals the rank of $\mathbf{A}|\mathbf{b}$ and both are equal to the number of unknowns (n) in the equations, then there is a unique solution to the equations.

2. If the rank of $\mathbf{A}$ equals the rank of $\mathbf{A}|\mathbf{b}$ but is less than the number of unknowns (n) in the equations, then there is an infinite number of solutions to the equations.
3. If the rank of $\mathbf{A}$ is less than the rank of $\mathbf{A}|\mathbf{b}$, then there is no solution to the equations.

Exercise 4.5.3 Calculate the ranks of the following matrices by putting them in row echelon form.

$$a. \begin{pmatrix} -1 & 2 & 3 \\ -5 & 2 & -3 \\ 3 & -1 & -4 \end{pmatrix} \quad b. \begin{pmatrix} -1 & 2 & 3 & 4 \\ -5 & 2 & 15 & 1 \\ 3 & -1 & 12 & 4 \end{pmatrix} \quad c. \begin{pmatrix} 1 & 1 & 1 & 6 \\ 2 & 3 & -1 & 5 \\ 5 & -1 & 2 & 9 \end{pmatrix}$$

Now, we can actually take Gaussian elimination one step further. In our examples so far, our final matrices had zeros below the main diagonal. We can continue using row operations to reduce this matrix to one where the only nonzero elements are the ones on that diagonal and all the values on the diagonal are 1. A matrix in this form is in *reduced row echelon form*, and the procedure to get there is called *Gauss–Jordan elimination.*[11] To see this in action, let us start with the matrix Equation (4.72) which is already in row echelon form. To put this in reduced row echelon form we start with the element in the bottom right-hand corner of the (3×3) matrix and make its value 1 (by multiplying the row by $-1/21$)

$$\left(\begin{array}{ccc|c} 1 & 1 & 1 & 6 \\ 0 & 1 & -3 & -7 \\ 0 & 0 & 1 & 3 \end{array}\right). \qquad \left(\mathrm{R}^{(3)} \times \frac{-1}{21}\right)$$

Now, we use that element as the pivot and use the third row to make the values of the elements above it all zero. We can do this using the row operation $\mathrm{R}^{(2)} + 3\mathrm{R}^{(3)}$ followed by $\mathrm{R}^{(1)} - \mathrm{R}^{(3)}$ to give

$$\left(\begin{array}{ccc|c} 1 & 1 & 0 & 3 \\ 0 & 1 & 0 & 2 \\ 0 & 0 & 1 & 3 \end{array}\right).$$

Finally, we make the middle element of the second row (the element on the diagonal) the pivot and use it to set all elements in the column above it to zero using the row operation $\mathrm{R}^{(1)} - \mathrm{R}^{(2)}$ to give

$$\left(\begin{array}{ccc|c} 1 & 0 & 0 & 1 \\ 0 & 1 & 0 & 2 \\ 0 & 0 & 1 & 3 \end{array}\right).$$

We can see that the values in the end column are now the solution to the original equations! So, by continuing row operations and putting the matrix in reduced row echelon form, we have found the solution we need without using back substitution.

It is useful to notice that when we add a multiple of one row (say row $\mathrm{R}^{(i)}$) to another row (row $\mathrm{R}^{(j)}$) to make an element zero, what we are doing is subtracting from each element in row $\mathrm{R}^{(j)}$ the quantity (a_{ji}/a_{ii}) multiplied by the corresponding element of row $\mathrm{R}^{(i)}$.

[11] This is named for Gauss (who we have already met) and Wilhelm Jordan (1842–1899) who studied geodesy, the shape and gravitational field of the Earth (Althoen and McLaughlin, 1987).

This means that the process of Gauss–Jordan elimination can easily be put into a computer program.

Exercise 4.5.4 Use Gauss–Jordan elimination to find a solution to the following equations:

$$2x + 3y + 2z = 0, \quad x - y + 2z = 1, \quad 3x - 2y - 4z = 2.$$

Gauss and Jordan have one more useful trick up their sleeves: we can also use Gauss–Jordan elimination to find the inverse of a matrix. To see why this is so, let us rethink what we are doing to a matrix when we use Gauss–Jordan elimination. Each of the three row operations we have been using can be performed by multiplying the matrix with another matrix of the appropriate form. For example, if we want to exchange the first and second rows of an arbitrary (3×3) matrix, then we can use the following multiplication:

$$\begin{pmatrix} 0 & 1 & 0 \\ 1 & 0 & 0 \\ 0 & 0 & 1 \end{pmatrix} \begin{pmatrix} a & b & c \\ d & e & f \\ g & h & i \end{pmatrix} = \begin{pmatrix} d & e & f \\ a & b & c \\ g & h & i \end{pmatrix}.$$

Exercise 4.5.5 Consider the general (3×3) matrix

$$\begin{pmatrix} a & b & c \\ d & e & f \\ g & h & i \end{pmatrix}.$$

What matrix would you have to multiply this with to perform the row operation $R^{(2)} - 3R^{(3)}$?

To get an augmented matrix into reduced row echelon form basically means that we have transformed the original matrix (**A**) into the unit matrix; each operation performed on **A** is also performed on the augmented part of the augmented matrix, the column vector **b**. So, a matrix that is in reduced row echelon form is equivalent to the identity matrix, and if it takes k row operations to get a matrix (**A**) into reduced row echelon form, then we have successively multiplied **A** by k transformation matrices, **M**, to get the unit matrix. In other words,

$$\mathbf{M}_k\mathbf{M}_{k-1}\mathbf{M}_{k-2}\cdots\mathbf{M}_2\mathbf{M}_1\mathbf{A} = \mathbf{I}.$$

But we know that $\mathbf{A}^{-1}\mathbf{A} = \mathbf{I}$, so $\mathbf{M}_k\mathbf{M}_{k-1}\mathbf{M}_{k-2}\cdots\mathbf{M}_2\mathbf{M}_1\mathbf{I} = \mathbf{A}^{-1}$. This is telling us that to calculate the inverse of **A** we need to apply the same transformations to the unit matrix that we make to the matrix **A**.

Example 4.18 Let us use Gauss–Jordan elimination to find the inverse of the matrix

$$\mathbf{A} = \begin{pmatrix} 1 & 1 & 0 \\ 0 & 1 & -1 \\ 1 & 2 & 1 \end{pmatrix}.$$

We will write **A** and **I** in two columns, applying the same transformation to each.

$$\mathrm{R}^{(3)} \to \mathrm{R}^{(3)} - \mathrm{R}^{(1)} \begin{pmatrix} 1 & 1 & 0 \\ 0 & 1 & -1 \\ 0 & 1 & 1 \end{pmatrix} \begin{pmatrix} 1 & 0 & 0 \\ 0 & 1 & 0 \\ -1 & 0 & 1 \end{pmatrix}$$

$$\mathrm{R}^{(3)} \to \mathrm{R}^{(3)} - \mathrm{R}^{(2)} \begin{pmatrix} 1 & 1 & 0 \\ 0 & 1 & -1 \\ 0 & 0 & 2 \end{pmatrix} \begin{pmatrix} 1 & 0 & 0 \\ 0 & 1 & 0 \\ -1 & -1 & 1 \end{pmatrix}$$

$$\mathrm{R}^{(3)} \to \frac{1}{2}\mathrm{R}^{(3)} \begin{pmatrix} 1 & 1 & 0 \\ 0 & 1 & -1 \\ 0 & 0 & 1 \end{pmatrix} \begin{pmatrix} 1 & 0 & 0 \\ 0 & 1 & 0 \\ -1/2 & -1/2 & 1/2 \end{pmatrix}$$

$$\mathrm{R}^{(2)} \to \mathrm{R}^{(2)} + \mathrm{R}^{(3)} \begin{pmatrix} 1 & 1 & 0 \\ 0 & 1 & 0 \\ 0 & 0 & 1 \end{pmatrix} \begin{pmatrix} 1 & 0 & 0 \\ -1/2 & 1/2 & 1/2 \\ -1/2 & -1/2 & 1/2 \end{pmatrix}$$

$$\mathrm{R}^{(1)} \to \mathrm{R}^{(1)} - \mathrm{R}^{(2)} \begin{pmatrix} 1 & 0 & 0 \\ 0 & 1 & 0 \\ 0 & 0 & 1 \end{pmatrix} \begin{pmatrix} 3/2 & -1/2 & -1/2 \\ -1/2 & 1/2 & 1/2 \\ -1/2 & -1/2 & 1/2 \end{pmatrix}$$

Gauss–Jordan elimination is an exceedingly good way to calculate inverse matrices and to solve systems of linear algebraic equations. This is because it is easily written as an algorithm, making it an ideal candidate for a computer program, and as we will see, it is also very efficient in the number of steps it needs to arrive at an answer.

4.5.1 Determinants

We met determinants when we calculated a vector product (Equation (4.42)), the scalar triple product (Equation (4.44)), and in calculating the inverse of a matrix (Equation (4.57)). It is now time that we looked at them in more detail. Determinants are only defined for square matrices, and for a (2×2) matrix **A**, the determinant can be written using different notations as

$$\det(\mathbf{A}) = \det \begin{pmatrix} a & b \\ c & d \end{pmatrix} = \begin{vmatrix} a & b \\ c & d \end{vmatrix} = ad - bc. \tag{4.73}$$

Calculating a determinant becomes a little more complicated for a (3×3) matrix:

$$\det \begin{pmatrix} a & b & c \\ d & e & f \\ g & h & i \end{pmatrix} = a \begin{vmatrix} e & f \\ h & i \end{vmatrix} - b \begin{vmatrix} d & f \\ g & i \end{vmatrix} + c \begin{vmatrix} d & e \\ g & h \end{vmatrix}. \tag{4.74}$$

Each element of the (3×3) matrix has an associated *minor*, which is the determinant formed from the elements that are not in the row or column of the chosen element. So, for example, in Equation (4.74), the minor of element b is obtained by neglecting all other elements in the first row and second column. Similarly, the minor of element d would be the determinant formed by discarding all elements in the second row and first column, i.e.,

$$\text{minor associated with element } d = \mathbf{M}_{21} = \begin{vmatrix} b & c \\ h & i \end{vmatrix} = bi - hc,$$

where d is the element on the second row of the first column of the matrix, so we have given the minor the same subscripts. The *cofactor* of the element on the ith row and jth column of the matrix $\mathbf{A}$ is then defined as

$$c_{ij} = (-1)^{i+j} M_{ij}. \tag{4.75}$$

Example 4.19 In this example, we will calculate the matrix of cofactors for the following matrix:

$$\mathbf{A} = \begin{pmatrix} 1 & 2 & 1 \\ -1 & 3 & 1 \\ 2 & -1 & 1 \end{pmatrix}.$$

Working element by element along the first row we find that

$$C_{11} = (-1)^2 \begin{vmatrix} 3 & 1 \\ -1 & 1 \end{vmatrix} = 4, \quad C_{12} = (-1)^3 \begin{vmatrix} -1 & 1 \\ 2 & 1 \end{vmatrix} = 3, \quad C_{13} = (-1)^4 \begin{vmatrix} -1 & 3 \\ 2 & -1 \end{vmatrix} = -5.$$

and similarly along the second row

$$C_{21} = (-1)^3 \begin{vmatrix} 2 & 1 \\ -1 & 1 \end{vmatrix} = -3, \quad C_{22} = (-1)^4 \begin{vmatrix} 1 & 1 \\ 2 & 1 \end{vmatrix} = -1, \quad C_{23} = (-1)^5 \begin{vmatrix} 1 & 2 \\ 2 & -1 \end{vmatrix} = 5,$$

and for the last row we get $C_{31} = -1$, $C_{32} = -2$, $C_{33} = 5$. So, the matrix of cofactors is

$$\mathbf{C} = \begin{pmatrix} 4 & 3 & -5 \\ -3 & -1 & 5 \\ -1 & -2 & 5 \end{pmatrix}.$$

Now we have a means to calculate the determinant of a square matrix of any size. We first choose a single row or column of the matrix, calculate the cofactors of each element along that row or column, and sum the products of the cofactors with the corresponding element of the original matrix. So, for a (3×3) matrix $\mathbf{A}$ whose elements are a_{ij} $(i, j = 1, 2, 3)$ and whose cofactors are C_{ij}, the determinant can be calculated as

$$\det(\mathbf{A}) = a_{11}C_{11} + a_{12}C_{12} + a_{13}C_{13}, \tag{4.76}$$

where we have performed what is called a *cofactor expansion* along the first row of $\mathbf{A}$. We can also expand along any other row, or even along any column, and the value of the determinant of $\mathbf{A}$ will be the same.

Example 4.20 We can use the cofactors that we calculated in Example 4.19 to calculate the determinant of the matrix

$$\mathbf{A} = \begin{pmatrix} 1 & 2 & 1 \\ -1 & 3 & 1 \\ 2 & -1 & 1 \end{pmatrix}.$$

Using Equation (4.76), we find that

$$\det(\mathbf{A}) = (1 \times 4) + (3 \times 2) + (-5 \times 1) = 5.$$

Exercise 4.5.6 Consider the matrix **A** and its corresponding matrix of cofactors **C** from Example 4.19. Show that the value of the determinant of **A** is the same if you perform a cofactor expansion along each row and each column.

The fact that the value of the determinant remains the same no matter which row or column you choose to expand along can be very helpful. This is because specific rows or columns in the matrix **A** may contain zeros, which will reduce the number of cofactors you have to calculate.

Exercise 4.5.7 Calculate the determinant of **A** by using a cofactor expansion along the third column, where

$$\mathbf{A} = \begin{pmatrix} 2 & 1 & 4 & 1 \\ 1 & 3 & 0 & 2 \\ 0 & 2 & 1 & 4 \\ 7 & 3 & 0 & 1 \end{pmatrix}.$$

The determinant has a lot of nice properties. For example, if we add a multiple of one row of a matrix to another row, the value of the determinant of the matrix does not change. Let us demonstrate this with a (3×3) matrix. We will work with the generic matrix

$$\mathbf{A} = \begin{pmatrix} a & b & c \\ d & e & f \\ g & h & i \end{pmatrix}, \tag{4.77}$$

and we start by multiplying the third row by β and adding the result to the first row. Using a cofactor expansion along the first row, the determinant of the resulting matrix is

$$\begin{vmatrix} a + \beta g & b + \beta h & c + \beta i \\ d & e & f \\ g & h & i \end{vmatrix} = (a + \beta g)(ie - fh) - (b + \beta h)(di - fg) + (c + \beta i)(dh - eg).$$

The terms that do not contain β are $a(ie - fh) - b(di - fg) + c(dh - eg) = \det(\mathbf{A})$, and those that do contain β are $\beta g(ie - fh) - \beta h(di - fg) + \beta i(dh - eg) = 0$. So, the value of the determinant remains the same.

Exercise 4.5.8 Show that the determinant of **A** from Equation (4.77) is unchanged if you multiply column 2 by β and add it to column 3.

What happens to the value of the determinant if we interchange two rows (or columns) in the matrix $\mathbf{A}$? In this case, the determinant gets multiplied by -1. Consider the matrix in Equation (4.77) again and swap the first and third rows. The determinant is

$$\begin{vmatrix} g & h & i \\ d & e & f \\ a & b & c \end{vmatrix} = g(ei - hf) - h(di - fa) + i(db - ae)$$

$$= -(a(ie - hf) - b(di - gf) + c(dh - ge)) = -\det(\mathbf{A}).$$

Exercise 4.5.9 Using the generic (3×3) matrix $\mathbf{A}$ from Equation (4.77), confirm that the following statements are true.

1. If all the elements of any one row or column of $\mathbf{A}$ are zero, then $\det(\mathbf{A}) = 0$.
2. If all the elements of any one row or column of $\mathbf{A}$ are multiplied by a constant β to obtain a new matrix, $\mathbf{B}$, then $\det(\mathbf{B}) = \beta \det(\mathbf{A})$.
3. If all the elements of a (3×3) matrix $\mathbf{A}$ are multiplied by a constant β, then $\det(\beta\mathbf{A}) = \beta^3 \det(\mathbf{A})$.
4. $\det(\mathbf{A}^{\mathrm{T}}) = \det(\mathbf{A})$.
5. If $\mathbf{A}$ and $\mathbf{B}$ are both (3×3) matrices, then $\det(\mathbf{AB}) = \det(\mathbf{A})\det(\mathbf{B})$, but in general $\det(\mathbf{A} + \mathbf{B}) \neq \det(\mathbf{A}) + \det(\mathbf{B})$.

There are two other important properties of determinants that we need to look at because they can help us provide an interpretation for the determinant. The first property is that if any two rows (or columns) of $\mathbf{A}$ are proportional to each other, then $\det(\mathbf{A}) = 0$. We will use our generic (3×3) determinant to demonstrate this by making the first row proportional to the second so that

$$\mathbf{A} = \begin{pmatrix} a & b & c \\ \beta a & \beta b & \beta c \\ g & h & i \end{pmatrix},$$

where β is a constant. The determinant is then

$$\det(\mathbf{A}) = a(\beta bi - \beta ch) - b(\beta ai - \beta cg) + c(\beta ah - \beta bg) = 0.$$

The last property we want to demonstrate is that if any row (or column) of $\mathbf{A}$ is a linear combination of other rows (or columns), then $\det(\mathbf{A}) = 0$. Let us take the third column in $\mathbf{A}$ to be a linear combination of the first and second, then the determinant is

$$\begin{vmatrix} a & b & \alpha a + \beta b \\ d & e & \alpha d + \beta e \\ g & h & \alpha g + \beta h \end{vmatrix} = a[e(\alpha g + \beta h) - h(\alpha d + \beta e)] - b[d(\alpha g + \beta h) - g(\alpha d + \beta e)] \tag{4.78}$$

$$+ (\alpha a + \beta b)(dh - eg) = 0.$$

These last two properties suggest that the determinant is telling us something about the linear independence of vectors, and indeed it is. To see this, recall that we can write a vector as a column vector and that we can create a matrix where each column of the matrix is one of these column vectors. If the determinant of the resulting matrix is nonzero, it is

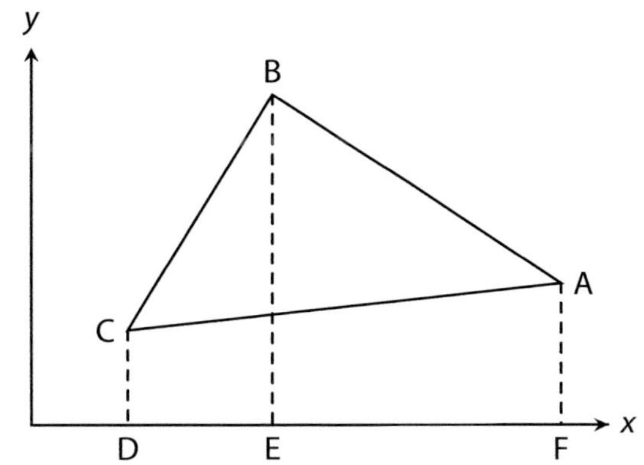

Figure 4.21 A triangle made up of three points, A (x_1, y_1), B (x_2, y_2), and C (x_3, y_3).

telling us that the vectors are all linearly independent. To see this in another way, consider two vectors $\mathbf{A} = a\hat{\mathbf{i}} + b\hat{\mathbf{j}}$ and $\mathbf{B} = c\hat{\mathbf{i}} + d\hat{\mathbf{j}}$ in two dimensions. If these vectors are linearly dependent, then the two vectors are proportional to each other, which means that they are parallel and their components differ by a constant factor. In other words,

$$\frac{a}{c} = \frac{b}{d} \implies ad - bc = 0;$$

that is, the determinant of the matrix formed from the components of the vectors is zero.

Exercise 4.5.10 Show, by calculating an appropriate determinant, that the vectors $\mathbf{a} = 3\hat{\mathbf{i}} - 2\hat{\mathbf{j}} + \hat{\mathbf{k}}$, $\mathbf{b} = \hat{\mathbf{i}} + 2\hat{\mathbf{j}} - \hat{\mathbf{k}}$, and $\mathbf{c} = -\hat{\mathbf{i}} - \hat{\mathbf{j}} - \hat{\mathbf{k}}$ are linearly independent.

Let us look a little bit more closely at the relationship between geometry and the determinant by calculating the area of a triangle (Figure 4.21). The area of the triangle is made up of the area of the three trapezoids ABEF, BCDE, and ACDF. If we work around the vertices starting from A and moving counterclockwise, then

$$\begin{aligned}\text{triangle area} &= \text{ABEF} + \text{BCDE} - \text{ACDF} \\ &= \frac{1}{2}\left((y_1 + y_2)(x_1 - x_2) + (y_2 + y_3)(x_2 - x_3) - (y_1 + y_3)(x1 - x_3)\right) \\ &= \frac{1}{2}(x_1y_2 - x_2y_1 + x_2y_3 - x_3y_2 - x_1y_3 + x_3y_1) \\ &= \frac{1}{2}\begin{vmatrix} x_1 & y_1 & 1 \\ x_2 & y_2 & 1 \\ x_3 & y_3 & 1 \end{vmatrix}.\end{aligned}$$

So, the determinant in two dimensions is the area of a triangle; in three dimensions, one can show in a similar manner than the determinant is the volume of a parallelepiped. Notice that the properties of the determinant tell us that if we swapped the order of two of the vertices (i.e., we went around the vertices of the triangle in the opposite direction), we would have obtained the negative of the above determinant. So, not only is the determinant the area of the triangle, it is a *signed area*. This should be familiar because we found that the vector product of two vectors can be calculated using determinants and can be interpreted as an area but with a direction (Section 4.3.2).

Exercise 4.5.11 Consider two vectors **A** and **B** in two dimensions. Show that the value of the determinant of the matrix whose columns are the vectors **A** and **B** is the area of the parallelogram formed when adding the two vectors.

This property of the determinant can be extended to three dimensions where, instead of representing an area, the determinant gives us the volume of a parallelepiped defined by three vectors. This should not be too surprising because we know from Section 4.3.3 that the value of the scalar triple product of three vectors can be calculated using a determinant, and the scalar triple product represents a volume.

The determinant is very useful for determining how an area or volume element changes under a coordinate transformation. We know that if we change coordinates from x to u in one dimension, then the line element dx changes according to the equation

$$dx = \frac{\partial x}{\partial u} du,$$

and we can use this to change variables in an integral, for example. What happens in two dimensions, where in general the new coordinates might not even be orthogonal (Figure 4.22)? Consider a coordinate change from rectangular Cartesian coordinates (x, y) to a new set of coordinates (u, v). We want to know what happens to vectors that define a small differential area $dxdy$. In the new coordinates, this area will be $dudv$, but what is the relationship between these two areas? Because we have chosen to look at vectors that lie along the coordinate lines, we can consider what happens to the components of a vector under a coordinate change in which we keep v constant and vary only u, or a coordinate change in which we keep u constant and vary v. In these cases, the vectors in the (u, v) coordinates have components

$$\left(\frac{\partial x}{\partial u} du, \frac{\partial y}{\partial u} du\right) \quad \text{and} \quad \left(\frac{\partial x}{\partial v} dv, \frac{\partial y}{\partial v} dv\right). \tag{4.79}$$

We take the vector product of these to find the area of the parallelogram defined by them, giving

$$dxdy = \left(\frac{\partial x}{\partial u} du, \frac{\partial y}{\partial u} du\right) \times \left(\frac{\partial x}{\partial v} dv, \frac{\partial y}{\partial v} dv\right) = \left(\frac{\partial x}{\partial u}\frac{\partial y}{\partial v} - \frac{\partial x}{\partial v}\frac{\partial y}{\partial u}\right) dudv.$$

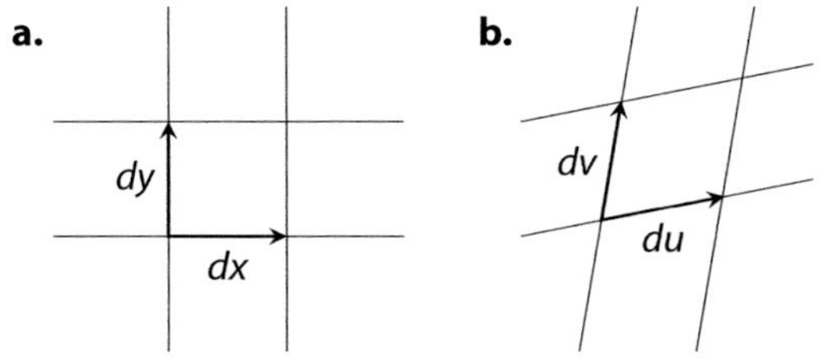

Figure 4.22 The area element $dxdy$ in Cartesian rectangular coordinates (a.) changes shape under a coordinate transformation to coordinates u and v which may not be orthogonal (b.).

This equation tells us how the area changes under the coordinate transformation, and we recognize the factor in front of the $dudv$ term as the determinant in two dimensions of the components of the vectors in Equation (4.79). We can extend this argument to three dimensions, but instead of taking the vector product to find the area of the parallelogram, we have to take the scalar triple product (Equation (4.44)) to calculate the area of the parallelepiped formed from the three vectors. If we transform the coordinates from (x_1, x_2, x_3) to (ξ_1, ξ_2, ξ_3), then the volume element transforms according to

$$dx_1dx_2dx_3 = J(\xi_1, \xi_2, \xi_3)d\xi_1d\xi_2d\xi_3 = \frac{\partial(x_1, x_2, x_3)}{\partial(\xi_1, \xi_2, \xi_3)}d\xi_1d\xi_2d\xi_3$$

$$= \begin{vmatrix} \frac{\partial x_1}{\partial \xi_1} & \frac{\partial x_1}{\partial \xi_2} & \frac{\partial x_1}{\partial \xi_3} \\ \frac{\partial x_2}{\partial \xi_1} & \frac{\partial x_2}{\partial \xi_2} & \frac{\partial x_2}{\partial \xi_3} \\ \frac{\partial x_3}{\partial \xi_1} & \frac{\partial x_3}{\partial \xi_2} & \frac{\partial x_3}{\partial \xi_3} \end{vmatrix} d\xi_1d\xi_2d\xi_3. \quad (4.80)$$

The determinant in Equation (4.80) is called the *Jacobian*.[12] We can use Equation (4.80) as a general means of transforming coordinates when we calculate two- and three-dimensional integrals.

Exercise 4.5.12 Calculate the Jacobian for the transformation between rectangular Cartesian coordinates in two dimensions and polar coordinates in two dimensions. Use this to calculate the integral of $f(x, y) = x^2 + y^2$ over the area of a disk of radius 1.

To recap, we have seen that the determinant of a matrix tells us something about the linear dependence of the columns (or rows) of the matrix, that a determinant in two dimensions is an area and in three dimensions is a volume, and that a matrix that has a determinant of zero has no inverse and is a singular matrix. We can see how this fits in with the solution of sets of linear simultaneous equations. In a linear system of two equations,

$$ax + by = e, \qquad cx + dy = f, \quad (4.81)$$

each equation represents a straight line in the (x, y) plane, and there are three possibilities for the solution of the equations (Figure 4.23).

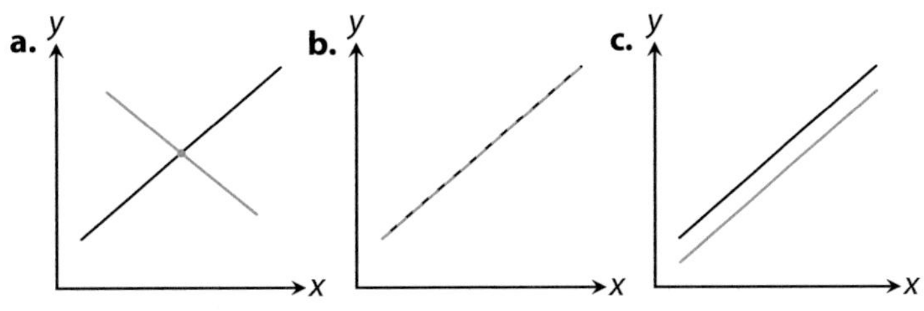

Figure 4.23 Three cases for the solution of Equation (4.81): a unique solution where the lines intersect each other (a.), the two lines are coincident, in which case there is an infinite number of possible solutions (b.), the two lines are parallel and not coincident and there is no solution (c.).

[12] Named after the German mathematician Carl Gustav Jacob Jacobi (1804–1851).

1. There is a unique solution, in which case the two straight lines represented by the equations intersect at a unique point (Figure 4.23a), and $a/c \neq b/d$, so the slopes of the lines are different. For example,

$$x + 2y = 5, \qquad 2x + 2y = 13.$$

2. There exists an infinite number solution, which occurs when the two straight lines are in fact the same, so the lines coincide (Figure 4.23b). For example, the two straight lines

$$x + y = 5, \qquad 2x + 2y = 10,$$

which are really the same (the second equation is simply twice the first equation) and in this case $a/c = b/d = e/f$.

3. No solution exists, in which case the two lines are parallel to each other but not coincident (Figure 4.23c). In other words, the slopes of the lines are constant, but their intercepts on the y axis are different. For example,

$$x + y = 5, \qquad 2x + 2y = 13.$$

In this case $a/c = b/d \neq e/f$.

We can see that what distinguishes the first case from the other two is that $a/c \neq b/d$, or that $ad - bc \neq 0$. The determinant is therefore a nice tool that allows us to determine if a unique solution to a set of linear simultaneous equations exists or not. If we write Equations (4.81) as a matrix equation $\mathbf{Ax} = \mathbf{B}$, where

$$\mathbf{A} = \begin{pmatrix} a & b \\ c & d \end{pmatrix}, \quad \mathbf{x} = \begin{pmatrix} x \\ y \end{pmatrix}, \quad \mathbf{B} = \begin{pmatrix} e \\ f, \end{pmatrix},$$

we have seen that we can think of the columns of $\mathbf{A}$ as being two column vectors

$$\begin{pmatrix} a \\ c \end{pmatrix} \quad \text{and} \quad \begin{pmatrix} b \\ d \end{pmatrix}.$$

The determinant is then telling us if these two vectors are linearly independent or not. In the cases shown in Figure 4.23b and Figure 4.23c, the two vectors are the same, so they are obviously not linearly independent. In higher dimensions, however, it becomes harder to tell if solutions exist just by looking at the equations.

If Equations (4.81) are a set of *homogeneous* equations, then the right-hand sides of the equations are zero and the equations are

$$ax + by = 0, \qquad cx + dy = 0.$$

These equations represent straight lines that pass through the origin, so there will always exist the solution $x = y = 0$. If the slopes of the lines are the same, then the lines are identical and the determinant of the matrix $\mathbf{A}$ is zero; if the determinant is not zero, then the slopes of the lines are different and the origin is the unique solution. This result holds for linear homogeneous equations with any number of variables.

We have seen two methods we can use to calculate the inverse of a matrix. We can use Gauss–Jordan elimination, or we can use a cofactor expansion of the matrix to calculate the determinant and use this to calculate the inverse matrix (e.g., Equation (4.57)). For small (e.g., (2×2) or (3×3) matrices), there is not too much difference in the computational effort involved in either method. However, we may often need to calculate the inverse of large (e.g., (10000×10000) or larger!) matrices. In these cases, **never** use a cofactor expansion. To see why, let us estimate the number of operations it takes to invert a matrix scaled with the size of the matrix for these different methods.

When we perform a Gaussian elimination, we use a combination of multiplications, additions, and subtractions to put the matrix into the correct form. Let us estimate the computational effort required to invert an $(n \times n)$ matrix. Starting with the first row of the matrix, we need to find out what multiple of the first row is needed to make the entry in the first column of the next row have a value of zero; that involves performing one division. Then, we need to multiply each element of the first row by this factor, and subtract it from each element in the second row; this will involve approximately n calculations (it will take at most n multiplications and n subtractions, but we are estimating here, so we are not worried about dropping a factor of 2). But, we need to apply this to $(n-1)$ rows, so we will need approximately $(n+1)(n-1) = n^2 - 1$ operations. We repeat this for the second row, but we do not perform any more calculations on the first row. So, we will need $(n-1)^1 - 1$ operations. For the third row, we will need approximately $(n-2)^2 - 1$ operations, and so on. In total, we will need

$$\sum_{k=1}^{n}(k^2 - 1) = \left(\sum_{k=1}^{n} k^2\right) - n = \frac{n(n-1)(2n-1)}{6} - n.$$

For large values of n, the number of operations will scale as n^3. Using a similar argument, the number of operations required for a cofactor expansion scales as n^5. So, for large matrices, using a cofactor expansion to calculate the determinant requires many more calculations and is a very inefficient method for obtaining the inverse of a matrix.

4.6 Kronecker Delta and Levi-Civita Symbol

The *Kronecker delta*[13] and *Levi-Civita symbol*[14] are tools that can help simplify calculations using the elements of matrices.

The Kronecker delta is defined by the equation

$$\delta_{ij} = \begin{cases} 1 & \text{when } i = j \\ 0 & \text{when } i \neq j \end{cases} \tag{4.82}$$

and is basically a way of singling out the diagonal elements of a matrix. As an example, we can write the scalar product of vectors $\mathbf{A} = a_1\hat{\mathbf{e}}_1 + a_2\hat{\mathbf{e}}_2 + a_3\hat{\mathbf{e}}_3$ and $\mathbf{B} = b_1\hat{\mathbf{e}}_1 + b_2\hat{\mathbf{e}}_2 + b_3\hat{\mathbf{e}}_3$ as

$$\mathbf{A}\cdot\mathbf{B} = a_1b_1 + a_2b_2 + a_3b_3 = \sum_{i=1}^{3} a_ib_i = \sum_{i=1}^{3}\sum_{j=1}^{3} a_ib_j\delta_{ij}$$

[13] Named for the German mathematician Leopold Kronecker (1823–1891).
[14] Named after Tullio Levi-Civita (1873–1941), an Italian mathematician and physicist.

because only those terms in the summation with $i = j$ remain. So, using the Kronecker delta we can write the equations for the scalar product of a set of orthonormal basis vectors as $\hat{\mathbf{e}}_i \cdot \hat{\mathbf{e}}_j = \delta_{ij}$.

If we are doing calculations with matrices, then the Kronecker delta acts as the unit matrix. For example, if we have an $(n \times n)$ matrix $\mathbf{A}$, then

$$\mathbf{AI} = \sum_{k=1}^{n} a_{ik}\delta_{kj},$$

so that, for a (3×3) matrix,

$$a_{12} = a_{11}\delta_{12} + a_{12}\delta_{22} + a_{13}\delta_{32} = a_{11} \times 0 + a_{12} \times 1 + a_{13} \times 0 = a_{12}.$$

Exercise 4.6.1 What is the value of $\sum_{i=1}^{3} \delta_{ik}\delta_{kj}$?

You will frequently see the *Einstein summation convention* being used to simplify the writing of equations involving indices. This convention states that repeated indices in an expression are summed over. Thus δ_{ii} is shorthand for

$$\delta_{ii} = \sum_{i=1}^{n}\sum_{j=1}^{n} \delta_{ij} = (\delta_{11} + \delta_{12} + \cdots + \delta_{1n}) + (\delta_{21} + \delta_{21} + \cdots$$
$$+ \delta_{2n}) + \cdots + (\delta_{n1} + \delta_{n2} + \cdots + \delta_{nn}) = n.$$

Exercise 4.6.2 If x_i are the components of an n-dimensional vector, what is $x_i x_j \delta_{ij}$?
Exercise 4.6.3 Show that $\delta_{ij}\delta_{jk} = \delta_{ik}$.

With all our work on determinants, we may be struck by the idea that a pattern emerges. For example, in the calculation of the determinant of an $(n \times n)$ matrix, the sign in front of the minors alternates between terms depending on whether or not $(i + j)$ is even or odd. We can think of this in terms of *permutations*. Let us think of the set of numbers 1234. We can arrive at any other ordering of these four number by successively interchanging two numbers at a time. For example, to obtain the sequence 1243 we simply swap the order of the last two digits. More permutations are needed to arrive at the sequence 3142: we can swap adjacent pairs of numbers, $1234 \to 1324 \to 3124 \to 3142$, or we can swap pairs that are not adjacent, $1234 \to 1243 \to 3241 \to 3142$. In both cases we have needed three permutations, and this will always be the case when we have multiple ways to get to the same permutation. In the above case we have used an odd number of interchanges to arrive at our desired permutation. We will require an even number of interchanges to obtain other sequences. Still other sequences will not be a permutation (e.g., if one of the digits is repeated). We can now define the Levi-Civita symbol:

$$\epsilon_{ijk\ldots} = \begin{cases} +1 & \text{for } ijk\ldots \text{ an even permutation} \\ -1 & \text{for } ijk\ldots \text{ an odd permutation} \\ 0 & \text{for } ijk\ldots \text{ not a permutation.} \end{cases} \tag{4.83}$$

This definition means that most of the values of ϵ_{ijk} are zero. In particular, any index that has a repeated value is zero; for example, $\epsilon_{121} = -\epsilon_{121}$, implying that $\epsilon_{121} = 0$ because it will take three permutations to swap the first and third indices. We can represent the Levi-Civita symbol as a matrix in two dimensions because $\epsilon_{11} = \epsilon_{22} = 0$ and $\epsilon_{12} = -\epsilon_{21} = 1$, but it is harder to depict a Levi-Civita symbol with more than two indices.

Exercise 4.6.4 Show that it takes one permutation to swap adjacent indices and three to swap the first and third indices in ϵ_{ijk} and hence $\epsilon_{ijk} = 0$ if any two indices have the same value.

So, in three dimensions, all the values of $\epsilon_{ijk} = 0$, except for

$$\epsilon_{123} = \epsilon_{231} = \epsilon_{312} = 1 \quad \text{and} \quad \epsilon_{132} = \epsilon_{213} = \epsilon_{321} = -1.$$

This allows us to write our equation for a determinant in an equivalent but more compact notation:

$$\det(\mathbf{A}) = \sum_{ij\ldots} \epsilon_{ij\ldots} a_{1i} a_{2j} \ldots . \tag{4.84}$$

For example, if $\mathbf{A}$ is a (2×2) matrix,

$$\det(\mathbf{A}) = \begin{vmatrix} a_{11} & a_{12} \\ a_{21} & a_{22} \end{vmatrix} = \epsilon_{12} a_{11} a_{22} + \epsilon_{21} a_{12} a_{21} = a_{11} a_{22} - a_{12} a_{21}.$$

This is also useful in writing vector equations. For example, we can write the components of the cross product of two vectors $\mathbf{A} = A_1\hat{\mathbf{\imath}} + A_2\hat{\mathbf{\jmath}} + A_3\hat{\mathbf{k}}$ and $\mathbf{B} = B_1\hat{\mathbf{\imath}} + B_2\hat{\mathbf{\jmath}} + B_3\hat{\mathbf{k}}$ as $(\mathbf{A} \times \mathbf{B})_i = \epsilon_{ijk} A_j B_k$. Writing this out in full for the $i = 1$ component, we see

$$(\mathbf{A} \times \mathbf{B})_1 = \epsilon_{1jk} A_j B_k = \epsilon_{123} A_2 B_3 + \epsilon_{132} A_3 B_2 = A_2 B_3 - A_3 B_2,$$

which is the expression we obtained before (Equation (4.39)).

There are some very useful identities that connect ϵ_{ijk} and $\delta_i j$. The starting point for proving many of these identities is the following relationship for the product of two Levi-Civita symbols:

$$\epsilon_{ijk}\epsilon_{lmn} = \begin{vmatrix} \delta_{il} & \delta_{im} & \delta_{in} \\ \delta_{jl} & \delta_{jm} & \delta_{jn} \\ \delta_{kl} & \delta_{km} & \delta_{kn} \end{vmatrix}, \tag{4.85}$$

which can be proven by tediously working through all the combinations of i, j, and k — which we will not do here! However, Equation (4.85) does lead us to a very important identity (note that the Einstein summation convention implies that $\delta_{kk} = 3$ in three dimensions):

$$\begin{aligned} \epsilon_{ijk}\epsilon_{lmk} &= 3\delta_{il}\delta_{jm} - 3\delta_{im}\delta_{jl} + \delta_{im}\delta_{jk}\delta_{kl} - \delta_{ik}\delta_{jm}\delta_{kl} + \delta_{ik}\delta_{jl}\delta_{km} - \delta_{il}\delta_{jk}\delta_{km} \\ &= 3\delta_{il}\delta_{jm} - 3\delta_{im}\delta_{jl} + \delta_{im}\delta_{jl} - \delta_{il}\delta_{jm} + \delta_{im}\delta_{jl}\delta_{il}\delta_{jm} \\ &= \delta_{il}\delta_{jm} - \delta_{im}\delta_{jl}. \end{aligned} \tag{4.86}$$

We can use this identity to prove the equation for the vector triple product (Equation (4.45)) of vectors **a**, **b**, and **c**,

$$\begin{aligned}\mathbf{a}\times(\mathbf{b}\times\mathbf{c}) &= a_i\hat{\mathbf{e}}_i(\epsilon_{jkl}b_jc_k\hat{\mathbf{e}}_l) = a_i\epsilon_{jkl}b_jc_k\epsilon_{ilh}\hat{\mathbf{e}}_h \\ &= \epsilon_{jkl}\epsilon_{hil}a_ib_jc_k\hat{\mathbf{e}}_h = (\delta_{jh}\delta_{ki} - \delta_{ji}\delta_{kh})a_ib_jc_k\hat{\mathbf{e}}_h \\ &= a_ib_jc_i\hat{\mathbf{e}}_j - a_ib_ic_k\hat{\mathbf{e}}_k = (a_ic_k\delta_{ik})b_j\hat{\mathbf{e}}_j - (a_ib_j\delta_{ij})(c_k\hat{\mathbf{e}}_k) \\ &= (\mathbf{a}\cdot\mathbf{c})\mathbf{b} - (\mathbf{a}\cdot\mathbf{b})\mathbf{c}\end{aligned}$$

Exercise 4.6.5 Show that $\epsilon_{ijk}\epsilon_{mjk} = 2\delta_{im}$ and that $\epsilon_{ijk}\epsilon_{ijk} = 6$.

Exercise 4.6.6 Show, using the Levi-Civita symbol, that $\mathbf{A}\cdot(\mathbf{B}\times\mathbf{C}) = \mathbf{C}\cdot(\mathbf{A}\times\mathbf{B}) = \mathbf{B}\cdot(\mathbf{C}\times\mathbf{A})$.

4.7 Eigenvalues and Eigenvectors

We have seen that multiplying a vector by a matrix performs a linear transformation on the vector; the vector can be stretched, rotated, sheared, and so on. We might wonder if, for any given matrix, there are certain vectors that remain unchanged, or *invariant*, when multiplied by a matrix. For example, if the matrix represents a rotation about the z axis, then any vector pointing along the z axis will not be rotated and will be invariant under that transformation. Similarly, if the transformation reflected vectors in any plane that contained the z axis, vectors pointing along the z axis would not be changed. But what about a matrix representing a more general transformation? Are there similar invariant directions, and if there are, how do we find them? The answers to these questions involve calculating the *eigenvectors* and *eigenvalues* of the matrix, and these are key components of many important multivariate data analyses that are commonly used in the Earth and environmental sciences.

We will restrict our discussion to square matrices,[15] **A**, and we want to ask if we can find vectors **v** such that when we evaluate the product **Av** we end up with a vector that points in the same direction as **v**. Note that we will allow the new vector to have a different length than **v**. This means that we are looking for a vector **v** such that

$$\mathbf{A}\mathbf{v} = \lambda\mathbf{v}, \tag{4.87}$$

where λ is a scalar. A vector **v** that solves this equation is called an *eigenvector* of the matrix **A**, and λ is the corresponding *eigenvalue*. We can rewrite Equation (4.87) as

$$(\mathbf{A} - \lambda\mathbf{I})\mathbf{v} = 0. \tag{4.88}$$

This is a homogeneous equation (the right-hand side of the equation is zero and all solutions will pass through the origin) and implies that either $(\mathbf{A} - \lambda\mathbf{I}) = 0$ or $\mathbf{v} = 0$. The solution $\mathbf{v} = 0$ is not really interesting—it is essentially telling us that multiplying something by zero gives the answer zero—so we want to find values of λ that satisfy $(\mathbf{A} - \lambda\mathbf{I}) = 0$. What

[15] The concepts we cover in the section can be extended to nonsquare matrices, but to do so involves concepts that are beyond the scope of this book; see Section 4.9.

is more, this matrix cannot have an inverse matrix. If it did, we could multiply Equation (4.88) on the left by the inverse matrix to get

$$(\mathbf{A} - \lambda\mathbf{I})^{-1}(\mathbf{A} - \lambda\mathbf{I})\mathbf{v} = \mathbf{I}\mathbf{v} = \mathbf{v} = 0,$$

which is our uninteresting solution again. This is telling us that the matrix $(\mathbf{A} - \lambda\mathbf{I})$ must be a singular matrix, which implies

$$\det(\mathbf{A} - \lambda\mathbf{I}) = 0. \tag{4.89}$$

Equation (4.89) will be polynomial in λ whose degree is the size of the square matrix. So, if $\mathbf{A}$ is a (2×2) matrix, the polynomial will be a quadratic; if $\mathbf{A}$ is a (3×3) matrix, the polynomial will be a cubic, and so on. Equation (4.89) is called the *characteristic equation*, and it will have n solutions for an $(n \times n)$ matrix. Once we have found these n values of λ, we can substitute them one at a time into Equation (4.88) and calculate the components of the eigenvector corresponding to each eigenvalue. However, because we have a homogeneous equation (Equation (4.88)) and the matrix $(\mathbf{A} - \lambda\mathbf{I})$ is singular, we will have an infinite number of solutions for each eigenvector. Let us look at an example to see how this works in practice.

Example 4.21 Finding the eigenvalues of (2×2) and (3×3) matrices is relatively easy. For example, let us find the eigenvalues and eigenvectors of the matrix

$$\mathbf{A} = \begin{pmatrix} 0 & 1 \\ -2 & 3 \end{pmatrix}. \tag{4.90}$$

The first thing we want to do is to find the eigenvalues. From Equation (4.89) we get

$$\begin{vmatrix} -\lambda & 1 \\ -2 & 3-\lambda \end{vmatrix} = \lambda^2 - 3\lambda + 2 = (\lambda - 1)(\lambda - 2) = 0,$$

so there are two distinct eigenvalues, $\lambda_1 = 1$ and $\lambda_2 = 2$. Now, let the components of the eigenvector $\mathbf{x}$ be (x_a, x_b). We take each eigenvalue in turn, substitute it into Equation (4.88) and solve for x_a and x_b. Starting with $\lambda_1 = 1$ we have

$$\left(\begin{pmatrix} 0 & 1 \\ -2 & 3 \end{pmatrix} - \begin{pmatrix} \lambda_1 & 0 \\ 0 & \lambda_1 \end{pmatrix}\right)\begin{pmatrix} x_a \\ x_b \end{pmatrix} = \begin{pmatrix} -1 & 1 \\ -2 & 2 \end{pmatrix}\begin{pmatrix} x_a \\ x_b \end{pmatrix} = \begin{pmatrix} -x_a + x_b \\ -2x_a + 2x_b \end{pmatrix} = \begin{pmatrix} 0 \\ 0 \end{pmatrix}.$$

This gives us the equations $x_a - x_b = 0$ and $2x_a - 2x_b = 0$, which are the same equation. We expected this because Equation (4.88) is a homogeneous equation and $\det(\mathbf{A} - \lambda\mathbf{I}) = 0$. This means that we can write one of the components of $\mathbf{x}$ in terms of the other, which is then a free parameter. The direction of the vector (i.e., the slope x_b/x_a) will always be the same, only the length of the vector will change. But since we are interested in the *direction* of the vector, this is sufficient. Since we are free to choose the value of one of x_a or x_b, we might as well choose a value that makes our lives easy; for example, we could choose $x_a = 1$ so that $x_b = 1$ too. Since we are only interested in the directions of the eigenvectors,

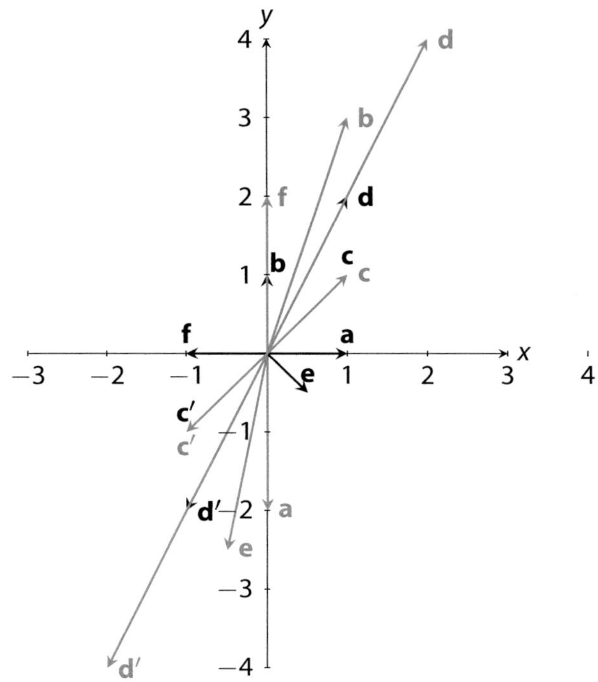

Figure 4.24 The effect of multiplying various vectors by the matrix **A** in Equation (4.90). The lettered gray vectors show the result of multiplying the corresponding black vectors by the matrix **A**. The directions of the eigenvectors **c**, **d**, **c**′, and **d**′ are the only ones that are unchanged by the multiplication, though their lengths are multiplied by the values of the corresponding eigenvalues.

we can make the eigenvector a unit vector using the length of the vector $\sqrt{2}$, so that finally we have the eigenvector

$$\mathbf{x}_1 = \frac{1}{\sqrt{2}}\begin{pmatrix}1\\1\end{pmatrix}.$$

We can perform similar calculations for the other eigenvalue, $\lambda_2 = 2$. In this case the equation for the components of the eigenvectors is $x_b = 2x_a$. If we choose $x_a = 1$, then $x_b = 2$, and dividing by the length of the vector we get the second eigenvector

$$\mathbf{x}_2 = \frac{1}{\sqrt{5}}\begin{pmatrix}1\\2\end{pmatrix}.$$

Notice that we have eigenvalue–eigenvector pairs: λ_1 and $\mathbf{x}_1$ form one pair and λ_2 and $\mathbf{x}_2$ form the other. Vectors that lie parallel to $\mathbf{x}_1$ do not change direction (that is the defining characteristic of an eigenvalue) or length ($\lambda_1 = 1$) when multiplied by **A**. However, vectors parallel to the eigenvector $\mathbf{x}_2$ are stretched by a factor of $\lambda_2 = 2$. All other vectors will change both their direction and magnitude (Figure 4.24).

For a (2×2) matrix, the characteristic equation is a quadratic equation. We can calculate this for any generic (2×2) matrix,

$$\mathbf{A} = \begin{pmatrix} a & b \\ c & d \end{pmatrix}, \tag{4.91}$$

giving a characteristic equation $\lambda^2 - (a+d)\lambda + (ad - bc) = 0$. The last term is simply the determinant of $\mathbf{A}$. The coefficient of λ is $(a+d)$, which is the sum of the diagonal terms of the matrix and is called the *trace*. The trace can be defined for any square $(n \times n)$ matrix as

$$\text{Tr}(\mathbf{A}) = \sum_{i=1}^{n} a_{ii}. \tag{4.92}$$

Now, if λ_1 and λ_2 are the eigenvalues of Equation (4.91), then they are solutions to a quadratic equation and we can write

$$(\lambda - \lambda_1)(\lambda - \lambda_2) = \lambda^2 - (\lambda_1 + \lambda_2)\lambda + \lambda_1\lambda_2 = 0,$$

so we must have that $\lambda_1 + \lambda_2 = \text{Tr}(\mathbf{A})$ and $\lambda_1\lambda_2 = \det(\mathbf{A})$. So, the trace of the matrix is equal to the sum of the eigenvalues, which we can see holds in Example 4.21. In fact, this holds for any square $(n \times n)$ matrix, not just for (2×2) matrices, and provides a quick and easy check on our calculations of eigenvalues. There are some useful properties of the trace. For example, if we have two square $(n \times n)$ matrices $\mathbf{A}$ and $\mathbf{B}$, then

$$\text{Tr}(\mathbf{A} + \mathbf{B}) = \sum_{i=1}^{n}(a_{ii} + b_{ii}) = \sum_{i=1}^{n} a_{ii} + \sum_{i=1}^{n} b_{ii} = \text{Tr}(\mathbf{A}) + \text{Tr}(\mathbf{B}).$$

Exercise 4.7.1 For a square $(n \times n)$ matrix $\mathbf{A}$, show that $\text{Tr}(c\mathbf{A}) = c\,\text{Tr}(\mathbf{A})$, where c is a scalar constant.

Exercise 4.7.2 If $\mathbf{A}$ is an $(n \times m)$ matrix and $\mathbf{B}$ is an $(m \times n)$ matrix, show that $\text{Tr}(\mathbf{AB}) = \text{Tr}(\mathbf{BA})$.

Exercise 4.7.3 For a square $(n \times n)$ matrix $\mathbf{A}$, show that $\text{Tr}(\mathbf{A}^{\text{T}}) = \text{Tr}(\mathbf{A})$.

So far we have only looked at cases where the eigenvalues of the matrix were real numbers. But this need not be the case. We can use the same methods to calculate complex eigenvalues and eigenvectors, but they have some additional properties that can make our calculations easier. Let us calculate the eigenvalues and eigenvectors of the matrix

$$\mathbf{X} = \begin{pmatrix} -1 & 4 \\ -2 & 3 \end{pmatrix}. \tag{4.93}$$

The characteristic equation for this matrix is $\lambda^2 - 2\lambda + 5 = (\lambda - (1+2i))(\lambda - (1-2i))$, so we have two eigenvalues, $\lambda_1 = 1 + 2i$ and $\lambda_2 = 1 - 2i$. The first thing to note is that the eigenvalues come in complex conjugate pairs, so a (3×3) matrix might have one real eigenvalue and two complex ones that are the complex conjugates of each other. We can calculate the eigenvectors as before, so that for $\lambda_1 = 1 + 2i$. If the eigenvector $\mathbf{v}$ has components (v_1, v_2), then

$$\begin{pmatrix} -2-2i & 4 \\ -2 & 2-2i \end{pmatrix}\begin{pmatrix} v_1 \\ v_2 \end{pmatrix} = 0,$$

so we have the equations

$$-2(i+1)v_1 + 4v_2 = 0, \qquad -2v_1 + 2(i-1)v_2 = 0.$$

After using a little manipulation we can see that these are both the same equation, $v_1 = v_2(1-i)$. Similarly, for the eigenvalue $\lambda_2 = 1 - 2i$, we have $v_1 = (1+i)v_2$. So, then the eigenvectors are

$$\mathbf{v}_1 = \begin{pmatrix} 1-i \\ 1 \end{pmatrix}, \qquad \mathbf{v}_1 = \begin{pmatrix} 1+i \\ 1 \end{pmatrix}.$$

These also form a complex conjugate pair. So, once we have found one complex eigenvalue and eigenvector, we can immediately write down the second pair.

We have seen that a matrix represents a linear transformation, so how do we interpret a real matrix that has complex eigenvalues and eigenvectors? Let us think about the matrix representing a rotation through an angle θ in two dimensions,

$$\mathbf{R} = \begin{pmatrix} \cos(\theta) & -\sin(\theta) \\ \sin(\theta) & \cos(\theta) \end{pmatrix}. \tag{4.94}$$

The characteristic equation for this matrix is $\lambda^2 - 2\lambda\cos(\theta) + 1 = 0$, and it has real roots only if $4\lambda^2(\cos^2(\theta) - 1) \geq 0$, or $\cos^2(\theta) \geq 1$. So, unless $\theta = 0, 2\pi, 4\pi \ldots$, the roots of this equation are complex. Therefore, we might suspect that complex eigenvectors and eigenvalues are related to rotations of vectors.

Before we go any further, we should remind ourselves of what multiplication of two complex numbers means geometrically. Let us take two complex numbers, $\lambda = a + ib = |\lambda|(\cos(\theta) + i\sin(\theta))$ and $z = u + iv$ (Appendix C). Then,

$$\mathrm{Re}(\lambda z) = |\lambda|(u\cos(\theta) - v\sin(\theta)), \quad \mathrm{Im}(\lambda z) = |\lambda|(u\sin(\theta) + v\cos(\theta)).$$

This is a pair of linear equations, and we know how to write them in matrix form as

$$\begin{pmatrix} \mathrm{Re}(\lambda z) \\ \mathrm{Im}(\lambda z) \end{pmatrix} = |\lambda| \begin{pmatrix} \cos(\theta) & -\sin(\theta) \\ \sin(\theta) & \cos(\theta) \end{pmatrix} \begin{pmatrix} u \\ v \end{pmatrix} = |\lambda| R_\theta \begin{pmatrix} u \\ v \end{pmatrix}.$$

In other words, we can think of the complex product λz as rotating the vector represented by z counterclockwise through the angle θ, followed by an expansion by a factor of $|\lambda|$.

To see how we translate these concepts into the world of eigenvalues and eigenvectors, let us concentrate on just (2×2) matrices. We write the complex eigenvalue λ and its corresponding eigenvector $\mathbf{v} = (v_1, v_2)$ as

$$\lambda = a + ib, \quad \mathbf{v} = \begin{pmatrix} v_1 \\ v_2 \end{pmatrix} = \begin{pmatrix} \mathrm{Re}(v_1) \\ \mathrm{Re}(v_2) \end{pmatrix} + i \begin{pmatrix} \mathrm{Im}(v_1) \\ \mathrm{Im}(v_2) \end{pmatrix},$$

and we will define a (2×2) matrix

$$\mathbf{V} = \begin{pmatrix} \mathrm{Re}(v_1) & \mathrm{Im}(v_1) \\ \mathrm{Re}(v_2) & \mathrm{Im}(v_2) \end{pmatrix} = \left(\mathrm{Re}(\mathbf{v}) \mid \mathrm{Im}(\mathbf{v})\right),$$

where we need to remember that $\mathrm{Re}(\mathbf{v})$ and $\mathrm{Im}(\mathbf{v})$ are actually column vectors. We can now write the eigenvalue equation as (remember that all the elements in $\mathbf{A}$ are real)

$$\mathbf{A}\mathrm{Re}(\mathbf{v}) = a\mathrm{Re}(\mathbf{v}) - b\mathrm{Im}(\mathbf{v}), \qquad \mathbf{A}\mathrm{Im}(\mathbf{v}) = b\mathrm{Re}(\mathbf{v}) + a\mathrm{Im}(\mathbf{v}).$$

So, we can write the right-hand side of these equations as

$$\mathbf{V}\begin{pmatrix} a \\ -b \end{pmatrix} \quad \text{and} \quad \mathbf{V}\begin{pmatrix} b \\ a \end{pmatrix}.$$

Therefore, if we define a new matrix

$$\mathbf{\Lambda} = \begin{pmatrix} a & b \\ -b & a \end{pmatrix},$$

we can write the eigenvalue equation as $\mathbf{AV} = \mathbf{V\Lambda}$. If we now write the complex eigenvalue as $\lambda = |\lambda|(\cos(\theta) + i\sin(\theta))$, we see that

$$\mathbf{\Lambda} = |\lambda| \begin{pmatrix} \cos(\theta) & \sin(\theta) \\ -\sin(\theta) & \cos(\theta) \end{pmatrix} = |\lambda| R_{-\theta},$$

which is a *clockwise* rotation through an angle θ followed by a scaling. For our matrix Equation (4.93), multiplication of a vector by λ_1 corresponds to a rotation of the vector through 63.435° followed by a scaling by a factor of $\sqrt{5}$.

We know that the eigenvectors of a matrix $\mathbf{A}$ are directions that remain unchanged when vectors are multiplied by $\mathbf{A}$, but they are more than that. Let us look at the eigenvectors from Example 4.21

$$\mathbf{x}_1 = \frac{1}{\sqrt{2}}\begin{pmatrix} 1 \\ 1 \end{pmatrix} \quad \text{and} \quad \mathbf{x}_2 = \frac{1}{\sqrt{5}}\begin{pmatrix} 1 \\ 2 \end{pmatrix}.$$

Vector $\mathbf{x}_2$ is not proportional to $\mathbf{x}_1$, so these vectors are linearly independent. In fact, the eigenvectors corresponding to different, distinct eigenvalues are always linearly independent. Let us show this is true for a (2×2) matrix $\mathbf{A}$, though it is true for any square matrix. We let $\mathbf{A}$ have two distinct eigenvalues λ_1 and λ_2 with $\lambda_1 \neq \lambda_2$, and we will let $\mathbf{v}_1$ and $\mathbf{v}_2$ be the corresponding eigenvectors. We want to show that $\mathbf{v}_1$ and $\mathbf{v}_2$ are linearly independent; in other words, we need to show that the equation $a_1\mathbf{v}_1 + a_2\mathbf{v}_2 = 0$ has only one solution, and that is $a_1 = a_2 = 0$. Operating on the left of this equation with $\mathbf{A}$ and using Equation (4.87) we get

$$\mathbf{A}(a_1\mathbf{v}_1 + a_2\mathbf{v}_2) = a_1\mathbf{A}\mathbf{v}_1 + a_2\mathbf{A}\mathbf{v}_2 = a_1\lambda_1\mathbf{v}_1 + a_2\lambda_2\mathbf{v}_2 = 0.$$

Now, let us take the equation of linear independence and multiply it by λ_1 to get

$$\lambda_1 a_1\mathbf{v}_1 + \lambda_1 a_2\mathbf{v}_2 = 0.$$

Subtracting these two equations gives

$$a_2\lambda_2\mathbf{v}_2 - \lambda_1 a_2\mathbf{v}_2 = a_2(\lambda_2 - \lambda_1)\mathbf{v}_2 = 0.$$

We know that $\mathbf{v}_2 \neq 0$ because it is an eigenvector. We also know that $\lambda_1 \neq \lambda_2$ because we stated that $\mathbf{A}$ had *distinct* eigenvalues. So, the only way this last equation can be true is if $a_2 = 0$. We can use a similar argument to show that $a_1 = 0$, so $\mathbf{v}_1$ and $\mathbf{v}_2$ are linearly independent. This condition holds in general, not just for (2×2) matrices. The fact that the eigenvectors are linearly independent shows that they form a basis. In the case of Example 4.21 this is a nonorthogonal basis.

Exercise 4.7.4 Show that the eigenvectors from Example 4.21 are nonorthogonal.

Exercise 4.7.5 Use the Gram–Schmidt algorithm to construct an orthogonal basis from the eigenvectors in Example 4.21.

This is all very nice, but what happens if the eigenvalues of the matrix are not distinct? To explore this case, consider the (3×3) matrix

$$\mathbf{A} = \begin{pmatrix} 0 & 1 & 1 \\ 1 & 0 & 1 \\ 1 & 1 & 0 \end{pmatrix}. \tag{4.95}$$

Exercise 4.7.6 Show that the eigenvalues of $\mathbf{A}$ in Equation (4.95) are given by the characteristic equation $(\lambda + 1)(\lambda + 1)(\lambda - 2) = 0$.

Two of the eigenvalues of $\mathbf{A}$ have the value -1, and these are called *degenerate eigenvalues*; there are two such eigenvalues in this case, so we say that the eigenvalue -1 has a multiplicity of 2. Calculating the eigenvector corresponding to $\lambda = 2$ is a generalization of what we did in Example 4.21. We substitute $\lambda = 2$ into $(\mathbf{A} - \lambda\mathbf{I})\mathbf{v} = 0$, giving

$$\begin{pmatrix} -2 & 1 & 1 \\ 1 & -2 & 1 \\ 1 & 1 & -2 \end{pmatrix} \begin{pmatrix} v_1 \\ v_2 \\ v_3 \end{pmatrix} \implies \begin{aligned} -2v_1 + v_2 + v_3 &= 0 \\ v_1 - 2v_2 + v_3 &= 0 \\ v_1 + v_2 - 2v_3 &= 0 \end{aligned} .$$

Any one of these three equations can be obtained from a linear combination of the other two showing us that we have three unknowns but only two independent equations. So, as in Example 4.21 we have a one-parameter family of solutions and, because we are only interested in the direction of the vector, we are free to give that parameter a value that makes our lives easier. Setting $v_1 = 1$, solving for v_2 and v_3, and normalizing the resulting vector by its length, we end up with the unit vector

$$\mathbf{v}_1 = \frac{1}{\sqrt{3}} \begin{pmatrix} 1 \\ 1 \\ 1 \end{pmatrix}.$$

Now, what about the other eigenvalue? If we follow the same procedure, we end up with just a single equation, $v + 1 + v_2 + v_3 = 0$. Now we have three unknowns but only one equation, so we have a two-parameter family of solutions as a result of the multiplicity of the eigenvalue. When this happens, we take a slightly different approach to finding $\mathbf{v}$. We have seen that, for distinct eigenvalues, the eigenvectors are all linearly independent, and this is a feature that we want to preserve. So, we should choose our free parameters such that we get two linearly independent eigenvectors corresponding to the same eigenvalue. For example, if we choose $v_1 = v_2 = 1$, then $v_3 = -2$, and the normalized eigenvector is

$$\mathbf{v}_2 = \frac{1}{\sqrt{6}} \begin{pmatrix} 1 \\ 1 \\ -2 \end{pmatrix}.$$

But if we choose $v_1 = v_3 = 1$, then $v_2 = -2$, and we end up with the eigenvector

$$\mathbf{v}_3 = \frac{1}{\sqrt{6}}\begin{pmatrix}1\\-2\\-1\end{pmatrix},$$

which is linearly independent from $\mathbf{v_2}$. So we still have a set of linearly independent vectors that can form a basis of three-dimensional space.

Recall that we can think of the columns of a matrix as being individual vectors. That means that we can form a matrix whose columns are the individual eigenvectors of the matrix $\mathbf{A}$. Let us look at the case in three dimensions.

Exercise 4.7.7 Consider a generic (3×3) matrix $\mathbf{A}$ (e.g., Equation (4.77)). Show that Equation (4.87) is equivalent to the following system of linear equations:

$$av_1 + bv_2 + cv_3 = \lambda v_1, \quad dv_1 + ev_2 + fv_3 = \lambda v_2, \quad gv_1 + hv_2 + iv_3 = \lambda v_3,$$

where λ is an eigenvalue of $\mathbf{A}$ and the corresponding eigenvector $\mathbf{v}$ has components (v_1, v_2, v_3).

For a three-dimensional system we have three eigenvalues (λ_1, λ_2, and λ_3) and three corresponding eigenvectors ($\mathbf{v}_1$, $\mathbf{v}_2$, $\mathbf{v}_3$). Let us write each eigenvector as a column of a new matrix $\mathbf{V}$ and the eigenvalues as elements of a diagonal matrix $\mathbf{\Lambda}$,

$$\mathbf{V} = \begin{pmatrix}(v_1)_1 & (v_2)_1 & (v_3)_1\\(v_1)_2 & (v_2)_2 & (v_3)_2\\(v_1)_3 & (v_2)_3 & (v_3)_3\end{pmatrix}, \qquad \mathbf{\Lambda} = \begin{pmatrix}\lambda_1 & 0 & 0\\0 & \lambda_2 & 0\\0 & 0 & \lambda_3\end{pmatrix}, \tag{4.96}$$

where $(v_i)_j$ is the jth element of the ith eigenvector. Then, the eigenvalue equation (Equation (4.87)) can be written as $\mathbf{AV} = \mathbf{V\Lambda}$.

Exercise 4.7.8 Show that the equation $\mathbf{AV} = \mathbf{V\Lambda}$ produces the same equations as those derived in Exercise 4.7.7, where $\mathbf{A}$ is a generic (3×3) matrix and $\mathbf{V}$ and $\mathbf{\Lambda}$ are defined in Equation (4.96).

If we multiply the equation $\mathbf{AV} = \mathbf{V\Lambda}$ on the right by $\mathbf{V}^{-1}$, we get

$$\mathbf{AVV}^{-1} = \mathbf{A} = \mathbf{V\Lambda V}^{-1}. \tag{4.97}$$

Equation (4.97) allows us to create a matrix $\mathbf{A}$ that has specified eigenvalues and eigenvectors.[16] Another interesting thing about this equation is that it tells us what an $(n \times n)$ matrix $\mathbf{A}$ does to a vector $\mathbf{x}$. Recall that the columns of $\mathbf{V}$ form a basis because they represent linearly independent vectors in n-dimensions, so multiplying $\mathbf{x}$ by $\mathbf{V}^{-1}$ transforms $\mathbf{x}$ to into a new coordinate system. Multiplying by the diagonal matrix $\mathbf{\Lambda}$ then expands or contracts the components of $\mathbf{x}$ in this coordinate system. Lastly, we multiply by $\mathbf{V}$ to return the deformed vector back to the original coordinate system (Figure 4.25). The matrices $\mathbf{A}$ and $\mathbf{\Lambda}$ are said to be *similar*, and Equation (4.99) is called a *similarity transformation*.

[16] This is how the author created the matrices for the examples and problems in this section!

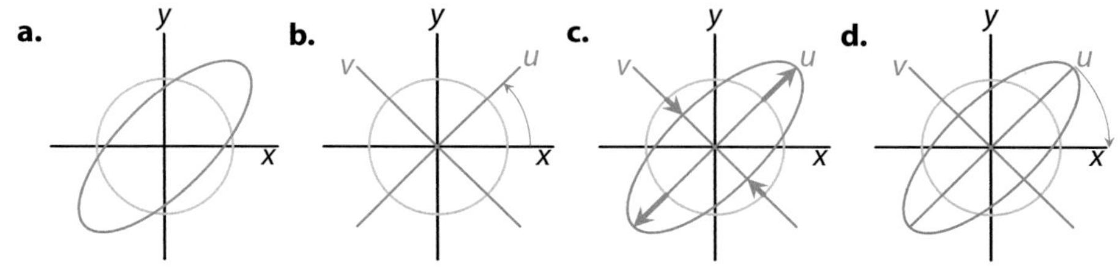

Figure 4.25 The similarity transform. A matrix **A** transforms the points that lie on a circle into a rotated ellipse (a.). The similarity transformation decomposes this into (b.) a rotation of coordinates to a new coordinate system (u, v) resulting from multiplication by $\mathbf{V}^{-1}$, followed by (c.) stretching and compressing along the axes of this new coordinate systems, and lastly (d.) undoing the initial coordinate transformation by multiplication by **V**.

Another advantage of the similarity transformation is that it helps us raise a matrix to a power. For example, if we wanted to calculate $\mathbf{A}^6$, we would have to do a lot of matrix multiplications. However, using the similarity transformation, we see that

$$\mathbf{A}^2 = (\mathbf{V\Lambda V}^{-1})(\mathbf{V\Lambda V}^{-1}) = \mathbf{V\Lambda}^2\mathbf{V}^{-1},$$

and because $\mathbf{\Lambda}$ is a diagonal matrix, $\mathbf{\Lambda}^2$ is also a diagonal matrix whose elements are the squares of the elements of $\mathbf{\Lambda}$. We can see that

$$\mathbf{A}^n = \mathbf{V\Lambda}^n\mathbf{V}^{-1},$$

so that raising a square matrix to an arbitrary power is relatively easy once we know the matrix of eigenvectors (**V**) and its inverse. Why would we want to do this? We have seen that the exponential function of a variable occurs frequently when we describe scientific phenomena. What do we do if we want to take the exponential of a matrix rather than a single variable? The exponential of a matrix **A** is *defined* by

$$e^{\mathbf{A}} = \sum_{n=0}^{\infty} \frac{\mathbf{A}^n}{n!} = \mathbf{I} + \mathbf{A} + \frac{1}{2!}\mathbf{A}^2 \cdots, \tag{4.98}$$

so being able to easily calculate the powers of **A** can be very useful.

Lastly, we can often use a similarity transformation to turn a square matrix into a diagonal matrix. If we start with the eigenvalue equation $\mathbf{AV} = \mathbf{V\Lambda}$ and multiply it on the left by $\mathbf{V}^{-1}$, we get

$$\mathbf{V}^{-1}\mathbf{AV} = \mathbf{V}^{-1}\mathbf{V\Lambda} = \mathbf{\Lambda}. \tag{4.99}$$

The square matrix **A** is said to be *diagonalizable* if it is similar to a diagonal matrix. Basically what Equation (4.99) is doing is performing a coordinate transformation to a new basis such that the matrix **A** becomes diagonal. So, in this new coordinate transformation, the matrix represents only expansions and compressions (Figure 4.25).

Example 4.22 To see how this works, let us diagonalize the following matrix,

$$\mathbf{A} = \begin{pmatrix} 4 & 2 \\ 1 & 3 \end{pmatrix}.$$

First, we have to find the eigenvalues and eigenvectors of **A**. The eigenvalues are $\lambda_1 = 5$ and $\lambda_2 = 2$, giving the normalized eigenvectors

$$\mathbf{v}_1 = \frac{1}{\sqrt{5}}\begin{pmatrix}2\\1\end{pmatrix}, \quad \mathbf{v}_2 = \frac{1}{\sqrt{2}}\begin{pmatrix}-1\\1\end{pmatrix}.$$

The matrix of eigenvectors and its inverse are then

$$\mathbf{V} = \begin{pmatrix}\frac{2}{\sqrt{5}} & \frac{-1}{\sqrt{2}}\\ \frac{1}{\sqrt{5}} & \frac{1}{\sqrt{2}}\end{pmatrix}, \qquad \mathbf{V}^{-1} = \begin{pmatrix}\frac{\sqrt{5}}{3} & \frac{\sqrt{5}}{3}\\ -\frac{\sqrt{2}}{3} & -\frac{2\sqrt{2}}{3}\end{pmatrix},$$

so that

$$\mathbf{V}^{-1}\mathbf{A}\mathbf{V} = \begin{pmatrix}5 & 0\\0 & 2\end{pmatrix}$$

which is the diagonal matrix formed from the eigenvalues of **A**.

Unfortunately, not all square matrices are diagonalizable. A *sufficient condition* that an $(n \times n)$ matrix **A** can be diagonalized is that it has n distinct eigenvalues and n linearly independent eigenvectors. However, there are some matrices that do not satisfy this condition but can still be diagonalized (the condition is a *sufficient* condition, so that any matrix satisfying it can be diagonalized, but not all matrices that can be diagonalized have to satisfy this condition). For example, consider the symmetric matrix in Equation (4.95). This has an eigenvalue of multiplicity 2, so it does not satisfy the condition, but we were still able to find three linearly independent eigenvectors. In this case, we can still diagonalize the original matrix

Exercise 4.7.9 Diagonalize the matrix in Equation (4.95).

Now, consider the matrix

$$\mathbf{A} = \begin{pmatrix}1 & 2\\0 & 1\end{pmatrix}.$$

This also has a repeated eigenvalue, $\lambda = 1$. Substituting this eigenvalue into the eigenvalue equation gives

$$\begin{pmatrix}0 & 2\\0 & 0\end{pmatrix}\begin{pmatrix}v_1\\v_2\end{pmatrix} = 0,$$

where v_1 and v_2 are the components of the eigenvector **v**. We can see that v_1 is a free parameter and $v_2 = 0$, so the eigenvector is

$$\mathbf{v} = \begin{pmatrix}1\\0\end{pmatrix}.$$

However, we can see that there are no other eigenvectors, so we cannot form the eigenvector matrix to perform the diagonalization.

Now that we have discussed determinants, eigenvectors, and eigenvalues, let us see if we can prove Euler's theorem. This is a little bit technical, but it shows the power of eigenvectors, determinants, and some of the techniques and methods that are common in using vectors. Recall that the theorem states that we can describe the motion of a rigid body on the surface of a sphere as a rotation about an axis that passes through the sphere. Since such an axis will not change when we apply the rotation, we might suspect that this has something to do with eigenvectors. So, we can prove the theorem by showing that any matrix describing motion of rigid objects on the surface of the sphere has eigenvectors.

Consider three points, P_1, P_2, and P_3, which define a rigid shape on the surface of a sphere. The vectors from the center of the sphere to these points are $\mathbf{R}_1$, $\mathbf{R}_2$, and $\mathbf{R}_3$. Moving the shape around on the surface of the sphere has the same effect as keeping the shape in one place and rotating the sphere underneath it. Doing this, the new coordinates of the points P_1, P_2, and P_3 are p_1, p_2, and p_3 with new position vectors $\mathbf{r}_1$, $\mathbf{r}_2$, and $\mathbf{r}_3$. We now form two matrices whose columns are the components of these vectors,

$$\mathbf{T} = \begin{pmatrix} X_1 & X_2 & X_3 \\ Y_1 & Y_2 & Y_3 \\ Z_1 & Z_2 & Z_3 \end{pmatrix}, \qquad \mathbf{t} = \begin{pmatrix} x_1 & x_2 & x_3 \\ y_1 & y_2 & y_3 \\ z_1 & z_2 & z_3 \end{pmatrix},$$

where $\mathbf{R}_1$ has components (X_1, Y_1, Z_1), $\mathbf{r}_1$ has components (x_1, y_1, z_1), and so on. Now,

$$\mathbf{T}^{-1}\mathbf{R}_1 = \begin{pmatrix} 1 \\ 0 \\ 0 \end{pmatrix}, \qquad \mathbf{T}^{-1}\mathbf{R}_2 = \begin{pmatrix} 0 \\ 1 \\ 0 \end{pmatrix}, \qquad \mathbf{T}^{-1}\mathbf{R}_3 = \begin{pmatrix} 0 \\ 0 \\ 1 \end{pmatrix}.$$

Exercise 4.7.10 Consider a (2×2) matrix $\mathbf{U}$ whose columns are the components of two position vectors $\mathbf{R}_1 = (x_1, y_1)$ and $\mathbf{R}_2 = (x_2, y_2)$, i.e.,

$$\mathbf{U} = \begin{pmatrix} x_1 & x_2 \\ y_1 & y_2 \end{pmatrix}.$$

Show that

$$\mathbf{U}^{-1}\mathbf{R}_1 = \begin{pmatrix} 1 \\ 0 \end{pmatrix}, \qquad \mathbf{U}^{-1}\mathbf{R}_2 = \begin{pmatrix} 0 \\ 1 \end{pmatrix}.$$

This means that we can define a new matrix $\mathbf{A} = \mathbf{t}\mathbf{T}^{-1}$ such that $\mathbf{A}\mathbf{T} = \mathbf{t}\mathbf{T}^{-1}\mathbf{T} = \mathbf{t}$ and $\mathbf{A}\mathbf{R}_i = \mathbf{r}_\mathbf{i}$.

The sphere is rigid, so any vector from the origin to the surface of the sphere will have the same length (i.e., the radius of the sphere). So, for any of these radial vectors

$$||\mathbf{r}|| = ||\mathbf{A}\mathbf{R}|| = ||\mathbf{R}||,$$

and using the transpose representation of the scalar product we see that

$$\mathbf{R}^\mathrm{T}\mathbf{A}^\mathrm{T}\mathbf{A}\mathbf{R} = \mathbf{R}^\mathrm{T}\mathbf{R} \implies \mathbf{A}^\mathrm{T}\mathbf{A} = \mathbf{I},$$

so $\mathbf{A}$ is an orthogonal matrix. What is more, we know that $\det(\mathbf{A}) = \pm 1$ for an orthogonal matrix. How can we decide on which sign to use? A good way to determine this is to start with a transformation that we know the sign of, and work from there. A good transformation to start with is one that does nothing (i.e., a null transformation), that is $\mathbf{A}\mathbf{R}_i = \mathbf{R}_i$. In this case $\mathbf{A} = \mathbf{I}$, so $\det(\mathbf{A}) = \det(\mathbf{I}) = +1$. We now choose a transformation that is infinitessimally different from this one, so that it rotates the sphere a small amount.

Recall that the determinant of a matrix can be thought of as a volume, and so as we move smoothly from the null transformation to this new one we want all the characteristics of the transformation to also change smoothly. But since the determinant of **A** can only be ± 1, it does not change smoothly; it can only switch between these two values without taking on any values in between. So, the determinant of **A** stays at a value of $+1$ always.

Now, let us see if we can show that $\det(\mathbf{A} - \mathbf{I}) = 0$, which would be a step toward showing that **A** has eigenvectors. First, using $\det(\mathbf{A}^T) = \det(\mathbf{A})$ we see that

$$\det(\mathbf{A} - \mathbf{I}) = \det((\mathbf{A} - \mathbf{I})^T = \det(\mathbf{A}^T - \mathbf{I}). \tag{4.100}$$

Because **A** is orthogonal, we also know that

$$\mathbf{A}\mathbf{A}^T - \mathbf{A} = (\mathbf{A}^T - \mathbf{I})\mathbf{A} = \mathbf{I} - \mathbf{A}. \tag{4.101}$$

So,

$$\begin{aligned} \det((\mathbf{A}^T - \mathbf{I})\mathbf{A}) &= \det(\mathbf{A}^T - \mathbf{I})\det(\mathbf{A}) && \text{because} \quad \det(\mathbf{XY}) = \det(\mathbf{X})\det(\mathbf{Y}) \\ &= \det(\mathbf{A}^T - \mathbf{I}) && \text{because} \quad \det(\mathbf{A}) = 1 \\ &= \det(\mathbf{A} - \mathbf{I}) && \text{from Equation (4.100).} \end{aligned}$$

But, we also know from Equation (4.101) that $\det(\mathbf{A}\mathbf{A}^T - \mathbf{A}) = \det(\mathbf{I} - \mathbf{A})$. Also, if **X** is a (3×3) matrix, we know that $\det(-\mathbf{X}) = (-1)^3 \det(\mathbf{X}) = -\det(\mathbf{X})$. So, we have shown that $\det(\mathbf{A} - \mathbf{I}) = \det(\mathbf{I} - \mathbf{A}) = -\det(\mathbf{A} - \mathbf{I})$. The only way this can happen is if $\det(\mathbf{A} - \mathbf{I}) = 0$, which in turn implies that there exists a vector **N** such that $\det(\mathbf{A} - \mathbf{I}) \cdot \mathbf{N} = 0$, so **N** is an eigenvector of the transformation matrix **A**.

Therefore, we can describe the motion of tectonic plates, for example, by rotations about different axes if we know the position vectors of the plates at the start and end of their path.

4.8 Vectors, Matrices, and Data

One area where we come across vectors and matrices a great deal is in handling data. Many data sets in the Earth and environmental sciences contain many variables collected in many locations around the globe over long periods of time, and matrices provide a convenient means for storing and analyzing these data. For example, say you want to analyze atmospheric aerosol data collected daily for a year at 100 different locations. The data at each location include temperature, pressure, humidity, wind speed, wind direction, and several components of aerosol composition (e.g., ammonium, hydrochloric acid, iron, lead, mercury, nitric acid, organic carbon, sulphur dioxide). Each station measures 13 variables, so the data could be represented as a $(13 \times 100 \times 365)$ matrix. This can be thought of as a cube consisting of 13 rows, 100 columns, and 365 "slices." Each slice is a (13×100) matrix representing a snapshot of all the measurements collected at all locations on that day. Each snapshot is then stacked, one behind the other, in sequence so that the first slice is the data from the first day, the second slice from the second day, and so on. Each column or row of the cube can be thought of as a vector, but now in a space that is impossible to visualize as arrows as we have.

Analysis of such data sets involves techniques of multivariate and time-series data analysis, and they are topics for other books. However, to properly understand these techniques requires a familiarity with vectors and matrices, particularly with eigenvectors and eigenvalues, as these concepts underlie many of the common techniques (such as principal component analysis) used to simplify large data sets and look for relationships between variables. Although we have only considered $(n \times m)$ matrices here, what we have learned translates directly to $(n \times m \times k)$ matrices and beyond. It is often impossible to visualize these large data sets as geometric vectors using arrows, but being able to picture in two and three dimensions what happens under a matrix multiplication or a vector product gives us a strong intuition as to what happens in these other cases.

4.9 Further Reading

There are many very good books that cover linear algebra, and many of them are more abstract than the approach taken here. A good book that is more applied is the text *Linear Algebra and its Applications* (Strang, 2006). Many books on mathematical methods in physics or engineering will also cover vectors and matrices with a more applied approach, but many of the problems will be physics based. A good textbook is *Mathematical Methods in the Physical Sciences* by Mary Boas (2006).

Vectors and matrices make a large appearance in fields such as structural geology and geochemistry. The textbooks *Fundamentals of Structural Geology* (Pollard and Fletcher, 2005) and *Structural Geology Algorithms* (Allmendinger et al., 2012) provide good examples of how vectors and matrices are used to understand geological structures. The book *Introduction to Geochemical Modeling* (Albarède, 1995), while more technically demanding than the book you are reading, shows how matrices are used to understand, analyze, and model geochemical systems.

Matrices and vectors appear in understanding and numerical modeling of the motions of fluids such as air and water, as explained in *Atmosphere, Ocean, and Climate Dynamics* (Marshall and Plumb, 2008). They also appear prominently in understanding the motions of tectonic plates and molten rocks (e.g., as seen in the book *Geodynamics* (Turcotte and Schubert, 2014)).

Very large matrices occur very often in modeling and understanding the Earth and environmental processes, and these matrices have to be solved and manipulated on computers. As a result, there a strong incentive in making numerical algorithms for handling vectors and matrices both fast and efficient. This is not always easy, but a good introduction to some of the techniques and issues can be found in the book *Numerical Recipes in C* (Press et al., 1992). This is a great book for starting to learn about specific numerical methods, how they work, and why they might fail. Numerical software packages such as R, Python, and Matlab come replete with many excellent routines for numerically handling matrices and vectors, but it is still useful to know when these routines can fail and give incorrect answers, and that requires an understanding of linear algebra.

Problems

4.1 Show that if two vectors **x** and **y** satisfy $||\mathbf{x}+\mathbf{y}|| = ||\mathbf{x}-\mathbf{y}||$, then **x** and **y** are orthogonal.

4.2 Use the rotation matrices to derive the transformation matrix for a rotation about the x axis through an angle θ followed by a rotation through an angle ϕ about the y axis.

4.3 In two dimensions, derive a matrix for a clockwise rotation by θ.

4.4 Two forces act on a body of mass m sitting on the surface of a spherical, rotating Earth of radius R. The force of gravity ($F_g = mg$, where g is the acceleration due to gravity) acts to pull the object toward the center of the planet. The *centrifugal force* arises from the rotation of the Earth and acts to pull the object away from the surface in a direction perpendicular to the axis of rotation and has a value $F_c = m\omega^2 r$, where ω is the angular velocity of the Earth (a constant) and r is the perpendicular distance of the body from the rotation axis. Derive an expression for the effective acceleration due to gravity by adding the two vectors. Derive an expression for the angle between the effective direction of gravity and the vertical at that latitude.

4.5 Consider the following matrices:

$$\mathbf{A} = \begin{pmatrix} 1 & 2 \\ 1 & 1 \end{pmatrix}, \quad \mathbf{B} = \begin{pmatrix} 1 & 2 \\ 2 & 1 \end{pmatrix}, \quad \mathbf{C} = \begin{pmatrix} 1 & -2 \\ 2 & 1 \end{pmatrix}.$$

1. Calculate and draw the action of the matrices $\mathbf{A}$, $\mathbf{A}^{\mathrm{T}}$, $\mathbf{B}$, $\mathbf{C}$, $\mathbf{C}^{\mathrm{T}}$ on the points $(x, y) = (0, 0)$, $(0, 1)$, $(1, 0)$, and $(1, 1)$, which form a square in the (x, y) plane.
2. Decompose **A** into a symmetric matrix ($\mathbf{A}_s$) and an antisymmetric matrix ($\mathbf{A}_a$), and calculate the action of these matrices on the four points in Question 1.

4.6 Consider a unit vector pointing along the x direction. Use the three-dimensional rotation matrices to calculate the effects on this vector of a rotation of 90° about the y axis followed by a rotation of 90° about the z axis. Compare this with the effect of doing the rotation about the z axis first, followed by the rotation about the y axis.

4.7 Calculate the products **AB** and **BA**, where

$$\mathbf{A} = \begin{pmatrix} a & b & c \\ d & e & f \\ g & h & k \end{pmatrix} \quad \text{and} \quad \mathbf{B} = \begin{pmatrix} 1 & 1 & 1 \\ 1 & 1 & 1 \\ 1 & 1 & 1 \end{pmatrix}.$$

4.8 Show that Equation (4.85) is true.

4.9 Consider the matrices

$$\mathbf{x} = \begin{pmatrix} x \\ y \end{pmatrix}, \quad \mathbf{A} = \begin{pmatrix} 2 & 1 \\ 3 & 4 \end{pmatrix}.$$

Evaluate the expression $\mathbf{x}^{\mathrm{T}}\mathbf{A}\mathbf{x} = 1$. This is the *quadratic form* associated with the matrix **A**. Perform a coordinate transformation to coordinates (X, Y) by diagonalizing $\mathbf{A}$, and show that the quadratic form is the equation of an ellipse, $X^2 + 5Y^2 = 1$. Calculate the quadratic form associated with the symmetric matrix

$$\mathbf{B} = \begin{pmatrix} 8 & 2 \\ 2 & 5 \end{pmatrix}.$$

By finding an appropriate coordinate transformation, show that this describes a rotated ellipse, and calculate the angle it is rotated by in the original coordinates.

4.10 Consider the matrix $\mathbf{R}_x(\theta)$ for a counterclockwise rotation in three dimensions by an angle θ about the x-axis. Show that $\mathbf{R}_x^{-1}(\theta) = \mathbf{R}_x(-\theta)$.

4.11 Calculate the rotation for a counterclockwise rotation through an angle θ about the x axis $\mathbf{R}_x(\theta)$, followed by a counterclockwise rotation through ϕ about the y axis $\mathbf{R}_y(\phi)$, followed by a counterclockwise rotation through ξ about the z axis $\mathbf{R}_z(\xi)$. Calculate the rotation matrix for the combined rotation $\mathbf{R}_x(\theta)\mathbf{R}_z(\xi)\mathbf{R}_y(\phi)$, and compare the two answers.

4.12 Determine the surface $S(x, y) = 2x^2 + 2xy + 2y^2 = 1$ by diagonalizing an appropriate matrix.

4.13 Consider the following chemical reaction involving calcium, carbon, oxygen, phosphorus, and hydrogen:

$$\nu_1 \mathrm{CaCO_3} + \nu_2 \mathrm{H_3PO_4} \rightarrow \nu_3 \mathrm{Ca_3(PO_4)_2} + \nu_4 \mathrm{H_2CO_3} + \nu_5 \mathrm{CO_2}.$$

Determine an algebraic equation for each element that balances the number of atoms on either side of the reaction. Write these equations in matrix form, and solve the matrix equation for ν_1 through ν_4.

4.14 Dimensional analysis calculations result in systems of linear equations that can be solved using matrices. Consider the drag force F acting on a sphere of radius r moving though a fluid of density ρ and dynamic viscosity μ with a velocity v.

1. Write down the dimensions of each of the variables in the problem.
2. Dimensional analysis tells us that $F \propto r^\alpha \rho^\beta \mu^\gamma v^\delta$. Complete the matrix $\mathbf{M}$ with each row corresponding to a dimension (i.e., [M], [L], etc.) and each column a dependent variable (i.e., r, ρ, μ, and v), with the entries being the power to which each dimension appears in that variable; that is

$$\mathbf{M} = \begin{array}{c|cccc} & r & v & \rho & \mu \\ \hline L & 1 & 1 & & \\ T & & & & \\ M & & & & \end{array}.$$

3. Form two column vectors: $\mathbf{A}$ containing the unknown powers (i.e. α, β etc.) in the dimensional analysis problem, and $\mathbf{B}$ containing the exponents of the dimensions of F.

4. Solve the matrix problem $\mathbf{MA} = \mathbf{B}$ by forming an appropriate augmented matrix and reducing it to row form to show that the solution is

$$\begin{pmatrix} 2 \\ 2 \\ 1 \\ 0 \end{pmatrix} + d \begin{pmatrix} -1 \\ -1 \\ -1 \\ 1 \end{pmatrix}.$$

5. Show that

$$F \propto r^2 v^2 \rho f\left(\frac{\mu}{v\rho r}\right).$$

4.15 The exponential of a square matrix can be defined using a power series expansion,

$$e^{\mathbf{A}} = \mathbf{I} + \mathbf{A} + \frac{1}{2!}\mathbf{A}^2 + \frac{1}{3!}\mathbf{A}^3 + \cdots, \qquad (4.102)$$

where $\mathbf{I}$ is the unit matrix.

1. Why does Equation (4.102) hold only for square matrices?
2. Show that in general $e^{\mathbf{A}}e^{\mathbf{B}} \neq e^{\mathbf{A}+\mathbf{B}}$.

4.16 The way in which rock deforms under force is characterized by the stress tensor, which can be written as a symmetric (3×3) matrix $\boldsymbol{œ}$. Consider a stress tensor

$$\boldsymbol{œ} = \begin{pmatrix} 20 & 5 & 1 \\ 5 & 10 & 40 \\ 1 & 40 & 30 \end{pmatrix}.$$

Diagonalize $\boldsymbol{œ}$. The components of the diagonalized matrix are called the *principal stresses*.

5 Probability

There are many situations in science where we need to make use of probability. In spite of our best efforts, all the measurements we make contain some degree of uncertainty. The digital readout of the instrument we use may give an answer to many decimal places, but that last digit is always uncertain. We quantify that uncertainty and its effects using statistical techniques that are based upon the mathematics of random variables. We will not delve into statistics and data analysis here, that is a topic for another book entirely (see Section 5.8). However, we will explore ideas in probability and random variables so that we can have have a better understanding of statistical tests and the assumptions that underlie them.

Probability is also useful for understanding processes that are either inherently random, or appear to be random. For example, the amount of rainfall at a given location on a given day can be thought of as being random, or the number of earthquakes in a region during a given length of time can be thought of as random. Events such as these are treated as the outcome of random processes that have a certain probability of occurring. This is not necessarily because the processes are really random, but because their complexity means that we cannot forecast them accurately. For example, to accurately and precisely predict when and where an earthquake will occur requires measurements and mathematical models of stress deep within the Earth, and at high resolutions to detect where stress is the highest. But, by treating earthquake occurrence as a random variable allows us to quantify and give some structure to our inability to accurately forecast all the processes involved in an earthquake.

5.1 What Is Probabililty?

We will often use simple, everyday examples such as tossing a coin or drawing cards from a deck of cards when explaining probability because probability can sometimes produce results that are counterintuitive, and using simple examples can sometimes make things clearer. First, we need to think a little bit about what we mean by "probability." When we toss a coin there are two possible outcomes for each toss: either heads or tails. The probability of getting either of these outcomes is the frequency with which we expect to get heads (or tails) if we toss the coin many, many times. We can formalize this by saying that, if we perform many experiments, the probability (P) of getting a given outcome (e.g., heads) is

$$P = \frac{\text{number of experiments that lead to a given outcome}}{\text{total number of experiments}}. \tag{5.1}$$

This is the basis of what is called the *frequentist* interpretation of probability, and it is the one that many find most intuitive.[1] But there is a problem here: how many experiments do we need? For example, flipping a coin four times has just given me the sequence HTHH (where H represents heads and T tails). Repeating this experiment gives TTTH. In the first case, the frequency of getting heads is 0.75 and the frequency of getting tails is 0.25, and these frequencies are reversed in the second experiment. Flipping a coin ten times gives me a sequence THHHTHTHTH, so that the frequency of getting heads or tails is 0.6 and 0.4, respectively. Repeating the experiment gives the sequence HHTHHTTHHH, giving frequencies of 0.7 and 0.3 for heads and tails. So, to obtain a consistent value for probability we have to modify Equation (5.1) to say something about how many experiments we need to do in order to get a consistent value for P. Mathematically, our definition of probability is consistent only in the limit of an infinite number of experiments so, the probability $P(x)$ of getting an outcome x (e.g., heads in a coin-flipping experiment) is

$$P(x) = \lim_{N\to\infty} \frac{N_x}{N}, \tag{5.2}$$

where N_x is the number of experiments with an outcome x and N is the total number of experiments. This definition implies that a probability must lie between 0 and 1 because $0 \leq N_x \leq N$. If $P(x) = 0$, then x will never be an outcome of the experiment, and if $P(x) = 1$, x will always be the only outcome. Having a definition of probability is one thing, but to be more useful we need to have rules for adding, multiplying, and combining probabilities. Before we look at these, we need to define some terms.

An *event* is defined as being some simple process that is well defined with a beginning and end. For example, the toss of a coin, drawing a card from a shuffled deck, the decay of an atom of ^{234}Th, the absorption of a photon by a water molecule in the ocean, or the occurrence of an earthquake at a given location during the next decade. The collection of all possible outcomes of an event is called the *sample space*. For the toss of a coin, the sample space has two elements: the coin lands heads up, or it lands tails up. For an earthquake, our sample space may be occurrence or nonoccurrence at a given location within a given time interval.

The first rule of probability is that the sum of the probabilities of all possible outcomes defined in the sample space is 1. If there are M possible outcomes and the probability of any given outcome x_i is $P(x_i)$, then

$$\sum_{i=1}^{M} P(x_i) = 1.0. \tag{5.3}$$

All this equation is saying is that *something* must happen. This is a very useful property because if the probability of getting a certain outcome X is $P(X)$, then Equation (5.3) implies that the probability of getting anything but that outcome is $1 - P(X)$, which is called the *complement* of X and is written $P(X^c)$.

[1] This is not the only interpretation of probability, nor was it the first, Another common interpretation is the Bayesian interpretation (Bertsch McGrayne, 2011) named after Thomas Bayes (1701–1761). In this interpretation, a probability is a measure of the amount of confidence one has in a statement. We shall not explore these different interpretations here, but some places to start are listed in Section 5.8.

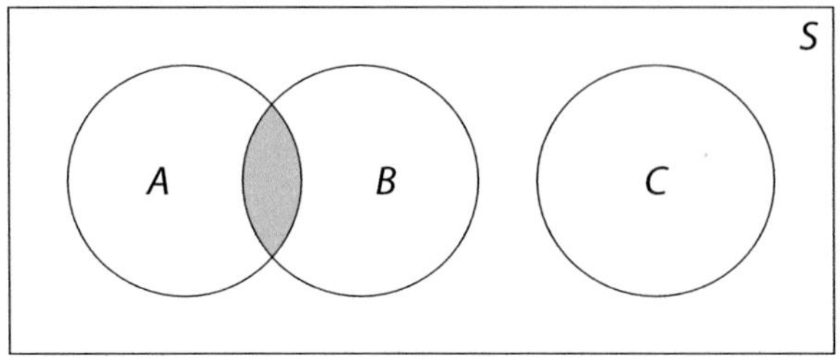

Figure 5.1 Using a Venn diagram to help with combining probabilities. S is the sample space, which contains three possible outcomes, A, B, and C. The shaded area represents cases where A and B occur together (i.e., $A \cap B$). C is disjoint from A and B, indicating that C cannot occur if either A or B occurs and vice versa.

Once we know the probability of a single event occurring, we can start to combine probabilities and ask for the probability of two earthquakes occurring in the same place within a year of each other, or the probability of rain occurring at a given location for three days in a row. Combining probabilities of multiple events can sometimes be confusing, and drawing a Venn diagram can be a useful tool to help us. Let us consider a sample space (S) with three possible outcomes, A, B, and C (Figure 5.1). Notice that the circles representing outcomes A and B overlap. This means that A and B can occur together. The circle representing C does not overlap with either A or B, showing that A and C are *mutually exclusive*, as are the events B and C—they do not occur together. For example, consider rolling a six-sided die with faces numbered 1 through 6 and coloured such that faces with even numbers are blue and those with odd numbers are green. The two outcomes, rolling the die and having a green face facing upward, and rolling the die and having a face with an even number facing upward, are mutually exclusive outcomes; they cannot occur together. This also shows us that mutually exclusive events are not independent; i.e., A and C in Figure 5.1 are not independent events. Two events are *independent* if having one of them occur does not affect the probability of the other one occurring. For example, if we separately toss two coins, the outcome of tossing one of the coins does not affect the outcome of the other.

Exercise 5.1.1 If you draw one card from a pack of cards and then immediately draw another without replacing the first, are the outcomes of these two events independent or not, and why?

Deciding whether or not events are independent or not is not always easy. For example, if it rained where you are yesterday, there might be a higher probability of it raining again today because a persistent weather pattern has occurred. This means that the probability of it raining today is not necessarily independent of the probability of it having rained yesterday. So, we should always give some thought as to whether or not the events we are interested in are truly independent or not.

The probability that both A *and* B occur is represented by the shaded area of Figure 5.1 and, if A and B are independent, is given by the product of the probability of A occurring and the probability of B occurring:

$$P(A \text{ and } B) = P(A \cap B) = P(A) \times P(B). \tag{5.4}$$

Why do we multiply the probabilities? If $P(A) = 0.5$ and $P(B) = 0.25$, then we expect that B will occur one quarter of the times that A occurs, so long as A and B are independent of each other; i.e., the probability of A occurring does not affect the probability of B occurring.

The probability that one outcome *or* another occurs is a little more involved. If the two outcomes are mutually exclusive (e.g., A and C in Figure 5.1), then

$$P(A \text{ or } C) = P(A \cup C) = P(A) + P(C). \tag{5.5}$$

For example, the probability of rolling a die and obtaining a 5 *or* a 6 would be $(1/6)+(1/6) = 1/3$. However, if the two outcomes are not mutually exclusive (e.g., A and B in Figure 5.1), then by simply adding the probabilities we are also including those cases where A *and* B occur together (the shaded area in Figure 5.1). So, we have to subtract the probability of $P(A \cap B)$, to give

$$P(A \text{ or } B) = P(A \cup B) = P(A) + P(B) - P(A) \times P(B). \tag{5.6}$$

Example 5.1 Let us look at this using the example of drawing a single card from a shuffled deck. What is the probability that the card you draw is a club *or* an ace? Because there is a card that is the ace of clubs, these two events are not mutually exclusive (they occur together as the ace of clubs). If A denotes that the card is a club and B that the card is an ace, then $P(A) = (13/52)$ and $P(B) = (4/52)$. There is only one ace of clubs in a deck of cards, so $P(A \cap B) = (1/52)$. The probability $P(A \cup B) = (13/52) + (4/52) - (1/52) = (4/13)$.

Exercise 5.1.2 What is the probability that a randomly selected person:

1. has a birthday in a month that ends with the letter "y"?
2. has a birthday in a month that has 31 days?
3. has a birthday in a month that has 31 days and ends in the letter "y"?
4. has a birthday in a month that has 31 days or ends in the letter "y"?

The probability of combined events and their complements are related by *De Morgan's law*:[2]

$$(A \cup B)^c = A^c \cap B^c \qquad \text{and} \qquad (A \cap B)^c = A^c \cup B^c. \tag{5.7}$$

Exercise 5.1.3 Use Venn diagrams to convince yourself that DeMorgan's law is true.

Frequently in science we find ourself working with *conditional probabilities*. These are situations where the probability of an outcome depends on another event already having occurred. For example, if we already know that a randomly selected person was born in a month with 31 days in it, what is the probability that they were born in a month with an "r" in the name? The fact that we already know they were born in a month with 31 days excludes February, April, June, September, and November from the list of possible months we should consider (i.e., excluding them from the sample space), leaving only those months having 31 days and whose names include the letter "r." This is the intersection of the two

[2] This law is named after the British logician Augustus De Morgan (1806–1871).

sets A = {January, March, May, July, August, October, December} and B = {January, February, March, April, September, October, November, December}, so that $A \cap B$ = {January, March, October, December} and the probability of B occurring given that A has already occurred is 4/7. We write conditional probabilities as $P(B \mid A)$, which we read as the "probability of B occurring *given that A has already occurred*." This is different from the probability of an "and" event (Equation (5.4)), where the second outcome is not conditional on the outcome of the first event. In our example, the probability $P(B \mid A) = 4/7$. This is the conditional probability of B given A, which is not the same thing as $P(B \cap A) = 1/3$. However, if we look at this result carefully we can see that $P(B \mid A)$ is related to $P(B \cap A)$ by

$$P(B \mid A) = \frac{P(B \cap A)}{P(A)} \tag{5.8}$$

for $P(A) \neq 0$. This shows that to calculate the conditional probability $P(B \mid A)$ we are basically finding the fraction of outcomes A that are also in B. For our month example we get

$$P(B \mid A) = \frac{P(B \cap A)}{P(A)} = \frac{4/12}{7/12} = 4/7,$$

which agrees with the answer we found before.

Example 5.2 The *residence time* is an important parameter for understanding pollution and the fate of substances in the environment. It is an estimate of the characteristic time that a substance stays within a region, such as a lake or an estuary, before being transformed or flushed out. Let us consider a lake that is well-mixed by winds and water flow. The probability $P(t)$ that a substance has a residence time $r(t)$ greater than t hours is given by $P(r(t)) = e^t$. If we know that the chemical pollutant has been in a lake for more than three hours, what is the probability that it will remain in the lake for more than four hours?

We need to calculate the conditional probability of obtaining a residence time greater than four hours given that the chemical has been in the lake for longer than three hours. Using Equation (5.8), we need to calculate

$$P(r(4) \mid r(3)) = \frac{P(r(4) \cap r(3))}{P(r(3))} = \frac{P(r(4))}{P(r(3))} = \frac{e^{-4}}{e^{-3}} \approx 0.37,$$

where we have used the fact that if the substance has been in the lake for four hours, it has definitely been there for longer than three hours. So there is about a 37% probability that the pollutant will be in the lake for more than four hours after it entered the lake.

Note that we can rearrange Equation (5.8) to give $P(A \cap C) = P(A \mid C)P(C)$, which provides another way to calculate $P(A \cap C)$ if finding the conditional probability is easy. We also know that $P(A) + P(A^c) = 1$, and a similar result also holds for conditional probabilities:

$$P(A \mid C) + P(A^c \mid C) = \frac{P(A \cap C)}{P(C)} + \frac{P(A^c \cap C)}{P(C)} = \frac{P(A \cap C) + P(A^c \cap C)}{P(C)} = \frac{P(C)}{P(C)} = 1,$$

where we have used the fact that C can be subdivided into two parts, the part that is also in A ($A \cap C$) and the part that is not in A ($A^c \cap C$).

Exercise 5.1.4 At a given location the probability that the maximum daily temperature $T_{\max} \leq 20°\text{C}$ on days that it rains is 0.3, and the probability that $T_{\max} \leq 20°\text{C}$ on days that it does not rain is 0.1. The probability that it will rain tomorrow is 0.7.

1. What is the probability that the maximum temperature tomorrow will be greater than 20°C?
2. Given that the maximum temperature tomorrow is less than or equal to 20°C, what is the probability that it will rain?

Note that an event A is independent of event B if $P(A \mid B) = P(A)$; i.e., the probability that A occurs given that B has occured is just the probability that A occurs. So, to show that two events are independent, we only need to show one of the following conditions is true, whichever is easiest,

$$P(A \mid B) = P(A), \qquad P(B \mid A) = P(B), \qquad P(A \cap B) = P(A) \times P(B).$$

Conditional probabilities are important for scientists because in many cases the results we are interested in are conditional on other factors. Let us look at an example to illustrate this. Assume that we are interested in the biogeochemical cycling of nitrogen in the environment. We have found a microbe that is able to convert nitrate to nitrogen using a novel biochemical pathway, and we have identified the genes responsible for this pathway in the microbe's DNA. From this, we have developed a test for the presence of these genes, but the test is not perfect. After doing a survey of many locations we find that the probability of finding these genes (G) in a sample is $P(G) = 0.02$, so they are quite rare. If the gene is present in the sample, then the probability of our test being positive (P) is $P(P \mid G) = 0.7$, but the probability of the test showing a positive result when the genes are absent (a false positive) is $P(P \mid G^c) = 0.1$.

Now, we collect samples from around the world and want to know if the gene is present in them or not. We know that we can get a positive test result whether the sample contains the genes or not, so we can ask what is the probability that an arbitrary sample tests positive? The tested sample either has the genes in it (G) or it does not (G^c), but either way we can get a positive test result (P). If the set of samples that test positive is P, then $P(P) = (P \cap G) \cup (P \cap G^c)$; i.e., the probability of a positive result is the probability of getting a positive result if the gene is in the sample *and* the probability of getting a false positive. But $(P \cap G)$ and $(P \cap G^c)$ have no elements in common, so they are mutually exclusive and we can write the probability of getting a positive test result as $P(P) = P(P \cap G) + P(P \cap G^c)$. Let us look at the first term on the right-hand side. Using Equation (5.4) we can write $P(P \cap G) = P(P) \times P(G)$. We know $P(G)$, but we do not yet know $P(P)$. However, we do know the conditional probability $P(P \mid G)$, so we can use Equation (5.8) to write $P(P \cap G) = P(P \mid G)P(G)$. Similarly, we can write the second term as $P(P \cap G^c) = P(P \mid G^c)P(G^c)$, so that the probability of getting a positive test result from an arbitrary sample is

$$P(P) = P(P \mid G)P(G) + P(P \mid G^c)P(G^c) = 0.7 \times 0.02 + 0.1 \times (1 - 0.02) = 0.112,$$

which is a little greater than the probability of getting a false positive.

We know that our test is not perfect and that we can get both false negative and false positive results. So, we want to know, if a test on a sample is positive, what is the probability that the gene really is in that sample? That is, we want to know the conditional probability $P(G \mid P)$, but we know $P(P \mid G)$. This is where we make use of Bayes' theorem, which tells us how to relate these two conditional probabilities. It is important to note that the logic of the two conditional probabilities is very different: $P(G \mid P)$ is the probability of the gene being present in a sample given that the test on that sample is positive, whereas $P(P \mid G)$ is the probability of the test being positive given that the gene is in the sample. Bayes' theorem states that

$$P(G \mid P) = \frac{P(P \cap G)P(G)}{P(P)} = \frac{P(P \mid G)P(G)}{P(P \mid G)P(G) + P(P \mid G^c)P(G^c)}. \tag{5.9}$$

To show that this is true we note that $P(G \mid P)P(P) = P(G \cap P)$. But we can also write $P(G \cap P) = P(P \mid G)P(G)$, so equating these expressions gives the first equality in Equation (5.9); the second equality comes from using our result above for $P(P)$ (you will see Bayes' theorem written in both ways). Now, we can calculate our conditional probability as $P(G \mid P) = 0.125$. So, if the test is positive, there is approximately 12% probability that the gene actually is in the sample. This is telling us that we should probably try and improve the accuracy of our test.

5.2 Random Variables, Expectation, and Variance

A *random variable* is a variable whose value is given by the outcome of a random process, and examples include the number of decays of an unstable isotope in a given amount of time, or the amount of daily precipitation at a given location. We will distinguish between the variable and its value by writing the variable using capital letters (e.g., X) and the value it takes using the corresponding lowercase letter (x). For example, we would write the daily precipitation at a given location as X, and if the precipitation at that location on January 31 was 1 cm, then $x = 1$ cm for that date. We will write the probability that a random variable X has a specific value x as $P(X = x)$.

If we know we are dealing with a random variable, it is reasonable to ask what is the most likely value that the variable can have? To find an answer to this, let us start by thinking about the arithmetic mean ($\bar{X}$) of a set of ten random numbers, e.g., $X = (1, 1, 5, 7, 6, 2, 9, 8, 2, 5)$:

$$\bar{X} = \frac{1+1+5+7+6+2+9+5+2+5}{10} = \frac{1}{N}\sum_{j=1}^{N} x_j = \sum_{j=1}^{N} \frac{x_j}{N}. \tag{5.10}$$

We can rearrange this sum by collecting together the occurrences of the same number,

$$\bar{X} = \left(\frac{2\times 1}{10}\right) + \left(\frac{3\times 5}{10}\right) + \left(\frac{2\times 2}{10}\right) + \left(\frac{1\times 7}{10}\right) + \left(\frac{1\times 6}{10}\right) + \left(\frac{1\times 9}{10}\right) = \sum_{j=1}^{10} x_j \left(\frac{n_{x_j}}{10}\right),$$

where n_{x_j} is the number of times that the number x_j appears in the sequence, so $n_{x_j}/10$ is the frequency that the number x_j appears in the sequence of random numbers. We define the *expected value* of the random variable as

$$E(X) = \sum_{j=1}^{M} x_j P(X = x_j), \tag{5.11}$$

where M is the number of possible values that X can take (in this case, ten). It is tempting to think of the expected value as being simply the mean value, but that is not necessarily the case, as Example 5.3 shows.

Example 5.3 Consider the outcome of rolling an unbiased, six-sided die with the sides labelled 1 through 6. The probability of obtaining any one of these values with a single roll of the die is 1/6. From Equation (5.11) we can calculate the *expected value* of a single roll of the die:

$$E(X) = 1 \times \frac{1}{6} + 2 \times \frac{1}{6} + 3 \times \frac{1}{6} + 4 \times \frac{1}{6} + 5 \times \frac{1}{6} + 6 \times \frac{1}{6} = 3.5.$$

If we roll the die ten times and get the values 6, 5, 1, 3, 1, 2, 2, 5, 1, 3, the mean value is 2.9. This is called the *sample mean* and is the mean value of a finite set of samples of a random variable. The difference between the mean and expected values is 0.6. The reason for this difference can be seen by looking at the frequencies of occurrence of each number in the sequence of rolls: e.g., the value 6 occurred one time, and the value 1 occurred three times. The sample of ten rolls is not sufficiently large to give us the real frequencies of the different outcomes. If we rolled the die many, many more times, we would expect that these frequencies would each converge to 1/6, and the mean and the expected value would be the same. This illustrates a problem with using finite samples to determine probabilities when Equation (5.2) requires a sample of infinite size. As the size of the sample increases, so the expected value and the sample mean start to converge, as can be seen in Figure 5.2. This problem of finite size samples arises repeatedly when analyzing real data.

Exercise 5.2.1 What is the value of $E(c)$ if c is a constant and not a random variable?

The expected value of a random number characterizes the value of the random variable, but it does not tell us anything about the range of values the random variable can take. For example, if the faces of the die in Example 5.3 were labelled 15, 12, 7, 0, −3, and −10, the expected value would still be 3.5, but the range of possible values that each roll can produce is much larger (−10 to 15 instead of 1 to 6). This spread in values can be characterized by the *variance*, and it is defined by

$$\text{Var}(X) = E[(X - E(X))^2] = \sum_{i=1}^{M} P(X = x_i)(x_i - \mu)^2, \tag{5.12}$$

where $\mu = E(X)$. Let us look at this equation a little more closely. First, it involves the square of the distance of each value $(X = x_i)$ from the expected value. This means that

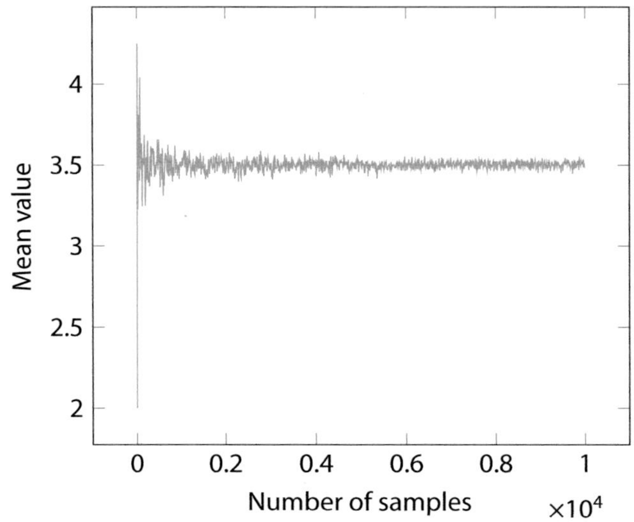

Figure 5.2 The behavior of the mean as the sample size increases. This figure shows the result of repeated simulations of rolling a single die, with each simulation using progressively more rolls (i.e., samples). As the number of times we roll the simulated die increases from 2 to 10^4, the mean of the simulated rolls gets closer to the expected value of 3.5, and the departures from that value decrease.

values of X less than $E(X)$ will not cancel with values greater than $E(X)$. Second, we multiply each term in the sum by the probability that the value of $X = x_i$ occurs; in other words, we are taking a weighted mean of the squared distances, with more probable values of X contributing more to the value of the sum. The variance is thus the expected value of the square of the differences between the values of the random variable and its expected value.

Any measurement that we make has some associated random uncertainty. This uncertainty might be very small, but it is there and it is a measure of the precision of the measurement. Measured values should be written as, for example, 2.45 ± 0.01, where the first value is the mean value of the measurements that were taken, and the second is the measure of uncertainty. We might be tempted to use the variance as this uncertainty, but there is a problem with this: the dimensions of the variance are the *square* of the dimensions of the measured value, so we cannot add the variance to the mean value. Instead of the variance, the *standard deviation* (σ) is used to express measurement error. This is defined by

$$\sigma = \sqrt{\text{Var}(X)}, \tag{5.13}$$

which overcomes the dimensional problems with the variance.

The expected value and variance of a random variable have useful properties. Consider two sequences of random numbers, $A = \{a_1, a_2, a_3 \ldots\}$ and $B = \{b_1, b_2, b_3 \ldots\}$. What is the expected value of $A + B$? We now have two random numbers to deal with, and we write the *joint probability* of A having a value a_i and B having a value b_j as $P(A = a_i, B = b_j)$. For example, in an experiment looking at spatial interactions between objects, we might want to place 1000 objects randomly in a plane by specifying their (x, y) coordinates; so the two random values are X and Y, and their values will be the x and y coordinates of the points. The joint probability is written as $P(X = x_i, Y = y_j)$ and it is the probability that the point will have coordinates (x_i, y_j). We can define the *marginal probability* as the probability that the x coordinate of a point is x_i, irrespective of the value of y_j. To calculate this we need to sum over the probabilities of all the possible values for Y, so

$$P(X = x_i) = \sum_y P(X = x_i, Y = y). \tag{5.14}$$

Now, using Equation (5.11) we can write the expected value of $A + B$ as

$$E(A + B) = \sum_i \sum_j (a_i + b_j) P(A = a_i, B = b_j), \tag{5.15}$$

which we can rearrange using Equation (5.14) to give

$$\begin{aligned} E(A + B) &= \sum_i \sum_j a_i P(A = a_i, B = b_j) + \sum_i \sum_j b_j P(A = a_i, B = b_j) \\ &= \sum_i a_i \sum_j P(A = a_i, B = b_j) + \sum_j b_j \sum_i P(A = a_i, B = b_j) \\ &= \sum_i a_i P(a_i) + \sum_j b_j P(b_j) = E(A) + A(B). \end{aligned}$$

So, the expected value of the sum of random variables is simply the sum of the expected values of the individual random variables.

Exercise 5.2.2 Show that $E(\beta A) = \beta E(A)$, where β is a constant.
Exercise 5.2.3 Show that if A and B are independent, then $E(AB) = E(A) \times E(B)$.

These results show us that calculating the expected value of a random variable is a linear operation. Calculating the variance, however, is not a linear operation. We can see that by looking at $\mathrm{Var}(cA)$,

$$\mathrm{Var}(cA) = E[(cA - cE(A))^2] = E[c^2(A - E(A))^2] = c^2 E[(A - E(A))^2] = c^2 \mathrm{Var}(A).$$

This is not a linear operation because the factor c gets squared in the operation of calculating the variance.

Excercise 5.2.4 Prove the following properties of the variance:

1. $\mathrm{Var}(X) = E(X^2) - (E(X))^2$.
2. If α and β are constants, then $\mathrm{Var}(\alpha X + \beta) = \alpha^2 \mathrm{Var}(X)$.
3. $\mathrm{Var}(X + Y) = \mathrm{Var}(X) + \mathrm{Var}(Y)$ for independent random variables X and Y.

The mathematical function that tells us the probability of obtaining a given outcome (x) for a random variable (X) is called the *probability distribution function*[3] or PDF, $p_X(x)$. For example, the PDF for a fair, six-sided die is $p_X(x) = 1/6$, where $x = 1, 2, 3, 4, 5, 6$. If

[3] Sometimes also called the *probability mass function*.

we had a dishonest die that preferentially gave us a value of 6 when rolled, with all other values having the same probability, then a PDF might be

$$P(X = x) = p_X(x) = \begin{cases} 2/15 & x = 1, 2, 3, 4, 5 \\ 1/3 & x = 6 \end{cases}.$$

The PDF has to describe a probability, so not all mathematical functions can be used as probability distribution functions. To start with, negative probabilities make little sense because of Equation (5.2), so this means that $p_X(x) \geq 0$. Also, because we require the sum of the probabilities of all possible outcomes of a random process to equal 1, the PDF must satisfy $\sum_i p_X(x_i) = 1$. What is interesting is that certain types of random process have very specific types of PDF. This can lead to a very powerful way of characterizing random variables in the environment, such as rainfall, without having to accurately represent the myriad processes that produce rainfall. There are many useful probability distribution functions, and we will look at some of the more common ones.

5.3 Discrete Random Variables

We will start by looking at some of the common distributions for *discrete random variables*, i.e., random variables that take specific, discrete values. For example, the result of flipping a coin is a discrete random variable because it can take one of only two values, heads or tails, which we could represent using the integers 0 and 1. A more interesting example might be the number of occurrences of an invasive species within a given area, a random number that also takes integer values.

5.3.1 Discrete Uniform Distribution

The simplest probability distribution is probably the discrete uniform distribution, where the probability of each possible outcome is the same (Figure 5.3). For example, for a fair, six-sided die, the probability of rolling the die and getting a value of 2 is 1/6, which is the same as the probability of rolling a value of 5. For an imaginary die with m sides, the probability of rolling the die and getting any specific integer k between 1 and m is

$$P(X = k) = p_X(k) = \frac{1}{m}. \tag{5.16}$$

Using Equation (5.11), the expected value for a uniform random variable (X) that can take integer values between 1 and m is

$$E(X) = \sum_{k=1}^{m} kP(X = k) = \sum_{k=1}^{m} \frac{k}{m} = \frac{1}{m}(1 + 2 + \cdots + m).$$

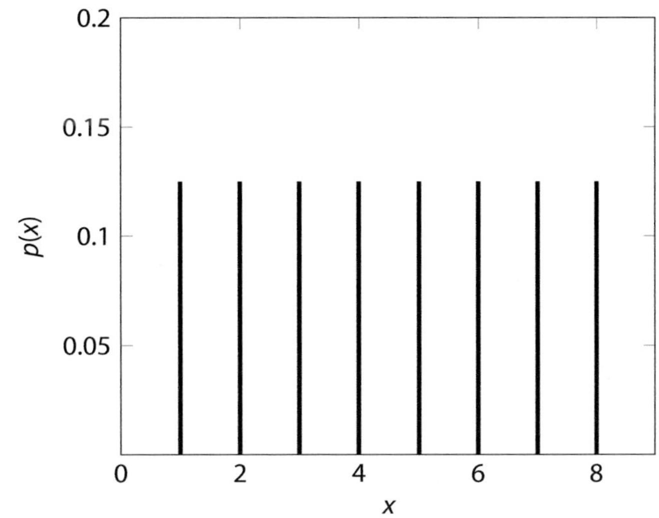

Figure 5.3 An example of a uniform random probability distribution where the random variable X can have integer values between 1 and 8, each occurring with an equal probability of $1/8 = 0.125$.

We know from Equation (3.3) that $(1 + 2 + \cdots + m) = (m(m + 1))/2$, so that

$$E(X) = \frac{1}{m}\sum_{k=1}^{m} k = \frac{1}{m}\frac{m(m+1)}{2} = \frac{m+1}{2}. \tag{5.17}$$

Similarly, the variance is

$$\sigma^2 = \frac{(m^2 - 1)}{12}. \tag{5.18}$$

Exercise 5.3.1 Prove Equation (5.18).

The random variable in Equation (5.16) took values from 1 to m. We can make this more general and consider a random variable X that can take integer values between a and b such that $a \leq x \leq b$, with the probability of X having any of these values being the same. Then, the PDF can be written as

$$P(X = x) = p_X(x) = \begin{cases} \frac{1}{b-a+1} & a \leq x \leq b \\ 0 & \text{otherwise} \end{cases}, \tag{5.19}$$

where x is an integer. In this case, the expected value and the variance can be written as

$$E(X) = \frac{a+b}{2}, \quad \text{Var}(X) = \frac{(b-a+1)^2 - 1}{12}. \tag{5.20}$$

Exercise 5.3.2 Show that the expected value and variance of the PDF given by Equation (5.19) are given by the expressions in Equation (5.20).

Another useful way to describe the probability distribution is the *cumulative distribution function* or CDF. This is defined as the probability distribution of getting an outcome less than or equal to a given value, i.e.,

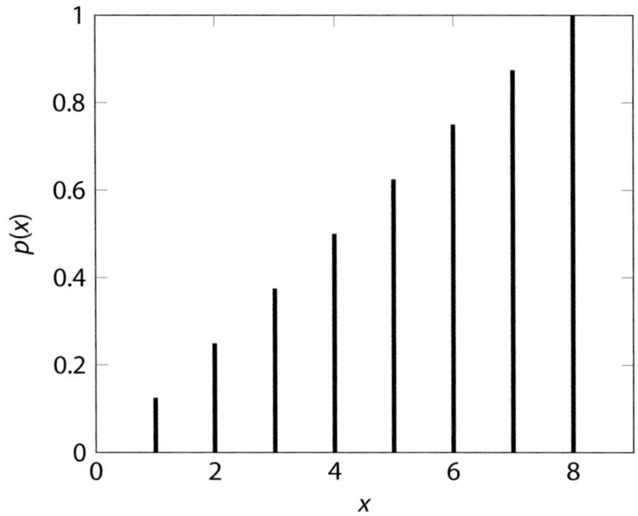

Figure 5.4 The cumulative distribution for the uniform random variable with probability distribution shown in Figure 5.3. In this case, the CDF is a sequence of uniform steps from 0.125 to 1.0. The probability of $x = 1$ is 0.125, the probability of $x = 1$ or $x = 2$ is $p(1) + p(2) = 0.25$, and so on.

$$F_X(x) = P(X \leq x) = \sum_{k \leq x} p_X(k), \tag{5.21}$$

where the sum is over all values of k less than or equal to x. For our uniform distribution in Figure 5.3, the cumulative distribution function is a series of uniform steps (Figure 5.4). For the uniform probability distribution function (Equation (5.19)), the CDF is

$$F_X(x) = \frac{x - a + 1}{b - a + 1}, \quad a \leq x \leq b. \tag{5.22}$$

We can see from Equation (5.21) that the CDF has some useful properties. First, if the random variable X can take values between $x = a$ and $x = b$, then the maximum value of the CDF is 1.0 and occurs for $x = b$ (Figure 5.4). Also, because the PDF cannot be negative, the CDF is monotonically increasing. If we wanted to find the probability that X had a value between x_1 and x_2 ($x_1 < x_2$), then $P(X = x; x_1 < x \leq x_2) = F_X(x_2) - F_X(x_1)$. This means that if we know the cumulative distribution, we can use it to calculate the probability for obtaining different ranges of values.

Exercise 5.3.3 Consider the function

$$f_X(x) = \begin{cases} \gamma x & x = 1, 2, 3, 4 \\ 0 & \text{otherwise} \end{cases}$$

for integer values of x.

1. Use the fact that the sum of a PDF must equal 1 to determine the value of γ needed to make $f_X(x)$ a PDF.
2. Calculate the CDF of $f_X(x)$.

3. Calculate the expected value, $E(X)$, of $f_X(x)$.
4. Calculate the variance of $f_X(x)$.

There are two other useful characteristics of a PDF that we need to know, the *median* and the *mode*. The median is the value of the random variable that lies at the middle of the distribution. Formally, we can write that the median m is the value of the random variable X such that

$$P(X \leq m) \geq \frac{1}{2} \quad \text{and} \quad P(X \geq m) \geq \frac{1}{2}. \tag{5.23}$$

For the uniform distribution given in Equation (5.19), the median $m = (a + b)/2$. The median is a useful measure because it is less affected by extreme values than the mean. For example, assume an experiment produced the integer numbers 5, 999, 1, 9, 6, 5, 3. Then to calculate the median we arrange them in ascending order (1, 3, 5, 5, 6, 9, 999) and choose the middle value; in this case, 5—half the values are less than 5 and half are larger than 5. However, the arithmetic mean of these numbers is 146.8. The mode is the most likely value of a random variable. For the uniform distribution, all values are equally likely so the distribution does not have a mode. In our imaginary data set, the mode is the value 5—it is the value that occurs the most, so it is the most likely one.

5.3.2 Binomial Distribution

A random variable that has only two possible outcomes (e.g., true or false, presence or absence, heads or tails) is called a *Bernoulli random variable*.[4] A *Bernoulli trial* is a single experiment or process that can have one of only two outcomes. A simple example is a single toss of a coin, which can give only heads or tails. Presence or absence studies are also Bernoulli trials where, for example, we are looking for the presence or absence of a specific mineral in a geological sample, or the presence of a specific organism in a geographic region. It is often useful to label the two outcomes of a Bernoulli trial as $x = 1$ and $x = 0$; for example, when tossing a coin we could label heads as $x = 1$ and tails as $x = 0$, or in a presence/absence experiment we could label presence by $x = 1$ and absence by $x = 0$. Then, if p is the probability of having a successful outcome (e.g., $x = 1$) we can write

$$P(X = 1) = 1, \qquad P(X = 0) = 1 - p.$$

You will often see Bernoulli variables denoted by $X \sim \text{Ber}(p)$ or $X \sim \text{Bernoulli}(p)$. The PDF of a Bernoulli random variable is

$$p_X(x) = \begin{cases} p & \text{for } x = 1 \\ 1 - p & \text{for } x = 0 \\ 0 & \text{otherwise} \end{cases}, \tag{5.24}$$

which is often written as $p_X(x) = p^x(1 - p)^{1-x}$, where $x = 0$ or 1.

[4] This probability distribution is named after Jacob Bernoulli (1655–1705), one of many mathematicians and scientists in the Swiss Bernoulli family.

Exercise 5.3.4 Show that for a Bernoulli random variable, $E(X) = p$, $E(X^2) = p$, and $\text{Var}(X) = p(1-p)$.

Usually we are interested in the results of multiple Bernoulli trials. For example, let us say that we conduct a one-time survey of 260 coastal salt marshes looking for the presence of an invasive new species.[5] Each marsh that we survey is a single Bernoulli trial, so our whole survey consists of 260 Bernoulli trials. For each trial we can use $x = 1$ to indicate that the invasive species is present, and $x = 0$ to indicate that it is absent. We would like to know the probability that the invasive species is found in any 10 sites, but *not* in the remaining 250.

To answer this, we need to make some assumptions. First, we assume that each Bernoulli trial (i.e., each survey of an individual marsh) is independent from all the others, so that finding the invasive species in one marsh does not affect the probability of finding it in any others. This assumption needs to be justified on a case-by-case basis and, in this example, we would have to use our understanding of the natural processes of how the plant spreads. But let us proceed assuming the separate trials are independent, so we can simply multiply probabilities. Second, we assume that the probability of finding our target species is the same for all the marshes. One way to do this is to survey all 260 marshes and then calculate the frequency with which the invasive species is found in any marsh. For example, say we find it in 13 of the 260 marshes, then we can set the probability of finding that species in any one marsh as $p(x = 1) = 0.05$. This means that the probability that it is absent in a marsh is 0.95.

We can use Equation (5.4) to write the probability that the species occurs in 10 marshes as $p(x = 1)^{10} = 0.05^{10} \approx 9.766 \times 10^{-14}$. But this says nothing about the remaining 250 sites, so we need to multiply (an "and" process) by the probability that it is absent from them, giving

$$p = (0.05)^{10} \times (1 - 0.05)^{250} \approx 2.634 \times 10^{-19}.$$

But the probability we have just calculated is for the probability of occurrence in 10 marshes; however, we want the probability that it occurs in **any** 10 marshes, so we need to calculate how many different sets of 10 marshes we can select from 260. There are 260 possible choices for our first marsh, 259 choices for the second one, 258 for the third, and so on, so the total number of ways we can select 10 marshes from 260 is

$$260 \times 259 \times \cdots \times 251 \approx 1.185 \times 10^{24}.$$

In general, the number of ways we can choose x objects from n possibilities is

$$n(n-1)(n-2)\dots(n-x+1).$$

This sort of looks like $n!$, but instead of the factors decreasing all the way to 1, they stop at $(n-x+1)$; in other words, it is $n!$, but missing all factors from $(n-x)$ to 1. So, we can write

$$n(n-1)\dots(n-x+1) = \frac{n(n-1)\dots(n-x+1)(n-x)(n-x-1)\dots 1}{(n-x)(n-x-1)\dots 1} = \frac{n!}{(n-x)!}.$$

[5] For example, the salt marsh grass *Spartina alterniflora* is native to the East and Gulf coasts of the United States, but it is an invasive species in other places such as the west coasts of the United States and China.

But now we have another counting problem—we could choose the same 10 sites but in a different order. That is, we have not distinguished between choosing marsh numbers 1, 2, 3, 4, 5, 6, 7, 8, 9, 10 and 10, 9, 8, 7, 6, 5, 4, 3, 2, 1, yet the two sets of marshes are clearly the same. To correct for this, we need to divide our answer by the number of ways of arranging ten objects (different arrangements imply different ordering). This is just 10!, or $x!$ in the general case. This means that the number of ways of selecting x objects from a total of n is

$$\frac{n!}{(n-x)!\,x!} = \binom{n}{x}, \tag{5.25}$$

which is called the *binomial coefficient*. Our final probability is then given by the *binomial distribution*,

$$P(x) = \frac{n!}{(n-x)!\,x!} p^x (1-p)^{n-x}. \tag{5.26}$$

A random variable X that obeys a binomial distribution is often written as $X \sim \text{Bin}(n, p)$, and we can use Equation (5.26) to calculate the probability of finding the invasive species in only 10 sites from the survey; the answer is 0.086, or 8.6%. This is the expected probability, so if we found an area with a higher probability than this, then we might suspect that the invasive species has begun to flourish there.

The binomial distribution has two parameters that we need to know, n and p. It is a discrete distribution, and if $p \neq 0.5$ the distribution is skewed and not symmetric about its maximum value (Figure 5.5).

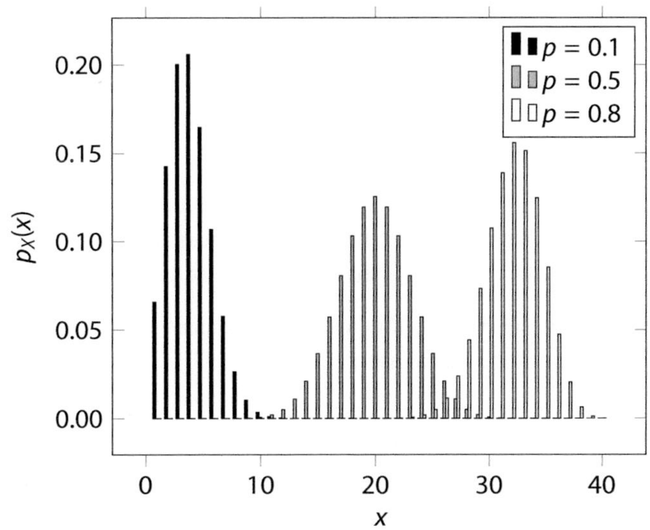

Figure 5.5 The binomial distribution for different values of p. The distribution is symmetric for $p = 0.5$, but it is not symmetric for other values.

We can calculate the expected value of a binomial distribution using Equation (5.11),

$$E(X) = \sum_{k=0}^{n} kP(X = k) = \sum_{k=0}^{n} k \binom{n}{k} p^k (1-p)^{n-k}.$$

This looks ungainly and not particularly helpful, so we want to simplify it a bit if possible. First of all, notice that we can get rid of some of the terms in the summation. For example, the term with $k = 0$ is zero, so we can neglect $k = 0$ in the summation, giving

$$E(X) = \sum_{k=1}^{n} k \binom{n}{k} p^k (1-p)^{n-k} = \sum_{k=1}^{n} \frac{kn!}{(n-k)!\,k!} p^k (1-p)^{n-k}.$$

Both the variables n and p are not affected by the summation index (e.g., $\sum_k npk = np\sum_k k$), so we can take a factor of np out of the summation:

$$E(X) = \sum_{k=1}^{n} \frac{kn(n-1)!}{(n-k)!\,k(k-1)!} pp^{k-1} (1-p)^{n-k} = np \sum_{k=1}^{n} \frac{(n-1)!}{(n-k)!\,(k-1)!} p^{k-1} (1-p)^{n-k}$$

$$= np \sum_{k=1}^{n} \binom{n-1}{k-1} p^{k-1} (1-p)^{n-k}.$$

Now, recalling the binomial theorem (Equation (3.9)) we see that

$$\sum_{k=1}^{n} \binom{n-1}{k-1} p^{k-1} (1-p)^{n-k} = (p + (1-p))^{n-1} = 1,$$

so we are left with something much more useful and insightful:

$$E(X) = \sum_{k=0}^{n} kP(X = k) = np. \tag{5.27}$$

Although we can write down a PDF for the binomial distribution using simple functions such as factorials and powers, we cannot do this for the CDF.[6]

Exercise 5.3.5 Show that the variance of the binomial distribution is $\sigma^2 = np(1-p)$.

Exercise 5.3.6 A biased coin has a probability $p_t = 0.6$ that it will land tails up. If the coin is flipped three times, calculate:

1. The mean and standard deviation of X, the number of tails.
2. The probability that you get at least two tails in the three flips.
3. The probability that you get an odd number of tails in three flips.

The binomial distribution has many applications in the Earth and environmental sciences. In general it applies to situations where

- there are two possible outcomes,
- the total number of observations (n) is fixed,
- each observation is independents of all the others,
- the probability of a successful outcome ($P(x = 1) = p$) is the same for all observations.

[6] The CDF for the binomial distribution is given in terms of what is called the regularized incomplete beta function, a function that is defined in terms of an integral (Abramowitz and Stegun, 1972).

For example, we could use a binomial distribution to determine the number of expected times that peaks in one variable occur at the same time as peaks in another, apparently unrelated variable, or we can calculate how many seismic hotspots we would expect to see at a given distance from the boundary of a tectonic plate (Julian et al., 2015) if their positions were completely random. If we found more than we expect from a random process, then this would indicate that hotspots are clustered near plate boundaries.

5.3.3 Poisson Distribution

The Poisson distribution is another commonly used discrete distribution.[7] It gives the probability of the number of occurrences of an event recorded in a fixed spatial extent or over a fixed interval of time. For example, if we are interested in the spread of vegetation over a landscape, we could use the Poisson distribution to examine the number of seeds counted in square meter areas over the landscape. Many observational techniques make use of radioactive isotopes, and the number of radioactive decays of a radioactive isotope detected per second follows a Poisson distribution. In fact, a simple rule of thumb is that if your variable counts something, then it will likely follow a Poisson distribution. So, the number of earthquakes per decade in a given region can be described by a Poisson distribution.

We can derive the Poisson distribution in several ways, but the simplest is to consider it as a limiting case of the binomial distribution as $n \to \infty$ but $p \to 0$, so that np remains finite. To see this, let us start with Equation (5.26),

$$P(x) = \left(\frac{n!}{(n-x)!\,x!}\right)(p^x)\left((1-p)^{n-x}\right),$$

and look at the behavior of each of the three terms in parentheses. If $n \to \infty$, then $n \gg x$ and the Binomial coefficient is approximately

$$\begin{aligned}\frac{n!}{(n-x)!\,x!} &= \frac{n(n-1)(n-2)\cdots(n-x+1)(n-x)(n-x-1)\cdots(2)(1)}{x!\,(n-x)(n-x-1)\cdots(2)(1)} \\ &= \frac{n(n-1)(n-2)\cdots(n-x+1)}{x!} \approx \frac{n^x}{x!},\end{aligned} \tag{5.28}$$

where, in the last step, we have used the fact that as n gets very large the dominant term in numerator will come from n multiplied by itself x times. Now, let us look at the last term in $P(x)$, $(1-p)^{n-x}$. If $n \gg x$, this will be approximately $(1-p)^n$, which we can expand using the binomial theorem:

$$\begin{aligned}(1-p)^n &\approx 1 - np + \frac{n(n-1)}{2!}p^2 - \frac{n(n-1)(n-2)}{3!}p^3 \cdots \\ &\approx 1 - np + \frac{(np)^2}{2!} - \frac{(np)^3}{3!} \cdots \approx e^{-np},\end{aligned} \tag{5.29}$$

[7] This is named for the French scientist Siméon Poisson (1781–1840) who made many contributions to mathematics and science, though found himself on the wrong side of the debate on the classical nature of light, believing light to be particles, not waves.

where we have used the power series expansion for the exponential function (Appendix B). So, using Equations (5.28) and (5.29) in Equation (5.26) we obtain the *Poisson distribution*,

$$P(x) \approx \left(\frac{n^x}{x!}\right) p^x e^{-np} = \frac{(np)^x}{x!} e^{-np} = \frac{\lambda^x e^{-\lambda}}{x!}, \tag{5.30}$$

where we have written $np = \lambda$, which is often called the *rate parameter* or *shape parameter* and gives the average number of events within a given interval. A random variable X that satisfies a Poisson distribution is often written as $X \sim \text{Poiss}(\lambda)$, reminding us that the Poisson distribution is a *one-parameter distribution*—it depends only on λ.

Example 5.4 On average there are six hurricanes that form in the Atlantic Ocean every year. If we assume that the number of hurricanes per year follows a Poisson distribution, we can calculate the probabilities that there will be a) exactly two hurricanes in a year, b) less than three hurricanes in a year, c) between two and five hurricanes in a year. The average number of hurricanes per year is six, so $\lambda = 6$ and

$$P(x) = e^{-6}\frac{6^x}{x!} \approx 2.479 \times 10^{-3}\frac{6^x}{x!}.$$

So, the probability that there will be exactly two hurricanes is

$$P(2) = e^{-6}\frac{6^2}{2!} \approx 0.045.$$

Equation (5.30) gives us the probability of an exact number of hurricanes. To determine the probability of there being less than three hurricanes we need to calculate the probability of there being ≤ 2 hurricanes; i.e., we need $P(0) + P(1) + P(2)$:

$$P(< 3) = P(\leq 2) = P(0) + P(1) + P(2) \approx 0.062.$$

To find the probability that there will be between two and five hurricanes per year we need to calculate the probability that there will be ≤ 2 hurricanes per year and the probability that there will be ≥ 5 per year so that

$$P(2 \leq x \leq 5) = P(\leq 5) - P(< 2) \approx 0.017.$$

Exercise 5.3.7 If X is a random variable with a mean value of 4 and follows a Poisson distribution, what is the probability that X has a value of a) 2, b) 4, c) 8?

Exercise 5.3.8 The probability that flooding will occur in a given region is 0.02 per year and follows a Poisson distribution. Calculate the probability of a) zero floods in a year, b) two floods in a year.

The Poisson distribution is not symmetric for small values of λ, but as λ increases in magnitude, it becomes more and more symmetric (Figure 5.6).

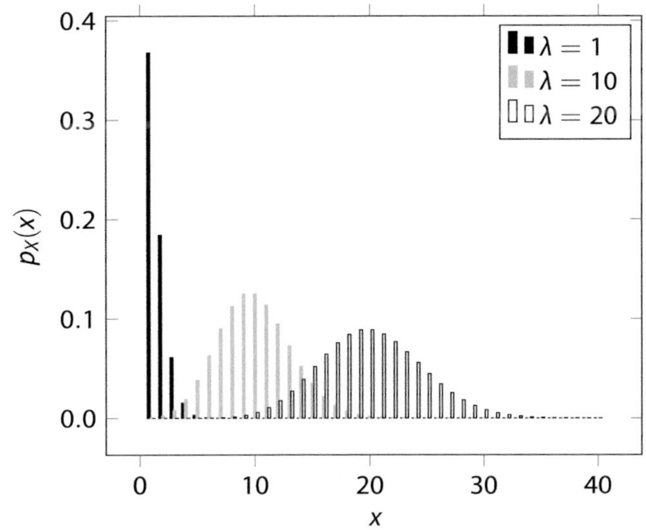

Figure 5.6 Plots of the Poisson distribution (Equation (5.30)) for different values of λ.

We can calculate the expectation and variance of the Poisson distribution; you should see some familiar techniques in these derivations:

$$E(X) = \sum_{x=0}^{\infty} x\frac{\lambda^x}{x!}e^{-\lambda} = \sum_{x=1}^{\infty} x\frac{\lambda^x}{x!}e^{-\lambda} = \sum_{x=1}^{\infty} \frac{\lambda^x}{(x-1)!}e^{-\lambda} = e^{-\lambda}\lambda\sum_{x=1}^{\infty}\frac{\lambda^{(x-1)}}{(x-1)!}$$

$$= e^{-\lambda}\lambda\left(\frac{\lambda^0}{0!} + \frac{\lambda^1}{1!} + \frac{\lambda^2}{2!} + \cdots\right) = e^{-\lambda}\lambda\sum_{x=0}^{\infty}\frac{\lambda^x}{(x)!} = \lambda e^{-\lambda}e^{\lambda} = \lambda. \quad (5.31)$$

Exercises 5.3.9 and 5.3.10 show that $Var(X) = \lambda$.

Exercise 5.3.9 Show that $Var(X) = E(X^2) - (E(X))^2 = E(X(X-1)) + \lambda - \lambda^2$.
Exercise 5.3.10 Show that $E(X(X-1)) = \lambda^2$ and hence that $Var(X) = \lambda$.

These results show that the expectation and variance of the Poisson distribution are the same. The CDF for the Poisson distribution is simply

$$F(y) = e^{-\lambda}\sum_{m=0}^{y}\frac{\lambda^m}{m!}. \quad (5.32)$$

The fact that $E(X) = \text{Var}(x) = \lambda$ for a Poisson distribution leads to a simple way to determine if a spatial or temporal process is truly random or if there is some process that clusters or disperses events. If the events are randomly distributed, then the distribution of the number of points in a given area, for example, will follow a Poisson distribution with $E(X) = \text{Var}(x) = \lambda$ (Figure 5.7). If the points are clustered more than they would be in a random distribution, then $\text{Var}(x) < E(X)$, and if they are dispersed more than in a random distribution, then $\text{Var}(x) > E(X)$.

There are many other discrete probability distributions that can be used to describe various processes, and many of these are described in some of the references listed in

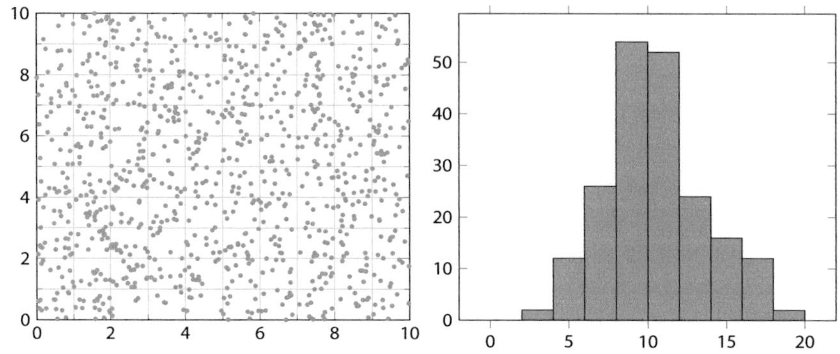

Figure 5.7 Randomly distributed points on a grid. The left-hand panel shows randomly distributed points on a (10×10) grid. The right-hand panel is a histogram of the number of points in a grid cell, and the histogram is a Poisson distribution with a mean value of 10.0 and variance of 10.0202, showing that the points are randomly distributed spatially. The difference between the mean and variance of the histogram arises because of the finite number of points in the example.

Section 5.8. For example, the *negative binomial distribution* can be used to calculate the probability of the *number* of successes in a string of Bernoulli trials.

5.4 Continuous Random Variables

The distributions we have looked at so far are discrete, i.e., the outcomes of the processes can be categorized by "yes or no," "presence or absence," or represent counts that can be represented by integers. However, they are not very helpful for describing random variables that can take noninteger values such as temperature, salinity, concentration, rainfall, etc. For this, we need distributions that are continuous so that we can represent numbers like 27.1, 28.005. But this poses a problem that forces us to think about what probability means for a continuous variable.

To see the problem let us start with a discrete, uniform distribution with the five integer values 1, 2, 3, 4, and 5 (Figure 5.8) and try and convert it into a continuous transformation by gradually increasing the number of points that lie between these integers. We know that the probability of obtaining any specific integer from 1 to 5 is 1/5. Let us add points halfway between these integer values—because our values are discrete, we can always relabel the values using integers. Now, in order to have a total probability of 1, the probability of getting any one of these nine outcomes is $1/9 \approx 0.111$. If we add further points that are halfway between each of the new points, the probability of obtaining any one value is $1/17 \approx 0.0588$. Continuing on this route until we have a continuous set of possible values between 1 and 5 will lead to an infinite number of possible outcomes, each of which has a probability of $1/\infty = 0$ of occurring. This is not particularly useful, but it points to the fact that for a continuous random variable, we cannot talk about the probability of getting

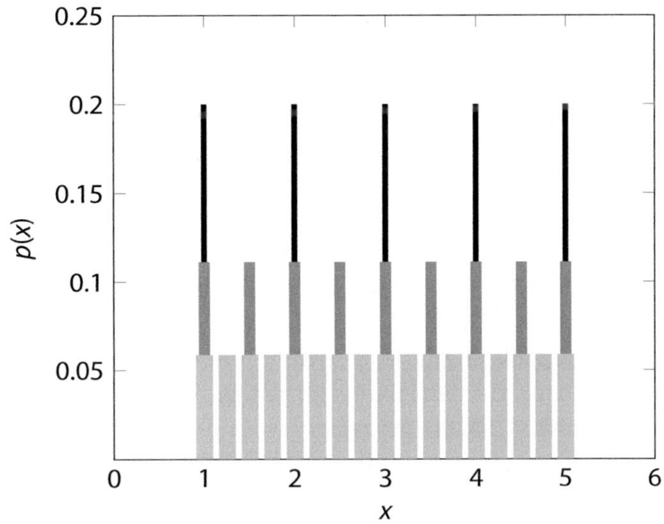

Figure 5.8 Trying to turn a discrete probability distribution into a continuous distribution. Starting with a discrete uniform distribution for the integers 1, 2, . . ., 5 (black bars), we first add points midway between each integer (dark gray bars). The height of the dark gray bars is lower than the black ones in order to ensure that the sum of the probabilities is 1. Adding further points halfway between each dark gray bar further decreases the probability of having a value as an outcome (light gray bars).

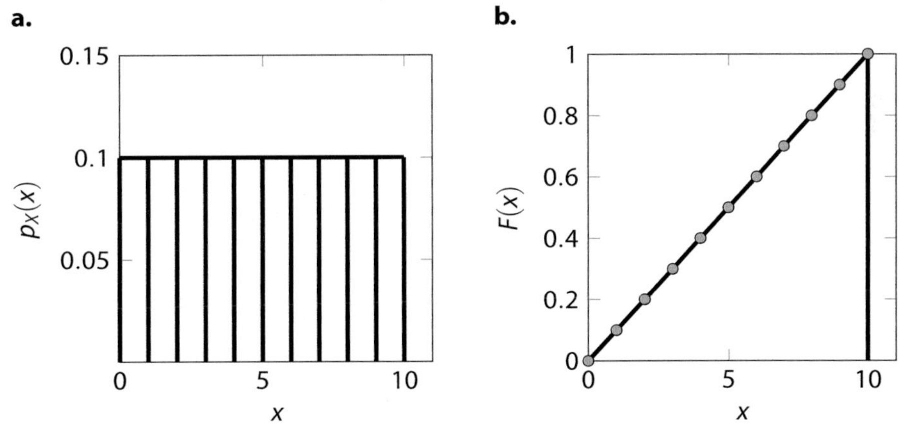

Figure 5.9 The uniform probability distribution for numbers between 0 and 10 (a.). The interval is divided into 10 equal-sized bins. The probability of having an outcome that lies between any two integer values is the area under the curve between those two integers. In (b.) we see the cumulative probability distribution corresponding to the uniform distribution in (a.).

a specific value. Instead, we have to talk about the probability of getting an outcome that lies within a given interval, say x to $x + \Delta x$.

To see how this works, let us take a continuous uniform distribution between 0 and 10 and divide it into ten equal intervals (Figure 5.9). Although, as we have just seen, the probability of getting a specific value is zero, the probability of getting a number that lies between 2 and 3 must be 1/10—the probability of getting *some* number between 0 and 10 must equal 1, so the probability of getting a value that lies in one of these ten equally sized bins must be 1/10. If instead we were to define our bins as being of width 2, then the probability of having an outcome in any bin is $1/5 = 0.2$. This makes sense; the larger the bin size, the greater the probability of randomly selecting a number within that bin. However, this probability value is the same as the area of each of the rectangular bins in Figure 5.9a. This suggests that the probability of getting a value between two numbers a and b for a continuous probability distribution is given by the area under the curve between the values a and b.

Let us look at this in a slightly different way that uses the cumulative probability distribution. The probability of getting a value in the interval 0 to 1 is, as we have seen above, 1/10; the probability of getting an outcome in the interval 0 to 2 is 1/5; the probability of getting a value in the interval 0 to 3 is 3/10, and so on. A plot of this looks like Figure 5.9b. In each case we are asking for the probability of getting an outcome less than or equal to the given value. For the uniform probability distribution, this cumulative probability distribution is a straight line through the origin reaching a maximum of 1—the probability of getting some number between 0 and 10 is 1.

Now, instead of a uniform distribution, assume our distribution is uniform except for numbers in one bin that have a much larger probability of occurring (Figure 5.10). Something interesting happens here. Where there is a greater probability of getting a value, the slope of the cumulative distribution increases, which makes sense because the increment in probability over that bin is greater but the width of the bin is still the same as all the others. In other words, the rate of increase of the CDF increases over that bin. This shows us that the derivative of the CDF is telling us something about how the PDF is

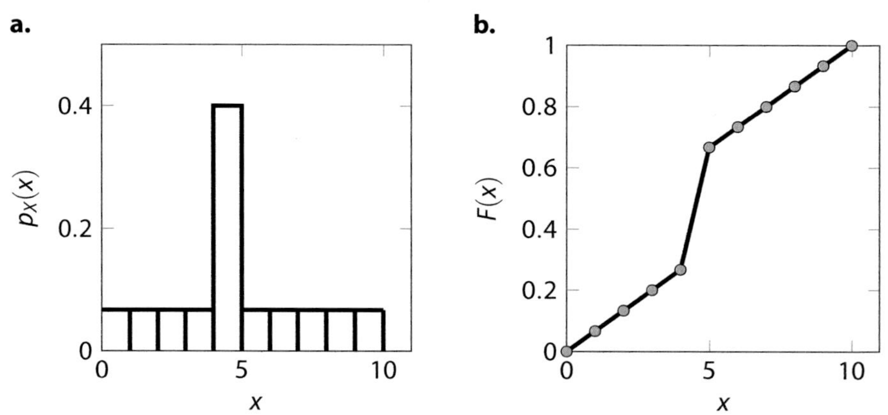

Figure 5.10 A nonuniform probability distribution for numbers between 0 and 10 (a.). Here, the probability of obtaining a number between 4 and 5 is four times the probability of obtaining a number between any other two consecutive integers. The cumulative distribution function (b.) between $x = 4$ and $x = 5$ rises at a steeper rate than between any other consecutive integers to account for the higher probability of obtaining a value in this range.

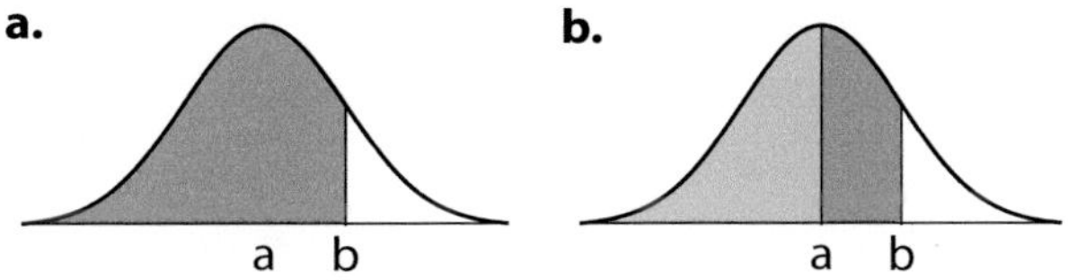

Figure 5.11 Equation (5.34) tells us that to calculate the probability of obtaining a value between $x = a$ and $x = b$ $(b > a)$ we first calculate the area under the curve up to $x = b$ (a.) and then subtract the area under the curve up to $x = a$ (b.).

changing. We can formalize this by saying that if $F(x) = P(X \leq x)$ is the CDF, then the PDF ($p_X(x)$) is given by

$$p_X(x) = \frac{dF}{dx}. \tag{5.33}$$

We can use the fundamental theorem of calculus from Chapter 2 to write the CDF as

$$F(a) = \int_{-\infty}^{a} p_X(x)\, dx,$$

and the probability of x lying between $x = a$ and $x = b$ $(b > a)$ is

$$P(a \leq x \leq b) = F(b) - F(a) = \int_{-\infty}^{b} p_X(x)\, dx - \int_{-\infty}^{a} p_X(x)\, dx = \int_{a}^{b} p_X(x)\, dx. \tag{5.34}$$

All this equation is saying is that to find the probability of obtaining a value of x between $x = a$ and $x = b$, we first calculate the area under the PDF ($p_X(x)$) up to the higher value of x ($x = b$), and subtract from it the area under the curve up to the lower value of x ($x = a$), as in Figure 5.11. The probability of the random variable x lying between a and b is then equal to the area under the PDF between $x = a$ and $x = b$, just as we suspected from Figure 5.9. How do we calculate the areas under these curves? Before computers and high-powered handheld calculators became common place, these were tabulated for specific values. However, most scientific computer languages such as MATLAB™ and Python, and many scientific calculators have functions that will calculate these for you.

We know that for the function $p_X(x)$ to actually represent a PDF it must satisfy certain conditions. The first condition is that $pX(x) \geq 0$ for all values of x. We also know that the sum of the probabilities of all possible values (i.e., summing all the probabilities in the sample space) must equal 1, so for a continuous variable we have

$$\int_{-\infty}^{\infty} p_X(x)\, dx = 1. \tag{5.35}$$

If the range of possible values of x does not extend to $\pm\infty$, but only ranges between $x = u$ and $x = v$, then Equation (5.35) becomes

$$\int_{u}^{v} p(x)\, dx = 1.$$

For example, if x represents the concentration of a substance we are interested in, the lowest value x can take is $x = 0$, rather than $x = -\infty$.

As an example, consider the continuous uniform distribution defined between $x = a$ and $x = b$. We have seen that the CDF for this distribution is a straight line $F(x) = mx + \beta$. The PDF is then

$$p_X(x) = \frac{dF}{dx} = m,$$

which is a uniform probability distribution. To satisfy Equation (5.35) we need

$$\int_a^b p_X(x)\,dx = \int_a^b m\,dx = m(b-a) = 1,$$

so the PDF is

$$p_X(x) = \frac{1}{b-a}. \tag{5.36}$$

We can define an expected value and a variance for continuous distributions by replacing the summations in Equations (5.11) and (5.12) with integrals (assuming that the integrals are finite) to give

$$E(x) = \mu = \int_{-\infty}^{\infty} xp(x)\,dx, \qquad \text{Var}(x) = \sigma^2 = \int_{-\infty}^{\infty} (x-\mu)^2 p(x). \tag{5.37}$$

Example 5.5 Let us calculate the expected value for a continuous uniform distribution defined between $x = a$ and $x = b$. Using Equation (5.36) in Equation (5.37) we get

$$E(x) = \int_a^b xp(x)\,dx = \int_a^b x\frac{1}{b-a}\,dx = \frac{1}{b-a}\left(\frac{b^2}{2} - \frac{a^2}{2}\right) = \frac{1}{2}\frac{(b-a)(b+a)}{b-a} = \frac{b+a}{2},$$

so the expected value is the midpoint of the interval, $(a + b)/2$, which is what we might have expected.

Exercise 5.4.1 Show that for a continuous uniform distribution defined between $x = a$ and $x = b$, the variance is

$$\sigma^2 = \frac{(b-a)^2}{12}.$$

Exercise 5.4.2 Calculate the expected value of the random variable X defined between $x = 0$ and $x = 2$ and that has a PDF of $p(x) = 3x^2$.

Exercise 5.4.3 Calculate the expected value and variance of X with a PDF $p(x) = \lambda e^{-\lambda x}$ and defined on $0 \leq x \leq \infty$.

5.4.1 Normal or Gaussian Distribution

The *normal*, or Gaussian distribution is one of the most important continuous probability distributions we will meet[8] and it is used in science to represent the distribution of many different continuous random variables. Its importance results from a theorem

[8] Named after Carl Friedrich Gauss (1777–1855), a German mathematician who made significant contributions in mathematics, geodesy, geophysics, astronomy, and physics.

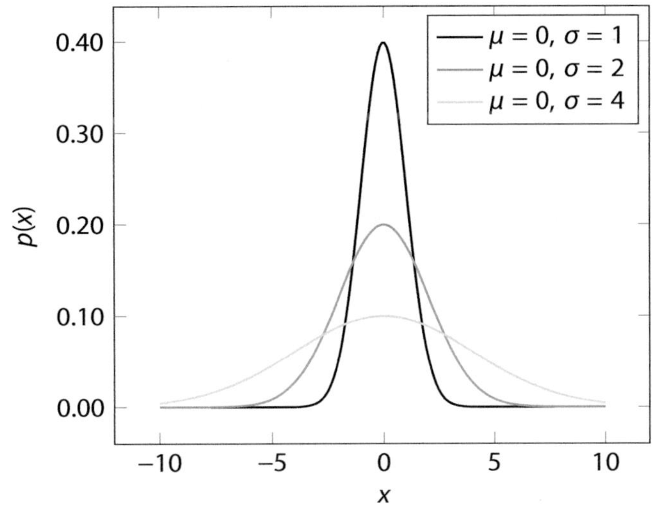

Figure 5.12 The Gaussian distribution for three different values of σ.

called the *central limit theorem* (Section 5.5), which allows us, under some very broad conditions, to approximate almost any continuous probability distribution function as a normal distribution. For example, any measurement we make will have some stochastic uncertainty making the measured quantity a random variable. However, in most cases we do not know the distribution function of that variable, but if the conditions of the central limit theorem are met, we can approximate it as a normal distribution. Consequently, the normal distribution lies at the heart of many data analysis and statistical techniques (see Section 5.8).

The normal distribution (Figure 5.12) is a two-parameter (μ and σ) distribution given by

$$p_X(x) = \frac{1}{\sigma\sqrt{2\pi}} \exp\left(-\frac{(x-\mu)^2}{2\sigma^2}\right). \tag{5.38}$$

This distribution is symmetric about $x = \mu$ (because of the term $(x - \mu)^2$) and is an even function. A random variable X that is described by a normal distribution is often written as $X \sim N(\mu, \sigma)$.

We can derive the normal distribution in many different ways. For example, we have mentioned that the normal distribution can be used to approximate many other PDFs, so we will derive it as a limiting case of the binomial distribution (Equation (5.26)) as $n \to \infty$ and $np \to \infty$ together. For this proof we will introduce a very important formula, *Stirling's formula*,[9] which gives an approximation to $n!$ as n becomes large:[10]

$$n! \approx \sqrt{2\pi n}\left(\frac{n}{e}\right)^n, \qquad n \to \infty. \tag{5.39}$$

[9] This is named after the Scottish mathematician James Stirling (1692–1770), but was first stated by Abraham de Moivre (1667–1754), who gave credit for its proof to Stirling.

[10] The proof of Stirling's formula can be a little involved, so we will not prove it here, but see Section 8.3.

Using Equation (5.39) on the three factorial terms in the binomial distribution and collecting up terms gives

$$p_X(x) = \frac{1}{\sqrt{2\pi n}}\left(\frac{x}{n}\right)^{-x-1/2}\left(\frac{n-x}{n}\right)^{-n+x-1/2} p^x(1-p)^{n-x}. \tag{5.40}$$

Exercise 5.4.4 Use Equation (5.39) to derive Equation (5.40) from Equation (5.26).

This does not look much like Equation (5.38); in particular there are no exponential functions in the expression. However, recall that $\exp(a\ln(b)) = \exp(\ln(b^a)) = b^a$ (see Appendix B), so we can rewrite the terms in parentheses to give

$$p_X(x) = \frac{1}{\sqrt{2\pi n}}\exp\left[-\left(x+\frac{1}{2}\right)\ln\left(\frac{x}{n}\right) - \left(n-x+\frac{1}{2}\right)\ln\left(\frac{n-x}{n}\right) + x\ln(p) + (n-x)\ln(1-p)\right]. \tag{5.41}$$

Now, we need to think about what happens to the binomial distribution as $n \to \infty$ and $np \to \infty$. We know that the expected value of the binomial distribution is np (Equation (5.27)), so we expect the distribution to have a peak at this point. If we let $x = np + \xi$, the variable ξ is then a measure of how far x is from the expected value of the binomial distribution. We can then use expansions for the logarithms (see Appendix B) to give us

$$p_X(x) = \frac{1}{\sqrt{2\pi n}}\frac{1}{\sqrt{p(1-p)}}\exp\left(-\frac{\xi^2}{2np(1-p)}\right). \tag{5.42}$$

Exercise 5.4.5 Fill in the steps leading to Equation (5.41) and Equation (5.42).

Recall that the variance of the binomial distribution is $\sigma^2 = np(1-p)$, so we can use this to write Equation (5.42) as

$$p_X(x) = \frac{1}{\sqrt{2\pi}\sigma}\exp\left(-\frac{\xi^2}{2\sigma^2}\right),$$

where ξ is a measure of how far x is from the expected value of the binomial. We can then write $\xi = x - \mu$, where μ is the expected value of the binomial to finally give us Equation (5.38). Thus we have shown that in the limit of large n and np, the binomial distribution is approximated by the normal distribution (Figure 5.13).

We can use Equation (5.37) to calculate the expected value and variance of the normal distribution. This is a useful exercise to do because it exposes us to some very useful techniques for evaluating difficult integrals, as well as presages what we will see in Chapter 8. The expectation is

$$E(X) = \int_{-\infty}^{\infty} x p_X(x)\,dx = \int_{-\infty}^{\infty} x\frac{1}{\sigma\sqrt{2\pi}}\exp\left(-\frac{(x-\mu)^2}{2\sigma^2}\right)\,dx.$$

The problem with evaluating this integral comes from the combination of the x and the quadratic term in the exponential. This makes it hard to find a substitution that allows us to

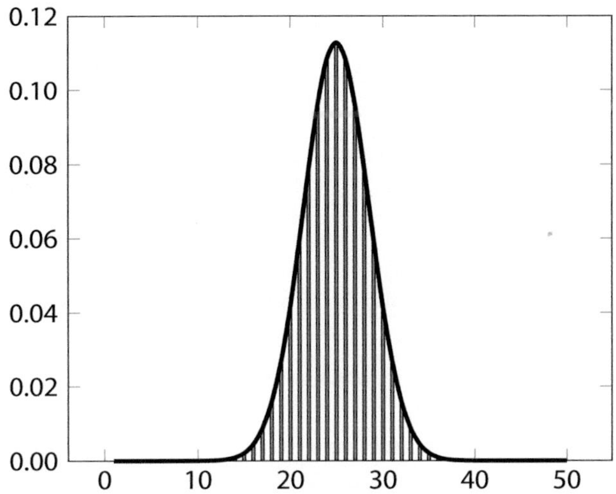

Figure 5.13 The binomial distribution (gray bars) for $p = 0.5$ and $np = 50$, compared with the normal distribution (solid line) with $\mu = 25$ and $\sigma = 3.535 = \sqrt{np(1-p)}$, showing that the normal is a good approximation to the binomial.

solve the integral. But, let us proceed anyway and use a substitution to remove the $(x - \mu)$ term. To do this we will define a new variable $y = x - \mu$, so that $dy = dx$ and the integral becomes

$$\begin{aligned} E(X) &= \frac{1}{\sigma\sqrt{2\pi}} \int_{-\infty}^{\infty} (y + \mu) \exp\left(-\frac{y^2}{2\sigma^2}\right) dy \\ &= \frac{1}{\sigma\sqrt{2\pi}} \left\{ \int_{-\infty}^{\infty} y \exp\left(-\frac{y^2}{2\sigma^2}\right) dy + \int_{-\infty}^{\infty} \mu \frac{1}{\sigma\sqrt{2\pi}} \exp\left(-\frac{y^2}{2\sigma^2}\right) dy \right\}. \end{aligned}$$

This looks like we have made things worse; we now have two integrals to contend with. However, we have removed the $(x - \mu)^2$ argument of the exponential, and we have also ended up with one integral that does not involve the combination of y multiplied by the exponential. Let us look at the first integral. The integrand is an odd function, and we know that an odd function will change signs as the variable y goes from positive to negative values — maybe there is a possibility that the integral over the negative values of y will cancel with the integral over the positive values of y. Let us divide the interval in the first integral into two equal halves:

$$\mathcal{I}_\infty = \frac{1}{\sigma\sqrt{2\pi}} \int_{-\infty}^{0} y \exp\left(-\frac{y^2}{2\sigma^2}\right) dy + \frac{1}{\sigma\sqrt{2\pi}} \int_{0}^{\infty} y \exp\left(-\frac{y^2}{2\sigma^2}\right) dy.$$

We want to try and rearrange things such that the limits of the two integrals are the same, and hopefully in doing so, one of these integrals will acquire a negative sign and so cancel with the other. We can swap the limits on the first integral and then use the fact that the integrand is an odd function to give

$$\begin{aligned}
I_\infty &= -\frac{1}{\sigma\sqrt{2\pi}}\int_0^{-\infty} y\exp\left(-\frac{y^2}{2\sigma^2}\right)\,dy + \frac{1}{\sigma\sqrt{2\pi}}\int_0^{\infty} y\exp\left(-\frac{y^2}{2\sigma^2}\right)\,dy\\
&= +\frac{1}{\sigma\sqrt{2\pi}}\int_0^{+\infty} (-y)\exp\left(-\frac{(-y)^2}{2\sigma^2}\right)\,dy + \frac{1}{\sigma\sqrt{2\pi}}\int_0^{\infty} y\exp\left(-\frac{y^2}{2\sigma^2}\right)\,dy\\
&= -\frac{1}{\sigma\sqrt{2\pi}}\int_0^{+\infty} y\exp\left(-\frac{(y)^2}{2\sigma^2}\right)\,dy + \frac{1}{\sigma\sqrt{2\pi}}\int_0^{\infty} y\exp\left(-\frac{y^2}{2\sigma^2}\right)\,dy\\
&= 0.
\end{aligned}$$

So, now the equation for the expected value simplifies to

$$E(X) = \frac{1}{\sigma\sqrt{2\pi}}\int_{-\infty}^{\infty} \mu\exp\left(-\frac{y^2}{2\sigma^2}\right)\,dy.$$

To evaluate this integral (note that the integrand is even) we first simplify the argument of the exponential by introducing a new change of variables $y = \sqrt{2}\sigma x$, so that $dy = \sqrt{2}\sigma dx$ and the integral becomes

$$E(X) = \frac{\mu}{\sqrt{\pi}}\int_{-\infty}^{\infty} e^{-x^2}\,dx. \tag{5.43}$$

Now, there are a couple of ways we can proceed from here. One involves recognizing that this integral is related to another integral—the *error function*—that we will meet in Chapter 8, and using the properties of that function to evaluate it. However, we will evaluate the integral with a more useful technique. Let us first square the integral and write

$$I^2 = \left(\int_{-\infty}^{\infty} e^{-x^2}\,dx\right)\left(\int_{-\infty}^{\infty} e^{-y^2}\,dy\right) = \int_{-\infty}^{\infty}\int_{-\infty}^{\infty} e^{-(x^2+y^2)}\,dx\,dy.$$

The argument of the exponential reminds us of the formula for the square of the radius in two dimensions ($r^2 = x^2 + y^2$). This suggests that it might be a good idea to switch to polar coordinates, so we let $x = r\cos(\theta)$, $y = r\sin(\theta)$, and the integral becomes

$$I^2 = \int_{\theta=0}^{\theta=2\pi}\int_{r=0}^{r=\infty} e^{-r^2} r\,dr\,d\theta = 2\pi\int_0^{\infty} re^{-r^2}\,dr,$$

where we have evaluated the θ integral directly because the integrand does not depend on θ. To evaluate the remaining integral we make the substitution $u = r^2$ and end up with an integral we can easily evaluate

$$I^2 = \pi\int_0^{\infty} e^{-u}\,du = \pi,$$

so that $I = \sqrt{\pi}$. Putting all these results together we find that the expected value for the normal distribution is

$$E(X) = \frac{\mu}{\sqrt{\pi}}\sqrt{\pi} = \mu. \tag{5.44}$$

Calculating the variance of the normal distribution starts using similar techniques to those we have just employed, but we end up with an integral that, unlike Equation (5.43), will require some more thought on our part to evaluation. Using similar techniques we can write the variance of the normal distribution as

$$\mathrm{Var}(X) = \int_{-\infty}^{\infty} (x-\mu)^2 \frac{1}{\sigma\sqrt{2\pi}} \exp\left(-\frac{(x-\mu)^2}{2\sigma^2}\right) dx = \frac{4\sigma^2}{\sqrt{\pi}} \int_0^{\infty} x^2 \mathrm{e}^{-x^2}\, dx. \quad (5.45)$$

Exercise 5.4.6 Derive Equation (5.45).

This integral is not so easy, but we will start by trying to simplify the integrand by using a change of variables to transform the argument of the exponential into something nicer. Notice that taking the square of this integral and converting to polar coordinates does not simplify the problem because of the presence of the x^2 term that multiplies the exponential.[11] Both terms in the integrand contain x^2, so we define $t = x^2$ so that $dx = (2\sqrt{t})^{-1} dt$ and Equation (5.45) becomes

$$\mathrm{Var}(X) = \frac{2\sigma^2}{\sqrt{\pi}} \int_0^{\infty} t^{3/2-1} \mathrm{e}^{-t}\, dt. \quad (5.46)$$

In this case our substitution has apparently made things a lot worse, and we appear to be stuck without a way to evaluate this integral. Let us generalize the integral in Equation (5.46) by defining the integral

$$\Gamma(n) = \int_0^{\infty} t^{n-1} \mathrm{e}^{-t}\, dt, \quad (5.47)$$

which is a function of n because t is a dummy variable, and is called the *Gamma function* and we will meet it again in Chapter 8. The integral in Equation (5.46) is then $\Gamma(3/2)$. Let us look to see if there are any values of n for which we can evaluate this integral. If $n = 1$ the integrand is just the negative exponential, so we can calculate $\Gamma(1)$. We can also evaluate the integral if $n = 1/2$; if we make the substitution $t = u^2$, we end up with an integral that looks like the one in Equation (5.43) which we know how to solve:

$$\Gamma(1/2) = \int_0^{\infty} t^{-1/2} \mathrm{e}^{-t}\, dt = 2\int_0^{\infty} \mathrm{e}^{-u^2}\, du = \sqrt{\pi}. \quad (5.48)$$

So, we can evaluate some of these integrals, but not the one we are interested in. However, we might be able to find a relationship between the values of $\Gamma(n)$ for different values of n. For example, if we know $\Gamma(n)$, can we also find $\Gamma(n+1)$. A relationship where we can write a function of $(n+1)$ in terms of the same function of n is called a *recurrence relationship*. Recall that if one of the factors in an integrand is the power of a variable, we can use integration by parts to lower the power by 1. So,

$$\Gamma(n+1) = \int_0^{\infty} t^n \mathrm{e}^{-t}\, dt = \left(-t^n \mathrm{e}^{-t}\right)\Big|_0^{\infty} + n\int_0^{\infty} t^{n-1}\mathrm{e}^{-t}\, dt = 0 + n\Gamma(n).$$

Therefore, for $n = 1/2$, $\Gamma(3/2) = (1/2)\Gamma(1/2) = \sqrt{\pi}/2$. And now we can complete our calculation of variance to give

$$\mathrm{Var}(X) = \sigma^2 \frac{4}{\sqrt{\pi}} \frac{1}{2} \frac{\sqrt{\pi}}{2} = \sigma^2. \quad (5.49)$$

[11] You should try such a transformation and see this for yourself.

We can now also give meaning to the two parameters of the normal distribution. The parameter μ is the mean or expected value of the distribution, and σ^2 is the variance.

Exercise 5.4.7 Show that the normal distribution is a valid PDF by a) showing $p(x)$ in Equation (5.38) is greater than or equal to zero and b) showing that

$$\int_{-\infty}^{\infty} p(x)\, dx = 1.$$

We often come across situations where we need to calculate a quantity that depends on a normally distributed random variable. For example, we might measure a variable $X \sim N(\mu, \sigma)$ but really be interested in the quantity $Y = aX + b$, where a and b are known constants. In such cases we might wonder what the probability distribution of Y is. To figure this out we are going to work not with the PDF itself, but rather with the CDF, which is related to the PDF by Equation (5.33). For a normal distribution, the probability that $x \leq u$ is given by the CDF

$$F(x \leq u) = \int_{-\infty}^{u} \frac{1}{\sigma\sqrt{2\pi}} \exp\left(-\frac{(x-\mu)^2}{2\sigma^2}\right) dx.$$

The probability that $ax + b \leq u$ is the probability that $x \leq (u - b)/a$, so that

$$F(ax + b \leq u) = F\left(x \leq \frac{u-b}{a}\right) = \int_{-\infty}^{(u-b)/a} \frac{1}{\sigma\sqrt{2\pi}} \exp\left(-\frac{(x-\mu)^2}{2\sigma^2}\right) dx.$$

Now, we want to transform variables to $y = ax + b$ so that $dy/dx = a$ and the limit $x = (u - b)/a$ become $y = u$. The integral becomes, after a little bit of algebra

$$F(ax + b \leq u) = F(y \leq u) = \int_{-\infty}^{u} \frac{1}{a\sigma\sqrt{2\pi}} \exp\left(-\frac{(y-(a\mu+b))^2}{2a^2\sigma^2}\right) dx,$$

which is the CDF of a random variable that obeys a normal distribution with $Y \sim N(a\mu + b, a^2\sigma^2)$. So, a linear transformation of a normally distributed random variable X gives a new, normally distributed random variable Y but with a different expected value and variance,

$$E(Y) = E(aX + b) = E(aX) + E(b) = aE(X) + E(b) = a\mu + b,$$
$$\text{Var}(Y) = \text{Var}(aX + b) = a^2\,\text{Var}(X) + \text{Var}(b) = a^2\sigma^2.$$

This is very useful, because if we choose $a = 1/\sigma$ and $b = -\mu/\sigma$, then $E(Y) = 0$ and $\text{Var}(Y) = 1$. This transformation is called a z transformation, and it produces a normal distribution in *standard form*. That is, if we have a random variable $X \sim N(\mu, \sigma)$ and apply the transformation $Z = (X - \mu)/\sigma$, then $Z \sim N(0, 1)$; i.e., Z is a normally distributed random variable with $\mu = 0$ and $\sigma = 1$ and is sometimes called the *z score*. For any value x of the random variable X, the Z score is the number of standard deviations that x is away from the mean (μ). Many statistical tests that you will come across require the variables in the data to be in standard form before you can apply the test.

We can use the cumulative probability distribution to calculate the probabilities of a random variable having a value within certain ranges. We know that the probability of the random variable X lying between $x = a$ and $x = b$ is the area under the PDF between

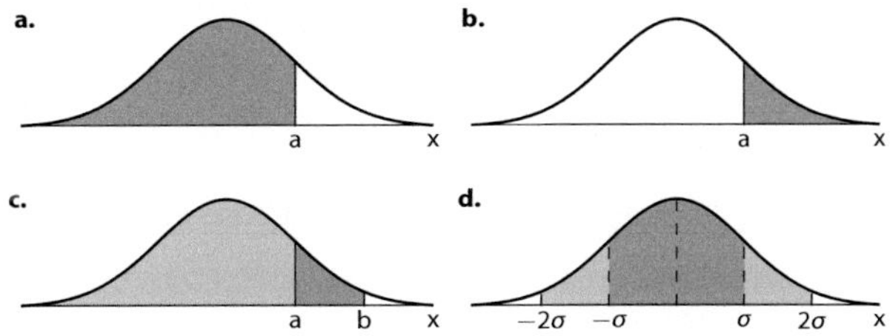

Figure 5.14 The relationship between a probability and the area under a Gaussian curve: (a.) the probability of obtaining a value of x from $-\infty$ to $x = a$; (b.) the probability of obtaining a value of $x \geq a$; (c.) the probability of obtaining a value of x lying between a and b (the area from $x = -\infty$ to a subtracted from the area from $x = -\infty$ to b; (d.) the area between $\pm 1\sigma$ and $\pm 2\sigma$ from the mean value.

$x = a$ and $x = b$ (Figure 5.14). To determine the probability of obtaining a value less than or equal to a, we calculate the area under the curve to the left of a (Figure 5.14a and d):

$$P(X \leq a) = \int_{-\infty}^{a} p(x)\, dx.$$

To calculate the probability of obtaining a value greater than a, we calculate the area to the right of $x = a$ (Figure 5.14b):

$$P(X \geq a) = \int_{a}^{\infty} p(x)\, dx = \int_{-\infty}^{\infty} p(x)\, dx - \int_{-\infty}^{a} p(x)\, dx.$$

To calculate the probability of obtaining a value that lies between $x = a$ and $x = b$ we subtract two integrals; i.e., subtract two cumulative probabilities (Figure 5.14c):

$$P(a \leq x \leq b) = \int_{-\infty}^{b} p(x)\, dx - \int_{-\infty}^{a} p(x)\, dx.$$

Such calculations lie at the basis of techniques of hypothesis testing in statistics and data analysis. Fortunately, we do not have to evaluate these integrals by hand every time we need to calculate a probability. Most scientific programming languages such as MATLAB™ and Python already contain programs that will make these calculations, but we still have to understand which integrals need to be evaluated. However, there are some useful numbers to know for a normal distribution:

- Approximately 68% of the area under the curve lies within $\pm 1\sigma$ of the mean. This implies that if we randomly choose a value from a normally distributed random variable, there is a 0.68 probability that it will lie within one standard deviation of the mean.
- Approximately 95% of the area under the curve lies within $\pm 2\sigma$ of the mean of the distribution.
- Approximately 99% of the area under the curve lies within $\pm 3\sigma$ of the mean of the distribution.

So, if we read that someone made a measurement whose average value was 10.34 and standard deviation was 0.01 (written as 10.34 ± 0.01), then this indicates a very narrowly peaked normal distribution because 99% of the area under the curve lies between 10.37

and 10.31. On the other hand, a value of 10.34 ± 3.0 would indicate a broad distribution and a measurement that is far less precise. Our job would then be to figure out why.

5.5 Law of Large Numbers and Central Limit Theorem

We mentioned earlier that the normal distribution is central to many techniques in statistics and data analysis and that this is because of something called the central limit theorem. In this section we will explore this theorem and why it is so important. Let us start by thinking of what happens when we take a measurement. When we do an experiment or take field samples we rarely if ever make a single measurement and stop there. Instead we take replicate measurements and use the average of these numbers as our best estimate. If the values of our replicates are very close to each other, then we have confidence that our measurements are precise. If the values of our replicates are spread over a large range, then our confidence in our measurements is low. In such a case, we might try and improve the precision of our measurements by taking more replicates.[12]

Why should the number of replicates affect the average of those replicates? To see why let us think of a series of N replicate measurements as a series of random numbers $x_1, x_2, x_3 \dots x_N$. Each random number is a single measurement and comes from a single, unknown probability distribution called the *population distribution*—this is the distribution that the variable actually follows, but we would require an infinite number of samples to determine it. What is more, we will assume that each measurement in the series is independent of the others. Such a sequence of random numbers forms what is called an *independent and identically distributed* (iid) set of random numbers. The average value of these measurements is simply

$$\bar{x} = \frac{1}{n}\sum_{i-1}^{n} x_i.$$

It is important to realize that this mean value may not be the expected value of the population distribution—we have seen this happen before in Example 5.3. However, we know that expectations are linear, so we can calculate the expected value of the mean,

$$E((\bar{X})) = \frac{1}{n}E(x_1 + x_2 + \cdots + x_n) = \frac{1}{n}\left(E(x_1) + E(x_2) + \cdots + E(x_n)\right) = \frac{1}{n}(n\mu) = \mu,$$

where μ is the expected value of the population distribution. Similarly, using the properties of the variance we can calculate the variance of the mean as

$$\text{Var}(\bar{X}) = \frac{1}{n^2}(n\sigma^2) = \frac{\sigma^2}{n},$$

where σ is the standard deviation of the measurements. This is telling us that the standard deviation of the mean (called the *standard error*) of the measured values is smaller than

[12] Accuracy and precision are not the same thing. Accurate measurements will have a mean value that is close to the "true" value, whereas precision is a measure of the random uncertainty in the measurement.

the standard deviation of the measurements themselves. We take advantage of this when we take replicate measurements; we basically exchange an improved precision for the additional hard work of making replicate measurements.

This leads us to an important inequality called *Chebyshev's inequality* (Theorem 5.1).[13]

Theorem 5.1 (Chebyshev's inequality) *For an arbitrary random variable X*

$$P(|X - E(X)| \geq a) \leq \frac{1}{a^2}\,\mathrm{Var}(X) \tag{5.50}$$

for any value of $a > 0$.

To show that this inequality is true assume that we have a continuous probability density distribution $p_X(x)$ for X and that $\mu = E(X)$. Then

$$\mathrm{Var}(X) = \int_{-\infty}^{\infty} (x-\mu)^2 p_X(x)\,dx \geq \int_{(x-\mu)\geq a} (x-\mu)^2 p_X(x)\,dx \tag{5.51}$$

$$\geq \int_{|x-\mu|\geq a} a^2 p_X(x)\,dx = a^2 P(|\ x - \mu\ | \geq a). \tag{5.52}$$

What is Chebyshev's inequality telling us? The left-hand side of the inequality is the probability that the value of the random variable X is greater than or equal to a distance a from the expected value of X. If we write $E(X) = \mu$ and $\mathrm{Var}(X) = \sigma^2$, then we can use the complement of the probability to write Equation (5.50) as

$$P(|X - \mu| < a) > 1 - \frac{\sigma^2}{a^2},$$

which says that the probability that the value of X is less than a distance a from the expected value is greater than $1 - (\sigma^2/a^2)$. Why is this useful? It is placing an upper limit on the probability of the value of a random variable lying a given distance from the mean. For example, let us choose a to be some multiple of the standard deviation, $a = n\sigma$, then the inequality says

$$P(|X - \mu| < n\sigma) > 1 - \frac{1}{n^2}.$$

Let us put some numbers in to clarify things a little. For example, if $n = 2$, this equation tells us that the probability that the value of X lies within two standard deviations of the mean value is greater than 0.75. That is nice to know, especially as we have not had to specify *what* the PDF of X is! In fact, Chebyshev's inequality says nothing about the probability distribution itself; it is a general rule that says that probability distributions are peaked around the mean value, and it gives us a constraint on how peaked the distribution is. However, it is only a lower bound, and it turns out to be quite a conservative one at that, because we know that for a normal distribution approximately 95% of the area under the curve lies within 2σ of the mean, which is certainly larger than the 75% given by Chebyshev's inequality. Where the inequality is useful is in giving us a very quick estimate of how peaked a distribution is.

[13] Named after the Russian mathematician Pafnuty Lvovich Chebyshev (1821–1894). You will see his name spelled in many ways, and another common spelling is Tchebysheff.

Example 5.6 We can use Chebyshev's inequality to show that if X is a random Poisson variable with an expected value of λ, then the probability that the value of X is within a distance $3\sqrt{\lambda}$ of the expected value is at most 1/9. Using Equation (5.50) we see that

$$P(|X-\lambda| \geq 3\sqrt{\lambda}) \leq \frac{\lambda}{9\lambda} = \frac{1}{9}.$$

As we have seen, we have a fundamental problem when we make a series of measurements. We know that the values we measure come from an as yet unknown probability distribution, the population distribution. Each measurement amounts to taking a single sample from that distribution, but it also has a random component to it that arises from measurement uncertainty, e.g., electronic noise in the detection equipment. We have seen that we can overcome this by taking replicate measurements (measuring the same thing over and over again) and taking the mean value of these replicates. We would like the mean and variance of these sample replicates to be a good approximation to the population mean and variance. Consider a series of measurements $X_1, X2, \ldots, X_n$ that have a mean value $E(\bar{X}_n) = \mu$ and variance $\text{Var}(\bar{X}_n) = \sigma^2/n$. As we make more and more measurements (i.e., n gets larger), what happens to the mean value of the samples? Applying Chebyshev's inequality we find

$$P(|\bar{X}_n - \mu| > \epsilon) = P(|\bar{X}_n - E(\bar{X}_n)| > \epsilon) \leq \frac{1}{\epsilon^2}\text{Var}(\bar{X}_n) = \frac{\sigma^2}{n\epsilon^2},$$

and as $n \to \infty$ (i.e., we have lots of measurements) the right-hand side of the equation tends to zero. This is called the *law of large numbers* (Theorem 5.2).

Theorem 5.2 (Law of Large Numbers) *If $\bar{X}_n$ is the average of n independent random variables with expectation μ and variance σ^2, then for any $\epsilon > 0$*

$$\lim_{n\to\infty} P(|\ \bar{X}_n - \mu\ | > \epsilon) = 0.$$

We have already seen this in Example 5.3, where the mean value of the roll of a die got closer to the expected value the more times we rolled the die. This is telling us that the more replicate measurements we can take, the better our estimate of the mean value will be (Figure 5.15). However, there is a very important caveat to the law of large numbers; it only applies if the expected value and variance of the population distribution are finite. This is generally true for most distributions that we deal with as scientists, but we should not be complacent. For example, a Cauchy distribution looks very similar to a normal distribution, but it does not approach zero as $x \to \infty$ quite as quickly as a Gaussian does (Figure 5.16). As a result, the Cauchy distribution has an expected value and a variance that are both infinite. Distributions like this are called "heavy-tailed" distributions because the tails of the distribution (the curve as $x \to \pm\infty$) approach zero slowly and so contribute a non-negligible amount to the area under the curve. We may be tempted to think that the Cauchy distribution is a pathological example, a mere curiosity, but that is not so because we get a Cauchy distribution when we divide two standardized normal distributions, implying that we need to be cautious when we divide two measurements because even though the measurements might be normally distributed, their ratio is not.

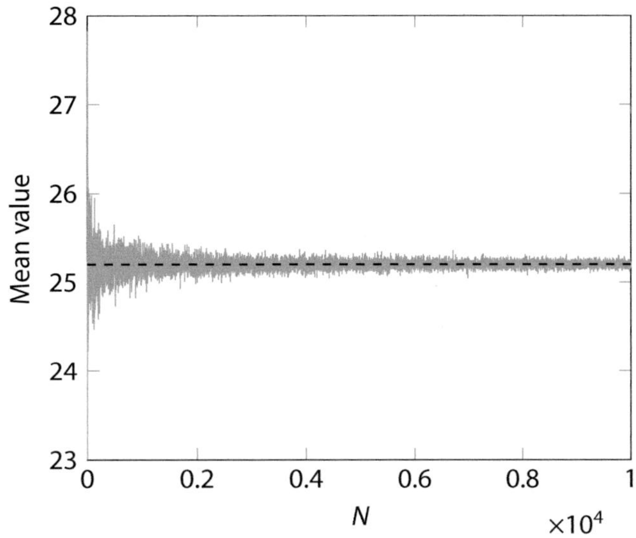

Figure 5.15 An illustration of the law of large numbers. Samples of different sizes ranging from $N = 2$ to $N = 10^4$ were taken from a normal distribution having $\mu = 25.2$. The plot shows the mean of each set of samples. As N increases, the mean of the replicate samples get closer and closer to the mean of the distribution they came from.

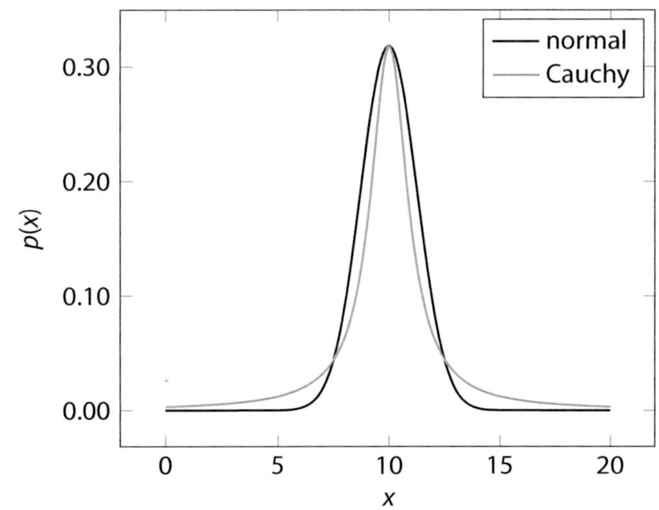

Figure 5.16 A comparison of the normal and Cauchy distributions. Notice that the values of the Cauchy distribution in the tails are greater than those of the normal distribution.

The law of large numbers leads us to the central limit theorem (Theorem 5.3), a very important theorem that underlies the importance of the Gaussian distribution for the way we analyze data. We have seen that the law of large numbers tells us that if we have n independent random numbers $(X_1, \ldots, X_n)$ taken from the same distribution with expectation μ and variance σ^2, then as $n \to \infty$ the mean of the random variables $\bar{X}_n$ will tend to μ. But, remarkably, we can also say something about the *distribution* of $\bar{X}_n$ itself.

Theorem 5.3 (Central limit theorem) *Let $X_1, X_2, \dots, X_n$ be any sequence of independent and identically distributed random variables taken from any probability distribution with a finite mean μ and finite variance σ^2. As $n \to \infty$, then $\bar{X}_n$ tends to a normal distribution with a mean value μ and a variance σ^2/n.*

The central limit theorem basically says that the mean of a large number of independent random variables is approximately a normal distribution, with the approximation getting better and better the more samples we have. This happens irrespective of the actual distribution that the random variables come from, so long as it has a finite mean and variance. There are, however, some important questions we should ask. For example, how many samples do we need for the distribution of means to look like a normal distribution? Such questions are easy to examine using computers, and some of the problems at the end of the chapter and online computer codes look at this.

5.6 Manipulating Random Variables

We have already seen in Section 5.4.1 that if X is a normally distributed variable, then $Y = aX + b$ (a and b constant) also follows a normal distribution. But what happens with more general manipulations? For example, what is the PDF of the sum of two random numbers?

5.6.1 Adding Continuous Random Variables

We know how to add two nonrandom numbers, but random numbers have, by their very nature, a distribution of values that they can take; which values should we choose to add? Let us think about this in a little more detail. Consider what happens if we add two continuous random variables, X and Y, that are both uniformly distributed between 0 and 1. In particular, we want to know what the PDF of $Z = X + Y$ is. The smallest value that $X + Y$ can be is 0, and this occurs when both random variables are 0. Similarly, the greatest value $X + Y$ can be is 2, which occurs when $X = Y = 1$. We would expect the probabilities of these two cases occurring to be small. This is because there is only one way in which we can get each of these values, whereas, for example, we can get a value of $Z = 1$ in many different ways ($X = 0.5$, $Y = 0.5$; $X = 0.9$, $Y = 0.1$; $X = 0.2$, $Y = 0.8$, etc.). This means that we would expect values close to $X + Y = 1$ to be more common, so the PDF should have a maximum value somewhere around $Z = 1$. What is more, the function $Z = X + Y$ is a linear function, so we expect that the PDF is a straight line from $Z = 0$ to the maximum value of the PDF, and a straight line with a negative slope from the maximum of the PDF to $Z = 2$.

We can generalize these arguments and make them more rigorous by again making use of the relationship between the PDF and the CDF of the function we are interested in. Let us write the probability distribution functions of X and Y as $p_X(x)$ and $p_Y(y)$, and assume that X and Y can take values from $-\infty$ to $+\infty$. The CDF of the random variable Z is

$$F_Z(z) = \mathrm{P}(Z \leq z) = \mathrm{P}(X + Y \leq z).$$

Now, if X and Y are independent random variables, we can use the basic rules of probability to write

$$\mathrm{P}(X + Y \leq z) = F_Z(z) = \int_{X+Y\leq z} p_X(x)p_Y(y)\,dx\,dy = \int_{y=-\infty}^{y=\infty}\int_{x=-\infty}^{x=z-y} p_X(x)p_Y(y)\,dx\,dy$$
$$= \int_{y=-\infty}^{y=\infty} F_X(z-y)p_Y(y)\,dy,$$

where F_X is the CDF of $p_X(x)$. But we want the PDF of Z ($p_Z(z)$), so using Equation (5.33) we find that

$$p_Z(z) = \frac{d}{dz}F_{X+Y}(z) = \int_{-\infty}^{\infty} p_X(z-y)p_Y(y)\,dy = \int_{-\infty}^{\infty} p_Y(z-x)p_X(x)\,dx, \tag{5.53}$$

where the last equality arises because we can switch the order of the two integrals. The integrals that appear in Equation (5.53) are called *convolutions*, and they appear often in probability and signal processing. Notice that the convolution is a function of z because it integrates over all values of x or y. So, how is this equivalent to an addition? Remember that when we add independent random variables (i.e., an "and" event) we multiply the probabilities (Equation (5.4)). But because there can be many ways in which we can add these two random variables and get the same answer, we have to sum (i.e., integrate for continuous random variables) over all the possibilities that give us the desired answer. To picture this process, consider the two pulse functions shown in Figure 5.17a. The PDF $f_X(x)$ is zero everywhere except between $0 \leq x \leq 1$, and the function $g_Y(x)$ is the same. We will write the convolution of these two functions as

$$H(c) = \int_{-\infty}^{\infty} g_Y(c-x)f_X(x)\,dx.$$

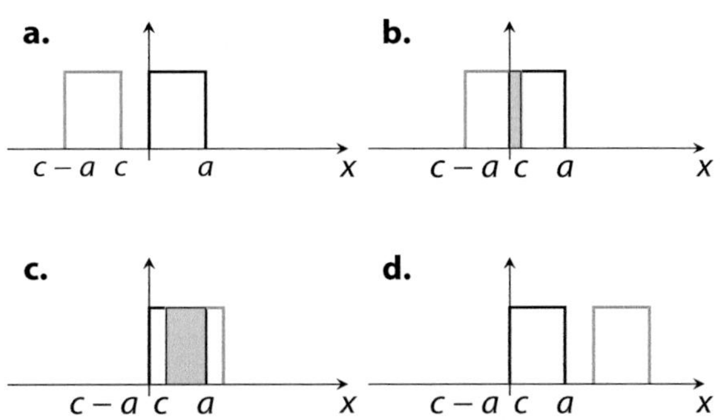

Figure 5.17 The operation of a convolution of the probability distribution functions, $f_X(x)$ (black curve) and $g_Y(x)$ (gray curve), of two random variables X and Y. The shaded areas represent the areas where the two curves overlap and so give a contribution to the convolution integral.

In performing the convolution, we first take the function $g_Y(x)$ and shift it to the left to get $g_Y(c-x)$ (Figure 5.17a) — remember c is fixed and we are integrating over x. For $c<0$, the functions $g_Y(c-x)$ and $f_X(x)$ do not overlap, so $H(c)=0$. As the value of c increases, the two functions start to overlap (Figure 5.17b) and $H(c)\neq 0$. As we continue to shift the function $g(x-c)$ along the x axis as c increases, the area of overlap of the two functions increases until we get a maximum overlap area when the two functions completely cover each other (which occurs when $c=a$). As we continue to increase c still further, the area of overlap decreases until we move $g_Y(c-x)$ far enough to the right that there is no longer any overlap between it and $f_X(x)$, in which case the value of $H(c)$ drops to zero again.

Now, let us return to the example we examined at the start of this section and see if our intuition was correct. Our two functions are

$$f_X(x)=\begin{cases}1 & 0\leq x\leq 1\\ 0 & \text{otherwise}\end{cases}\quad \text{and}\quad g_Y(x)=\begin{cases}1 & 0\leq x\leq 1\\ 0 & \text{otherwise}\end{cases}.$$

From Equation (5.53) we have that, in general,

$$H(z)=\int_{-\infty}^{\infty} f_Y(z-x)g_Y(x)\,dx,$$

but $g(x)=0$ for values of x that are outside of the range $x=0$ to $x=1$, and $g(x)=1$ for all values of x inside of that interval, so we can write this integral as

$$H(z)=\int_0^1 f_Y(z-x)\,dx.$$

By a similar argument, $f_Y(z-x)=1$ if $0\leq z-x\leq 1$ and 0 otherwise, so we can break down the integral into two pieces:

$$p(z)=\int_0^z f_Y(z-x)\,dx=\int_0^z dx=z,\qquad \text{for } 0\leq z\leq 1$$

and

$$p(z)=\int_{z-1}^1 f(z-x)\,dx=\int_{z-1}^1 dx=2-z,\qquad \text{for } 1<z\leq 2,$$

and $p(z)=0$ for $z<0$ and $z>2$. So, we end up with

$$p(z)=\begin{cases}z & 0\leq z\leq 1\\ 2-z & 1<z\leq 2,\\ 0 & \text{otherwise}\end{cases}$$

which is the triangular function shown in Figure 5.18. Therefore, our intuition from the start of this section was correct.

Example 5.7 We can use these results to show that the sum of two standardized normal random variables, X and Y, is itself a normalized random variable. We are dealing with random variables as X and Y that are in standard form, so $\mu=0$ and $\sigma=1$ and they both have the same PDF,

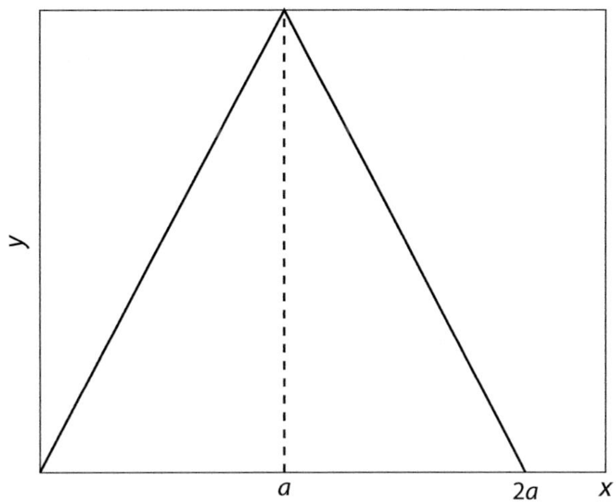

Figure 5.18 The result of the convolution of the two pulse functions in Figure 5.17.

$$f_X(x) = \frac{1}{\sqrt{2\pi}} \exp\left(-\frac{x^2}{2}\right), \qquad g_Y(y) = \frac{1}{\sqrt{2\pi}} \exp\left(-\frac{y^2}{2}\right).$$

Then, using the convolution equation,

$$p(z) = \frac{1}{2\pi} \int_{-\infty}^{\infty} \exp\left(-\frac{(z-y)^2}{2}\right) \exp\left(-\frac{y^2}{2}\right) dy = \frac{e^{-z^2/4}}{2\pi} \int_{-\infty}^{\infty} \exp\left(-\left(y - \frac{z}{2}\right)^2\right) dy$$
$$= \frac{e^{-z^2/4}}{2\pi} \sqrt{\pi} = \frac{e^{-z^2/4}}{2\sqrt{\pi}},$$

where we have made use of the integration techniques we saw in Section 5.4.1.

Exercise 5.6.1 Use the convolution to show that the sum of two random variables X and Y having an exponential distribution

$$f(x) = g(y) = \begin{cases} \lambda e^{-\lambda x} & x \geq 0 \\ 0 & \text{otherwise} \end{cases}$$

has the form

$$p_{Z=X+Y}(z) = \begin{cases} z\lambda^2 e^{-\lambda z} & z \geq 0 \\ 0 & \text{otherwise} \end{cases}.$$

Exercise 5.6.2 If X and Y are independent random variables following normal distributions with mean and standard deviations (μ_X, σ_X) and (μ_Y, σ_Y) respectively, then show that the sum $Z = X + Y$ is a normal distribution with mean $\mu_Z = \mu_X + \mu_Y$ and standard deviation $\sigma_Z^2 = \sigma_X^2 + \sigma_Y^2$.

The convolution may seem to be a rather abstract construct concerned with adding random variables, but this is not the case, and convolutions find practical application in understanding data. For example, a watershed (or drainage basin) is a region where precipitation drains through soils to rivers and ends up in a single outlet, such as a river, estuary, or lake. Watersheds vary in size from small areas that drain into a single lake to large areas such as the Mississippi watershed in the United States or the Amazon watershed.[14] When rain falls on a watershed it does not instantaneously appear at the river mouth. Instead some of the water travels through the soils, some travels overland, and some might be evaporated and fall again as rain somewhere else in the watershed. So, the relationship between the frequency of rainfall in the watershed and the river discharge is a complicated one, but it can be represented as a convolution. For example, if $f_R(\tau)$ represents rainfall over the watershed and $g_W(t-\tau)$ is the *basin response function* (i.e., the function that describes how water travels though the watershed), then the runoff over time $q(t)$ can be written as a convolution (Karamouz et al., 2012):

$$q(t) = \int_0^t g_W(t-\tau) f_R(\tau)\, d\tau.$$

You will often hear such integrals described as *filters*—the function $g_W(t-\tau)$ filters the input ($f_R(\tau)$) to produce an output signal $q(t)$. Similarly, observed seismic signals represent a convolution of the original signal (e.g., from an earthquake or deliberate explosion) with a filter representing the propagation of that signal through different layers of the Earth (Zhou, 2014). Because they are connected to filters, convolutions are often dealt with in detail in texts on data analysis and signal processing (see Section 5.8).

5.6.2 Transforming Random Variables

Sometimes it is necessary to transform a random variable. For example, we might know the PDF of a random variable X but actually be interested in the quantity $\exp(-kX)$, and there are several techniques we can use. The first one we are going to look at uses the cumulative distribution function in a way we have seen before, so we will only give an example of how this works.

Example 5.8 Let us use the cumulative distribution method to calculate the PDF of the continuous random variable $Y = -\ln(X)$, given that X is a continuous random variable with a uniform probability distribution defined on $0 \le y \le 1$. The first thing we need to do is to determine the CDF of X. This is

$$F_X(x) = \mathrm{P}(X \le x) = \begin{cases} 0 & x \le 0 \\ x & 0 \le x \le 1 \\ 1 & x \ge 1 \end{cases}.$$

Because $0 \le y \le 1$, we have that $y = -\ln(x) > 0$, so y takes on positive, nonzero values. Now, we need to determine the CDF of Y in terms of the CDF of X:

$$\begin{aligned} F_Y(y) &= \mathrm{P}(Y \le y) = \mathrm{P}(-\ln(X) \le y) = \mathrm{P}(\ln(X) > -y) \\ &= \mathrm{P}(X > \mathrm{e}^{-y}) = 1 - \mathrm{P}(X \le \mathrm{e}^{-y}) = 1 - F_X(\mathrm{e}^{-y}). \end{aligned}$$

[14] The Mississippi River drains most of the land in the United States between the Rocky Mountains in the west and the Appalachian Mountains in the east. The Amazon watershed is the largest in the world and covers almost two fifths of the South American landmass.

We have already determined that $F_X(x) = x$ for $0 \le x \le 1$, so

$$F_Y(y) = 1 - F_X(e^{-y}) = 1 - e^{-y}.$$

And now that we have the CDF of Y, we can calculate the PDF by taking its derivative:

$$f_Y(y) = \frac{d}{dy}F_Y(y) = \frac{d}{dy}(1 - e^{-y}) = e^{-y}.$$

Exercise 5.6.3 Use the cumulative distribution method to find the PDF of $Y = -\ln(1 - X)$, where X is a continuous random variable uniformly distributed on the interval $0 \le x \le 1$.

Exercise 5.6.4 If X is a continuous random variable with the PDF

$$f_X(x) = \begin{cases} 4x & 0 < x < 1 \\ 0 & \text{otherwise} \end{cases},$$

find the PDF of $Y = 4X^4$.

Another way of funding the distribution of a transformed variable is the so-called PDF-*method*. In Chapter 2 we learned that some integrals can be evaluated by making a suitable transformation of variables. For example, to evaluate the integral

$$\int_{x=a}^{x=b} f(x)\,dx$$

using a transformation of variables $y = y(x)$, we would first substitute for x in the integrand using the inverse function $y^{-1}(x)$, then replace the limits with $y(a)$ and $y(b)$, and finally substitute dx with $(dx/dy)dy$. When we transform random variables using the PDF, we follow the same path, though there are some additional things that we need to take into account.

Consider a random variable X with a PDF $f_X(x)$ so that the probability that $a \le x < b$ is

$$P(a \le x < b) = \int_{x=a}^{x=b} f_X(x)\,dx.$$

We want to calculate the PDF $g_Y(y)$ of the random variable $Y = y(x)$. If $a \le x < b$, then $y(a) \le Y < y(b)$ and $P(y(a) \le Y < y(b)) = P(a \le x < b)$. In other words

$$P(y(a) \le Y < y(b)) \int_{x=a}^{x=b} f_X(x)\,dx = \int_{y(a)}^{y(b)} f_X(x(y))\frac{dy}{dx}\,dy,$$

so that $f_X(x(y))\frac{dy}{dx} = g_Y(y)$ is the PDF of Y.

Example 5.9 We can use the probability distribution method to determine the PDF in Example 5.8. We already know that

$$f_X(x) = \begin{cases} 1 & 0 \le x < 1 \\ 0 & \text{otherwise} \end{cases}.$$

Also $y = -\ln(x)$, so the inverse function is $x = e^{-y}$. Therefore, the PDF of Y is

$$g_Y(y) = f_X(x(y)) \left|\frac{dx}{dy}\right| = 1\left|-e^{-y}\right| = e^{-y}.$$

Exercise 5.6.5 If X is a random variable with a PDF

$$f(x) = \begin{cases} 4x & 0 \leq x < 1 \\ 0 & \text{otherwise} \end{cases},$$

what is the PDF of the random variable $Y = \sqrt{X}$?

Exercise 5.6.6 If X is a normally distributed random variable with PDF

$$f_X(x) = \frac{1}{\sigma\sqrt{2\pi}} \exp\left(-\frac{(x-\mu)^2}{\sigma^2}\right),$$

what is the distribution of the random variable

$$Z = \frac{X-\mu}{\sigma}?$$

Both the cumulative or PDF approaches can be used, but sometimes one is easier than the other.

5.7 Monte Carlo Methods

Monte Carlo techniques are computational methods that rely on sampling and manipulating random numbers.[15] They can be thought of as experimental probability, and they have become very popular since high-powered personal computers have become ubiquitous. Monte Carlo simulation techniques can be used to tackle a wide range of scientific problems from numerically integrating functions, performing error propagation in observational results, and simulating stochastic processes such as the propagation of light through the atmosphere, ocean (Mobley, 1994), and vegetation canopies.

5.7.1 Monte Carlo Error Propagation

Let us look at a simple example of error propagation to get an idea of the basics and some of the things we need to be concerned about. Assume that we have measured two variables (e.g., temperature and salinity in the oceans) and determined their uncertainties, and we want to calculate the value and uncertainty of a third variable (e.g., density) that depends on the first two. We have seen in Section 2.9.1 how to do this using calculus, but that method assumes that the uncertainties are small so that we can linearize the function. This might

[15] Modern Monte Carlo techniques were developed in the 1930s and 1940s by a group of physicists and mathematicians working on the Los Alamos project. The method was invented by Stanislaw Ulam and developed by John von Neumann, Nicolas Metropolis, and others (Cooper, 1989). It is named after a famous casino in Monaco.

not be the case. The idea behind Monte Carlo error propagation is to use our knowledge of the uncertainties in the observed values to *simulate* the uncertainty in the derived variable. To do this, we use a *pseudorandom number generator* to produce a very large number of *realizations* or samples from these distributions (see Box 5.1).[16] Then, for each realization we calculate the quantity we are interested in. We now have many realizations of this quantity and we can look at its distribution and calculate its mean and variance.

Box 5.1 Random Number Generators

Monte Carlo methods involve calculating many (frequently tens of thousands) random numbers. This is because the methods rely on having a good statistical representation of the simulation. In our error propagation example we calculated 10,000 values of each variable (P, V, and T). Other applications of the Monte Carlo technique may require far more; e.g., simulating the propagation of light through the atmosphere or ocean where photons can be scattered a vast number of times before being absorbed. Using computers is the only practical way of calculating the required numbered of random numbers. However, generating truly random numbers is very difficult. In practice, computers use algorithms to generate *pseudorandom numbers*. These algorithms require an input number (called a *seed*) from which they generate sequences of apparently random numbers. This sequence will eventually repeat itself, showing that it is not truly random — a good modern random number generator will repeat itself after approximately 10^9 numbers have been generated, poorly written random number generators repeat themselves much sooner. What is more, the algorithm will always produce the same sequence of random numbers given the same value for the seed. This can be a good thing, because it allows scientists to reproduce simulations. Many computer packages and languages (e.g., Python, R, MATLAB, Fortran, C) have good random number generators either built in or readily available. This makes Monte Carlo techniques easy to use.

As a simple example, we can look at the error we get when using the ideal gas law

$$N = \frac{PV}{RT} \tag{5.54}$$

to calculate the number of moles of a gas (N) given its pressure (P), volume (V), and temperature (T); the constant $R = 0.082$ L atm K^{-1} mol^{-1} is the *universal gas constant*. Assume that we measure $P = 0.268 \pm 0.012$ atmospheres, $V = 1.26 \pm 0.05$ L, and $T = 294.2 \pm 0.3$ K, where the uncertainties are the standard deviation of the respective measurement. If we assume that all the measured uncertainties are normally distributed, we can use a pseudorandom number generator on a computer to create many (e.g., 10000) samples of the values of P, V, and T and use these values in Equation (5.54) to calculate 10000 values of N, from which we can calculate a mean and standard deviation for N (Figure 5.19).

Exercise 5.7.1 Use the propagation of errors formula in Section 2.9.1 to calculate the error in N and compare the value with the one obtained by Monte Carlo error propagation.

For small uncertainties in the measured values, the Monte Carlo method gives uncertainties that agree well with the linearized error propagation we met in Section 2.9.1. The Monte

[16] You can see why this method was not popular before computers became ubiquitous and powerful.

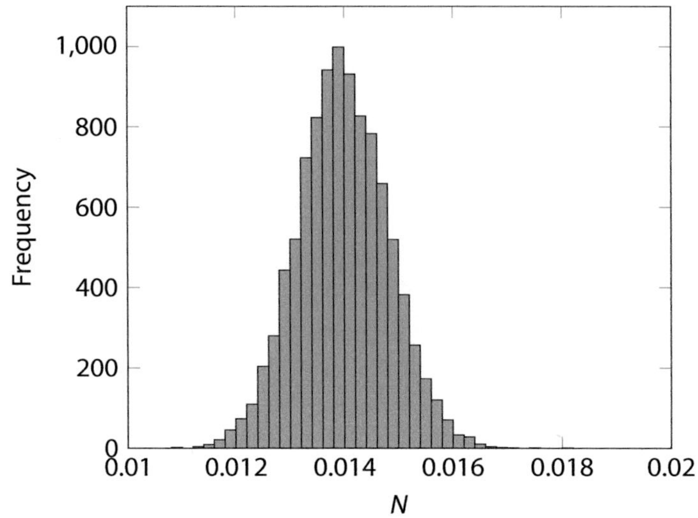

Figure 5.19 The result of a Monte Carlo error analysis using Equation (5.54). The distribution has a mean value of $\bar{N} = 0.0140$ moles and standard deviation of $\sigma_N = 0.0008$.

Carlo method comes into its own, however, when we have a complicated equation that is difficult to differentiate, or if the equations are sufficiently nonlinear and the measurement uncertainties sufficiently large that the linearization assumption does not hold, or if the measured values have a skewed probability distribution function.

Monte Carlo techniques are inherently probabilistic in that we have a distribution of numbers that we *randomly* sample many times. Each sample can then be used to make further calculations. The key point is that, with a computer, we can produce many thousands of samples and so obtain reliable statistics on the results of our calculations.

5.7.2 Monte Carlo Integration

We have seen in Chapter 2 that it is often not possible to evaluate an integral analytically, and in such cases we have to resort to numerical methods. The methods described in Section 2.16 involve finding an approximation to the function we are integrating and then calculating the area under the curve given by that approximation. Monte Carlo techniques provide another means of evaluating an integral numerically, and they can be particularly efficient for either multidimensional integrals or integrals of very complicated functions.

There are several different ways of evaluating an integral using Monte Carlo techniques. The simplest Monte Carlo integration technique basically involves filling an area with random points and calculating the fraction of them that fall within the desired region—this is sometimes referred to as the hit-or-miss, or rejection sampling method. For example, to evaluate the integral

$$\int_a^b f(x)\,dx = \int_{-1.5}^{2.5} (x^3 - x^2 - 3x + 4)\,dx \tag{5.55}$$

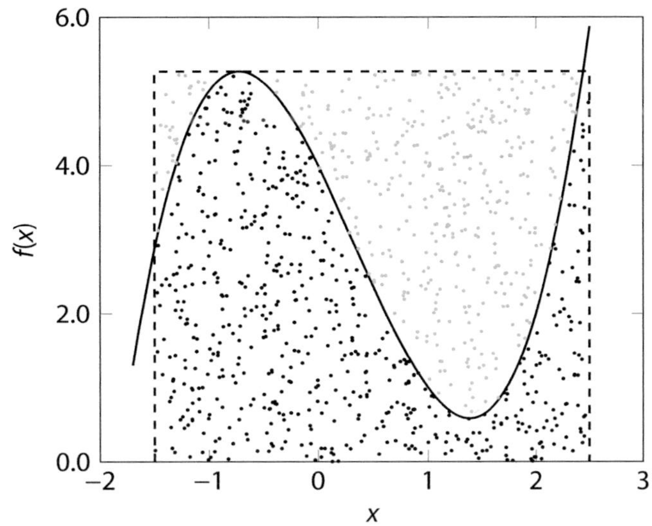

Figure 5.20 Monte Carlo integration of the function $f(x)$ given in Equation (5.55) using a "hit-or-miss" method. We create a large number of random points (x_i, y_i), where both x_i and y_i are sampled from a uniform distribution. Points where $y_i \leq f(x_i)$ are coloured black, and those where $y_i > f(x_i)$ are coloured grey. The value of the integral is given by the proportion of points $y_i \leq f(x_i)$ multiplied by the area of the box shown by the dashed lines. The value of the integral in Equation (5.55) is approximately 12.167 and the value from sampling 10^5 points is 12.164.

we first find the maximum of the function $f(x)$ between $x = a$ and $x = b$ and draw a box from the x axis to the maximum value of $f(x)$ and between $x = a$ and $x = b$ (Figure 5.20). To calculate the integral, we generate a very large number of uniformly distributed random points and keep track of the number of points that fall below the curve $y = f(x)$; i.e., we keep track of the number of points (x_i, y_i) such that $y_i \leq f(x_i)$. The value of the integral is then given by the area of the box multiplied by the fraction of random points that fall under the curve. This is a very simple method, but requires a very large number of random points to get an accurate answer. However, for two- and three-dimensional integrals, it can be more computationally efficient than other techniques.

An alternative Monte Carlo method is the sample-mean method (Figure 5.21). Let us say that we want to numerically evaluate the integral

$$I = \int_a^b f(x)\, dx, \tag{5.56}$$

where $f(x)$ is some function that we know. If we pick N random numbers (x_i) that are uniformly distributed between $x = a$ and $x = b$, then an estimate of I is

$$I(N) = \frac{(b-a)}{N} \sum_{i=1}^{N} f(x_i). \tag{5.57}$$

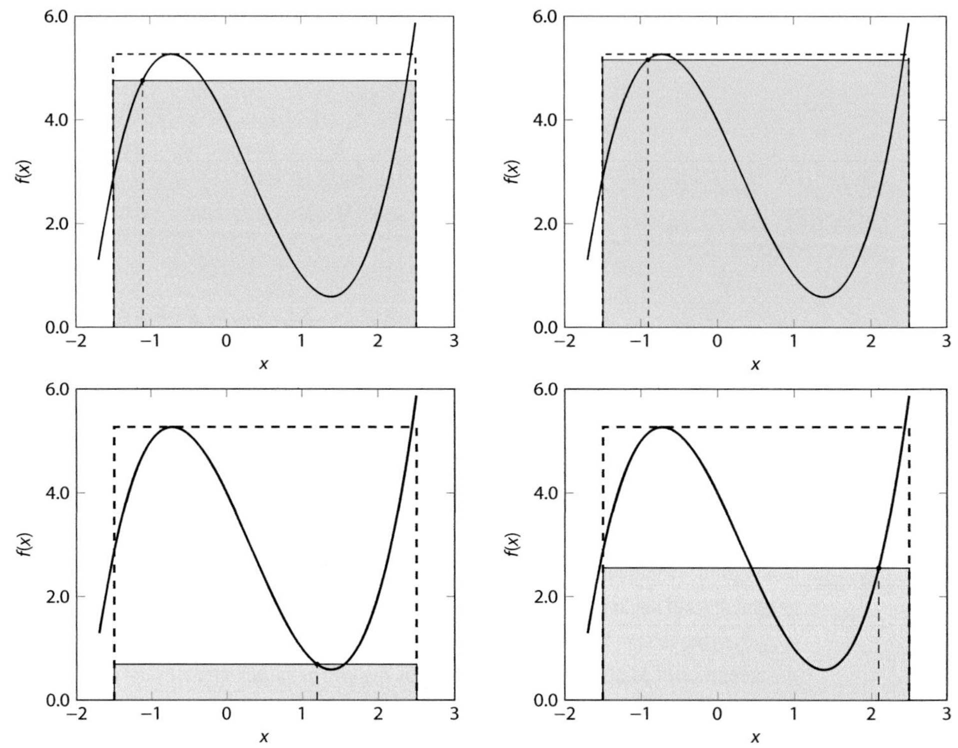

Figure 5.21 A graphical depiction of the sample-mean method of Monte Carlo integration. The function $f(x)$ being integrated is the same function as shown in Figure 5.20. Each panel shows a random value of x (gray dashed line) chosen between the limits of integration. The corresponding value of $f(x)$ is calculated and a rectangle (gray area) constructed whose height is the value of $f(x)$ and width is the distance between the limits of integration and the area of the rectangle is constructed. This procedure is repeated many times.

What is happening here and why does this work? The x_i are random values of x, so

$$\frac{1}{N}\sum_{i=1}^{N} f(x_i) = \langle f(x)\rangle$$

is simply the average value of the function $f(x)$ evaluated at the N random values x_i, This is an estimate of the average value of the function $f(x)$ above the x-axis. Multiplying this average value by $(b - a)$ gives us an estimate of the area under the curve. We can think of this in another way by rewriting Equation (5.57) to put the factor $(b - a)$ inside the summation,

$$I(N) = \frac{1}{N}\sum_{i=1}^{N}(b-a)f(x_i).$$

This equation is the *average* area of rectangles whose base is $(b - a)$—the interval over which we are integrating—and whose height is randomly chosen, though note that the heights all lie on the curve $f(x)$. Now, because x_i is a random number, the value $I(N)$ is also a random number, so we can calculate its expected value as:

$$E[I(N)] = E\left[\frac{(b-a)}{N}\sum_{i=1}^{N} f(x_i)\right] = \frac{(b-a)}{N}E\left[\sum_{i=1}^{N} f(x_i)\right] = \frac{(b-a)}{N}\sum_{i=1}^{N} E[f(x_i)]$$
$$= \frac{(b-a)}{N}\sum_{i=1}^{N}\left(\int_a^b f(x)p(x)\,dx\right) = \frac{1}{N}\sum_{i=1}^{N}\left(\int_a^b (b-a)f(x)\frac{1}{(b-a)}\,dx\right)$$
$$= \frac{1}{N}\sum_{i=1}^{N}\left(\int_a^b f(x)\,dx\right) = \int_a^b f(x)\,dx,$$

where we have used the PDF for a uniform distribution (Equation (5.36)). So, the expected value of our Monte Carlo estimate is indeed the value of the integral we are interested in. For the numerical methods we explored in Section 2.16 we found that we could improve the estimates of our numerical solution by increasing the number of rectangles we used. In an analogous way we can ask how the accuracy of the Monte Carlo estimate changes as we change the value of N.

$$\text{Var}[I(N)] = \text{Var}\left[\frac{1}{N}\sum_{i=1}^{N}\frac{f(x_i)}{p(x_i)}\right] = \frac{1}{N^2}\sum_{i=1}^{N}\text{Var}\left[\frac{f(x_i)}{p(x_i)}\right] = \frac{1}{N^2}\sum_{i=1}^{N}\text{Var}[Y]$$
$$= \frac{1}{N}\text{Var}[Y],$$

so that the standard deviation of the value of the integral is

$$\sigma_{I(N)} = \frac{1}{\sqrt{N}}\sigma_Y. \tag{5.58}$$

This equation tells us that if we want to improve the accuracy of our estimate by a factor of 10, we need to increase the value of N by a factor of 100.

Having seen these two methods of Monte Carlo integration, we might wonder which is the more accurate. To explore this we can evaluate the integral

$$I = \int_0^1 \sin^2\left(\frac{1}{x}\right)\,dx \tag{5.59}$$

using both techniques; the relative errors in both cases are shown in Figure 5.22, which indicates that for this integral there is not a great deal of difference between the two, though the sample-mean method has the potential to produce smaller relative errors (the gray line drops to values of approximately 2×10^{-6}).

5.8 Further Reading

There are many good books on probability and its application in the Earth and environmental sciences. *Chance in Biology* (Denny and Gaines, 2002) shows how probability can be used in understanding many biological phenomena of interest to Earth scientists. The relationships between probability, statistics, data analysis, and science are explored in a very practical and accessible way in *The Ecological Detective* (Hilborn and Mangel, 1997). The history of Bayes' theorem and its uses is fascinating and full of interesting characters,

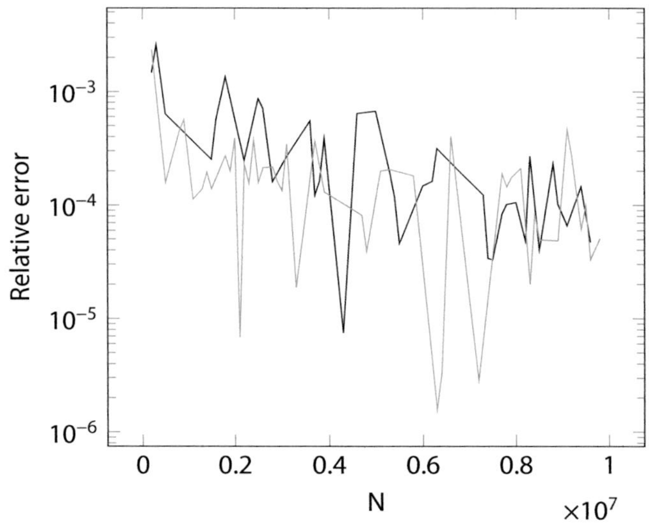

Figure 5.22 The relative errors for the Monte Carlo evaluation of Equation (5.59) using the simple hit-or-miss algorithm (black line) and the sample-mean algorithm (gray line) using different numbers of samples (*N*).

and is well described in *From Cosmos to Chaos* (Coles, 2006) and *The Theory That Would Not Die* (Bertsch McGrayne, 2011). More sophisticated Monte Carlo methods, as well as a discussion of random number generators, can be found in *Numerical Recipes in C* (Press et al., 1992).

Problems

5.1 According to the United States Geological Survey, there were 1346 earthquakes of magnitude ≥ 7.0 worldwide between 1900 and 2016, and there were 92 earthquakes of magnitude ≥ 8.0 worldwide between 1905 and 2015. Assuming that the probability of an earthquake occurring somewhere on the Earth during one year follows a Poisson distribution, calculate:

1. The probability of there being at least one earthquake of magnitude ≥ 7.0 in a year.
2. The probability of there being at least one earthquake of magnitude ≥ 8.0 in a year.
3. The probability of there being no earthquake of magnitude ≥ 7.0 in a year.
4. The probability of there being no earthquake of magnitude ≥ 8.0 in a year.

5.2 Consider a random variable A with $\mathrm{Var}(A) \neq 0$. Now consider a random variable $B = -A$.

1. Show that $\text{Var}(A) + \text{Var}(B) = 2\text{Var}(A)$.
2. Show that $\text{Var}(A + B) \neq \text{Var}(A) + \text{Var}(B)$.

5.3 Zircons are small (~0.1 mm) crystals that are found in many rocks. They are chemically inert and can be used to date samples of rock because they trap uranium and lead when they form. A geologist has a large number of rock samples that have all be crushed into small grains, and she needs to date the samples. She knows that, in general there is a probability of 0.01 of finding a zircon crystal in a typical grain. What is the probability of finding 1 zircon crystal in 100 rock grains? What is the probability of finding 5 crystals in 200 grains?

5.4 In the Earth sciences we sometimes need a uniformly random distribution of points on the surface of a sphere. Here we run into problems because the areas of lines of latitude and longitude converge at the north and south poles.

1. Use the facts that we require a uniform probability and that the total area of a sphere is 4π steradians to show that the probability density for a point on the sphere is $f(x) = 1/(4\pi)$.
2. Use your answer from Question 1 to show that the probability of finding a point in an area dA on a sphere of radius 1 is

$$f(\theta, \phi) = \frac{1}{4\pi} \sin(\phi).$$

5.5 Mean wind speed (v) is often represented using a *Weibull distribution* with $v > 0$:

$$P(v) = \alpha \frac{v^{\alpha-1}}{v_0^{\alpha}} \exp\left[-\left(\frac{v}{v_0}\right)^{\alpha}\right],$$

where v_0 is called the *location* parameter and is proportional to the mean wind speed, and α determines the width of the distribution.

1. Show that the cumulative distribution function is

$$F_V(v) = P(V \leq v) = 1 - \exp\left[-\left(\frac{v}{v_0}\right)^{\alpha}\right].$$

2. Use the transformation $Y = V^{\alpha}$ to show that

$$F_Y(y) = 1 - e^{-y/v_0^{\alpha}} \quad y > 0$$

and that the PDF of Y is an exponential PDF

$$f(y) = \frac{1}{v_0^{\alpha}} e^{-y/v_0^{\alpha}} \quad y > 0.$$

5.6 The distribution of wave heights on the ocean can be described by the *Rayleigh distribution*, which has a CDF

$$F(h) = 1 - \exp\left(-2\left(\frac{h}{H_s}\right)^2\right),$$

where H_s is a parameter called the *significant wave height*, a measure of the average wave height. In the early morning of February 4, 2013, a buoy in the North Atlantic measured a significant wave height of 19 m. What is the probability that the buoy experienced

1. a wave of height greater than 20 m?
2. a wave of height greater than 40 m?

5.7 Monthly rainfall at a given location can often be modeled using a gamma PDF

$$p(x) = \frac{\lambda^a x^{a-1}}{\Gamma(a)} e^{-\lambda x},$$

where the function $\Gamma(a)$ is the one we met in Equation (5.47).

1. Using the fact that $\Gamma(n+1) = n\Gamma(n)$, show that the expected value of the gamma distribution is

$$E(X) = \frac{a}{\lambda}.$$

2. At a certain location, the mean rainfall during March is 11.3 cm with a standard deviation of 2.2 cm. If the monthly rainfall can be represented by a gamma distribution, what is the probability of receiving more than 15 cm of rain during March? (Use the fact that the variance of the gamma distribution is $\sigma^2 = a/\lambda^2$.)

5.8 The number of major earthquakes in a region follows a Poisson distribution with an average of four earthquakes every 100 years. Calculate the probability that there will be at least one major earthquake in the region in the next five years.

5.9 This problem is inspired by a story told to the author by his PhD supervisor, Dr. John Barrow. A book of this length will undoubtedly contain typographical errors. Given that two of the author's colleagues have read the whole manuscript, one of them found α errors, the other found ω errors, and out of these there were μ errors that they both found. However, there are likely to be some errors that remain unfound. Show that the total number of errors (found and unfound) is $T = \alpha\omega/\mu$ and that the number of errors remaining to be found is $(\alpha - \mu)(\omega - \mu)/\mu$.

5.10 Global warming can affect temperatures by increasing the mean, increasing the variance, or both. Assume the annual high temperature at a location is given by a normal distribution with a mean 17°C with a standard deviation of 8°C. What is the probability of having a high temperature greater than 30°C? If the annual mean high temperature increases to 19°C but the standard deviation remains the same, what is the probability of having a high temperature in excess of 30°C? If the mean high temperature is 17°C but the standard deviation increases to 10°C, what is the probability of obtaining a high temperature greater than 30°C? If the mean increases to 19°C and the standard deviation increases to 10°C, what is the probability of having a high temperature greater than 30°C? (You should use a calculator or computer to calculate the probabilities.)

5.11 When rain falls on a watershed we can think of it moving, via runoff, through a series of lakes (or reservoirs) such that the output from the nth lake is the input to the $(n+1)$th lake. The output $q(t)$ from a single lake at time t after the rain fell can be given by the convolution integral

$$q(t) = \int_0^t I(t)\frac{1}{k}\exp\left(-\frac{t-\tau}{k}\right)\,d\tau,$$

where $I(t)$ is input to the lake and k is a constant.

1. If the output from the first lake flows into a second lake, show that the output from the second lake is

$$q(t) = \frac{t}{k^2}\mathrm{e}^{-t/k}.$$

2. If there are n lakes in a chain, with the output from the nth lake flowing into the $(n+1)$th lake, show that the output from the nth lake is

$$q(t) = \frac{1}{k\Gamma(n)}\left(\frac{t}{k}\right)^{n-1}\mathrm{e}^{-t/k}.$$

5.12 Calculate the convolution of $f(x) = \mathrm{e}^{-x}$ and $g(x) = \sin(x)$.

5.13 Consider an exponential PDF

$$p_X(x) = \frac{1}{k}\mathrm{e}^{-x/k}.$$

Calculate the transformed PDF under the transformation $Y = \mathrm{e}^X$; the result is known as the *Pareto distribution*.

5.14 Computers can be very useful in helping to hone our intuition about probability. Scientific software packages such as MATLAB™ and Python have routines that will randomly sample different probability distributions. This problem uses these to examine the central limit theorem. Consider the following PDFs:

1. A uniform distribution between $x = 0$ and $x = 1$.
2. A Poisson distribution with $\lambda = 24.7$.
3. The sum of two normal distributions, one with $\mu = 23.5$ and $\sigma = 8.7$, and the other with $\mu = 18.6$ and $\sigma = 7.3$.

For each distribution, use a computer to generate 500 sets of two samples and take the mean of each set of two samples. Calculate the mean and standard deviation of these 500 mean values and compare them with the mean and standard deviation of the original distribution. Repeat this procedure using three samples, then four samples, and so on. How many samples do you need before the mean *and* standard deviation of the samples you calculate agree with reasonable accuracy with the means and standard deviations of the original distributions?

5.15 Write a computer code to use a Monte Carlo method to calculate the value of π by evaluating

$$\int_0^1 \sqrt{1 - x^2}\, dx.$$

5.16 Consider two independent, random variables X and Y that both have standardized normal probability distributions.

1. Construct the joint probability distribution function $p_{X,Y}(x, y)$.
2. Use the transformation $U = f(X, Y) = X/Y$, $V = g(X, Y) = Y$ to show that

$$p_{U,V}(u, v) = \frac{v}{2\pi} \exp\left(\frac{-v^2(u^2 + 1)}{2}\right).$$

3. Use integration by parts to show that

$$p_U(u) = \frac{1}{\pi(u^2 + 1)},$$

which is a Cauchy distribution.

6 Ordinary Differential Equations

Many of the questions we want to ask as scientists involve understanding how fast something is changing, either with respect to time, space, or some other variable. For example, we might be interested in how fast plant biomass increases as temperature increases, or we might have an equation for the rate of change of the concentration of a pollutant in a lake in terms of the rates of input (the sources) and output (the sinks) of the pollutant. We have already seen that rates of change are described mathematically by a derivative, so we need to ask if we can solve an equation for a function $y(x)$ that contains derivatives of y with respect to x. This is the realm of differential equations. Unlike solving algebraic equations where we want to find a number that satisfies a given equation, solving a differential equation involves trying to find a *function* that satisfies the equation. The equation contains derivatives of the function, so solving the equation will require integrating it; and as we have seen in Chapter 2, we cannot evaluate every integral in terms of elementary functions, so it is likely that we cannot solve all differential equations in terms of elementary functions. This is true, and we will discuss both *analytical* and *numerical* methods for solving differential equations, as well as *qualitative* methods that give us insight into the behavior of the solution without having to explicitly solve the equation. In this chapter we will restrict our attention to functions of a single variable so the equation will contain ordinary derivatives, hence these equations are called *ordinary differential equations* (ODEs). Equations that involve derivatives of functions of more than one variable are called *partial differential equations* (PDEs), and we will look at those in Chapter 10.

How do differential equations arise from the problems we are studying? Differential equations describe the rate at which one variable is changing with respect to another.[1] For example, the rate of change of atmospheric pressure with height, the rate of change in the number of atoms of a radioactive substance over time, the change in temperature of fresh magma as it cools. Many ODEs that we come across represent the difference between rates of input (i.e., sources) and rates of output (i.e., sinks). As a simple example, the rate of change with respect to time of money in a bank account is the difference between the rate of input (how much money is deposited per month) and the rate of output (the amount of money being spent in a month).

Example 6.1 The number of atoms ($N(t)$) of a radioactive element changes over time (t) as the atoms decay at a rate proportional to the number of atoms present at that time. If we write the proportionality constants as λ with dimensions of [T]−1, we can derive a

[1] Even though we talk about how *fast* one variable changes with respect to another, the changes need not be with respect to time.

differential equation for the rate of change of $N(t)$ with time. If our sample has $N(t_0)$ atoms at time t_0, then over a time interval Δt, the number of atoms that decay will be $\lambda N(t)\Delta t$. This should have dimensions of "number of atoms," which it does. So, assuming that no new atoms of the element are created (i.e., there is no source of new atoms, only a sink), the change ΔN in the number of atoms in the time interval Δt is

$$\Delta N = (\text{Number of atoms that decay in time interval } \Delta t) = -\lambda N(t)\Delta t,$$

where the minus sign is used because there is a *loss* of atoms. We can rearrange this equation by dividing both sides by Δt, and taking the limit as $\Delta t \to 0$, giving us the equation

$$\lim_{\Delta t \to \infty} \frac{\Delta N(t)}{\Delta t} = \frac{dN(t)}{dt} = -\lambda N(t). \tag{6.1}$$

Does this equation make sense? The derivative is negative (so long as $\lambda > 0$), so $N(t)$ is decreasing, which is what we expect. Also, the dimensions of the left-hand side of the equation are atoms per time, which is the same as the dimensions of the right-hand side. So, the equation is dimensionally consistent and agrees with the fact that radioactive decay causes the number of atoms in the sample to decrease.

Example 6.2 The concentration of a pollutant in a lake depends on the balance between how fast the pollutant enters the lake via the river flowing into it, and the rate of loss of the pollutant from a single river that flows out of the lake, assuming there are no sources or sinks of pollutant in the lake itself. Consider a lake (Figure 6.1) with an inflow (river A) and and an outflow (river B). The water entering the lake contains a pollutant, which is mixed within the lake and exits the lake through river B. We want a differential equation for the rate of change of pollutant concentration in the lake. Problems that deal with concentrations of a substance are best formulated initially in terms of the total mass of that substance.

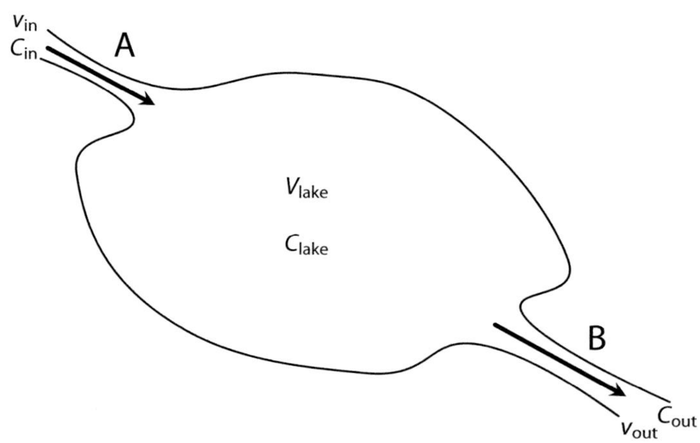

Figure 6.1 Water with a pollutant concentration C_{in} enters the lake from the river A at a rate v_{in}. The lake has a volume V_{lake} and concentration of pollutant C_{lake}. Water leaves the lake at a rate v_{out} via river B and has a pollutant concentration of C_{out}.

This is because the concentration of a substance depends on its mass and the volume it occupies, and both might be changing with time. Mass, however, is conserved, making it easier to deal with. In a time interval Δt, the change in mass (M) of pollutant in the lake is

$$\Delta M = (\text{Gain of pollutant mass in time } \Delta t) - (\text{Loss of pollutant mass in time } \Delta t).$$

To calculate the *rate* at which the pollutant enters the lake, we need to know the flow rate of river A entering the lake (v_{in}, in units of volume per time) and the concentration of pollutant in the river water (C_{in} in units of grams per volume). The mass of pollutant flowing into the lake during a time Δt is then $v_{\text{in}} \times C_{\text{in}} \times \Delta t$; this expression has dimensions of $([\text{L}]^3[\text{T}]^{-1})([\text{M}][\text{L}]^{-3})[\text{T}] = [\text{M}]$, i.e., a mass, which is the correct dimension for ΔM.

We are going to have to use some simplifying assumptions to obtain an expression for the loss of pollutant from the lake. We will assume that water from the input river mixes instantaneously with the water already in the lake, so that there is always a uniform concentration of pollutant across the lake. This means that the concentration of pollutant in water leaving the lake is the same as the concentration C_{lake} in the lake itself. We can now develop an expression for the loss of pollutant over the time interval Δt: $v_{\text{out}} \times C_{\text{lake}} \times \Delta t$. Since there are no other gains or losses, mass balance of the pollutant within the lake implies that the change in pollutant mass in the lake over the time interval Δt is $\Delta M = v_{\text{in}} C_{\text{in}} \Delta t - v_{\text{out}} C_{\text{lake}} \Delta t$. Dividing by Δt and taking the limit as $\Delta t \to 0$ gives

$$\frac{dM}{dt} = v_{\text{in}} C_{\text{in}} - v_{\text{out}} C_{\text{lake}}.$$

To obtain an expression for the rate of change in C_{lake}, we assume that the volume of the lake (V_{lake}) is constant (which implies $v_{\text{in}} = v_{\text{out}}$), so

$$\frac{dM}{dt} = V_{\text{lake}} \frac{dC_{\text{lake}}}{dt} = v_{\text{in}} C_{\text{in}} - v_{\text{in}} \times C_{\text{lake}},$$

giving finally

$$\frac{dC_{\text{lake}}}{dt} = \frac{v_{\text{in}} C_{\text{in}} - v_{\text{in}} C_{\text{lake}}}{V_{\text{lake}}}. \tag{6.2}$$

The dimensions of the left-hand side of the equation are $[\text{M}][\text{L}]^{-3}[\text{T}]^{-1}$, or mass per volume per time, and the dimensions of the right-hand side are also moles per volume per time, so the equation is dimensionally consistent.

Example 6.3 As a last example, let us derive an equation for the rate of change of atmospheric pressure with height in the atmosphere, assuming that the atmosphere is in hydrostatic equilibrium (Figure 6.2). This problem requires a little more thought because there are no gain or loss terms as there were in the previous examples. Atmospheric pressure at the surface of the Earth comes about from the weight of the atmosphere above. The condition of hydrostatic equilibrium is a simplification that says that the atmosphere is static and not moving vertically. This implies that there are no upward or downward motions of air and the pressure at a given height in the atmosphere is uniform. We can therefore consider a vertical column of air as being representative of the atmosphere at any location, and we choose the column to be in the shape of a cylinder. When we have to choose a shape to solve a problem, it is often a good idea to pick a simple one so that we

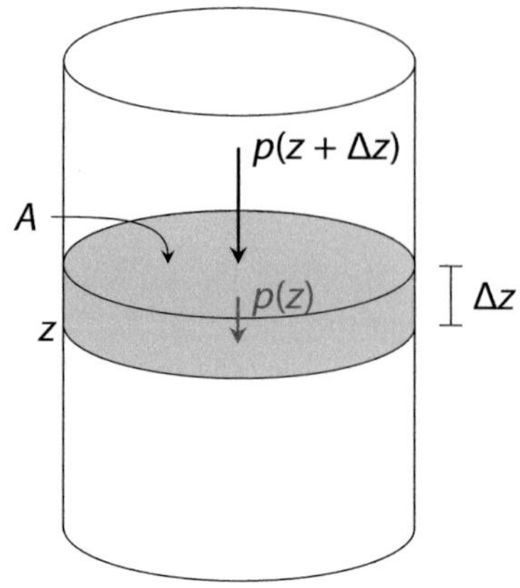

Figure 6.2 A disk formed from two horizontal slices in a cylindrical column of air. The lower disk is at a height z in the atmosphere and experiences a pressure (a force per unit area) $p(z)$ due to the weight of the air in the column above it. The disk at height $z + \Delta z$ experiences a lower pressure because the weight (gM) of the air between z and $z + \Delta z$ (the gray volume) is not acting on the upper layer.

can easily calculate areas and volumes. Frequently, because terms cancel out, the actual shape does not really matter in the end.

We want to derive an equation for the rate of change of pressure with height, so consider two slices through the cylinder at heights z and $z + \Delta z$. Recall from Chapter 1 that pressure has dimensions of a force per unit area. The slice at height z is supported from below by the pressure ($p(z)$) of the atmosphere below it, so the upward force acting on the surface is $p(z)A$, where A is the cross-sectional area of the cylinder. The forces from above pushing the slice down include the weight of the atmosphere between z and $z + \Delta z$, and the weight of the atmosphere above height $z + \Delta z$. Newton's second laws tells us that the first of these is $gM = g\rho A \Delta z$,[2] where ρ is the density of air, M is the mass of atmosphere in the layer between heights z and $z + \Delta z$, and g is the acceleration due to gravity. The second downward force is just $p(z + \Delta z)A$. For this slice of the column not to move, the upward directed forces must balance the downward ones; that is, if we take the upward direction as positive, then

$$p(z)A - p(z + \Delta z)A - g\rho \Delta z = 0 \quad \Longrightarrow \quad \frac{p(z + \Delta z) - p(z)}{\Delta z} = -g\rho,$$

or, by taking the limit as $\Delta z \to 0$,

$$\frac{dp}{dz} = -g\rho.$$

This equation is dimensionally correct, and the left-hand side is the derivative of pressure with height in the atmosphere (which is what we want), but the right-hand side contains a different variable, the density ρ. So, to get any further we need to find a relationship

[2] We are using the law that $F = ma$, where F is the force acting, m is the mass of the body, and a the acceleration of the body caused by the force.

that will allow us to write the density as a function of pressure. If the atmosphere can be considered as an *ideal gas*,[3] then the ideal gas law relates the density to the pressure via

$$p = \frac{\rho R_g T}{M_a},$$

where $R_g = 8.3143\ \mathrm{J\,K^{-1}\,mol^{-1}}$, is the universal gas constant; M_a is the mean molecular weight of dry air, which we also treat as a constant; T is the temperature (in degrees Kelvin), which will depend on z, as will ρ. We can then substitute for ρ to get

$$\frac{dp}{dz} = -\frac{gM_a}{R_g T(z)} p(z). \tag{6.3}$$

A good strategy for solving complicated problems is to simplify the problem first, as we have seen in Examples 6.2 and 6.3. Once we have a solution to the simplified version, we can start to add back the complexity by relaxing assumptions one at a time, and trying to solve the more complicated problem. Doing this helps us develop an intuition for the problem. However, some of these assumptions may appear unphysical or unrealistic, but may be reasonable under some circumstances. For example, our assumption in Example 6.2 that water entering the lake is instantaneously mixed throughout the lake volume may be a reasonable approximation for a small lake with a large or fast river flowing into it. In such cases, the time for water to circulate and mix in the lake can be shorter than the time it takes water entering the lake from river A to cross it and exit through river B, so we can assume that the pollutant is uniformly distributed throughout the lake.

Exercise 6.0.1 Show that the hydrostatic equation derived in Example 6.2 is dimensionally consistent.

Exercise 6.0.2 The radioactive isotope ^{234}Th has a half-life of 24.1 days and is often used in oceanography to track processes that occur on timescales of weeks to a few months. ^{234}Th decays to ^{234}Pa at a rate of $\lambda_{\mathrm{T}}\ \mathrm{d}^{-1}$ and is formed by the decay of U^{238}, which decays at a rate of $\lambda_{\mathrm{U}}\ \mathrm{d}^{-1}$. Derive a differential equation for the rate of change of ^{234}Th in the surface ocean, treating the surface ocean as a single, uniform layer of thickness L.

Exercise 6.0.3 Particles in the atmosphere are created at a constant rate of $\kappa\ \mathrm{s}^{-1}$ and destroyed by binary collisions (i.e., collisions between two, and only two, particles) at a rate of $k\ \mathrm{m}^3\ \mathrm{s}^{-1}$. N is the number of particles m^{-3}.

1. Derive an equation for the number of new particles formed in the time interval Δt, making sure that the equation is dimensionally consistent.

[3] An ideal gas consists of pointlike particles that do not exert any forces on each other and bounce off each other without any loss of energy when they collide. Although it is an idealized, theoretical construction, there are many cases where a real gas approximates an ideal gas quite well. One advantage of assuming the atmosphere is an ideal gas is that it allows us to relate the density of the gas to its pressure using the ideal gas law.

2. Derive an equation for the number of particles lost by collisions with other particles in the time interval Δt, again making sure that the equation is dimensionally consistent.
3. Use the equations from Questions 1 and 2 to derive a differential equation for the rate of change of N.

6.1 Terminology and Classification

The method we use to solve a particular differential equation depends largely on the type of equation we are dealing with. Consequently, we need to learn some terminology before we can proceed. The most general ODE for a function $y(x)$ can be written as

$$a_0(x,y)\frac{d^{(n)}y(x)}{dx^{(n)}} + a_1(x,y)\frac{d^{(n-1)}y(x)}{dx^{(n-1)}} + \cdots + a_{n-1}(x,y)\frac{dy(x)}{dx} + a_n(x,y)y(x) = f(x,y). \tag{6.4}$$

In Equation (6.4), $y(x)$ is the unknown function of x that we want to find, the coefficients $a_i(x, y)$ can be constants, functions of x, functions of y, or functions of both, and $f(x, y)$ is a known function, often called the *forcing function*, which depends on the problem at hand. The *order* of the differential equation is the order of the highest derivative that appears in the equation; the following ODEs are all examples of second order differential equations:

$$a.\ 2y(x)\frac{d^2y}{dx^2} + x\frac{dy}{dx} = 0, \qquad b.\ 3x\frac{d^2y}{dx^2} + 5\frac{dy}{dx} + 6y = e^{x^2}, \qquad c.\ \frac{d^2y}{dx^2} = 6.$$

An ODE is *linear* if the unknown function appears linearly in the equation, otherwise it is a *nonlinear* ODE. So, example equation (a) is a nonlinear ODE because the first term contains $y(x)$ multiplied by its second derivative. Example equations (b) and (c) are both linear ODEs. If $f(x, y) = 0$ in Equation (6.4), then the equation is called a *homogeneous* equation, otherwise it is an *inhomogeneous* equation. Lastly, if the independent variable (x in Equation (6.4)) appears only in the derivatives (i.e., f and a_i are either constants or functions of only y), then the equation is called *autonomous*. As in Chapter 2, we will use a variety of common, space-saving notations for the derivative including $y'(x)$ for the derivative of y with respect to x, and $\dot{y}(t)$ for the derivative with respect to time. We will write derivatives more compactly and simply write y' and $\dot{y}$ for $y'(x)$ and $\dot{y}(t)$.

Example 6.4 Let us classify the following ODE using these categories:

$$a.\ y(x)\frac{dy}{dx} = x^2\sin(x), \quad b.\ x\frac{d^3y}{dx^3} + \frac{dy}{dx} = 0, \quad c.\ y^2\frac{d^2y}{dx^2} + y\frac{dy}{dx} + y = 0.$$

Equation (a) is a nonlinear (because it contains the term $y(x)y'(x)$), first order (it contains only a first order derivative), inhomogeneous (because of the presence of the $x^2\sin(x)$ term) equation. Equation (b) is a third order, linear, homogeneous equation, and (c) is a second order, nonlinear, homogeneous, autonomous equation.

Once we have classified a differential equation we have a better idea of which techniques we can use to solve it. There are many ODEs for which we can find analytic solutions (i.e., solutions using a paper and pen). However, finding a solution involves evaluating an integral, which is not always possible, so we often have to resort to numerical methods to solve the ODE. But before we do, there is a great deal we can, *and should*, learn about the solution of a differential equation using various simplifications and qualitative techniques, and we will explore some of these techniques later in this chapter. There are several reasons for not being too hasty in pursuing a numerical solution to an ODE. First, what we learn analytically can provide a check of our numerical solution. Second, we gain an understanding of how the solution behaves that can guide how we interpret a numerical solution.[4] We will concentrate our explorations on first and second order differential equations because these are the most common types of ODE that occur in science. This is because we are generally interested in equations for the rate of change of a variable (the first derivative) or the acceleration or deceleration of a variable (the second derivative).

6.2 First Order Differential Equations

We will start by examining the simpler ODEs first. If $y(x)$ is a function of x, then the most general first order ODE we can write is

$$a(x, y)\frac{dy}{dx} + b(x, y) = 0. \tag{6.5}$$

This equation is already too general for us to make any progress in solving it. For example, we do not know the form of the functions $a(x, y)$ and $b(x, y)$, both of which may be functions of $y(x)$, the unknown function we want to find. So, we shall have to simplify further.

6.2.1 First Order Linear Differential Equations

We can make our lives a little easier by recognizing that nonlinear equations are harder to solve than linear ones, though often the nonlinear equations are more interesting! So, to start with, we shall restrict ourselves to linear first order ODEs. This means that the function $y(x)$ appears linearly in the ODE, which implies that $a(x, y)$ must be a function only of x and $b(x, y) = b(x)y + c(x)$. The most general equation of this type is

$$a(x)\frac{dy}{dx} + b(x)y = c(x), \tag{6.6}$$

which we can simplify by dividing through by $a(x)$ to get

$$\frac{dy}{dx} + p(x)y = q(x), \qquad \text{where} \qquad p(x) = \frac{b(x)}{a(x)} \quad \text{and} \quad q(x) = \frac{c(x)}{a(x)}. \tag{6.7}$$

[4] An analogy is to think about taking a long hike in an unfamiliar area; you do not want to embark on the hike without any knowledge of what might lie ahead. Ideally you would like to have an accurate and detailed map of the area—this is like having the full analytical solution to the ODE. Failing that, a rough sketch on a piece of paper can be very useful in preventing you getting lost — this is analogous to having a qualitative understanding of the solutions without a full analytical solution.

Equation (6.7) is often called the *standard form*. We might be tempted to go straight ahead and integrate Equation (6.7), giving

$$\int \frac{dy}{dx}\,dx + \int p(x)y(x)\,dx = \int q(x)\,dx.$$

The first term on the left-hand side is just $y(x)$, which is what we want, so this is promising. If we can evaluate the integral on the right-hand side of the equation, then we are close to finding our solution. However, the problem lies in evaluating the second term on the left-hand side; we do not know what $y(x)$ is yet, so we cannot evaluate this integral, and unfortunately, this straightforward approach leads us nowhere. When our initial ideas lead to a dead end we can try to simplify the problem further.

We could reduce the number of terms in the equation by making either $q(x) = 0$ or $p(x) = 0$. Putting them both equal to zero gives us an equation with a solution y = constant, which is not particularly interesting. If $p(x) = 0$, the differential equation becomes

$$\frac{dy}{dx} = q(x).$$

We can formally integrate both sides of the equation with respect to x,[5]

$$\int \frac{dy}{dx}\,dx = \int dy = y(x) = \int q(x)\,dx,$$

and we can find the function $y(x)$ so long as we can evaluate the integral on the right-hand side of the equation. Let us pause for a minute to understand what we have done. Removing the $p(x)y(x)$ term has allowed us to separate the y and x parts of the equation; the left-hand side of the equation is just y and the right-hand side depends only on x. This separation of variables is a powerful technique that we will meet many times.

What happens if we make $q(x) = 0$ instead? In this case, Equation (6.7) becomes a first order, linear, homogeneous equation that is also separable:

$$\frac{dy}{dx} = -p(x)y.$$

The terms on the right-hand side of the equation are neatly factored into something that is a function solely of x, $p(x)$, multiplied by something that is solely a function of y, y itself. As a result, we can rearrange the equation and integrate over x:

$$\int \frac{dy}{y} = -\int p(x)\,dx.$$

We can immediately integrate the left-hand side of the equation to give us

$$\ln|y| + c_1 = -\int p(x)\,dx,$$

or

$$y(x) = C\exp\left(-\int p(x)\,dx\right) \qquad \text{for} \qquad C \geq 0,$$

[5] It is worth noting that replacing $(dy/dx)dx$ by dy is strictly a sleight of hand. This step can be made rigorous, but you should never think of this process as "cancelling the dx terms".

and we can find y so long as we can evaluate the integral on the right-hand side. We have already seen an example of this type of equation (Equation (6.1)),

$$\frac{dN(t)}{dt} = \pm\lambda N(t), \tag{6.8}$$

where $N(t)$ is a function of t and λ is a constant. If $\lambda > 0$ and the right-hand side of Equation (6.8) is $-\lambda N(t)$, then N is a decreasing function of time; but if the right-hand side is $+\lambda N(t)$, then $N(t)$ is an increasing function of time. The constant λ tells us how fast $N(t)$ is changing. Equation (6.8) describes many natural phenomena, such as the rate of change of the amount of a radioactive element, or the rate of change of a population of bacteria in a pond, or the attenuation of light as it passes from the surface to the deep ocean. We can now solve Equation (6.8) by dividing both sides of the equation by $N(t)$ and integrating with respect to t:

$$\begin{aligned}
\int \frac{1}{N}\frac{dN}{dt}\,dt &= \pm\lambda \int dt \\
\int \frac{dN}{N} &= \pm\lambda t + C_1 \\
\ln(|N|) + C_2 &= \pm\lambda t + C_1 \\
\ln(|N|) &= \pm\lambda t + C_1 - C_2 = \pm\lambda t + C \\
N &= \exp(\pm\lambda t + C) = \exp(C)\exp(\pm\lambda t) = N_0\exp(\pm\lambda t),
\end{aligned}$$

where we have chosen $N \geq 0$ because for most cases we are interested in (e.g., radioactive decay), so having $N < 0$ does not make physical sense. Notice that a first order ODE will require us to integrate the left- and right-hand sides of the equation, and we can combine the two constants of integration (C_1 and C_2) into a single constant (C). In the example we have written the constant $\exp(C)$ as N_0. The reason for this is that when $t = 0$, $N(t = 0) = \exp(C)$, so writing $\exp(C)$ as N_0 reminds us that the constant factor is just the value of N when $t = 0$.

Example 6.5 To show how separation of variables works, let us integrate the ODE

$$\frac{dy}{dx} = -2x^2 y.$$

Dividing the equation by y separates the variables, giving

$$\frac{1}{y}\frac{dy}{dx} = -2x^2,$$

and integrating both sides with respect to x produces

$$y(x) = C\exp\left(-2\int x^2\,dx\right) = C\exp\left(-\frac{2}{3}x^3\right).$$

What do we do if $q(x) \neq 0$? If this is the case, we have a first order, linear, inhomogeneous ODE, Equation (6.7). Let us multiply Equation (6.7) by a function $R(x)$ that we do not yet know,

$$R(x)\frac{dy}{dx} + R(x)p(x)y = R(x)q(x). \tag{6.9}$$

This seems like a crazy thing to do, because we now have two functions that we do not know, $y(x)$ and $R(x)$. But, we will find that we can choose $R(x)$ in such a way that simplifies the equation and allows us to solve it. Recall the product rule for derivatives (Equation (2.10)),

$$\frac{d}{dx}u(x)y(x) = u(x)\frac{dy}{dx} + y(x)\frac{du}{dx}.$$

If we can find $R(x)$ such that

$$R(x)\frac{dy}{dx} + R(x)p(x)y = \frac{d}{dx}\left(R(x)y(x)\right) = R(x)\frac{dy}{dx} + y(x)\frac{dR}{dx}, \tag{6.10}$$

then we can replace the left-hand side of Equation (6.9) with

$$\frac{d}{dx}(R(x)y(x)),$$

which we can immediately integrate. Equation (6.10) tells us that for us to be able to do this, $R(x)$ must satisfy the equation

$$\frac{dR}{dx} = R(x)p(x).$$

But this is a first order linear homogeneous ODE that we now know how to solve:

$$R(x) = \exp\left(\int p(x)\, dx\right). \tag{6.11}$$

The function $R(x)$ is called an *integrating factor* of the ODE. So, to solve our inhomogeneous equation, we calculate the integrating factor (if the function $p(x)$ is integrable), multiply the whole equation by it, then integrate. If we can evaluate the integral of the right-hand side of Equation (6.9), then we have solved the ODE.

Example 6.6 We can use an integrating factor to solve the ODE

$$\frac{dy}{dx} + \frac{2}{x}y = \frac{e^{2x}}{x^2}.$$

Using Equation (6.11), the integrating factor is given by

$$R(x) = \exp\left(2\int \frac{1}{x}\, dx\right) = e^{2\ln(x)} = e^{\ln(x^2)} = x^2.$$

Multiplying the ODE by the integrating factor gives

$$x^2\frac{dy}{dx} + 2xy = \frac{d}{dx}(x^2y(x)) = e^{2x}$$

$$\int \frac{d}{dx}(x^2y(x))\, dx = x^2y = \int e^{2x} = \frac{1}{2}e^{2x} + C$$

$$y(x) = \frac{e^{2x} + C_1}{2x^2},$$

where $C_1 = 2C$ is a constant.

Exercise 6.2.1 Use an integrating factor to find solutions to the following ODEs; in each case, start by writing the differential equation in the form of Equation (6.7).

1. $(x^3 - 1)\dfrac{dy}{dx} + 3xy = x^3$
2. $\dfrac{dy}{dx} + 4y = e^x$
3. $\dfrac{dy}{dx} + y\cot(x) = \sin(x)$
4. $\dfrac{dy}{dx} + xy = xe^{x^2/2}$

Exercise 6.2.2 Use an integrating factor to find the solution to Equation (6.2) assuming that v_{in}, v_{out}, V_{lake}, and C_{in} are all constant in time.

Another method for solving general first order linear inhomogeneous equations is called *variation of parameters*. Let us assume that we can write a solution to Equation (6.7) as the sum of two parts: $y(x) = y_h(x) + y_p(x)$. Substituting this into Equation (6.7) gives

$$y_h' + y_p' + p(x)(y_h(x) + y_p(x)) = (y_h' + p(x)y_h(x)) + (y_p' + p(x)y_p(x)) = q(x).$$

We now make the assumption that $y_h(x)$ satisfies the homogeneous equation

$$y_h' + p(x)y_h(x) = 0,$$

which is something that we know how to solve:

$$y_h(x) = Ae^{P(x)}, \qquad \text{where} \qquad P(x) = -\int p(x)\,dx. \tag{6.12}$$

Now we make a guess. We guess that $y_p(x)$ has a similar form, but instead of a constant A, we multiply the exponential by an unknown function $v(x)$, so that

$$y_p(x) = v(x)e^{P(x)}.$$

It may seem that we have just replaced one unknown function ($y_p(x)$) with another ($v(x)$). We have, but we have also done a little more than that; our guess for $y_p(x)$ implies that it can be factored into two parts, one that we know, and one we have yet to find. The "magic" in the technique is that this produces an equation for $v(x)$ that can be nicer to deal with,

$$\begin{aligned} y_p' + p(x)y_p(x) &= \frac{d}{dx}\left(v(x)e^{P(x)}\right) + p(x)v(x)e^{P(x)} \\ &= v'e^{P(x)} + v(x)P'(x)e^{P(x)} + p(x)v(x)e^{P(x)} \\ &= v(x)\left(P'(x)e^{P(x)} + p(x)e^{P(x)}\right) + v'e^{P(x)}. \end{aligned}$$

We know that y_h solves the homogeneous equation (we set things up that way), so

$$\frac{dy_h}{dx} = AP'e^{P(x)} + Ap(x)e^{P(x)} = A(P'e^{P(x)} + p(x)e^{P(x)}) = 0,$$

and therefore either $A = 0$, which is not interesting, or $P'\exp(P(x)) + p(x)\exp(P(x)) = 0$, which is satisfied because of the definition of $P(x)$ (Equation (6.12)). Our equation for y_p now becomes

$$y_p' + p(x)y_p(x) = v'e^{P(x)} = q(x), \tag{6.13}$$

which we can solve for $v(x)$ by integration. The final solution to the differential equation can then be written:

$$y(x) = y_h(x) + y_p(x) = Ae^{P(x)} + v(x)e^{P(x)}, \quad P(x) = \int p(x)\,dx, \quad v(x) = \int q(x)e^{-P(x)}\,dx. \tag{6.14}$$

Example 6.7 We can use variation of parameters to solve the differential equation

$$\frac{dy}{dx} + 3xy = x^3. \tag{6.15}$$

First, we solve the homogeneous equation

$$\frac{dy}{dx} + 3xy = 0$$

to get the function $y_h(x)$. Rearranging the equation and integrating with respect to x gives

$$\int \frac{dy_h}{y_h} = \ln(y_h) = -3\int x\,dx = -\frac{3}{2}x^2 + \text{constant},$$

so that

$$y_h(x) = Ae^{P(x)} = A\exp\left(-\frac{3}{2}x^2\right).$$

We can now find the function $v(x)$ using Equation (6.13):

$$\frac{dv}{dx} = x^3 e^{-3x^2/2}, \quad \text{so that } v(x) = \int x^3 \exp\left(\frac{3}{2}x^2\right)dx,$$

which can be evaluated using integration by parts to give

$$v(x) = \frac{3x^2 - 2}{9}\exp\left(\frac{3}{2}x^2\right).$$

The full solution is then

$$y(x) = y_h(x) + y_p(x) = y_h(x) + v(x)e^{P(x)} = A\exp\left(-\frac{3}{2}x^2\right) + \frac{3x^2 - 2}{9}. \tag{6.16}$$

Exercise 6.2.3 Confirm that Equation (6.16) is a solution to Equation (6.15) by differentiating Equation (6.16).

The solutions we have found so far have been *general solutions*—they contain an unspecified constant that arises from the integration (e.g., A in Equation (6.16)). This means that Equation (6.16) actually represents an infinite number of solutions, each differing from the others by the value of A. To pick a *specific solution*, i.e., to find a value for the constant, requires more information. For example, if we knew the value of $y(x)$ at a single, specific value of x, say x_0, then we could substitute these values into Equation (6.16) and solve for the value of A. The pair of numbers $(x_0, y(x_0))$ is called the *initial condition*.[6] An *initial value problem* consists of specifying an ODE together with a set of

[6] The term *initial conditions* may suggest something to do with time, but they do not have to be. They simply express information about a known value of the specific solution.

initial conditions that allow us to find a specific solution. As we will see later, there are other conditions, *boundary conditions*, that we can use to determine a specific solution for certain types of problems.

6.2.2 Direction Fields

We may not always be able to solve a differential equation, but we can still learn a great deal about the behavior of the solutions by using *direction fields*. Direction fields show us the directions of the tangents to the solutions of the ODE; they can be sketched by hand, but are also easy to plot using a computer. As an example, let us start with a simple equation and analyze

$$\frac{dy}{dx} = 10 - 2y. \tag{6.17}$$

Although we can solve this specific equation, we want to obtain a good idea of what solutions to the equation look like without actually solving it. We have not specified an initial condition, so we are going to be looking at the general solution of the equation. Plotting direction fields relies on the fact that the derivative is the slope of the tangent to the curve $y(x)$, so the solutions to the equation will be curves that are tangent to the derivative. To see how this works in practice, let us look at Equation (6.17) and start by looking for the (x, y) values that give specific values of the derivative. Let us first look for places where the derivative is zero; this will be where the curve $y(x)$ is parallel to the x axis. These points occur when $y' = 0$, which implies $y = 5$. If $y > 5$, then $y' < 0$ and y decreases as x increases. Alternatively, if $y > 5$, then $y' > 0$ and y increases as x increases. What is more, in both cases, as y moves further from $y = 5$, the slopes of the curves get steeper and steeper (Figure 6.3). Thus, we have a general understanding of the behavior of the solutions to Equation (6.17); if $y > 5$, solutions decrease with a slope that flattens out as the solution approaches the line $y = 5$; and if $y < 5$, the solutions increase, and as they approach $y = 5$, the solutions flatten out again. We can get a computer to do this systematically by setting up a grid of points in the (x, y)-plane and at each grid point calculate the value of the derivative (e.g., from Equation (6.17)). Recall that the derivative tells us the slope of the tangent to the curve at that location, so at each grid point we can draw an arrow whose direction is given by the slope of the solution at that point. We can also use Pythagoras' theorem to represent the magnitude of the slope by the length of the arrow (Figure 6.3).

For a more complicated example, let us examine Equation (6.15). First, we look for the values of x and y where $dy/dx = 0$. Curves where $dy/dx =$ constant are called *isoclines*, and if $dy/dx = 0$, they are called *nullclines*. Putting $dy/dx = 0$ in Equation (6.15) gives $x(x^2 - 3y) = 0$, so $x = 0$ and $x = \pm\sqrt{3y}$. The first solution ($x = 0$) tells us that the slope of all solutions is parallel to the x axis when $x = 0$ (Figure 6.4). A solution also has a zero slope at the point where it intersects the curve $x = \pm\sqrt{3y}$. Plotting the direction field using a computer we can see that as x becomes large, the directions all line up along a curve. This makes sense, because we can see from Equation (6.16) that as x gets large, the exponential term becomes small and the solution is approximately $y \approx (3x^2 - 2)/9$, which is the equation of the dashed curve in Figure 6.4.

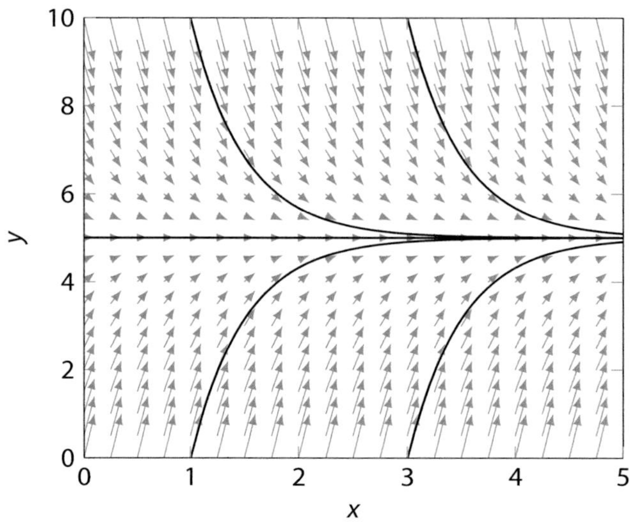

Figure 6.3 A direction field for the differential equation (6.17). The black lines are specific solutions to the differential equation, showing that the direction field given by the gray arrows represent the slopes of the solutions. Each arrow shows the direction of the slope of the solution at the point where the base of the arrow is located with the magnitude of each arrow representing the magnitude of the rate of change at that point. The collection of arrows is called the direction field.

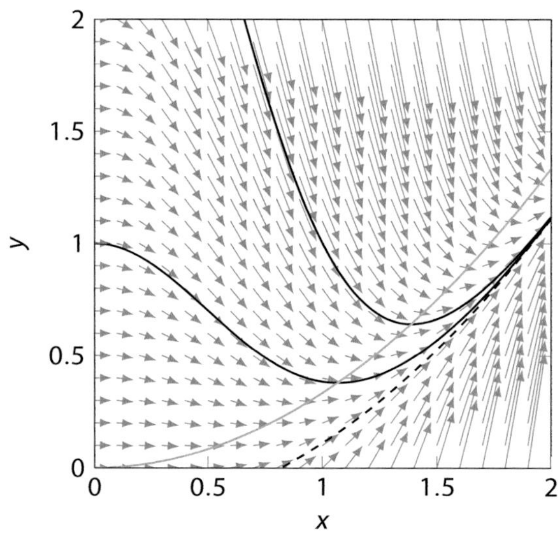

Figure 6.4 The direction field for Equation (6.15) for $x \geq 0$. The gray line is the curve $y = x^2/3$, where the slopes of the curves are zero, and the dashed line is the curve $y = x^2/3 - 2/9$. The two solid black curves are two specific solutions to the differential equation, showing that the curves asymptotically approach the dashed curve as x increases.

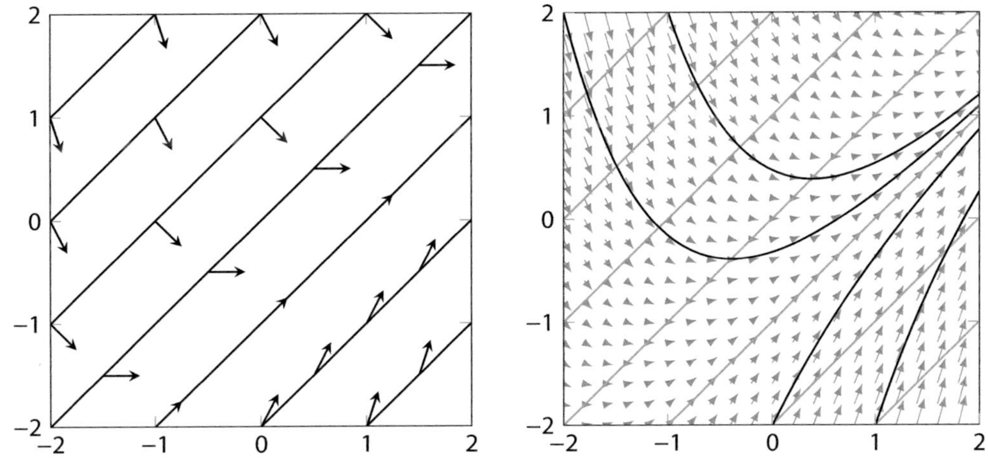

Figure 6.5 The left-hand panel shows the nullcline and isoclines for the differential equation $y' = x - y$ for $y' = c = -2, -1, 1, 2$, as well as the directions of the slopes of solution curves when they intersect those lines. The right-hand panel shows a computer-generated direction field and four solution curves.

We do not actually need a computer to sketch a direction field, we can quite easily do it by hand. As an example, let us see how we would sketch the direction field for the ODE $y' = x - y$. The first thing we want to do is find the nullclines. These are curves along which $y' = 0$, and for this equation it is the curve $y = x$, i.e., a straight line through the origin. So, at the point where any solution curve of the ODE intersects $y = x$, the slope of the curve will be zero and the solution curves will be parallel to the x axis (Figure 6.5). The isolines are given by the equation $y' = c =$ constant or $y = x - c$. For $c = 0$, this is the nullcline. To obtain other isoclines, we pick values of c. For example, for the isocline given by $c = 1$, the direction field is at a constant value of $\tan^{-1}(1) = 45°$. This means that the slope of each solution curve is 45° when it intersects the line $y = x - 1$. Figure 6.5 shows several isoclines and the slopes of the solution curves as they cross each isocline. We can see that for solutions above the line $y = x - 1$, the slopes of the solution curves are initially negative, become zero as they cross the line $y = x$, and then become positive, getting asymptotically closer to the line $y = x - 1$. The slopes of solution curves lying below the line $y = x - 1$ are positive, and the solution curves also tend to get closer to the line $y = x - 1$. From this sketch, we can easily draw in the shape of the solution curves by recalling that the solution curves will be tangent to the direction field.

Exercise 6.2.4 Using pen and paper, find the nullclines and isoclines and sketch the direction field and some typical solutions for the equation $y' = (y^2 - 2)(1 - y)^2$.

6.2.3 First Order Nonlinear Equations

So far we have considered first order linear equations. We can make things more complicated by making the equation nonlinear. First order nonlinear equations are harder to solve than linear ones, but there are three broad strategies that we can follow in pursuit of a solution.

6.2.3.1 Separation of Variables

We have already seen an example of separation of variables. The general idea is that if we have a differential equation

$$\frac{dy}{dx} = f(x, y),$$

we see if we can rearrange the equation such that $f(x, y) = X(x)Y(y)$; that is, $f(x, y)$ can be written as a function of x multiplied by a function of y. Then the ODE becomes

$$\frac{dy}{dx} = X(x)Y(y),$$

which we can write as

$$\int \frac{1}{Y(y)}\frac{dy}{dx}\, dx = \int \frac{dy}{Y(y)} = \int X(x)\, dx.$$

What we have done is split the equation so that all the y dependency is on one side of the equals sign, with all the terms depending on x on the other. If we can evaluate the two integrals, we can solve the differential equation.

Example 6.8 We can use separation of variables to find the general solution to the differential equation

$$\frac{dy}{dx} = \frac{\cos(x)}{\sin(y)}.$$

First, we rearrange the equation such that

$$\sin(y)\frac{dy}{dx} = \cos(x),$$

and integrating both sides of the equation with respect to x gives

$$\int \sin(y)\frac{dy}{dx}\, dx = \int \sin(y)\, dy = \int \cos(x)\, dx.$$

Evaluating the integrals we find

$$\cos(y) = -\sin(x) + C,$$

where C is a constant so that

$$y = \cos^{-1}(C - \sin(x)).$$

Exercise 6.2.5 Identify which of the following equations can, in principle, be solved using separation of variables. Note that you may have to make use of identities in Appendix B.

1. $\dfrac{dy}{dx} = y^2 e^{-y}\sin(x)$
2. $\dfrac{dy}{d\theta} = \sin(y(\theta)+\theta) - \sin(y(\theta)-\theta)$
3. $\dfrac{dy}{dx} + y\cot(x) = 0$
4. $\dfrac{dy}{dx} = xe^{x^2/2} + 2y(x)$

6.2.3.2 Exact Equations

Not all nonlinear equations are separable; however, some of them may be *exact equations*. Consider an ODE of the form

$$\frac{dy(x)}{dx} = -\frac{P(x,y)}{Q(x,y)}. \tag{6.18}$$

The fact that both $P(x, y)$ and $Q(x, y)$ depend on both x and y means that we cannot use separation of variables to solve the equation unless $P(x, y)$ and $Q(x, y)$ can be written as $P(x, y) = U(x)Y(y)$ and $Q(x, y) = W(x)Z(y)$. However, we can write Equation (6.18) in differential form

$$P(x,y)\,dx + Q(x,y)\,dy = 0.$$

Now, if we can find a function $\phi(x, y)$ such that

$$d\phi = P(x,y)\,dx + Q(x,y)\,dy, \tag{6.19}$$

then our differential equation has become $d\phi(x, y) = 0$, which we can easily integrate to give $\phi(x, y) = \text{constant}$. Under what conditions does such a function, $\phi(x, y)$, exist, and how can we find it? If we take the differential of $\phi(x, y)$, we get

$$d\phi = \frac{\partial\phi}{\partial x}\,dx + \frac{\partial\phi}{\partial y}\,dy, \tag{6.20}$$

and equating Equation (6.20) with Equation (6.19) we find that $P(x, y)$ and $Q(x, y)$ must satisfy the equations

$$P(x,y) = \frac{\partial\phi}{\partial x}, \qquad Q(x,y) = \frac{\partial\phi}{\partial y}. \tag{6.21}$$

Taking second derivatives of Equation (6.21) we see that

$$\frac{\partial^2\phi}{\partial y\partial x} = \frac{\partial P}{\partial y}, \qquad \frac{\partial^2\phi}{\partial x\partial y} = \frac{\partial Q}{\partial x},$$

but we know from the properties of partial derivatives that these two expressions must be equal, so we must have

$$\frac{\partial P}{\partial y} = \frac{\partial Q}{\partial x}. \tag{6.22}$$

Equation (6.22) provides us with a condition that must be satisfied for the ODE to be an exact equation. The solution to the differential equation is then, from Equation (6.19),

$$\phi(x,y) = \int_{x_0}^{x} P(x,y)\,dx + \int_{y_0}^{y} Q(x,y)\,dy + \text{constant}. \tag{6.23}$$

Example 6.9 As an example, Let us solve the differential equation

$$\frac{dy}{dx} = -\frac{2x + y^3}{3y(xy - 2)}.$$

This equation is nonlinear and not separable. We can write it as a differential,

$$(2x + y^3)dx + 3y(xy - 2)dy = 0,$$

and set

$$P(x, y) = 2x + y^3, \qquad Q(x, y) = 3y(xy - 2).$$

For the differential equation to be an exact equation, it must satisfy Equation (6.22), which in this case becomes

$$\frac{\partial P}{\partial y} = 3y^2, \qquad \frac{\partial Q}{\partial x} = 3y^2,$$

so the condition is satisfied and the equation is exact. We can now use either of the two equations in Equation (6.21) to start calculating the function ϕ; sometimes starting with one equation is easier than the other. We will choose to start first with the differential equation for $P(x, y)$ so that

$$\frac{\partial \phi}{\partial x} = P(x, y) = 2x + y^3.$$

When we integrate this equation we will have an unknown function of y instead of a constant of integration. This is because $P(x, y)$ is given by the partial derivative of ϕ with respect to x, so any term in $P(x, y)$ that contains just y (e.g., $3y$, $2y^2 \sin(y)$) is treated as a constant when we take that derivative. So, when we integrate with respect to x, we will have an unknown function of y instead of a constant. So,

$$\phi(x, y) = \int (2x + y^3)\, dx = x^2 + xy^3 + g(y). \tag{6.24}$$

We find the function $g(y)$ by using the other equation from the pair in Equation (6.21). Differentiating Equation (6.24) with respect to y we get, using Equation (6.21),

$$\frac{\partial \phi}{\partial y} = 3xy^2 + \frac{dg}{dy} = Q(x, y) = 3y(xy - 2).$$

Cancelling terms gives the equation

$$\frac{dg}{dy} = -6y, \qquad \text{so that} \qquad g(y) = -3y^2 + c,$$

so

$$\phi(x, y) = x^2 + xy^3 - 3y^2 + c.$$

We can now assemble the solution to the differential equation using Equation (6.23) to give a one-parameter family of solutions $\phi(x, y) = k$, where k is constant:

$$x^2 + xy^3 - 3y^2 = k - c = s.$$

To calculate the value of the constant s we would need an initial or boundary condition.

Exercise 6.2.6 Determine if the following equations are exact, and if they are, find the solution using the method of exact differentials.

1. $\dfrac{dy}{dx} = \dfrac{\cos(y) + y\cos(x)}{\sin(x) - x\sin(y)}$
2. $\dfrac{dy}{dx} = \dfrac{2x^2y + x}{x^2}$
3. $\dfrac{dy}{dx} = -\dfrac{x}{2y}$
4. $\dfrac{dy}{dx} = -\dfrac{y^3 + 2x}{3xy^2}$

6.2.3.3 Transforming a Nonlinear to a Linear Equation

The methods we have described so far for solving ODEs are neither infallible nor comprehensive, and finding solutions for some ODEs requires a certain amount of experience and guile. For example, we can sometimes use a transformation of the variables in a nonlinear ODE to convert it into a linear ODE that we can solve. For example, nonlinear ODEs that have the form

$$\frac{dy}{dx} + P(x)y(x) = Q(x)y^n, \qquad n \neq 1, \tag{6.25}$$

are called *Bernoulli differential equations*.[7] It is the presence of the y^n term on the right-hand side of the equation that causes the problems in finding a solution; if $n = 0$ or 1, the equation is linear and we can solve it using the methods we have already discussed. It would be nice if we could find a transformation that factored out the y^n term from the right-hand side of Equation (6.25). It would be even nicer if we could find a transformation that did this but also made the whole differential equation linear. But what transformations will work? We might be tempted to try the transformation $z = y^n$, but this turns out to not be helpful.

Exercise 6.2.7 Use the transformation $z = y^n$ in Equation (6.25) and show that the transformed equation is

$$\frac{dz}{dx} + nP(x)z = nQ(x)z^{(2n-1)/n)}. \tag{6.26}$$

Exercise 6.2.8 Notice that the left-hand side of Equation (6.26) has the same form as the left-hand side of Equation (6.25), but with y replaced by z. Why does this happen?

Exercise 6.2.8 shows us that if we use a power law transformation, then the left-hand side of Equation (6.25) keeps its linear form, which is nice because linear equations are easier to solve. So, what power law transformation do we need to cancel out the y^n factor?

Exercise 6.2.9 Use the transformation $z = y^m$ in Equation (6.25) to show that if $m = 1 - n$, then the ODE becomes

$$\frac{dz}{dx} + (1 - n)P(x)z = (1 - n)Q(x).$$

[7] The Bernoulli family produced many great mathematicians and scientists. This equation is named after Jacob Bernoulli (1655–1705), who also made significant contributions to the field of probability.

This is now a first order linear equation that we can try and solve. We can also see how the transformation is chosen specifically to remove the nonlinear terms without introducing any other nasty terms.

Example 6.10 The key to successfully using a transformation of variables is finding the right transformation to use. For Bernoulli equations, the choice is straight-forward. Let us find the general solution of the differential equation

$$\frac{dy}{dx} + y(x) = y^{2/3}.$$

The right-hand side of the equation contains a term $y^{2/3}$, which suggests using the transformation $z = y^{1/3}$. This gives the equation

$$\frac{dz}{dx} = \frac{1}{3}y^{-2/3}\frac{dy}{dx} = +\frac{1}{3} - \frac{1}{3}z,$$

which we can solve for $z(x)$ by finding an integrating factor to give the solution

$$z = 1 + Ce^{-x/3}.$$

Transforming back into the original variables, we find

$$y(x) = (1 + Ce^{-x/3})^3.$$

Example 6.11 Material in a landslide, on land or underwater, starts at a given point on a slope where a fracture occurs. The material then accelerates down the slope until it reaches a constant velocity (V_∞) at a distance L downslope from the fracture. The velocity $V(l)$ of material in the landslide at a distance l downslope from the fracture point can be represented by the ODE

$$\frac{dV}{dl} + \frac{1+\eta}{l}V = \frac{\eta}{l}\frac{V^2}{V_\infty}, \tag{6.27}$$

where $\eta > 0$ is a parameter. Let us solve the equation given that $V = V_\infty$ for $l = L$. Equation (6.27) is a Bernoulli equation with $n = 2$. To solve this equation we will first *nondimensionalize* the variables. This is a useful technique for reducing the number of parameters in a differential equation and can often result in a simpler equation. In this case, we have two parameters, η and V_∞. We can nondimensionalize the velocity $V(l)$ by dividing it by V_∞ and defining a new dimensionless velocity $v = V/V_\infty$; notice that v varies from 0 to 1. To make the velocity dimensionless, we have scaled the variable velocity (V) with a constant velocity (V_∞) that is specific to the problem. This new dimensionless velocity equals 1 when $l = L$, so it makes sense to also define a new dependent variable $\xi = l/L$. Then we know that the solution to our equation must satisfy $v = 1$ when $\xi = 1$. Substituting the new variables into Equation (6.27) gives us

$$\frac{dv}{d\xi} + \frac{1+\eta}{\xi}v = \frac{\eta}{\xi}v^2. \tag{6.28}$$

This equation may look to be dimensionally incorrect because two terms have a v and the third has a v^2 in it. But remember that these are now dimensionless velocities, so each term in the equation is dimensionless. To solve the equation, we make the substitution $u = v^{1-n} = v^{-1}$ ($n = 2$ for Equation (6.28)), giving us

$$u' = -\frac{v'}{v^2} = -\frac{\eta}{\xi} + \frac{(1+\eta)}{\xi}u.$$

This is a first order linear equation in u that we can solve by calculating the integrating factor

$$I = \exp\left(-(1+\eta)\int \frac{d\xi}{\xi}\right) = \xi^{-(1+\eta)}$$

to give the general solution

$$u = \frac{\eta}{(1+\eta)} + c\xi^{(1+\eta)}.$$

Substituting the values for the boundary condition $u = 1$ at $\xi = 1$ tells us that $c = 1 - \eta/(1+\eta)$. Substituting this into the equation and converting back to the original variables results in the solution

$$V(l) = V_\infty\left\{\frac{\eta}{1+\eta}\left(1 - \left(\frac{l}{L}\right)^{1+\eta}\right) + \left(\frac{l}{L}\right)^{1+\eta}\right\}^{-1}.$$

The key to determining which transformations to use to solve a differential equation is to understand why and how different transformations work. We use transformations to eliminate those terms that are causing us trouble, and hopefully end up with a simpler equation that we can solve. For example, for a Bernoulli equation (Equation (6.25)), the troublesome term is $Q(x)y^n$. By looking at how a general power law transformation worked with that equation, we were able to find a general transformation that could factor out the unwanted term in the equation. So, the success of these transformations requires us to first recognize which terms are causing us trouble, and how differentiation affects different types of function. Discovering a successful transformation can often take some guesswork and a few attempts in order to get it right.

Even small changes to the form of a differential equation can render a soluble ODE insoluble. Consider, for example, an ODE of the form

$$\frac{dy}{dx} = Q(x)y^2(x) + P(x)y(x) + R(x). \tag{6.29}$$

This slight generalization of the Bernoulli equation is called a *Riccati equation*.[8] If $Q(x) = 0$, then the equation is a linear one and we have a good chance of solving it analytically. If $R(x) = 0$, the equation is a Bernoulli equation, and we have seen how to tackle those. For the full equation, however, there is no standard set of techniques for finding an analytical solution. The reason for this takes us ahead of ourselves a little bit. If we make the substitution to a new function $w(x)$ such that

$$y(x) = -\frac{1}{Q(x)w(x)}\frac{dw}{dx},$$

[8] This is named after Jacopo Riccati (1676–1754), who was a mathematician and lawyer in Venice.

then the Riccati equation becomes

$$\frac{d^2w}{dx^2} - \left(\frac{1}{Q(x)}\frac{dQ}{dx} + P(x)\right)\frac{dw}{dx} + Q(x)R(x)w(x) = 0,$$

which is a second order linear homogeneous ODE and there are no known techniques for solving a general equation of this type.[9] However, there is some hope. If we find, by any means (even guesswork), one solution of a Riccati equation, we can reduce the Riccati equation to a Bernoulli equation. To see how this works, let us assume that $y_a(x)$ is a solution of a Riccati equation, then we assume that the full solution has the form $y(x) = y_a(x) + u(x)$, where $u(x)$ is an unknown function. Substituting this into Equation (6.29) gives

$$\frac{du}{dx} = [P(x) + 2Q(x)y_a(x)]u(x) + Q(x)u^2(x), \tag{6.30}$$

which is a Bernoulli equation.

Exercise 6.2.10 Derive Equation (6.30) from Equation (6.29).

Example 6.12 Let us find the general solution to the ODE

$$\frac{dy}{dx} = y^2 - \frac{y}{x} - \frac{1}{x^2}.$$

The first thing to recognize here is the pattern in the terms on the right-hand side of the equation; there is a steady progression in powers of y from y^2 to y^0 with a simultaneous progression of powers of $1/x$. This progression in $1/x$ is suggestive of the derivative of a power law. If we substitute $y = x^n$ into the equation, then we get

$$nx^{n-1} = x^{2n} - x^{n-1} - x^{-2},$$

which is satisfied by $n = -1$. So, a solution to the equation is $y_a = 1/x$. Letting $y(x) = y_a(x) + u(x)$ and substituting this into the ODE gives

$$\frac{du}{dx} = u^2 - \frac{u}{x},$$

which is a Bernoulli equation. Making the substitution $u = v^{-1}$ gives the linear equation

$$\frac{dv}{dx} + \frac{v}{x} = -1,$$

which has a solution

$$v(x) = -\frac{x}{2} + \frac{A}{x},$$

so

$$y(x) = \frac{2A + x^2}{x(2A - x^2)}.$$

---

[9] There are, however, techniques for specific cases of the functions $Q(x)$, $P(x)$, and $R(x)$.

6.2.3.4 Nondimensionalization and Scaling

In Example 6.11, we nondimensionalized the variables in the problem, thereby reducing the number of parameters in the equation. Differential equations that represent real-world problems often contain many different parameters, and these equations can often be simplified by introducing *nondimensional* variables. In Example 6.11, we transformed variables from those that had dimensions (e.g., V with dimensions $[\mathrm{L}][\mathrm{T}]^{-1}$) to ones without dimensions by scaling the variables V and l by parameters having the same dimensions as V and l and which characterized the problem (V_∞ and L respectively). This reduced the problem from one with three parameters (η, V_∞, and L; L appeared in the boundary conditions) to a one-parameter (η) problem with nondimensional boundary conditions.

Let us look at another example. The Verhulst equation is a first order nonlinear ODE that describes changes in a population of organisms arising from growth and mortality.[10] For example, we might be interested in changes in $N(t)$, the number of microbial cells per unit volume of soil, which has dimensions $[\mathrm{L}]^{-3}$. The differential equation is

$$\frac{dN}{dt} = (\alpha - \beta N)N, \qquad \alpha, \beta > 0. \tag{6.31}$$

Since N represents the number of organisms per unit volume of soil, it is always greater than or equal to zero. The term αN is therefore positive and represents the growth of the population. To be dimensionally consistent, α must have dimensions $[\mathrm{T}]^{-1}$ and it is the *growth rate* per organism; this is often called the *specific growth rate*. The term $-\beta N$ is negative and represents the loss of organisms, for example by mortality. For Equation (6.31) to be dimensionally consistent, β must have dimensions $[\mathrm{L}]^3[\mathrm{T}]^{-1}$. We can choose to scale the time variable by α (because α has only one dimension, time, and does not include length) to create a new, dimensionless time variable $\tau = \alpha t$. Substituting into Equation (6.31) and dividing through by α gives

$$\frac{dN}{d\tau} = \left(1 - \frac{\beta}{\alpha}N\right)N.$$

Notice that β/α has dimensions $[\mathrm{L}]^3$, so $\beta/\alpha N$ is dimensionless. We can create a new dimensionless variable

$$\tilde{N} = \frac{\beta}{\alpha}N,$$

and Equation (6.31) becomes

$$\frac{d\tilde{N}}{d\tau} = (1 - \tilde{N})\tilde{N}.$$

It would appear that this equation has no parameters, but that is not quite right because we still need to specify the initial condition, $N(t = 0) = N_0$. So, we have reduced the problem from an equation with three parameters (α, β, N_0) to a problem with one free parameter (N_0).

[10] This equation, also known as the *Logistic equation*, was developed by the Belgian mathematician Pierre François Verhulst (1804–1849) after studying Thomas Malthus's work *An Essay on the Principle of Population*.

Could we have chosen a different nondimensionalization for the equation? Yes. For example, instead of scaling N by β/α, we could have scaled it by the initial value, N_0, so that

$$N^* = \frac{N}{N_0},$$

and we would end up with an ODE

$$\frac{dN^*}{d\tau} = \left(1 - \left(\frac{\beta N_0}{\alpha}\right) N^*\right) N^* = (1 - \xi N^*)N^*, \quad \xi = \frac{\beta}{\alpha}N.$$

This equation also has one parameter, ξ, but the initial condition is now $N^*(\tau = 0) = 1$.

Nondimensionalizing the variables in an ODE can have other useful consequences. In particular, it can reduce the range of values that variables take. For example, in our landslide problem (Example 6.11), velocities might have values of a few meters per second and x values could be hundreds or thousands of meters (Chaytor et al., 2009), giving a factor of 10^2–10^3 between these variables. By nondimensionalizing the variables, we effectively normalize these numbers so that the scaled distance now only varies between 0 and 1, and the velocities are between 0 and approximately 1 times V_∞. The *rescaling* that we get with nondimensionalization can be very helpful if we have to numerically solve ODEs. This is because numerical methods can sometimes have problems dealing with very large and very small numbers at the same time, so scaling the variables so they all have similar ranges can be useful.

Exercise 6.2.11 The spruce budworm is an extremely destructive insect that can wreak severe damage on conifer forests. An equation for the population abundance of spruce budworm is (Ludwig et al., 1978):

$$\frac{dN}{dt} = \gamma N\left(1 - \frac{N}{\kappa}\right) - \frac{\beta N^2}{\alpha^2 + N^2},$$

where N is the number of budworms per unit area, γ is the budworm birth rate, and κ is called the *carrying capacity* and depends on the habitat.

1. What are the dimensions of γ, κ, α, and β?
2. Choose new, scaled variables and nondimensionalize the ODE; how many parameters remain in the equation?

6.2.4 A Question of Uniqueness

Once we have found a general solution to a differential equation, it is reasonable to wonder if it is the only solution; in other words, is the solution to the equation unique? Let us consider our general first order linear differential equation, Equation (6.7), with the condition that $y(a) = b$ for some value $x = a$. We are going to examine the question of uniqueness using a commonly found mathematical argument: we first assume that there are in fact two solutions, and then try to show that the equation implies that these two solutions are in fact identical, thereby showing that the solution is unique. Let us write our two solutions as $y_1(x)$ and $y_2(x)$. Because we want to show that these are identical, it us more convenient to look at the function $v(x) = y_1(x) - y_2(x)$. Because we know that $y_1(x)$ and $y_2(x)$ are both solutions to the differential equation, we have that

$$\frac{dv(x)}{dx} = \frac{dy_1(x)}{dx} - \frac{dy_2(x)}{dx} = -p(x)v(x) \tag{6.32}$$

and $v(a) = y_1(a) - y_2(a) = b - b = 0$. But equation (6.32) is separable and has a solution

$$v(x) = A\exp\left(-\int p(\xi)\,d\xi\right).$$

We also know that $v(a) = 0$, and because the exponential function is only zero at $-\infty$, this means that $A = 0$, so $v = 0$ for all values of x. Therefore we must conclude that $y_1(x) = y_2(x)$ and the solution is unique.

Showing that a solution actually exists is a little harder. We start with Theorem 6.1, which

Theorem 6.1 *If $F(x, y)$ is continuous, then the initial value problem*

$$\frac{dy}{dx} = F(x, y), \qquad y(x = x_0) = y_0$$

has a solution $y = f(x)$ that is defined for a neighborhood of x_0 such that $|x - x_0| < \delta$, where $\delta > 0$.

states that so long as $F(x, y)$ is a continuous function, then there exists a solution to the initial value problem in a region of x sufficiently close to the initial point, x_0. How close is sufficiently close? That depends on the initial value problem itself. We may also ask if this solution is unique, a question answered by Theorem 6.2.

Theorem 6.2 *If $F(x, y)$ is continuously differentiable,*[11] *then the solution is also unique.*

So continuity ensures the existence of a solution, but continuous differentiability ensures that is unique. As an example, let us consider the initial value problem

$$\frac{dy}{dx} = y^{2/3}, \qquad y(x = 0) = 0.$$

The function $y^{2/3}$ is continuous, so Theorem 6.1 tells us that a solution exists. The initial conditions tell us that $y = 0$ at $x = 0$, but the ODE itself tells us that $y' = 0$ at $x = 0$, therefore $y = 0$ for all values of x is one solution. However, we can solve this ODE to get the general solution

$$y(x) = \frac{1}{3}(x + c)^3.$$

Substituting $y(x = 0) = 0$ tells us that $c = 0$, resulting in a solution $y(x) = x^3/3$. But this is obviously a different solution from the one we just found, the trivial zero solution. We have apparently found two solutions to the same initial value problem; the solution is not unique! The reason for this is that $F(y) = y^{2/3}$ is not differentiable at $x = 0$. If we differentiate $F(y)$, we find

$$\frac{dF}{dy} = \frac{2}{3}\frac{1}{y^{1/3}},$$

[11] Mathematicians write this as $F(x, y) \in C^1$.

which is not defined at $y = 0$. If we had initial conditions of $y = 0$ at a value of $x \neq 0$, then the conditions of the theorem would be satisfied and the ODE would have a unique solution.

The concept of uniqueness for the solution of a differential equation is very important. It implies that if we have a differential equation that satisfies the existence and uniqueness theorems, then only *one* solution of the equation can pass through a given point; if solutions of an ODE could intersect, then the solutions would not be unique at that point and violate the uniqueness theorem.

6.3 Solving Differential Equations in Practice

We want to use differential equations to help us describe and understand phenomena that we see around us. To do this, we first have to derive an ODE that does this, as we did in the examples at the start of this chapter. We then have to solve the equation and interpret the solution. In practice, this can be hard to do, especially for complicated equations. So, in practice we often follow a more cautious process and first derive an equation for a simpler or idealized situation with the hope that we can solve it. After having done so, we can understand the behavior of the solution to the equation and use that knowledge to make the ODE more realistic by judiciously introducing more complexity. Hopefully we can solve this new equation, learn from the behavior of its solution, and repeat the process until we have an equation that describes the natural phenomenon to our satisfaction.

Let us look at a couple of examples in detail to see how this process works. We shall start by returning to the polluted lake we met in Example 6.2. We are going to follow our maxim of keeping a problem simple to start with (i.e., examining an idealized problem), understanding this simple problem first before making the problem more complicated (i.e., more realistic).

In our earlier derivation we had assumed that the volume of the lake was constant, but here we will allow the volume of the lake to change with changing outflow (but with F_{in} constant), and we will start by looking at how the volume of the lake changes over time. The rate of change of the volume (V) of the lake is a balance between how fast water enters the lake (F_{in} in units of $\text{m}^3\ \text{h}^{-1}$) and how fast it leaves (F_{out} in units of $\text{m}^3\ \text{h}^{-1}$):

$$\frac{dV}{dt} = F_{\text{in}} - F_{\text{out}}. \tag{6.33}$$

We cannot proceed much further without knowing something more about F_{in} and F_{out}. A simple, probably realistic assumption is that F_{in} is imposed by processes external to the lake itself (e.g., rainfall further upstream), so F_{in} is likely independent of the volume of the lake itself. What can we assume about F_{out}? The simplest (albeit unrealistic) assumption is that it is constant: $F_{\text{out}} = F = \text{constant}$, but this assumption can run into problems. For example, if $F_{\text{in}} < F$, then $dV/dt < 0$ and the volume of the lake will decrease over time. But when $V = 0$, F_{out} still equals F, so the outflow from the lake will still occur even though there is no water remaining in the lake. This assumption is not very realistic, though it may apply in situations when V can never be zero.

A somewhat more realistic assumption is that $F_{\text{out}} \propto h(t)$, where $h(t)$ is the time varying depth of the lake. In this case, the deeper the lake, the more it will overflow into the outflow river. Equation (6.33) then becomes

$$\frac{dV}{dt} = F_{\text{in}} - ah(t), \tag{6.34}$$

where a is a constant.

Exercise 6.3.1 Critique the assumptions behind Equation (6.34).

The problem now is that we have two unknown (but related) functions of time: $V(t)$, which we want to know, and $h(t)$. We need to write the equation in terms of one or the other, and we will choose to use V:

$$\frac{dV}{dt} = F_{\text{in}} - ah(t) = F_{\text{in}} - a\frac{V(t)}{A} = F_{\text{in}} - kV(t), \quad k = \frac{a}{A}, \tag{6.35}$$

where A is the surface area of the lake, which we assume to be constant, implying that the sides of the lake are purely vertical so that there is no change in lake area as the volume of the lake changes.

Exercise 6.3.2 What are the dimensions of k in Equation (6.35)?

Before we solve Equation (6.35), we will look at the steady state solution. This is *always* a good thing to do once you have derived a differential equation because it helps us to see if the equation makes sense, and frequently knowing the steady state solution helps us to understand the behavior of the nonsteady state solution. The steady state volume ($\tilde{V}$) is given by

$$\frac{d\tilde{V}}{dt} = F_{\text{in}} - k\tilde{V} = 0 \implies \tilde{V} = \frac{F_{\text{in}}}{k} = \frac{AF_{\text{in}}}{a}. \tag{6.36}$$

What does Equation (6.36) tell us? First, the larger the rate of water flowing into the lake, the larger the volume of the lake. Also, a is the proportionality constant that determines the outflow rate, so decreasing a decreases F_{out}, which leads to an increase in V, which, all other things being equal, also makes sense.

For a steady state solution we can calculate something called the *residence time* of the system. The residence time of a quantity that is in steady state is the value of that variable divided by the input rate (or output rate) of the same variable.[12] This is an important concept in many problems because it gives an estimate of the average amount of time spent in the system. For our lake equation, the residence time of the water is an estimate of the average time that a given parcel of water spends in the lake and is given by the total amount of water in the lake ($\tilde{V}$) divided by the inflow, or outflow, rate:

$$\tau = \frac{\tilde{V}}{F_{\text{in}}}.$$

[12] The system is in steady state, so the rate of input must equal the rate of output.

If two lakes have the same value for F_{in} but one has a larger steady state volume, then that lake will have the longer residence time. Similarly, if two lakes have the same steady state volume, the one with the greater inflow rate (or outflow rate, since they are equal for a steady state volume) has the shorter residence time. Why is the residence time important? Processes within the lake may act to alter the nature of the water passing through it. For example, organisms in the lake may consume oxygen from the water, so the longer a parcel of water stays in the lake, the more oxygen will be consumed and oxygen concentrations in the water will become lower and lower—possibly a critical problem for animals that require oxygen.

Exercise 6.3.3 Check that τ has the dimension of time.

Now we can proceed to solve Equation (6.35). Before we do, we should notice that the equation still contains the constant input rate, F_{in}. But we know that we can relate F_{in} to the steady state volume $\tilde{V}$, and because we are interested in the volume of the lake, it makes more sense to write Equation (6.35) as

$$\frac{dV}{dt} = k\tilde{V} - kV(t) = k(\tilde{V} - V(t)), \tag{6.37}$$

which we can solve by making the substitution $y = \tilde{V} - V(t)$ to give a solution

$$V(t) = \tilde{V} + ce^{-kt},$$

where c is a constant. If we know that the volume of the lake at time $t = 0$ (before any changes happen to the lake) is V_0, then we can solve for c and write the solution as

$$V(t) = \tilde{V} + (V_0 - \tilde{V})e^{-kt}. \tag{6.38}$$

What does this equation tell us? If the volume of the lake starts at its steady state concentration, i.e., $V_0 = \tilde{V}$, then the second term in Equation (6.38) is zero, and the volume of the lake stays at the steady state value. However, if there has been a sudden heavy rainfall and the volume of the lake after the storm is V_0, then the difference between V_0 and $\tilde{V}$ decays exponentially with time and $V \to \tilde{V}$ as $t \to \infty$. In fact, any perturbation from the steady state volume will decay exponentially over time, so the volume of the lake is stable.

Now let us make the problem a little more realistic (and more complicated) by making F_{in} a function of time as well. For example, we could assume that the input varies periodically about some constant, average value,

$$F_{in}(t) = F_0 + A\cos(\nu t), \tag{6.39}$$

where F_0 and A are constants and ν is the frequency of changes to the input.[13] The differential equation for $V(t)$ becomes

$$\frac{dV}{dt} + kV(t) = F_0 + A\cos(\nu t), \tag{6.40}$$

which can be solved by calculating the integrating factor.

[13] The variable ν must be a frequency because, recall from Chapter 1, the dimensions of the argument of a function such as sine must be dimensionless, and t has dimensions of a time.

Exercise 6.3.4 What are the dimensions of k in Equation (6.40)?

Exercise 6.3.5 Show that the general solution to Equation (6.40) is

$$V(t) = \frac{F_0}{k} + Be^{-kt} + \frac{C}{k^2 + \nu^2}(k\sin(\nu t) - \nu\cos(\nu t)), \tag{6.41}$$

where B and C are constants. [*Hint:* you will have to use integration by parts twice to evaluate an integral.]

Exercise 6.3.6 Use the substitution $k = D\cos(\phi)$, $\nu = D\sin(\phi)$, and the formulae in Appendix B to write Equation (6.41) as

$$V(t) = V_0 + Be^{-kt} + \frac{C}{k}\left(1 + \frac{\nu^2}{k^2}\right)^{-1/2}\sin(\nu t - \phi). \tag{6.42}$$

Let us look and see what Equation (6.42) is telling us (Figure 6.6). Recall that ν is the frequency with which the input flow is varying and the volume of the lake also varies with the same frequency as the inflow. The oscillation of the inflow and the oscillation of the lake volume do not vary together but are *out of phase*, with the phase difference given by ϕ. The natural timescale of response of the lake volume is k^{-1}, and the variable ϕ compares ν and k because $\phi = \tan^{-1}(\nu/k)$. If $\nu \gg k$, then the lake responds to changes in volume slower than the changes in the inflow. In this case, ν/k will be large, so that $\phi \approx \pi/2$ and the lake volume is 90° out of phase with the inflow; i.e., the lake volume follows a cosine curve, whereas the inflow follows a sine curve. If the lake responds more rapidly than changes in the inflow ($\nu \ll k$), then $\phi \approx 0$ and the lake volume tracks the changes in the inflow.

Now let us add yet more complexity (i.e., reality) and start to tackle the problem we had originally. How does the concentration of a substance (e.g., a pollutant) in the lake vary if

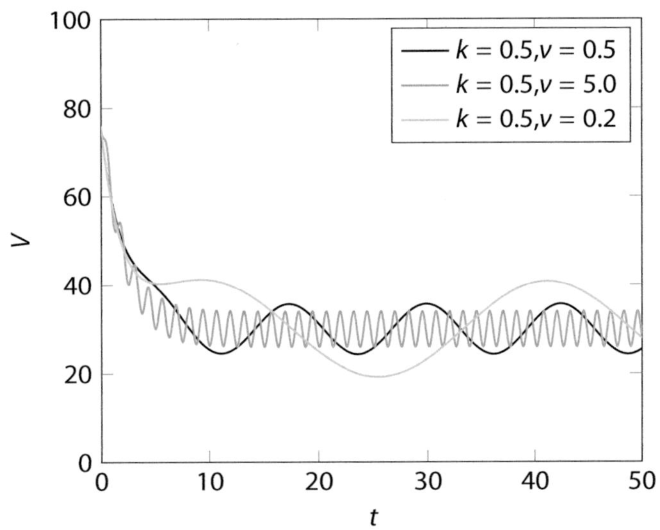

Figure 6.6 A plot of Equation (6.42) with $V_0 = 30$, $B = 50$, $C = 2$, $\phi/\pi/2$, showing the effects of different values of ν. The solutions show the exponential decrease from the initial value followed by the oscillations in V as t increases.

the inflow bringing it into the lake varies over time? We will again employ the strategy of starting with very simple (even though they might be unrealistic) scenarios and gradually building up the complexity. Let us assume that the concentration of the substance in the lake is C moles L^{-3}, the concentration in the inflow is C_{in} moles L^{-3}, and the concentration in the outflow is the same as that in the lake. We also have to take into account any reactions that change the amount of the substance within the lake. As we have found out, it is easier to develop an equation using mass as the dependent variable, so we get

$$\frac{d(CV)}{dt} = C\frac{dV}{dt} + V\frac{dC}{dt} = F_{in}C_{in} - F_{out}C - R(t, C, V), \tag{6.43}$$

where we have used $R(t, C, V)$ to represent the reactions that can also remove (hence the negative sign) the substance from the lake waters.

The first solution we should look for is the steady state with F_{in} and F_{out} constant in time; we will also assume for the moment that there are no reactions taking place (i.e., $R = 0$). Equation (6.43) then tells us that

$$C = C_{in}\frac{F_{in}}{F_{out}}.$$

Notice that we have assumed that CV is in steady state. If, in addition, V is not changing and $F_{in} = F_{out}$, then $C = C_{in}$. This is the statement of conservation of mass and is the behavior of a *conservative tracer*, i.e., changes in the pollutant concentration follow changes in the water flow.

Exercise 6.3.7 Write an equation for the residence time of the substance in the lake.

If the volume of the lake is in steady state, then $V = \tilde{V}$ and $F_{in} = F_{out}$, and Equation (6.43) becomes (assuming no reactions)

$$\frac{dC}{dt} = \frac{F_{in}}{\tilde{V}}(C_{in} - C) = \frac{F_{in}C_{in}}{\tilde{V}}\left(1 - \frac{C}{C_{in}}\right),$$

which we already know how to solve (the equation has the same form as Equation (6.37)):

$$C(t) = C_{in} + (C_0 - C_{in})e^{-k_1 t}, \tag{6.44}$$

where $C_0 = C(t = 0)$ and $k_1 = F_{in}/\tilde{V}$. This equation tells us that the concentration of the substance varies in the same way as the volume of water in the lake. A sudden pulse in the concentration of the pollutant in the input waters will show as an exponentially decreasing pollutant concentration in the lake.

What happens if we keep the assumption of no reactions, but make the flow nonsteady state? To simplify the problem a little, we will assume that the lake is initially at steady state, and then flooding caused an accident upstream that spilled a large amount of a pollutant into the river flowing into the lake. The ODE for the mass of pollutant in the lake is

$$\frac{d(CV)}{dt} = C\frac{dV}{dt} + V\frac{dC}{dt} = F_{in}C_{in} - F_{out}C,$$

where we assume that C_{in} is constant. But we know that

$$\frac{dV}{dt} = F_{in} - F_{out},$$

so our ODE becomes

$$\frac{dC}{dt} = \frac{F_{\text{in}}}{V(t)}(C_{\text{in}} - C).$$

We can use Equation (6.38) to substitute for $V(t)$, and if we define $x = C_{\text{in}} - C$, we end up with the equation

$$\frac{dx}{dt} = -\frac{kx}{1 - \beta e^{-kt}}. \tag{6.45}$$

Exercise 6.3.8 Use the substitution $u = e^{mx}$ and partial fractions to show that

$$\int \frac{dx}{a + be^{mx}} = \frac{1}{am}(mx - \ln(a + be^{mx})), \tag{6.46}$$

where a, b, and m are constants.

Exercise 6.3.9 Use the integral from Equation (6.46) and the initial conditions $C = C_0$ at $t = 0$ to show that the solution to Equation (6.45) is

$$C(t) = C_{\text{in}} + (C_0 - C_{\text{in}})\frac{V_0}{(V_0 - \tilde{V}) + \tilde{V}e^{kt}}.$$

Lastly, we will add a reaction that removes the substance from the lake waters. To make things a little easier, we will keep the assumption of steady state flow and assume that the rate at which the substance is removed from the lake is γCV. Our ODE now becomes

$$\frac{dC}{dt} = kC_{\text{in}}\left(1 - \frac{k+\gamma}{k}\frac{C}{C_{\text{in}}}\right),$$

which has a steady state concentration of

$$\tilde{C} = \frac{k}{k+\gamma}C_{\text{in}}.$$

We can then write our ODE in the form

$$\frac{dC}{dt} = kC_{\text{in}}\left(1 - \frac{C}{\tilde{C}}\right), \qquad \text{which has a solution} \qquad C(t) = \tilde{C}\left(1 - Ae^{-(k+\gamma)t}\right).$$

This is similar to the equation for the change in volume over time (as it must be because the ODEs are the same), but the rate of exponential decline is greater because of the additional way that pollutant can be removed from the lake.

These examples demonstrate how we start with a simple (though possibly unrealistic) mathematical description of the system we are interested in, and gradually add complexity until we have an equation that describes the system to our satisfaction. At each step along the way, we solve our equations and try to understand the behavior of the solutions, using this understanding to guide any modifications we need to make to the equations.

6.4 Second Order Differential Equations

Second order ODEs are characterized by the presence of a second derivative, which adds a new level of complexity to the equation. The presence of the second derivative means that to solve these equations we will have to perform two integrations, thereby introducing another constant to the problem. These equations tend to arise when we are interested in the rate of change of a rate of change of a variable, in other words, the acceleration or deceleration of a variable if we are differentiating with respect to time, for example.

To start thinking about some of these issues, let us look at some very simple second order ODEs. The first is

$$\frac{d^2y}{dt^2} = 0. \tag{6.47}$$

If we interpret y as a distance and t as time, then this is the equation for an object moving with zero acceleration. We can integrate both sides of this equation once

$$\int \frac{d}{dt}\frac{dy}{dt}\,dt = \int 0\,dt, \qquad \text{giving} \qquad \frac{dy}{dt} = a = \text{constant},$$

which tells us that if the acceleration is zero, then the magnitude of the velocity is constant. Integrating once more we get $y(t) = at + b$, where b is a second constant. We now need two conditions on the solution of the equation in order to find values for these constants, and there is a choice of possible conditions we can use. If we specify the value of $y(t)$ and its first derivative at some point $t = t_0$, then we have what is called an *initial value problem* (IVP). We could also specify the value of $y(t)$ at two points, $t = t_0$ and $t = t_1$, and this type of problem is called a *boundary value problem* because we usually use t_0 and t_1 to represent the end points or boundary points that we are interested in. There are other types of boundary values we could specify: for example, instead of the values of $y(t)$ at $t = t_0$ and $t = t_1$, we could specify the values of the first derivative $y'(t)$ at $t = t_0$ and $t = t_1$, or we could specify $y(t_0)$ and $y'(t_1)$, and so on. We will have more to say about BVP later.

A slightly more complicated second order ODE is

$$\frac{d^2y}{dt^2} = a = \text{constant}, \tag{6.48}$$

which, by integrating twice, produces $y(t) = \frac{1}{2}at^2 + bt + c$, where b and c are the constants we need to find. We can add more complexity by looking at an equation where the right-hand side is not constant:

$$\frac{d^2y}{dt^2} = y. \tag{6.49}$$

Finding a solution to this equation is a little harder because we cannot just integrate both sides of this equation. However, we are looking for a function that, when differentiated twice, gives us the same function back. The exponential function $y(t) = e^t$ is such a function, but so is $y(t) = e^{-t}$ (you should differentiate these two functions to show that they both satisfy Equation (6.49)). This looks like we have two very different solutions to the same ODE, one solution that grows over time and another that decays. But it gets worse, because $y(t) = 12e^t$ is also a solution, as is $y(t) = 0.05e^{-t}$, and even $y(t) = 12e^t + 0.05e^{-t}$. In fact, there is an infinite number of solutions of the form $y(t) = Ae^t + Be^{-t}$ with the initial or boundary conditions telling us the values of the constants A and B.

Let us make a slight change to Equation (6.49) and look at

$$\frac{d^2y}{dt^2} = -y. \tag{6.50}$$

Now we are looking for a function that changes sign when you differentiate it every other time. The functions $\sin(t)$ and $\cos(t)$ both have that property, so $y(t) = A\sin(t)$ and $y(t) = B\cos(t)$ are both solutions to Equation (6.50).

Exercise 6.4.1 Calculate the second derivative of $y(t) = A\sin(t) + B\cos(t)$, and show that $y(t)$ is also a solution to Equation (6.50).

So far we have been able to directly integrate these second order ODEs, or use our knowledge of elementary functions to deduce what the solution is. But these methods will not get us much further. Fortunately, there are some systematic methods for solving certain classes of second order ODE.

6.4.1 Second Order Linear Differential Equations

We will again start our exploration with linear equations. A general second order linear ODE has the form

$$P(x)\frac{d^2y}{dx^2} + Q(x)\frac{dy}{dx} + R(x)y = S(x), \tag{6.51}$$

with the *standard form* obtained by dividing by $P(x)$:

$$\frac{d^2y}{dx^2} + q(x)\frac{dy}{dx} + r(x)y = s(x), \quad q(x) = \frac{Q(x)}{P(x)}, \; r(x) = \frac{R(x)}{P(x)}, \; s(x) = \frac{S(x)}{P(x)}. \tag{6.52}$$

We can simplify things further by considering *homogeneous* equations, i.e., equations with $S(x) = 0$. These equations have several nice features, the first being that if $y_1(x)$ and $y_2(x)$ are both solutions of the equation, then so is $y(x) = a_1y_1(x) + a_2y_2(x)$. This is called the *principle of superposition*, and we can easily demonstrate it to be true. Consider a general second order linear homogeneous ODE in standard form,

$$\frac{d^2y}{dx^2} + q(x)\frac{dy}{dx} + r(x)y = 0, \tag{6.53}$$

and let us assume that both $y_1(x)$ and $y_2(x)$ are solutions. We claim that $y(x) = a_1y_1(x) + a_2y_2(x)$ is also a solution. To see this, substitute $y(x)$ into the left-hand side of the differential equation

$$\begin{aligned} a_1y_1'' + a_2y_2'' + q(x)a_1y_1' + q(x)a_2y_2' + r(x)a_1y_1 + r(x)a_2y_2 \\ = a_1[y_1'' + q(x)y_1' + r(x)y_1] + a_2[y_2'' + q(x)y_2' + r(x)y_2)] \\ = a_1(0) + a_2(0) = 0, \end{aligned}$$

so $y(x)$ satisfies the ODE as long as the functions $y_1(x)$ and $y_2(x)$ also satisfy it. By following the same line of argument, we can see that the principle of superposition does not work for inhomogeneous equations ($s(x) \neq 0$ in Equation (6.52)). In that case, substituting $y(x) = a_1 y_1(x) + a_2 y_2(x)$ would give

$$a_1 y_1'' + a_2 y_2'' + q(x) a_1 y_1' + q(x) a_2 y_2' + r(x) a_1 y_1 + r(x) a_2 y_2 = (a_1 + a_2) s(x) \neq s(x),$$

unless $a_1 + a_2 = 1$, but in general $y(x)$ will not be a solution of the inhomogeneous equation.

How do we find solutions to Equation (6.53)? To start with, let us consider the equation

$$ay'' + by' + cy = 0, \tag{6.54}$$

where a, b, and c are all constants. If we assume $y(x) \sim x^n$, where n is constant, then each term on the left-hand side of the equation contains a different power of x, and the only solution to the equation has $a = b = c = 0$. In order to preserve the fact that a, b, and c are all constants, $y(x)$ must be a function whose derivative is proportional to itself; i.e., an exponential. In order for $y = \exp(mx)$ to be a solution, we require (by substituting this solution into Equation (6.54))

$$am^2 e^{mx} + bme^{mx} + ce^{mx} = (am^2 + bm + c)e^{mx} = 0.$$

But we know that $y = \exp(mx) \neq 0$ for any real value of x, so for the exponential to be a solution, we require that m is the solution of the quadratic equation

$$am^2 + bm + c = 0. \tag{6.55}$$

This equation is called the *characteristic equation* or *characteristic polynomial* of the differential equation. The solution of the characteristic equation is so

$$m_\pm = \frac{-b \pm \sqrt{b^2 - 4ac}}{2a}. \tag{6.56}$$

The values of $m_\pm$ will fall into three classes depending on the value of $b^2 - 4ac$. If $b^2 - 4ac > 0$, then the two roots (m_+ and m_-) of the quadratic equation are real and distinct. In this case we have two solutions, $y_1(x) = \exp(m_+ x)$ and $y_2(x) = \exp(m_- x)$. Using the principle of superposition, we can write a more general solution as

$$y(x) = c_1 e^{m_+ x} + c_2 e^{m_- x}, \tag{6.57}$$

where c_1 and c_2 are constants.

If $b^2 - 4ac = 0$, then we have only one root to the characteristic equation, so we have one solution, $y(x) = \exp(mx)$. Can we find another solution? Yes, and it turns out that $y(x) = x\exp(mx)$ is also a solution, so we can write down the general solution as

$$y(x) = c_1 e^{mx} + c_2 x e^{mx}. \tag{6.58}$$

Exercise 6.4.2 Show that $y(x) = x\exp(mx)$ is also a solution of Equation (6.54) if $b^2 - 4ac = 0$.

The last case to consider is when $b^2 - 4ac < 0$. Then the characteristic equation has two complex roots that form a complex conjugate pair, $m_+ = (\alpha + i\beta)$ and $m_- = (\alpha - i\beta)$. Using Euler's equation (Appendix B), we can write a general solution in the form

$$
\begin{aligned}
y(x) &= c_1 e^{(\alpha+i\beta)x} + c_2 e^{(\alpha-i\beta)x} \\
&= c_1 e^{\alpha x}\left(\cos(\beta x) + i\sin(\beta x)\right) + c_2 e^{\alpha x}\left(\cos(\beta x) - i\sin(\beta x)\right) \\
&= e^{\alpha x}\left([c_1 + c_2]\cos(\beta x) + i[c_1 - c_2]\sin(\beta x)\right) \\
&= e^{\alpha x}\left(A\cos(\beta x) + iB\sin(\beta x)\right). \qquad (6.59)
\end{aligned}
$$

Notice that in this case the solution has a real part and an imaginary part:

$$\mathrm{Re}(y(x)) = Ae^{\alpha x}\cos(\beta x), \quad \mathrm{Im}(y(x)) = Be^{\alpha x}\sin(\beta x).$$

This may still seem a little messy, so we can form a general solution to the ODE in a slightly different way. Let

$$y_1(x) = e^{\alpha x}\left(\cos(\beta x) + i\sin(\beta x)\right) \quad \text{and} \quad y_2(x) = e^{\alpha x}\left(\cos(\beta x) - i\sin(\beta x)\right).$$

We know that both of these solve the ODE, so by the principle of superposition the following combinations are also solutions to the equation:

$$\frac{1}{2}(y_1(x) + y_2(x)) = e^{\alpha x}\cos(\beta x), \quad \frac{1}{2i}(y_1(x) - y_2(x)) = e^{\alpha x}\sin(\beta x).$$

Therefore, if the roots of the characteristic equations are complex, we can also write the general solution in the form

$$y(x) = Ae^{\alpha x}\cos(\beta x) + Be^{\alpha x}\sin(\beta x). \qquad (6.60)$$

If we also have initial or boundary conditions for the problem, then we can find the values of the constants that appear and obtain specific solutions from the general one.

Example 6.13 We can now solve the following equations:

1. $\dfrac{d^2y}{dx^2} + 5\dfrac{dy}{dx} + 6 = 0, \quad y(0) = 1,\ y'(0) = 2.$
2. $\dfrac{d^2y}{dx^2} + 5\dfrac{dy}{dx} + 6 = 0, \quad y(0) = 1,\ y(1) = 0.$
3. $\dfrac{d^2y}{dx^2} - 4\dfrac{dy}{dx} + 4 = 0, \quad y(0) = 1,\ y'(0) = 0.$
4. $\dfrac{d^2y}{dx^2} - 4\dfrac{dy}{dx} + 7 = 0, \quad y(0) = 1,\ y'(0) = 0.$

Each of the equations is a second order linear homogeneous ODE with constant coefficients, so to solve them we need to calculate the characteristic equation. Using a solution of the form $y = e^{mx}$, the characteristic equation for the first equation is

$$m^2 + 5m + 6 = (m+2)(m+3) = 0,$$

so the general solution has the form $y(x) = Ae^{-2x} + Be^{-3x}$, with a first derivative $y'(x) = -2Ae^{-2x} - 3Be^{-3x}$. Substituting in the initial values $y(0) = 1$, $y'(0) = 2$ and solving for A and B gives the final solution

$$y(x) = 5e^{-2x} - 4e^{-3x}.$$

The second equation is the same ODE as the first, but with a set of boundary conditions instead of initial conditions. The general solution $y(x) = Ae^{-2x} + Be^{-3x}$ is the same, but substituting the values $y(0) = 1$, $y(1) = 0$ gives the final solution

$$y(x) = \frac{1}{1-e}e^{-2x} - \frac{e}{1-e}e^{-3x}.$$

The third ODE has a characteristic equation $m^2 - 4m + 4 = (m-2)^2 = 0$, and so we have a repeated root, $m = 2$. The solution to the ODE is then $y(x) = e^{2x}(A + Bx)$. Substituting in the initial conditions gives $A = 1$, $B = -2$, so the solution is

$$y(x) = e^{2x}(1 - 2x).$$

The last ODE has a characteristic equation with complex roots $2 \pm i\sqrt{3}$. The general solution to the equation is

$$y(x) = Ae^{2x}\cos(\sqrt{3}x) + Be^{2x}\sin(\sqrt{3}x).$$

Substituting in the initial conditions gives that $A = 1$, $B = -2/\sqrt{3}$, so the solution is

$$y(x) = e^{2x}\cos(\sqrt{3}x) - \frac{2e^{2x}}{\sqrt{3}}\sin(\sqrt{3}x).$$

We have been using the term "general solution" quite a lot without really having said what it means for a second order linear homogeneous ODE. It is time to look into this in more detail, and in doing so we will see what conditions a general solution must satisfy to earn that name. Let us look again at Equation (6.49). We have seen that both $y_1(t) = \sin(t)$ and $y_2(t) = \cos(t)$ are solutions to the equation, and from the principle of superposition, we know that $y_a(t) = A\sin(t) + B\sin(t)$ is also a solution. We claim that this is the *general* form of solution, i.e., all solutions of the ODE can be derived from $y_a(t)$ using suitable choices of the constants A and B. Why is this claim true? Why is not $y(t) = 4A\sin(t) + 3B\sin(t)$ not the general solution? The reason is that there is something special about $y_1(t) = \sin(t)$ and $y_2(t) = \cos(t)$; they are *linearly independent*. This means that there is no value of the constant k for which $y_1(t) = ky_2(t)$; if we can find such a value of k, then the two functions are *linearly dependent*.[14]

If $y_1(t)$ and $y_2(t)$ are solutions to a second order linear homogeneous ODE and are *linearly independent* functions, then the general solution of the ODE is $y(t) = Ay_1(t) + By_2(t)$, where A and B are constants. Now we can see why $y_1(t) = 4\sin(t)$ and $y_2(t) = 3\sin(t)$ cannot be combined to form a general solution; the functions $y_1(t)$ and $y_2(t)$ are related by $y_1(t) = (4/3)y_2(t)$ and are linearly dependent. However, there is no constant k such that $\sin(t) = k\cos(t)$, so these are linearly independent functions. This can be formalized using a quantity called the *Wronskian*.[15]

The Wronskian of two functions $y_1(x)$ and $y_2(x)$ is defined by

$$W(y_1, y_2) = y_1(x)\frac{dy_2}{dx} - y_2(x)\frac{dy_1}{dx}. \tag{6.61}$$

[14] This concept is very similar to linear dependence of vectors, which we have seen in Chapter 4.
[15] Named after the Polish mathematician Józef Maria Hoene-Wroński (1777–1853).

The Wronskian is useful because $W(y_1, y_2) \neq 0$ if, and only if, the functions y_1 and y_2 are linearly independent. We can see this quite easily by assuming that the two functions are linearly dependent, so that $y_1(x) = ky_2(x)$. Then

$$W(y_1, y_2) = ky_2(x)\frac{dy_2}{dx} - y_2(x)\frac{d(ky_2(x))}{dx} = 0.$$

If $y_1(x)$ and $y_2(x)$ are not linearly dependent, then $y_1(x) \neq ky_2(x)$, and $W(y_1, y_2) \neq 0$. What is more, if $y_1(x)$ and $y_2(x)$ are solutions to the second order linear homogeneous ODE

$$\frac{d^2y}{dx^2} + a(x)\frac{dy}{dx} + b(x)y(x) = 0,$$

then

$$y_1'' + a(x)y_1' + b(x)y_1(x) = 0, \qquad y_2'' + a(x)y_2' + b(x)y_2(x) = 0.$$

If we multiply the first equation by $-y_2(x)$ and multiply the second by $y_1(x)$, then subtracting the two equations results in

$$(y_1y_2'' - y_2y_1'') + a(x)(y_1y_2' - y_2y_1') = 0.$$

The second term in this equation is just $W(y_1, y_2)$, and the first term is the derivative of $W(y_1, y_2)$ with respect to x.

Exercise 6.4.3 Given two functions $y_1(x)$ and $y_2(x)$, show that the derivative of the Wronskian of these functions is given by $y_1y_2'' - y_2y_1''$.

So, we can write

$$\frac{dW}{dx} + a(x)W = 0,$$

which we can solve to give

$$W = C\exp\left(-\int a(x)\,dx\right).$$

Because the function e^x is never zero, this tells us that if the constant $C \neq 0$, then $W(y_1, y_2)$ can never be zero for finite values of x. If, on the other hand, $C = 0$, then $W(y_1, y_2) = 0$. This result is called *Abel's theorem*,[16] and it shows that if $y_1(x)$ and $y_2(x)$ are linearly independent and satisfy a second order linear homogeneous ODE, they cannot become linearly dependent for some specific value or values of x, and vice versa. So, once we have found our linearly independent solutions $y_1(x)$ and $y_2(x)$, we can form a general solution that always holds.

How do we solve a second order linear nonhomogeneous ODE? If the equation has constant coefficients, then we can still, in principle, solve the equation, though we have to work harder to get the solution. Let us assume we have the following differential equation for the function $y(x)$,

$$ay'' + by' + cy = g(x), \tag{6.62}$$

[16] Named after the Norwegian mathematician Niels Henrick Abel (1802–1829).

where $g(x)$ is some specified function of x and a, b, and c are constants. If we have two solutions, $y_1(x)$ and $y_2(x)$, to Equation (6.62), then the function $y(x) = y_1(x) - y_2(x)$ satisfies the equation

$$ay'' + by' + cy = (ay_1'' + by_1' + cy_1) - (ay_2'' + by_2' + cy_2) = g(x) - g(x) = 0.$$

In other words, $y(x)$ satisfies a second order linear homogeneous ODE with constant coefficients, which we know how to solve. So, if we can find, by any means, *one* solution (say $y = y_1(x)$) to Equation (6.62), then to find any other solution we only have to solve the corresponding homogeneous ODE. In fact, if $y_p(x)$ is any solution of Equation (6.62) and y_a and y_b are linearly independent solutions of the corresponding homogeneous equation, then the general solution of Equation (6.62) is

$$y(x) = y_p(x) + Ay_a(x) + By_b(x), \tag{6.63}$$

where A and B are constants. The solution $y_p(x)$ is called the *particular solution* of Equation (6.62), and the solution to the corresponding homogeneous equation is called the *complementary function*. But the question remains, how do we find $y_p(x)$? There are two common approaches. The first is called the *method of undetermined coefficients* and works if $g(x)$ is made up of exponentials, polynomials, sines or cosines, or mixtures of these functions; it is based on informed guesswork. The second method is called the *variation parameters* and, although more involved, can be used when $g(x)$ is not suitable for using the method of undetermined coefficients.

6.4.1.1 Undetermined Coefficients

The method of undetermined coefficients gets its name because we assume a form of the particular solution based on the form of $g(x)$ and with constant coefficients that we have to determine. This method depends on $g(x)$ containing sines, cosines, polynomials, or exponentials because of the patterns that come from taking derivatives of these functions. For example, the derivative of an exponential is always an exponential, so if $g(x) = \alpha e^{\beta x}$, then we can be quite sure that $y(x)$ must also contain an exponential. Therefore, we can assume a solution of the form $y(x) = Ae^{\beta x}$, where A is a constant, substitute it into the ODE, and find the specific value of A that makes this function a solution to the differential equation.

Example 6.14 Let us use the method of undetermined coefficients to find particular solutions for the following differential equations:

$$a.\ y'' - y' + 2y = 3e^{2x}, \quad b.\ y'' - y' + 2y = 3e^{2x} + 6.$$

Equation (a) has an exponential function for $g(x)$, so we *assume* a particular solution of the form $y_p(x) = Ae^{2x}$. To find the value of A, we substitute this solution into the equation to find

$$4Ae^{2x} - 2Ae^{2x} + 2Ae^{2x} = 4Ae^{2x} = 3e^{2x}.$$

Therefore $A = 3/4$, giving a particular solution $y_p(x) = (3/4)e^{2x}$.

We need to modify our guess of the particular solution for equation (b); we need to include a constant in $y_p(x)$. This will vanish when we take its derivative. So, assuming $y_p(x) = Ae^{2x} + B$ and substituting this into equation (b) we find

$$4Ae^{2x} + 2B = 3e^{2x} + 6,$$

and for this equation to hold we must have $A = 3/4$ as before and $B = 3$, giving a particular solution of $y_p(x) = (3/4)e^{2x} + 3$.

Similarly, we know that the derivative of a sine is a cosine, and the second derivative of a sine is the negative sine function. This means that if $g(x)$ contains either a sine function, a cosine function, or both, then $y(x)$ must contain a sum of sines of cosines. Why a sum of sines and cosines? Equation (6.62) contains a first derivative of $y(x)$, so we need to have both in the solution.

Example 6.15 As an example, let us find the particular solution of the equation

$$y'' - y' + 2y = 4\sin(2x).$$

Let us first try a solution of the form $y_p(x) = A\sin(2x)$, where A is a constant. Differentiating this twice and substituting into the ODE, we find

$$-2A\sin(2x) - 2A\cos(2x) = 4\sin(2x).$$

We can equate the sine terms on the two sides of the equation, but this leaves the cosine term. So $y_p(x) = A\sin(2x)$ alone cannot be a solution to the ODE; differentiating has introduced a cosine term. Therefore, we have to try $y_p(x) = A\sin(2x) + B\cos(2x)$. Differentiating this expression twice and substituting into the ODE gives

$$(-2A + 2B)\sin(2x) - (2A + 2B)\cos(2x) = 4\sin(2x),$$

and we can now equate the sine and cosine terms on both sides of the equation to get

$$-2A + 2B = 4, \quad 2A + 2B = 0,$$

which implies $A = -1$, $B = 1$. So, our particular solution is $y_p(x) = -\sin(2x) + \cos(2x)$.

Lastly, let us consider the case when $g(x)$ is a polynomial. In this case we assume that $y_p(x)$ is also a polynomial. But, what order of polynomial should we choose? We know that each time we differentiate a polynomial, we get a polynomial of one less order (e.g., differentiating a cubic gives a quadratic), so the highest order term will come from the nondifferentiated term (cy in Equation (6.62)), and this has to cancel the highest order term on the right-hand side of the ODE.

Example 6.16 Let us try to find the particular solution of the equation

$$y'' - y' + 2y = 4x^3.$$

We choose as a trial solution $y_p(x) = ax^3 + bx^2 + cx + d$. Differentiating twice and substituting into the ODE gives

$$2ax^3 + (2b - 3a)x^2 + (6a - 2b + 2c)x + (2b - c + 2d) = 4x^3.$$

Equating powers of x on both sides of the equation gives: $2a = 4$, implying $a = 2$; $2b - 3a = 0$, which gives $b = 3$; $6a - 2b + 2 = 0$, giving $c = -3$; and lastly $2b - c + 2d = 0$, yielding $d = -9/2$. So, our particular solution is

$$y_p(x) = 2ax^3 + 3x^2 - 3x - 9/2.$$

Exercise 6.4.4 Show that the particular solutions derived in Examples 6.14–6.16 satisfy their respective ODEs.

Exercise 6.4.5 Find the particular solution of the ODE $y'' - 3y' + 2y = e^{3x} + 2x^2$. Note that the particular solution will be a sum of two terms.

6.4.1.2 Variation of Parameters

If the right-hand side of the ODE is not a combination of sines, cosines, exponentials, and polynomials, then we have to resort to the method of variation of parameters. We have met this method already in Section 6.2.1, but with second order differential equations, things become a little messier. We want to solve the equation

$$ay'' + by' + cy = g(x).$$

The idea here is that if $y_1(x)$ and $y_2(x)$ are two linearly independent solutions of the corresponding homogeneous equation (i.e., $ay'' + by' + cy = 0$), then we know that the general solution of the homogeneous equation is $y(x) = c_1 y_1(x) + c_2 y_2(x)$, where c_1 and c_2 are constants. Now we assume that the particular solution has the same form but with the constant replaced by unknown functions, so that

$$y_p(x) = v_1(x)y_1(x) + v_2(x)y_2(x).$$

Substituting this into the inhomogeneous equation we get that

$$g(x) = a(v_1''y_1 + v_2''y_2 + 2(v_1'y_1' + v_2'y_2')) + b(v_1'y_1 + v_2'y_2). \tag{6.64}$$

But now we have two unknown functions ($v_1(x)$ and $v_2(x)$) and only one equation. Remember that we are finding a *single* solution of the ODE, so we can try and see what happens if we choose $v_1(x)$ and $v_2(x)$ such that one of the terms in Equation (6.64) is zero. For example, if we choose v_1 and v_2 such that

$$v_1'y_1 + v_2'y_2 = 0, \tag{6.65}$$

then the term multiplying b is zero, and we also simplify the term multiplying the constant a because differentiating Equation (6.65) gives

$$v_1''y_1 + v_2''y_2 + v_1'y_1' + v_2'y_2' = 0, \quad \text{so that} \quad g(x) = a(v_1'y_1' - v_2'y_2').$$

We now end up with two equations,

$$v_1' y_1' + v_2' y_2' = \frac{g(x)}{a} \quad \text{and} \quad v_1' y_1 + v_2' y_2 = 0, \tag{6.66}$$

for two unknown functions that, recalling our results from Chapter 4, will have a unique solution if $y_1(x)$ and $y_2(x)$ are linearly independent; which they are because they form the general solution of the homogeneous ODE. Equations (6.66) can be solved to yield

$$v_1' = -\frac{g(x)y_2(x)}{a(y_1(x)y_2'(x) - y_1(x)'y_2(x))}, \qquad v_2' = \frac{g(x)y_1(x)}{a(y_1(x)y_2'(x) - y_1(x)'y_2(x))}, \tag{6.67}$$

where we recognize that the denominator in these two equations is the Wronksian of the solutions of the homogeneous equation. Equation (6.67) allows us to calculate the functions v_1 and v_2:

$$v_1 = -\int \frac{g(x)y_2(x)}{a(y_1(x)y_2'(x) - y_1(x)'y_2(x))}\, dx, \qquad v_2 = \int \frac{g(x)y_1(x)}{a(y_1(x)y_2'(x) - y_1(x)'y_2(x))}\, dx, \tag{6.68}$$

so long as we can evaluate the two integrals in Equation (6.68).

Exercise 6.4.6 Show that Equation (6.67) is the solution to Equation (6.66).

Notice that for variation of parameters to work we need to have already found the general solution to the corresponding homogeneous equation, so if we can find the particular solution using this method, we will have found the general solution of Equation (6.62).

Example 6.17 We can use the method of variation of parameters to find the general solution of the differential equation

$$y'' - 3y' + 2y = \left(1 + \mathrm{e}^{-x}\right)^{-1}.$$

To use the method of variation of parameters, we first need to solve the corresponding homogeneous problem, $y''-3y' + 2y = 0$. The characteristic equation for the homogeneous equation is $m^2 - 3m + 2 = (m - 2)(m - 1) = 0$, so two solutions are $y_1(x) = \mathrm{e}^x$ and $y_2(x) = \mathrm{e}^{2x}$. Using these solutions, Equation (6.67) becomes

$$\frac{dv_1}{dx} = -\frac{\mathrm{e}^{-x}}{1 + \mathrm{e}^{-x}}, \quad \frac{dv_2}{dx} = \frac{\mathrm{e}^{-2x}}{1 + \mathrm{e}^{-x}}.$$

We can integrate the first equation using the substitution $w(x) = 1 + \mathrm{e}^{-x}$, which gives

$$v_1(x) = \int \frac{1}{w}\, dw = \ln(|w|) + C = \ln(1 + \mathrm{e}^{-x}) + C_1,$$

where we have dropped the absolute value because $1 + \mathrm{e}^{-x} > 0$. We can integrate the second equation using the substitution $w(x) = \mathrm{e}^{-x}$, giving

$$v_2(x) = -\int \frac{w}{1 + w}\, dw = -\int \left(1 - \frac{1}{1 + w}\right) dw = -\mathrm{e}^{-x} + \ln(1 + \mathrm{e}^{-x}) + C_2.$$

The particular solution of the ODE is then

$$y_p(x) = v_1(x)y_1(x) + v_2(x)y_2(x) = \mathrm{e}^x\left(\ln(1+\mathrm{e}^{-x}) + C_1\right) + \mathrm{e}^{2x}\left(\ln(1+\mathrm{e}^{-x}) - \mathrm{e}^{-x} + C_2\right),$$

and the general solution to the differential equation is

$$\begin{aligned}
y(x) &= a_1 y_1(x) + a_2 y_2(x) + y_p(x) \\
&= a_1\mathrm{e}^x + a_2\mathrm{e}^{2x} + \mathrm{e}^x\left(\ln(1+\mathrm{e}^{-x}) + C_1\right) + \mathrm{e}^{2x}\left(\ln(1+\mathrm{e}^{-x}) - \mathrm{e}^{-x} + C_2\right) \\
&= (a_1 + C_1)\mathrm{e}^x + (a_2 + C_2)\mathrm{e}^{2x} + \mathrm{e}^x\left(\ln(1+\mathrm{e}^{-x})\right) + \mathrm{e}^{2x}\left(\ln(1+\mathrm{e}^{-x}) - \mathrm{e}^{-x}\right) \\
&= A\mathrm{e}^x + B\mathrm{e}^{2x} + \mathrm{e}^x\left(\ln(1+\mathrm{e}^{-x})\right) + \mathrm{e}^{2x}\left(\ln(1+\mathrm{e}^{-x}) - \mathrm{e}^{-x}\right),
\end{aligned}$$

which shows that we could have neglected the integration constants C_1 and C_2 when calculating $v_1(x)$ and $v_2(x)$ because they can just be incorporated into the complementary function.

Exercise 6.4.7 Work through the details of Example 6.17.

The method of variation of parameters also works for some second order linear equations that have variable coefficients; e.g., a, b, and c in Equation (6.62) are functions of x. However, the integrals involved in calculating $v_1(x)$ and $v_2(x)$ become more complicated and may not be able to be evaluated analytically.

The solution of second order nonlinear differential equations is more a case of trial and error (Murphy, 1960; Zwillinger, 1997). In Section 6.5 we will introduce some techniques for getting approximate analytical solutions, and later we will look at how we can find qualitative information about the solution of certain classes of nonlinear equations. If none of these methods work, then we have to resort to numerical methods to solve the equations (Section 6.10). However, for the time being, we will explore some more aspects of, second order linear ODEs.

6.4.2 Oscillations and Waves

Second order ODEs are important because they can be used to represent oscillations. Oscillations and vibrations are ubiquitous in the Earth and environmental sciences, from the daily cycle of day and night, to the motion of the Earth during an earthquake. A full description of waves requires an understanding of *partial differential equations* (Chapter 10), but we can use ODE to examine the motion of a body moving up and down or from side to side as a wave passes.

Let us start by looking at the ODE

$$\ddot{x}(t) + \omega_0^2 x = 0, \tag{6.69}$$

where ω_0 is a constant, $x(t)$ is a displacement from a mean level, and we are using a dot over a letter to represent a derivative with respect to time. The characteristic polynomial for Equation (6.69) is $m^2 + \omega_0^2 = 0$, which implies that $m = \pm i\omega_0$ and the solution to the equation is $x(t) = a\sin(\omega_0 t) + b\cos(\omega_0 t)$, where a and b are constants. We can simplify this solution a little to make it easier to interpret. To do this we will replace a and b with

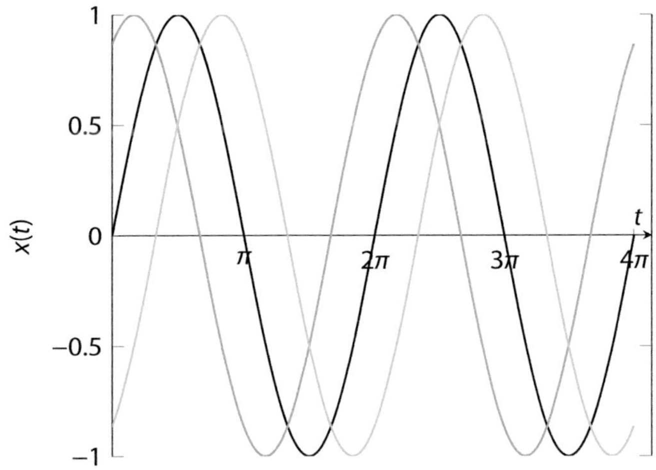

Figure 6.7 The solution (Equation (6.70)) to Equation (6.69) for $\phi_0 = 0$ (black), $\phi_0/\pi/3$ (dark gray), and $\phi_0 = -\pi/3$ (light gray). At $t = 0$, each curve is at a different point in its oscillation.

two new constants ϕ_0 and A such that $\tan(\phi_0) = b/a$, $a = A\cos(\phi_0)$, and $b = A\sin(\phi_0)$. Then we can rewrite the solution in the form

$$x(t) = A\cos(\phi_0)\sin(\omega_0 t) + A\sin(\phi_0)\cos(\omega_0 t) = A\sin(\omega_0 t + \phi_0). \tag{6.70}$$

This equation represents what is called *simple harmonic motion*, i.e., a simple sinusoidal oscillation. How do we interpret the constants A, ϕ_0, and ω_0? The constant ϕ_0 is called the *initial phase* of the oscillation. If we set $t = 0$, then the initial amplitude of the oscillation is $A\sin(\phi_0)$, so ϕ_0 is basically telling us how far along in one period of the oscillation we are at $t = 0$ (Figure 6.7). The constant ω_0 is the *angular frequency* of the oscillation; if τ is the period of the oscillation, then angular frequency is $\omega_0 = 2\pi/\tau$.

Example 6.18 Consider a spring attached to a support at one end with a mass m at the other (Figure 6.8). The force of gravity will pull the mass down, extending the spring. The spring will exert a force in the opposite direction that is proportional to the extension; this is called Hooke's law.[17] If the spring is allowed to reach equilibrium (so that it is not moving) and then the mass is pulled down a small amount then released, can we find the angular frequency and period of the resulting oscillation? We let y be the distance below the support. Newton's second law tells us that the force of gravity pulling the mass down is mg, where g is the acceleration due to gravity.[18] The spring exerts a force $F_r = ky$ in the opposite direction, where k is a constant. So, invoking Newton's second law again, we can write

$$m\frac{d^2y}{dt^2} = mg - ky.$$

[17] Named after the irascible physicist and polymath Robert Hooke (1635–1703) who, though brilliant, was involved in major disputes with many notable scientists of his day, including Isaac Newton and Christopher Wren.

[18] The famous $F = ma$, force equals mass times acceleration.

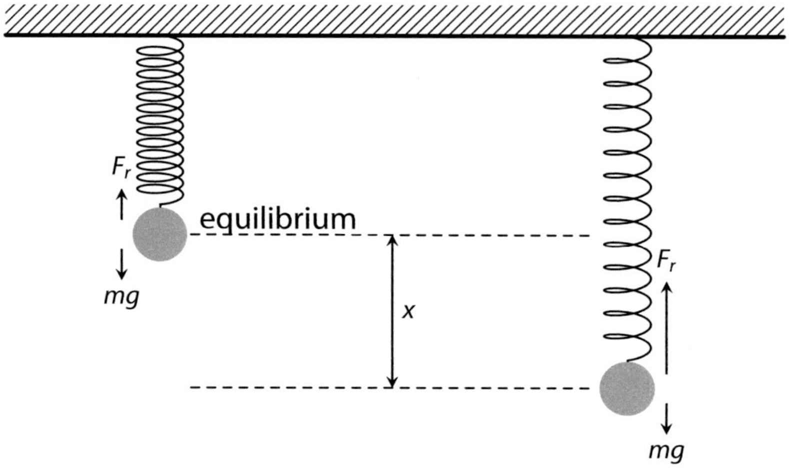

Figure 6.8 A spring with a weight of mass m is suspended from a support. The left-hand picture shows the equilibrium configuration of the spring where the force of gravity (mg) is balanced by the spring's restoring force F_r; the arrows representing mg and F_r are of the same length, showing that the magnitudes of the two forces are the same but that they act in opposite directions. In the right-hand picture, the weight has been pulled down, stretching the spring beyond its equilibrium length, so that the restoring force F_r is greater than mg and there is a net upward force on the mass at the end of the spring.

Under equilibrium, the force of gravity and the restoring force must balance, so the equilibrium distance (y_{equil}) that m hangs below the support given by

$$m\frac{d^2y}{dt^2} = 0 \implies y_{\text{equil}} = \frac{mg}{k}.$$

We want to look at a displacement from y_{equil}, so define a new distance $x = y - y_{\text{equil}}$, and the ODE becomes

$$\frac{d^2x}{dt^2} = -\frac{k}{m}x.$$

This is the equation for simple harmonic motion, and by comparison with Equation (6.69) the angular frequency (ω) and period (τ) of the oscillation are

$$\omega = \sqrt{\frac{k}{m}}, \qquad \tau = 2\pi\sqrt{\frac{m}{k}}.$$

The motion described by Equation (6.69) is rather unrealistic because there is no force damping the motion; the oscillation described by the sine curve will go on forever without changing. Real systems have forces such as friction and drag that act to dampen the motion. In many cases, the damping force is proportional to the velocity of the moving object so that $F_d = -c\dot{x}(t)$. The ODE describing the motion is

$$m\ddot{x} + c\dot{x} + kx = 0, \quad \text{i.e.,} \quad \ddot{x} + 2b\dot{x} + \omega_0^2 x = 0, \tag{6.71}$$

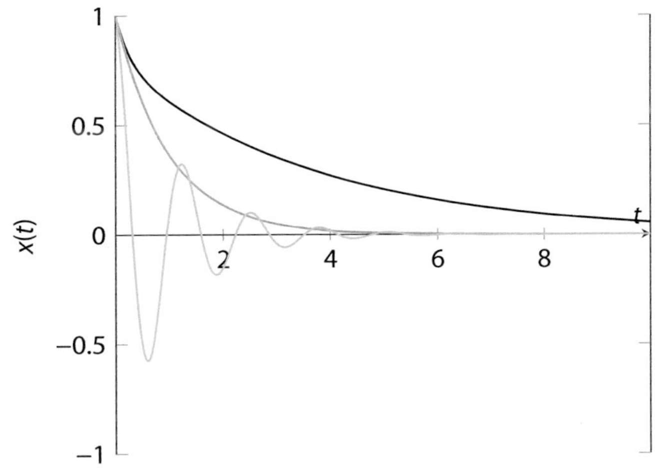

Figure 6.9 The different types of solution for the damped simple harmonic oscillator: over-damped (black), critically damped (gray), underdamped (light gray).

where we have defined $2b = c/m$ and $\omega_0^2 = k/m$.[19] The characteristic equation of Equation (6.71) is $\lambda^2 + 2b\lambda + \omega_0^2 = 0$, which has roots

$$\lambda_\pm = -b \pm \sqrt{b^2 - \omega_0^2}.$$

There are three possibilities for λ, and each gives us a different type of motion (Figure 6.9).

1. If $b > \omega_0$, then there are two real, distinct roots to the characteristic equation. What is more, because $(b^2 - \omega_0^2) < b^2$, these roots will both be negative and the solution to Equation (6.71) is

$$x(t) = A\mathrm{e}^{\lambda_+ t} + B\mathrm{e}^{\lambda_- t},$$

where A and B are constants, $\lambda_+ < 0$, and $\lambda_- < 0$. This describes a situation where the amplitude of the oscillation decays exponentially over time, there is no oscillation, and the motion is called *overdamped* (Figure 6.9).
2. If $b = \omega_0$, then there is only a single root ($\lambda = -b = -\omega_0$) to the characteristic equation, and the solution to the differential equation is

$$x(t) = (At + B)\mathrm{e}^{-\omega_0 t},$$

which also decays as t gets large, though depending on the value of A, the amplitude may initially increase for a short time. This is the *critically damped* case, and the solution decays faster than the overdamped solution (Figure 6.9). To see why this is the case, let us look again at the overdamped solution. The rate at which the solution decays

[19] You may be wondering why we have chosen to add a factor of 2 here, and used ω_0^2 instead of ω_0. We are cheating; we already know the answer, and these definitions mean we will not have square roots and factors of 1/2 littering our answer.

is determined by the term with the *smallest* absolute value of λ, and since $\lambda_\pm$ have dimensions of $[\mathrm{T}]^{-1}$, we can define a timescale for the decay as

$$t_d = \frac{1}{\mid \lambda_+ \mid} = \frac{1}{b - \sqrt{b^2 - \omega_0^2}} = \frac{1}{\omega_0}(b + \sqrt{b^2 - \omega_0^2}) > \frac{1}{\omega_0}.$$

Similarly, we can define a timescale for decay for the critically damped case as $1/\omega_0$ which is smaller than t_d, so the amplitude decays faster in the critically damped case.

3. Lastly, what happens if $b < \omega_0$? In this case, the roots of the characteristic equation are complex and we have a solution that we can write in the form

$$x(t) = A\mathrm{e}^{-bt}\cos(\omega t + \phi_0), \quad \text{where} \quad \omega = \sqrt{\omega_0^2 - b^2}.$$

This equation describes an oscillation with an exponentially decaying amplitude and is called the *underdamped* case (Figure 6.9).

All the oscillations we have looked at so far have been unforced; there have been no external forces driving the oscillation and the resulting ODEs have been homogeneous. We shall now look at the case of *forced oscillations*. These occur where an external force acts to drive the motion of the oscillator. For example, in Figure 6.8 the support could also be moving up and down with some frequency. Similarly, seismographs that measure motions of the Earth's crust are examples of forced oscillators. With that example in mind, let us consider the case where the driving force obeys a cosine function $F_d(t) = F\cos(\omega t)$. To start with, we will consider the undamped situation so that our ODE is

$$\ddot{x} + \omega_0^2 = \frac{F}{m}\cos(\omega t). \tag{6.72}$$

Although this is an inhomogeneous equation, we already know the general solution of the corresponding homogeneous equation, so all we have to do is find the particular solution.

Exercise 6.4.8 Use the method of undetermined coefficients to show that the particular solution of Equation (6.72) is $F\cos(\omega t)/(m(\omega_0^2 - \omega^2))$.

The general solution of Equation (6.72) is

$$x(t) = A\sin(\omega_0 t + \phi_0) + \frac{F}{m}\frac{\cos(\omega t)}{(\omega_0^2 - \omega^2)}. \tag{6.73}$$

We know what the first term looks like, but the second term contains a factor $(\omega_0^2 - \omega^2)$ in the denominator, so this term becomes very large if ω_0 is very close to ω, and becomes infinite when $\omega_0 = \omega$. This means that as ω approaches ω_0, the second term on the right-hand side of Equation (6.73) dominates and the amplitude of the oscillation increases (Figure 6.10). This phenomenon is called *resonance*, and ω is called the *driving frequency* of the system.

An interesting thing happens under certain conditions. Let us consider the solution with initial conditions $\dot{x}(0) = x(0) = 0$, so that Equation (6.73) becomes

$$x(t) = \frac{F}{m(\omega_0^2 - \omega^2)}(\cos(\omega t) - \cos(\omega_0 t)).$$

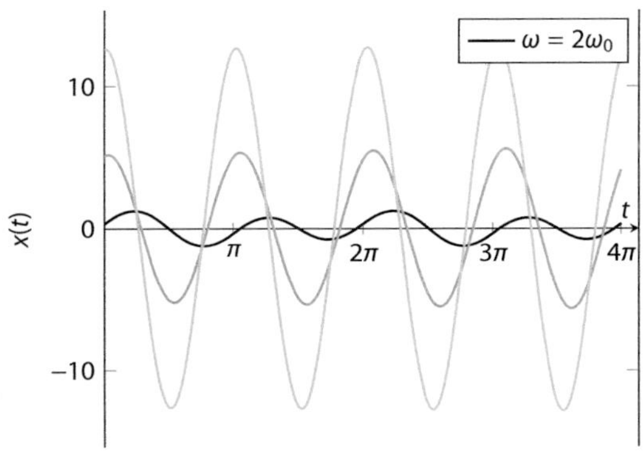

Figure 6.10 Resonance for the undamped simple harmonic oscillator (Equation (6.73): $\omega = 2\omega_0$ (black), $\omega = 0.975\omega_0$ (gray), $\omega = 0.99\omega_0$ (light gray)).

We can now use the fact that (see Appendix B) $2\sin(\theta)\sin(\phi) = \cos(\theta - \phi) - \cos(\theta + \phi)$ to rewrite this equation as

$$x(t) = \frac{2F}{m(\omega_0^2 - \omega^2)} \sin\left(\frac{1}{2}(\omega_0 - \omega)t\right) \sin\left(\frac{1}{2}(\omega_0 + \omega)t\right) = G(t)\sin\left(\frac{1}{2}(\omega_0 + \omega)t\right), \tag{6.74}$$

where we have defined

$$G(t) = \frac{2F}{m(\omega_0^2 - \omega^2)} \sin\left(\frac{1}{2}(\omega_0 - \omega)t\right).$$

Equation (6.74) represents two oscillations with different frequencies. If $\omega \approx \omega_0$, then $(\omega_0 - \omega) \ll (\omega_0 + \omega)$ and the equation represents a high frequency oscillation whose amplitude is *modulated* by a second, lower frequency oscillation (Figure 6.11). This is referred to as the *beat phenomenon* and can be heard when two musical instruments simultaneously play notes that are slightly different.[20]

What happens if we add a damping term to the equation for a forced oscillation? In this case our ODE is

$$\ddot{x} + 2b\dot{x} + \omega_0^2 x = \frac{F}{m}\cos(\omega t). \tag{6.75}$$

Exercise 6.4.9 Use the method of undetermined coefficients to show that a particular solution of Equation (6.75) is

$$y_p(t) = \frac{F}{m[(\omega_0^2 - \omega^2) + 4b^2\omega^2]^{1/2}}((\omega_0^2 - \omega^2)\cos(\omega t) + 2b\omega\sin(\omega t)). \tag{6.76}$$

As we have seen, all the solutions to the homogeneous equation (Equation (6.71)) decay with time, whereas the particular solution (Equation (6.76)) does not. The complementary function for Equation (6.75) is called the *transient solution* because after a sufficient length of time, the transient solution has decayed away and the solution to the ODE

[20] For example, when musicians are tuning up before a performance.

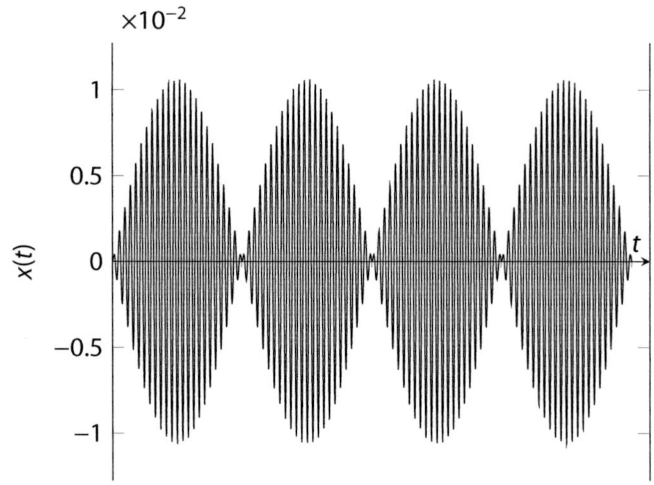

Figure 6.11 The beat phenomenon that occurs when $\omega_0 \approx \omega$ for an undamped, forced oscillation. This plot is for the oscillation given by Equation (6.74) with $\omega_0 = 48$ and $\omega = 46$, and shows the high frequency oscillation with an amplitude modulated by an oscillation at a lower frequency.

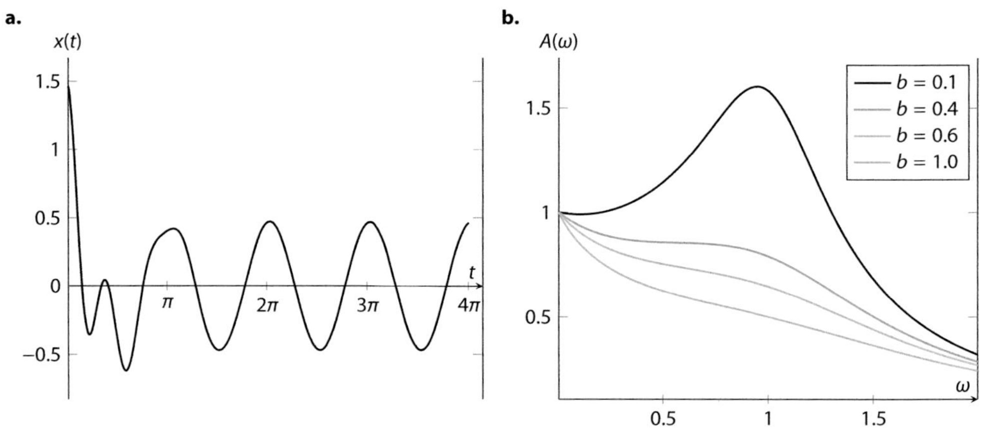

Figure 6.12 A plot of a solution to Equation (6.75), showing the decay of the transient solution leaving only a oscillating solution (a.) The amplitude (Equation (6.77)) of the particular solution of Equation (6.75) as a function of ω for different values of *b* with $\omega_0 = 1$; (b.) as *b* becomes small, the function becomes more strongly peaked about $\omega = \omega_0$. This is an example of *resonance*.

is dominated by the particular solution (Figure 6.12). The amplitude of the particular solution

$$A(\omega) = \frac{F}{m[(\omega_0^2 - \omega^2) + 4b^2\omega^2]^{1/2}} \tag{6.77}$$

has a maximum when $\omega = \omega_0$ (Figure 6.12), and the amplitude of oscillations at that frequency increases as the size of the damping term (i.e., the value of b in Equation (6.75)) decreases. Resonance is an important concept in science. Many systems have a natural frequency of oscillation (ω_0), but if the system is forced at a frequency close to, or at, ω_0, then the amplitude of the oscillations can increase dramatically. The classic example of this is pushing on a child's swing. If we push at the same frequency of the oscillations of the swing itself, then the amplitude of the swing gets larger and larger. But resonance is important for many other situations, such as the ability of molecules in the atmosphere to absorb radiation at specific frequencies.

6.5 Series Solutions and Singular Solutions

The second order differential equations we have looked at so far have had constant coefficients. However, a general second order linear homogeneous ODE has a standard form

$$\frac{d^2y(x)}{dx^2} + p(x)\frac{dy(x)}{dx} + q(x)y(x) = 0,$$

where $p(x)$ and $q(x)$ are not constants. There are no general techniques for finding general solutions to these equations. However, we can often find approximate solutions in the neighborhood of a given point, $x = x_0$, by assuming that they can be written as a power series. To see how this works, let us find a power series solution to an equation with constant coefficients that we have already solved:

$$\frac{d^2y}{dx^2} + y(x) = 0. \tag{6.78}$$

We are going to look for a solution of Equation (6.78) in the neighborhood of $x = 0$, so we first assume that the solution to the equation has the form

$$y(x) = a_0 + a_1x + a_2x^2 + \cdots = \sum_{n=0}^{\infty} a_nx^n,$$

from which we can calculate the derivatives

$$\frac{dy}{dx} = a_1 + 2a_2x + 3a_3x^2 + \cdots = \sum_{n=1}^{\infty} a_nnx^{n-1} \text{ and}$$

$$\frac{d^2y}{dx^2} = 2a_2 + 6a_3x + 12a_4x^2 + \cdots = \sum_{n=2}^{\infty} a_nn(n-1)x^{n-2}.$$

Substituting these expressions into Equation (6.78) gives

$$\sum_{n=2}^{\infty} a_nn(n-1)x^{n-2} + \sum_{n=0}^{\infty} a_nx^n = 0.$$

It would be nice if we could combine these terms, but they appear to be quite different. However, we can rewrite the first term as

$$\sum_{n=2}^{\infty} a_nn(n-1)x^{n-2} = \sum_{n=0}^{\infty} a_{n+2}(n+2)(n+1)x^n,$$

so that our equation becomes

$$\sum_{n=0}^{\infty}[a_{n+2}(n+2)(n+1)+a_n]x^n = 0.$$

For this equation to be true, the coefficients for each power of x must be the same on both sides of the equation, so

$$a_{n+2}(n+2)(n+1)+a_n = 0 \implies a_{n+2} = -\frac{a_n}{(n+2)(n+1)}. \tag{6.79}$$

Equation (6.79) is called a *recurrence relationship* because we can use it to calculate the value of any coefficient a_n if we know the values of a_0 and a_1; we need both values because Equation (6.79) relates a_{n+2} to a_n, not a_{n+1}. So,

$$\begin{aligned}
a_2 &= -\frac{a_0}{2}, & a_3 &= -\frac{a_1}{3\times 2},\\
a_4 &= -\frac{a_2}{4\times 3} = \frac{a_0}{4!}, & a_5 &= -\frac{a_3}{5\times 4} = \frac{a_1}{5!},\\
a_6 &= -\frac{a_4}{6\times 5} = -\frac{a_0}{6!}, & a_7 &= -\frac{a_5}{7\times 6} = -\frac{a_1}{7!}, \quad \text{etc.,}
\end{aligned}$$

from which we can discern the patterns

$$a_{2n} = (-1)^n\frac{a_0}{(2n)!}, \qquad a_{2n+1} = (-1)^n\frac{a_1}{(2n+1)!}.$$

Substituting these back into the series solution we get the solution

$$y(x) = a_0\sum_{n=0}^{\infty}(-1)^n\frac{x^{2n}}{(2n)!} + a_1\sum_{n=0}^{\infty}(-1)^n\frac{x^{2n+1}}{(2n+1)!}. \tag{6.80}$$

We can see from Equation (6.80) that we still have two constants (a_0 and a_1) that we do not have values for, and we will need to use initial or boundary conditions to find them. If we look at Equation (6.80) we can recognize the two series as being the Maclaurin series for $\cos(x)$ and $\sin(x)$.[21] This agrees with the solution we found to Equation (6.49), which is good news. We also know that the power series for cosine and sine both converge for all values of x, not just in the neighborhood of $x = 0$; however, this will not always be the case for power series solutions of ODEs. If we want to find a power series solution about a point other than $x = 0$, say $x = x_0$, then we simply use a power series expansion about that point, i.e., $y(x) = \sum_{n=0}^{\infty} a_n(x-x_0)^n$.

To be able to develop power series solutions in general we need to know a little more about the ODE itself. Let us start with a general second order, linear homogeneous ODE:

$$P(x)\frac{d^2y}{dx^2} + Q(x)\frac{dy}{dx} + R(x)y(x) = 0. \tag{6.81}$$

To get the standard form of the equation we divide through by $P(x)$:

$$\frac{d^2y}{dx^2} + q(x)\frac{dy}{dx} + r(x)y(x) = 0, \quad \text{where} \quad q(x) = \frac{Q(x)}{P(x)},\ r(x) = \frac{R(x)}{P(x)}. \tag{6.82}$$

[21] The Maclaurin series is a Taylor series expanded about $x = 0$ and is named after the Scottish mathematician Colin Maclaurin (1698–1746).

But, we can only do this at those values of x for which $P(x) \neq 0$. Any point $x = x_0$ for which $q(x)$ and $r(x)$ are finite is called an *ordinary point*. In fact an ordinary point is *analytic*, which means that it is infinitely differentiable at that point, so that we can find a Taylor series for the function $y(x)$ about the point and that Taylor series converges to the value of the function. Remember that a Taylor series is a power series where the coefficients depend on the derivatives of the function; a polynomial has a finite number of terms, but a function that is not a polynomial has an infinite number of terms, so it requires the function to have derivatives of all orders. If the point $x = x_0$ is not analytic, then it is called a *singular point*.

If $x = x_0$ is an ordinary point of the equation $y'' + p(x)y' + q(x)y = 0$, then we can always find two linearly independent power series solutions at $x = x_0$. What is more, the series solutions converge for $|x - x_0| < D$, where D is the distance from x_0 to the nearest singular point of the ODE; i.e., the series solutions converge for all values of x between $x = x_0$ and the value of x at the nearest singular point. For example, the equation

$$\frac{d^2y}{dx^2} - 2x\frac{dy}{dx} + y = 0$$

does not have a singular point for any finite values of x and the nearest singular points are at $x = \pm\infty$. So, for this equation, we can find two linearly independent series solutions that are valid for $-\infty < x < \infty$.

The solution, $y(x)$, of an ODE near a singular point is often a very important part of the solution because $y(x)$ changes very rapidly there. For example, the differential equation

$$t^2\frac{d^2x}{dt^2} - 12x(t) = 0 \implies \frac{d^2x}{dt^2} - \frac{12x(t)}{t^2} = 0 \tag{6.83}$$

has a singular point at $t = 0$; the second term becomes infinite there.

Exercise 6.5.1 Verify by direct substitution that $x_1(t) = t^2$ and $x_2(t) = t^{-1}$ are two linearly independent solutions of Equation (6.83) for $t \neq 0$.

The general solution of Equation (6.83) for $t \neq 0$ is $x(t) = At^4 + Bt^{-3}$, where A and B are constants. As $t \to 0$, the second term in the solution becomes large and is changing like $\sim t^{-3}$, whereas the first term becomes small and changes like $\sim t^4$. The solution is therefore changing very rapidly and becoming very large near the singular point. But we have a problem if we want to use our power series method to find a solution. The method we used to derive Equation (6.80) for an ordinary point will not work for a singular point because we will not be able to find a Taylor series for $x(t)$ at the singular point; in our example, the function $x_2(t) = Bt^{-3}$ does not have a Taylor series at $x = 0$ because all the derivatives of that function are infinite there. This can be a significant problem for any attempt to numerically solve the equation near $x = 0$, and in such cases a combination of numerical and analytic approaches is often needed.

It may seem that we are unable to use the series solution technique near a singular point. However, not all singular points of an ODE are the same. If the singularity is not too strong—that is, if the singular terms in the ODE increase to infinity slowly enough—then we can still make progress. But how slowly is slowly enough? A point $x = x_0$ is called a *regular singular point* of the second order linear homogeneous ODE

$$\frac{d^2y(x)}{dx^2} + p(x)\frac{dy(x)}{dx} + q(x)y(x) = 0 \tag{6.84}$$

if

$$\lim_{x\to x_0}(x - x_0)p(x) \quad \text{and} \quad \lim_{x\to x_0}(x - x_0)^2 q(x) \quad \text{are both analytic.} \tag{6.85}$$

A singular point that does not satisfy these conditions is called an *irregular singular point* or *essential singularity*.

We are sometimes interested in solutions to differential equations as the independent variable (e.g., x) tends to infinity. If the singular point is at $x = \infty$, how do we study the solution there? The answer is to transform the variables using $x = 1/z$, and then look at the nature of the point $z = 0$. If we make this transformation, then writing $w(z) = y(z^{-1})$ we have

$$\frac{dy(x)}{dx} = \frac{dy(z^{-1})}{dz}\frac{dz}{dx} = \frac{dw(z)}{dz}\left(-\frac{1}{x^2}\right) = -z^2\frac{dw}{dz}$$

$$\frac{d^2y}{dx^2} = \frac{d}{dz}\frac{dy}{dx}\frac{dz}{dx} = -z^2\frac{d}{dz}\left(-z^2\frac{dw}{dz}\right) = z^4\frac{d^2w}{dz^2} + 2z^3\frac{dw}{dz}.$$

With these substitutions, Equation (6.84) becomes

$$z^4\frac{d^2w}{dz^2} + (2z^3 - z^2p(z^{-1}))\frac{dw}{dz} + q(z^{-1})w = 0, \tag{6.86}$$

and if we put it into standard form, we can examine the nature of any singular points by looking at

$$\frac{2z - p(z^{-1})}{z^2} \quad \text{and} \quad \frac{q(z^{-1})}{z^4}.$$

If these remain finite as $z \to 0$, then the point $x = \infty$ in the original equation is an ordinary point. If they diverge no more rapidly than $1/z$ and $1/z^2$ respectively, then $x = \infty$ is a regular singular point, otherwise it is an irregular singular point.

Example 6.19 Let us find and classify the singular points of the ODE

$$x^2\frac{d^2y}{dx^2} + x\frac{dy}{dx} + (x^2 - n^2)y = 0.$$

For finite values of x we first put the equation into standard form,

$$\frac{d^2y}{dx^2} + \frac{1}{x}\frac{dy}{dx} + \frac{(x^2 - n^2)}{x^2}y = 0,$$

from which we see that $x = 0$ is a singular point. To determine the nature of the singular point we need to evaluate Equation (6.85):

$$\lim_{x\to 0}\frac{x}{x} = 1 < \infty \quad \text{and} \quad \lim_{x\to 0}\frac{x^2(x^2 - n^2)}{x^2} = -n^2,$$

which is finite, so the point $x = 0$ is a regular singular point. To examine the points as $x \to \infty$, we make the transformation $x = 1/z$ and $w(z) = y(z^{-1})$, which results in the following equation (in standard form):

$$\frac{d^2w}{dz^2} + \frac{1}{z}\frac{dw}{dz} + \frac{1-n^2z^2}{z^2} = 0.$$

The point $z = 0$ is a singularity, and to determine its nature we need to evaluate the limits

$$\lim_{z\to 0}\frac{z}{z} = 1 \quad \text{and} \quad \lim_{z\to 0}\frac{z^2(1-n^2z^2)}{z^4} = \infty,$$

so that $x = \infty$ is an irregular singular point.

Exercise 6.5.2 Identify and classify any singular points of the following equations:

1. $x^2y'' - 2xy' + y = 0$
2. $(x^2-4)^2y'' - (x+2)y' + y = 0$
3. $x^2y'' + 4e^xy' + 2\cos(x)y = 0$
4. $x^2(1-x)^2y'' + x(9-x^2)y' + (1+2x)y = 0$.

We are sometimes interested in the solution of an ODE near a singular point. It turns out that we can only do this if the singular point is a regular singular point, and even then we must modify the method we used for an ordinary point where we looked for a solution of the form

$$y(x) = \sum_{n=0}^{\infty} a_n(x-x_0)^n.$$

We must modify this if $x = x_0$ is a regular singular point and instead look for a power law solution of the form

$$y(x) = (x-x_0)^r \sum_{n=0}^{\infty} a_n(x-x_0)^n = \sum_{n=0}^{\infty} a_n(x-x_0)^{n+r}, \tag{6.87}$$

where we have to also determine the value of r. This additional factor of $(x-x_0)^r$ allows us to handle the singularity at $x = x_0$. As an example of the method, Let us look for a power series solution to the equation from Example 6.19,

$$x^2\frac{d^2y}{dx^2} + x\frac{dy}{dx} + (x^2-n^2)y = 0. \tag{6.88}$$

This is called *Bessel's equation*,[22] and we will meet it again in later chapters because it is an equation that keeps appearing when we solve real-world problems. Writing the equation in standard form shows us that $x = 0$ is a regular singular point. So, we use Equation (6.87) and look for a power law solution about the point $x_0 = 0$,

$$y(x) = \sum_{j=0}^{\infty} a_j x^{r+j}, \qquad a_0 \neq 0,$$

[22] Named after Friedrich Bessel (1784–1846), though the equation was first studied by Daniel Bernoulli (1700–1782).

where we have used a slightly different notation to avoid confusion with the n in Equation (6.88). Calculating the derivatives and substituting back into the differential equation gives us

$$\sum_{j=0}^{\infty} a_j(r+j)(r+j-1)x^{r+j} + \sum_{j=0}^{\infty} a_j(r+j)x^{r+j} + \sum_{j=0}^{\infty} a_j x^{r+j+2} - n^2 \sum_{j=0}^{\infty} a_j x^{r+j} = 0,$$

or, combining terms and factoring out the common factor of x^r,

$$\sum_{j=0}^{\infty} a_j[(r+j)^2 - n^2]x^j + \sum_{j=0}^{\infty} a_j x^{j+2} = 0.$$

This is a power series:

$$\begin{aligned} &\left\{a_0(r^2 - n^2) + a_1[(r+1)^2 - n^2]x + a_2[(r+2)^2 - n^2] + \cdots\right\} \\ &\qquad + \left\{a_0 x^2 + a_1 x^3 + a_2 x^4\right\} \cdots = 0, \end{aligned} \tag{6.89}$$

and as such the powers of x on each side of the equals sign must be the same. Setting $j = 0$ gives us an equation for r, called the *indical equation*, $a_0[r^2 - n^2] = 0$, so $r = \pm n$ because we have set $a_0 \neq 0$. The indical equation allows us to calculate the values of the constant r, and we have two cases to consider.

If $r = n$, then comparing x^1 terms in Equation (6.89) gives

$$a_1(2n + 1) = 0,$$

which implies that, for general values of n, $a_1 = 0$. Comparing the x^j terms in Equation (6.89) gives

$$a_j[(r+j)^2 - n^2] + a_{j-2} = 0.$$

This equation is a little awkward because we have to remember that j in this equation starts at $j = 2$; we have already considered the cases when $j = 0$ and $j = 1$. We can make this a little more explicit by shifting j by 2 so that j becomes $j + 2$ and $j - 2$ becomes j. The equation then becomes

$$a_{j+2} = -\frac{a_j}{(j+2)(2n+j+2)},$$

which is a recurrence relationship telling us how to calculate every other value of a_j. We know that $a_1 = 0$, so this recurrence relationship tells us that $a_3 = a_5 = a_7 = \cdots = 0$. For even values of j, we have

$$a_2 = -\frac{a_0 n!}{2^2 1!\,(n+1)!}, \qquad a_4 = \frac{a_0 n!}{2^4 2!\,(n+2)!}, \qquad a_6 = -\frac{a_0 n!}{2^6 3!\,(n+3)!} \cdots$$

from which we can see the general relationship

$$a_{2p} = (-1)^p \frac{a_0 n!}{2^{2p} p!\,(n+p)!}.$$

Putting this all together, for $r = n$, the series solution to Equation (6.88) is

$$y(x) = a_0 x^n \left(1 - \frac{n!\,x^2}{2^2 1!\,(n+1)!} + \frac{n!\,x^4}{2^4 2!\,(n+2)!} + \cdots\right)$$
$$= a_0 \sum_{j=0}^{\infty} (-1)^j \frac{n!\,x^{n+2j}}{2^{2j} j!\,(n+j)!} = a_0 2^n n! \sum_{j=0}^{\infty} (-1)^j \frac{1}{j!\,(n+j)!} \left(\frac{x}{2}\right)^{n+2j}.$$

If we choose $a_0 = 1/(2^n n!)$, we arrive at the standard form of a *Bessel function*, $J_n(x)$, which is a function *defined by its power series*

$$J_n(x) = \sum_{j=0}^{\infty} (-1)^j \frac{1}{j!\,(n+j)!} \left(\frac{x}{2}\right)^{n+2j}. \tag{6.90}$$

Bessel functions often occur in solutions to problems that have a cylindrical symmetry, and Equation (6.90) is our first glimpse at a useful mathematical function that can only be expressed as an infinite series; we shall meet other such useful functions in Chapter 8.

How do we know if we will be able to find a power series solution? The answer is provided by an important theorem called *Fuchs' theorem*[23] which states that it is always possible to find at least one power series solution to a second order ODE so long as the expansion is about either an ordinary point or a regular singularity. If we expand about an irregular singularity, then the method may fail.

6.6 Higher Order Equations

Solutions for higher order differential equations become harder to find and are more complicated than the ones we have examined here. Fortunately, they do not appear very often in the Earth and environmental sciences. However, there are some types of higher order equation that we can hope to solve. The simplest case occurs when the unknown function only appears in the highest order derivative, i.e., for a function $x(t)$

$$\frac{d^n x}{dt^n} = g(t),$$

where $g(t)$ may be a constant. We can solve this equation if we can integrate it n times.

Example 6.20 Some higher order equations are easy to solve, requiring only repeated integration. For example, the differential equation

$$\frac{d^4 y}{dx^4} = \gamma,$$

where γ is a constant, can be solved by simply integrating four times to give

$$y(x) = \frac{1}{24}\gamma x^4 + \frac{1}{6}Ax^3 + \frac{1}{2}Bx^2 + Cx + D.$$

---

[23] Named after Lazarus Immanuel Fuchs (1833–1902).

We can also solve higher order equations if we can find a judicious substitution to reduce it to a lower order ODE that we can solve. For example, we can solve

$$\frac{d^4y}{dx^4} + \frac{d^2y}{dx^2} = g(x)$$

by making the substitution

$$u = \frac{d^2y}{dx^2} \implies \frac{d^2u}{dx^2} + u = g(x).$$

If we can solve this equation for u, then we stand a chance of solving the second order equation for y.

Higher order linear homogeneous equations with constant coefficients can be solved using similar methods to those for second order equations. The general form of the equation is

$$a_n\frac{d^ny}{dx^n} + a_{n-1}\frac{d^{n-1}y}{dx^{n-1}} + \cdots + \frac{d^2y}{dx^2} + \frac{dy}{dx} + y(x) = 0, \qquad a_n \neq 0,$$

and, by analogy with a second order equation, will have a general solution of the form

$$y(x) = c_1y_1(x) + c_2y_2(x) + \cdots + c_{n-1}y_{n-1}(x) + c_ny_n(x),$$

where c_i are constants and $y_1(x)$, $y_2(x)$, $\ldots y_n(x)$ are n linearly independent solutions of the equation. Finding a specific solution requires knowing n initial or boundary conditions so that the constants c_i can be found. Linearly independent solutions can be found by assuming a solution of the form $y(x) = e^{mx}$, substituting it into the equation, and solving the resulting characteristic equation. The problem though is that we will have to solve an nth order polynomial in m, and that might not always be possible to do analytically. If we can find the roots of the polynomial, the solutions then fall into different categories:

- If m is real and a distinct, nonrepeated root of the characteristic equation, then $y(x) = Ae^{mx}$ is a solution, where A is a constant. If m is a real root that is repeated k times, then $y(x) = A_0e^{mx}$, $y(x) = A_1xe^{mx}$, $y(x) = A_2x^2e^{mx}$, up to $y(x) = A_{k-1}x^{k-1}e^{mx}$ are all solutions.
- If m is a complex conjugate pair, $m = \alpha \pm i\beta$, then the solutions are $y(x) = Ae^{\alpha x}\cos(\beta x)$ and $y(x) = Be^{\alpha x}\sin(\beta x)$, where A and B are constants. If m is a repeated complex root, then the pairs $y(x) = A_0e^{\alpha x}\cos(\beta x)$, $y(x) = B_0e^{\alpha x}\sin(\beta x)$, up to $y(x) = A_{n-1}x^{n-1}e^{\alpha x}\cos(\beta x)$, $y(x) = B_{n-1}x^{n-1}e^{\alpha x}\sin(\beta x)$ are all solutions.

6.7 Differential Equations in Practice

If we can find the solution to a differential equation using the methods we have discussed so far, then it is worth the effort to do so. This is because we then have access to all the information we need to determine how the solutions of the equation behave; though we still have some work to do in analyzing the solutions and interpreting the results. However, there are many equations which cannot be solved by the techniques we have examined. Fortunately there are techniques we can use to understand how the solutions to these equations behave without having to solve the equations. A good strategy to pursue is to start by looking at the steady state solutions and then trying to find approximate solutions to the ODE in the neighborhood of the steady state. These approximate solutions often provide considerable insight into the behavior of the general solutions and can also provide us with valuable information to help us determine the validity of a numerical solution to the equation.

Let us consider as an example a simple climate model that balances the energy entering the climate system from the Sun and the energy leaving the Earth via radiation into space. The simplest such model is the zero-dimensional energy balance model. This model considers the average surface temperature (T) of the Earth (Kaper and Engler, 2013) and leads to the equation

$$c\frac{dT}{dt} = \frac{1}{4}(1-\alpha)Q - \sigma\gamma T^4, \tag{6.91}$$

where c is the average heat capacity of the Earth, Q is the amount of solar radiation striking the Earth (called the *solar constant*), α is the albedo of the Earth (the fraction of incident energy from the Sun that is reflected back into space), σ is Stefan's constant, and γ is a parameter called the *emissivity* of the Earth. Equation (6.91) actually has a general solution, but it is not very helpful because it involves an equation for T that cannot be solved without a computer. However, we can learn a lot about the solution of the equation if we are not concerned with the most general of solutions.

As we have mentioned, a good strategy for understanding the behavior of any differential equation is to look for the steady state solution (i.e., $dT/dt = 0$). The steady state solution for Equation (6.91) is

$$T_0 = \left[\frac{(1-\alpha)Q}{4\gamma\sigma}\right]^{1/4}. \tag{6.92}$$

You might argue that we have cheated because we have deliberately neglected the feature we were interested in, the fact that the temperature can change with time. However, if we assume that in most physical, realistic situations of interest, the system we are interested in is close to the steady state solution, then we can start to look for time-dependent solutions of the equation that are also close to steady state; we will see what we mean by "close" in a short while. To do this, we define a new variable, $\theta(t)$, that measures the difference between the actual, time dependent solution ($T(t)$) and the steady state solution, $\theta(t) = T(t) - T_0$. Then we substitute this into Equation (6.91), realizing that T_0 is a constant, so its derivative is zero, to get

$$\begin{aligned}\frac{d\theta}{dt} &= \frac{1}{4}(1-\alpha)Q - \sigma\gamma(T_0+\theta)^4\\ &= \frac{1}{4}(1-\alpha)Q - \sigma\gamma(T_0^4 + 4T_0^3\theta + 6T_0^2\theta^2 + 4T_0\theta^3 + \theta^4).\end{aligned} \tag{6.93}$$

Equation (6.92) tells us that $\frac{1}{4}(1-\alpha)Q - \sigma\gamma T_0^4 = 0$, and substituting this into Equation (6.93) leaves us with the equation

$$\frac{d\theta}{dt} = -\sigma\gamma(4T_0^3\theta + 6T_0^2\theta^2 + 4T_0\theta^3 + \theta^4). \tag{6.94}$$

We can simplify this equation further by recalling that we are interested in solutions that are close to T_0, in other words, θ is small (<1). This means that $\theta \gg \theta^2 \gg \theta^3$, etc. Since we are looking at an approximation already, we can concentrate on only the dominant terms and neglect everything else. Therefore, we neglect all terms containing θ^2 or higher powers. This process is called *linearization* because we are keeping only the terms that are linear in the variable of interest, and results in the equation

$$c\frac{d\theta}{dt} = -4\sigma\gamma T_0^3\theta. \tag{6.95}$$

This equation gives us important information about the behavior of the solution of the original differential equation. To start with, if $\theta > 0$ (that is, we increase the temperature a little bit beyond the steady state), the derivative is negative, so θ decreases with time and the temperature moves back toward the steady state temperature T_0 (i.e., $\theta = 0$). If, on the other hand, we decrease the temperature from the steady state (i.e., $\theta < 0$), then the derivative is positive and we push the temperature $T(t)$ back toward the steady state again. In other words, the steady state solution is *stable* for small perturbations. Because the equation is linear (we constructed it that way through the linearization process), we can solve it to get

$$\theta(t) = \theta(t = 0)\exp\left(-\frac{4\sigma\gamma}{c}T_0^3 t\right), \tag{6.96}$$

so we even have a typical timescale for the time that perturbations in the temperature take to fade away,

$$\tau_\theta = \frac{c}{4\sigma\gamma T_0^3}. \tag{6.97}$$

As we will see later, having this kind of understanding of the behavior of the solution of a differential equation is very useful when we use numerical techniques.

6.7.1 Phase Plane

The *phase plane* is a useful tool for visualizing the behavior of solutions of autonomous ODEs. To see how this works, let us consider a system of coupled first order autonomous differential equations (notice that the equations can be linear or nonlinear),

$$\dot{x} = f(x, y), \qquad \dot{y} = g(x, y), \tag{6.98}$$

where the derivative is with respect to t. We can find the steady state solutions of this system by finding the (x, y) values that solve the equations $f(x, y) = g(x, y) = 0$. The solutions to Equations (6.98) are curves in the (x, y) plane. The uniqueness properties of the solutions tells us that these curves do not cross each other, so for each initial condition $((x_0, y_0))$ there is a unique curve that passes through that point; these curves are sometimes called *trajectories*. We can construct the phase plane by hand or by using a computer, and it gives us a qualitative understanding of the behavior of the solutions to the equations.

We can extend this idea to second order autonomous differential equations,

$$\ddot{x} + g(x, \dot{x}) = 0,$$

if we define a new variable, $y = \dot{x}$. We can then write this single equation as a pair of coupled first order autonomous differential equations,

$$\dot{x} = y, \qquad \dot{y} = -g(x, y),$$

from which we can construct the phase plane. In the next section we shall see how the phase plane can help us understand the behavior of systems of linear ODEs.

6.8 Systems of Linear Differential Equations

So far we have considered problems that involve only a single ODE. However, many problems involve systems of coupled differential equations, where the solution of one ODE depends on the solution of another. For example, our polluted lake problem could be generalized to involve multiple lakes along a river so that the pollutant gradually makes its way down the chain of rivers and lakes. The change of pollutant with time in the last lake will depend on how the pollutant changes in all the preceding lakes.

If $y_1(x), \dots, y_n(x)$ are functions (e.g., the concentration of pollutant in n connected lakes) related by a system of first order linear ODEs with constant coefficients, then we can write

$$\begin{aligned}
y_1' &= a_{11}y_1 + a_{12}y_2 + \cdots + a_{1n}y_n + f_1(x)\\
y_2' &= a_{21}y_1 + a_{22}y_2 + \cdots + a_{2n}y_n + f_2(x)\\
\vdots &= \ \vdots \ + \ \vdots \ + \cdots + \ \vdots \ + \ \vdots\\
y_n' &= a_{n1}y_1 + a_{n2}y_2 + \cdots + a_{nn}y_n + f_n(x),
\end{aligned}$$

where a_{ij} are constants. We can write this system of equations as a matrix equation in the form

$$\mathbf{Y}' = \mathbf{A}\mathbf{Y} + \mathbf{F}(x), \tag{6.99}$$

where

$$\mathbf{Y} = \begin{pmatrix} y_1 \\ y_2 \\ \vdots \\ y_n \end{pmatrix}, \quad \mathbf{A} = \begin{pmatrix} a_{11} & a_{12} & \cdots & a_{nn} \\ a_{21} & a_{22} & \cdots & a_{2n} \\ \vdots & \vdots & & \vdots \\ a_{n1} & a_{n2} & \cdots & a_{nn} \end{pmatrix}, \quad \mathbf{F} = \begin{pmatrix} f_1(x) \\ f_2(c) \\ \vdots\ f_n(x) \end{pmatrix}.$$

Let us start by looking at the homogeneous case, i.e., $f_i(x) = 0$. Equation (6.99) then becomes the matrix differential equation $\mathbf{Y}' = \mathbf{A}\mathbf{Y}$. This is a system of linear equations with constant coefficients, and given our experience with such equations, a good solution to try is $\mathbf{Y} = \mathbf{K}\exp(rx)$. If we substitute this solution into the matrix differential equation, we find that $r\mathbf{K} = \mathbf{A}\mathbf{K}$, which we recognize as an eigenvalue equation where r is an eigenvalue of the matrix $\mathbf{A}$ and $\mathbf{K}$ is the corresponding eigenvector. This is nice because we already know that eigenvectors corresponding to different eigenvalues will always be linearly independent, so the corresponding solutions of the ODE will also be linearly independent.

We will concentrate on two-dimensional systems because it is easier to see what is happening (and do the calculations!), but much of what we will discuss holds for three-dimensional and higher systems, though there are some important complications (Arnold, 1978). For a system of two equations there are three possibilities for the eigenvalues of **A**: the eigenvalues are real and distinct, the eigenvalues form a complex conjugate pair, or there is a real, repeated eigenvalue.

An important aspect of the behavior of differential equations is the existence of a steady state solution where the derivatives are all zero. For systems of linear equations, the origin is the only steady state solution. However, this is not the case for nonlinear systems of equations, as we shall see later. We will refer to this as a steady state, even if the derivatives are not with respect to time. If the equation does have derivatives with respect to time, then the steady state solutions are especially important. Deviations from the steady state can grow, in which case the steady state is an unstable one, or can decay leading to a stable steady state. For example, if we take a normal wooden pencil and lay it flat on its side on a table top, this is a stable steady state. If we tap the pencil gently, it will move horizontally along the table, but it will remain flat on the surface. If we stand the pencil on its blunt end on the table, then we can normally (so long as the end has not been chewed!) balance the pencil upright in this way. If we tap the pencil very gently, it might waver a little, but will remain upright. If we give the pencil a slightly harder push, it will fall. This shows that when balanced like this, the pencil is stable to small perturbations, but larger perturbations can cause it to become unstable. However, balancing the pencil on its sharp point is almost impossible to do because any slight deviation from the pencil being absolutely upright will cause it to fall, and this is an unstable state.

We have seen that straight lines are important in the analysis of these systems, so let us look for straight line solutions. A straight line will have a constant direction in the $(\tilde{x}, \tilde{y})$ plane, but can vary in length along that direction as time changes. In other words, we are looking for a solution that has the form $\tilde{\mathbf{X}}(t) = f(t)\mathbf{U}$, where $f(t)$ is a function of time, and $\mathbf{U}$ is a constant vector. To see under what conditions this is a solution of the linear system, we differentiate it and compare it with the linear system:

$$\frac{d}{dt}\tilde{\mathbf{X}}(t) = \frac{df}{dt}\mathbf{U} = \mathbf{A}\tilde{\mathbf{X}} = f(t)\mathbf{A}\tilde{\mathbf{U}}.$$

We know that $\mathbf{U} \neq 0$; if it were, there would be no direction for the straight line. So this vector equation tells us that $\mathbf{A}\tilde{\mathbf{U}}$ has to be in the same direction as $\tilde{\mathbf{U}}$; in other words, $\mathbf{A}\tilde{\mathbf{U}} = \lambda\tilde{\mathbf{U}}$, which is an eigenvalue equation. The equation for $f(t)$ then becomes $\dot{f} = \lambda f(t)$, which has the familiar solution $f(t) = ce^{\lambda t}$, so $\tilde{\mathbf{X}}(t) = ce^{\lambda t}\mathbf{U}$ along this straight line direction. If $\lambda > 0$, then the move away from the critical point along the direction $\mathbf{u}$, and if $\lambda < 0$, we move toward the critical point, which is what we saw in Figures 6.14 and 6.18.

So, we have characterized the direction that trajectories move along the straight line solution, but what about the nature of the critical point? To answer this, let us try and generalize a little bit and look at the eigenvalue equation $\mathbf{A}\tilde{\mathbf{U}} = \lambda\tilde{\mathbf{U}}$. Remember that the Jacobian matrix is just a matrix of numbers, so for this matrix equation to have a nontrivial solution (i.e., a solution other than $\tilde{\mathbf{U}} = 0$) we need $\det(\mathbf{A} - \lambda\mathbf{I}) = 0$. Writing

$$\mathbf{A} = \begin{pmatrix} a & b \\ c & d \end{pmatrix}$$

the characteristic eigenvalue equation is $\lambda^2-(a+d)\lambda+(ad-bc)=0$. Now we can see that $a+d=\text{Tr}\mathbf{A}=T$ and $(ad-bc)=(det)\mathbf{A}=D$, so we can write the characteristic eigenvalue equation in terms of the trace and determinant of the Jacobian matrix: $\lambda^2-T\lambda+D=0$. If λ_1 and λ_2 are the two solutions of this equation, then

$$(\lambda-\lambda_1)(\lambda-\lambda_2)=\lambda^2-(\lambda_1+\lambda_2)\lambda+\lambda_1\lambda_2=0,$$

and by comparison, $T=\lambda_1+\lambda_2$ and $D=\lambda_1\lambda_2$. What is more, the solutions to $\lambda^2-T\lambda+D=0$ are

$$\lambda_1=\frac{1}{2}\left(T+\sqrt{T^2-4D}\right),\quad \lambda_2=\frac{1}{2}\left(T-\sqrt{T^2-4D}\right).$$

To investigate the nature of the critical point, we can now work through the different cases.

- If $T^2-4D>0$, then the roots λ_1 and λ_2 are real and the solution has the general form $x(t)=c_1e^{\lambda_1 t}\mathbf{u_1}+c_2e^{\lambda_2 t}\mathbf{u_2}$. The signs of λ_1 and λ_2 tell us about the nature of the critical point:
  - If $D=\lambda_1\lambda_2<0$ (i.e., λ_1 and λ_2 have opposite signs), then $T^2-4D>T^2$ and $\lambda_1<0<\lambda_2$ and the critical point is a saddle.
  - If $0<D<T^2/4$, then $0<T^2-4D<T^2$ and both λ_1 and λ_2 have the same sign. If $T>0$, then λ_1 and λ_2 are positive and the critical point is an unstable node, and if $T<0$, then they are both negative and the critical point is a stable node.
- If $T^2-4D>0$, then λ_1 and λ_2 are complex conjugates: $\lambda_1=(\alpha+i\beta)/2$ and $\lambda_2=(\alpha-i\beta)/2$, where $\alpha=T$ and $\beta=\sqrt{4D-T^2}$. If $T<0$, the critical point is a stable spiral; if $T>0$, it is an unstable spiral; if $T=0$, the critical point is a center.

So, the stability and nature of the critical points can be determined from the trace and determinant of the Jacobian matrix. These results are nicely summarized in Figure 6.13.

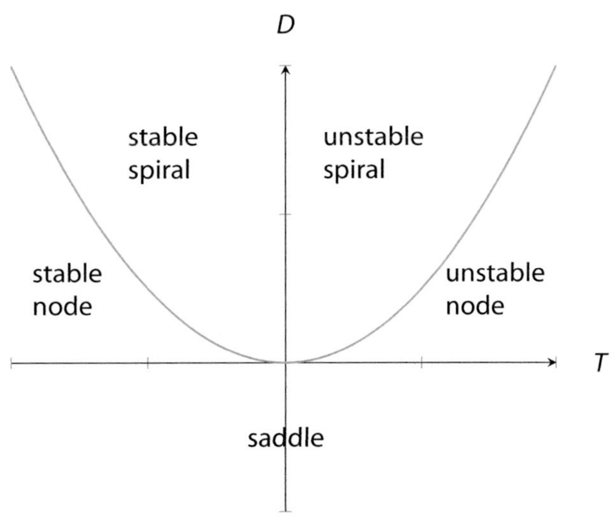

Figure 6.13 The trace-determinant diagram showing the conditions for the stability and nature of critical points in a two-dimensional linear system of odes. The curve is the parabola $D=T^2/4$.

Exercise 6.8.1 Find the natures of the two critical points of the Lotka–Volterra system using the trace and determinant of the Jacobian matrix.

6.8.1 Real, Distinct Eigenvalues

Let us look first at the simplest case, where **A** has real, distinct eigenvalues.

Example 6.21 Find the general solution of the linear system

$$u' = -2u + 2v, \qquad v' = 2u + v.$$

First, we note that $u(x) = v(x) = 0$ is a steady state solution, or *critical point* of the equations. We can write the equations in matrix form $\mathbf{U}' = \mathbf{AU}$, where

$$\mathbf{U} = \begin{pmatrix} u(x) \\ v(x) \end{pmatrix}, \qquad \mathbf{A} = \begin{pmatrix} -2 & 2 \\ 2 & 1 \end{pmatrix}.$$

To solve this equation we can use our experience with linear ODEs so far and look for a solution of the form $\mathbf{U} = \mathbf{K}\mathrm{e}^{\lambda x}$, where $\mathbf{K}$ is a constant vector and λ is a constant. Substituting this into the matrix form of the differential equations we find that $\mathbf{AK} = \lambda\mathbf{K}$. So, for our guess to be a solution of the equations, $\mathbf{K}$ must be an eigenvector of $\mathbf{A}$, and λ must be an eigenvalue of $\mathbf{A}$, and we know how to find these. The characteristic equation for the matrix $\mathbf{A}$ is $\lambda^2 + \lambda - 6 = (\lambda + 3)(\lambda - 2) = 0$, so the eigenvalues of $\mathbf{A}$ are $\lambda_1 = -3$ and $\lambda_2 = 2$. The corresponding eigenvectors are

$$\mathbf{K}_1 = \begin{pmatrix} -2 \\ 1 \end{pmatrix}, \qquad \mathbf{K}_2 = \begin{pmatrix} 1 \\ 2 \end{pmatrix},$$

so the general solution of the ODE is

$$\mathbf{U} = C_1 \begin{pmatrix} -2 \\ 1 \end{pmatrix} \mathrm{e}^{-3x} + C_2 \begin{pmatrix} 1 \\ 2 \end{pmatrix} \mathrm{e}^{2x}. \tag{6.100}$$

What is the significance of the eigenvalues and eigenvectors for a system of ODEs? Let us look at Equation (6.100) in a bit more detail and examine what happens when $C_1 = 0$. The eigenvector $(1, 2)$ is a vector that defines a direction in the (u, v) plane. The exponential factor (e^{2x}) is always positive and affects only the magnitude of the eigenvector, which will change as x changes, but the direction will stay the same. The constant C_1 can alter the overall magnitude of the vector (this will not change with x) and its direction (C_1 can be positive or negative). So, this solution will be a straight line, parallel to the direction given by the vector $(1, 2)$, that will increase in length as x becomes large. Similarly, for the other solution, except that it will be defined by the other eigenvector $((-2, 1))$ and its magnitude will be decreasing as x increases. In this case, the trajectories of the solutions form a saddle (Figure 6.14). The directions defined by the eigenvectors are *separatrices* because they separate the different behaviors of the trajectories. In this case, the origin

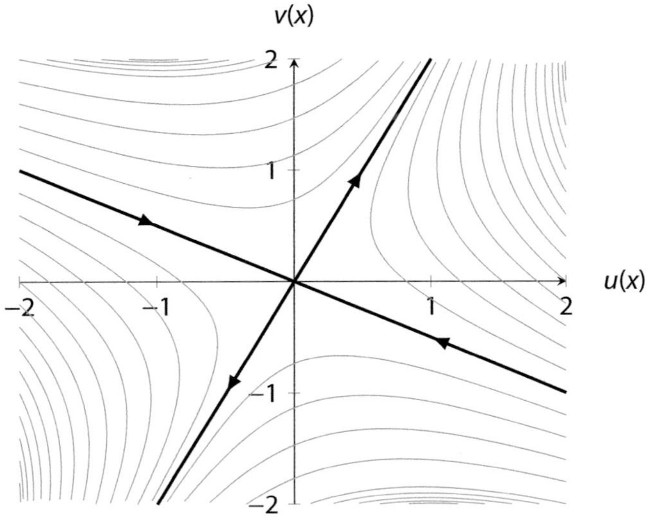

Figure 6.14 A phase plane of the system of differential equations in Example 6.21. The thick black lines show the directions specified by the two eigenvectors, with the arrows showing the direction of the trajectories.

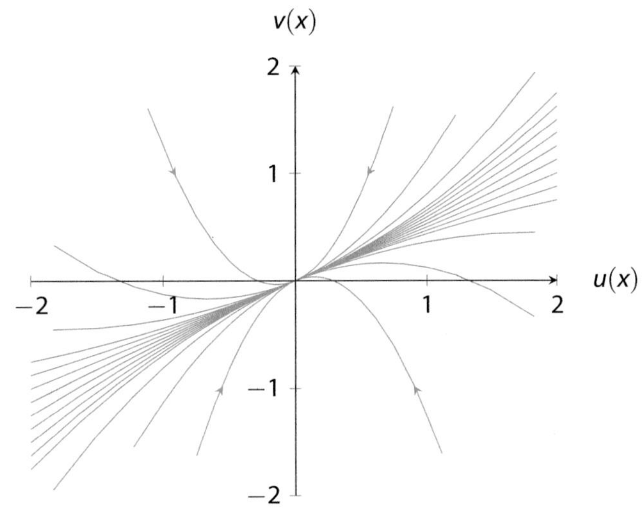

Figure 6.15 A phase plane of the system of differential equations showing a stable node.

is an unstable point and the steady state of the system; it is unstable because there are trajectories that move away from it.

In Example 6.21, the signs of the two eigenvalues were different. If the signs of the eigenvalues are the same, then the steady state is a *node*. If the eigenvalues are positive the node is an *unstable node*, and trajectories move away from the steady state. If the eigenvalues are all negative, the node is a *stable node* (Figure 6.15) and all trajectories move toward it.

6.8.2 Complex Conjugate Eigenvalues

We get a different type of behavior when the eigenvalues are a complex conjugate pair. We should expect from our previous explorations that the solutions will be composed of sine and cosine functions, so they represent some kind of periodic behavior.

Example 6.22 Find the solution of the initial value problem

$$u' = -\frac{u}{2} + v, \qquad v' = -u - \frac{v}{2}.$$

Writing the system of equations as a matrix equation $\mathbf{U}' = \mathbf{AU}$, the eigenvalues of the $\mathbf{A}$ are the complex conjugate pair $\lambda = -1/2 + i$ and $\bar{\lambda} = -1/2 - i$. We need only find the eigenvector for λ because, if all the elements of the matrix $\mathbf{A}$ are real and $\mathbf{w}$ satisfies the eigenvector equation $\mathbf{Aw} = \lambda\mathbf{w}$, then $\mathbf{A}\bar{\mathbf{w}} = \bar{\lambda}\bar{\mathbf{w}}$. So, knowing λ and $\mathbf{v}$ allows us to immediately write down $\bar{\lambda}$ and $\bar{\mathbf{w}}$. The two eigenvectors are then

$$\mathbf{w} = \begin{pmatrix} 1 \\ i \end{pmatrix}, \quad \bar{\mathbf{w}} = \begin{pmatrix} 1 \\ -i \end{pmatrix},$$

so the general solution of the ODEs is a linear combination of $\mathbf{w}e^{\lambda t}$ and $\bar{\mathbf{w}}e^{\bar{\lambda}t}$. We know from Appendix C that we can form two real functions from a complex conjugate pair, so our two real solutions are, using Euler's formula,

$$\frac{\mathbf{w} + \bar{\mathbf{w}}}{2} = \mathrm{Re}(\mathbf{w}) = e^{-t/2}\begin{pmatrix} \cos(t) \\ -\sin(t) \end{pmatrix}, \qquad \frac{\mathbf{w} - \bar{\mathbf{w}}}{2i} = \mathrm{Im}(\mathbf{w}) = e^{-t/2}\begin{pmatrix} \sin(t) \\ \cos(t) \end{pmatrix},$$

and the general solution can be written

$$\mathbf{U} = \begin{pmatrix} u(t) \\ v(t) \end{pmatrix} = e^{-t/2}\left[A\begin{pmatrix} \sin(t) \\ \cos(t) \end{pmatrix} + B\begin{pmatrix} \cos(t) \\ -\sin(t) \end{pmatrix}\right],$$

which describes trajectories that spiral *inward* to the origin because of the negative exponential (Figure 6.16).

In Example 6.22 the eigenvalues had real and imaginary parts. If the real part of the eigenvalues is zero, then the phase portrait becomes a *center* where the trajectories are closed ellipses about the origin (Figure 6.17).

Exercise 6.8.2 Find the general, real solution of the system of equations

$$u' = \frac{u}{2} + v,$$
$$v' = -u + \frac{v}{2},$$

and sketch the phase plane. By considering the signs of u' and v' at the point $u = 1, v = 0$, determine the direction of the trajectories.

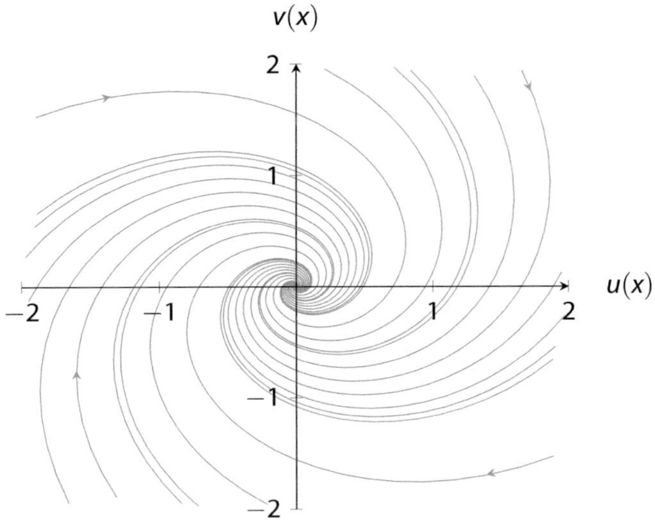

Figure 6.16 A phase plane of the system of differential equations showing a stable spiral.

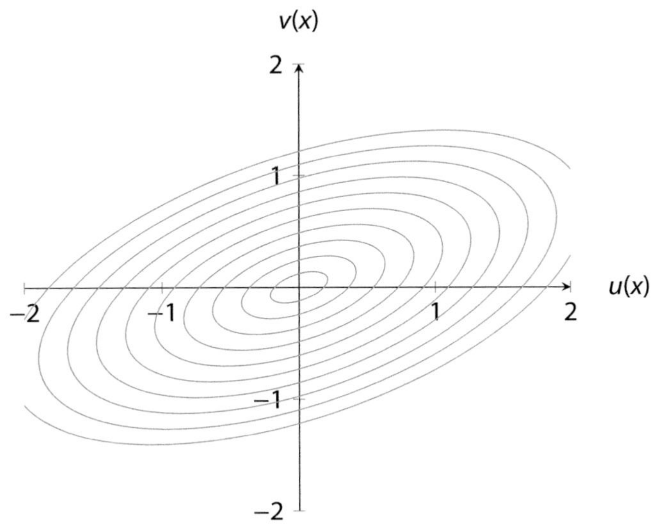

Figure 6.17 A phase plane of the system of differential equations showing a center.

6.8.3 Repeated Roots

The last case we are going to consider here is when the characteristic equation has repeated roots.

Example 6.23 Find the general solution of the linear system

$$u' = u + v, \qquad v' = -u + 3v.$$

The characteristic equation for this system is $\lambda^2 - 4\lambda + 4 = (\lambda - 2)^2 = 0$, so there is only a single eigenvalue, $\lambda = 2$. The corresponding eigenvector equation is

$$\begin{pmatrix} -1 & 1 \\ -1 & 1 \end{pmatrix} \begin{pmatrix} \beta_1 \\ \beta_2 \end{pmatrix} = 0,$$

giving an eigenvector $(1, 1)$. The problem here is that we have found only one solution, and we need another to get the general solution. It turns out that a second solution can be found that has the form

$$te^{\lambda x} \begin{pmatrix} \beta_1 \\ \beta_2 \end{pmatrix} + e^{\lambda x} \begin{pmatrix} \gamma_1 \\ \gamma_2 \end{pmatrix},$$

where

$$(\mathbf{A} - \lambda \mathbf{I}) \begin{pmatrix} \gamma_1 \\ \gamma_2 \end{pmatrix} = \begin{pmatrix} \beta_1 \\ \beta_2 \end{pmatrix}.$$

This means that, in our case,

$$\begin{pmatrix} -1 & 1 \\ -1 & 1 \end{pmatrix} \begin{pmatrix} \gamma_1 \\ \gamma_2 \end{pmatrix} = \begin{pmatrix} 1 \\ 1 \end{pmatrix},$$

so $\gamma_2 = 1 + \gamma_1$, giving a vector $(\gamma_1, (1 + \gamma_1))$. We can choose γ_1 however we like, so we might as well choose something to make our lives easier. So, setting $\gamma_1 = 0$ gives the general solution to the linear system as

$$\begin{pmatrix} u \\ v \end{pmatrix} = C_1 e^{\lambda x} \begin{pmatrix} 1 \\ 1 \end{pmatrix} + C_2 \left(x e^{\lambda x} \begin{pmatrix} 1 \\ 1 \end{pmatrix} + e^{\lambda x} \begin{pmatrix} 0 \\ 1 \end{pmatrix} \right).$$

A system like this that has only a single linearly independent eigenvector is called an *improper node*.

The system of equations in Example 6.23 has a phase portrait shown in Figure 6.18. If the system of equations has repeated eigenvalues but with two linearly independent eigenvectors, then the phase portrait is called a *proper node*; this can only occur for a very specific form of equations where the matrix of coefficients, $\mathbf{A}$, is proportional to the identity matrix, i.e.,

$$\mathbf{A} = \begin{pmatrix} a & 0 \\ 0 & a \end{pmatrix} = a\mathbf{I}.$$

Exercise 6.8.3 Consider a system of linear differential equations $\mathbf{y}' = \mathbf{A}\mathbf{y}$, where $\mathbf{A}$ has a single, real, repeated eigenvalue (λ) with a corresponding eigenvector $\mathbf{v}$. Then we know that a solution is $\mathbf{v}e^{\lambda x}$.

1. Assume that $\mathbf{u} = x\mathbf{v}e^{\lambda x}$ is also a solution. Show that this implies that the eigenvector $\mathbf{v} = 0$, which contradicts the fact that it is an eigenvector of $\mathbf{A}$.
2. Assume that $\mathbf{u} = x\mathbf{v}e^{\lambda x} + \mathbf{u}e^{\lambda x}$ is a solution of the system of linear differential equations. Show that this implies $(\mathbf{A} - \lambda\mathbf{I})\mathbf{v} = 0$ and $(\mathbf{A} - \lambda\mathbf{I})\mathbf{u} = \mathbf{v}$.

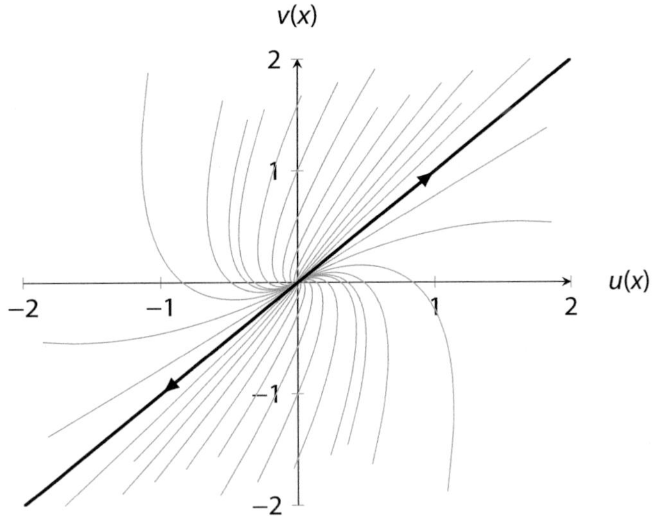

Figure 6.18 A phase plane of the system of differential equations in Example 6.23. The thick black lines show the directions specified by the two eigenvectors, with the arrows showing the direction of the trajectories.

6.9 Systems of Autonomous Nonlinear Equations

Our explorations in Section 6.8 provided us with a nice classification of the types of behavior we can see in systems of linear ODEs, but what about systems of nonlinear equations? Let us look at a classic example, the *predator–prey equation* or *Lotka–Volterra Equation*,[24]

$$\frac{dx}{dt} = ax - bxy, \quad \frac{dy}{dt} = -cy + dxy, \tag{6.101}$$

where a, b, c, and d are all positive constants. This system of autonomous equations is often used to represent the change in population over time of a prey ($x(t)$) and its predator ($y(t)$). The constant a represents the growth rate of the prey population and b the rate of population decline by being consumed by the predator, c represents the rate of mortality of the predator, and d is a measure of how much the predator population grows from consuming its prey.

Exercise 6.9.1 If x and y have dimensions of number of organisms per unit area, what are the dimensions of the constants a, b, c, and d?

We cannot solve the Lotka–Volterra equations analytically, but before resorting to numerical solutions, we can examine what the solutions look like qualitatively. This is very useful for understanding the roles that parameter values take in determining the nature of

[24] These equations were developed independently by the American mathematician Alfred Lotka (1880–1949), who developed them when looking at autocatalytic chemical reactions, and Vito Volterra, an Italian mathematician who was interested in mathematical biology.

the solutions, but also helps provide something to compare our numerical solutions with, thereby providing a check on our numerical algorithm and programming acumen.

The equations are nonlinear, so we cannot directly use the methods in Section 6.8 to examine the qualitative behavior of the solutions. But let us start by looking at Equation (6.101) in more detail. First, if there are no predators (i.e., $y = 0$), then $\dot{x} = ax$ and the prey population grows exponentially. The interaction between the predator and prey arises from a multiplicative term, which captures the idea that the more predators or prey there are, the more they will interact, so prey mortality will increase. If the prey become extinct ($x = 0$), then the predator population will decline exponentially because $\dot{x} = 0$ and $\dot{y} = -cy$.

The first thing to do when analyzing a system of nonlinear autonomous equations is to look for steady state solutions or critical points. Setting $\dot{x} = \dot{y} = 0$, Equation (6.101) gives

$$x(a - by) = 0, \quad y(dx - c) = 0.$$

Solving these equations for x and y gives two steady state solutions, $(x, y) = (0, 0)$ and $(x, y) = (c/d, a/b)$.

Once we have found the steady states, the next task is to examine the nature of the solutions near these points. To do so, we are going to *linearize* Equation (6.101) about each critical point. This will leave us with a linear system that we know how to analyze. To see how this works, let us denote a steady state solution as the point (x_0, y_0) and define new variables $\epsilon = x - x_0$ and $\eta = y - y_0$, which measure the distance away from the steady state in the x and y directions. By substituting these definitions into Equation (6.101) we obtain the following differential equations for ϵ and η:

$$\dot{\epsilon} = (ax_0 - bx_0y_0) + \epsilon(a - by_0) - \eta bx_0 - b\epsilon\eta, \tag{6.102}$$

$$\dot{\eta} = (-cy_0 + dx_0y_0) - \eta(c - dx_0) + \epsilon dy_0 + d\epsilon\eta. \tag{6.103}$$

The first term in parentheses on the right-hand side of each equation is zero because, by comparison with Equation (6.101), we see that these are just the derivatives evaluated at the steady state point (x_0, y_0). If we also assume that ϵ and η are small (i.e., we are looking at a region close to the steady state), then the terms containing $\epsilon\eta$ will be smaller than those with just ϵ or η and can be neglected. As a result, we end up with a set of linearized equations for ϵ and η,

$$\dot{\epsilon} = \epsilon(a - by_0) - \eta bx_0, \qquad \dot{\eta} = -\eta(c - dx_0) + \epsilon dy_0, \tag{6.104}$$

which is a two-dimensional linear system in ϵ and η. We have just seen how to analyze systems of equations like this, so we can determine the nature of each steady state in turn. Substituting the values for the steady state $(x_0, y_0) = (0, 0)$ into Equation (6.104) gives the equations

$$\dot{\epsilon} = \epsilon a, \quad \dot{\eta} = -\eta c,$$

showing that ϵ grows and η decays, showing that the origin is a saddle. Substituting the values for the second steady state into Equation (6.104) gives

$$\dot{\epsilon} = -\eta\frac{bc}{d}, \quad \dot{\eta} = \epsilon\frac{da}{b}, \tag{6.105}$$

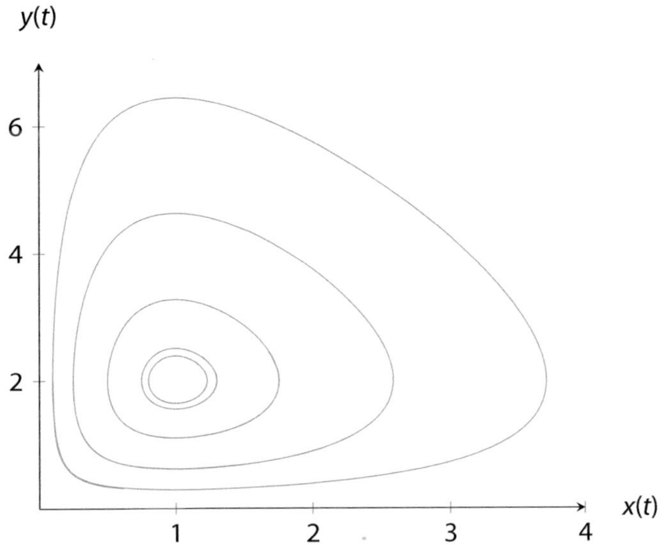

Figure 6.19 A phase plane for the Lotka–Volterra equations.

which describes a center with the trajectories in the phase plane being ellipses (Figure 6.19). We can see this from the equation

$$\frac{d\epsilon}{d\eta} = -\omega^2 \frac{\eta}{\epsilon}, \quad \text{where} \quad \omega^2 = \frac{b^2 c}{d^2 a},$$

which we can solve by separation of variables to give

$$\epsilon^2 + \omega^2 \eta^2 = c.$$

This is the equation of an ellipse. It is important to appreciate the consequences of the assumptions we have made. In particular, the linearized equations (Equation (6.104)) only apply close to the steady states. We can see this in Figure 6.19, where close to the steady state $(x_0, y_0) = (1, 2)$, the trajectories in the phase plane are close to being elliptical. However, as we move further from the steady state, the trajectories become less like ellipses because the nonlinear terms ($b\epsilon\eta$ and $d\epsilon\eta$) that we neglected as being small are no longer small and have an effect.

Exercise 6.9.2 Write the linearized Lotka–Volterra equations in matrix form ($\dot{\mathbf{x}} = \mathbf{A}\mathbf{x}$) and confirm the nature of the two critical points by calculating the characteristic equation of **A**.

Exercise 6.9.3 By differentiating the equation for $\dot{\epsilon}$ and solving the resulting second order differential equation, find explicit solutions for $\epsilon(t)$ and $\eta(t)$ in the vicinity of the steady state.

We can approach the analysis of a nonlinear system in a slightly different way. Let us start with a generic two-dimensional, nonlinear autonomous system of differential equations:

$$\frac{dx}{dt} = F(x, y), \quad \frac{dy}{dt} = G(x, y), \tag{6.106}$$

where $F(x, y)$ and $G(x, y)$ are nonlinear functions of x and y only. Just as with the linear system, we are interested in the behavior of this system near a steady state solution. The steady state solutions are given by the solutions to the equation $F(x_0, y_0) = G(x_0, y_0) = 0$; but, unlike the case of linear systems, the steady state points might not be located at the origin (which was the case for the Lotka–Volterra system). Because we are interested in the behavior close to (x_0, y_0), we can expand the functions $F(x, y)$ and $G(x, y)$ in Taylor series about the critical point,

$$F(x, y) = F(x_0, y_0) + (x - x_0) \left.\frac{\partial F}{\partial x}\right|_{(x_0,y_0)} + (y - y_0) \left.\frac{\partial F}{\partial y}\right|_{(x_0,y_0)} + \cdots$$

$$G(x, y) = G(x_0, y_0) + (x - x_0) \left.\frac{\partial G}{\partial x}\right|_{(x_0,y_0)} + (y - y_0) \left.\frac{\partial G}{\partial y}\right|_{(x_0,y_0)} + \cdots$$

where we have ignored all higher order, nonlinear terms, that is we have linearized the equations; note that if the derivatives of F and G are zero at the critical point, then linearization does not work and we have to consider the effects of the higher order terms. We know that $F(x_0, y_0) = G(x_0, y_0) = 0$, so defining new variables $\tilde{x} = x - x_0$ and $\tilde{y} = y - y_0$, we end up with a linear system of differential equations for $\tilde{x}$ and $\tilde{y}$ that we can write in matrix form as

$$\frac{d}{dt}\begin{pmatrix}\tilde{x}\\ \tilde{y}\end{pmatrix} = \begin{pmatrix} \left.\frac{\partial F}{\partial x}\right|_{(x_0,y_0)} & \left.\frac{\partial F}{\partial y}\right|_{(x_0,y_0)} \\ \left.\frac{\partial G}{\partial x}\right|_{(x_0,y_0)} & \left.\frac{\partial G}{\partial y}\right|_{(x_0,y_0)} \end{pmatrix} \begin{pmatrix}\tilde{x}\\ \tilde{y}\end{pmatrix}. \tag{6.107}$$

The matrix of partial derivatives is just the Jacobian matrix, and because each derivative is evaluated at the critical point, this is just a matrix of numbers. Equation (6.107) is a two-dimension linear system of equations that we can write as $\tilde{\mathbf{X}} = \mathbf{A}\tilde{\mathbf{X}}$, where $\tilde{\mathbf{X}} = (\tilde{x}, \tilde{y})$, and we know how to analyze this to find the nature of the solutions near the critical points.

Exercise 6.9.4 Find the natures of the two critical points of the Lotka–Volterra system using the trace and determinant of the Jacobian matrix.

6.10 Numerical Solution

As we have mentioned, there are many advantages to obtaining a solution to a differential equation analytically. Once we have a solution we are able to find the general behavior of the solutions to the equation. The problem is that to obtain such a solution we have to be able to evaluate one or more integrals, and we know from Chapter 2 that not all integrals can be evaluated in terms of elementary functions. So, we might suspect that we are unable to solve to every set of ODEs analytically. In these cases, we may have to resort to numerical solutions. There are many numerical techniques for obtaining accurate numerical approximations to the solutions of ODEs, and many scientific computing systems (e.g., MATLAB, Python) contain routines for doing this. In this section we are going to look at some of the common numerical routines for solving ODEs.

We may legitimately ask why we should spend time trying to find solutions to ODEs analytically when we can simply get accurate solutions using appropriate numerical methods. Numerical solutions have limitations and dangers, and it is worth getting to know what these are. Numerical solutions are, first and foremost, approximations to the actual solution. How good those approximations are depends on the algorithm used, its implementation, and the equations themselves. In fact, it is important to appreciate that arithmetic operations performed by a computer can be inaccurate. This is because computers can typically store only a finite, set number of digits for any number, so numbers that have binary representations that contain more digits than the computer can store will be rounded. This is not too much of a problem for most computations because the number of digits a computer sets aside to store each number is quite large. However, it is something we need to be aware of. In addition, a numerical solution requires definite values for the parameters of the equation, and the behavior of the solutions can change dramatically as these parameters vary. Determining the parameter values that produce these different behaviors requires a large number of computer simulations. Having said this, we have to recognize that many differential equations we come across in the Earth and environmental sciences cannot be solved analytically, and we have to use numerical methods to solve them. But it is still worth spending some time looking for approximate solutions, special case solutions (e.g., steady states, or solutions when certain terms are zero), and the qualitative behavior of the solutions. If nothing else, comparing them with the numerical solutions will give us confidence that our methods are correct!

There are many algorithms for solving ODEs numerically (Acton, 1990; Press et al., 1992; Shampine, 1994), and choosing the best algorithm for a specific problem requires understanding how these algorithms work and what type of problem we are dealing with. For example, most problems we will come across are initial value problems, where the differential equations are given along with the value of the unknown function and its derivatives at a certain time; the idea is that we want to know the behavior of the system moving forward from that time. However, we may also have a boundary value problem, where the equation is specified along with the value of the function at discrete points, and we need to know the behavior of the solution between these points.

We will start by looking at algorithms for initial value problems. These are largely based on making approximations for the slope of the unknown functions, and then using the slope to take small, discrete steps in time, thereby advancing the solution.

6.10.1 Euler Method and Its Relations

The simplest numerical algorithm for solving ODEs is called the Euler method.[25] The basic idea is that the differential equation

$$\frac{dy}{dx} = f(x, y)$$

gives us information about the slope of the unknown function $y(x)$ at a point. So, if we also have some initial data, i.e., the value of $y(x_0)$ at the point x_0, then we can calculate the

[25] Named after the Swiss mathematician Leonhard Euler (1707–1783).

slope at x_0 and use it to find the value of $y(x_0 + \Delta x)$ at a nearby point $x + \Delta x$. Recalling Taylor's theorem, the value of y at $x_0 + \Delta x$ can be obtained knowing $y(x_0)$ by

$$y(x_0 + \Delta x) = y(x_0) + \Delta x \left.\frac{dy}{dx}\right|_{x_0} + \frac{1}{2}\Delta x^2 \left.\frac{d^2y}{dx^2}\right|_{x_0} + \cdots \tag{6.108}$$

If we neglect all the terms in the Taylor expansion that are of second order or higher in Δx, we are left with the equation of a straight line, the tangent (i.e., slope) of the curve at the point x_0:

$$y(x_0 + \Delta x) = y(x_0) + \Delta x \left.\frac{dy}{dx}\right|_{x_0}. \tag{6.109}$$

We have basically linearized the differential equation (i.e., we have approximated the curve by a straight line over the interval Δx). We can also think of this as a *finite difference* approximation to the derivative

$$\left.\frac{dy}{dx}\right|_{x=x_0} \approx \frac{y(x_0 + \Delta x) - y(x_0)}{\Delta x} = \frac{\Delta y}{\Delta x}. \tag{6.110}$$

The basic idea behind numerically solving an initial value problem is that we start from the initial conditions. We then take a step (Δx) using the initial conditions to approximate the solution, giving us an approximation to a new point, $(x_0 + \Delta x, y_1)$. We then use this value to take another step and so on until we reach the value of x that we want to know. As an example of how we can implement this algorithm, let us say we want to numerically integrate the following differential equation,

$$\frac{dy}{dx} = f(x, y), \qquad \text{given the initial condition} \qquad y(x_0) = y_0, \tag{6.111}$$

and we want to find the value of y at some point $x = x_b$. We start by choosing either a step size (Δx) or the number of steps (n) that we want to use to get from $x = x_0$ to $x = x_b$; these are related by $\Delta x = (x_b - x_0)/n$. We then calculate $f(x_0, y_0)$ and use Equation (6.109) to calculate the value of $y_1(x_1)$ at $x_1 = x_0 + \Delta x$. We then use this new value of y to calculate the derivative $f(x_0 + \Delta x, y(x_0 + \Delta x)) = f(x_1, y_1)$ and move to the next step, $y_2(x + 2\Delta x)$, and so on. So, at the nth step, we have

$$x_n = x_0 + n\Delta x, \qquad y_{n+1} = y_n + \Delta x f(x_n, y_n). \tag{6.112}$$

As we can see from Figure 6.20, the accuracy of the method depends in part on the size of Δx. A smaller value of Δx means that the departure of the straight line approximation from the curve will be smaller and we will hopefully obtain a more accurate answer. However, we can also see that any small difference between the value of y that we calculate and the real value of y will potentially increase as we perform more and more steps.

Let us look more closely at the sources of error. The *global truncation error* at the nth step is defined as

$$E_n = y(x_n) - y_n, \tag{6.113}$$

where $y(x_n)$ is the exact value of $y(x)$ at $x = x_n$ and y_n is the approximate value we get from the numerical method. The obvious source of error is that Equation (6.112)

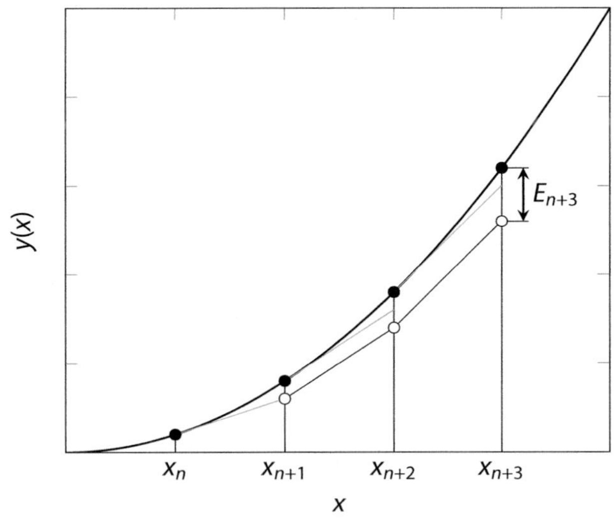

Figure 6.20 The Euler method. The exact solution to the equation is represented by the black curve, and we want to find an approximate, numerical solution at the points x_i, x_{i+1} etc. The gray lines are the tangents to the curve at these locations. The Euler method consists of taking steps of size $\Delta x = x_{n+1} - x_n$, using the tangent to the curve at the point x_n as the slope. The resulting solution, shown by the open circles, will diverge from the real solution; E_{n+3} is the error at x_{n+3}, i.e., the difference between the real solution and the numerical approximation. However, the smaller we can make the step size Δx, the better the approximation will be.

truncates the Taylor series expansion at the second term; that is, we are not not considering derivatives of second order or higher. Without those higher order terms we lose all knowledge of the curvature of $y(x)$. To understand the error in the Euler method, we need to investigate the effects of this truncation. Recall that from Box 2.1 we can write (expanding $y(x)$ about the point x_n)

$$y(x_{n+1}) = y(x_n) + \Delta x f(x_n, y(x_n)) + \frac{1}{2}(\Delta x)^2 y''(\xi), \tag{6.114}$$

where ξ is some value of x that lies between x_n and x_{n+1}. However, we cannot evaluate this because to do so we would have to know the exact solution ($y(x_n)$) of the differential equation in order to evaluate $f(x_n, y(x_n))$. But, we know from Theorem 2.4 that there exists a value of y, call it η, that lies between the Euler calculated value y_n and the exact value $y(x_n)$ such that

$$f(x_n, y(x_n)) = f(x_n, y_n) + \left.\frac{\partial f}{\partial y}\right|_{x_n,\eta} [y(x_n) - y_n] = f(x_n, y_n) + \left.\frac{\partial f}{\partial y}\right|_{x_n,\eta} E_n.$$

We can substitute this expression for $f(x_n, y(x_n))$ into Equation (6.114) to get

$$y(x_{n+1}) = y(x_n) + \Delta x \left(f(x_n, y_n) + \left.\frac{\partial f}{\partial y}\right|_{x_n,\eta} E_n \right) + \frac{1}{2}(\Delta x)^2 y''(\xi).$$

Using Equation (6.112) gives us that the difference between the numerical solution and actual solution at the point x_{n+1} is

$$y_{n+1} - y(x_{n+1}) = y_n - y(x_n) + \Delta x \left.\frac{\partial f}{\partial y}\right|_{x_n,\eta} E_n - \frac{1}{2}(\Delta x)^2 y''(\xi),$$

and using Equation (6.113) we can write

$$E_{n+1} = \left[1 + \Delta x \left.\frac{\partial f}{\partial y}\right|_{x_n,\eta}\right] E_n - \frac{1}{2}(\Delta x)^2 y''(\xi). \qquad (6.115)$$

Equation (6.115) tells us how the error changes between successive steps for the Euler method and that the error is made up of two parts. The *local truncation error* (or LTE) is defined as the error that is incurred at step $(n + 1)$ if there is zero error at step n. In other words, if we had perfect knowledge of the exact solution to Equation (6.111) at x_n, and then took a single Euler step to find y at $x_{n+1} = x_n + \Delta x$, the difference between the Euler value y_{n+1} and the exact value at $y(x_{n+1})$ is the local truncation error. We can calculate the local truncation error for the Euler method by setting $E_n = 0$ in Equation (6.115),

$$\text{local truncation error} = -\frac{1}{2}(\Delta x)^2 y''(\xi),$$

which varies with the step size as $(\Delta x)^2$. So, for a single step, the smaller we make the step size, the better. The term

$$\left[1 + \Delta x \left.\frac{\partial f}{\partial y}\right|_{x_n,\eta}\right] E_n$$

in Equation (6.115) tells us the error at step $(n + 1)$ that results from the error that has already occurred at step n. That is, if we do not know $y(x_n)$ perfectly (e.g., if we approximate $y(x_n)$ by y_n using the Euler method), then this factor tells us how the error at x_n contributes to the error at x_{n+1}. The actual value of E_{n+1} will depend on the values and signs of the various factors in Equation (6.115). In analyzing methods such as the Euler method, it is often more useful to know what the *maximum* possible error is. The largest value of E_{n+1} will occur when $\partial f/\partial y$ is positive and E_n and $y''(\xi)$ have opposite signs. So, we can write

$$|E_{n+1}| \leq \left[1 + \Delta x \left|\frac{\partial f}{\partial y}\right|_{x_n,\eta}\right] |E_n| + \frac{1}{2}(\Delta x)^2 |y''(\xi)|,$$

which is nice, but we have two terms we cannot evaluate because we do not know the values of η and ξ. However, we know that a function that is continuous between $x_n \leq x \leq x_{n+1}$ has maximum and minimum values either at $x = x_n$ or $x = x_{n+1}$ or somewhere in between. So, between $x_n \leq x \leq x_{n+1}$ we know that there exist numbers K and M such that

$$\left|\frac{\partial f}{\partial y}\right|_{x_n,\eta} \leq K, \quad \text{and} \quad |y''(\xi)| \leq M,$$

so we can write

$$|E_{n+1}| \leq (1 + K\Delta x)|E_n| + \frac{1}{2}M(\Delta x)^2, \qquad (6.116)$$

which is often written using the "big-O" notation as $|E_{n+1}| \leq (1 + K\Delta x)|E_n| + O((\Delta x)^2)$ (see Box 6.1).[26] Equation (6.116) gives us an upper bound on the value of the

[26] This notation is sometimes called *Landau notation*, named after the German mathematician Edmund Landau (1877–1938).

Box 6.1 **The Big-O Notation**

The big-O notation is often used in mathematics and numerical analysis in several different ways. For our purposes, we can think of it intuitively in the following way. If we have a power series expansion of a function

$$f(x) = a_1x + a_2x^2 + a_3x^3 + \cdots$$

then if x is small, we might approximate the power series by neglecting all terms containing powers of x higher than x^2 (if $0 < x < 1$ then $x^2 \ll x, x^3 \ll x$, etc.) and we can write $f(x) = a_1x + O(x^2)$ to remind us that, in this limit as $x \to 0$, we have neglected all these higher order terms. We can make this slightly more formal in the following way. If we have two functions $f(x)$ and $g(x)$ that are defined on $a \leq x \leq b$ and we have a point x_0 that lies between $x = a$ and $x = b$, then $f(x) = O(g(x))$ as $x \to x_0$ means that for x close to x_0, $|f(x)| \leq N|g(x)|$ for some positive constant N. This basically tells us that $f(x)$ behaves like $g(x)$ multiplied by a constant as $x \to x_0$. In our previous example, this means that

$$|f(x) - a_1x| \leq Nx^2 \quad \text{as } x \to 0,$$

which tells us approximately how fast $(f(x) - a_1x)$ approaches zero.

truncation error in terms of global properties of the function $f(x, y)$ in the interval $x_n \leq x \leq x_{n+1}$.

Can we obtain a similar inequality to Equation (6.116) but for the global truncation error? Let us assume that we start with perfect knowledge of the initial conditions; i.e., the initial condition is obtained from the exact solution of Equation (6.111). Since this implies $|E_0| = 0$, we have

$$\begin{aligned}
|E_1| &\leq \frac{1}{2}M(\Delta x)^2, \\
|E_2| &\leq (1 + K\Delta x)\frac{1}{2}M(\Delta x)^2 + \frac{1}{2}M(\Delta x)^2, \\
|E_3| &\leq (1 + K\Delta x)^2\frac{1}{2}M(\Delta x)^2 + (1 + K\Delta x)\frac{1}{2}M(\Delta x)^2 + \frac{1}{2}M(\Delta x)^2,
\end{aligned}$$

and so on. We can see a pattern building here and infer that

$$|E_n| \leq \frac{1}{2}M(\Delta x)^2 \sum_{k=0}^{n-1}(1 + K\Delta x)^k.$$

The summation is a geometric series, and we have seen how to calculate the sum of such a series in Section 3.2. So, we can write

$$|E_n| \leq \frac{1}{2}\frac{M\Delta x}{2K}\left[(1 + K\Delta x)^{x_n/\Delta x} - 1\right]. \tag{6.117}$$

If we make Δx small, so that $K\Delta x < 1$, then

$$\lim_{\Delta x \to 0}(1 + K\Delta x)^{x_n/\Delta x} = \mathrm{e}^{Kx_n},$$

and for small step sizes

$$|E_n| \leq \frac{1}{2}\frac{M\Delta x}{2K}(\mathrm{e}^{Kx_n} - 1). \tag{6.118}$$

Table 6.1 Forward Euler solutions of growing exponential

Δx	y_{exact}	y_{euler}	Abs. error	Rel. error
2.0000e-01	1.4841e+02	9.5396e+01	5.3017e+01	3.5723e-01
1.0000e-01	1.4841e+02	1.1739e+02	3.1022e+01	2.0903e-01
5.0000e-02	1.4841e+02	1.3150e+02	1.6912e+01	1.1395e-01
2.0000e-02	1.4841e+02	1.4127e+02	7.1454e+00	4.8146e-02
1.0000e-02	1.4841e+02	1.4477e+02	3.6404e+00	2.4529e-02
1.0000e-03	1.4841e+02	1.4804e+02	3.7032e-01	2.4952e-03

Equation (6.118) tells us that the truncation error changes linearly with Δx, and as a result, the Euler method is called an *order one method*. So, if we halve the step size, we halve the truncation error. The order of the method is equal to the number of factors of Δx that we have in the estimate for the upper bound on the global truncation error. Later we will come across some higher order methods where, for example, the truncation error varies according to $(\Delta x)^2$, and this will be a second order method. Table 6.1 shows results from using the forward Euler method to numerically integrate the initial value problem

$$\frac{dy}{dx} = y, \quad y(x = 0) = 1,$$

from $x = 0$ to $x = 5$ for various values of the step size Δx. The numerical solution is y_{euler}, and because we know the exact solution y_{exact} to this ODE we can calculate the absolute error $|y_{exact} - y_{euler}|$ and the relative error $(y_{exact} - y_{euler})/y_{exact}$. As we expect, as the step size is decreased, both the absolute and the relative errors decrease, but the number of steps we have to take increases, thereby increasing the computational time.

The *stability* of the numerical method is another factor we have to take into account. We would like to know if the difference between the numerical and the exact solution grows or decays away, and if so, how quickly? To examine this, we look at a simple initial value problem,

$$\frac{dy}{dx} = \mu y, \qquad y(x = 0) = y_0 \neq 0.$$

The exact solution to this problem is $y(x) = y_0 \exp(\mu x)$. The constant μ can be a complex number, and if $\mathrm{Re}(\mu) < 0$, the solution decays to zero as x tends toward infinity. If we were to solve this problem using the Euler method with a step size $\Delta x = h$, we would have

$$y_{k+1} = y_k + h\mu y_k = (1 + h\mu)y_k,$$

so

$$y_k = (1 + h\mu)^k y_0.$$

If this is to have the same behavior as the actual solution, i.e., $y \to 0$ as $x \to \infty$, then we require that

$$|1 + h\mu| < 1.$$

We can visualize this stability criterion by plotting it on the complex plane (Figure 6.21a). This tells us that if $\mathrm{Re}(\mu) > 0$, then no matter how small a value of h we choose, any errors

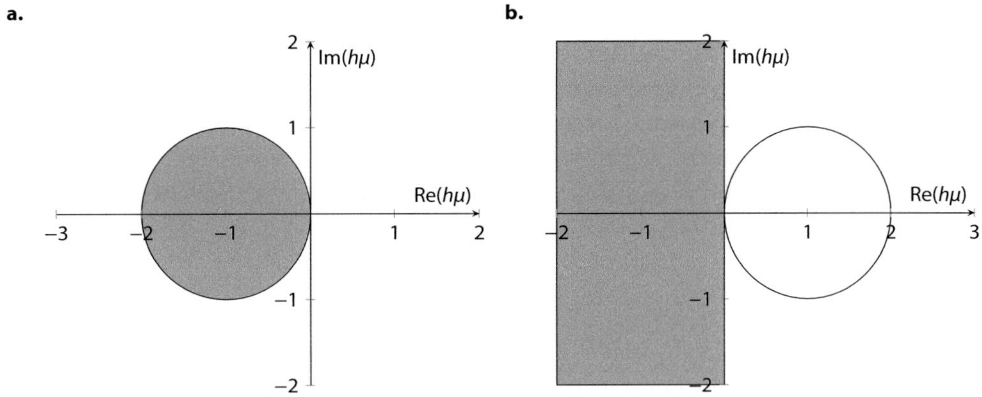

Figure 6.21 Stability diagrams for (a.) the forwards and (b) the backward Euler methods. This is plotted on axes of real and imaginary parts of $h\mu$ with the gray areas showing the values of $h\mu$ where the methods are stable.

will grow and the numerical approximation will be *unstable*, and what is worse, the errors will grow exponentially fast.

Exercise 6.10.1 Show that if μ is a real number, then the stability criterion for the Euler method becomes $0 < h < -2/\mu$.

It is important to realize that the question of stability depends on both the method being used *and* the equation being integrated.

The numerical method given by Equation (6.112) is called the *explicit Euler method*, or the *forward Euler method*. It is called "explicit" because, given the value of $y(x)$ at a given point, we can straightforwardly calculate the value of y at the next point. We do this by basically marching forward, taking a single step in the direction we want to go; to calculate the solution at one point we need only have the information at the previous point. In other words, we evaluate the slope at the point we know and go from there. However, we could think of evaluating the slope at the point we want to find and working backward. In this case, we would use the finite-difference formula

$$\left.\frac{dy}{dx}\right|_{x+h} \approx \frac{y(x+h) - y(x)}{h} \tag{6.119}$$

instead of Equation (6.110) to approximate the derivative, so that

$$y(x+h) \approx y(x) + h\left.\frac{dy}{dx}\right|_{x+h}. \tag{6.120}$$

The problem with this approach is that we have to evaluate the derivative on the right-hand side of the equation at a point we do not yet know—we have not yet calculated a value of y at $x + h$, so we cannot calculate the derivative at that point. However, we can solve this equation numerically as an algebraic equation using a method such as Newton's method (Section 2.7.2). Unlike the forward Euler method, this new method does not give an explicit equation for $y(x + h)$, but rather an implicit equation for $y(x + h)$ and consequently this is called an *implicit* method. Because it uses a backward finite difference approximation for the derivative, it is sometimes called the *backward Euler method* or *implicit Euler method.*

Jut as with the forward Euler method, the local truncation error and rate of convergence for the backward Euler method are $O(h^2)$, so we might wonder why we should even contemplate using it, it has the same LTE but requires more computation to solve. The answer lies in its stability. If we examine the same problem as before, we find that

$$y_n = \frac{1}{(1 - h\mu)^n} y_0, \tag{6.121}$$

and if λ is a real number, this method is stable for all choices of $h = \Delta x$, i.e., it is unconditionally stable (Figure 6.21b). To see this, note that for the backward Euler method to be stable, Equation (6.121) implies that $| 1/(1 - h\mu) | < 1$, so that $| (1 - h\mu) | > 1$. This tells us that either $(1 - h\mu) > 1$ or $(1 - h\mu) < -1$, so for the method to be stable, $h\mu < 0$ or $h\mu > 2$. Now, the step size (h) is positive and we have specified that $\mu < 0$, so $h\mu < 0$ and the method is unconditionally stable.

We can now see that the reason for using the backward Euler method is that it allows us to accurately solve problems that the forward Euler method would have difficulties with.

Both the forward and backward Euler methods use the derivative of the function at a single point to approximate the actual derivative of the function over an interval Δx. The forward Euler method uses the derivative at x, and the backward Euler method uses the derivative at $x + \Delta x$. But the actual derivative of $y(x)$ changes over the interval Δx (unless the solution to the ODE is a straight line) and the linear approximation neglects the curvature of the curve. Can we find a method that somehow accounts for any changes in the slope of the curve? One possibility is to take the average of the slopes we calculate at x and $x + \Delta x$. This might give a better approximation to the slope given the fact the actual slope changes over the interval (Figure 6.22). In this case, to numerically solve Equation (6.111) we would write

$$\frac{\Delta x}{\Delta y} = \frac{y_{n+1} - y_n}{\Delta x} = \frac{1}{2}\left(f(x_n, y_n) + f(x_{n+1}, y_{n+1})\right), \tag{6.122}$$

which gives an implicit method because we need to know the value of the function y at x_{n+1} to evaluate $f(x_{n+1}, y_{n+1})$. In practice, to evaluate Equation (6.122) we first take a forward Euler step to get an approximation for y_{n+1} and use that value to calculate the derivative at x_{n+1}, so

$$\frac{y_{n+1} - y_n}{\Delta x} = \frac{1}{2}(f(x_n, y_n) + f(x_{n+1}, (y_n + \Delta x f(x_n, y_n)))). \tag{6.123}$$

This method is called Heun's method,[27] or the *improved Euler method*, and is a member of a general family of numerical methods for solving ODEs called *predictor–corrector* methods. The name arises because we first make a prediction step (in this case, the forward Euler step) and then use that to make a correction to our answer. Heun's method requires more calculations for each step, and therefore more computational effort, but it is a second order scheme with an error 1/4 of that of the other methods. So, the additional computational effort might be worth it.

[27] Named after the German mathematician Karl Heun (1859–1929).

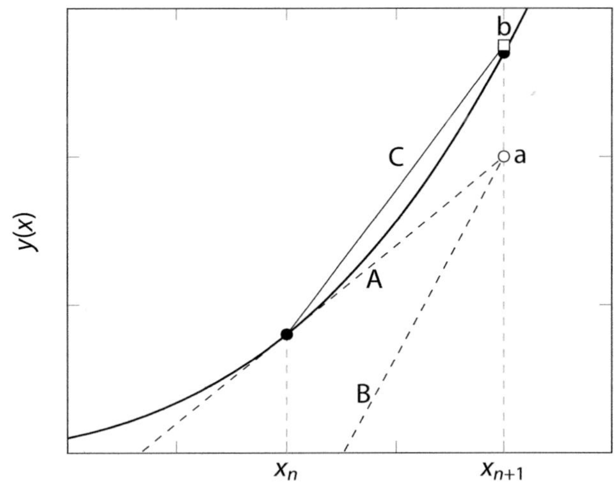

Figure 6.22 Heun's method for the numerical solution of an ode. The exact solutions are shown by the black circles. We first take a single forward Euler step (the dashed line A) starting at x_n, giving the point shown by the open circle at *a*. We then use Equation (6.111) to calculate the slope at the point (x_{n+1}, y_{n+1}), which is shown by the dashed line B. We take the average of the slopes of lines A and B and use this to take a single step (line C) from x_n to x_{n+1}, giving the final, corrected solution point shown by the open square at *b*.

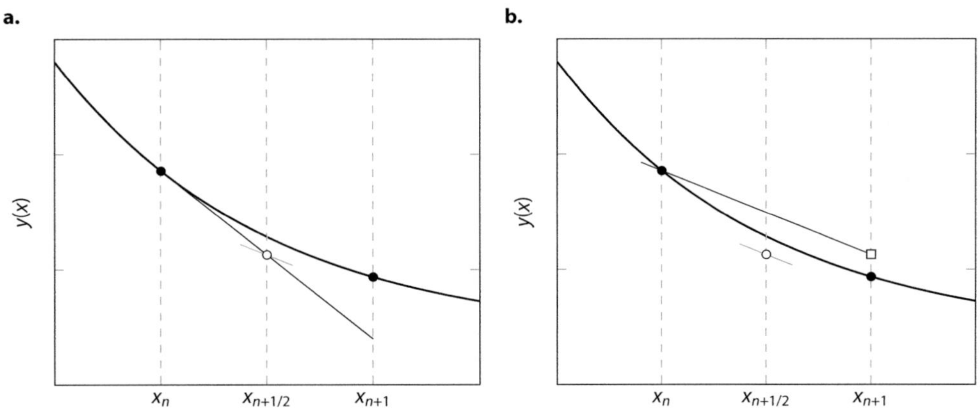

Figure 6.23 The steps of the modified Euler method. First (a.) we use the slope at the point x_n to take a forward Euler step to find $y(x_{n+1/2})$ (open circle). We use this approximation to calculate the slope at the midpoint (the gray line through the open circle). Lastly (b.), we use this new slope to take a full forward Euler step from x_n to $xn + 1$, giving the new approximation shown by the open square.

The last variant of the Euler method we want to look at is called the *modified Euler method* (Figure 6.23). In this case, instead of taking the average of the slopes at both end points of the interval x to $x + \Delta x$, this method uses a point in the middle of the step to calculate the slope. To see why this might be a good idea, let us consider a forward Euler step that uses the slope at the midpoint of the interval Δx,

$$y_{n+1} = y(x_n + \Delta x) = y(x_n) + \Delta x \left.\frac{dy}{dx}\right|_{x+\Delta x/2} .$$

We can now use a Taylor series to expand the derivative in this expression:

$$\left.\frac{dy}{dx}\right|_{x+\Delta x/2} = \left.\frac{dy}{dx}\right|_x + \frac{1}{2}\Delta x \left.\frac{d^2y}{dx^2}\right|_x + \frac{1}{4}(\Delta x)^2 \left.\frac{d^3y}{dx^3}\right|_x + \cdots$$

so that

$$y(x_n + \Delta x) = y(x_n) + \Delta x \left.\frac{dy}{dx}\right|_{x_n} + \frac{1}{2}(\Delta x)^2 \left.\frac{d^2y}{dx^2}\right|_{x_n} + \frac{1}{4}(\Delta x)^3 \left.\frac{d^3y}{dx^3}\right|_{x_n} + \cdots$$

The first three terms on the right-hand side of this equation are just the first three terms of the Taylor series expansion of $y(x_n + \Delta x)$. Recall that the standard forward Euler method truncates the expansion at the Δx term, but by considering a single, forward Euler half step, we get an approximation that agrees with the Taylor expansion up to the $(\Delta x)^2$ term. The $(\Delta x)^3$ term is almost the same as the next term in the Taylor series, but the coefficient is 1/4 instead of 1/6. So, by taking a half step and using the derivative at that midway point ($\Delta x/2$) to approximate the slope over the whole interval Δx, we gain a lot of accuracy in the solution.

A standard notation used in describing numerical methods is to represent the slope of the function evaluated at the different points by $k_1, k_2, \ldots$, etc. For the modified Euler method, the first slope is calculated at (x_n, y_n), so we write $k_1 = f(x_n, y_n)$. The second slope is calculated at the midpoint of the interval $h = \Delta x$ using the slope calculated at (x_n, y_n), i.e., $k_2 = f(x_n + 0.5h, y_n + 0.5hk_1)$. Lastly, the new point uses this new slope to take an Euler step from (x_n, y_n). Putting this all together, we can write the whole method compactly as

$$\begin{aligned} k_1 &= f(x_n, y_n), \\ k_2 &= f(x_n + 0.5h, y_n + 0.5hk_1), \\ y_{n+1} &= y_n + hk_2. \end{aligned} \tag{6.124}$$

Exercise 6.10.2 Numerically solve the ODE

$$\frac{dx}{dt} = \frac{x^2}{1+t}, \quad x(t=0) = 1,$$

from $t = 0$ to $t = 2$ using the forward Euler method with a step size $\Delta t = 0.1$, the backward Euler method, Heun's method, and the modified Euler method, and compare the values at $t = 2$ from all four methods.

6.10.2 Higher Order Methods: Runge–Kutta

Runge–Kutta methods are the workhorses of solving initial value problems involving ODEs.[28] The basic idea builds on the modified and improved Euler method in that Runge–Kutta Methods use slopes calculated at the starting point, the midpoint and the end point of the step, and then combine them in a weighted average to calculate the estimate of y_{n+1} at x_{n+1}. This is illustrated schematically in Figure 6.24. To derive the required equations

[28] Named after German mathematicians Carl Runge (1856–1927) and Martin Kutta (1867–1944).

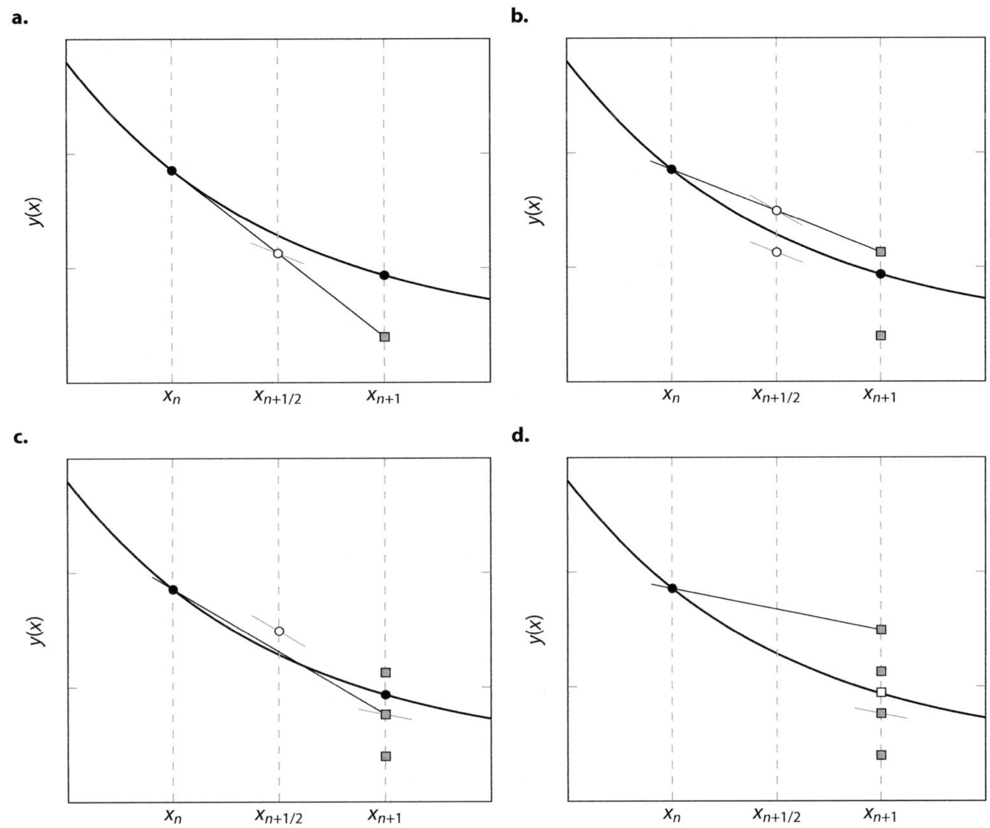

Figure 6.24 The four slopes used in the fourth order Runge–Kutta method. The first slope (a.) is the one calculated at x_n. We keep the point this predicts at x_{n+1} (the gray square) and calculate the y value at the midpoint $x_{n+1/2}$. Using the slope calculated at the midpoint (b.), we calculate a new point (gray square) at x_{n+1} starting from x_n. We again calculate the slope at the midpoint of this line (open circle). We use this new slope (c.) to calculate another approximation of y_{n+1} (gray square), but this time calculate the slope at (x_{n+1}, y_{n_1}). We use this last slope (d.) to calculate one more approximation to y_{n+1} starting at x_n. We now have four estimates of y_{n+1} (the gray squares) and we combine them in a weighted average to get our final estimate, the white square.

formally we can start looking again at the improved Euler method. The Euler methods approximate the derivative by considering only the linear terms in a Taylor expansion

$$y(x+h) = y(x) + hx\left.\frac{dy}{dx}\right|_x,$$

where h is the step size. Heun's method works by averaging the slopes at the beginning and end of the step,

$$\begin{aligned} y_{n+1} &= y_n + \frac{h}{2}\left[f(x_n, y_n) + f(x_{n+1}, y_{n+1})\right] \\ &= y_n + \frac{h}{2}f(x_n, y_n) + \frac{h}{2}f(x_{n+1}, y_{n+1}) \\ &= y_n + \frac{1}{2}k_1 + \frac{1}{2}k_2, \end{aligned} \tag{6.125}$$

where we have replaced $hf(x_n, y_n)$ and $hf(x_{n+1}, y_{n+1})$ with the constants k_1 and k_2 respectively. We know that $y_{n+1} \approx y_n + hf(x_n, y_n)$, so we can write

$$k_2 = hf(x_{n+1}, y_{n+1}) = hf(x_{n+1}, y_n + hf(x_n, y_n)) = hf(x_{n+1}, y_n + k_1).$$

We can write the whole algorithm in a similar way to Equation (6.124):

$$k_1 = hf(x_n, y_n), \qquad k_2 = hf(x_{n+1}, y_n + k_1), \qquad y_{n+1} = y_n + \frac{1}{2}(k_1 + k_2). \tag{6.126}$$

What if we started with the higher order derivatives in the initial Taylor expansion? Can we generalize this result? Keeping terms up to the third derivative, we have

$$y(x + h) = y(x) + h \left.\frac{dy}{dx}\right|_x + \frac{1}{2}h^2 \left.\frac{d^2y}{dx^2}\right|_x + \frac{1}{3!}h^3 \left.\frac{d^3y}{dx^3}\right|_x + \cdots$$

and if we write

$$\frac{dy}{dx} = f(x, y(x)),$$

then we can write the Taylor expansion as

$$\begin{aligned} y(x + h) = y(x) + hf(x, y(x)) &+ \frac{1}{2}h^2 \left(\frac{\partial f}{\partial x} + \frac{\partial f}{\partial y}\frac{dy}{dx}\right) \\ &+ \frac{h^3}{3!}\left(\frac{\partial^2 f}{\partial x^2} + 2\frac{\partial^2 f}{\partial x \partial y}\frac{dy}{dx} + \left(\frac{dy}{dx}\right)^2 \frac{\partial^2 f}{\partial y} + \frac{dy}{dx}\left(\frac{\partial f}{\partial y}\right)^2 + \frac{\partial f}{\partial x}\frac{\partial f}{\partial y}\right) + \cdots \end{aligned} \tag{6.127}$$

However, by analogy with Equation (6.126), we can also write

$$y(x + h) = y(x) + \alpha_1 k_1 + \alpha_2 k_2 + \cdots + \alpha_n k_n,$$

where

$$\begin{aligned} k_1 &= hf(x, y) \\ k_2 &= hf(x + \beta_{21}h, y + \beta_{21}k_1) \\ k_3 &= hf(x + \beta_{31}h + \beta_{32}h, y + \beta_{31}k_1 + \beta_{32}k_2) \\ &\vdots \qquad\qquad \vdots \\ k_n &= hf\left(x + h\sum_{m=1}^{n-1} \beta_{nm}, y + h\sum_{m=1}^{n-1} \beta_{nm}k_m\right). \end{aligned} \tag{6.128}$$

Let us look at the case $n = 2$ in detail to see how we find the values of the constants α_i and β_{ij}. Starting with Equation (6.127), we keep terms up to the second order derivative:

$$y(x + h) = y(x) + hf(x, y(x)) + \frac{1}{2}h^2 \left(\frac{\partial f}{\partial x} + \frac{\partial f}{\partial y}\frac{dy}{dx}\right), \tag{6.129}$$

and since we want our other equation to the same order, we take

$$y(x+h) = y(x) + \alpha_1 k_1 + \alpha_2 k_2 \tag{6.130}$$
$$k_1 = hf(x,y) \tag{6.131}$$
$$k_2 = hf(x + \beta_{21}h, y + \beta_{21}k_1). \tag{6.132}$$

Now Taylor expand Equation (6.132) up to $O(h^2)$,

$$k_2 = hf + \beta_{21}h^2\left(\frac{\partial f}{\partial x} + f\frac{\partial f}{\partial y}\right),$$

and substitute the values of k_1 and k_2 back into Equation (6.130), giving

$$y(x+h) = y(x) + (\alpha_1 + \alpha_2)hf + \alpha_2 h^2 \beta_{21}\left(\frac{\partial f}{\partial x} + f\frac{\partial f}{\partial y}\right).$$

Comparing this with Equation (6.129) gives

$$\alpha_1 + \alpha_2 = 1, \qquad \alpha_2\beta_{21} = 0.5$$

These equations do not have a unique solution (we have two equations and three unknowns), so we have the freedom to choose the value of one of the parameters for our convenience, so long as it leads to a solution for the other two that is consistent. For example, if we choose $\beta_{21} = 1$, then $\alpha_1 = \alpha_2 = \frac{1}{2}$ and we recover Heun's method. But we could also have chosen $\beta_{21} = 0.75$, giving $\alpha_1 = 1/3$, $\alpha_2 = 2/3$.

The most commonly used form of Runge–Kutta takes the expansion up to $O(h^4)$, giving a fourth order Runge–Kutta algorithm

$$y(x+h) = y(x) + \frac{1}{6}(k_1 + k_2 + k_3 + k_4) \tag{6.133}$$
$$k_1 = hf(x,y)$$
$$k_2 = hf(x + \frac{1}{2}h, y + \frac{1}{2}k_1)$$
$$k_3 = hf(x + \frac{1}{2}h, y + \frac{1}{2}k_2)$$
$$k_4 = hf(x + h, y + k_3).$$

Table 6.2 compares the results of using different methods to numerically integrate the ode $y' = -y$ with initial condition $y(x=0) = 1$ from $x = 0$ to $x = 10$ using a constant step size of $\Delta x = 0.1$; only every tenth point is shown in the table. We can see that the Runge–Kutta method performs the best, producing results that agree with the exact solution to three or four decimal places. Choosing a smaller step size would produce more accurate results for all the methods.

The fourth order Runge–Kutta routine is a good choice for integrating single or systems of ODEs, and it is often the first choice that many people will use. It does require more computations than the other techniques we have looked at (we have to evaluate $f(x,y)$ four times per step), but this is balanced by the fact that the method allows us to use larger step sizes to achieve the same accuracy as methods such as the modified Euler method. Other advantages of the Runge–Kutta method are that it is easy to code on a

Table 6.2 A comparison of the Euler, implicit Euler, and Runge–Kutta methods for a decaying exponential

x	y_{exact}	Euler	Backward Euler	Runge–Kutta
0	1.0000	1.0000	1.0000	1.0000
1.0	3.6788×10^{-1}	3.4868×10^{-1}	3.8554×10^{-1}	3.6788×10^{-1}
2.0	1.3534×10^{-1}	1.2158×10^{-1}	1.4865×10^{-1}	1.3534×10^{-1}
3.0	4.9787×10^{-2}	4.2391×10^{-2}	5.7272×10^{-2}	4.9787×10^{-2}
4.0	1.8316×10^{-2}	1.4781×10^{-2}	2.2061×10^{-2}	1.8316×10^{-2}
5.0	6.7379×10^{-3}	5.1538×10^{-3}	8.5066×10^{-3}	6.7380×10^{-3}
6.0	2.4788×10^{-3}	1.7970×10^{-3}	3.3122×10^{-3}	2.4788×10^{-3}
7.0	9.1188×10^{-4}	6.2658×10^{-4}	1.2898×10^{-3}	9.1189×10^{-4}
8.0	3.3546×10^{-4}	2.1847×10^{-4}	4.6409×10^{-4}	3.3547×10^{-4}
9.0	1.2341×10^{-4}	7.6177×10^{-5}	1.6210×10^{-4}	1.2341×10^{-4}
10.0	4.5400×10^{-5}	2.6561×10^{-5}	5.6523×10^{-5}	4.5400×10^{-5}

computer, and that it is a stable method. However, it is not a panacea and does have some disadvantages (and potential pitfalls). Unlike the Euler method, it is not easy to estimate the errors incurred in using a Runge–Kutta problem, and the simplest way to assess the accuracy of a Runge–Kutta computer code is to extensively test it using ODEs that have known analytical solutions.

As with all numerical methods for solving ODEs, the choice of step size (Δx) is very important. If we choose a step size that is too small, then we might end up performing too many unnecessary computations; if we choose too large a step size we might jump over crucial features of the solution. One way to get around this is to use an adaptive step size algorithm (Press et al., 1992). A fourth order Runge–Kutta with an adaptive step size will allow the step size to increase when the solution is smooth and the numerical solution has a high accuracy, and it will automatically reduce the step size when the accuracy requires it. Most modern implementations of ordinary differential solvers will use adaptive step size algorithms.

There are classes of differential equation that are not well suited to being integrated using straightforward Runge–Kutta methods. One common class of such equations is so-called *stiff* equations (Press et al., 1992; Shampine, 1994). For example, consider the initial value problem

$$\frac{dy}{dx} = a - by - ce^{-x}, \quad y(x=0) = 0,$$

where a, b, and c are constants. The solution of this initial value problem is

$$y(x) = \frac{a}{b} - \frac{c}{(b-1)}e^{-x} + \left(\frac{c}{(b-1)} - \frac{a}{b}\right)e^{-bx}.$$

If b is very large (e.g., 1000), then the third term in the solution decays to zero very fast compared to the other exponential term. But, if we were interested in this transient behavior, we would have to use an extremely small time step to capture it. Stiff systems of equations occur where the different terms of an ODE have very different scales over which they change. For example, many systems of chemical reactions involve reactions that occur very rapidly coupled to other reactions that occur very slowly and the ODE describing such a system can be very stiff. Stiff differential equations are explored more in Problem 6.25.

One important thing to remember about computer programs and numerical solutions is that the computer always does what you tell it to do, not necessarily what you meant to tell it to do. So, there is plenty of opportunity for human errors to creep in to a program. Even without these, numerical methods have weaknesses and shortcomings, and it is useful to have an understanding of these and to devise checks on your numerical solutions.

All of the methods we have talked about so far have considered only first order ODEs. How do we deal with higher order differential equations? The answer is to write the higher order equation as a system of first order equations by defining new variables. For example, let us say we wanted to numerically integrate the following initial value problem:

$$x\frac{d^2y}{dx^2} + 2\frac{dy}{dx} - 6y = x + e^{2x}, \qquad y(x = 0) = 0, \qquad y'(x = 0) = 0.$$

We first define a new variable

$$u(x) = \frac{dy}{dx}$$

and write our ODE as the system of coupled first order equations:

$$\frac{dy}{dx} = u(x), \qquad x\frac{du}{dx} + 2u(x) - 6y(x) = x + e^{2x}, \qquad y(x = 0) = 0, \qquad u(x = 0) = 0.$$

We can now use any of the numerical methods we have talked about to find an approximate solution to the problem.

Exercise 6.10.3 Write the following fourth order differential equation as a system of first order differential equations:

$$D\frac{d^4w}{dx^4} + g(\rho_m - \rho_w)w(x) = V(x),$$

where ρ_m, g, and ρ_w are constants. This equation describes the deflection of a tectonic plate under a load described by the function $V(x)$.

6.10.3 Boundary Value Problems

Boundary value problems BVPs cannot be solved numerically in the same way that initial value problems are solved. The main reason for this is that the we do not have all the information available at one boundary to fully initiate a stepping numerical algorithm. A common method of numerically solving a two-point BVP is the *shooting method*. This is basically a smart way of homing in on a good approximation to the solution. The way we do this is to make the BVP look like an IVP. For example, let us say our problem specifies the value of the unknown function $y(x)$ at both the boundaries ($x = x_a$ and $x = x_b$). We start by *assuming* a value for the first derivative at $x = x_a$, say $y'(x = a) = \alpha$. We now have an IVP and we can use the methods we already know to integrate the equations to $x = b$. Unless we are impossibly lucky, the value of our numerical solution at $x = b$,

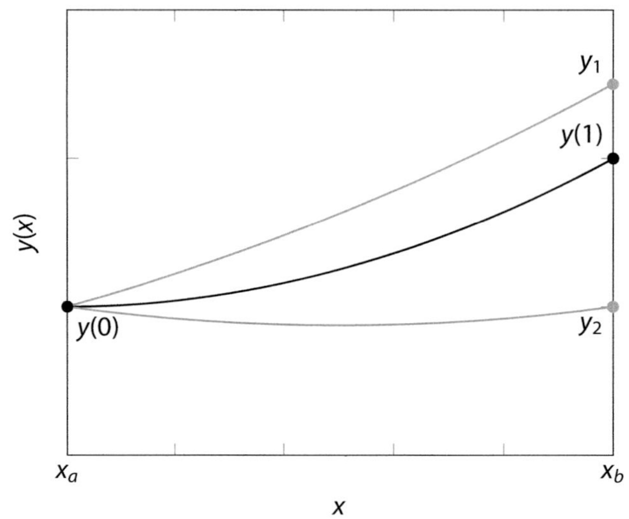

Figure 6.25 The shooting method for solving a boundary value problem. The idea is to repeatedly choose different values of the first derivative at x_a and solve the resulting initial value problems (gray curves). We then have a set of values for the initial slope and the solution at $x = x_b$ that we can solve to obtain the desired solution.

$y_1(x = b)$, will not be the same as our specified boundary condition $y(x = b)$. If we chose a different value of α, we would end up with a different value of y at x_b, $y_2(x = b)$. So, we can think of the difference between the numerical value $(y_i(x = b))$ and the actual value of $y(x = b)$ as being a function of α: i.e., $y_i(x = b) - y(x = b) = h(\alpha)$. We can then use a root-finding algorithm such as Newton's method to find the value of α that makes $y_i(x = b) - y(x = b) = 0$ and gives us the solution we need (Figure 6.25). So, the shooting method is basically an initial value problem wrapped inside a root-finding problem.

Shooting methods are a good starting point for numerically solving BVPs (Press et al., 1992), but other methods also exist. For example, we can convert the BVP into a matrix problem by replacing the derivatives with *finite difference* approximations. Let us see how this works with a simple example. Consider the simple BVP

$$\frac{d^2y}{dx^2} + y(x) = 2, \quad y(x = 0) = 1, \ y(x = 1) = 0, \tag{6.134}$$

where we have specified the value of the function $y(x)$ on two boundaries. With a finite difference approach, the first thing we do is subdivide the interval between the boundaries into a set of equally distant points (Figure 6.26). If the boundaries are located at x_a and x_b, then we can create N intervals of size

$$\Delta x = \frac{x_b - x_a}{N}.$$

Just as with the Euler method, we are going to use a Taylor series to approximate the derivatives at the grid points x_i, but we now have several choices. When we used finite differences for the Euler method we used only *forward differences*,

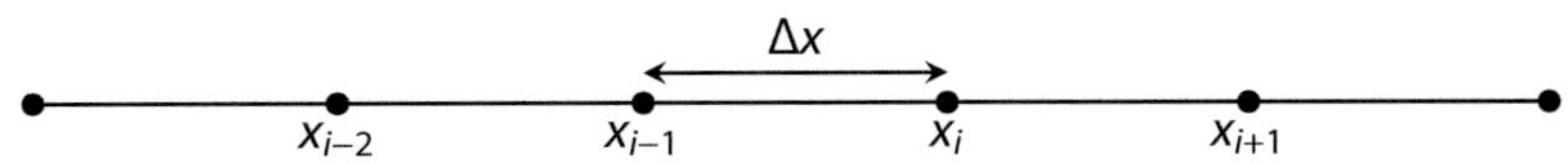

Figure 6.26 Setting up a grid along the *x* axis for using a finite difference scheme.

$$\left.\frac{dy}{dx}\right|_{x_i} \approx \frac{y(x_{i+1}) - y(x_i)}{\Delta x} = \frac{y_{i+1} - y_i}{\Delta x}, \tag{6.135}$$

because the initial values gave us the value of y_i at x_i, and we wanted to march forward to find the value of y_{i+1}. But for a BVP we do not have all the information to allow us to march along the x axis one step at a time. Instead, we could discretize a first derivative using *backward differences*,

$$\left.\frac{dy}{dx}\right|_{x_i} \approx \frac{y(x_i) - y(x_{i-1})}{\Delta x} = \frac{y_i - y_{i-1}}{\Delta x}, \tag{6.136}$$

or *central differences*,

$$\left.\frac{dy}{dx}\right|_{x_i} \approx \frac{y(x_{i+1}) - y(x_{i-1})}{2\Delta x} = \frac{y_{i+1} - y_{i-1}}{2\Delta x}. \tag{6.137}$$

But Equation (6.134) contains a second derivative. To form the finite difference approximation to a second derivative we make use of the two Taylor series

$$y(x + \Delta x) = y(x) + \Delta(x) \left.\frac{dy}{dx}\right|_x + \frac{(\Delta x)^2}{2}\left.\frac{d^2y}{dx^2}\right|_x + \frac{(\Delta x)^3}{6}\left.\frac{d^3y}{dx^3}\right|_x + \cdots$$

$$y(x - \Delta x) = y(x) - \Delta(x) \left.\frac{dy}{dx}\right|_x + \frac{(\Delta x)^2}{2}\left.\frac{d^2y}{dx^2}\right|_x - \frac{(\Delta x)^3}{6}\left.\frac{d^3y}{dx^3}\right|_x + \cdots .$$

Subtracting one of these equations from the other and rearranging gives

$$\left.\frac{d^2y}{dx^2}\right|_x \approx \frac{y(x+\Delta x) - 2y(x) + y(x - \Delta x)}{(\Delta x)^2}. \tag{6.138}$$

We can now write Equation (6.134) as

$$\frac{y_{i+1} - 2y_i + y_{i-1}}{(\Delta x)^2} + y_i = 2 \implies y_{i+1} - (2 - (\Delta x)^2)y_i + y_{i-1} = 2(\Delta x)^2.$$

This is a large number of simultaneous, linear equations for the unknown quantities y_i. We can write these equations explicitly as

$$\begin{aligned}
y_0 &= 1 & (i = 0)\\
y_2 - (2 - (\Delta x)^2)y_1 + y_0 &= 2(\Delta x)^2 & (i = 1)\\
y_3 - (2 - (\Delta x)^2)y_2 + y_1 &= 2(\Delta x)^2 & (i = 2)\\
y_4 - (2 - (\Delta x)^2)y_3 + y_2 &= 2(\Delta x)^2 & (i = 3)\\
\vdots &= \vdots & \\
y_N - (2 - (\Delta x)^2)y_{N-1} + y_{N-2} &= 2(\Delta x)^2 & (i = N - 1)\\
Y_N &= 0 & (i = N + 1),
\end{aligned}$$

where the first and last equations are the boundary conditions $y_0 = 1$ and $y_N = 0$. These equations can be written as a matrix equation

$$\begin{pmatrix} 1 & 0 & 0 & 0 & \cdots & 0 & 0 & 0 \\ 1 & -(2-(\Delta x)^2) & 1 & 0 & \cdots & 0 & 0 & 0 \\ 0 & 1 & -(2-(\Delta x)^2) & 1 & \cdots & 0 & 0 & 0 \\ \vdots & & \vdots & & \vdots & & \vdots & \\ 0 & 0 & 0 & 0 & \cdots & 1 & -(2-(\Delta x)^2) & 1 \\ 0 & 0 & 0 & 0 & \cdots & 0 & 0 & 1 \end{pmatrix} \begin{pmatrix} y_1 \\ y_2 \\ y_3 \\ \vdots \\ y_{N-1} \\ y_N \end{pmatrix} = \begin{pmatrix} 1 \\ (\Delta x)^2 \\ (\Delta x)^2 \\ \vdots \\ (\Delta x)^2 \\ 0 \end{pmatrix},$$

which we can solve numerically using any one of a number of techniques (Press et al., 1992).

In solving Equation (6.134) we had a BVP where the value of the function $y(x)$ was given on the boundaries. However, there are different types of boundary conditions: if the boundaries are at x_a and x_b, then

$$y(x_a) = \alpha,\ y(x_b) = \beta \quad \text{Dirichlet boundary conditions,} \tag{6.139}$$

$$y'(x_a) = \alpha,\ y'(x_b) = \beta \quad \text{Neumann boundary conditions,} \tag{6.140}$$

$$\left.\begin{aligned} A_1 y(x_a) + A_2 y'(x_a) = \alpha \\ B_1 y(x_b) + B_2 y'(x_b) = \beta \end{aligned}\right\} \quad \text{Mixed boundary conditions,} \tag{6.141}$$

$$y(x_a) = y(x_b),\ y'(x_a) = y'(x_b) \quad \text{Periodic boundary conditions.} \tag{6.142}$$

How would we incorporate Neumann boundary conditions into our method, for example? Let us consider the boundary condition at x_b as being $y'(x_b) = 0$. Using a centered finite difference approximation for the first derivative at that boundary tells us that $y_{N-1} = y_{N+1}$. Our last equation then becomes $2y_{N-1} - (2 - (\Delta x)^2)y_N = 0$ instead of $y_{N+1} - (2 - (\Delta x)^2)y_N + y_{N+1} = 0$.

Exercise 6.10.4 If we wanted to solve the BVP on the interval $x_a \le x \le x_b$

$$\frac{d^2y}{dx^2} + y(x) = 2$$

with periodic boundary conditions, what would the corresponding matrix of finite difference equations be?

6.10.4 Computer Algebra Systems

It would be remiss not to mention the use of *computer algebra* or *symbolic computing* systems such as Maple, Mathematica, and Sage.[29] In addition, computer languages such as Python and MATLAB have modules for symbolic computing.[30] These packages are very powerful and useful for removing the toil from our calculations, but they should not be

[29] Sage is an Open Source system and you can learn more about it at www.sagemath.org.
[30] The Python module, SymPy, and the MATLAB Symbolic Math Toolbox are very useful, but the Symbolic Math Toolbox has to be purchased as an additional module to the standard MATLAB package.

treated as a substitute for knowing how to make these calculations ourselves. One reason for this is that to use these packages effectively to solve ODEs requires knowledge of the effects of different substitutions and transformations of variables. When used with knowledge and understanding, computer algebra systems can be very useful tools.

6.11 Dynamical Systems and Chaos

We are frequently unable to solve an ODE or system of ODEs analytically. We have seen that in such cases we can use numerical methods to obtain an approximate solution to the equations. We have also seen that there is often a lot we can learn about the behavior of the system before we launch into a numerical solution.

As we have seen, one of the useful things to know about a dynamical system is its steady state solution. Let us look at a simple example. Newton's law of cooling says that the rate of change of temperature (T) of a body is proportional to the difference between the temperature of the body and that of its surroundings (T_0). In other words,

$$\frac{dT}{dt} = -k(T - T_0), \tag{6.143}$$

where k is a positive constant. If the body is warmer than its surroundings ($T > T_0$), then $T - T_0 > 0$ and the temperature of the body cools. If the body is cooler than its surroundings, then $T - T_0 < 0$ and the right-hand side of the equation is positive, indicating that the body warms up. We can easily find the solution to this equation,

$$T(t) = T_0 + (T(0) - T_0)e^{-kt}, \tag{6.144}$$

and Figure 6.27 shows solutions for two cases, when $T(0) > T_0$ and $T(0) < T_0$. In both cases we see that the temperature of the body approaches T_0 asymptotically, i.e., it will only

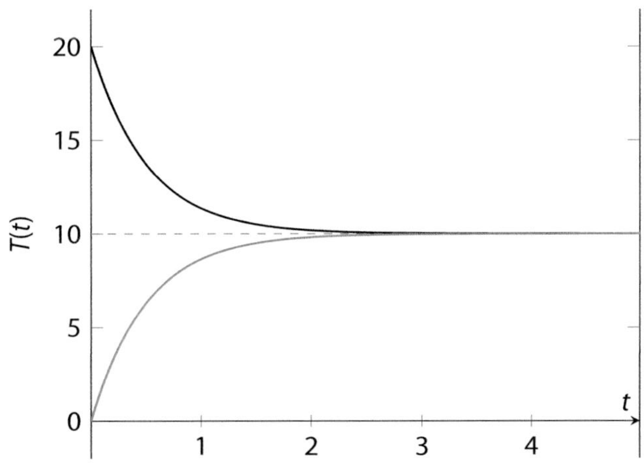

Figure 6.27 Two solutions of Equation (6.143) for $T_0 = 10$: $T(t = 0) > T_0$ (black curve) and $T(t = 0) < T_0$ (gray curve).

actually equal T_0 when $t = \infty$ (see Equation (6.144)). What is more, once the temperature has reached T_0, it stays there forever because when $T = T_0$, $dT/dt = 0$, so $T = T_0$ is the steady state solution.

To find the steady state solution, we set the derivative in Equation (6.143) to zero and solve the resulting equation

$$-k(T - T_0) = 0, \qquad \text{i.e., at steady state,} \qquad T = T_0.$$

So, this system has a single steady state, $T = T_0$; this is something we have derived mathematically, but in this case we were also able to determine the steady state from considering the physics (this might not always be the case). For such a simple equation, we can determine the stability of this steady state by looking at the sign of dT/dt either side of the steady state temperature. For $T > T_0$, $dT/dt < 0$, so the temperature will decrease to the value T_0. For $T < T_0$, $dT/dt > 0$, so the temperature will increase to the value T_0 (Figure 6.27). In this case it is easy to see that the direction of change for T is always toward the steady state, making the steady state a *stable* one; any perturbation away from the steady state in either direction will cause the system to move back toward the steady state.

Example 6.24 An equation describing a simple model for the global temperature of a planet is (Kaper and Engler, 2013):

$$c\frac{dT}{dt} = \frac{1}{4}(1 - \alpha)S_0 - \sigma\epsilon T^4, \tag{6.145}$$

where c is a constant representing the average *heat capacity* of the planet, α is the albedo, $S_0 = 1386$ W m^{-2} is the rate of solar radiation received per square meter at the top of the atmosphere (called the *solar constant*), σ is the Stefan–Boltzmann constant,[31] and ϵ is a constant parameter called the *emissivity*. The terms on the right-hand side of Equation (6.145) represents the rate energy input from incident solar radiation and the rate of energy leaving the planet by thermal radiation. Let us calculate the steady states and determine their stability (a) when α is constant, and (b) when α is given by the following equation of temperature,

$$\alpha(T) = 0.5 - 0.2\tanh\left(\frac{T - 265}{R}\right), \tag{6.146}$$

where R is a constant. In case (a), where α is constant, the steady state is

$$T^* = \left(\frac{(1 - \alpha)S_0}{4\sigma\epsilon}\right)^{1/4}.$$

To examine the stability of this solution we look at the behavior of a small deviation from the steady state, $\tau = T - T^*$. Substituting this into Equation (6.145) and linearizing by keeping only terms that are linear in τ, we find that

$$\frac{d\tau}{dt} \approx -4\sigma\epsilon\tau(T^*)^3.$$

[31] This is named after Austrian mathematicians Josef Stefan (1835–1893) and Ludwig Eduard Boltzmann (1844–1906).

If $\tau > 0$ (i.e., the planet heats up), then $d\tau/dt < 0$, so the perturbation to the temperature decreases and the temperature of the planet moves back toward the steady state temperature. If $\tau < 0$ (i.e., the planet cools), then $d\tau/dt > 0$ and the temperature increases back toward T^*. Therefore the steady state is a stable one.

In case (b), where α is a function of temperature given by Equation (6.146), the steady state temperature is given by the equation

$$\frac{4\sigma\epsilon T^4}{S_0} = 0.2\tanh\left(\frac{T-265}{R}\right) + 0.5. \tag{6.147}$$

This is not an equation that we can solve analytically, but it can be solved numerically or graphically by plotting the left-hand side and the right-hand side and seeing where the two curves intersect. Let us write

$$F_1(T) = 4\sigma\epsilon T^4, \quad F_2(T) = S_0\left[0.2\tanh\left(\frac{T-265}{R}\right) + 0.5\right],$$

so that the steady state temperatures are those points where $F_1(T) = F_2(T)$ (Figure 6.28). For the values of the constants that we have chosen, we see that there are three equilibrium points, A, B, and C. We can look at the relative magnitudes of F_1 and F_2 to determine the stability of each point. If the planet is in the steady state A and we decrease the temperature slightly, then $F_2 > F_1$ and $dT/dt > 0$, so the temperature of the planet increases back toward the steady state. If we increase the temperature slightly, then $F_1 > F_2$, $dT/dt < 0$ and the temperature decreases back to the steady state. So, A is a stable steady state solution. In a similar manner we can see that B is an unstable steady state and C is another stable steady state.

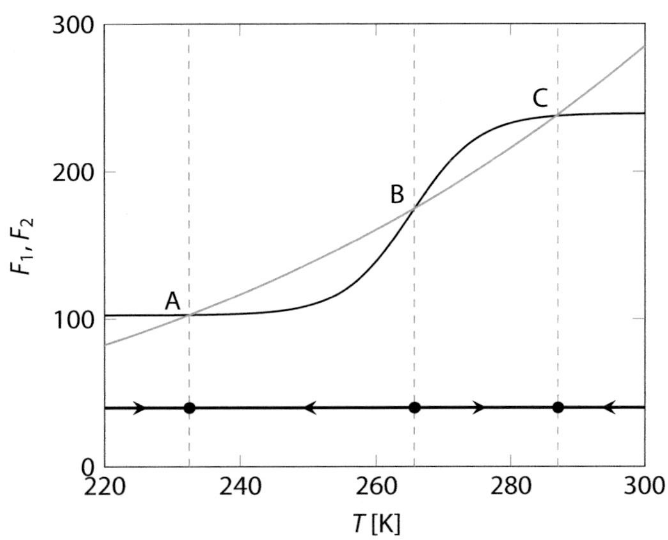

Figure 6.28 Finding the solution to Equation (6.147). The thin gray curve shows the function $F_1(T)$, and the thin black curve shows $F_2(T)$. The thick black line indicates the stability of the three steady state solutions A, B, and C.

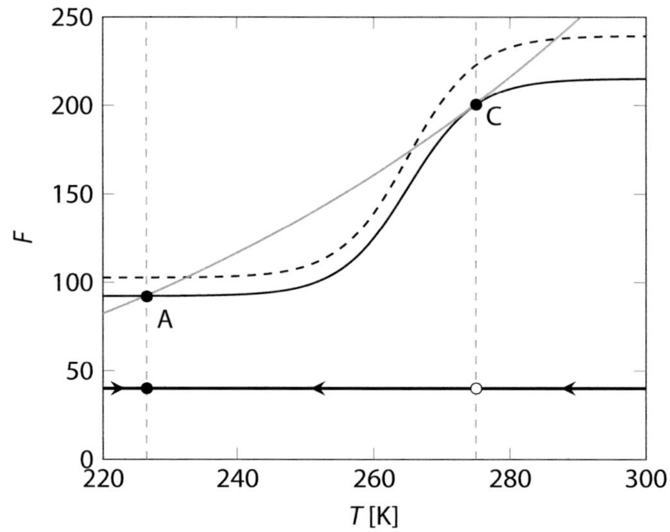

Figure 6.29 The effect of decreasing S_0 on the number and nature of the steady state solutions for a simple climate model. The dashed black curve uses $S_0 = 1368\ \mathrm{W\,m^{-2}}$, where the solid black curve uses $S_0 = 1230\ \mathrm{W\,m^{-2}}$. The steady states *B* and *C* from Figure 6.28 have merged in this figure to produce the single steady state *C*.

Example 6.24 has a very interesting feature: the number and nature of the steady states depends on the values of the constants in Equations 6.145 and 6.146. For example, let us look at what happens if we change the amount of incoming solar radiation by changing S_0.[32] The solar constant is a proportionality constant for the amount of energy the Earth receives, so decreasing S_0 simply shifts curves vertically, which not only changes the number of steady state solutions, it also changes their nature (Figure 6.29). As we lower the value of S_0, the steady state points B and C in Figure 6.28 get closer together until eventually they merge (Figure 6.29). The stability of the merged steady state takes on characteristics of both previous points. For example, if we increase the temperature slightly, then $dT/dt < 0$ and the temperature moves back to the steady state. But if we decrease T, then $dT/dt < 0$ as well, so the temperature continues to decrease until it reaches the (stable) steady state A.

This phenomenon of merging and disappearing of equilibrium points is called a *bifurcation*. Bifurcations are important features in understanding the nature of nonlinear systems, and there are several different types of bifurcation that each have different properties (see e.g., Guckenheimer and Holmes, 1983).

[32] Models of the evolution of stars indicate that the solar radiation received by the Earth in the distant past was approximately a quarter of what it is today. This would have resulted in an Earth that was completely frozen, but there is strong evidence that liquid water was present on the planet at that time. This problem is called the *faint young Sun paradox* (Feulner, 2012).

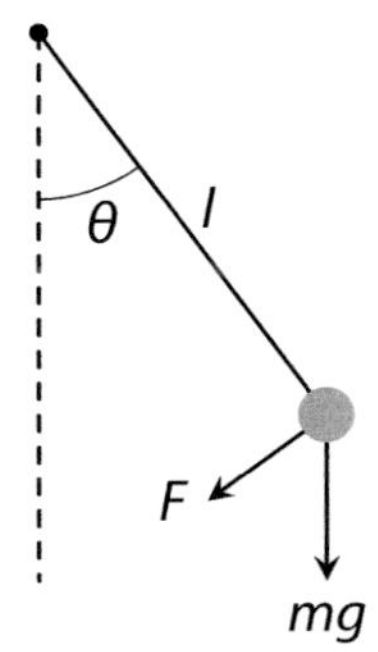

Figure 6.30 A pendulum of mass m attached to the end of a string of length l and oscillating about the vertical. Gravity acts vertically downward on the mass with a force mg and the force pulling the pendulum back to the vertical is $F = (mg/l)\sin(\theta)$.

6.11.1 Chaos

As systems of differential equations become more complicated, so their solutions can behave in more complicated, and sometimes unexpected and alarming, ways. This is important because we use differential equations to describe how systems behave in the real world and we would like to know if the behavior exhibited by the solution reflects that of the real-world system it represents.

Let us start with a simple system that is easy to understand, the pendulum. We will look at a pendulum with a mass m at the end of a string of length l (Figure 6.30). We will imagine a situation where the pendulum is initially stationary and hanging vertically down. We then pull it to one side through an angle θ and let it go so that the pendulum swings back and forth. The equation that governs the change in θ over time (t) is

$$\frac{d^2\theta}{dt^2} = -\frac{g}{l}\sin(\theta), \tag{6.148}$$

where g is the gravitational acceleration. This is a nonlinear equation (because of the $\sin(\theta)$ term), but if we consider only small values of the angle, then $\sin(\theta) \approx \theta$ and the equation is a linear one, the familiar equation for simple harmonic motion (Equation (6.69)) with $\omega^2 = g/l$. We know that the solution to this equation is a sine or a cosine function. We can write Equation (6.148) as two coupled, first order ODEs

$$\frac{d\theta}{dt} = \psi, \quad \frac{d\psi}{dt} = -\frac{g}{l}\sin(\theta),$$

and we can look at the solution and the phase plane, the plot of $\dot{\theta} = \psi$ against θ (Figure 6.31). The phase portrait consists of concentric ellipses showing the oscillating motion of the pendulum.

What happens if we do not restrict ourselves to small angles, but instead integrate the full nonlinear equation (Equation (6.148))? Because we are no longer restricting θ to be small, we have the possibility that the pendulum can swing all the way to the top and make a complete revolution. The resulting phase plane is shown in Figure 6.32. For small

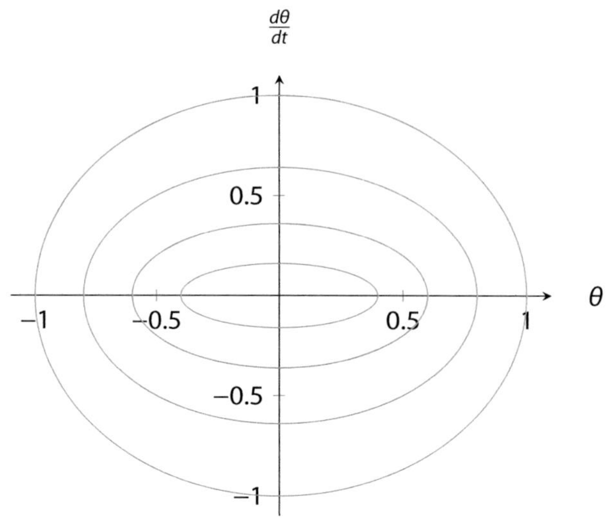

Figure 6.31 The phase plane for the linear oscillator $\ddot{\theta} = -(g/l)\theta$ showing the closed trajectories representing the pendulum oscillating around the vertical.

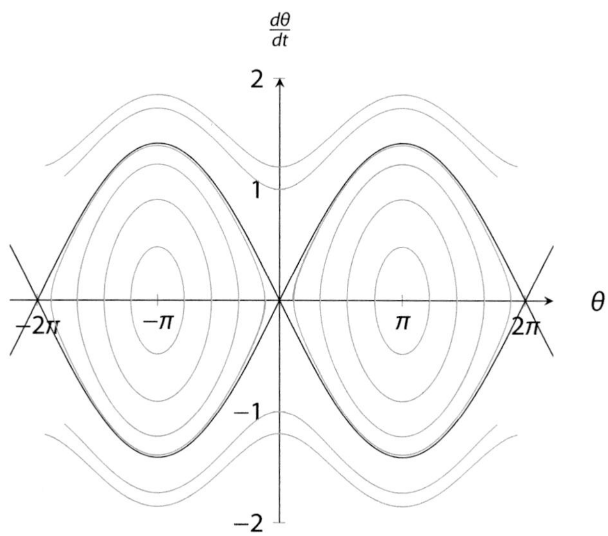

Figure 6.32 The phase plane for the nonlinear pendulum Equation (6.148). The black curve is the *separatrix* that separates small oscillations about the vertical from motion where the pendulum moves in a complete circle.

oscillations, we still have the closed periodic orbits as before. However, there are new trajectories in the phase plane that move smoothly between all angles. These represent the cases when the pendulum starts off with enough energy that it makes complete revolutions; remember that there is no friction in this example, so the pendulum will keep going for ever. There are also some special trajectories that separate these two kinds of motion.

These represent motion where the pendulum just reaches the top of the circle at zero velocity and then tips over to continue.

We have seen in Section 6.4.2 that adding either a forcing term, a damping term, or both can affect the nature of the solutions of the linear oscillator, but what happens for a nonlinear oscillator?

Let us look at a different system, the *van der Pol oscillator*.[33] This is described by the equation

$$\ddot{x} - \epsilon(1 - x^2)\dot{x} + x = 0, \tag{6.149}$$

where $\epsilon > 0$ is a constant parameter that defines the amount of damping. The equation was first developed to describe the behavior of certain types of electrical circuits, but has since found use describing many natural phenomena, including the motion of geological plates in earthquakes (Cartwright et al., 1999). As usual, to look at the phase plane we define a new variable to create a two-dimensional system of autonomous equations,

$$\dot{x} = y, \quad \dot{y} = \epsilon(1 - x^2)y - x.$$

Exercise 6.11.1 Show that the origin is the only steady state solution of the van der Pol oscillator and that it is an unstable focus for $\epsilon < 2$ and an unstable node for $\epsilon > 2$.

The linearized form of Equation (6.149) near the origin is

$$\ddot{x} - \epsilon\dot{x} + x = 0,$$

which describes a damped harmonic oscillator but with *negative* damping for $\epsilon > 0$, so we can guess that any perturbation from the steady state solution is likely to be unstable.

Although the van der Pol oscillator has only one steady state solution, it turns out that this equation has another, rather special type of solution that occurs when we are far enough from the steady state that nonlinear terms in the equations become important. To see this, let us look at the full damping term $(-\epsilon(1-x^2))$. If $|x| < 1$, then this term is always negative and the amplitude of the oscillation will continue to grow. In this case, the trajectory will move away from the critical point as a growing spiral—a spiral because the amplitude of the oscillation is always increasing. Once these amplitudes are such that $|x| > 1$, then the damping term switches sign and starts to act like a normal damping term, in which case we might expect to see some periodic behavior arise. We cannot solve Equation (6.149) analytically, so we have to use a numerical solution to show the existence of the *limit cycle* (Figure 6.33). A stable limit cycle is a closed trajectory in the phase plane such that all trajectories spiral to asymptotically meet it. We can see this in Figure 6.33 where trajectories from inside and outside the limit cycle asymptotically approach it. Proving limit cycles exists for a given dynamical system is not easy, and there is no standard way to do it, but they do occur in many types of oscillatory phenomena.

Another equation that has unusual behavior is the *Duffing equation*.[34] This is a nonlinear, damped, forced oscillator

$$\ddot{x} + r\dot{x} - x + x^3 = A\cos(\Omega t), \tag{6.150}$$

[33] Named after Balthasar van der Pol (1889–1959), a Dutch physicist and electrical engineer.
[34] Named after the German engineer Georg Duffing (1861–1944).

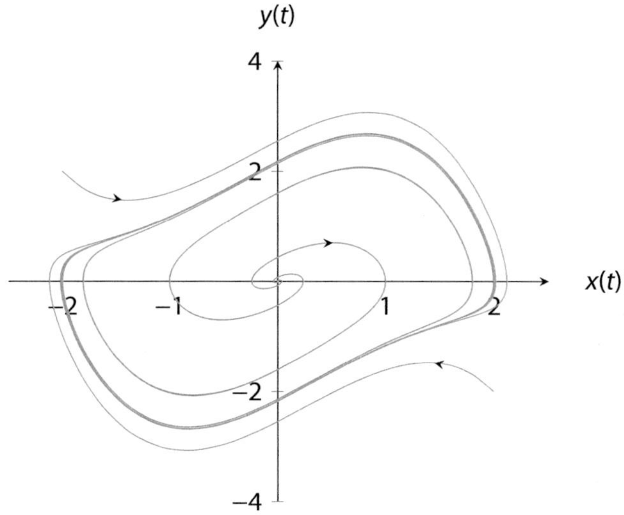

Figure 6.33 The phase plane for the van der Pol oscillator (Equation (6.149)) for $\epsilon = 1.0$ showing the existence of a limit cycle.

where r is a damping term, A describes the amplitude of the forcing, and Ω is the forcing frequency.

Exercise 6.11.2 Show that for the Duffing equation with no driving force (i.e., $A = 0$), the critical points are (a) the origin, which is a saddle, and (b) the points $(\pm 1, 0)$, which are both stable points with each point being a stable spiral for $r < \sqrt{8}$ and a stable node for $r > \sqrt{8}$.

Something very curious and interesting happens as we increase the amplitude of the forcing A (Figure 6.34). For $A = 0.7$ there appears to be an oscillation, but it takes the system two cycles to return to its starting point (Figure 6.34a). If we increase A slightly to $A = 0.75$, this two-cycle oscillation appears to have split into a four-cycle oscillation (Figure 6.34b). Increasing A again to 0.8 produces a much more complicated figure where there are numerous trajectories with complicated behaviors described as *chaos* (Figure 6.34c). It is important to realize that there are no random or stochastic variables in the equation; it is purely deterministic and yet can give rise to extremely complicated behaviors. This process where changing a parameter value in the equation leads to a repeated doubling of the period of an oscillation is called the *period doubling route to chaos* (Strogatz, 2001). If we increase A further still to $A = 0.9$, then the chaotic behavior seems to disappear (Figure 6.34d).

We have seen that with both bifurcations and chaos, changing the value of a parameter in a system of differential equations can lead to remarkably different types of behavior. This is important because many systems we want to study are described by equations we cannot solve analytically, and we have to rely on numerical methods to study them. However, we have to take care because small changes in a parameter can lead to very different results. In fact, chaotic systems are characterized by a phenomenon called *sensitive dependence on initial conditions*. This means that if the system is chaotic, any small change in the initial

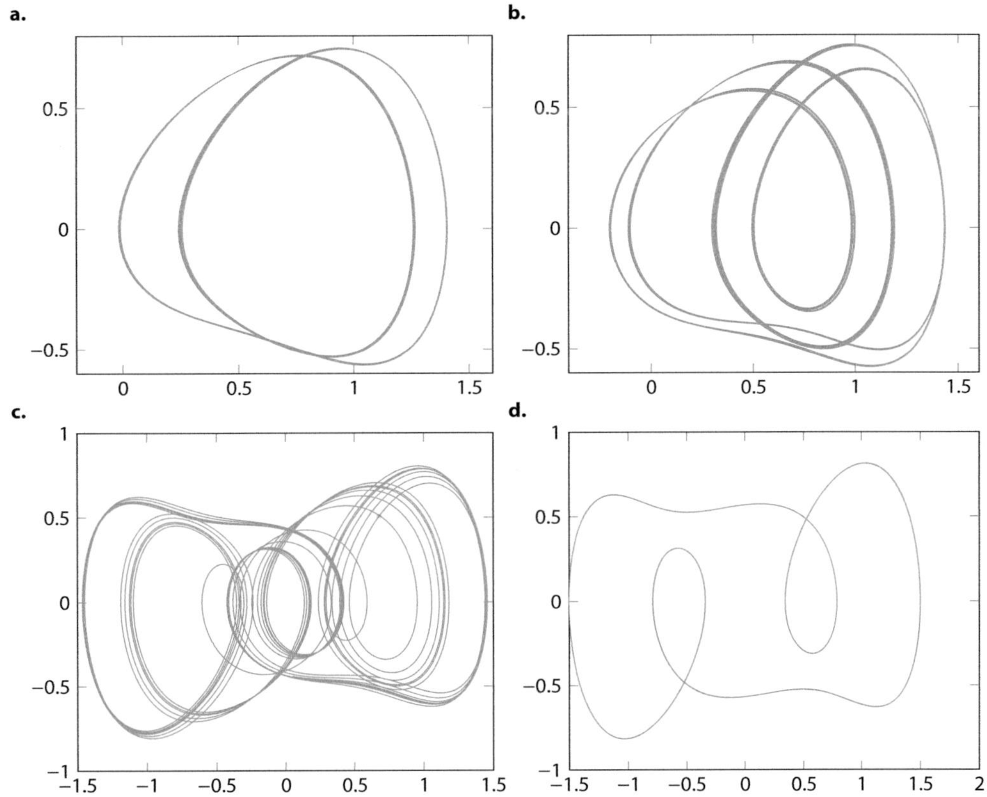

Figure 6.34 The phase planes for the Duffing equation (Equation (6.150)) for $r = 1.0$, $\Omega = 1.0$, and $A = 0.7$ (a.), $A = 0.75$ (b.), $A = 0.8$ (c.), $A = 0.9$ (d.).

conditions used to numerically integrate the equations can lead to dramatically different solutions. There are some other canonical systems of ODEs that exhibit chaotic behavior; for example, the Lorenz system,[35]

$$\dot{x} = \sigma(y - x), \quad \dot{y} = rx - y - xz, \quad \dot{z} = xy - bz, \tag{6.151}$$

which was derived from the equations for the convection of a fluid in a shallow layer of fluid (Lorenz, 1963). A simpler system is the Rössler system of equations,[36]

$$\dot{x} = -(y + z), \quad \dot{y} = x + \frac{1}{5}y, \quad \dot{z} = \frac{1}{5} + z(x - \mu). \tag{6.152}$$

These are studied more in the problems for this chapter. The existence of deterministic chaos in real-world systems is not always easy to demonstrate. However, the dynamics of various tectonic plate systems are possibly chaotic (Huang and Turcotte, 1990). Similarly, models of the reversals in the Earth's magnetic field exhibit chaotic behavior that mimics the observed timing of reversals inferred from volcanic rocks (Cortini and Barton, 1994;

[35] Named after the American mathematician and meteorologist Edward Lorenz (1917–2008).
[36] Discovered by the German biochemist Otto Rössler.

Chillingworth and Holmes, 1980; Ito, 1980). Chaotic behavior is also seen in equations describing the dynamics of populations (May et al., 1987; Solé and Bascompte, 2006).

6.12 Boundary Value Problems, Sturm–Liouville Problems, and Green's Functions

Most of the differential equations we have looked at so far have been initial value problems. However, boundary value problems are also common. For example, we may have a differential equation that describes the flow of heat within the Earth. If we have information about the temperature at the core of the planet and at the surface, we can solve the equation as a BVP and calculate the distribution of heat within the planet. BVPs also arise in studying hydrodynamic flows in the atmosphere and oceans, as well as in problems concerning groundwater flow. The remaining sections of this chapter are a little more technical than the preceding ones, but they set the stage for dealing with partial differential equations later.

To solve an IVP we need to have information about the solution and its derivatives at the initial point. To solve a BVP we need conditions on the solution or its derivatives on the boundary of the domain we are interested in. This means we are missing information that would allow us to march forward from one boundary to the other; for example, as we saw in Section 6.10.3, without knowing the value of the function and its derivative at the same point, we cannot take incremental steps.

It is important to realize that, although an IVP and a BVP might be described by the same differential equation, the properties of the solutions to the two problems can be very different indeed. For example, we have seen that under a broad set of conditions of continuity and differentiability, an IVP has a unique solution. However, this is not necessarily the case for a BVP. Let us consider the simple ODE

$$\frac{d^2y(x)}{dx^2} = -y(x).$$

We know that this equation represents an oscillation of y as x changes, and has a general solution $y(x) = a_1 \cos(x) + a_2 \sin(x)$, where a_1 and a_2 are real constants. For an initial value problem we specify $y(x = 0)$ and $y'(x = 0)$, allowing us to calculate values for a_1 and a_2 and obtain a specific solution. For example, if $y(x = 0) = 0$ and $y'(x = 0) = 1$, then $a_1 = 0$ and $a_2 = 1$, giving the solution $y(x) = \sin(x)$. Now, instead of initial conditions, let us give boundary conditions at $x = 0$ and $x = \pi$. Substituting the first boundary condition into the solution tells us again that $a_1 = 0$, leaving us with a solution $y(x) = a_2 \sin(x)$. We need the other boundary condition to find the value of a_2. Let us consider a couple of possibilities:

- If $y(\pi) = b \neq 0$, then we can see that there is no solution to the ODE because $\sin(\pi) = 0$, giving us $y(\pi) = a_2 \sin(\pi) = 0 = b \neq 0$, which contradicts the assumption that $b \neq 0$.
- If $y(\pi) = 0$, then the constant a_2 is not determined and we have an infinite number of possible solutions, one for each possible value of a_2.

So, we have a unique solution for the IVP, but not for the BVP.

Let us generalize our ODE a little and look at the equation

$$\frac{d^2y(x)}{dx^2} = -\lambda y(x), \quad \lambda > 0, \qquad y(x=0)=0,\ y(x=\ell)=0. \tag{6.153}$$

If $\lambda > 0$, then the general solution is $y(x) = A\cos(\sqrt{\lambda}x) + B\sin(\sqrt{\lambda}x)$, where A and B are constants, and if $\lambda = 0$, the general solution is $y(x) = C + Dx$. Now let us impose the boundary value conditions $y(0) = 0$ and $y(\ell) = 0$, where $\ell > 0$. If we substitute these into the solution for $\lambda = 0$, we find that $C = D = 0$ and the only solution is $y = 0$, which is not particularly useful. If $\lambda > 0$, then substituting the boundary condition at $x = 0$ into the general solution tells us that $A = 0$. Substituting the boundary condition at $x = l$ tells us that $B\sin(\sqrt{\lambda}l) = 0$, so either $B = 0$ or $\sin(\sqrt{\lambda}l) = 0$, with B being undetermined. The solution to $\sin(\sqrt{\lambda}l) = 0$ is $\sqrt{\lambda}l = n\pi$, where $n = 0, \pm1, \pm2, \ldots$ We can discard the case when $n = 0$ because this implies that $\lambda = 0$, which is just the case we had before. So, the solution to the BVP is

$$y(x) = B\sin\left(\frac{n\pi x}{L}\right), \qquad n = \pm1,\ \pm2,\ \pm3\ldots,$$

where B is an arbitrary constant.

The result of all this is that, for some values of λ Equation (6.153) has no solutions, and for other values of λ the equation has infinitely many solutions. Equation (6.153) is an *eigenvalue problem* and is analogous to the eigenvalue equations ($\mathbf{Av} = \lambda\mathbf{v}$) we met in Chapter 4, but instead of having a matrix $\mathbf{A}$ acting on a vector $\mathbf{v}$, we have a derivative acting on a function. Eigenvalue problems appear very often when we are dealing with BVPs and there is a special class of such problems, *Sturm–Liouville problems*, which are very important.

For the remainder of this chapter we will focus our attention on BVPs associated with second order linear inhomogeneous ODEs for $y(x)$ that have the form

$$p(x)y'' + q(x)y' + r(x)y = f(x). \tag{6.154}$$

Recall that in general we cannot solve such an equation for arbitrary functions $p(x)$, $q(x)$, $r(x)$, and $f(x)$. However, to start with, let us examine the conditions that these functions would have to satisfy in order to make Equation (6.154) an exact equation. The reason for doing this is that if we can make Equation (6.154) an exact equation then we can immediately perform one integration, giving a first order equation that we might have a better chance of solving. If we add and subtract the terms $p''y$, $p'y'$, and $q'y$ to the Equation (6.154) (i.e., we are adding zero to both sides), then we can rewrite it as

$$(py' - p'y + qy)' + (p'' - q' + r)y = f(x). \tag{6.155}$$

If $p(x)$, $q(x)$, and $r(x)$ are such that $(p'' - q' + r) = 0$, then Equation (6.155) is an exact equation and can be formally integrated to obtain a first order differential equation,

$$py' - p'y + qy = \int f(x)\,dx + C,$$

which we may be able to solve.

Exercise 6.12.1 Derive Equation (6.155).

Exercise 6.12.2 Why did we choose to add and subtract the terms $p''y$, $p'y'$, and $q'y$ to Equation (6.154) and not terms like $q''y$, $r'y'$ etc?

Exercise 6.12.3 Show that the differential equation

$$\frac{2}{x}y'' - \frac{3}{x^2}y' + \frac{2}{x^3}y = e^x$$

satisfies $(p'' - q' + r) = 0$ and has a solution

$$y(x) = x^{1/2}e^x(x-1) + \frac{1}{2}C_1x^{5/2} + C_2x^{1/2},$$

where C_1 and C_2 are constants.

This is helpful, but there are many differential equations that do not satisfy $(p'' - q' + r) = 0$. We are going to once more do one of those apparently crazy things that appear at first to make our lives harder. We are going to multiply Equation (6.154) by a new, unknown function $v(x)$:

$$v(x)p(x)y'' + v(x)q(x)y' + v(x)r(x)y = v(x)f(x), \tag{6.156}$$

and then add and subtract $(pv)''y$, $(pv)'y$, and $(qv)'y$. Doing this, and collecting up terms, our differential equation becomes

$$(vpy' - (vp)'y + vqy)' + ((pv)'' - (qv)' + rv)y = v(x)f(x),$$

so, arguing as before, if we can find a function $v(x)$ that satisfies

$$(pv)'' - (qv)' + rv = pv'' + (2p' - q)v' + (p'' - q' + r)v = 0, \tag{6.157}$$

then we can multiply Equation (6.154) by the function $v(x)$ and obtain the exact equation

$$(pv)y' - (pv)'y + (qv)y = \int f(x)v(x)\,dx + C.$$

If we can evaluate the integral on the right-hand side, then we have a first order differential equation for y that we might be able to solve. The function $v(x)$ is called an *integrating factor* (cf. Equation (6.11)), and Equation (6.157) is called the *adjoint equation* for Equation (6.154).

We still have to find the function $v(x)$ that satisfies the adjoint equation. However, some forms of ODE do not even require us to do that. Let us consider the differential equation Equation (6.154) again, but write it in standard form,

$$y'' + \frac{q(x)}{p(x)}y' + \frac{r(x)}{p(x)}y = \frac{f(x)}{p(x)}.$$

We can multiply the equation by a new unknown function $\eta(x)$, so that

$$\eta(x)y'' + \eta(x)\frac{q(x)}{p(x)}y' + \eta(x)\frac{r(x)}{p(x)}y = \eta(x)\frac{f(x)}{p(x)}.$$

Thinking back to our derivation of integrating factors, we can see that if

$$\eta' = \eta\frac{q(x)}{p(x)},$$

then the first two terms of the differential equation can be written as $(\eta(x)y')'$. For this to be the case we need

$$\eta = \exp\left(\int \frac{q(x)}{p(x)}\, dx\right).$$

Why is this useful? Because it means that we can, in principle, write Equation (6.154) in the form

$$\frac{d}{dx}\left(P(x)\frac{dy}{dx}\right) + Q(x)y = P(x)\frac{d^2y}{dx^2} + \frac{dP}{dx}\frac{dy}{dx} + Q(x)y = F(x), \tag{6.158}$$

where

$$P(x) = \eta x, \qquad Q(x) = \frac{\eta(x)r(x)}{p(x)}, \qquad F(x) = \frac{\eta(x)f(x)}{p(x)}.$$

Example 6.25 Let us write the differential equation

$$x^2y'' + xy' + 6y = 0$$

in the form of Equation (6.158). First, we check to see that the equation is not already in the required form. Comparing with Equation (6.158) we have

$$P(x) = x^2, \quad P'(x) = 2x \neq x,$$

so the equation is not in the required form. Putting the equation into standard form

$$y'' + \frac{1}{x}y' + \frac{6}{x^2}y = 0,$$

we need to calculate

$$\eta = \exp\left(\int \frac{dx}{x}\right) = x,$$

so the differential equation becomes

$$\frac{1}{x}x^2y'' + \frac{1}{x}xy' + \frac{6}{x}y = (xy')' + \frac{6}{x}y = 0.$$

Why is it important, or even interesting, to write a differential equation in the form of Equation (6.158)? To answer this, let us find the adjoint of

$$\frac{d}{dx}\left(p(x)\frac{dy}{dx}\right) + q(x)y(x) = p(x)y'' + p'(x)y' + q(x)y = 0. \tag{6.159}$$

The adjoint of Equation (6.159) is

$$p(x)v'' + (2p'(x) - p')v' + (p'' - p'' + q(x))y = 0$$
$$p(x)v'' + p'(x)v' + q(x)v = 0;$$

in other words, the same equation we started with. A differential equation that can be written in the form of Equation (6.159) such that the original equation and its adjoint are the same is called *self-adjoint*.

Self-adjoint equations may appear to be too restrictive, but the derivation we just did shows that we can transform any second order linear ODE into a self-adjoint equation by finding the function $\eta(x)$ that transforms Equation (6.154) into Equation (6.158).

Many textbooks on differential equations make use of what is called an *operator notation*. You can think of this as a shorthand notation, but it can be written in different ways; for example,

$$L[y] = \mathcal{L}y = \left(p(x)\frac{d^2}{dx^2} + q(x)\frac{d}{dx} + r(x) \right) y(x) = p(x)y'' + q(x)y' + r(x)y,$$

and we say that the *operator* L (or $\mathcal{L}$; you will see both notations used) operates on the function $y(x)$; here we will use the notation $L[y]$ because it is explicit that the operator L operates on the function y. We will often write the operator itself as $[\cdot]$, so that in this example,

$$L[\cdot] = p(x)\frac{d^2}{dx^2} + q(x)\frac{d}{dx} + r(x).$$

If $L[\cdot]$ is a self-adjoint operator and we have two functions $u(x)$ and $v(x)$, what is $uL[v] - vL[u]$?

$$\begin{aligned} uL[v] - vL[u] &= u(x)\left(\frac{d}{dx}\left(p(x)\frac{dv}{dx} \right) + q(x)v \right) - v(x)\left(\frac{d}{dx}\left(p(x)\frac{du}{dx} \right) + q(x)u \right) \\ &= u(x)\frac{d}{dx}\left(p(x)\frac{dv}{dx} \right) - v(x)\frac{d}{dx}\left(p(x)\frac{du}{dx} \right) \\ &= f derx \left(p(x)u(x)\frac{dv}{dx} - p(x)v(x)\frac{du}{dx} \right) \\ &= [puv' - pvu']', \end{aligned} \tag{6.160}$$

which is called *Lagrange's identity*.[37] If we integrate Equation (6.160), we arrive at *Green's identity*,[38]

$$\int_a^b (uL[v] - vL[u])\, dx = \int_a^b \frac{d}{dx}(puv' - pvu')\, dx = (puv' - pvu')\big|_a^b. \tag{6.161}$$

Having done some groundwork, we are now in a position to think about the following class of BVPs called *Sturm–Liouville* problems:[39]

$$\left. \begin{aligned} \frac{d}{dx}\left(p(x)\frac{dy}{dx} \right) + q(x)y + \lambda r(x)y(x) = L[y] + \lambda r(x)y(x) = 0 \\ a_1 y(a) + a_2 y'(a) = 0 \\ b_1 y(b) + b_2 y'(b) = 0 \end{aligned} \right\}, \tag{6.162}$$

where $a \leq x \leq b$ and the boundary conditions are called *homogeneous boundary conditions*. If we look at the operator form of the differential equation, we see a similarity to

[37] Named after Italian mathematician Joseph Lagrange (1736–1813), who also made significant advances in physics and astronomy.

[38] Named after George Green (1793–1841), whose famous theorem we met in Chapter 2.

[39] These are named after the French mathematicians Jacques Sturm (1803–1855) and Joseph Liouville (1809–1882).

a matrix eigenvalue equation $\mathbf{A}\mathbf{v} = \lambda\mathbf{v}$, and this leads to calling the parameter λ in Equation (6.162) an eigenvalue and the functions that satisfy Equation (6.162) *eigenfunctions*. For example, Equation (6.153) has the form of a Sturm–Liouville problem, with $r(x) = 1$, $q(x) = 0$, and $p(x) = 1$, and we saw from that example that for some values of λ the equation admits an infinite number of solutions. We can ask if there is any relationship between these solutions. To answer this, let us assume that we have two solutions $y_m(x)$ and $y_n(x)$ that correspond to two different eigenvalues λ_m and λ_n with $\lambda_m \neq \lambda_n$, so that

$$\frac{d}{dx}\left(p(x)\frac{dy_m(x)}{dx}\right) + q(x)y_m(x) + \lambda_m r(x)y_m(x) = 0,$$
$$\frac{d}{dx}\left(p(x)\frac{dy_n(x)}{dx}\right) + q(x)y_n(x) + \lambda_n r(x)y_n(x) = 0.$$

If we multiply the first equation by $y_n(x)$ and the second by $y_m(x)$, subtract the equations and integrate the result over the interval $a \leq x \leq b$, we get

$$\int_a^b \left(y_n\frac{d}{dx}(py'_m) - y_m\frac{d}{dx}(py'_n)\right)dx + (\lambda_m - \lambda_n)\int_a^b r(x)y_m(x)y_n(x)\,dx = 0.$$

Let us look at the term

$$\begin{aligned}\int_a^b y_n\frac{d}{dx}(py'_m)\,dx &= \int_a^b \frac{d}{dx}(y_n p(x)y'_m)\,dx - \int_a^b p(x)y'_m\frac{dy_n}{dx}\,dx \\ &= (y_n p(x)y'_m)\Big|_a^b - \int_a^b p(x)y'_m\frac{dy_n}{dx}\,dx,\end{aligned}$$

where we have used the method of integration by parts. Similarly,

$$\begin{aligned}\int_a^b y_m\frac{d}{dx}(py'_n)\,dx &= \int_a^b \frac{d}{dx}(y_m p(x)y'_n)\,dx - \int_a^b p(x)y'_n\frac{dy_m}{dx}\,dx \\ &= (y_m p(x)y'_n)\Big|_a^b - \int_a^b p(x)y'_n\frac{dy_m}{dx}\,dx.\end{aligned}$$

Subtracting these terms gives us

$$\begin{aligned}\int_a^b \left(y_n\frac{d}{dx}(py'_m) - y_m\frac{d}{dx}(py'_n)\right)dx &= (y_n p(x)y'_m)\Big|_a^b - (y_m p(x)y'_n)\Big|_a^b \\ &\quad - \int_a^b p(x)y'_n y'_m\,dx + \int_a^b p(x)y'_n y'_m\,dx\end{aligned}$$

The two integrals are identical, so they cancel each other, leaving only the two terms evaluated at the boundaries. We can use the homogeneous boundary conditions (Equation (6.162)) to write these terms as

$$p(a)\left(-y_m(a)\frac{a_1}{a_2}y_n(a) + y_m(a)\frac{a_1}{a_2}y_n(a)\right) + p(b)\left(-y_m(b)\frac{b_1}{b_2}y_n(b) + y_m(b)\frac{b_1}{b_2}y_n(b)\right) = 0,$$

and we are left with

$$(\lambda_m - \lambda_n)\int_a^b r(x)y_m(x)y_n(x)\,dx = 0,$$

which implies that either $\lambda_m = \lambda_n$, which is counter to our original assumption, or

$$\int_a^b r(x) y_n(x) y_m(x)\, dx = 0. \tag{6.163}$$

When the functions $y_m(x)$ and $y_n(x)$ satisfy Equation (6.163), they are said to be *orthogonal*. As an example, consider the case when $r(x) = 1$, $y_n(x) = \sin(x)$, and $y_m(x) = \cos(x)$ on the interval $-\pi \le x \le \pi$. The integral in Equation (6.163) is then

$$\int_{-\pi}^{\pi} \sin(x)\cos(x)\, dx = 0,$$

so the functions $\sin(x)$ and $\cos(x)$ are orthogonal on the interval $-\pi \le x \le \pi$.

If we think back to Chapter 4 we can see an analogy here between basis vectors and orthogonal functions. Indeed, the eigenfunctions $y_m(x)$ form what is called a complete orthogonal set on the interval $a \le x \le b$. Again, by analogy with basis vectors, this allows us to represent any well-behaved function $u(x)$ defined on the interval $a \le x \le b$ as a linear combination of the eigenfunctions

$$u(x) = \sum_{n=0}^{\infty} v_n y_n(x). \tag{6.164}$$

We can find the coefficients v_n by multiplying both sides of Equation (6.164) by $r(x)y_m(x)$ and integrating over the interval $a \le x \le b$. Because of the orthogonality condition, the only term that survives is the one for which $m = n$, so

$$v_m = \frac{\int_a^b u(x) y_m(x) r(x)\, dx}{\int_a^b [y_m(x)]^2 r(x)\, dx}. \tag{6.165}$$

How does this help us solve our BVP? Let us assume that we have a BVP $L[y] = f(x)$, with homogeneous boundary conditions; i.e., we have a differential equation such as Equation (6.154) with homogeneous boundary conditions. We know that we can rewrite this as a Sturm–Liouville problem,

$$L[\psi_n(x)] = -\lambda_n \psi_n(x) r(x), \tag{6.166}$$

where $\psi_n(x)$ are the eigenfunctions of the operator $L[\cdot]$. We now make use of the fact that the eigenfunctions form a basis to write the solution as an infinite series

$$y(x) = \sum_{n=1}^{\infty} a_n \psi_n(x).$$

Substituting this expansion into the BVP and using Equation (6.166) we get

$$L[y] = L\left[\sum_{n=1}^{\infty} a_n \psi_n(x)\right] = -\sum_{n=1}^{\infty} a_n \lambda_n r(x)\psi(x) = f(x).$$

We can multiply this equation by $\psi_m(x)$ and use the orthogonality property of the functions $\psi_m(x)$ to write

$$\int_a^b f(x)\psi_m(x)\,dx = -\sum_{n=1}^{\infty} a_n\lambda_n \int_a^b \psi_n(x)\psi_m(x)r(x)\,dx$$
$$= -a_m\lambda_m \int_a^b \psi_m^2(x)r(x)\,dx,$$

which we can solve to find the coefficients a_m. We then have the complete solution to the BVP.

Example 6.26 Let us use these techniques to solve the BVP

$$x\frac{d^2y}{dx^2} + \frac{d}{dy}x + \frac{y}{x} = \frac{1}{x}, \qquad y(1) = y(\mathrm{e}) = 0.$$

We first note that the equation is already a self-adjoint equation and the self-adjoint operator is

$$L[\psi] = \frac{d}{dx}\left(x\frac{d\psi}{dx}\right) + \frac{\psi}{x},$$

so the associated Sturm–Liouville problem is $L[\psi] = -\lambda r(x)\psi(x)$ with boundary conditions $\psi(1) = \psi(\mathrm{e}) = 0$. This is the differential equation

$$x^2\psi'' + x\psi' + (1 + \lambda r(x)x)\psi(x) = 0.$$

If we choose the function $r(x) = 1/x$ to match the right-hand side of the original equation, then this ODE has solutions (i.e., the eigenfunctions)

$$\psi_n(x) = A\sin(n\pi\ln(x)), \qquad n = 1, 2, \ldots$$

with the eigenvalues $\lambda = 1 - n^2\pi^2$. It is conventional to normalize the eigenfunctions such that

$$\int_1^{\mathrm{e}} \psi_n^2 r(x)\,dx = 1,$$

which gives us that $A = \sqrt{2}$. Now that we have found the eigenfunctions, we can write the solution to our original problem as

$$y(x) = \sqrt{2}\sum_{n=1}^{\infty} a_n \sin(n\pi\ln(x)).$$

Substituting this into the original equation, multiplying by $\psi_m(x)$, and using the orthogonality condition gives us that

$$a_n = \frac{\sqrt{2}}{n\pi}\frac{(-1)^n - 1}{n^2\pi^2 - 1},$$

so we have our solution.

Exercise 6.12.4 Fill in the details in Example 6.26.

6.12.1 Green's Functions

Many real-world systems that vary in time experience an external forcing and, as we have seen in Section 6.4.2 and Section 6.11, this can dramatically affect the solutions of the ODEs that describe these systems. Let us look briefly at second order inhomogeneous linear equations again. A general equation of this type can be written

$$\frac{d^2u(t)}{dt^2} + k(t)\frac{du(t)}{dt} + p(t)u(t) = f(t). \tag{6.167}$$

We know from Section 6.4.1 that this equation has a general form of $u(t) = u_h(t) + u_p(t)$, where $u_h(t)$ is the general solution to the corresponding homogeneous equation (i.e., Equation (6.167) with $f(t) = 0$) and $u_p(t)$ is a particular solution of the inhomogeneous equation.

Let us use the technique of variation of parameters to find a particular solution, which we will assume has the form

$$u_p(t) = v_1(t)\phi_1(t) + v_2(t)\phi_2(t), \tag{6.168}$$

where $v_1(t)$ and $v_2(t)$ are functions we need to find and $\phi_1(t)$ and $\phi_2(t)$ are any two linearly independent solutions of the homogeneous equation. If we substitute Equation (6.168) into equation Equation (6.167), we obtain

$$\begin{aligned} &v_1[\ddot{\phi}_1 + k(t)\dot{\phi}_1 + p(t)\phi_1] + v_2[\ddot{\phi}_2 + k(t)\dot{\phi}_2 + p(t)\phi_2] \\ &\quad + 2(\dot{\phi}_1\dot{v}_1 + \dot{\phi}_2\dot{v}_2) + \phi_1\ddot{v}_1 + \phi_2\ddot{v}_2 + k(t)[\dot{v}_1\phi_1 + \dot{v}_2\phi_2] = f(t). \end{aligned}$$

The first two terms in square brackets vanish because ϕ_1 and ϕ_2 are solutions to the homogeneous equation. If we now choose $(\dot{v}_1\phi_1 + \dot{v}_2\phi_2) = 0$, then the term multiplying $k(t)$ vanishes. Because $(\dot{v}_1\phi_1 + \dot{v}_2\phi_2)$ is zero, its derivative

$$\phi_1\ddot{v}_1 + \phi_2\ddot{v}_2 + \dot{\phi}_1\dot{v}_1 + \dot{\phi}_2\dot{v}_2 = 0,$$

so we are left with

$$\dot{\phi}_1\dot{v}_1 + \dot{\phi}_2\dot{v}_2 = f(t).$$

Now we have two equations for two unknowns, $\dot{v}_1$ and $\dot{v}_2$. We can write these two equations in terms of a matrix equation $\mathbf{Ax} = \mathbf{b}$:

$$\begin{pmatrix} \phi_1 & \phi_2 \\ \dot{\phi}_1 & \dot{\phi}_2 \end{pmatrix} \begin{pmatrix} \dot{v}_1 \\ \dot{v}_2 \end{pmatrix} = \begin{pmatrix} 0 \\ f \end{pmatrix}. \tag{6.169}$$

Exercise 6.12.5 Show that the determinant of $\mathbf{A}$ is nonzero.

Because the determinant of $\mathbf{A} \neq 0$, we have

$$x = \begin{pmatrix} \dot{v}_1 \\ \dot{v}_2 \end{pmatrix} = \frac{1}{|\mathbf{A}|}\begin{pmatrix} \dot{\phi}_2 & \phi_2 \\ -\dot{\phi}_1 & \phi_1 \end{pmatrix}\begin{pmatrix} 0 \\ f(t) \end{pmatrix},$$

so that

$$\dot{v}_1 = \frac{-\phi_2 f}{W(\phi_1, \phi_2)}, \qquad \dot{v}_2 = \frac{\phi_1 f}{W(\phi_1, \phi_2)},$$

where $W(\phi_1, \phi_2$ is the Wronskian (cf. Equation (6.61)). Therefore, substituting into Equation (6.168) we get

$$u_p(t) = \int_{t_0}^{t} \frac{[\phi_1(z)\phi_2(t) - \phi_2(z)\phi_1(t)]}{W(z)} f(z)\, dz = \int_{t_0}^{t} G(t, z) f(z)\, dz, \tag{6.170}$$

where the function $G(t, z)$ is called the *Green's function*.[40] The general solution to Equation (6.167) can then be written as

$$u(t) = A\phi_1(t) + B\phi_2(t) + \int_{t_0}^{t} G(t, z) f(z)\, dz, \tag{6.171}$$

where A and B are constants. If we have suitable initial values for Equation (6.167), then we can find the values of A and B.

Now let us see what happens for the BVP

$$\frac{d^2 y(x)}{dx^2} + k(x)\frac{dy}{dx} + p(x)y(x) = f(x) \tag{6.172}$$

for $x_l < x < x_r$ with the boundary conditions at the left-hand (at $x = x_l$) and right-hand (at $x = x_r$) boundaries:

$$\alpha_1 y(x_l) - \beta_1 \left.\frac{dy}{dx}\right|_{x_l} = 0, \qquad \alpha_2 y(x_r) - \beta_2 \left.\frac{dy}{dx}\right|_{x_r} = 0. \tag{6.173}$$

We are going to use a similar line of reasoning as before, and start with two linearly independent solutions ($\psi_1(x)$ and $\psi_2(x)$) to the homogeneous equation (i.e., Equation (6.172) with $f(x) = 0$). Notice that knowing the solutions are linearly independent helps us cancel all sorts of terms. For simplicity, we are going to impose the left-hand boundary condition on $\psi_1(x)$ and the right-hand boundary condition on $\psi_2(x)$. From Equation (6.171) we know that the general solution of the inhomogeneous equation can be written as

$$y(x) = c_1\psi_1(x) + c_2\psi_2(x) + \int_{x_l}^{x_r} \left(\frac{\psi_1(z)\psi_2(x) - \psi_1(x)\psi_2(z)}{W(z)}\right) f(z)\, dz.$$

We can differentiate this equation using Leibniz's rule (Section 2.11.4) to get

$$\frac{dy}{dx} = c_1\frac{d\psi_1(x)}{dx} + c_2\frac{d\psi_2(x)}{dx} + \int_{x_l}^{x_r} \left(\frac{\psi_1(z)\psi_2'(x) - \psi_1'(x)\psi_2(z)}{W(z)}\right) f(z)\, dz.$$

Now we can apply the boundary conditions to the general solution. First, the left-hand boundary condition says that (note that the limits on the integral are both x_l, so the integral vanishes)

$$\alpha_1 y(x_l) - \beta_1 y'(x_l) = c_1[\alpha_1\psi_1(x_l) - \beta_1\psi_1'(x_l)] + c_2[\alpha_1\psi_2(x_l) - \beta_1\psi_2'(x_l)] = 0.$$

[40] This is the same George Green of Green's theorem and Green's identity.

However, because we have set up the problem so that ψ_1 satisfies the boundary condition at $x = x_l$, we are left with

$$c_2[\alpha_1 \psi_2(x_l) - \beta_1 \psi_2'(x_l)] = 0,$$

which implies that $c_2 = 0$.

The right-hand boundary condition is a little more involved because the integral does not vanish there. But, by similar reasoning we get

$$c_1 = \int_{x_l}^{x_r} \frac{\psi_2(z) f(z)}{W(z)}\, dz,$$

so that our solution becomes

$$y(x) = \left\{ \int_{x_l}^{x_r} \frac{\psi_2(z) f(z)}{W(z)}\, dz \right\} \psi_1(x) + \int_{x_l}^{x_r} \left(\frac{\psi_1(z)\psi_2(x) - \psi_1(x)\psi_2(z)}{W(z)} \right) f(z)\, dz.$$

After some cancellations we are left with

$$y(x) = \int_{x_l}^{x} \frac{\psi_1(z)\psi_2(x)}{W(z)} f(z)\, dz + \int_{x}^{x_r} \frac{\psi_1(x)\psi_2(z)}{W(z)} f(z)\, dz.$$

So, we define the Green's functions for the BVP as

$$G(x, z) = \begin{cases} \frac{\psi_1(z)\psi_2(x)}{W(z)} & x_l < z \le x \\ \frac{\psi_1(x)\psi_2(z)}{W(z)} & x \le z < r \end{cases}, \tag{6.174}$$

and the solution is

$$y(x) = \int_{x_l}^{x_r} G(x, z) f(z)\, dz.$$

Green's functions are important because, as we will see later, they can be thought of as the response of the differential equation to a point source. If we then know the Green's function for a given differential operator $L[]$, we can then construct the solution to an arbitrary source. Let us look at a simple example. Consider the first order inhomogeneous equation

$$L[y] = \frac{dy}{dx} + 2xy = f(x), \qquad y(x = a) = 0. \tag{6.175}$$

The integrating factor for the equation is e^{x^2}, so we can write Equation (6.175) as

$$\frac{d}{dx}(e^{x^2} y(x) = f(x) e^{x^2},$$

which we can integrate to give

$$y(x) = \int_a^x e^{x'^2 - x^2} f(x')\, dx' = \int_a^x G(x; x') f(x')\, dx',$$

where we have used a dummy variable (x') in the integral. Once we have the Green's function, then we have a solution to the ODE for any $f(x)$ we need; so long as we evaluate the integral.

6.13 Further Reading

Although this chapter on differential equations has been a long one, we have only scratched the surface of the subject. It is a vast, fascinating, and immensely useful field of study, and much of what we study as the Earth and environmental scientists relies upon it. There are many excellent textbooks that go deeper and further into the topic, a standard one being *Elementary Differential Equations and Boundary Value Problems* by Boyce and DiPrima (2012).

Finding the analytical solutions for differential equations is a skill that takes practice. However, there are some good books that list solutions and techniques for solutions. A particularly nice, if somewhat older one is the collection *Ordinary Differential Equations and Their Solutions* by Murphy (1960). This lists many ODEs, their solutions, and the techniques used to solve them. A more modern collection is the *Handbook of Differential Equations* by Zwillinger (1997). Although these collections are very useful, we will often have to use our understanding of transformations to get the equation we are stuck on into a form that is listed in the book.

Nonlinear dynamics, chaos, and the study of bifurcations is a large field, but a good entry-level book is *Nonlinear Dynamics and Chaos* by Strogatz (2001). A more advanced and mathematically sophisticated treatise is *Nonlinear Oscillations, Dynamical Systems, and Bifurcations of Vector Fields* by Guckenheimer and Holmes (1983). An older, somewhat more advanced text that looks at the nature of critical points in three dimensions is the book *Ordinary Differential Equations* by Arnold (1978). A book that explores chaos in geological and geophysical processes at a similar mathematical level to this one is *Fractals and Chaos in Geology and Geophysics* by Turcotte (1997).

Well-written and accurate codes for numerically solving differential equations are readily available in many common programming languages such as C, C++, and FORTRAN. Common scripting languages (such as Python and MATLAB) contain excellent packages for numerically integrating ODEs. However, it can be dangerous to treat these computer codes as foolproof, and a broad understanding of how they work is essential for using them properly. The book *Numerical Recipes in C* by Press et al. (1992) is a standard reference for getting an introduction to a wide variety of numerical methods and algorithms, and it provides a lot of straightforward, practical advice. What is more, the book is available with computer code examples in different programming languages including FORTRAN and C++. The two books by Forman Acton, *Numerical Methods That (Usually) Work* (1990) and *Real Computing Made Real* (1996) are both excellent texts covering a wide range of numerical methods, but from a very practical viewpoint—these are excellent books to read if your work involves substantial numerical computing. More detailed and technical books that specifically cover the numerical solution of differential equations include *Numerical Solution of Ordinary Differential Equations* by Shampine (1994), which covers many different algorithms, and *A First Course in the Numerical Analysis of Differential Equations* by Iserles (2008), which is considerably more technical than any of the books already mentioned, but worth the effort to read.

Problems

6.1 Classify the following differential equations:

1. $\frac{dx}{dt} + 2x = 4 + \sin^2(\omega t)$
2. $\frac{d^2x}{dt^2} + 2\frac{dx}{dt} - x = 0$
3. $\frac{d^2x}{dt^2} + 3\left(\frac{dx}{dt}\right)^2 + x = e^{-\kappa t}$
4. $\frac{dx}{dt} + x^2 = 0$
5. $\frac{dy}{dx} = x^2 + 2xy + x^4y^2$
6. $\frac{dy}{dx} = \frac{y}{x}$

6.2 Find the general solution to the following first order equations:

1. $t\frac{dy}{dt} = 2t^2 + 3y, \quad t > 0$
2. $\sin(\phi)\frac{dr}{d\phi} = \tan(\phi) - r\cos(\phi)$, $0 < \phi < \pi/2$.
3. $\frac{dv}{dt} + vt = v$
4. $\frac{du}{dx} - u = u^2 e^{-x}$
5. $(\sin(x) + 2xy)dx + (y\cos(x) + y^2)dy = 0$
6. $(x^2 + 2)\frac{dy}{dx} = -2xy$

6.3 Find the general solution to the following second order equations:

1. $y'' + 3y' - 2y = e^{2x}$
2. $y'' + y' - 2y = \sin(x)$
3. $y'' + y' + y = xe^{-x}$
4. $y'' - y = \tan(x)$
5. $y'' + 4y' + 4y = x^2$
6. $y'' - 4y' + 4y = e^x\left(\frac{3 + 2x^2}{12x}\right)$

6.4 Ecological models of population growth tend to result in nonlinear differential equations. An example is the Verhulst population model

$$\frac{dN}{dt} = (\alpha - \beta N(t))N(t),$$

where $N(t)$ is the population number, α is the constant *specific growth rate* (i.e., the growth rate per unit population number), and β is a constant mortality rate. Solve this differential equation with the initial condition $N(t = 0) = N_0 =$ constant; note that because the problem concerns populations, $N(t) \geq 0$ because having a negative population makes no sense.

6.5 A BVP that expresses mass balance of ice in a circular ice sheet is

$$\frac{1}{r}\frac{d}{dr}\left(rh^5\left(\frac{dh}{dr}\right)^3\right) + 1 = 0,$$

where h is a dimensionless measure of the height of the ice sheet and r is the dimensionless radius of the ice sheet, which varies from 0 to 1. The boundary conditions are

$$\left.\frac{dh}{dr}\right|_{r=0} = 0 \quad \text{and} \quad h(r = 1) = 0.$$

Integrate the equation to show that

$$h(r) = \left[\frac{4}{2^{4/3}}\left(1 - r^{4/3}\right)\right]^{3/8}.$$

6.6 Radioactive isotopes are frequently used to learn more about specific Earth and environment processes as well as to trace material through a given system. Consider a radioisotope R_1 that decays to a radioisotope R_2, which in turn decays to a stable isotope R_3 (e.g., the uranium–thorium sequence). A system of equations representing the changes in the number of atoms (N_i) of each isotope is

$$\frac{dN_1}{dt} = -\lambda_1 N_1, \quad \frac{dN_2}{dt} = \lambda_1 N_1 - \lambda_2 N_2, \quad \frac{dN_3}{dt} = \lambda_2 N_2,$$

where λ_1, λ_2, and λ_3 are the decay constants for R_1, R_2, and R_3 respectively.

1. Solve the equations for N_1, N_2, and N_3.
2. Show that if $\lambda_1 \ll \lambda_2$ (the system is said to be in *secular equilibrium*), then

$$N_2(t) \approx \frac{\lambda_1 N_1(t = 0)}{\lambda_2}(1 - e^{-\lambda_2 t}).$$

6.7 The radioisotope ^{234}Th is often used to trace the sinking flux of particulate matter from the surface ocean. ^{234}Th is produced by the decay of ^{238}U, which has an approximately constant concentration in the ocean. ^{238}U is dissolved, whereas ^{234}Th can exist in both dissolved form (Th_d) and absorbed onto particles (Th_p) that can sink through the water column. Consider a surface layer of the ocean of thickness Z,

$$\frac{dTh_d}{ddt} = \lambda_U \mathrm{U} - \lambda_{Th} Th_d - kTh_d P$$

$$\frac{dTh_p}{ddt} = kTh_d P - \lambda_{Th} Th_p - \frac{v}{Z} Th_p,$$

where λ_U, λ_{Th}, and k are constants, P is the constant particle concentration, and v is the constant particle sinking velocity. Solve the equations for the dissolved and particulate thorium assuming that $Th_p(t = 0) = 0$ and Th_d is in secular equilibrium with U at $t = 0$.

6.8 A falling spherical particle of mass m experiences a gravitational force pulling it down, a buoyancy force, and a drag force. A differential equation for the velocity of the particle is

$$m\frac{dv}{dt} = -mg + \frac{4}{3}\pi r^3 \rho_w g - 6\pi\mu r v,$$

where g is the acceleration due to gravity, r is the radius of the particle, ρ_w is the density and μ the viscosity of water, and v the particle sinking speed. Solve this equation for v and show that the terminal velocity is (where ρ_p is the particle density)

$$v_t = \frac{2}{9}\frac{g}{\mu}(\rho_p - \rho_w)r^2.$$

6.9 Three lakes have different concentrations (x_i) of pollutant:

$$\dot{x_1} = -k_1 x_1, \quad \dot{x_2} = k_1 x_1 - k_2 x_2, \quad \dot{x_3} = k_2 x_2 - k_3 x_3,$$

where k_i are constants. Write the system of equations in matrix form (see Section 6.8) and solve the equations.

6.10 Many volcanic islands lie on submarine ridges that stretch across the oceans. These ridges act as an additional loading over and above that of the water of the oceans, causing the Earth's lithosphere to flex under the extra weight. The lithosphere lies on top of the denser asthenosphere, which provides a buoyancy force upward. A simple model for the vertical displacement (z) of the lithosphere as a function of distance from the ridge (x) is given by

$$D\frac{d^4 z}{dx^4} + g\Delta\rho z = F, \tag{6.176}$$

where D is a parameter called the *flexural rigidity*, g is the acceleration due to gravity, $\Delta\rho$ is the difference between the density of the asthenosphere and the overlying water, and F is the downward vertical force per unit length caused by the additional loading. Assume that D, $\Delta\rho$, and F are constants and solve Equation (6.176).

6.11 Lakes and other small bodies of water are often approximated as *continuously stirred tank reactors* (CSTR). This can be a useful approximation when looking at the fate of chemicals in a lake because it provides a useful approximation to the residence times of material in the lake. A fundamental property of a CSTR is that the concentration of any substance in the outflow is the same as that in the container (i.e., the lake). Consider a lake that has a single inflow and single outflow, and assume that there is a sudden inflow pulse of concentration C_0 of a chemical that is not normally present in the lake. If the rate of output flow is v, write down a first order differential equation for the rate of change of concentration (C) of the chemical in the lake with initial conditions $C = C_0$ at $t = 0$, and solve the equation. Calculate the distribution of residence times,

$$E(t) = C(t)\left(\int_0^\infty C(t)\,dt\right)^{-1}.$$

6.12 A spherical raindrop falls through the air. As it does so, it evaporates at a rate proportional to its surface area. The equation for the rate of change of velocity (v) of the droplet over time, t, is

$$\frac{dv}{dt} - \frac{\alpha}{\rho}\frac{3}{r_0 - (\alpha/\rho)t}v = g,$$

where ρ is the density of water, r_0 is the initial radius of the raindrop, g is the acceleration due to gravity, and $\alpha > 0$ is a constant. Solve the equation for $v(t)$.

6.13 Many situations occur where a moving particle accumulates mass: a raindrop falling through a cloud, or a particle of marine detritus falling through the ocean. Assume that the rate at which a spherical, falling particle accumulates mass is proportional to its surface area, and it always retains a spherical shape.

1. Write down a differential equation for the rate of change of the radius of the particle with time and solve it, assuming that the radius of the particle is r_0 at time $t = 0$.
2. Newton's laws of motion imply that

$$\frac{d}{dt}(mv) = mg,$$

where m is the mass of the particle, v is its velocity, and g is the acceleration due to gravity. Use this to find a differential equation for the rate of change of the particle velocity with time and solve it, assuming that $v = 0$ at time $t = 0$.

6.14 Aerosols and other small, submicron-sized particles such as colloids in aquatic systems can collide and aggregate to form larger particles. As a result, the number of particles per unit volume changes with particle size. An initial population of submicron, monodisperse particles (i.e., they all the have the same size) will coagulate to form larger particles, so that the size distribution will change over time. We can divide the size of particles into classes such that particles in the first size class all have the size of the initial particle, particles in the second size class have twice the size of those in the first, and so on. If n_k is the number of particles of size class k in unit volume, then the rate of change of n_k is given by

$$\frac{dn_k}{dt} = \frac{K}{2}\sum_{i+j=k} n_i n_j - K n_k \sum_{i=1}^{\infty} n_i, \tag{6.177}$$

where K = constant is a measure of the frequency of collisions.

1. What are the dimensions of K?
2. Define $N_\infty = \sum_{i=1}^{\infty}$ as the total number of particles present and show that

$$\frac{dN_\infty}{dt} = \frac{K}{2}\sum_{k=1}^{\infty}\sum_{i+j=k} n_i n_j - K N_\infty^2.$$

3. Write out the first three or four terms in summation in the first term on the right-hand side and show that

$$\frac{K}{2}\sum_{k=1}^{\infty}\sum_{i+j=k} n_i n_j = \frac{K}{2} N_\infty^2.$$

4. Show that the total number of particles present varies according to

$$N_\infty = \frac{N_\infty(t=0)}{1 + (K/2)N_\infty(t=0)t},$$

where $N_\infty(t = 0)$ is the value of N_∞ at time $t = 0$.
5. Write down differential equations for the rate of change of n_1 and n_2 and solve them.
6. Use the expressions for n_1 and n_2 as functions of time to show that

$$n_k(t) = N_\infty(t=0)\frac{(t/\tau)^{k-1}}{(1 + (t/\tau))^{k+1}},$$

where $\tau = (2/K)N_\infty(t = 0)$.

6.15 Photochemistry is chemistry driven by light and is an important process in many environments. It plays a role in the chemistry of the upper layers of bodies of water and in the chemistry of the atmosphere. Of particular importance to humans is its role in the formation of smog. In particular, concentrations of ozone in city air varies with the time of day, being highest during the middle of the day and lowest at night.

1. Assume that the sunlight over a day varies according to time t (measured in fractions of a day) as $S = 1 - \cos(t/T)$, where T is a constant. What is the value of T?
2. Assume that the rate of formation of ozone is proportional to the sunlight present, and that the rate of loss of ozone occurs with a time scale of τ, measured in units of days. Write down a differential equation for the rate of change of ozone (O) in the city air.
3. Solve the equation you have just derived and show that for $\tau < T$ (i.e., rapid loss of ozone), the amount of ozone in the air closely tracks the available sunlight.

6.16 Consider a general Riccati equation for $y(x)$ of the form

$$y' = a(x) + b(x)y(x) + c(x)y^2(x).$$

Show that if we multiply the equation by

$$g = \exp\left(-\int b(x)\,dx\right)$$

and make the substitution $v(x) = gy(x)$, we can remove the $b(x)y(x)$ term from the equation. Further, show that if we then make the substitution

$$w(x) = -\frac{1}{c(x)w(x)}\frac{dw}{dx},$$

the ODE becomes

$$c(x)\frac{d^2w}{dx^2} - \frac{dc}{dx}\frac{dw}{dx} + a(x)c^2(x)w(x) = 0.$$

6.17 Using the online code for the forward Euler method accompanying this chapter as a template, write computer code to solve a given differential equation using Heun's method and the modified Euler method. Use your code to solve the IVP

$$\frac{dy}{dx} = y, \quad y(x = 0) = 1$$

between $x = 0$ and $x = 5$ for a variety of step sizes and compare your results with those in Table 6.1.

6.18 Stratified fluids, where the density varies monotonically with depth, are common in the environment. If we displace a parcel of fluid vertically in a stratified fluid, then the density of the parcel of fluid will be different from the fluid around it, and this can lead to some interesting dynamics. Let us consider a stratified ocean (but we could also consider a stratified atmosphere). The equation of motion of the parcel of fluid is

$$\frac{d^2z}{dt^2} = \frac{gz}{\rho_0}\frac{d\rho}{dz}, \tag{6.178}$$

where z is the depth in the ocean, ρ_0 is the density of the parcel of water, g is the acceleration due to gravity, and $d\rho/dz$ is the density gradient of the surrounding water. Assuming that the density gradient is constant, define

$$N^2 = -\frac{g}{\rho_0}\frac{d\rho}{dz}$$

and solve Equation (6.178) for the cases when $N^2 > 0$ and $N^2 < 0$, describing the motion of the parcel of fluid in each case.

6.19 The concentrations of many chemical compounds in the environment are very dynamic, but how they change with time depends on the type of reactions they experience. In this problem you will consider various types of reactions between the three compounds A, B, and C.

1. A first order irreversible chemical reaction can be written as $A \rightarrow B$. The reaction is called a first order reaction because the reaction of A involves only itself. If the reaction has a rate constant κ, then the concentrations of A and B vary according to

$$\frac{dA}{dt} = -\kappa \mathrm{A}, \quad \text{and} \quad \frac{dB}{dt} = \kappa \mathrm{A}, \tag{6.179}$$

so that

$$\frac{dA}{dt} + \frac{dB}{dt} = 0,$$

which is an expression of *mass balance* (i.e., no material is created or destroyed, only converted from one form (A) to another (B)). Solve Equation (6.179) for A and B, assuming that the initial concentrations are such that at $t = 0$, $\mathrm{A} = \mathrm{A}_0$, and $\mathrm{B} = 0$, and sketch (or plot) the solutions.

2. A first order reversible reaction can be written $A \rightleftharpoons B$, indicating that the chemical A can form the chemical B and vice versa. We now have two rate constants, κ_1 for the reaction from A to B, and κ_{-1} for the reaction from B to A. So,

$$\frac{d}{d\mathrm{A}}t = -\kappa_1\mathrm{A} + \kappa_{-1}\mathrm{B}.$$

If the initial conditions are such that at $t = 0$, $\mathrm{A} = \mathrm{A}_0$, and $\mathrm{B} = 0$, then mass balance implies that $\mathrm{B} = \mathrm{A}_0 - \mathrm{A}$. Solve the differential equation for A and sketch (or plot) the time evolution of A and B.

3. A second order reaction involves two reactants. This can occur in several ways. Consider first the irreversible reaction $\mathrm{A} + \mathrm{A} \rightarrow C$. If the rate constant for the reaction is κ, then

$$\frac{1}{2}\frac{dA}{dt} = -\kappa A^2.$$

Solve this equation for A, given that at $t = 0$, $\mathrm{A} = \mathrm{A}_0$, and sketch (or plot) the solution.

4. Second order reactions can also involve two reactants, for example A+B $\rightarrow C$. If κ is the rate constant for the reaction, then A and B satisfy the following differential equations

$$\frac{dA}{dt} = -\kappa AB, \qquad \frac{dB}{dt} = -\kappa AB.$$

Use the substitution $x = (A_0 - A) = (B_0 - B)$, where A_0 and B_0 are the initial values of A and B at $t = 0$ to solve the differential equation for the ratio $A(t)/B(t)$.

6.20 We have seen that not all ODEs can be solved analytically. This applies to some deceptively simple ODEs such as Airy's equation

$$\frac{d^2y}{dx^2} = xy. \tag{6.180}$$

1. Classify the point $x = 0$ and find a power series solution for Equation (6.180) about $x = 0$.
2. Numerically solve Equation (6.180) from $x = -8$ to $x = +2$ using a Runge–Kutta algorithm with the two sets of initial conditions, $y(x = 0) = 1$, $y'(x = 0) = 0$ and $y(x = 0) = 0$, $y'(x = 0) = 1$, and compare the numerical and power series solutions.

6.21 The Lorenz equations (Equation (6.151)) show a range of behaviors as the values of the parameters change.

1. Show that the origin is the only steady state solution if $r < 1$ and that it is stable.
2. Show that for $r = 1$ there are two additional steady states indicating that a possible bifurcation has occurred.
3. Use a Runge–Kutta algorithm to numerically integrate the Lorenz equations (Equation (6.151)) with the parameter values $\sigma = 10$, $r = 28$, $b = 8/3$.

6.22 It is not always possible to find a power series solution to an ODE. Consider the equation

$$\frac{d^2y}{dx^2} - \frac{2y}{x^2} = 0 \tag{6.181}$$

and show that $y = x^2$ and $y = x^{-1}$ are both solutions, but that a power series solution about $x = 0$ does not exist, and determine why.

6.23 Show that there are no power law solutions about the point $x = 0$ for the equation

$$\frac{d^2y}{dx^2} - \frac{3y}{x^3} = 0$$

and say why.

6.24 Consider the equation

$$\frac{d^2y}{dx^2} + \frac{1}{x^2}\frac{dy}{dx} - \frac{\xi^2 y}{x^2} = 0.$$

Show that $x = 0$ is an irregular singular point, but that a power law solution exists with the recurrence relationship

$$a_{j+1} = \frac{\xi^2 - j(j-1)}{j+1} a_j.$$

Show that this series converges for only special values of ξ where the series has only a finite number of terms.

6.25 Consider the system of linear ODEs

$$u' = 500u + 1500v, \qquad v' = -502u - 1502v, \qquad u(0) = 1,\ v(0) = 0.$$

1. Solve the system of equations analytically.
2. Use different numerical schemes (i.e., Euler methods, Runge–Kutta) to solve these equations and explain any differences with the analytical solution.

7 Vectors and Calculus

We have seen that many environmental quantities we are interested in can be described mathematically using vectors. To understand how these vectors change spatially and temporally we need to combine what we know of vectors with our knowledge of calculus—a subject creatively called *vector calculus*. In this chapter we will largely be dealing with *vector fields*, where there is a vector (a fluid velocity, for example) that has a value at every point in the space we are considering (e.g., the ocean, the atmosphere, or a lake). This means that a vector field ($\mathbf{F}$) is a function of the spatial coordinates, $\mathbf{F} = \mathbf{F}(x, y)$, so that as x and y change, so do the magnitude and the direction of the vector. Some examples of vector fields are shown in Figure 7.1. In this chapter we will learn to calculate how these vectors change as the values of x, y, and z change (the "calculus" of vector calculus) and discover some useful theorems that connect integrals of vectors along paths and over surfaces.

Vector calculus lies at the heart of understanding many processes in the Earth and environmental sciences. To start with, it provides a framework for describing how moving fluids transport material in the environment and how heat and chemical compounds diffuse. For example, it provides a mathematical framework for understanding how heat moves from the Earth's core to the surface, how pollutants are transported through groundwater flows, how vortices are formed and move, how the Earth's magnetic field behaves, how chemicals move through the natural environment, and a myriad other natural processes. This makes vector calculus a very powerful tool.

7.1 Differentiating a Vector

Let us start by thinking about how we differentiate a vector. The basic idea is very similar to that of differentiating a function, except that we have the added complication that a vector can be described in terms of components and basis vectors. We will start by differentiating a vector whose components in a Cartesian basis are functions of time, i.e., $\mathbf{x} = a(t)\hat{\mathbf{i}} + b(t)\hat{\mathbf{j}} + c(t)\hat{\mathbf{k}}$. For example, we might imagine that the vector describes the (x, y, z) coordinates of a particle that is moving through the atmosphere (Figure 7.2). The basis vectors in Cartesian coordinates are constant, they do not change with space or time, so the derivative of $\mathbf{x}$ with respect to time is

$$\frac{d\mathbf{x}}{dt} = \dot{\mathbf{x}} = \frac{da}{dt}\hat{\mathbf{i}} + \frac{db}{dt}\hat{\mathbf{j}} + \frac{dc}{dt}\hat{\mathbf{k}}. \tag{7.1}$$

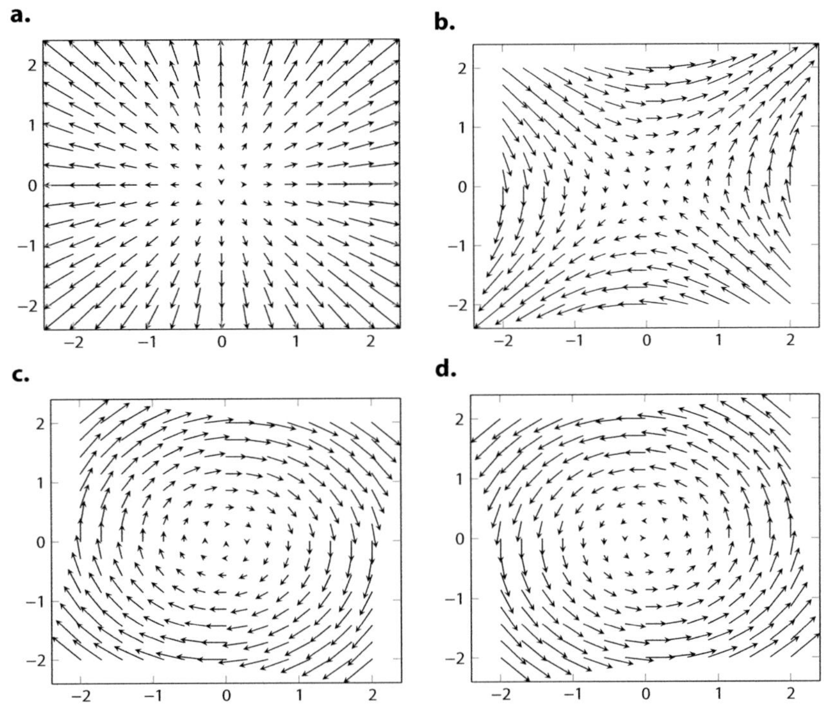

Figure 7.1 Examples of vector fields: (a.) $\mathbf{F}(x, y) = x\hat{\mathbf{i}} + y\hat{\mathbf{j}}$; (b.) $\mathbf{F}(x, y) = y\hat{\mathbf{i}} + x\hat{\mathbf{j}}$; (c.) $\mathbf{F}(x, y) = y\hat{\mathbf{i}} + -x\hat{\mathbf{j}}$; (d.) $\mathbf{F}(x, y) = -y\hat{\mathbf{i}} + x\hat{\mathbf{j}}$.

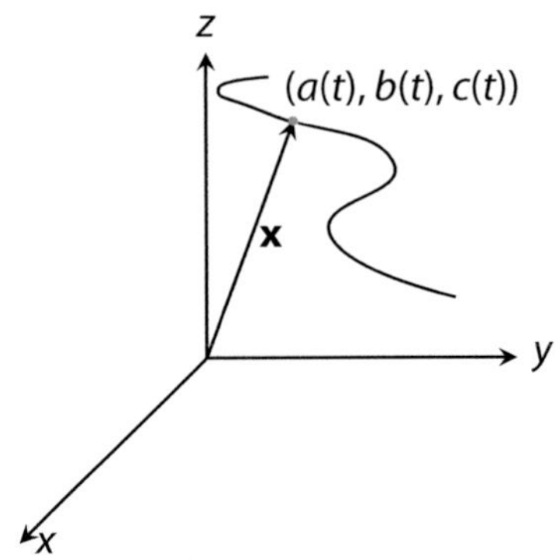

Figure 7.2 The trajectory of a particle moving through space. The parameter t parameterizes the curve and the vector $\mathbf{x}$ points from the origin to the particle as it moves along the trajectory.

For example, the derivative of $\mathbf{x} = 4t^2\hat{\mathbf{i}} + 2t\hat{\mathbf{j}} + 6\hat{\mathbf{k}}$ with respect to time is $\dot{\mathbf{x}} = 8t\hat{\mathbf{i}} + 2\hat{\mathbf{j}}$. Because the components are simply functions, the usual rules of differentiation that we met in Chapter 2 still apply.

Exercise 7.1.1 Calculate the derivatives with respect to time of the following vectors:

$$\text{a. } \mathbf{x} = \sin(4t)\hat{\mathbf{i}} + 2\cos(3t)\hat{\mathbf{j}}, \quad \text{b. } \mathbf{x} = e^{-2t}\sin(2t)\hat{\mathbf{i}} + t^2\tan(t)\hat{\mathbf{j}} - e^{-t^2}\sin(2t^2+3)\hat{\mathbf{k}}.$$

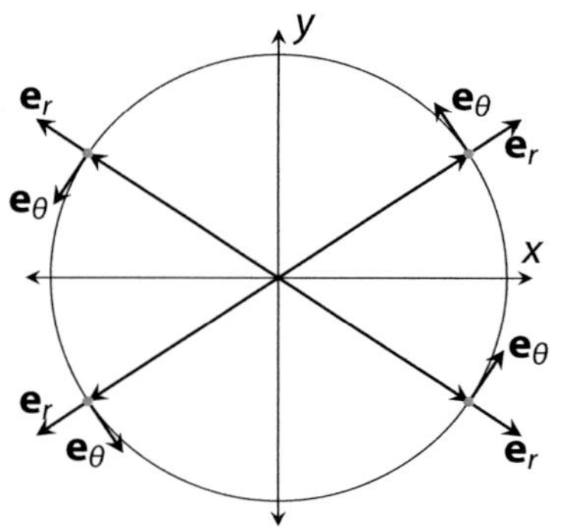

Figure 7.3 The trajectory of a particle moving along a circle in two dimensions. The position of the particle is shown at four points, and the basis vectors $\mathbf{e}_\theta$ and $\mathbf{e}_r$ have different directions at those four points.

However, as we know from Chapter 4, not all basis vectors are constant. We can see this by looking at the trajectory of a particle moving counterclockwise in a circle in two dimensions (Figure 7.3). We can describe motion in terms of Cartesian (i.e., (x, y) coordinates) and basis vectors $\hat{\mathbf{i}}$ and $\hat{\mathbf{j}}$. Alternatively, we can use polar coordinates (r, θ) with basis vectors $\hat{\mathbf{e}}_r$ and $\hat{\mathbf{e}}_\theta$ that have constant magnitude (they are orthonormal basis vectors), but their direction changes as θ changes (Figure 7.3). This means that the derivative of a vector written using polar coordinates has to take into account the changes in the directions of the vectors. The easiest way to see how this works is to write the polar basis vectors in terms of Cartesian basis vectors. In polar coordinates, the basis vector $\hat{\mathbf{e}}_r$ points radially outward from the origin, and the vector $\hat{\mathbf{e}}_\theta$ is orthogonal to $\hat{\mathbf{e}}_r$ pointing in the direction of motion and is tangent to the circle. In terms of Cartesian coordinates, we can write these vectors as

$$\hat{\mathbf{e}}_r = \cos(\theta)\hat{\mathbf{i}} + \sin(\theta)\hat{\mathbf{j}}, \quad \hat{\mathbf{e}}_\theta = -\sin(\theta)\hat{\mathbf{i}} + \cos(\theta)\hat{\mathbf{j}},$$

where θ is a function of time because the particle is moving in a circle. Therefore,

$$\frac{d\hat{\mathbf{e}}_r}{dt} = -\sin(\theta)\frac{d\theta}{dt}\hat{\mathbf{i}} + \cos(\theta)\frac{d\theta}{dt}\hat{\mathbf{j}} = \frac{d\theta}{dt}\hat{\mathbf{e}}_\theta,$$
$$\frac{d\hat{\mathbf{e}}_\theta}{dt} = -\cos(\theta)\frac{d\theta}{dt}\hat{\mathbf{i}} - \sin(\theta)\frac{d\theta}{dt}\hat{\mathbf{j}} = -\frac{d\theta}{dt}\hat{\mathbf{e}}_r.$$

Knowing how the basis vectors change with time allows us to differentiate any vector given in terms of polar coordinates. If $\mathbf{W} = w_r(t)\hat{\mathbf{e}}_r + w_\theta(t)\hat{\mathbf{e}}_\theta$, then

$$\frac{d\mathbf{W}}{dt} = \frac{dw_r}{dt}\hat{\mathbf{e}}_r + w_r\frac{d\hat{\mathbf{e}}_r}{dt} + \frac{dw_\theta}{dt}\hat{\mathbf{e}}_\theta + w_\theta\frac{d\hat{\mathbf{e}}_\theta}{dt} = \left(\frac{dw_r}{dt} - w_\theta\frac{d\theta}{dt}\right)\hat{\mathbf{e}}_r + \left(\frac{dw_\theta}{dt} + w_r\frac{d\theta}{dt}\right)\hat{\mathbf{e}}_\theta.$$

For example, let us consider a point P moving along an arbitrary trajectory in two dimensions. We can write the position vector for P as $\mathbf{R} = r(t)\hat{\mathbf{e}}_r$. This describes how the length and direction of the vector from the origin to the particle change as the particle moves along its path. This might seem strange because there appears to be no dependence of $\mathbf{R}$ on the angle θ, but the variation with θ is hidden in $\hat{\mathbf{e}}_r = \cos(\theta)\hat{\mathbf{i}} + \sin(\theta)\hat{\mathbf{j}}$; remember that the direction of $\hat{\mathbf{e}}_r$ changes as the particle moves along its trajectory. The velocity of P is then the derivative of $\mathbf{R}$ with respect to time,

$$\mathbf{v} = \dot{\mathbf{R}} = \frac{d}{dt}(r(t)\hat{\mathbf{e}}_r) = \dot{r}\hat{\mathbf{e}}_r + r\frac{d\hat{\mathbf{e}}_r}{dt} = \dot{r}\hat{\mathbf{e}}_r + r\dot{\theta}\hat{\mathbf{e}}_\theta,$$

and we see that although the position vector seems to be just a function of r and t, the fact that $\hat{\mathbf{e}}_r$ depends upon θ introduces an $\hat{\mathbf{e}}_\theta$ component to the velocity. For motion in a circle, the distance of the particle from the origin is constant and $\dot{r} = 0$, and the velocity is at right angles to the radius and has a magnitude $r\dot{\theta}$. The acceleration of the particle is

$$\mathbf{a} = \dot{\mathbf{v}} = (\ddot{r} - r\dot{\theta}^2)\hat{\mathbf{e}}_r + (r\ddot{\theta} + 2\dot{r}\dot{\theta})\hat{\mathbf{e}}_\theta.$$

Exercise 7.1.2 Derive expressions for the velocity and acceleration of a particle moving along an arbitrary trajectory in three dimensions using (a) cylindrical coordinates and (b) spherical coordinates.

We know how to differentiate the product of two functions, but how do we differentiate the product of two vectors? We need to take some care here because there are different types of product that we can have. If $\mathbf{V}$ and $\mathbf{U}$ are vectors whose components are functions of time, and $\alpha(t)$ is a function of time, then

$$\frac{d(\alpha\mathbf{V})}{dt} = \dot{\alpha}\mathbf{V} + \alpha\dot{\mathbf{V}}, \quad \frac{d(\mathbf{U}\cdot\mathbf{V})}{dt} = \dot{\mathbf{U}}\cdot\mathbf{V} + \mathbf{U}\cdot\dot{\mathbf{V}}, \quad \frac{d(\mathbf{U}\times\mathbf{V})}{dt} = \dot{\mathbf{U}}\times\mathbf{V} + \mathbf{U}\times\dot{\mathbf{V}}, \tag{7.2}$$

where we have a dot over a letter to indicate differentiation with respect to time. These expressions look very similar to the rules for differentiating the product of two functions. There is one catch, however, and that is we need to be careful to preserve the order of vectors when we differentiate a vector product because $\mathbf{U}\times\mathbf{V} = -\mathbf{V}\times\mathbf{U}$.

Let us look at an example of motion of an object in a circle at a constant speed. The vector $\mathbf{r}(t)$ points from the center to the object and the components of the vector are functions of time. Because the point is moving in a circle, the length of the vector $\mathbf{r}$ is constant, which means that $r^2 = \mathbf{r}\cdot\mathbf{r} =$ constant. If we differentiate this equation with respect to time, we find

$$0 = \frac{d(\mathbf{r}\cdot\mathbf{r})}{dt} = \mathbf{r}\cdot\frac{d\mathbf{r}}{dt} + \frac{d\mathbf{r}}{dt}\cdot\mathbf{r} = 2\mathbf{r}\cdot\frac{d\mathbf{r}}{dt},$$

in other words $\mathbf{r}\cdot\dot{\mathbf{r}} = \mathbf{r}\cdot\mathbf{v} = 0$, which tells us that the velocity vector is always orthogonal to the radius vector. We also specified that the object moved with a constant speed; i.e., the length of the velocity vector is constant. Applying the same reasoning again we find that $\mathbf{v}\cdot\dot{\mathbf{v}} = \mathbf{v}\cdot\mathbf{a} = 0$, where $\mathbf{a}$ is the acceleration vector. This tells us that the velocity vector is orthogonal to the acceleration vector. Since the motion is two dimensional, this means that the acceleration vector is parallel to the radius vector and either they point in the same direction or they point in opposite directions. To determine which it is we need to know the angle between $\mathbf{r}$ and $\mathbf{a}$, so we need an equation that contains the scalar product $\mathbf{r}\cdot\mathbf{a}$. Let us take the derivative of $\mathbf{r}\cdot\mathbf{v} = 0$, which gives $\mathbf{r}\cdot\mathbf{a} + \mathbf{v}\cdot\mathbf{v} = 0$, or rearranging, $\mathbf{r}\cdot\mathbf{a} = |\mathbf{r}||\mathbf{a}|\cos(\phi) = -\mathbf{v}\cdot\mathbf{v} = -v^2$, where ϕ is the angle between the vectors $\mathbf{r}$ and $\mathbf{a}$.

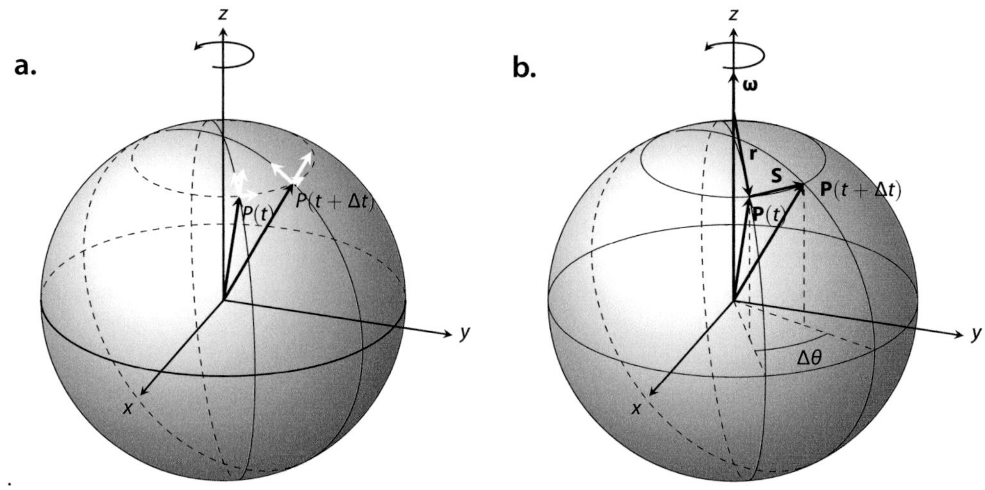

Figure 7.4 The relationship between a fixed coordinate system (x, y, z) and a rotating coordinate system fixed at a point on the sphere (white arrows). As the sphere rotates counterclockwise, the point $P(t)$ moves to the point $R(t + \Delta t)$ (a.). To an observer moving with the sphere, a basis (white arrows) defined at $P(t)$ will not change as the sphere (and observer) rotate. However, the directions of the basis vectors will change to an observer who is not rotating with the sphere and who uses the (x, y, z) coordinate system. The sphere rotates with an angular velocity ω. The vector $\mathbf{P}(t)$ connecting the origin of the sphere to the point P rotates to the vector $\mathbf{P}(t + \Delta t)$ connecting the origin to point $P(t + \Delta t)$ in a time Δt (b.) with $\mathbf{S}$ being the vector connecting point $P(t)$ to point $P(t + \Delta t)$.

This tells us that $\cos(\phi) < 0$, so $\phi = 180°$, and that $ar = v^2$. Therefore the acceleration, $a = v^2/r$ (called the *centripetal acceleration*), points toward the center of the circle (the vector $\mathbf{r}$ points away from the center).

Now, let us extend these calculations and look at the motion of a point moving on the surface of a rotating sphere, a situation with considerable relevance to understanding motions on the Earth. We need to be careful when we consider the motions of the atmosphere and ocean from the perspective of someone standing on the Earth. The atmosphere and the oceans are not fixed to the surface of the Earth, so they do not corotate with it like we do.[1] In addition, the movement of air and water is governed by the forces acting on them, and we use Newton's laws to relate these forces to how the air and water move. But Newton's laws only apply in what is called an *inertial frame of reference*, which is a set of coordinates that are not experiencing a force. A coordinate system centered on an observer standing on the Earth is experiencing a force, the centripetal force, because it is moving in a circle as the Earth rotates (Figure 7.4).

We can define a coordinate system that is fixed in space and not rotating that will serve as the coordinate system in our inertial reference frame—the (x, y, z) coordinate system in Figure 7.4. From the point of view of this coordinate system, the (x, y, z) coordinates

[1] This is true except for the microscopic layer of air or water that is immediately next to the solid Earth, giving rise to the so-called *no-slip* boundary conditions.

of someone standing on the Earth are continuously changing as the Earth rotates. We can also define a set of coordinates on the surface of the planet using latitude, longitude, and altitude above the surface. If we remain fixed at one location on the surface, our earthbound coordinates do not change as the Earth rotates (the white vectors in Figure 7.4a). However, as far as someone using the fixed, inertial coordinate system is concerned, the directions of the basis vectors of our earthbound coordinates do change as the Earth rotates (Figure 7.4a). What is the relationship between these coordinate systems? How does a vector **P** describing a point on the surface of the sphere in the (x, y, z) coordinates change as the sphere rotates? The vector $\mathbf{P}(t)$ connects the origin of the sphere to the point P at time t (Figure 7.4b). After a time interval Δt the Earth has rotated through an angle $\Delta\theta$, so in the (x, y, z) coordinates the vector has moved to $\mathbf{P}(t + \Delta t)$. The vector **S** connects the two points in the (x, y, z) coordinates.[2] Therefore, we can write

$$\mathbf{S} = \mathbf{P}(t + \Delta t) - \mathbf{P}(t). \tag{7.3}$$

If the angle $\Delta\theta$ is small, then we can write $|\mathbf{S}| = |\mathbf{r}|\Delta\theta = |\mathbf{r}|\omega\Delta t$, where the angular velocity is given by $\boldsymbol{\omega} = \omega\hat{\boldsymbol{\omega}}$. Now the vector **S** is orthogonal to both **P** and $\boldsymbol{\omega}$ (Section 4.3.2.1), so we can write a unit vector in the direction of **S** as

$$\hat{\mathbf{S}} = \frac{\boldsymbol{\omega} \times \mathbf{P}}{|\boldsymbol{\omega} \times \mathbf{P}|}. \tag{7.4}$$

In addition, the projection of **P** onto the z axis is given by $\hat{\boldsymbol{\omega}} \cdot \mathbf{P}$. Pythagoras' theorem then tells us that

$$|\mathbf{r}|^2 = |\mathbf{P}|^2 - |\hat{\boldsymbol{\omega}} \cdot \mathbf{P}|^2 = |\hat{\boldsymbol{\omega}}|^2|\mathbf{P}|^2 - |\hat{\boldsymbol{\omega}} \cdot \mathbf{P}|^2 = |\hat{\boldsymbol{\omega}} \times \mathbf{P}|^2,$$

where we have used the fact that $|\hat{\boldsymbol{\omega}}| = 1$ (because $\hat{\boldsymbol{\omega}}$ is a unit vector) and the result from Exercise 4.3.10. Because the length of a vector is a positive quantity, we have that $|\mathbf{r}| = |\hat{\boldsymbol{\omega}} \times \mathbf{P}|$, so $|\mathbf{S}| = |\mathbf{r}|\omega\Delta t = |\boldsymbol{\omega} \times \mathbf{P}|\Delta t$. Therefore, the vector **S** connecting points $P(t)$ and $P(t + \Delta t)$ is

$$\mathbf{S} = |\mathbf{S}|\hat{\mathbf{S}} = |\boldsymbol{\omega} \times \mathbf{P}|\Delta t \left(\frac{\boldsymbol{\omega} \times \mathbf{P}}{|\boldsymbol{\omega} \times \mathbf{P}|}\right) = (\boldsymbol{\omega} \times \mathbf{P})\Delta t.$$

If we now let $\Delta t \to 0$, then

$$\left(\frac{d\mathbf{r}}{dt}\right)_{\text{inertial}} = \left(\frac{d\mathbf{r}}{dt}\right)_{\text{rotating}} + (\boldsymbol{\omega} \times \mathbf{P}), \tag{7.5}$$

where we have allowed for any additional changes in **P** that occur within the rotating coordinates (e.g., the object might change latitude).

Equation (7.5) shows us how to transform any vector between the two coordinate systems. Let us apply it to the acceleration vector. If $\mathbf{v}_{\text{inertial}}$ and $\mathbf{v}_{\text{rotating}}$ are the velocity vectors in the inertial and rotating coordinates, then Equation (7.5) gives us that $\mathbf{v}_{\text{inertial}} = \mathbf{v}_{\text{rotating}} + (\boldsymbol{\omega} \times \mathbf{R})$, where we have renamed **P** as **R** to signify a radius. We can now apply Equation (7.5) again to the acceleration vector to get

[2] Note that from the viewpoint of someone standing stationary on the Earth, the points $P(t)$ and $P(t + \Delta t)$ are the same; the person has not moved.

$$\left(\frac{d\mathbf{v}_{\text{inertial}}}{dt}\right)_{\text{inertial}} = \left(\frac{d\mathbf{v}_{\text{inertial}}}{dt}\right)_{\text{rotating}} + (\boldsymbol{\omega} \times \mathbf{v}_{\text{inertial}})$$
$$= \left(\frac{d}{dt}(\mathbf{v}_{\text{rotating}} + \boldsymbol{\omega} \times \mathbf{r})\right)_{\text{rotating}} + \boldsymbol{\omega} \times (\mathbf{v}_{\text{rotating}} + \boldsymbol{\omega} \times \mathbf{r})$$
$$= \left(\frac{d}{dt}\mathbf{v}_{\text{rotating}}\right)_{\text{rotating}} + 2\boldsymbol{\omega} \times \mathbf{v}_{\text{rotating}} + \boldsymbol{\omega} \times \boldsymbol{\omega} \times \mathbf{r}. \tag{7.6}$$

Equation (7.6) shows us that when we move between the inertial and rotating coordinates, the acceleration of an object picks up two additional terms. The term $2\boldsymbol{\omega} \times \mathbf{v}_{\text{rotating}}$ is called the *Coriolis force*,[3] which affects motions within the rotating coordinates, and the term $\boldsymbol{\omega} \times \boldsymbol{\omega} \times \mathbf{r}$ is the *centrifugal force*, which concerns the rotation of the coordinate frame. These relationships are fundamental to understanding the large-scale motion of the atmosphere and oceans on a rotating planet (Vallis, 2017).

7.2 Gradient

Many environmental quantities vary continuously in space. For example, air temperature varies with height in the atmosphere and geographical location, and the density of seawater varies with depth and location in the ocean. As a result, we can define a function $\rho(x, y, z)$ that represents the seawater density at each point in the ocean. Such a function is called a *scalar field* because the function $\rho(x, y, z)$ is a scalar and it has a value at every point, so it is called a field. The equation $\rho(x, y, z) = \rho_0 =$ constant defines a surface called a *level set* on which the density is constant at all points; these are the three-dimensional versions of the contours on a map. Level sets are important for understanding processes in the environment. For example, in oceanography surfaces of constant seawater density are called *isopycnals* and can be used to understand how changes in density affect the movement of water.

Once we have a scalar field we would like to know how fast it changes as we move in any direction in space. For example, consider the two-dimensional contours shown in Figure 7.5. The function ϕ changes more rapidly with distance along the path AB than it does along the path CD—the level sets are more tightly spaced—so ϕ changes more

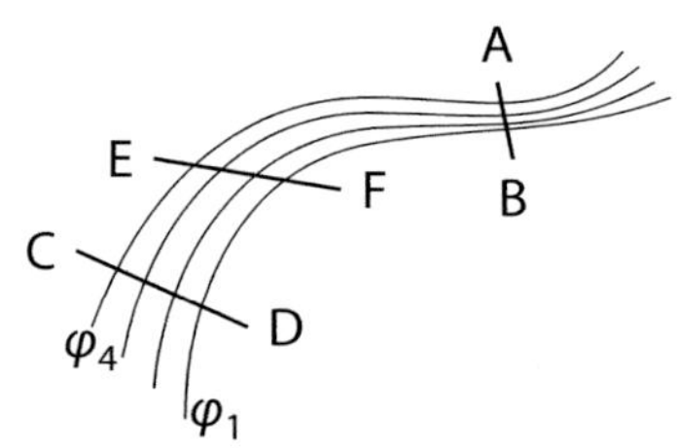

Figure 7.5 Four contours (i.e., level curves) of a function $\phi(x, y)$ with values ϕ_1 to ϕ_4. The gradient of ϕ is different along the three paths AB, CD, and EF.

[3] Named after the French scientist and engineer Gaspard-Gustave de Coriolis (1792–1843).

rapidly along AB. How does ϕ change along a straight line path such as EF? Let us write a unit vector in the direction along EF as $\hat{\mathbf{w}} = a\hat{\mathbf{\imath}} + b\hat{\mathbf{\jmath}} + c\hat{\mathbf{k}}$ where a, b, and c are constants. If the distance we travel along the path is ℓ, then the straight line connecting points E and F is described by the equations

$$x = x_E + a\ell, \quad y = y_E + b\ell, \quad z = z_E + c\ell,$$

where the coordinates of the point E are (x_E, y_E, z_E). The derivative of $\phi(x, y, z)$ with respect to ℓ, the distance along the path from E to F, is then

$$\frac{d\phi}{d\ell} = \frac{\partial\phi}{\partial x}\frac{dx}{d\ell} + \frac{\partial\phi}{\partial y}\frac{dy}{d\ell} + \frac{\partial\phi}{\partial z}\frac{dz}{d\ell} = a\frac{\partial\phi}{\partial x} + b\frac{\partial\phi}{\partial y} + c\frac{\partial\phi}{\partial z}. \tag{7.7}$$

But, a, b, and c are the x, y, and z components of $\hat{\mathbf{w}}$, so Equation (7.7) looks like the dot product of two vectors if we define a vector

$$\nabla\phi = \text{grad}\,\phi = \frac{\partial\phi}{\partial x}\hat{\mathbf{\imath}} + \frac{\partial\phi}{\partial y}\hat{\mathbf{\jmath}} + \frac{\partial\phi}{\partial z}\hat{\mathbf{k}}. \tag{7.8}$$

The vector defined in Equation (7.8) is called the *gradient* of the scalar ϕ. The symbol ∇ is often called *del*, *grad*, or *nabla*.[4] We can now write Equation (7.7) as

$$\frac{d\phi}{d\ell} = \nabla\phi \cdot \hat{\mathbf{w}} = |\nabla\phi||\hat{\mathbf{w}}|\cos(\theta) = |\nabla\phi|\cos(\theta),$$

where θ is the angle between the vectors $\nabla\phi$ and $\hat{\mathbf{w}}$. The derivative $d\phi/d\ell$ has its greatest value when $\cos(\theta) = 1$. This tells us that the magnitude of $\nabla\phi$ represents the greatest value of the spatial gradient of the function ϕ, and the quantity $\nabla\phi \cdot \hat{\mathbf{u}}$ is the projection of that gradient onto the vector $\hat{\mathbf{u}}$; in other words, how much of the gradient of ϕ is in the direction given by $\hat{\mathbf{u}}$.

The object ∇ is a *vector valued function*, and in Cartesian coordinates it is

$$\nabla = \hat{\mathbf{\imath}}\frac{\partial}{\partial x} + \hat{\mathbf{\jmath}}\frac{\partial}{\partial y}\hat{\mathbf{\jmath}} + \hat{\mathbf{k}}\frac{\partial}{\partial z}. \tag{7.9}$$

It is not a proper vector because we cannot calculate its length, but it is a function in that it takes an input (in this case, a scalar function) and returns the gradient of the input. It can be a good idea to write the components of ∇ as in Equation (7.9) with the basis vectors preceding the derivatives to remind ourselves that the derivatives do not act on the basis vectors.

As an example, we can calculate the gradient of the scalar field $\phi(x, y, z) = 3x^2 e^{-z}\sin(y) + 2xy$,

$$\nabla\phi = \hat{\mathbf{\imath}}(6xe^{-z}\sin(y) + 2y) + \hat{\mathbf{\jmath}}(3x^2e^{-z}\cos(y) + 2x) - \hat{\mathbf{k}}(3x^2e^{-z}\sin(y).$$

We can take the dot product of this vector with another to find the gradient of ϕ in different directions. For example, if $\mathbf{u} = 3\hat{\mathbf{\imath}} + 4\hat{\mathbf{\jmath}}$, then a unit vector in that direction is $\hat{\mathbf{u}} = (3/5)\hat{\mathbf{\imath}} + (4/5)(\hat{\mathbf{\jmath}})$ and

$$\nabla\phi \cdot \hat{\mathbf{u}} = \frac{2xe^{-z}}{5}(9\sin(y) + 2x\cos(y)) + \frac{2}{5}(3y + 2x).$$

[4] This strange name derives from the ancient Greek word for a harp and is used because of the general upside down triangular shape of a harp. The name was suggested by William Smith, a scholar of the Old Testament, to the physicist Peter Tait in a letter to Tait in November 1870.

Exercise 7.2.1 Calculate the vector gradient of the following scalar functions:

$$\text{a. } \phi(x, y) = x^2 + y^2, \quad \text{b. } \phi(x, y, z) = e^{x^2+y^2} \sin(z), \quad \text{c. } \phi(x, y, z) = x^2y^2z^2.$$

Exercise 7.2.2 Calculate the gradient of the function $\phi(x, y, z) = x^2y^2z^2$ in the directions given by the following vectors: a. $\mathbf{u} = \hat{\mathbf{k}}$, b. $2\hat{\imath} - 3\hat{\jmath} + \hat{\mathbf{k}}$.

Let us see what happens if we choose to take the dot product of $\nabla\phi$ with a unit vector $\hat{\mathbf{u}}$ that is tangent to a level set. The function ϕ is constant on a level set, so $\nabla\phi = 0$ and $\nabla\phi \cdot \hat{\mathbf{u}} = 0$, which tells us that the gradient of ϕ is orthogonal to the level set, so the direction of the largest gradient of ϕ is perpendicular to the surface describing the level set.

So far we have considered the scalar field to be a function of only space, but what if it is a function of time as well? What is more, let us assume that we move along a path with a velocity $\mathbf{u}$ so that our coordinates $(x(t), y(t), z(t))$ along the trajectory are also functions of time. In that case, we can write the scalar as $\phi = \phi(t, x(t), y(t), z(t)) = \phi(t, \mathbf{r}(t))$, where $\mathbf{r}(t)$ is the position vector along the path we are traveling. For example, we could think of a small balloon being moved with air. The velocity ($\mathbf{u}$) of the balloon will vary with time as it is swept along by the movements of the air. Now let us assume that the balloon is carrying a thermometer so that it can measure the air temperature (ϕ) as it moves. How does the temperature change with time? Taking the derivative of ϕ with respect to t we get

$$\frac{d\phi}{dt} = \frac{\partial\phi}{\partial t} + \frac{\partial\phi}{\partial x}\frac{dx}{dt} + \frac{\partial\phi}{\partial y}\frac{dy}{dt} + \frac{\partial\phi}{\partial z}\frac{dz}{dt} = \frac{\partial\phi}{\partial t} + \mathbf{u} \cdot \nabla\phi. \tag{7.10}$$

The derivative contains two parts. The first part represents the local rate of change of temperature of the air where the balloon is; for example, the balloon could also be carrying a heater, which would heat up the air in the vicinity of the balloon. The second term on the right-hand side ($\mathbf{u} \cdot \nabla\phi$) represents the rate of change of temperature because the balloon is moving to different locations that might have a different local temperature. Equation (7.10) is called a *material derivative* and is used to represent the rate of change of environmental variables (e.g., temperature, density, chemical concentration) within moving fluids. You will often see the material derivative written as

$$\frac{D\phi}{Dt} = \frac{\partial\phi}{\partial t} + \mathbf{u} \cdot \nabla\phi.$$

As we shall see later, some vector fields can be written as the gradient of a scalar field. If for a given vector field $\mathbf{V}$ we can find a function ϕ such that $\mathbf{V} = \nabla\phi$, then ϕ is called a *potential* and the vector field $\mathbf{V}$ is said to be a *gradient vector field.*. It turns out that every well-behaved function has a gradient vector field — we just have to be able to differentiate the function ϕ—but the converse is not true. That is, not all vector fields $\mathbf{V}$ can be written as the gradient of a function. Vector fields that *can* be written as the gradient of a function are called *conservative vector fields*. Conservative vector fields have important properties that make them useful for describing forces such as the gravitational field; for example, we will see later that the work a conservative force performs on an object as it moves is independent of the path it takes.

7.3 Divergence and Curl

We have seen that we can use the operator ∇ to find the spatial gradients of scalar fields, but what about vector fields? Taking the gradient of a vector yields a new mathematical object that will have to wait until Chapter 11 for us to explore. However, we can take vector products (the dot and cross products) of ∇ with vectors. The first of these is the *divergence*. For a vector field $\mathbf{V}(x, y, z) = V_x(x, y, z)\hat{\mathbf{\imath}} + V_y(x, y, z)\hat{\mathbf{\jmath}} + V_z(x, y, z)\hat{\mathbf{k}}$, the divergence is defined as the dot product of ∇ with $\mathbf{V}$,

$$\nabla \cdot \mathbf{V} = \operatorname{div} \mathbf{V} = \left(\hat{\mathbf{\imath}}\frac{\partial}{\partial x} + \hat{\mathbf{\jmath}}\frac{\partial}{\partial y} + \hat{\mathbf{k}}\frac{\partial}{\partial z}\right) \cdot (V_x\hat{\mathbf{\imath}} + V_y\hat{\mathbf{\jmath}} + V_z\hat{\mathbf{k}})$$
$$= \hat{\mathbf{\imath}} \cdot \hat{\mathbf{\imath}}\frac{\partial V_x}{\partial x} + \hat{\mathbf{\jmath}} \cdot \hat{\mathbf{\jmath}}\frac{\partial V_y}{\partial y} + \hat{\mathbf{k}} \cdot \hat{\mathbf{k}}\frac{\partial V_z}{\partial z} = \frac{\partial V_x}{\partial x} + \frac{\partial V_y}{\partial y} + \frac{\partial V_z}{\partial z}. \qquad (7.11)$$

The divergence of a vector field is a scalar quantity, just like the dot product of two vectors, and you will sometimes see it written as div($\mathbf{V}$).

Now, we can determine the meaning of the divergence in a nonrigorous way as follows. Let us consider a vector field $\mathbf{V}$ that represents the velocity of water. Imagine that we hold an infinitesimally small wire frame in the shape of a parallelepiped steady at a given location in the flow of water, and we want to know the net flow through the parallelepiped (Figure 7.6). We can set up coordinates (x, y, z) such that the coordinate axes are parallel to the faces of the cube. Let us first concentrate on the flow parallel to the side of length Δx. The component of the velocity in this direction that crosses the face of the parallelepiped at $x = x_a$ is $\mathbf{V}(x_a, y, z) \cdot \mathbf{n}$, where $\mathbf{n}$ is the unit normal to the face at $x = x_a$. But, we have set up coordinates such that two sides of the cube are parallel to the coordinate axes, so this component is simply $\mathbf{V}(x_a, y, z) \cdot \hat{\mathbf{\imath}} = V_x(x_a, y, z)$, where V_x is the x component of $\mathbf{V}$. Similarly, the component of the flow crossing the face of the parallelepiped at $x = x_b$ is $V_x(x_b, y, z)$. So, the total flows crossing each face are $V_x(x_a, y, z)dydz$ and $V_x(x_b, y, z)dydz$; the dimensions of a velocity are $[\mathrm{L}][\mathrm{T}]^{-1}$ and the dimensions of $V_x dydz$ are $[\mathrm{L}]^3[\mathrm{T}]^{-1}$; i.e., they represent a volume of water flowing per unit time across each face. The net flow into (or out of) the parallelepiped in the x direction is then

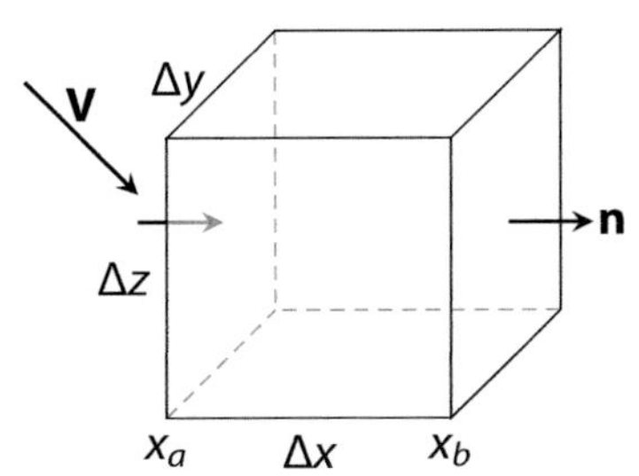

Figure 7.6 A parallelepiped with sides of length Δx, Δy, and Δz is put into a fluid flowing with a velocity vector $\mathbf{V}$. Each face of the parallelepiped has a unit vector $\mathbf{n}$ that is normal to that face.

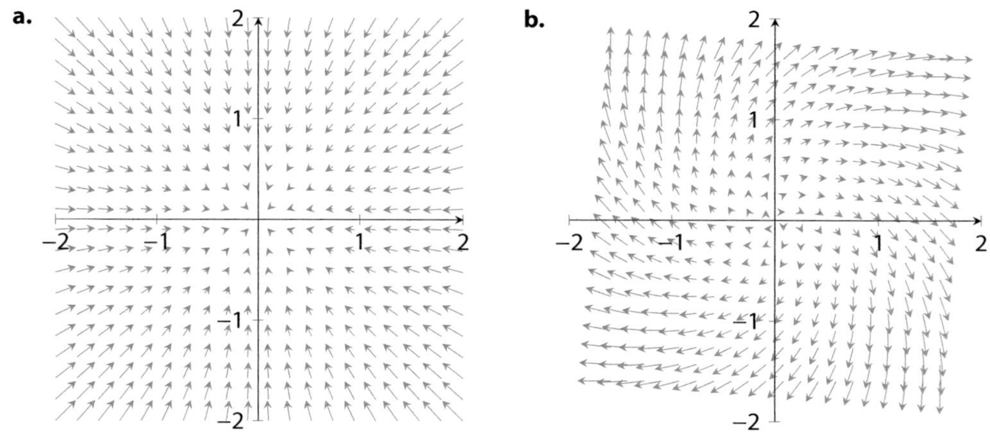

Figure 7.7 Examples of vector fields with nonzero divergence: (a.) the vector field $-x\hat{\imath} - y\hat{\jmath}$ and (b.) the vector field $(x+y)\hat{\imath} - (x-y)\hat{\jmath}$.

$$(V_x(x_a) - V_x(x_b))dydz = \left(\frac{\partial V_x}{\partial x}dx\right) dydz.$$

We can repeat this argument for the other two directions and arrive at similar expressions that, when combined, give the total net flow as

$$\left(\frac{\partial V_x}{\partial x} + \frac{\partial V_y}{\partial y} + \frac{\partial V_z}{\partial z}\right) dxdydz = \nabla \cdot \mathbf{V}dxdydz.$$

So, the divergence of the vector $\mathbf{V}$ is the net loss (or gain) rate per unit volume at a point in the fluid. For example, consider the two vector fields $\mathbf{v}_1 = -x\hat{\imath} - y\hat{\jmath}$ and $\mathbf{v}_2 = (x+y)\hat{\imath} + (-x+y)\hat{\jmath}$ (Figure 7.7). The divergence of $\mathbf{v}_1$ is

$$\nabla \cdot \mathbf{v}_1 = \frac{\partial(-x)}{\partial x} + \frac{\partial(-y)}{\partial y} = -2;$$

the vector field has a negative constant divergence and, as can be seen from Figure 7.7a, is converging toward the origin. For the vector field $\mathbf{v}_2$, we have

$$\nabla \cdot \mathbf{v}_2 = \frac{\partial}{\partial x}(x+y) + \frac{\partial}{\partial y}(-x+y) = 2,$$

so $\mathbf{v}_2$ is diverging from the origin (Figure 7.7b).

Exercise 7.3.1 Calculate the divergence of the vector field $\mathbf{v} = -(x^2 - x)\hat{\imath} - (y^2 - y)\hat{\jmath}$ and find the regions where the vector field is divergent and convergent.

Vector fields with a positive divergence are often called *sources* and those with a negative divergence are often called *sinks*. The reason for this can be seen from Figure 7.7, where it appears that a field with a negative divergence (i.e., a convergent field) is disappearing and vice versa for a field with a positive divergence. If the vector field represents the flow of a physical variable such as the mass of a pollutant carried by water, then a positive divergence indicates the presence of a source of that pollutant and a negative divergence indicates the presence of a sink. Such sources and sinks indicate the presence of reactions that create or remove the pollutant within the region of positive or negative divergence.

For example, let us consider a conserved quantity such as the mass M of a pollutant in water. We can define a vector $\mathbf{u} = \rho(x, y, z)\mathbf{v}(x, y, z)$, where ρ is the density of the pollutant as a function of position and $\mathbf{v}$ is the velocity of the fluid. The vector $\mathbf{u}$ has dimensions of $[\mathrm{M}][\mathrm{L}]^{-2}[\mathrm{T}]^{-1}$ and is called a *flux*. We have seen that the divergence represents a rate of change of a vector per unit volume V, so we might suspect that there exists a relationship that looks something like

$$\frac{\partial M}{\partial t} + S(x, y, z, t)V + \nabla \cdot \mathbf{u}V = 0 \implies \frac{\partial \rho}{\partial t} + S(x, y, z, t) + \nabla \cdot (\rho \mathbf{v}) = 0, \tag{7.12}$$

where $S(x, y, z, t)$ represents the rate per unit volume of formation or removal of the pollutant; i.e., the net combination of sources and sinks.[5] This is a *continuity* or *conservation equation* and represents the temporal and spatial changes that occur to the pollutant. This will in general be a partial differential equation that will require techniques from Chapter 10 to solve. If ρ is constant in time and uniform over space, then Equation (7.12) tells us that $\nabla \cdot \mathbf{v} = 0$, and a fluid obeying this equation is called an *incompressible* fluid.

Exercise 7.3.2 Show that Equation (7.12) is dimensionally correct.

We have seen that if $\phi(x, y, z)$ is a scalar field, then its gradient $\nabla\phi$ is a vector, so we might wonder what we get if take the divergence of $\nabla\phi$. In Cartesian coordinates, we get

$$\nabla \cdot \nabla\phi = \nabla^2\phi = \frac{\partial^2\phi}{\partial x^2} + \frac{\partial^2\phi}{\partial y^2} + \frac{\partial^2\phi}{\partial z^2}, \tag{7.13}$$

which is a scalar quantity. The operator ∇^2 is called the *Laplacian*,[6] and you will often see it written as Δ instead of ∇^2. The Laplacian appears in many partial differential equations that describe many interesting physical phenomena in the Earth and environmental sciences, as we shall see in Chapter 10.

Exercise 7.3.3 Calculate the Laplacian of the vector fields $\mathbf{v}_1 = -x\hat{\imath} - y\hat{\jmath}$, $\mathbf{v}_2 = (x + y)\hat{\imath}(-x + y)\hat{\jmath}$, and $\mathbf{v}_3 = -(x^2 - x)\hat{\imath} - (y^2 - y)\hat{\jmath}$.

Now that we have seen how to take the dot product of ∇ with a vector, let us look at what we get when we take the vector product of ∇ with a vector in Cartesian coordinates,

$$\nabla \times \mathbf{V} = \text{curl}\,\mathbf{V} = \left(\frac{\partial V_z}{\partial y} - \frac{\partial V_y}{\partial z}\right)\hat{\imath} + \left(\frac{\partial V_x}{\partial z} - \frac{\partial V_z}{\partial x}\right)\hat{\jmath} + \left(\frac{\partial V_y}{\partial x} - \frac{\partial V_x}{\partial y}\right)\hat{\mathbf{k}}$$

$$= \begin{vmatrix} \hat{\imath} & \hat{\jmath} & \hat{\mathbf{k}} \\ \frac{\partial}{\partial x} & \frac{\partial}{\partial y} & \frac{\partial}{\partial z} \\ V_x & V_y & V_z \end{vmatrix}. \tag{7.14}$$

This is called the *curl* of the vector $\mathbf{V}$, and you will often see it written as curl($\mathbf{V}$). We have to be a little careful about vector fields in two dimensions because strictly speaking they do not have a curl. This is because the only component of Equation (7.14) that is nonzero

[5] We will show a bit more rigorously that this is the case a little later in this chapter.
[6] Named after French scholar Pierre-Simon Laplace (1749–1827).

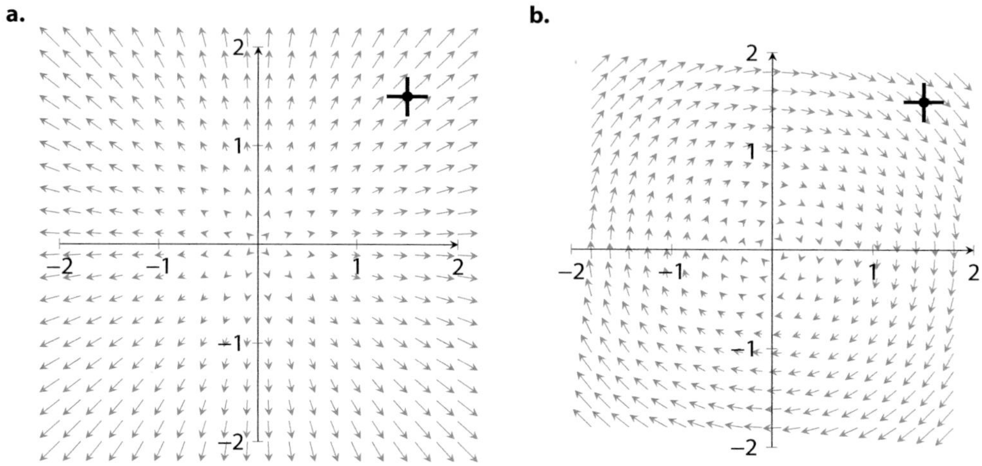

Figure 7.8 Examples of vector fields: (a.) the vector field $\mathbf{v}_1 = x\hat{\mathbf{i}} + y\hat{\mathbf{j}}$, which has a zero curl, and (b.) the vector field $\mathbf{v}_2 = y\hat{\mathbf{i}} - x\hat{\mathbf{j}}$, which has a nonzero curl. Both figures have representations of a small paddle wheel in them to show if the field has a nonzero curl.

in two dimensions is the $\hat{\mathbf{k}}$ component, and we strictly need three dimensions to calculate a curl. However, when we need to take the curl of a vector field in two dimensions, we create the fiction that there exists a third dimension and calculate the $\hat{\mathbf{k}}$ component.

Exercise 7.3.4 Show that we can write the components of curl of $\mathbf{V}$ using the Levi-Civita symbol (Equation (4.83)) so that the ith component of the curl of $\mathbf{V}$ is $(\nabla \times \mathbf{V})_i = \epsilon_{ijk}\nabla_j V_k$.

Exercise 7.3.5 Calculate the curl of the following vector fields: a. $\mathbf{v}_1 = (x + y)\hat{\mathbf{i}} - (x - y)\hat{\mathbf{j}}$, b. $\mathbf{v}_2 = x^2\hat{\mathbf{i}} + y^2\hat{\mathbf{j}} + z^2\hat{\mathbf{k}}$, c) $\mathbf{v}_3 = x^2 z\hat{\mathbf{i}} + z^2 x^2\hat{\mathbf{j}} + xy\hat{\mathbf{k}}$.

Let us look at the curl of some vector fields to try and gain some intuition into its physical interpretation. We will restrict our attention to two-dimensional vector fields for the sake of simplicity (and to make the figures easier to draw and understand!). Let us start by considering the two vector fields $\mathbf{v}_1 = x\hat{\mathbf{i}} + y\hat{\mathbf{j}}$ and $\mathbf{v}_2 = y\hat{\mathbf{i}} - x\hat{\mathbf{j}}$ (Figure 7.8). We can easily calculate the curl of these two vector fields: $\nabla \times \mathbf{v}_1 = 0$ and $\nabla \times \mathbf{v}_2 = -2\hat{\mathbf{k}}$. It would appear from this calculation that the curl of a vector field has something to do with rotation. Indeed, if we were to think of the vector field as representing the velocity of water and we placed a small paddle wheel into the vector field in Figure 7.8a, the paddle wheel would not rotate, but it would if placed anywhere in the vector field shown in Figure 7.8b. This is because in Figure 7.8a the water velocity is just pushing the paddle wheel outward, but in Figure 7.8b the water velocity is greater on one side of the paddle wheel than the other, creating a torque that causes the paddle wheel to spin.

However, we need to be a little cautious. We might think that any vector field showing a circulation has a nonzero curl, but this would be wrong, as Figure 7.9 shows. Figure 7.9a shows a vector field that seems to have a circulation, but if we calculate its curl we find that

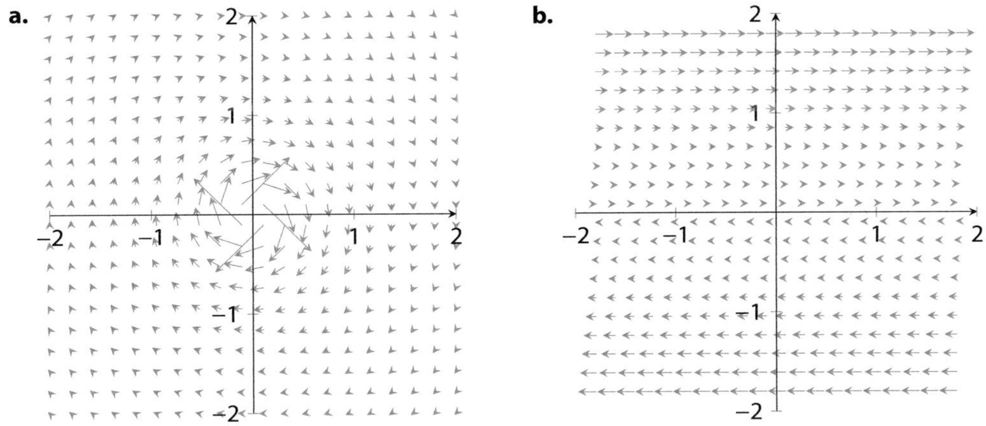

Figure 7.9 Examples of vector fields: (a.) the vector field $\mathbf{v}_1 = y/(x^2 + y^2)\hat{\imath} - x/(x^2 + y^2)\hat{\jmath}$, which has a zero curl, and (b.) the vector field $\mathbf{v}_2 = y\hat{\imath}$, which has a nonzero curl.

$$\nabla \times \mathbf{v_1} = \frac{\partial}{\partial x}\left(\frac{-x}{x^2 + y^2}\right) - \frac{\partial}{\partial y}\left(\frac{y}{x^2 + y^2}\right) = -\frac{2}{x^2 + y^2} + \frac{2(x^2 + y^2)}{(x^2 + y^2)^2} = 0,$$

whereas the curl of the vector field shown in Figure 7.9b is

$$\nabla \times \mathbf{v_2} = -\frac{\partial y}{\partial y} = -1.$$

Exercise 7.3.6 Using the paddle wheel analogy, determine why the vector field in Figure 7.9a has zero curl while the vector field in Figure 7.9b has a nonzero curl.

If the vector field shown in Figure 7.9b represents the velocity of a fluid, then it is an example of what is called a *shear* velocity field; these are ubiquitous in the environment. But such flow fields can produce a rotation, and this is a very important process in geophysical fluid dynamics where it affects the formation and evolution of cyclones and tornados in the atmosphere and currents in the ocean.

What we have seen is that the curl of a vector field is a measure of the rotation the movement of the fluid causes about a single point. It can be a good idea to always think of the curl in terms of an imaginary microscopic paddle wheel that can be inserted at any point in the vector field; if the vector field causes the paddle wheel to spin, then the flow has a nonzero curl. The curl of a vector field is frequently called the *vorticity*, which can be confusing because that makes us think of the large-scale circulation of vortices. To make matters even more confusing, a vector field that has a zero curl (e.g., the one shown in Figure 7.9a) is called an *irrotational vector field*.

7.3.1 Vector Identities

Manipulating gradients, divergences, and curls of vector fields requires some familiarity with the relationships between them. First, let us look at the curl of the gradient of a scalar:

$$\nabla \times (\nabla \phi) = \begin{vmatrix} \hat{\mathbf{i}} & \hat{\mathbf{j}} & \hat{\mathbf{k}} \\ \frac{\partial}{\partial x} & \frac{\partial}{\partial y} & \frac{\partial}{\partial z} \\ \frac{\partial \phi}{\partial x} & \frac{\partial \phi}{\partial y} & \frac{\partial \phi}{\partial z} \end{vmatrix}$$
$$= \hat{\mathbf{i}}\left(\frac{\partial^2 \phi}{\partial y \partial z} - \frac{\partial^2 \phi}{\partial z \partial y}\right) - \hat{\mathbf{j}}\left(\frac{\partial^2 \phi}{\partial x \partial z} - \frac{\partial^2 \phi}{\partial z \partial x}\right) + \hat{\mathbf{k}}\left(\frac{\partial^2 \phi}{\partial x \partial y} - \frac{\partial^2 \phi}{\partial y \partial x}\right) = 0. \tag{7.15}$$

This shows us that the curl of the gradient of a scalar field is zero. This means that if a vector field $\mathbf{V}$ can be written as the gradient of a scalar field ϕ, then the vector field is irrotational; for example, the gravitational force of a body can be written as the gradient of a scalar (the gravitation potential we met in Section 3.3), so it is an irrotational vector field.

Exercise 7.3.7 Show that $\nabla \cdot (\nabla \times \mathbf{A}) = 0$.
Exercise 7.3.8 Show that $\nabla \times (\nabla \times \mathbf{A}) = \nabla(\nabla \cdot \mathbf{A}) - \nabla^2 \mathbf{A}$.
Exercise 7.3.9 Show that $\nabla \cdot (\nabla\phi \times \nabla\psi) = 0$, where ϕ and ψ are functions.
Exercise 7.3.10 Show that $\nabla \cdot (\psi\nabla\phi - \phi\nabla\psi) = \psi\nabla^2\phi - \phi\nabla^2\psi$ for functions ϕ and ψ.

These identities are important, because they can be used to help simplify problems involving vector fields. For example, we have seen that an irrotational vector field has $\boldsymbol{\omega} = \nabla\times\mathbf{v} = 0$. Thus, Equation (7.15) then tells us that for an irrotational flow we can write $\mathbf{v} = \nabla\phi$ for some scalar function ϕ. If the flow is also incompressible, then $\nabla \cdot \mathbf{v} = 0$, so $\nabla^2\phi = 0$, which shows that if we can find ϕ by solving $\nabla^2\phi = 0$ then we know the full velocity field.

Recall that an incompressible vector field $\mathbf{v}$ has a zero divergence.[7] Exercise 7.3.7 then shows that we can find a vector $\mathbf{A}$ such that $\mathbf{v} = \nabla \times \mathbf{A}$; the vector $\mathbf{A}$ is called a *vector potential*. It turns out that we can decompose any vector field into an irrotational and an incompressible field. For example, there are two kinds of seismic wave that travel through the Earth. *Primary waves* are compression waves and are the faster moving;[8] these are described by $\nabla \times \mathbf{x} = 0$, where $\mathbf{x}(t)$ is the time dependent displacement vector of the material from its original position. The vector identities then tell us that we can write $\mathbf{x} = \nabla\phi$. *Secondary waves* are slower than the primary waves and are *shear waves* that cause displacements of the material perpendicular to the direction the wave is traveling. These waves are described by $\nabla\cdot\mathbf{x} = 0$, so we can write $\mathbf{x} = \nabla\times\mathbf{A}$ for some *vector potential* $\mathbf{A}$. The actual motion of the Earth during a seismic event will be a combination of these motions. This decomposition of a vector field $\mathbf{x}$ into a part that has $\nabla\cdot\mathbf{x} = 0$ (the *solenoidal* part) and a part that has $\nabla\times\mathbf{x} = 0$ (the *irrotational* part) is called *Helmholtz's theorem*.[9]

7.4 Curvilinear Coordinate Systems

So far in our exploration of the calculus of vectors we have mostly worked in Cartesian coordinates (x, y, z), but for many problems these may not be the most convenient coor-

[7] Incompressible vector fields are sometimes also called *solenoidal* vector fields
[8] These are like sound waves in that they compress and extend the material that they travel through.
[9] Named after the German scientist Hermann von Helmholtz (1821–1894), who also made significant contributions to our understanding of how we perceive sound and light.

dinates to use. For example, for problems that have a spherical symmetry (e.g., a problem about the Earth) where quantities depend only on the distance from the center of a sphere, working in Cartesian coordinates would involve quantities such as $r = \sqrt{x^2 + y^2 + z^2}$. This means our derivatives and integrals would become quite complicated, as we saw in Section 2.15.2. However, we can define new coordinates that take advantage of the symmetries of the sphere, making the mathematics easier. Such a coordinate system is called a *curvilinear coordinate system*, and although using such coordinate systems simplifies the formulation and solution of a problem, we lose some of the nice features of Cartesian coordinates.

In Cartesian coordinates, a point (x_0, y_0, z_0) in space is defined by the intersection of three planes: $x = x_0 =$ constant, $y = y_0 =$ constant, and $z = z_0 =$ constant (Figure 7.10). Each of theses planes is orthogonal to the others, and this is always the case no matter what the values of x_0, y_0, and z_0 are, and small differences (dx, dy, or dz) in coordinate values are real, physical distances. In addition, a normal vector to any one of these planes always points in the same direction; for example, the normal to the $x = x_0$ plane always points along the x axis in the direction of increasing values of x. We also know that we can write any vector ($\mathbf{A}$) in terms of the basis vectors $\hat{\mathbf{e}}_x = \hat{\mathbf{\imath}}$, $\hat{\mathbf{e}}_y = \hat{\mathbf{\jmath}}$, $\hat{\mathbf{e}}_z = \hat{\mathbf{k}}$, so that $\mathbf{A} = A_x\hat{\mathbf{e}}_x + A_y\hat{\mathbf{e}}_y + A_z\hat{\mathbf{e}}_z = A_x\hat{\mathbf{\imath}} + A_y\hat{\mathbf{\jmath}} + A_z\hat{\mathbf{k}}$, where A_x, A_y, and A_z are the components of $\mathbf{A}$ with respect to the basis vectors. One of the convenient aspects of Cartesian coordinates is that the basis vectors are always orthogonal to each other, so that the Cartesian basis vectors at one point in space are orthogonal to the Cartesian basis vectors at every other point. However,

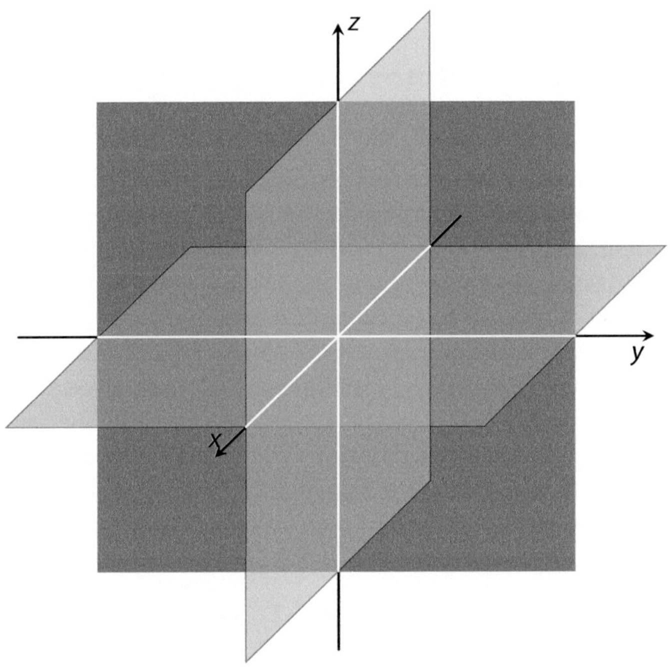

Figure 7.10 The familiar Cartesian coordinates (x, y, z) of a point defined by the intersection of the three coordinate planes.

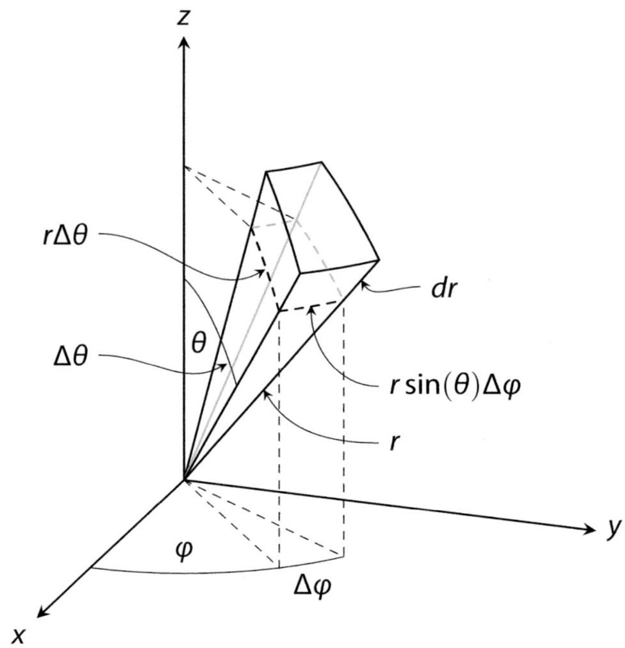

Figure 7.11 The definition of spherical coordinates (r, θ, ϕ).

we have seen that is not true in other coordinate systems such as polar coordinates (e.g., Figure 7.3)—the basis vectors are orthogonal to each other at a point, but the basis vectors at one point are not necessarily orthogonal to those at another point. This means that we have to be careful when we take the dot product of two vectors for example. In Cartesian coordinates we always have $\mathbf{A} \cdot \mathbf{B} = A_x B_x + A_y B_y + A_z B_z$, even if the vectors **A** and **B** are not defined at the same point.

Spherical coordinates are very useful for solving problems in the Earth sciences. A point in space in these coordinates is specified by a distance, the radius (r) of a point from the origin, and two angles (ϕ and θ), which we can think of as being similar to longitude and latitude on the surface of the Earth (Figure 7.11). The problem we want to address is how do we calculate spatial derivatives in such a coordinate system? To take spatial derivatives of a vector or scalar field at a point in real space we need to be able to refer to distances. Two of the coordinates in the spherical systems are not distances, but angles. So, we need to find a means of working entirely in distances. To see how to do this, let us make our mathematics simpler by working first in two-dimensional polar coordinates ((r, θ)) (Figure 7.12) before generalizing the results to any curvilinear coordinate system. We know that we can relate Cartesian coordinates to polar coordinates by $x = r\cos(\theta)$ and $y = r\sin(\theta)$ so that

$$dx = \frac{\partial x}{\partial r}dr + \frac{\partial x}{\partial \theta}d\theta = \cos(\theta)dr - r\sin(\theta)d\theta,$$
$$dy = \frac{\partial y}{\partial r}dr + \frac{\partial y}{\partial \theta}d\theta = \sin(\theta)dr + r\cos(\theta)d\theta.$$

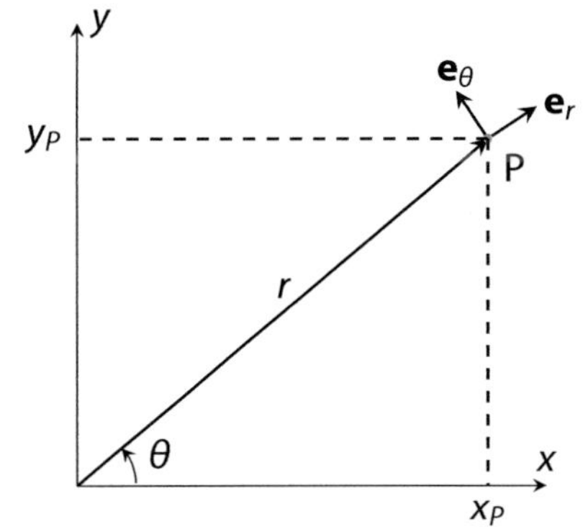

Figure 7.12 Cartesian ((x, y)) and polar ((r, θ)) coordinates in two dimensions.

Now, let us consider a small displacement $d\rho$ from the point P in Figure 7.12. By Pythagoras' theorem, we know that in Cartesian coordinates we can write $(d\rho)^2 = (dx)^2 + (dy)^2$, and because dx and dy are lengths, $d\rho$ will also be a length. What is the corresponding measurable length in polar coordinates? To determine this, we can calculate $d\rho^2$ in terms of r and θ. First, we calculate $(dx)^2$ and $(dy)^2$,

$$(dx)^2 = \left(\frac{\partial x}{\partial r}dr + \frac{\partial x}{\partial \theta}d\theta\right)^2 = \cos^2(\theta)(dr)^2 + r^2\sin^2(\theta)(d\theta)^2 - 2r\cos(\theta)\sin(\theta)drd\theta,$$

$$(dy)^2 = \left(\frac{\partial y}{\partial r}dr + \frac{\partial y}{\partial \theta}d\theta\right)^2 = \sin^2(\theta)(dr)^2 + r^2\cos^2(\theta)(d\theta)^2 + 2r\cos(\theta)\sin(\theta)drd\theta.$$

so that

$$(dx)^2 + (dy)^2 = \left[\left(\frac{\partial x}{\partial r}\right)^2 + \left(\frac{\partial y}{\partial r}\right)^2\right](dr)^2 + \left[\left(\frac{\partial x}{\partial \theta}\right)^2 + \left(\frac{\partial y}{\partial \theta}\right)^2\right](d\theta)^2,$$

because the cross terms

$$2\left(\frac{\partial x}{\partial r}\frac{\partial x}{\partial \theta} + \frac{\partial y}{\partial r}\frac{\partial y}{\partial \theta}\right)drd\theta$$

cancel out, leaving us with

$$(dx)^2 + (dy)^2 = (dr)^2 + r^2(d\theta)^2 = h_r^2(dr)^2 + h_\theta^2(d\theta)^2,$$

where $h_r = 1$ and $h_\theta = r$ are called the *scale factors*. The quantities dr and $rd\theta$ are actual distances, and we can use them to calculate derivatives. For example, the vector gradient of a scalar function in two-dimensional polar coordinates is

$$\nabla\Phi(r,\theta) = \frac{\partial \Phi}{\partial r}\hat{\mathbf{e}}_r + \frac{1}{r}\frac{\partial \Phi}{\partial \theta}\hat{\mathbf{e}}_\theta.$$

Let us generalize this to any curvilinear coordinate system. We will work in three dimensions and set the coordinates of a point P as (p_1, p_2, p_3) in the curvilinear coordinates and (x, y, z) in Cartesian coordinates. Then x, y, and z are functions of p_1, p_2, and p_3, and we can write

$$dx = \frac{\partial x}{\partial p_1} dp_1 + \frac{\partial x}{\partial p_2} dp_2 + \frac{\partial x}{\partial p_3} dp_3,$$

with similar expressions for dy and dz. Then

$$(dx)^2 = \sum_i^3 \sum_j^3 \frac{\partial x}{\partial p_i} \frac{\partial x}{\partial p_j} dp_i \, dp_j = \sum_{i,j} \frac{\partial x}{\partial p_i} \frac{\partial x}{\partial p_j} dp_i \, dp_j,$$

where we have written $\sum_{i,j}$ as shorthand for the double summation. We can write down similar expressions for $(dy)^2$ and $(dz)^2$, and combining these gives

$$\begin{aligned}(dr)^2 &= (dx)^2 + (dy)^2 + (dz)^2 \\ &= \sum_{i,j} \frac{\partial x}{\partial p_i} \frac{\partial x}{\partial p_j} dp_i \, dp_j + \sum_{i,j} \frac{\partial y}{\partial p_i} \frac{\partial y}{\partial p_j} dp_i \, dp_j + \sum_{i,j} \frac{\partial z}{\partial p_i} \frac{\partial z}{\partial p_j} dp_i \, dp_j \\ &= \sum_{i,j} g_{ij} \, dp_i \, dp_j,\end{aligned}$$

where the matrix g_{ij} is called the *metric* and is defined by

$$g_{ij} = \frac{\partial x}{\partial p_i} \frac{\partial x}{\partial p_j} + \frac{\partial y}{\partial p_i} \frac{\partial y}{\partial p_j} + \frac{\partial z}{\partial p_i} \frac{\partial z}{\partial p_j}.$$

If the curvilinear coordinates (p_1, p_2, p_3) are orthogonal (i.e., the lines of constant coordinate values are orthogonal to each other), then the off-diagonal terms are zero (i.e., $g_{ij} = 0$ if $i \neq j$). So, for an orthogonal curvilinear system (i.e., the only nonzero terms in g_{ij} are those with $i = j$) we can define scale factors

$$h_i^2 = g_{ii} = \left(\frac{\partial x}{\partial q_i}\right)^2 + \left(\frac{\partial y}{\partial q_i}\right)^2 + \left(\frac{\partial z}{\partial q_i}\right)^2$$

such that $(dr)^2 = (h_1 dp_1)^2 + (h_2 dp_2)^2 + (h_3 dp_3)^2$. The vector line element in the curvilinear coordinates is then $d\mathbf{r} = h_1 dp_1 \hat{\mathbf{e}}_1 + h_2 dp_2 \hat{\mathbf{e}}_2 + h_3 dp_3 \hat{\mathbf{e}}_3$, where the quantities $h_i dp_i$ are actual lengths in the direction given by the basis vectors $\hat{\mathbf{e}}_i$. Notice that for a Cartesian coordinate system, $h_x = h_y = h_z = 1$ do not depend on the location of the point.

We can now define the gradient of a scalar in a general orthogonal curvilinear coordinate system:

$$\nabla\phi(p_1, p_2, p_3) = \hat{\mathbf{e}}_1 \frac{1}{h_1} \frac{\partial \phi}{\partial p_1} + \hat{\mathbf{e}}_2 \frac{1}{h_2} \frac{\partial \phi}{\partial p_2} + \hat{\mathbf{e}}_3 \frac{1}{h_3} \frac{\partial \phi}{\partial p_3}. \tag{7.16}$$

When we defined the divergence of a vector field we looked at the flows in and out of a rectangular box. In a curvilinear coordinate system, the box is no longer rectangular. However, the same principle applies and the logic of the derivation is the same, so for a vector $\mathbf{A}$ with components (A_1, A_2, A_3) in a curvilinear coordinate system with coordinates (p_1, p_2, p_3) the divergence is

$$\nabla \cdot \mathbf{A} = \frac{1}{h_1 h_2 h_3} \left(\frac{\partial}{\partial p_1}(A_1 h_2 h_3) + \frac{\partial}{\partial p_2}(A_2 h_3 h_1) + \frac{\partial}{\partial p_3}(A_3 h_1 h_2) \right) \tag{7.17}$$

and the curl is

$$\nabla \times \mathbf{A} = \frac{1}{h_1 h_2 h_3} \begin{vmatrix} h_1 \hat{\mathbf{e}}_1 & h_2 \hat{\mathbf{e}}_2 & h_3 \hat{\mathbf{e}}_3 \\ \frac{\partial}{\partial p_1} & \frac{\partial}{\partial p_2} & \frac{\partial}{\partial p_3} \\ h_1 A_1 & h_2 A_2 & h_3 A_3 \end{vmatrix}, \tag{7.18}$$

and lastly, the Laplacian is

$$\nabla^2 \phi = \frac{1}{h_1 h_2 h_3} \left(\frac{\partial}{\partial p_1} \left(\frac{h_2 h_3}{h_1} \frac{\partial \phi}{\partial p_1} \right) + \frac{\partial}{\partial p_2} \left(\frac{h_3 h_1}{h_2} \frac{\partial \phi}{\partial p_2} \right) + \frac{\partial}{\partial p_3} \left(\frac{h_1 h_2}{h_3} \frac{\partial \phi}{\partial p_3} \right) \right). \tag{7.19}$$

Now we can answer our original question as to what the spatial derivatives look like in three-dimensional spherical coordinates.

Example 7.1 Let us calculate expressions for the gradient, divergence, curl, and Laplacian in spherical coordinates. First, we need to calculate the scale factors. The relationships between Cartesian $((x, y, z))$ and spherical $((r, \phi, \theta))$ coordinates are

$$x = r\cos(\phi)\sin(\theta), \quad y = r\sin(\phi)\sin(\theta), \quad z = r\cos(\theta),$$

so the scale factors are given by

$$h_r^2 = \left(\frac{\partial x}{\partial r}\right)^2 + \left(\frac{\partial y}{\partial r}\right)^2 + \left(\frac{\partial z}{\partial r}\right)^2 = 1$$

$$h_\phi^2 = \left(\frac{\partial x}{\partial \phi}\right)^2 + \left(\frac{\partial y}{\partial \phi}\right)^2 + \left(\frac{\partial z}{\partial \phi}\right)^2 = r^2 \sin^2(\theta)$$

$$h_\theta^2 = \left(\frac{\partial x}{\partial \theta}\right)^2 + \left(\frac{\partial y}{\partial \theta}\right)^2 + \left(\frac{\partial z}{\partial \theta}\right)^2 = r^2.$$

Therefore,

$$\nabla f(r, \phi, \theta) = \hat{\mathbf{e}}_r \frac{\partial f}{\partial r} + \hat{\mathbf{e}}_\phi \frac{1}{r\sin(\theta)} \frac{\partial f}{\partial \phi} + \hat{\mathbf{e}}_\theta \frac{1}{r} \frac{\partial f}{\partial \theta},$$

from which we can write that the gradient operator in spherical coordinates is

$$\nabla = \hat{\mathbf{e}}_r \frac{\partial}{\partial r} + \hat{\mathbf{e}}_\phi \frac{1}{r\sin(\theta)} \frac{\partial}{\partial \phi} + \hat{\mathbf{e}}_\theta \frac{1}{r} \frac{\partial}{\partial \theta}.$$

Similarly, the divergence of a vector $\mathbf{u} = (u_r, u_\theta, u_\phi)$ is

$$\nabla \cdot \mathbf{u} = \frac{1}{r^2} \frac{\partial}{\partial r}(r^2 u_r) + \frac{1}{r\sin(\theta)} \frac{\partial}{\partial \theta}(\sin(\theta) u_\theta) + \frac{1}{r\sin(\theta)} \frac{\partial u_\phi}{\partial \phi},$$

the curl is

$$\nabla \times \mathbf{u} = \frac{1}{r\sin(\theta)} \left(\frac{\partial}{\partial \theta}(u_\phi \sin(\theta)) - \frac{\partial u_\theta}{\partial \phi} \right) \hat{\mathbf{e}}_r + \frac{1}{r\sin(\theta)} \left(\frac{\partial u_r}{\partial \phi} - \sin(\theta) \frac{\partial r u_\phi}{\partial r} \right) \hat{\mathbf{e}}_\theta + \frac{1}{r} \left(\frac{\partial r u_\theta}{\partial r} - \frac{\partial u_r}{\partial \theta} \right) \hat{\mathbf{e}}_\phi,$$

and the Laplacian is

$$\nabla^2\phi = \frac{1}{r^2}\frac{\partial}{\partial r}\left(r^2\frac{\partial\phi}{\partial r}\right) + \frac{1}{r^2\sin(\theta)}\frac{\partial}{\partial\theta}\left(\sin(\theta)\frac{\partial\phi}{\partial\theta}\right) + \frac{1}{r^2\sin^2(\theta)}\frac{\partial^2\phi}{\partial\phi^2}. \tag{7.20}$$

Exercise 7.4.1 Show that Equations 7.16–7.19 are correct.

Exercise 7.4.2 Calculate the gradient and Laplacian of the scalar function $(x^2 + y^2 + z^2)^{3/2}$ using Cartesian coordinates, and then calculate the gradient and Laplacian of r^3 in spherical polar coordinates.

Exercise 7.4.3 For a vector $\mathbf{A} = A_r\hat{\mathbf{e}}_r + A_\phi\hat{\mathbf{e}}_\phi + A_\theta\hat{\mathbf{e}}_\theta$ in spherical coordinates, calculate expressions for the divergence and curl of $\mathbf{A}$.

Exercise 7.4.4 Calculate the Laplacian of $1/r$ in spherical coordinates.

7.5 Integrals and Vectors

We know what it means to integrate a function, but how do we interpret the integral of a vector field? Let us consider the specific case of the force $\mathbf{F}(x, y, z)$ acting on a particle as it moves through the atmosphere or the ocean. If this force causes the particle to move through a distance $d\mathbf{r}$, then the *instantaneous work* done by the force on the particle is $\mathbf{F}\cdot d\mathbf{r}$. As the particle moves to a new position, the magnitude and direction of the force may change, so the work will change as the particle moves through space (Figure 7.13). If

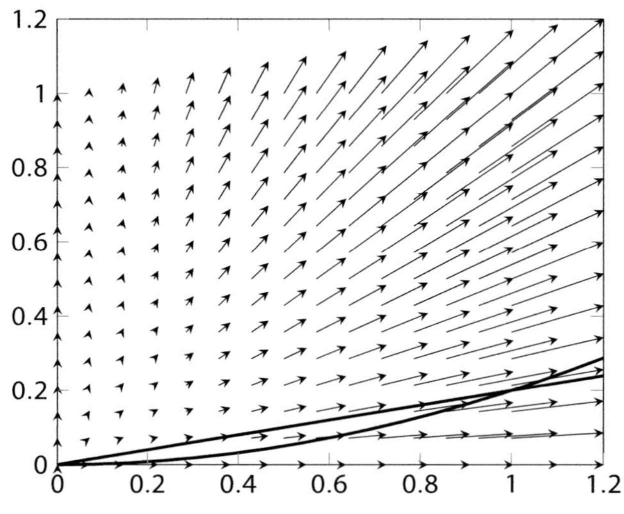

Figure 7.13 The vector field ($\mathbf{F}$) and two paths ($y = x/5$ and $y = x^2/5$) from Example 7.2, showing that particles following different paths will experience different magnitudes and directions of the vector field $\mathbf{F}$.

we know the trajectory that the particle follows, then we can calculate the total work done by the force on the particle by evaluating the *line integral*

$$\int \mathbf{F}\cdot d\mathbf{r}$$

over the whole trajectory. The integrand is a scalar function, so we can use the methods we developed in Section 2.15.1 to evaluate the integral.

Example 7.2 A particle moves in a force field given by $\mathbf{F} = x^2\hat{\imath} + xy\hat{\jmath}$ along the paths (a) $y = x/5$ and (b) $y = x^2/5$ between $x = 0$ and $x = 1$. Let us calculate the total work done by the force on the particle along the two different paths. The general position vector in two-dimensions is $\mathbf{r} = x\hat{\imath} + y\hat{\jmath}$, so $d\mathbf{r} = dx\hat{\imath} + dy\hat{\jmath}$ and $\mathbf{F}\cdot d\mathbf{r} = x^2\,dx + xy\,dy$. For this example, we will use the equation of the paths to substitute for one of the variables in the integral. For path (a), $y = x/5$ and $dy = dx/5$, so

$$\int \mathbf{F}\cdot d\mathbf{r} = \int x^2\,dx + xy\,dy = \int_{x=0}^{x=1}\left(x^2 + \frac{x^2}{25}\right)dx = \int_{x=0}^{x=1}\frac{26}{25}x^2\,dx = \frac{26}{75}.$$

For path (b), $y = x^2/5$ and $dy = 2x^2/5$, so

$$\int \mathbf{F}\cdot d\mathbf{r} = \int_0^1\left(x^2 + \frac{2}{25}x^4\right)dx = \frac{131}{375}.$$

There are some interesting aspects to Example 7.2. The first is that even though both paths start and end at the same points ($(x, y) = (0, 0)$ and $(x, y) = (0.2, 0.2)$), the work done by the force is different for the two paths. This is understandable because the vector field is different for the two paths. But, this need not always be the case, and vector fields for which the line integral is the same irrespective of the path taken are called *conservative fields*—the vector field in Example 7.2 is a *nonconservative* vector field. Conservative fields play an important role in many real-world problems that we come across in the Earth and environmental sciences. For example, if a vector field representing a force is a conservative field, then it means that the work done by that force on a particle moving between two points A and B is the same irrespective of the path that the particle takes between A and B. One important example of a conservative field is the gravitational field, where work done by the gravitational field of a perfect sphere depends on only the change in distance from the center of the sphere.

We can tell if a vector field $\mathbf{F}$ is conservative by evaluating the integral of $\mathbf{F}\cdot d\mathbf{r}$ over different paths, but there is another way that is often simpler. Let us assume that we can write $\mathbf{F}$ as the gradient of some scalar function ϕ; i.e., $\mathbf{F} = \nabla\phi$. We can parameterize the position vector $\mathbf{r}(s) = x(s)\hat{\imath} + y(s)\hat{\jmath} + z(s)\hat{\mathbf{k}}$ with the parameter s so that

$$\mathbf{F}\cdot d\mathbf{r} = \nabla\phi\cdot d\mathbf{r} = \frac{\partial\phi}{\partial x}\frac{dx}{ds}\,ds + \frac{\partial\phi}{\partial y}\frac{dy}{ds}\,ds + \frac{\partial\phi}{\partial z}\frac{dz}{ds}\,ds.$$

If we now calculate the line integral of this quantity from point A to point B, we get

$$\int_A^B \mathbf{F}\cdot d\mathbf{r} = \int_A^B \left(\frac{\partial\phi}{\partial x}\frac{dx}{ds} + \frac{\partial\phi}{\partial y}\frac{dy}{ds} + \frac{\partial\phi}{\partial z}\frac{dz}{ds}\right) ds = \int_A^B \frac{d}{ds}\phi\, ds = \phi(B) - \phi(A);$$

in other words, the integral just depends on the value of ϕ at the two end points and not on how we got from A to B. This is a form of the fundamental theorem of calculus (Section 2.10), but applied to line integrals. We have already seen (Equation (7.15)) that $\nabla \times (\nabla\phi) = 0$; therefore if $\mathbf{F} = \nabla\phi$, then $\nabla \times \mathbf{F} = 0$; therefore we have shown that a conservative vector field must also be irrotational, as we have stated before, so we can write down the *Gradient theorem*

$$\int_A^B \nabla\phi \cdot d\mathbf{r} = \phi(B) - \phi(A). \tag{7.21}$$

In other words, the line integral of the vector field along a path is given by the difference in the value of the potential function at the start and end of the path.

Example 7.3 Let us use what we have found out to show that the vector field $\mathbf{F} = x^2\hat{\mathbf{\imath}} + xy\hat{\mathbf{\jmath}}$ is conservative. The easiest way to show this is to calculate the curl of the vector field,

$$\nabla \times \mathbf{F} = \left(\hat{\mathbf{\imath}}\frac{\partial}{\partial x} + \hat{\mathbf{\jmath}}\frac{\partial}{\partial y}\right) \times (x^2\hat{\mathbf{\imath}} + xy\hat{\mathbf{\jmath}}) = y\hat{\mathbf{k}} \neq 0,$$

so the vector field is nonconservative.

Exercise 7.5.1 If the vector $\mathbf{v}(\ell)$ represents the velocity of a particle moving along a path parameterized by the parameter ℓ, what is the value of

$$\int_{\ell=a}^{\ell=b} \mathbf{v}(\ell)\, d\ell,$$

and how would you interpret this value?

This raises the question, if we already have a conservative vector field $\mathbf{F}$, how do we find the potential function ϕ? Let us consider a vector field $\mathbf{F} = y^2/2\hat{\mathbf{\imath}} + (xy+z)\hat{\mathbf{\jmath}} + (y)\hat{\mathbf{k}}$. We know that if $\phi(x, y, z)$ is a potential function for $\mathbf{F}$, then $\mathbf{F} = \nabla\phi$, so ϕ must satisfy the equations

$$\frac{\partial\phi}{\partial x} = \frac{1}{2}y^2, \quad \frac{\partial\phi}{\partial y} = xy + z, \quad \frac{\partial\phi}{\partial k} = y. \tag{7.22}$$

We start by choosing one of these equations and integrating it. Starting with the first equation we get

$$\phi(x, y, z) = \int \left(\frac{1}{2}y^2\, dx\right) = \frac{1}{2}y^2 x + C_1(y, z),$$

where instead of a constant of integration we have a function $C_1(y, z)$, because such a function will be zero when we differentiate $\phi(x, y, z)$ with respect to x. We can now differentiate this function with respect to y and compare the result with the second equation in Equation (7.22):

$$\frac{\partial\phi}{\partial y} = xy + \frac{\partial C_1}{\partial y} = xy + z.$$

We can solve this equation for the derivative and integrate to get

$$C_1 = \int z\,dy = zy + C_2(z),$$

so the potential becomes

$$\phi(x, y, z) = \frac{1}{2}y^2x + zy + C_2(z).$$

We can now differentiate this with respect to z and use the last equation in Equation (7.22) to get $C_2(z) = 0$, so our potential function is

$$\phi(x, y, z) = \frac{1}{2}y^2x + zy.$$

Strictly, $\phi(x, y, z)$ should involve an integration constant, but in practice this is often neglected because with conservative vector fields we are mostly interested in the differences in the potential at different locations (Equation (7.21)), so the constant will cancel out.

Exercise 7.5.2 Find the potential function for the vector field $\mathbf{F} = (yz + 2)\hat{\mathbf{i}} + (xz + 4y)\hat{\mathbf{j}} + (xy + 2)\hat{\mathbf{k}}$.

This seems all very neat and tidy, but we have neglected a few crucial details. To examine these, let us consider a two-dimensional vector field

$$\mathbf{F} = -\frac{\sin(\theta)}{r}\hat{\mathbf{i}} + \frac{\cos(\theta)}{r}\hat{\mathbf{j}} = -\frac{y}{x^2 + y^2}\hat{\mathbf{i}} + \frac{x}{x^2 + y^2}\hat{\mathbf{j}} = A(x, y)\hat{\mathbf{i}} + B(x, y)\hat{\mathbf{j}}, \tag{7.23}$$

where $r = \sqrt{x^2 + y^2}$ is the length of the position vector and (r, θ) are polar coordinates. Now, if $\phi(x, y)$ is a potential function for the vector field, then

$$\frac{\partial \phi}{\partial x} = A(x, y) = -\frac{y}{x^2 + y^2}, \quad \frac{\partial \phi}{\partial y} = B(x, y) = \frac{x}{x^2 + y^2}.$$

We can start by integrating either equation and we arrive at the potential function

$$\phi(x, y) = \tan^{-1}\left(\frac{y}{x}\right), \tag{7.24}$$

which satisfies both equations. This looks great, but there is a problem—a very serious problem. Let us consider what happens if $x = 0$. In that case,

$$\frac{\partial \phi}{\partial y} = B(x, y) = 0,$$

which implies that when $x = 0$, ϕ = constant = C. But what is the value of C? Well, if we let $x \to 0$ from above in Equation (7.24), then $\phi(x, y) \to (\pi/2)$. This looks all right. But, if we let $x \to 0$ from below, then $\phi(x, y) \to -(\pi/2)$, and we can see that we have a problem—the potential function has a discontinuity at $x = 0$. As a result, we have to exclude that line from our calculations; in effect we "cut it out" of the plane so that the (x, y) plane effectively has a tear in it. In mathematical terms, for the potential function to exist we require that the region in which we define the function to be *simply connected*, which loosely means that there are no holes or tears in the region of the plane we are considering.

How do we integrate a vector field along a more complicated path, such as a helical path for example? The answer is that we parameterize the path, just as we did when we integrated scalar functions along a path in Section 2.15.1. We can represent a path in space by coordinates that change as a parameter varies. For example, if a point P is represented by the coordinates (x, y, z), then if each of these coordinates is a function of a parameter λ, the point will move in space as λ changes because the values of x, y, and z will change; the point P will trace out a path, $P(\lambda)$. So we can describe the path using three one-parameter functions, $x(\lambda)$, $y(\lambda)$, and $z(\lambda)$, where $\lambda = a$ at the start of the path and $\lambda = b$ at the end of the path; if $P(\lambda = a) = P(\lambda = b)$, then the path is a closed path. We can write this path using a position vector $\mathbf{r}(\lambda) = x(\lambda)\hat{\mathbf{\imath}} + y(\lambda)\hat{\mathbf{\jmath}} + z(\lambda)\hat{\mathbf{k}}$ and the tangent vector to the curve is given by the derivative

$$\frac{d\mathbf{r}}{d\lambda} = \frac{dx}{d\lambda}\hat{\mathbf{\imath}} + \frac{dy}{d\lambda}\hat{\mathbf{\jmath}} + \frac{dz}{d\lambda}\hat{\mathbf{k}}.$$

To see how we find a suitable parameterization of a curve, let us look at some examples. The equation $x^2 + y^2 = 4$ describes a circle of radius 2 and centered on the origin. We know that we can use polar coordinates to write $x(\lambda) = 2\cos(\lambda)$ and $y(\lambda) = 2\sin(\lambda)$ so that as λ varies from 0 to 2π, the position vector $\mathbf{r} = 2\cos(\lambda)\hat{\mathbf{\imath}} + 2\sin(\lambda)\hat{\mathbf{\jmath}}$ describes the circle. The tangent vector to the circle is then (Figure 7.14)

$$\frac{d\mathbf{r}}{d\lambda} = -2\sin(\lambda)\hat{\mathbf{\imath}} + 2\cos(\lambda)\hat{\mathbf{k}}.$$

Exercise 7.5.3 The planes described by the equations $2x - y + z = 5$ and $x + y - z = 1$ meet in a line. Solve the two equations describing the planes by setting z as a free, undetermined parameter and show that the tangent vector of the line is $\hat{\mathbf{\jmath}} + \hat{\mathbf{k}}$.

Finding a suitable parameterization might not always be easy, but we found in Section 2.15.1 that we could parameterize curves using the path length. Similarly, reformulating this concept in terms of vectors will also provide a natural parameterization

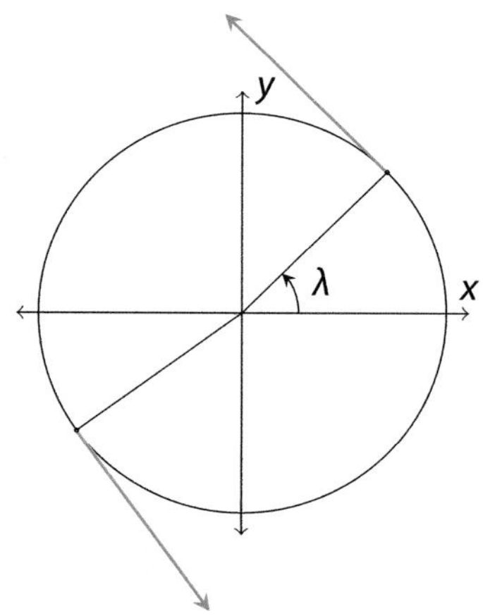

Figure 7.14 The circle $x^2 + y^2 = 4$ being traced out by the position vector $\mathbf{r} = 2\cos(\lambda)\hat{\mathbf{\imath}} + 2\sin(\lambda)\hat{\mathbf{\jmath}}$ with tangent vectors shown at two locations.

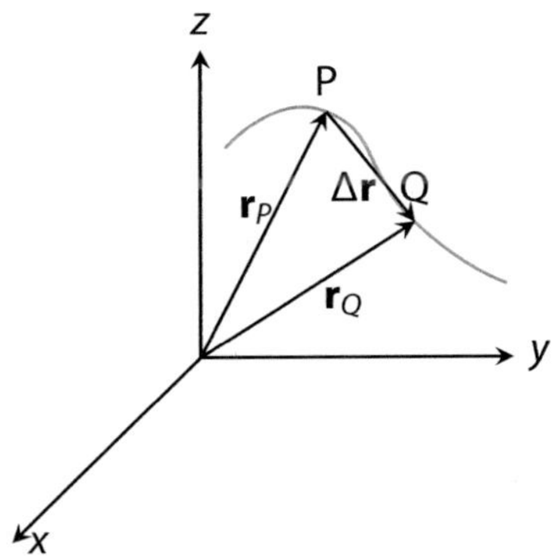

Figure 7.15 The derivation of path length using vectors. The path (in gray) connects points P, with position vector $\mathbf{r}_P$, and Q, with position vector $\mathbf{r}_Q$. The vector $\Delta\mathbf{r}$ connects the points P and Q.

of a path. Let us consider two points, P and Q, on a path in three dimensions (Figure 7.15). The position vectors $\mathbf{r}_P$ and $\mathbf{r}_Q$ connect the origin to the two points, and the vector $\Delta\mathbf{r} = \mathbf{r}_Q - \mathbf{r}_P$ connects the two points so that

$$|\Delta\mathbf{r}| = |\mathbf{r}_Q - \mathbf{r}_P| = \sqrt{\Delta\mathbf{r}\cdot\Delta\mathbf{r}}.$$

If the path is parameterized with a parameter λ, then $\mathbf{r} = \mathbf{r}(\lambda)$ and

$$|\Delta\mathbf{r}| = \sqrt{\Delta\mathbf{r}\cdot\Delta\mathbf{r}} = \left(\frac{\Delta\mathbf{r}}{\Delta\lambda}\cdot\frac{\Delta\mathbf{r}}{\Delta\lambda}\right)^{1/2}\Delta\lambda,$$

where $\Delta\lambda = \lambda_Q - \lambda_P$. If we now take the limit as the distance between P and Q tends to zero, the distance $|\Delta\mathbf{r}|$ will tend to the path length ds given by

$$ds = \left(\frac{d\mathbf{r}}{d\lambda}\cdot\frac{d\mathbf{r}}{d\lambda}\right)^{1/2} d\lambda = (\mathbf{r}'\cdot\mathbf{r}')^{1/2}d\lambda,$$

where a prime represents a derivative with respect to the parameter λ. This means that the path length between two points, P and Q, on a path is

$$s = \int_{\lambda_P}^{\lambda_Q} ds = \int_{\lambda_P}^{\lambda_Q} (\mathbf{r}'\cdot\mathbf{r}')^{1/2}d\lambda,$$

where $\lambda_P < \lambda_Q$, because otherwise we would have a negative path length.

Exercise 7.5.4 Calculate the path length for the path given by $\mathbf{r}(\lambda) = 2\cos(\lambda)\hat{\mathbf{i}} + 2\sin(\lambda)\hat{\mathbf{j}}$ for the parameter λ varying between $\lambda = 0$ to $\lambda = 2\pi$.

Now, let us see how we can apply this to evaluating the line integral of the dot product $\mathbf{F}\cdot d\mathbf{r}$ along a parameterized path. We first need to know how to write $d\mathbf{r}$ (not $|d\mathbf{r}|$) in terms of $d\lambda$; we need to have a vector that we can use to take the dot product with $\mathbf{F}$. Consider an object moving along the path given by the position vector $\mathbf{r}$. At any instant the object is moving in the *direction* of the tangent to the curve at that point on the curve. So, we can define a unit vector that points in the direction of the tangent vector as

$$\frac{\mathbf{r}'(\lambda)}{|\mathbf{r}'(\lambda)|},$$

so, writing $d\mathbf{r}$ as a direction (given by the unit vector) multiplied by a magnitude gives

$$d\mathbf{r} = \left(\frac{\mathbf{r}'(\lambda)}{|\mathbf{r}'(\lambda)|}\right) |\mathbf{r}'(\lambda)|\, d\lambda = \mathbf{r}'(\lambda)\, d\lambda.$$

We can now write the line integral of $\mathbf{F} \cdot d\mathbf{r}$ along a curve C from point P to point Q as

$$\int_C \mathbf{F} \cdot d\mathbf{r} = \int_{\lambda_P}^{\lambda_Q} \mathbf{F}(\mathbf{r}(\lambda)) \cdot \mathbf{r}'(\lambda)\, d\lambda, \tag{7.25}$$

where λ_P and λ_Q are the values of the parameter λ at points P and Q on C. For example, let us calculate the line integral of the vector $\mathbf{F} = x^2\hat{\mathbf{\imath}} + xy\hat{\mathbf{\jmath}}$ on the parameterized path (which should be recognizable) $\mathbf{r}(\lambda) = \cos(\lambda)\hat{\mathbf{\imath}} + \sin(\lambda)\hat{\mathbf{\jmath}}$ from $\lambda = 0$ to $\lambda = \pi$. We have $\mathbf{F}(\mathbf{r}(\lambda)) = \cos^2(\lambda)\hat{\mathbf{\imath}} + \cos(\lambda)\sin(\lambda)\hat{\mathbf{\jmath}}$ and $\mathbf{r}'(\lambda) = -\sin(\lambda)\hat{\mathbf{\imath}} + \cos(\lambda)\hat{\mathbf{\jmath}}$, giving

$$\int_C \mathbf{F} \cdot d\mathbf{r} = \int_0^{\pi} (-\cos^2(\lambda)\sin(\lambda) + \cos^2(\lambda)\sin(\lambda)\, d\lambda) = 0.$$

We also need to know how to integrate a vector field over a surface. For example, we can represent the flux of energy from the Sun as a vector that strikes a surface such as the Earth at an angle (e.g., Figure 4.12). To find the total amount of energy incident on the whole, or part, of the Earth, we need to integrate the dot product of the flux vector and the normal to surface as the vectors vary over the surface. We have seen that we can parameterize a curve with a one-parameter family of functions, so perhaps we can parameterize a surface with a two-parameter family of functions such that the position vector $\mathbf{r}$ of any point on the surface can be written as

$$\mathbf{r}(\lambda\mu) = x(\lambda, \mu)\hat{\mathbf{\imath}} + y(\lambda, \mu)\hat{\mathbf{\jmath}} + z(\lambda, \mu)\hat{\mathbf{k}}.$$

For example, we know that we can represent any point on the surface of a sphere of radius R by two coordinates, latitude and longitude, so a suitable parameterization in this case would be $\mathbf{r}(\lambda\mu) = R\cos(\lambda)\sin(\mu)\hat{\mathbf{\imath}} + R\sin(\lambda)\sin(\mu)\hat{\mathbf{\jmath}} + R\cos(\lambda)\hat{\mathbf{k}}$. When looking at curves, we calculated a small element ($d\mathbf{r}$) of the curve by looking at the tangent vector to the curve and this allowed us to relate it to the parameter interval $d\lambda$. Surfaces are a little more complicated because we have two coordinates and two tangent vectors, one for each coordinate curve, defining a plane (Figure 7.16).[10] If the position vector $\mathbf{r}(\lambda, \mu)$ defines a position in the surface $\mathcal{S}$, then the tangent vector along the coordinate lines μ = constant is given by the partial derivative of $\mathbf{r}(\lambda, \mu)$ with respect to λ ($\mathbf{r}_\lambda$) and vice versa. We know from Section 4.3.2 that, so long as the two vectors $\mathbf{r}_\lambda$ and $\mathbf{r}_\mu$ are not parallel, $\mathbf{r}_\lambda(P)$ and $\mathbf{r}_\mu(P)$ define a plane at point P, and this plane is called the *tangent plane*. We also know from Section 4.3.2 that the unit normal vector to this plane is given by

$$\hat{\mathbf{n}} = \frac{\mathbf{r}_\lambda \times \mathbf{r}_\mu}{|\mathbf{r}_\lambda \times \mathbf{r}_\mu|}. \tag{7.26}$$

[10] Recall from Chapter 4 that any two vectors define a plane.

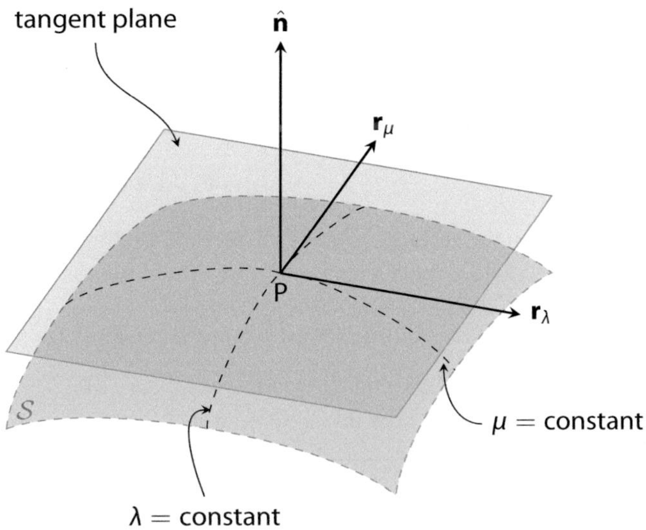

Figure 7.16 The tangent plane to a surface $\mathcal{S}$ at a point P. The vectors $\mathbf{r}_\lambda$ and $\mathbf{r}_\mu$ are tangent to the coordinate curves μ = constant and λ = constant in the surface $\mathcal{S}$ and lie in the tangent plane. The vector $\hat{\mathbf{n}}$ is normal to both $\mathbf{r}_\lambda$ and $\mathbf{r}_\mu$.

There is an ambiguity in sign here because $\mathbf{r}_\lambda \times \mathbf{r}_\mu = -\mathbf{r}_\mu \times \mathbf{r}_\lambda$, so we always have to be aware of the direction of the normal to a surface. Mostly, this is defined as being positive for an outward-pointing normal, but you should always check to make sure what convention an author is using.

We now want the position vector $\mathbf{r}(\lambda, \mu)$ to move over the surface, which means that we need coordinates λ and μ to be functions of a parameter τ (i.e., we have $\lambda = \lambda(\tau)$ and $\mu = \mu(\tau)$), and as τ varies, the position vector will draw out a curve C on the surface $\mathcal{S}$. The tangent to that curve is

$$\frac{d\mathbf{r}}{d\tau} = \frac{d}{d\tau}(\mathbf{r}(\lambda(\tau), \mu(\tau))) = \mathbf{r}_\lambda \frac{d\lambda}{d\tau} + \mathbf{r}_\mu \frac{d\mu}{d\tau},$$

showing us that this tangent vector is a linear combination of the vectors $\mathbf{r}_\lambda$ and $\mathbf{r}_\mu$, so it also lies in the tangent plane to the surface.

We can calculate the equation of the tangent plane at any point in the surface using what we learned in Section 4.3.2.1. Briefly, if P is a given point in the surface with a position vector $\mathbf{r}(P) = x(P)\hat{\mathbf{\imath}} + y(P)\hat{\mathbf{\jmath}} + z(P)\hat{\mathbf{k}}$ and $\mathbf{r} = x\hat{\mathbf{\imath}} + y\hat{\mathbf{\jmath}} + z\hat{\mathbf{k}}$ is the position vector of an arbitrary point in the tangent plane, then $(\mathbf{r} - \mathbf{r}(P)) \cdot \hat{\mathbf{n}} = 0$, where $\hat{\mathbf{n}}$ is the unit normal vector to the tangent plane at point P.

Exercise 7.5.5 Calculate the equation of the tangent plane to a sphere of unit radius at the point $\theta = 0$.

We now have the vector machinery to integrate over the area of a surface. First, we represent the position vector of an arbitrary point in the surface as $\mathbf{r}(\lambda, \mu) = x(\lambda, \mu)\hat{\mathbf{\imath}} + y(\lambda, \mu)\hat{\mathbf{\jmath}} + z(\lambda, \mu)\hat{\mathbf{k}}$, where λ and μ are coordinates in the surface. We also have to assume that the functions $x(\lambda, \mu)$, $y(\lambda, \mu)$, and $z(\lambda, \mu)$ and their derivatives are smooth so that all the appropriate limits converge properly, but this is usually the case in the Earth and environmental sciences. We now need to find an expression for an element of the surface area (dA) in terms of λ and μ. As we have seen, the differential

$$d\mathbf{r} = \mathbf{r}_\lambda d\lambda + \mathbf{r}_\mu d\mu$$

at a point P in the surface defines two vectors, $\mathbf{r}_\lambda d\lambda$ and $\mathbf{r}_\mu d\mu$, that lie in the tangent plane at P. These vectors form a parallelogram whose area is (Section 4.3.2)

$$dA = |\mathbf{r}_\lambda d\lambda \times \mathbf{r}_\mu d\mu| = |\mathbf{r}_\lambda \times \mathbf{r}_\mu|\, d\lambda\, d\mu,$$

and in the limit that $d\lambda$ and $d\mu$ tend to zero, dA (which is an area element in the tangent plane) will tend to the area of an element in the surface. So

$$A = \iint_S dA = \iint_{\mathcal{D}} |\mathbf{r}_\lambda \times \mathbf{r}_\mu|\, d\lambda\, d\mu, \tag{7.27}$$

where the first double integral is over the surface S in three dimensions and the second integral is over the two-dimensional space $\mathcal{D}$ defined by the coordinates in the surface itself (e.g., λ and μ).

Example 7.4 To calculate the area of a sphere of radius R, we first write the three-dimensional coordinates of the sphere in terms of coordinates in the surface itself, so that

$$x = R\cos(\lambda)\sin(\mu), \quad y = R\sin(\lambda)\sin(\mu), \quad z = R\cos(\mu),$$

where $0 \leq \lambda \leq 2\pi$ and $0 \leq \mu \leq \pi$. The position vector of a point in the surface is then

$$\mathbf{r}(\lambda, \mu) = R\cos(\lambda)\sin(\mu)\hat{\mathbf{i}} + R\sin(\lambda)\sin(\mu)\hat{\mathbf{j}} + R\cos(\mu)\hat{\mathbf{k}}$$

and the tangent vectors $\mathbf{r}_\lambda$ and $\mathbf{r}_\mu$ are given by

$$\mathbf{r}_\lambda = -R\sin(\lambda)\sin(\mu)\hat{\mathbf{i}} + R\cos(\lambda)\sin(\mu)\hat{\mathbf{j}} + 0\hat{\mathbf{k}},$$
$$\mathbf{r}_\mu = R\cos(\lambda)\cos(\mu)\hat{\mathbf{i}} + R\sin(\lambda)\cos(\mu)\hat{\mathbf{j}} - R\sin(\mu)\hat{\mathbf{k}},$$

and $dA = |\mathbf{r}_\lambda \times \mathbf{r}_\mu|\, d\lambda\, d\mu = R^2\sin(\mu)\, d\lambda\, d\mu$. The integral over the whole surface is then

$$A = \int_0^{2\pi}\int_0^{\pi} R^2\sin(\mu)\, d\mu, d\lambda = 4\pi R^2.$$

Now that we know how to calculate dA, we can calculate the integral of a scalar function $f(x, y, z)$ over a surface,

$$\iint_S f(x, y, z)\, dA = \iint_{\mathcal{D}} f(x(\lambda, \mu), y(\lambda, \mu), z(\lambda, \mu))|\mathbf{r}_\lambda \times \mathbf{r}_\mu|\, d\lambda\, d\mu.$$

But we really want to know how to integrate a vector field $\mathbf{F}$ over a surface, and for this we need to know the component of the vector field parallel to the normal to the surface element (see Figure 4.12). The vector area element of the surface is $d\mathbf{A} = \hat{\mathbf{n}}dA$, where $\hat{\mathbf{n}}$ is the unit normal to the area element (Figure 7.16), which is given by Equation (7.26) so that

$$d\mathbf{A} = \frac{\mathbf{r}_\lambda \times \mathbf{r}_\mu}{|\mathbf{r}_\lambda \times \mathbf{r}_\mu|}|\mathbf{r}_\lambda \times \mathbf{r}_\mu|\, d\lambda\, d\mu = (\mathbf{r}_\lambda \times \mathbf{r}_\mu)\, d\lambda\, d\mu$$

and

$$\iint_S \mathbf{F} \cdot d\mathbf{A} = \iint_S \mathbf{F} \cdot \hat{\mathbf{n}}\, dA = \iint_{\mathcal{D}} \mathbf{F}(\mathbf{r}(\lambda, \mu)) \cdot (\mathbf{r}_\lambda \times \mathbf{r}_\mu)\, d\lambda\, d\mu. \tag{7.28}$$

Exercise 7.5.6 Calculate the integral of the vector field $\mathbf{F} = x\hat{\mathbf{\imath}} + y\hat{\mathbf{\jmath}} + z\hat{\mathbf{k}}$ over the surface of a sphere of radius $R = 1$.

We can calculate volume integrals in a similar way except that now a position vector has to be written as a three-parameter family of vector functions,

$$\mathbf{r}(\lambda, \mu, \nu) = x(\lambda, \mu, \nu)\hat{\mathbf{\imath}} + y(\lambda, \mu, \nu)\hat{\mathbf{\jmath}} + z(\lambda, \mu, \nu)\hat{\mathbf{k}},$$

and we have to use a scalar triple product (Equation (4.44)) to obtain the volume element

$$dV = |\mathbf{r}_\lambda \cdot \mathbf{r}_\mu \times \mathbf{r}_\nu|\, d\lambda\, d\mu\, d\nu = \begin{vmatrix} x_\lambda & y_\lambda & z_\lambda \\ x_\mu & y_\mu & z_\mu \\ x_\nu & y_\nu & z_\nu \end{vmatrix} d\lambda\, d\mu\, d\nu.$$

Example 7.5 One way of detecting differences in the density of rock below the surface of the Earth is to measure changes in the gravitational field at the surface. Consider a conical volume of rock of depth ξ and angle ϵ with the apex of the cone at the surface of the Earth (Figure 7.17). The density of the cone is $\rho_E + \delta\rho$, where ρ_E is the density of the surrounding rock and $\delta\rho > 0$. Let us calculate the change in the gravitational field experienced by a unit mass located at the apex of the cone. If we set up coordinates such that the origin is located at the apex of the cone, and the unit vector in the z direction is positive, pointing upward, then the change in gravitational field at the apex of the cone is

$$\Delta\mathbf{F} = -G\delta\rho\hat{\mathbf{k}} \iiint_{\text{cone}} \frac{z}{(x^2 + y^z + z^2)^{3/2}} dV,$$

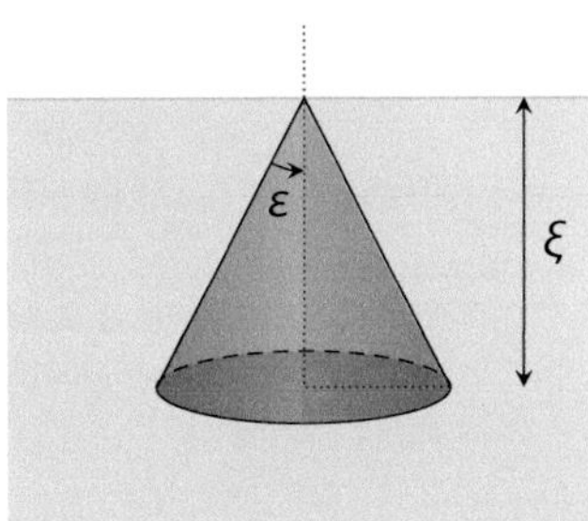

Figure 7.17 A cone of dense material is embedded in the Earth. The cone has a height of ξ and an angle of ϵ.

where dV is a volume element and the integral is evaluated over the volume of the cone and we have used the fact that the cone is symmetric about the z axis to realize that the components of $\mathbf{F}$ in the $\hat{\imath}$ and $\hat{\jmath}$ directions will cancel out once we integrate over the cone. Let us evaluate the integral in terms of spherical polar coordinates (we could choose to use a different set of coordinates, such as cylindrical coordinates) so that, using the formula to convert from Cartesian to spherical polar coordinates,

$$r = (x^2 + y^2 + z^2)^{1/2}, \quad z = r\cos(\theta), \quad dV = r^2 \sin(\theta)\, dr\, d\theta\, d\phi,$$

where r is the radius, θ is the angle with the z axis, and ϕ is the azimuthal angle. Then

$$\begin{aligned}\Delta\mathbf{F} &= -G\delta\rho\hat{\mathbf{k}} \int_0^{2\pi} \int_0^{\epsilon} \int_0^{\xi\sec(\theta)} \cos(\theta)\sin(\theta)\, dr\, d\theta\, d\phi \\ &= -2\pi G\delta\rho\xi\hat{\mathbf{k}} \int_0^{\epsilon} \sin(\theta)\, d\theta = -2\pi G\delta\rho\xi(1-\cos(\epsilon))\hat{\mathbf{k}}.\end{aligned}$$

7.5.1 Divergence Theorem

The *divergence theorem* relates an integral over a volume to one over the surface encompassing that volume. This turns out to be quite a useful thing to do, as we will see shortly. Earlier (Section 7.3), we considered the flow of a fluid through a volume and showed that it was linked to the divergence of a vector field. Let us look at this in more detail. If ρ is the mass density of the fluid and $\mathbf{v}$ its velocity, then the vector $\mathbf{F} = \rho\mathbf{v}$ is the flux of fluid mass with dimensions $[\mathrm{M}][\mathrm{L}]^{-2}[\mathrm{T}]^{-1}$. Now let us consider the flux of this fluid through a small element $d\sigma$ of a surface Σ (Figure 7.18); for example, in Cartesian (x, y, z) coordinates, we might have $d\sigma = dx\, dy$, and in spherical coordinates (r, θ, ϕ), $d\sigma = r^2 \sin^2(\theta) d\theta\, d\phi$. The outward-facing normal vector for the area element is $\mathbf{n}$, so that

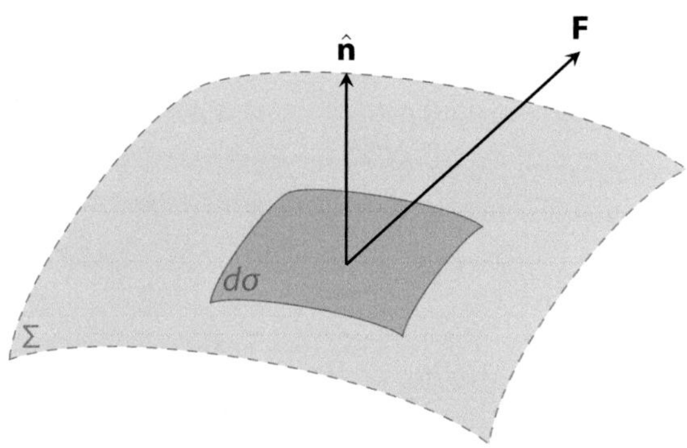

Figure 7.18 A surface Σ with a small surface element $d\sigma$ that has an outward-facing normal $\hat{\mathbf{n}}$. The vector $\mathbf{F}$ represents the flux of material flowing through the element of area $d\sigma$.

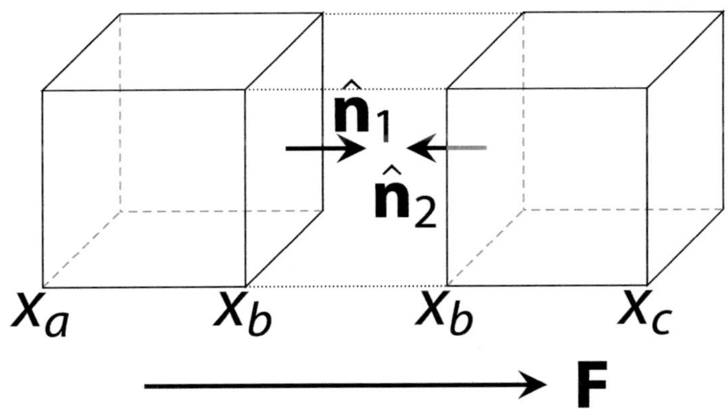

Figure 7.19 Two adjacent cubic volumes that have been separated at their common interface. The vector **F** represents a flux of material moving from the left-hand cube into the right-hand cube. The dot product of the unit normal $\hat{\mathbf{n}}_2$ of the right-hand cube with **F** will be the negative of the dot product of $\hat{\mathbf{n}}_1$ with **F** and the net flux across the faces will cancel.

the flow, or transport, of fluid through $d\sigma$ is $(\mathbf{F} \cdot \mathbf{n})d\sigma$ and the transport out of the whole surface Σ is the integral of this quantity over the whole surface

$$\iint_{\Sigma} \mathbf{F} \cdot \mathbf{n}\, d\sigma,$$

where we have used the double integral sign to remind ourselves that this is a double integral over the surface Σ.[11]

Now let us subdivide the volume inside the surface Σ into small volume elements of size $d\tau$; for example, if the volume was a rectangular box in Cartesian coordinates we could subdivide into smaller rectangular boxes, each with a volume $d\tau = dx\, dy\, dz$. We know from our exploration of the divergence that the flow of mass out of the volume element $d\tau$ is $\nabla \cdot \mathbf{F}\, d\tau$. So, if there are N small element volumes within Σ, the total out of Σ is the sum of all the outward flows from each volume element $d\tau$, i.e.,

$$\sum_{i=1}^{N} \nabla \cdot \mathbf{F}_i\, d\tau_i,$$

where the index i labels each elemental volume. However, when we add up all the flows in and out of each of these small volume elements we see that many of these flows cancel out. To see this, consider the small volume elements as being cubes (Figure 7.19). Each face of each cube has its own outward-facing normal vector so that the flow out of a face of one cube equals the *negative* of the flow into the neighboring cube. When we add the flows across all the faces of all these cubes, these interior flows will cancel each other. The result of adding the flows across the faces of all the small elements is that we are left with only those flows across the outer surfaces of the exterior volume elements; i.e., across the surface Σ. If we now take the limit as $d\tau \to 0$, we get

11 Some authors use only a single integral but use a symbol, such as Σ, at the base of the integral to tell the reader whether the integral is over a surface or a volume.

$$\iiint_\tau \nabla \cdot \mathbf{F}\, d\tau = \iint_\Sigma \mathbf{F} \cdot \mathbf{n}\, d\sigma, \tag{7.29}$$

which is a statement of the divergence theorem, also sometimes called the *Gauss' theorem*.[12] This is a useful theorem because it is sometimes easier to evaluate a surface integral than to evaluate a volume integral (and sometimes vice versa). For example, let us integrate the vector field $\mathbf{F} = (2x^2 + y)\hat{\mathbf{i}} + 3y^2\hat{\mathbf{j}} - 2z\hat{\mathbf{k}}$ over the surface of the cube defined by $0 \leq x \leq 1, 0 \leq y \leq 1, 0 \leq z \leq 1$. We could parameterize each surface of the cube and evaluate the integral that way, but it is far easier to use Equation (7.29). First, we calculate

$$\nabla \cdot \mathbf{F} = \frac{\partial}{\partial x}(2x^2 + y) + \frac{\partial}{\partial y}(3y^2) + \frac{\partial}{\partial z}(-2z) = 4x + 6y - 2.$$

We then use the divergence theorem to write the surface integral as a volume integral and integrate over the volume

$$\int_{x=0}^{x=1} \int_{y=0}^{y=1} \int_{z=0}^{z=1} (4x + 6y - 2)\, dx\, dy\, dz = 3.$$

Exercise 7.5.7 Evaluate the integral of $\mathbf{F} = (2x + y)\hat{\mathbf{i}} + 3y\hat{\mathbf{j}} + (z - x)\hat{\mathbf{k}}$ over the surface of the sphere $x^2 + y^2 + z^2 = 4$.

The divergence theorem plays an important role in developing conservation equations. These are equations that are used to describe how material is transformed and transported in the real world, and they are fundamental for describing processes in the Earth and environmental sciences. For example, we might want to describe how nitrate concentrations change in the oceans as it is consumed by algal cells, produced by microbial processes, and transported by the movements of water, or we might want to model how atmospheric ozone concentrations change. Let us consider a region of space within which we have a volume V of any shape that is enclosed by a surface Σ (Figure 7.20) and a scalar field $C(x, y, z, t)$ that is defined throughout the region. To make things concrete, let us say that C represents the amount of a specific substance per unit volume in the atmosphere (so the dimensions will be $[N][L]^{-3}$)—ozone, for example. Within the volume V, this substance can be formed

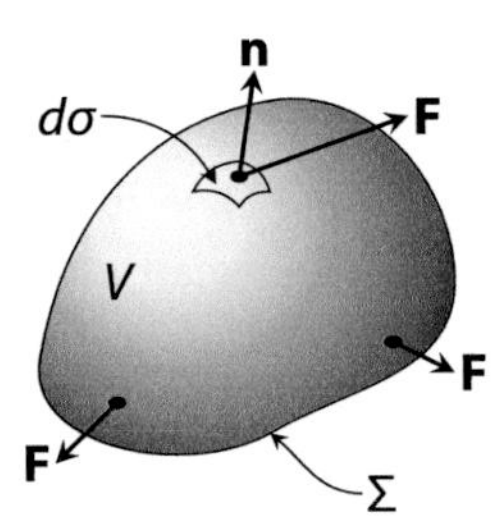

Figure 7.20 A volume V bounded by a surface Σ with a flux vector $\mathbf{F}$ varying in both direction and magnitude over the surface. A small element of the surface $d\sigma$ has a unit normal $\mathbf{n}$.

[12] Named after Johann Carl Friedrich Gauss (1777–1855), who provided the first general proof of the theorem, though Joseph Louis Lagrange (1736–1813) was the first to discover it.

or destroyed by chemical or photochemical reactions, or it can be transported into or out of V by the movements of the air, or it can diffuse into or out of V. We want to derive an equation that tells us how C changes with time within that volume. At any instant of time, the total amount of the substance within the volume V is given by the integral of C over the volume. So, the total rate of change of C within V is the time derivative of this integral, and it is made up of terms representing chemical transformation, transport, or diffusion. Let us look at them in turn. We can combine the rates of formation and destruction per unit volume into a single, net rate of formation per unit volume S; the total net rate of formation within the volume is then the integral of S over the volume V.

We can use the concept of flux to represent the process that moves C across the boundary of the volume. The flux of a substance through a surface is the amount of the substance that passes through the surface per unit time per unit area. So, for transport by the motions of the air, the flux across the surface is $C\mathbf{v}$; this has dimensions $[\mathrm{N}][\mathrm{L}]^{-2}[\mathrm{T}]^{-1}$, which is correct for a flux, and this is called an *advective process* because the motion of the fluid advects the material with it. We have to integrate this over the surface Σ to obtain the net transport out of the volume V. Similarly, we can represent transport by means other than fluid flow (e.g., diffusion) by a flux $\mathbf{F}$ that has to be integrated over the surface. It is worth reminding ourselves that a small element $d\boldsymbol{\sigma}$ of the surface Σ has a unit normal that by convention points out of the surface.[13] Because C, $\mathbf{v}$, and $\mathbf{F}$ all vary with space, we have to take the dot product of $C\mathbf{v}$ and F with $d\boldsymbol{\sigma}$ to get the component of the flow in the direction of the normal (see Figure 4.12 and the discussion preceding it).

Given all of this, we can write a general conservation equation for C as

$$\frac{d}{dt}\int_V C\,dV = \int_V S\,dV - \iint_\Sigma C\mathbf{v}\cdot d\sigma - \iint_\Sigma \mathbf{F}\cdot d\sigma = \int_V S\,dV - \iint_\Sigma (C\mathbf{v}+\mathbf{F})\cdot d\sigma, \quad (7.30)$$

where the negative signs on the transport terms are a result of the fact that $d\sigma$ is outward pointing. The term involving the integral over the surface Σ represents the total net transport of the substance out of the volume V (remember that the vector $d\sigma$ is pointing out of the volume), in other words the net rate of change of mass of the substance within the volume V resulting from transport across the boundary Σ. So, if we divide Equation (7.30) by the volume V (which we assume to be constant with time), we can write

$$\frac{d\bar{C}}{dt} = \bar{S} + \frac{1}{V}\left(\frac{dM_{\text{in}}}{dt} - \frac{dM_{\text{out}}}{dt}\right), \quad (7.31)$$

where $\bar{C}$ is the average concentration of the substance within the volume, $\bar{S}$ is the average net rate per unit volume of its formation or destruction within V, and M_{in} and M_{out} are the total masses of the substance flowing in and out of V. Equation (7.31) is called a conservation equation because it expresses the conservation of the substance as it is transported and transformed in the environment.

Exercise 7.5.8 Convince yourself that the signs on the right-hand side of Equation (7.31) are correct.

[13] The vector $d\sigma$ here represents the vector $|d\sigma|\mathbf{n}$, where $\mathbf{n}$ is the normal vector to the element of area.

The equation is an ordinary differential equation in terms of the average concentration, so we have lost all information about the spatial variability of C. Such equations are useful in formulating simple *box models* of environmental systems, and being ODEs, they can be relatively simple to solve. In fact, we have already come across such an equation in Section 6.3, where we looked at changes of the concentration of a pollutant in a lake over time.

Can we use a similar technique to construct equations that also retain the spatial variability of C? Yes, but we have to exercise a little caution. For example, in our lake example from Section 6.3 we have to think about how the concentration of the pollutant varies across the lake and with depth in the lake. One way to do this is to follow a similar argument to the derivation of Equation (7.31) but to choose very small volumes V such that the spatial variability of C is represented by the change in C from volume to volume, though the concentration within each small volume is spatially uniform; such volumes are called *representative volumes*. We will assume that an equation similar to Equation (7.30) holds for each representative volume. This may be reasonable if the representative volumes are sufficiently small and homogeneous, but it is not always the case. For example, the representative volume in an ocean circulation model might be tens of kilometers on a side in the horizontal and tens of meters deep. As we have seen in Equation (1.14), processes such as turbulence occur on many scales, and these can affect the distribution of substances within a representative volume.[14] The problem with Equation (7.30) is that it contains both surface and volume integrals. We would like to have only one type of integral, and because we are integrating over the volume to get the rate of change of total mass, it makes sense to try and convert the surface integrals into a volume integral. To do this, we apply the divergence theorem to the surface integrals in Equation (7.30), to give

$$\frac{d}{dt}\int_V C\,dV = \int_V S\,dV - \int_V \nabla\cdot(C\mathbf{v}+\mathbf{F})\,dV.$$

The volume and surface of our representative volume is assumed fixed, so we can take the derivative with respect to time inside the integral sign to give us

$$\int_V \left\{\frac{\partial C}{\partial t} + \nabla\cdot(C\mathbf{v}+\mathbf{F}) - S\right\} dV = 0.$$

We have not said anything specific about the volume (except that it is sufficiently small and constant), so this equation must hold for any volume, and the only way this can be true is if

$$\frac{\partial C}{\partial t} + \nabla\cdot(C\mathbf{v}+\mathbf{F}) - S = 0. \tag{7.32}$$

Equation (7.32) is a general conservation equation for C as a function of space and time. Let us apply it to two important cases. First, the conservation of mass. Recall that C is a concentration, so we let $C(x, y, z, t) = \rho(x, y, z, t)$, where ρ is the mass per unit volume of fluid (e.g., water or air). We know that we cannot create or destroy mass, so $S = 0$. If we also assume that the only transport of mass is by fluid motions with a velocity $\mathbf{v}$ (i.e., there

[14] In large computer simulations, this problem is overcome by using what are called *subgrid parameterizations* that represent important processes occurring on scales smaller than the representative volume (Stensrud, 2009).

is no diffusion), then we end up with a general mass conservation equation, or *advection equation*,

$$\frac{\partial \rho}{\partial t} + \nabla \cdot (\rho \mathbf{v}) = 0. \tag{7.33}$$

If ρ is spatially uniform, then we have

$$\frac{\partial \rho}{\partial t} + \rho \nabla \cdot \mathbf{v} = 0.$$

Equation (7.33) is one of the basic equations of fluid dynamics, and it is part of the system of equations that are used in geophysical fluid dynamics to determine the motion of the ocean and atmosphere.

As another example, let us derive an equation that tells us how the temperature of a substance varies spatially and temporally. This is going to be a little tricky because although we want an equation for temperature (something we measure), we actually need to be thinking in terms of heat, which is a form of energy. We are going to end up with an equation for how heat is transported by a fluid; for example, we might want to have an equation that tells us how heat moves within molten rock. To start with, we can write $C(x, y, z, t) = c_p \rho(x, y, z, t) T(x, y, z, t)$, where ρ is the fluid density, T is its temperature, and c_p is a constant called the *specific heat* at constant pressure and temperature and it has units of energy per unit mass of fluid per degree Kelvin.[15] Heat can be moved around by fluid motion ($C\mathbf{v} = c_p \rho(x, y, z, t) T(x, y, z, t)\mathbf{v}$) and by conduction. When heat flows by conduction it moves from regions of high temperature to regions of low temperature at a rate that is proportional to the magnitude of the temperature gradient: $\mathbf{F} = -k\nabla T$. The constant of proportionality (k) is called the *thermal conductivity* and varies from substance to substance. Note that the negative sign in the expression for $\mathbf{F}$ indicates that heat flows from high temperatures to low temperatures. Lastly, there can be sources of heat within each volume. For example, the volume may contain chemicals that react and give off heat. Putting all of this together in Equation (7.32) we get

$$\frac{\partial}{\partial t}(c_p \rho T) + \nabla \cdot (c_p \rho T \mathbf{v}) - \nabla \cdot (k \nabla T) - S = 0,$$

and if we assume that ρ, c_p, and $\mathbf{v}$ are constants, then we can write this equation as

$$\frac{\partial T}{\partial t} + \mathbf{v} \cdot \nabla T = \frac{DT}{dt} = \frac{k}{\rho c_p} \nabla^2 T + \frac{S}{\rho c_p}, \tag{7.34}$$

where we have used Equation (7.10) and the quantity $k/(\rho c_p)$ is called the thermal diffusivity.

Exercise 7.5.9 What are the dimensions of the thermal diffusivity?

Exercise 7.5.10 A chemical species has a concentration $C(x, y, z, t)$ in the ocean. It is formed by chemical reactions at a constant rate k_1 and consumed by other reactions at a rate $k_2 C$. The species is carried by motions of the water and also diffuses through the water with a flux $\mathbf{F} = -\mathcal{D}\nabla C$. Develop a conservation equation for C; such an equation is called an *advection–diffusion–reaction equation*.

[15] Air at surface conditions has a value of $c_p \approx 1\ \mathrm{J\ g^{-1}\ K^{-1}}$, whereas water at room temperature has a value of $c_p \approx 4.2\ \mathrm{J\ g^{-1}\ K^{-1}}$.

Equations such as (7.33) and (7.34) are used in understanding how heat and chemical substances are moved through the environment (Vallis, 2017; Glover et al., 2011), and they form the basis of numerical simulations of environmental and geophysical fluid flows, such as computer models of lakes and rivers, the atmosphere, oceans (Miller, 2007), and global climate (Griffies, 2004).

7.5.2 Stokes' Theorem

Another important theorem is *Stokes' theorem*, which relates the integral of the curl of a vector field to the integral around a closed path and is a three-dimensional equivalent of Green's theorem (Equation (2.58)). Let us first consider what happens if we take the line integral of a vector field $\mathbf{v}$ around a closed path

$$\Gamma = \oint_C \mathbf{v} \cdot d\mathbf{r}. \tag{7.35}$$

If we do this for a scalar function of one variable, then we know that

$$\int_a^a f(x)\, dx = 0,$$

but this need not be the case for vectors, and the quantity Γ in Equation (7.35) is often called the *circulation* of the vector field. As you might imagine from its name, this quantity is inspired by the properties of fluids in motion. To see what this quantity means, let us consider a fluid moving with a velocity $\mathbf{v} = \gamma y\hat{\mathbf{i}}$, where $\gamma > 0$ is a constant (Figure 7.21). A velocity field like this is called a *shear flow*, and they are very common in the natural environment. Let us consider a generic closed path C that is the boundary of a region S (Figure 7.21) with a unit normal to S of $\hat{\mathbf{n}} = -\hat{\mathbf{k}}$, which is into the paper. The circulation of $\mathbf{v}$ is

$$\Gamma = \oint_C \gamma y\hat{\mathbf{i}} \cdot d\mathbf{r} = \gamma A, \tag{7.36}$$

where A is the area of S enclosed by the curve C.

Exercise 7.5.11 Show that Equation (7.36) is correct by splitting the curve C into two parts.

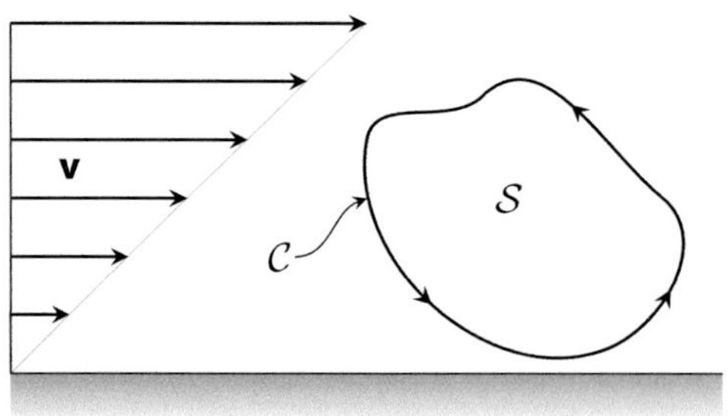

Figure 7.21 The shear velocity $\mathbf{v} = \gamma y\hat{\mathbf{i}}$ with a closed path C enclosing a region S. The circulation of the velocity field is calculated in a counterclockwise direction around C.

We know that the vorticity is the curl of the velocity field, $\boldsymbol{\omega} = \nabla \times \mathbf{v} = -\gamma \hat{\mathbf{k}}$, which in this case is a uniform vector field with a constant magnitude flowing into the paper. The *flux of vorticity* flowing through the surface $\mathcal{S}$ is

$$\iint_S \hat{\mathbf{n}} \cdot \nabla \times \mathbf{v}\, dA = \iint_S \hat{\mathbf{n}} \cdot \boldsymbol{\omega}\, dA = \iint_S -\hat{\mathbf{k}} \cdot (-\gamma \hat{\mathbf{k}}) dA = \gamma A = \Gamma.$$

This is an example of Stokes' theorem: if C is a closed path that is the boundary of a surface $\mathcal{S}$ with normal vector $\hat{\mathbf{n}}$, and $\mathbf{v}$ is a vector field, then

$$\oint_C \mathbf{v} \cdot d\mathbf{r} = \iint_S \hat{\mathbf{n}} \cdot \nabla \times \mathbf{v}\, dA. \tag{7.37}$$

For a fluid, Stokes' theorem tells us that the circulation of the fluid around a closed curve C equals the flux of vorticity of the fluid through the surface $\mathcal{S}$ that is bounded by C.

We have seen that the vorticity is at first sight a strange quantity; it is the rotation of a fluid at a point (recall the visualization of putting an infinitesimally small paddle wheel into the flow to see if it would rotate). For solid body rotation, where all the parts of the body rotate with the same angular velocity, the vorticity can be related to the angular velocity. For solid body rotation, the linear velocity of any point is $\mathbf{v} = \boldsymbol{\Omega} \times \mathbf{r}$, where $\boldsymbol{\Omega}$ is the angular velocity and $\mathbf{r}$ is the distance from the axis of rotation (Section 4.3.2.1). If we align the coordinates such that the angular velocity vector points along the z axis, then $\boldsymbol{\Omega} = \Omega \hat{\mathbf{k}}$, where Ω is the magnitude of $\boldsymbol{\Omega}$. Then we have that $\mathbf{u} = -\Omega y \hat{\mathbf{i}} + \Omega x \hat{\mathbf{j}}$ and the vorticity is $\boldsymbol{\omega} = \nabla \times \mathbf{u} = 2\Omega \hat{\mathbf{k}}$. So in this special case, the vorticity is twice the angular velocity, though this is not true for all flows.

Stokes theorem is very interesting and powerful, because we have not said anything about the shape of the surface $\mathcal{S}$, requiring only that it be bounded by the curve C. This means that for the same curve C we can have an infinite number of different surfaces bounded by that curve, and Stokes' theorem will hold for all of them. Let us explore this a little bit by looking at a vector field $\mathbf{v} = y\hat{\mathbf{i}} + xz\hat{\mathbf{j}} + \hat{\mathbf{k}}$ and considering the closed curve to be the simplest thing we can think of, a circle of radius $r = 1$ in the (x, y) plane centered on the origin (Figure 7.22). The vorticity of the vector field is $\nabla \times \mathbf{v} = -x\hat{\mathbf{i}} + (z - 1)\hat{\mathbf{k}}$.

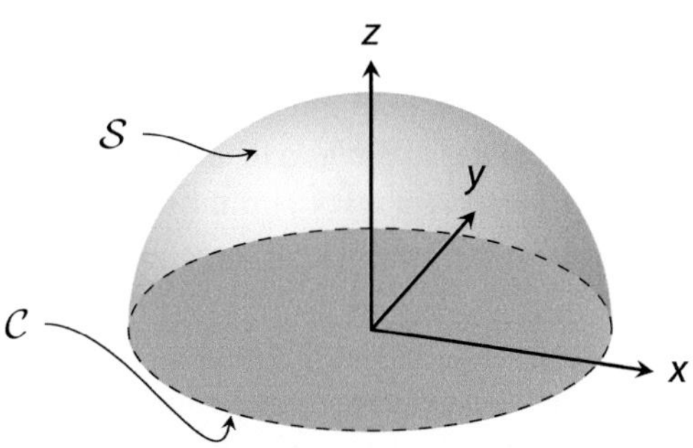

Figure 7.22 The surface $\mathcal{S}$ bounded by a curve C. Notice that there can be many surfaces all bounded by the same curve. It is useful to think of the analogy of blowing soap bubbles through a circular hoop; the curve (the hoop) is always the same, but there can be many different shapes of the surface of the soap bubble before it detaches from the hoop.

Exercise 7.5.12 Show that, for the vector field $\mathbf{v} = y\hat{\mathbf{\imath}} + xz\hat{\mathbf{\jmath}} + \hat{\mathbf{k}}$,

$$\int_C \mathbf{v} \cdot d\mathbf{r} = -\pi,$$

where C is the unit circle in the (x, y) plane and the path along the circle is traversed in a counterclockwise direction.

To check the validity of Stokes' theorem, let us consider a surface that is a hemisphere of radius $r = 1$ such that $z \geq 0$ for the whole hemisphere (Figure 7.22). The equation of the hemisphere is then $x^2 + y^2 + z^2 = 1$ with $z \geq 0$ and the unit normal to the surface of the sphere points outward, away from the origin.

Exercise 7.5.13 Show that the unit normal vector to the surface of the hemisphere is

$$\hat{\mathbf{n}} = x\hat{\mathbf{\imath}} + y\hat{\mathbf{\jmath}} + z\hat{\mathbf{k}}$$

and use this to show that

$$\iint_S \nabla \times \mathbf{v} \cdot \hat{\mathbf{n}} dA = -\pi$$

and that Stokes' theorem holds.

Exercise 7.5.14 Show that Stokes' theorem also holds if S is a disk of radius $r = 1$ in the (x, y) plane.

The remarkable thing about Stokes' theorem is that it relates a line integral to a surface integral. In doing so, it allows us to choose how we evaluate certain integrals, and it also provides a firm mathematical underpinning for many fluid and transport processes we are interested in.

7.6 Further Reading

Many books on mathematical methods in physics and engineering cover vector calculus in considerable detail. *Mathematical Methods in the Physical Sciences* by Boas (2006) provides a lot of detail at a level similar to this text. A book that can help you to build a good intuition about the gradient, divergence, and curl is *Div, Grad, Curl, And All That* by Schey (2004). Applications of vector calculus occur in fluid dynamics, rock deformation, and transport processes. *Atmospheric and Oceanic Fluid Dynamics* by Vallis (2017) is a comprehensive text on of geophysical fluid dynamics that requires knowledge of vector calculus. A good place to read in detail about the Coriolis effect is *An Introduction to the Coriolis Force* by Stommel and Moore (1989), which although older was written by one of the foremost oceanographers of the twentieth century (Henry Stommel) and is very accessible.

Problems

7.1 Consider a vector $\mathbf{r} = x\hat{\mathbf{i}} + y\hat{\mathbf{j}} + z\hat{\mathbf{k}}$. Calculate:
1. $\nabla(|\mathbf{r}|)$, 2. $\nabla \cdot \mathbf{r}$, 3. $\nabla \times \mathbf{r}$, 4. $\nabla^2(1/|\mathbf{r}|)$.

7.2 Show that $\nabla \times (\mathbf{u} \times \mathbf{v}) = (\nabla \cdot \mathbf{v})\mathbf{u} - (\nabla \cdot \mathbf{u})\mathbf{v} + (\mathbf{v}\nabla)\mathbf{u} - (\mathbf{u}\nabla)\mathbf{v}$, where for a vector $\mathbf{u} = u_x\hat{\mathbf{i}} + u_y\hat{\mathbf{j}} + u_z\hat{\mathbf{k}}$,

$$(\mathbf{u}\nabla) = u_x\frac{\partial}{\partial x} + u_y\frac{\partial}{\partial y} + u_z\frac{\partial}{\partial z}.$$

7.3 Sketch the vector field

$$\mathbf{V} = \frac{1}{(x^2 + y^2 + z^2)^{3/2}}(x\hat{\mathbf{i}} + y\hat{\mathbf{j}} + z\hat{\mathbf{k}})$$

and guess from your sketch if the $\nabla \cdot \mathbf{V}$ is zero or nonzero. Confirm your guess by calculating the divergence of $\mathbf{V}$ at an arbitrary point that is not $(x, y, z) = (0, 0, 0)$. Why is it important to not include the origin when calculating the divergence?

7.4 Cylindrical coordinates can also be useful for solving problems. For example, we might have an equation that represents the flow of magma through a lava tube. Cylindrical coordinates are defined by two lengths and an angle. These are the usual two-dimensional polar coordinates ((r, θ)) in a plane, and the linear distance along the z axis:

$$x = r\cos(\theta), \quad y = r\sin(\theta), \quad z = z.$$

Derive expressions for the gradient and Laplacian of a scalar as well as the divergence and curl of a vector field in cylindrical coordinates.

7.5 The gravitational field at a point A due to a body B of mass M is given by

$$\mathbf{F} = -\frac{GM}{r^3}\mathbf{r},$$

where $\mathbf{r}$ is the radial position vector connecting the center of B to point A, G is a constant, and $r = ||\mathbf{r}||$. Show that $\mathbf{F}$ is a conservative force and that the potential is

$$\phi = -\frac{GM}{r}.$$

7.6 Consider two surfaces S_1 and S_2 that share a common boundary C such that the orientation of the boundary is the same for the two surfaces. If $\mathbf{v}$ is a continuous vector field that passes through both surfaces, what is the relationship between

$$\iint \nabla \times \mathbf{v} \cdot d\mathbf{S}$$

for the two surfaces?

7.7 Geostrophic flow occurs when the Coriolis force affecting the motion of a fluid balances the pressure gradient

$$2\boldsymbol{\Omega} \times \mathbf{u} = -\frac{1}{\rho}\nabla p,$$

where $\boldsymbol{\Omega}$ is the angular velocity, $\mathbf{u}$ the fluid velocity, ρ the fluid density, and p the pressure. If, in addition, $\nabla \cdot \mathbf{u} = 0$, show that $(\boldsymbol{\Omega} \cdot \nabla)\mathbf{u} = 0$. If $\boldsymbol{\Omega} = \Omega\hat{\mathbf{k}}$, show that the velocity $\mathbf{u}$ is constant in a direction parallel to $\hat{\mathbf{k}}$ (this result is called the *Taylor–Proudman theorem*).

7.8 Use Cartesian coordinates to calculate the divergence of the vector field $\mathbf{u}_1 = (1/r)\hat{\mathbf{r}}$, where $r = \sqrt{x^2 + y^2 + z^2}$ and $\hat{\mathbf{r}}$ is a unit vector in the radial direction. Calculate the divergence of the vector field $\mathbf{u}_2 = x\hat{\mathbf{i}} + y\hat{\mathbf{j}} + z\hat{\mathbf{k}}$ and compare your result with that for $\mathbf{u}_1$.

7.9 Consider a function $h(r)$ that is a scalar function of radius only. Calculate the Laplacian of h.

7.10 Consider the vector field $\mathbf{F} = xy^2\hat{\mathbf{i}} + yz\hat{\mathbf{j}} + (x + y + z)\hat{\mathbf{k}}$ and calculate $\nabla \times \mathbf{F}$, $\nabla \times \nabla \times \mathbf{F}$, $\nabla \cdot \mathbf{F}$, and $\nabla(\nabla \cdot \mathbf{F})$.

7.11 Show that $\nabla \cdot (kr^n\hat{\mathbf{r}}) = (n + 2)kr^{n-1}$, where r is the radial distance and $\hat{\mathbf{r}}$ is a unit vector in the radial direction.

7.12 Calculate the direction derivative of $\phi(x, y, z) = x^2y^3 + 2z$ in the direction $\mathbf{v} = \hat{\mathbf{i}} - \hat{\mathbf{j}} + \hat{\mathbf{k}}$.

7.13 A curve is described by the equation in Cartesian coordinates $z = 4\sqrt{x^2 + y^2}$. What is the equation of the curve in (a) cylindrical coordinates and (b) spherical coordinates? What shape does the equation describe?

7.14 Consider a cylinder given by the equation $x^2 + y^2 = 9$ with z taking any value between 0 and 1. Find a parameterization of the cylinder in terms of a single angle (θ) and a length (u), and show that the area element $dA = 3\,d\theta\,du$.

7.15 A function whose Laplacian is zero is called a *harmonic function*. Which of the following functions are harmonic?
1. $f(x, y) = x^2 + y^2$, 2. $f(x, y) = y\cos(x) - x\sin(y)$, 3. $f(x, y) = 12xy$.

7.16 Consider the vector field $\mathbf{F} = x\hat{\mathbf{i}} + y\hat{\mathbf{j}} + z\hat{\mathbf{k}}$. If S is a surface that encloses a volume V, show that the flux of $\mathbf{F}$ through S is $3V$.

7.17 Is the vector field $\mathbf{v} = y\hat{\mathbf{i}} + x^2z\hat{\mathbf{j}} + z\hat{\mathbf{k}}$ a conservative field?

7.18 Sketch the vector field $\mathbf{v} = -y\hat{\mathbf{i}} + x\hat{\mathbf{j}}$ and calculate its line integral around a circle of radius $r = 1$ in the (x, y) plane oriented counterclockwise.

7.19 In Figure 7.19 we showed that the net flux across the boundary between two adjacent cubes is zero when the vector representing the flux, $\mathbf{F}$, is parallel to the unit normal vectors. Show that this also holds if $\mathbf{F}$ is at an angle to the unit normal $(\hat{\mathbf{n}}_1)$ of the left-hand cube.

7.20 The material derivative of the velocity $\mathbf{V}$ of a fluid moving in a gravitational field can be written as

$$\frac{D\mathbf{V}}{Dt} = -\frac{1}{\rho}\nabla p - \nabla \Phi,$$

where t is time, ρ is the fluid density, p is the fluid pressure, and Φ is the potential of the gravitational field. Consider the motion of the fluid along a closed path with line element $d\boldsymbol{\ell}$.

1. Show that

$$\oint \frac{D\mathbf{V}}{Dt} \cdot d\boldsymbol{\ell} = \oint \frac{D}{Dt}(\mathbf{V} \cdot d\boldsymbol{\ell}) - \oint \mathbf{V} \cdot d\mathbf{V}.$$

2. Show that

$$\oint \mathbf{V} \cdot d\mathbf{V} = 0, \qquad \text{and} \oint \nabla \Phi \cdot d\boldsymbol{\ell} = 0.$$

3. Use these results to show that

$$\frac{D}{Dt}\Gamma = -\oint \frac{1}{\rho} dp,$$

where Γ is the circulation of the velocity vector field. This result is called the *Bjerknes circulation theorem*.

4. Show that

$$-\oint \frac{1}{\rho} dp = \iint_A \left(\frac{\nabla \rho \times \nabla p}{\rho^2} \right) \cdot \mathbf{n}\, dA,$$

where A is the area enclosed by the closed path and $\mathbf{n}$ is the unit normal vector to the area.

8 Special Functions

By now we should be very familiar with what are called *elementary functions*. These are functions such as e^x, $\sin(\theta)$, and $\log_{10}(x)$, and we have seen that they are very useful in representing phenomena that we are interested in. For example, the sine and cosine functions can describe oscillatory and wavelike behaviors, the exponential function can describe radioactive decay or the growth of bacterial colonies, or the decrease in light intensity as it passes through the atmosphere, the ocean, or a dense plant canopy. Many of these functions describe solutions to differential equations that describe phenomena we are interested in, but we have seen that not all differential equations can be solved in terms of elementary functions. This creates a problem because some of these equations appear very frequently. In these cases, the solutions to these equations are given specific names—we have already come across some of them, such as the Bessel function (Equation (6.90)). In other cases, we find that certain types of integral arise over and over again, and these are also given specific names. So, in this chapter we are going to expand our repertoire of functions and delve a little into their properties; the reward for doing so is that we will be able to use mathematics to examine more complicated and interesting phenomena.

8.1 Heaviside Function

We have already seen examples where we have analyzed how a system responds to a sudden change in the forcing or external driver. For example, in Section 6.3 we examined a polluted lake where a sudden pulse of water or pollutant flowing into the lake changed its volume and the concentration of the pollutant in it. In Chapter 6 we handled this by using the perturbed conditions (i.e., the conditions immediately after the sudden change) as our initial condition and found the solution describing how the system recovered from the perturbation (i.e., the pulse of water). But how would we deal with a situation where a second pulse occurred shortly after the first so that the lake would not have enough time to return to its normal state before the second pulse occurred? A general method for representing sudden shifts in systems makes use of the *Heaviside function*.[1]

[1] This function is named after the scientist Oliver Heaviside (1850–1925), who made significant advances in mathematics, physics, and electrical engineering, paving the way for much of modern telecommunications (Hunt, 2012).

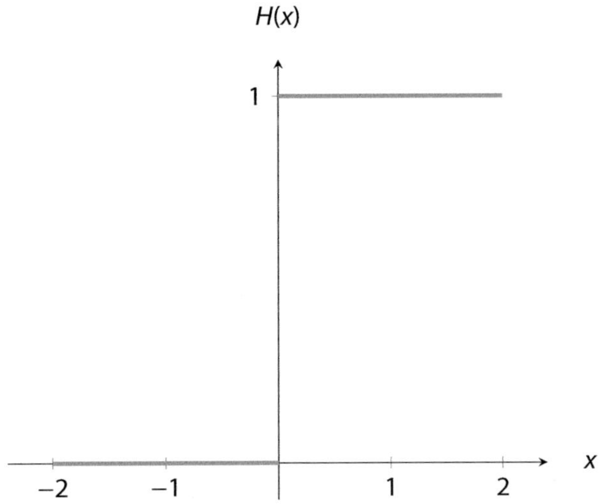

Figure 8.1 A plot of the Heaviside function, Equation (8.1).

The Heaviside function (also called the *step function*) is defined by

$$H(x) = \begin{cases} 0 & x < 0 \\ 1 & x \geq 0 \end{cases} \tag{8.1}$$

and describes a jump, or step discontinuity, at $x = 0$ (Figure 8.1). We can place the jump at any value of x by shifting the origin of the x axis. For example, if we need to have a jump at $x = a$, then we shift the origin of the x axis such that the argument to the Heaviside function has a value of zero when $x = a$, i.e.,

$$H(x - a) = \begin{cases} 0 & x < a \\ 1 & x \geq a \end{cases}. \tag{8.2}$$

We can also easily represent piecewise functions using the Heaviside function. For example, if

$$g(x) = \begin{cases} \alpha(x) & x < a \\ \beta(x) & x \geq a \end{cases},$$

then we can write $g(x) = \alpha(x) + (\beta(x) - \alpha(x))H(x - a)$. This works because for $x < a$, $H(x-a) = 0$, so $g(x) = \alpha(x)$; and for $x \geq a$, $H(x-a) = 1$, so $g(x) = \alpha(x)+\beta(x)-\alpha(x) = \beta(x)$. We can easily add more pieces to the function, though we have to be careful that we always subtract the correct pieces.

Exercise 8.1.1 Write the function

$$g(x) = \begin{cases} x & x < 1 \\ x^2 & 1 \leq x < 2 \\ x^4 & x > 2 \end{cases}$$

using the Heaviside function.

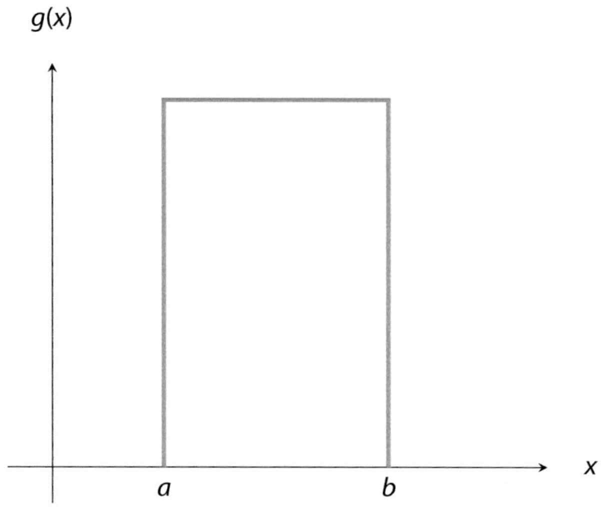

Figure 8.2 A plot of the box function, Equation (8.3).

We can also use the Heaviside function to create a *box function*

$$g(x) = \begin{cases} 0 & x < a \\ 1 & a \le x < b \\ 0 & x \ge b \end{cases}, \tag{8.3}$$

where $a < b$ (Figure 8.2). Using the Heaviside function, we can rewrite this function as $g(x) = H(x - a) - H(x - b)$, so that if $x < a$, then $H(x - a) = H(x - b) = 0$ and $g(x) = 0$; and if $x \ge b$, then $H(x - a) = H(x - b) = 1$ and $g(x) = 0$ again, but if $a \le x < b$, then $H(x - a) = 1$ and $H(x - b) = 0$, so that $g(x) = 1$. The box function is very useful for representing phenomena that switch on and off. We will see how to solve ODEs that involve the Heaviside function in Chapter 9.

Exercise 8.1.2 Show that

$$\int_{-\infty}^{x} H(x)\,dx = \begin{cases} 0 & x < 0 \\ x & x > 0 \end{cases},$$

which is sometimes called the *ramp function*.

8.2 Delta Function

What happens if we make the box function in Figure 8.2 thinner and thinner, but keep the area of the box constant as we do it (Figure 8.3)? As the box (or rectangular pulse) gets narrower and narrower, the height of the pulse must become taller so that the area under the curve remains the same. Let us see if we can translate this into the language of

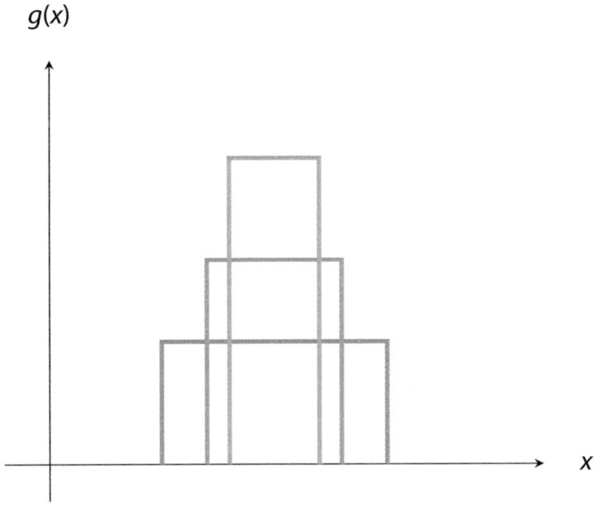

Figure 8.3 Making a rectangular box thinner but keeping the area under the box function constant.

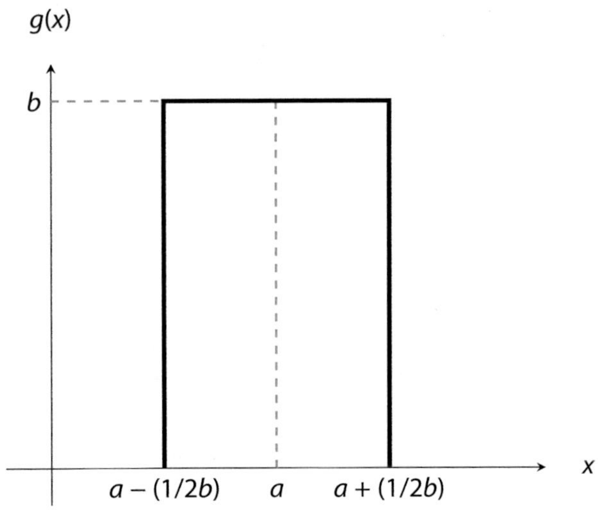

Figure 8.4 A rectangular pulse of width $1/b$ and height b centered on $x = a$.

mathematics. To start with, let us consider a box centered on $x = a$ with a width of $1/b$ and height b, so that the area under the curve is 1 (Figure 8.4), and the function is given by

$$\Delta(x, a) = \begin{cases} b & a - \frac{1}{2b} \leq x \leq a + \frac{1}{2b} \\ 0 & \text{otherwise} \end{cases}$$

and

$$\int_{-\infty}^{\infty} \Delta(x, a)\, dx = 1,$$

where we have extended the limits on the integral to $\pm\infty$ because the function is zero outside of the range $a - 1/2b \leq x \leq a + 1/2b$. Now let us consider a continuous function $f(x)$ that is defined for $-\infty < x < \infty$ and consider what happens to the value of the integral

$$\int_{a-1/2b}^{a+1/2b} f(x)\Delta(x,a)\,dx \qquad \text{as } b \to \infty;$$

in other words, as the rectangular pulse gets narrower but the area under the pulse remains constant. The mean value theorem (Equation (2.53)) tells that there exists a number ξ that lies between $(a - 1/2b)$ and $(1 + 1/2b)$ such that

$$f(\xi)\Delta(\xi,a) = \frac{1}{(1 + 1/2b) - (1 - 1/2b)} \int_{a-1/2b}^{a+1/2b} f(x)\Delta(x,a)\,dx,$$

and rearranging this equation gives us

$$\int_{a-1/2b}^{a+1/2b} f(x)\Delta(x,a)\,dx = \left[\left(a + \frac{1}{2b}\right) - \left(a - \frac{1}{2b}\right)\right] f(\xi)\Delta(\xi,a) = \frac{1}{b} f(\xi)b = f(\xi).$$

Now, if we take the limit as $b \to 0$, then both limits of the integral become a, but since $\Delta(x,a)$ is zero outside of the rectangular box, we can extend the limits of the integral to $\pm\infty$. In this limit, the function $\Delta(x,a)$ becomes an infinitely thin spike located at $x = a$; and in this limit, it is customary to write $\Delta(x,a) = \delta(x - a)$. Lastly, in the limit $b \to 0$, the value $x = \xi$ becomes $x = a$, so the right-hand side of the equation becomes $f(a)$. The equation then becomes

$$\int_{-\infty}^{\infty} f(x)\delta(x - a)\,dx = f(a), \tag{8.4}$$

and the function $\delta(x - a)$ is called the *Dirac delta function*, or just the delta function.[2] Equation (8.4) tells us something that is rather curious; if we multiply a function $f(x)$ by the Dirac delta function $\delta(x - a)$ and integrate from minus infinity to plus infinity, the result is the value of the function at $x = a$. We can understand this intuitively by thinking about what happens to $f(x)\delta(x - a)$. The delta function can be loosely thought of as an infinitely thin spike located at $x = a$ (ignoring that the spike is infinitely tall) and is zero everywhere else. So, when we multiply $f(x)$ by $\delta(x-a)$ and integrate, the only contribution we get to the integral is when $x = a$. But why can we not just multiply $f(x)$ by $\delta(x - a)$? Why do we need the integral? The reason is that we have been extremely cavalier in our calculation leading up to Equation (8.4). The delta function is not a normal function like a sine or exponential, but is what is called a *generalized function* or *distribution*. Although we arrived at Equation (8.4) by appealing to our intuition, a derivation of this equation can be made mathematically rigorous, but to do so would take us far beyond the scope of this book.

Exercise 8.2.1 What is the value of the following integrals?

$$\text{a. } \int_{-\infty}^{\infty} f(x)\delta(x)\,dx, \quad \text{b. } \int_{-\infty}^{\infty} \sin(x)\delta(x)\,dx, \quad \text{c. } \int_{-infty}^{infty} (3x^2 - 2)\delta(x - 4)\,dx.$$

[2] This function is named after the English theoretical physicist Paul Adrien Maurice Dirac (1902–1984), widely regarded as one of the most important theoretical physicists of the twentieth century.

The delta function is such an unusual mathematical object that we might legitimately wonder why it is useful. Delta functions are frequently used to approximate situations where a natural system is subject to a large forcing that acts over a very brief time. For example, let us look at the first order linear differential equation

$$a\frac{dx}{dt} + bx = \delta(t-\tau),$$

where a, b, and τ are constants. The solution to the homogeneous equation is an exponential function, but how does this change when the right-hand side contains a delta function? If we integrate both sides of the equation from $\tau - t_0$ to $\tau + t_0$, we find that

$$a(x(\tau+t_0) - x(\tau-t_0)) + b\int_{\tau-t_0}^{\tau+t_0} x\, dt = \int_{\tau-t_0}^{\tau+t_0} \delta(t-\tau)\, dt = 1.$$

Now, if we make t_0 smaller and smaller, then the second term on the left-hand side of the equation gets smaller and smaller, becoming zero when $t_0 = 0$, but what about the first term? As $t_0 \to 0$, $x(\tau + t_0)$ tends to $x(\tau)$ from above (i.e., $x(\tau_+)$) and similarly $x(\tau - t_0)$ tends to $x(\tau)$ from below (i.e., $x(\tau_-)$). Note that we have assumed that the solution $x(t)$ is piecewise continuous. So, we get that

$$x(\tau_+) - x(\tau_-)) = \frac{1}{a}$$

and the solution has a discontinuity at $t = \tau$. This is showing us that a sudden impulse to the forcing of a differential equation produces a discontinuity in the solution, as we would expect.

Delta functions might appear to be too esoteric to be of any use, but that is not the case. For example, the *point source* is a useful fiction that is frequently used to mathematically model situations where a source of chemical has a much smaller physical size than the region we are modeling. For example, we might want to develop equations describing the release of methane by microbes in soils. The size of a microbe is much smaller than the depth of soil we are interested in, so we could represent the positions of the microbes using delta functions.

The delta function has some useful properties. For example,

$$\int_{-\infty}^{x} \delta(y-a)\, dy = \begin{cases} 0 & x < a \\ 1 & x > 1 \end{cases},$$

which is just the Heaviside function $H(x-a)$ (Equation (8.1)), so that

$$\frac{d}{dx}H(x-a) = \delta(x-a).$$

The delta function is a symmetric function, so $\delta(x-a) = \delta(a-x)$, and if we scale the argument of the delta function by a nonzero constant (a), then

$$\delta(ax) = \frac{1}{|a|}\delta(x).$$

Lastly, the derivative of the delta function is

$$\frac{d}{dx}\delta(x) = -\frac{1}{x}\delta(x).$$

We motivated our discussion of the delta function by looking at what happens to the integral of a rectangular pulse in the limit that we make the pulse have zero width and infinite height. We need not have used a rectangular pulse, and indeed there are other useful representations of the delta function that involve taking limits of other functions. One particularly useful representation is that of a Gaussian that gets narrower and narrower:

$$\delta(x) = \lim_{\sigma\to 0}\frac{1}{\sqrt{2\pi}\sigma}\exp\left(-\frac{x^2}{2\sigma^2}\right). \tag{8.5}$$

So far we have looked at the delta function in one dimension, but what happens if we want to use the delta function with vector arguments? If $\mathbf{x} = x\hat{\mathbf{i}} + y\hat{\mathbf{j}} + z\hat{\mathbf{k}}$ and $\mathbf{x}_0 = x_0\hat{\mathbf{i}} + y_0\hat{\mathbf{j}} + z_0\hat{\mathbf{k}}$ are vectors in three dimensions, then

$$\int_{-\infty}^{\infty}\delta(\mathbf{x}-\mathbf{x}_0)\,d\mathbf{x} = \int_{-\infty}^{\infty}\int_{-\infty}^{\infty}\int_{-\infty}^{\infty}\delta(x-x_0)\delta(y-y_0)\delta(z-z_0)\,dx\,dy\,dz.$$

If we are working in a coordinate system other than rectangular coordinates, then we need to take the Jacobian of the coordinate transformation into account. For example, if $\mathbf{r}$ and $\mathbf{r}_0$ are vectors in spherical polar coordinates (r, θ, ϕ), then

$$\int \delta(\mathbf{r}-\mathbf{r}_0)\,d\mathbf{r} = \int_{-\infty}^{\infty}\int_0^{2\pi}\int_0^{\pi}\frac{1}{r^2\sin(\phi)}\delta(r-r_0)\delta(\theta-\theta_0)\delta(\phi-\phi_0)r^2\sin(\phi)\,dr\,d\theta\,d\phi,$$

so that in spherical polar coordinates

$$\delta(\mathbf{r}-\mathbf{r}_0) = \frac{1}{r^2\sin(\phi)}\delta(r-r_0)\delta(\theta-\theta_0)\delta(\phi-\phi_0).$$

Exercise 8.2.2 What is the Dirac delta function in three-dimensional cylindrical coordinates?

8.3 Gamma and Error Functions

Some functions are defined by integrals that cannot be evaluated in terms of elementary functions. We have already met one of these in Chapter 5, the *Gamma function* (Equation (5.47)). The Gamma function has many applications in probability and statistics; for example, it is used to model the probability distribution of rainfall amounts, to represent the variation of aerosol concentration with particle size, and it allows us to generalize the concept of the factorial.

To motivate the connection with the factorial, let us start with an integral we know. If $\alpha > 0$, then

$$\int_0^{\infty} e^{-\alpha x}\,dx = -\frac{1}{\alpha}e^{-\alpha x}\bigg|_0^{\infty} = -\frac{1}{\alpha}.$$

Now, let us repeatably differentiate both sides of this equation with respect to α, giving

$$\int_0^\infty x\mathrm{e}^{-\alpha x}\,dx = \frac{1}{\alpha^2}, \qquad \int_0^\infty x^2\mathrm{e}^{-\alpha x}\,dx = \frac{2}{\alpha^3}, \qquad \int_0^\infty x^3\mathrm{e}^{-\alpha x}\,dx = \frac{3!}{\alpha^4}, \cdots$$

We can see a pattern emerging, and that is

$$\int_0^\infty x^n\mathrm{e}^{-\alpha x}\,dx = \frac{n!}{\alpha^{n+1}}.$$

This is an interesting formula because, if we put $\alpha = 1$, we can write the factorial of an integer in terms of an integral,

$$\int_0^\infty x^n\mathrm{e}^{-x}\,dx = n! \qquad n = 0, 1, 2\ldots \tag{8.6}$$

Why should we want to do this when we already know how to calculate factorials? It is true that evaluating the factorial of an integer n is easy enough, though if n is very large it becomes a very lengthy and tedious calculation. For large values of n it might be easier, and computationally faster, to evaluate the integral instead. We can also use Equation (8.6) to calculate the factorial of 0; if we set $n = 0$ in Equation (8.6), then

$$0! = \int_0^\infty \mathrm{e}^{-x}\,dx = 1.$$

In addition, in arriving at Equation (8.6) we have not specified that n has to be an integer, and this allows us to extend the concept of a factorial to nonintegers. By convention, the symbol ! is reserved for factorials of integers, and for nonintegers (p) we use $\Gamma(p)$,

$$\Gamma(p) = \int_0^\infty x^{p-1}\mathrm{e}^{-x}\,dx, \tag{8.7}$$

which converges for all $p > 0$, so that we can write

$$\Gamma(p+1) = \int_0^\infty x^p\mathrm{e}^{-x}\,dx = p! \qquad p > -1. \tag{8.8}$$

Equation (8.7) is sometimes referred to as *Euler's formula*. The Gamma function has some nice properties that help us with calculations. For example, we have already seen in Section 5.4.1 that if we integrate Equation (8.8) using integration by parts, we can derive a recursion relationship,

$$\Gamma(p) = \frac{1}{p}\Gamma(p+1). \tag{8.9}$$

Exercise 8.3.1 Use Equation (8.9) to show that the Gamma function is undefined for $p \leq 0$.

We mentioned that if n has a large integer value, then it may be easier to calculate the factorial using Equation (8.6) to calculate $n!$ rather than calculating it directly. Let us see if we can find an approximate value of Equation (8.6) for large values of n. First, let us think a little about the integrand, $g(x) = x^n\mathrm{e}^{-x}$. As x increases from 0 to ∞, the term x^n will increase, and will do so quite fast if n is large. Conversely, e^{-x} decreases as x increases, and also does so quite rapidly. We can see for small values of x that the integrand will increase like x^n, but for large values of x it will decrease like e^{-x}. So, the function must have a maximum somewhere. To find the location of the maximum, let us write the integrand as $x^n e^{-x} = \exp(-x + n\ln(x))$, and by differentiating with respect to x we see that $(n\ln(x) - x)$ has a maximum at $x = n$.

Exercise 8.3.2 Plot the function $g(x) = x^n e^{-x}$ between $x = 0$ and $x = 10$ for three values of n between 2 and 8 and confirm the description of $g(x)$ in the preceding paragraph.

Now, since the integral is the area under the curve, we might expect that the biggest contribution to the integral comes from values of x in the neighborhood of the maximum of the integrand. We can then use a Taylor expansion to approximate the function in the neighborhood of its maximum.

Exercise 8.3.3 If $g(x) = x^n e^{-x}$, use a Taylor expansion to show that, to second order

$$g(x-n) = -n + n\ln(n) - \frac{1}{2n}(x-n)^2 + \cdots \tag{8.10}$$

Using Equation (8.10) we can write

$$n! \approx \int_0^\infty \exp\left(-n + n\ln(n) - \frac{1}{2n}(x-n)^2\right) dx = e^{-n+n\ln(n)} \int_0^\infty \exp\left(-\frac{1}{2n}(x-n)^2\right) dx. \tag{8.11}$$

The integral now looks like the integral of the Gaussian probability distribution (Equation (5.38)) with a mean value of n and a variance of n. We know that the integral of the probability distribution from $x = -\infty$ to $x = +\infty$ must be equal to one, but the lower limit of the integral in Equation (8.11) is $x = 0$. But we are looking for an approximation to $n!$ when n is large, and we know that most of the area under a Gaussian distribution is contained within two or three standard deviations. So, we can hopefully extend the lower limit of the integral to $x = -\infty$ without incurring too much of an error. Therefore, we can write

$$n! \approx e^{-n+n\ln(n)}\sqrt{2\pi n} = \sqrt{2\pi n}\, n^n e^{-n}, \tag{8.12}$$

which is *Stirling's formula* (Equation (5.39)).

The *error function* is another function defined by an integral that cannot be evaluated in terms of elementary functions. As its name suggests, the error function often appears in statistics and data analysis, but it also appears when studying diffusive processes in the environment (see Chapter 10). The error function is defined by

$$\operatorname{erf}(x) = \frac{2}{\sqrt{\pi}} \int_0^x e^{-t^2}\, dt, \tag{8.13}$$

and it is an odd function, i.e., $\operatorname{erf}(-x) = -\operatorname{erf}(x)$.

Exercise 8.3.4 Using the fact that $\Gamma(1/2) = \sqrt{(\pi)}$, show that

$$\operatorname{erf}(\infty) = \frac{2}{\sqrt{\pi}} \int_0^\infty e^{-t^2}\, dt = \frac{2}{\sqrt{\pi}} \frac{1}{2} \Gamma(1/2) = 1.$$

Instead of the error function, you will often see the *complementary error function*, which is defined as

$$\operatorname{erfc}(x) = 1 - \operatorname{erf}(x) = \frac{2}{\sqrt{\pi}} \int_x^\infty e^{-t^2}\, dt.$$

8.4 Orthogonal Functions and Orthogonal Polynomials

There are other special functions that arise, especially as solutions to Sturm–Liouville problems. Recall from Section 6.12 that the solutions to Sturm–Liouville problems are orthogonal eigenfunctions and that they form a complete set so that they can be used as a basis. Similar concepts of orthogonality and eigenvectors appeared in Chapter 4, where we could picture what orthogonal vectors looked like; it is harder to envisage what orthogonal functions look like, so let us explore this connection a little bit more.

Recall that two vectors **x** and **y** are orthogonal when

$$\mathbf{x}\cdot\mathbf{y} = \sum_{i=1}^{n} = x_1y_1 + x_2y_2 + \cdots x_ny_n = 0,$$

where n is the dimensionality of the space. How do we interpret two functions that are orthogonal? Let us consider the two functions

$$f(x) = \sin\left(\frac{\pi x}{2}\right) \qquad \text{and} \qquad g(x) = \cos\left(\frac{\pi x}{2}\right)$$

and evaluate them at $x = -1$, $x = 0$, and $x = +1$ so that the interval or step size between different points is $\Delta x = 1$. We will write these as three vectors,

$$\mathbf{x} = \begin{pmatrix}-1\\0\\1\end{pmatrix}, \qquad \mathbf{f}(x) = \sin\left(\frac{\pi\mathbf{x}}{2}\right) = \begin{pmatrix}-1\\0\\1\end{pmatrix}, \qquad \mathbf{g}(x) = \cos\left(\frac{\pi\mathbf{x}}{2}\right) = \begin{pmatrix}0\\1\\0\end{pmatrix},$$

from which we can calculate the dot product between the vectors **f** and **g**, which we will call U:

$$U = \sum_{i=1}^{3}\mathbf{f}(x_i)\cdot\mathbf{g}(x_i) = (-1\times 0) + (0\times 1) + (1\times 0) = 0.$$

This suggests that if we do indeed think of these objects as vectors, then they are orthogonal to each other. In this particular case, the vectors are three-dimensional and we can easily visualize them. What happens if we add more points in between $x = -1$ and $x - 1$ so that n gets larger? Doing this increases the number of the components of the vectors and increases the dimension of the space we are looking at.

Exercise 8.4.1 Repeat these calculations, but with $x = -1$, $x = -0.5$, $x = 0$, $x = 0.5$, $x = 1$ (i.e., $\Delta x = 0.5$). Note that the vectors are now five-dimensional (i.e., $n = 5$), which is not easy to visualize. You may find it useful to know that $\cos(\pi/4) = \sin(\pi/4) = 1/\sqrt{2}$.

Exercise 8.4.2 Repeat Exercise 8.4.2 $x = -1, -2/3, -1/2, -1/3, 0, 1/3, 1/2, 2/3, 1$. Note that $\sin(\pi/6) = \cos(\pi/3) = 1/2$, $\cos(\pi/6) = \sin(\pi/3) = \sqrt{3/2}$.

As we add more x values between $x = -1$ and $x = 1$, we increase the dimensionality of the vectors (i.e., n increases) and decrease the value of Δx, but the quantity $\sum_{i=1}^{n}\mathbf{f}(x_i)\cdot\mathbf{g}(x_i)$ stubbornly remains zero. As we let $\Delta x \to 0$, the number of components of the vectors becomes infinite, suggesting that we have an infinite dimensional space even though we are still looking at points between $x = -1$ and $x = 1$. Second, we move from functions evaluated at a discrete number of points to continuous functions, and we have seen that when we move from a summation of discrete points to a continuum of points we replace the summation by an integral. So, in this limit of $\Delta x \to 0$, U becomes

$$U = \int_{-1}^{1} f(x)g(x)\,dx = \int_{-1}^{1} \sin\left(\frac{\pi x}{2}\right)\cos\left(\frac{\pi x}{2}\right)\,dx = 0, \tag{8.14}$$

and we interpret this equation to mean that the two functions $f(x)$ and $g(x)$ are orthogonal to each other on the interval $-1 \leq x \leq 1$. Also, recall that taking the dot product of a vector with itself tells us the magnitude, or length, of the vector. Similarly,

$$\int_{-1}^{1} f(x)f(x)\,dx$$

is a quantity that we can think of as the magnitude of the function $f(x)$. You will frequently see an additional function $r(x)$ in the integrand in Equation (8.14), and this is called the *weighting function* and has to be strictly positive.

Recall from Chapter 4 that in n dimensions, a set of n orthogonal eigenvectors are linearly independent and we can write any other vector as a linear combination of the eigenvectors. This is similar for orthogonal functions. So, if $f(x)$ and $g(x)$ are orthogonal on the interval $a \leq x \leq b$,

$$f(x)\cdot g(x) = \int_{a}^{b} r(x)f(x)g(x)\,dx = 0 \tag{8.15}$$

and any well-behaved function on the interval $a \leq x \leq b$ can be written as a linear combination of $f(x)$ and $g(x)$. Note that we have written $f(x)\cdot g(x)$ to represent the integral in the Equation (8.15) because this is a common notation. This is a very powerful concept, as we shall see in Chapter 10.

Many differential equations that describe natural phenomena can be described by second order ODEs that can be written as Sturm–Liouville problems which have solutions that can be written as orthogonal polynomials. Recall that in Section 6.12 we saw that we could rewrite a second order ODE in a self-adjoint form by multiplying by a suitable function. So, for example, if the differential operator has the form

$$\mathcal{L}(x) = p(x)\frac{d^2}{dx^2} + q(x)\frac{d}{dx} + r(x),$$

then multiplying by the function

$$w(x) = \frac{1}{p(x)}\exp\left(\int \frac{q(x)}{p(x)}\,dx\right)$$

allows us to write the eigenvalue problem $\mathcal{L}\psi x = \lambda\psi(x)$ in self-adjoint form because

$$w(x)\mathcal{L}\psi(x) = P(x)\frac{d^2}{dx^2} + Q(x)\frac{d}{dx} + w(x)r(x),$$

where

$$\frac{dP}{dx} = \frac{d}{dx}\exp\left(\int \frac{q(x)}{p(x)}\,dx\right) = \frac{q(x)}{p(x)}\exp\left(\int \frac{q(x)}{p(x)}\,dx\right) = Q(x),$$

which is the requirement for $\mathcal{L}$ to be in self-adjoint form. It turns out that if nth order polynomial solutions to the Sturm–Liouville problem exist, then they can be written in the form

$$y_n(x) = \frac{1}{w(x)} \frac{d^n}{dx^n}(w(x)p(x))^n. \tag{8.16}$$

Equation (8.16) is known as *Rodrigues' formula*.[3] This formula provides another, more compact way to write the eigenfunctions of the differential operator.

8.5 Legendre Polynomials

Legendre polynomials are solutions of a second order ODE called *Legendre's differential equation*,[4]

$$(1 - x^2)\frac{d^2y}{dx^2} - 2x\frac{dy}{dx} + \mu y = 0, \tag{8.17}$$

where μ is a constant. This ODE appears frequently when we examine problems that have an axial symmetry, such as a cylindrical or spherical symmetry.

To solve Equation (8.17) in general, we have to resort to finding a power series expansion. Let us look for a power series solution in the neighborhood of $x = 0$. Recalling our classification of singular points, we see that Equation (8.17) has regular singular points at $x = \pm 1$ and $x = \infty$. This implies that there exists a series solution about the point $x = 0$ that will converge for all values of $|x| < 1$ for any value of μ.

Exercise 8.5.1 Show that the recursion relationship for the power series expansion of Equation (8.17) about the ordinary point $x = 0$ is

$$a_{m+2} = \frac{m(m+1) - \mu}{(m+1)(m+2)} a_m, \qquad m = 0, 1, 2, \ldots$$

Exercise 8.5.2 Use the ratio test to show that the radius of convergence of the power series solution is 1.

If we try to develop a power series solution at the regular singular point $x = 1$, we find that we get a term containing $\ln(1 - x)$, which is infinite at $x = 1$. Similarly, at $x = -1$ we get a term containing $\ln(1 + x)$, which is also infinite at $x = 1$. But, if the constant $\mu = m(m+1)$ for integer values of m, then the series will terminate with a finite number of terms because the coefficients in the power series will be zero for all higher order terms. In other words, our infinite series solutions will become polynomials of order m. These polynomials are called the *Legendre polynomials*, $P_n(x)$ (Table 8.1), and we have already met them in Equation (3.15). By convention, the Legendre polynomials are scaled such that $P_n(x) = 1$ at $x = 1$.

[3] Named after French mathematician Benjamin Olinde Rodrigues (1795–1851).

[4] This equation is named after the mathematician Adrien-Marie Legendre (1752–1833), whose portrait has been confused with that of the politician Louis Legendre in many textbooks (Duren, 2009).

Table 8.1 The first six Legendre polynomials

i	$\mu = i(i+1)$	Legendre polynomial
0	0	$P_0(x) = 1$
1	2	$P_1(x) = x$
2	6	$P_2(x) = \frac{1}{2}(3x^2 - 1)$
3	12	$P_3(x) = \frac{1}{2}(5x^3 - 3x)$
4	20	$P_4(x) = \frac{1}{8}(35x^4 - 30x^2 + 3)$
5	30	$P_5(x) = \frac{1}{8}(63x^5 - 70x^3 + 15x)$

Exercise 8.5.3 Use Rodrigues' formula to show that

$$P_n(x) = \frac{1}{s^n n!}\frac{d^n}{dx^n}(x^2 - 1)^n.$$

Many of the properties of the Legendre polynomials can be derived using a function called a *generating function*. Let us consider the function

$$\Phi(x, h) = (1 - 2xh + h^2)^{-1/2}, \qquad |\, h \,| < 1. \tag{8.18}$$

Now, for simplicity, let us define $y = 2xh - h^2$ so that $\Phi = (1 - y)^{-1/2}$. We can now easily expand Φ in terms of y (this was the reason for making the definition, just to make our lives easier) and, once we have substituted back for h, we can collect terms:

$$\begin{aligned}\Phi(y) = (1 - y)^{-1/2} &= 1 + \frac{1}{2}y + \frac{3}{8}y^2 \ldots = 1 + \frac{1}{2}(2xh - h^2) + \frac{3}{8}(2xh - h^2)^2 + \cdots \\ &= 1 + hx + h^2\left(\frac{3}{2}x^2 - \frac{1}{2}\right) + \cdots\end{aligned}$$

and we see that each coefficient of h^n looks suspiciously like one of the Legendre polynomials from Table 8.1. Indeed, if we continue this further, we find that

$$\Phi(x, h) = P_0(x) + hP_1(x) + h^2P_2(x) + \cdots = \sum_{n=0}^{\infty} h^n P_n(x). \tag{8.19}$$

The function $\Phi(x, h)$ is the generating function.

Example 8.1 Let us see how we can use the generating function to find some properties of the Legendre polynomials. First, let us put $x = -x$ in Equation (8.18) and use Equation (8.19) to find

$$\begin{aligned}\Phi(-x, h) = (1 + 2xh + h^2)^{-1/2} &= \sum_{n=0}^{\infty} h^n P_n(-x) \\ = (1 - 2x(-h) + (-h)^2)^{-1/2} &= \sum_{n=0}^{\infty} (-h)^n P_n(x),\end{aligned}$$

which implies that $P_n(-x)h^n = P_n(x)(-h)^n = P_n(x)(-1)^n h^n$. This gives us a convenient formula for calculating the Legendre polynomial of a negative number,

$P_n(-x) = (-1)^n P_n(x)$. This equation tells us that $P_n(x)$ is an even function (i.e., symmetric about $x = 0$) if n is an even integer and it is an odd function if n is an odd integer. Next, if we differentiate Equation (8.18) with respect to h, we find a nice recurrence formula

$$nP_n(x) = (2n-1)P_{n-1}(x) - (n-1)P_{n-2}(x), \qquad n = 2, 3, \ldots$$

and if we differentiate Equation (8.18) with respect to x, we find

$$P_n'(x) - 2xP_{n-1}'(x) + P_{n-2}' = P_{n-1}(x), \qquad n = 2, 3, \ldots$$

We can also integrate both sides of Equation (8.18) from $x = -1$ to $x = +1$ and use the orthogonality property to see that

$$\int_{-1}^{1} (P_n(x))^2 \, dx = \frac{2}{2n+1}, \qquad n = 0, 1, 2 \ldots$$

We have come across Legendre polynomials before when we used the binomial theorem to derive a multipole expansion of a gravitational field (Equation (3.15)). In fact, Legendre polynomials occur frequently in multipole expansions of fields that have an axial symmetry. They also appear when we solve partial differential equations that also have an axial symmetry, as we shall see in Chapter 10. The shapes of the Legendre polynomials (Figure 8.5) also make them ideal for use in approximating more complex functions. For example, they are frequently in climate models used to represent climate variables that vary latitudinally (North, 1975; North and Kim, 2017).

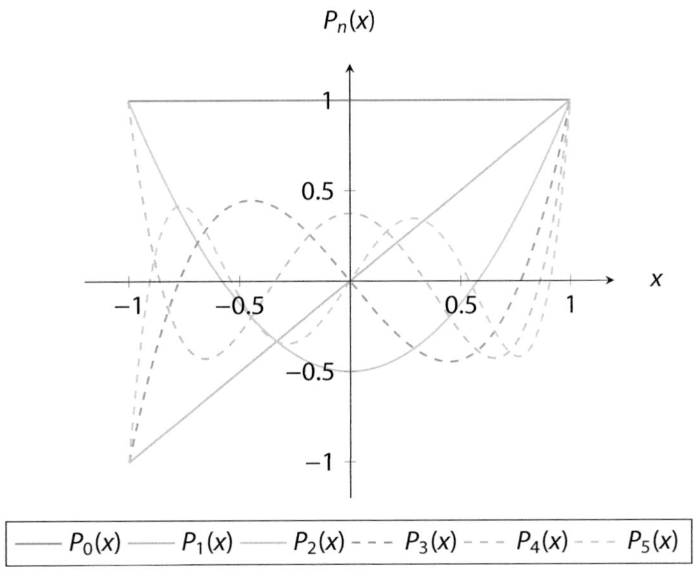

Figure 8.5 Plots of the first six Legendre polynomials (Table 8.1).

8.5.1 Associated Legendre Functions and Spherical Harmonics

Legendre's differential equation (Equation (8.17)) is a special case of the *associated Legendre equation*

$$(1-x^2)\frac{d^2y}{dx^2} - 2x\frac{dy}{dx} + \left(m(m+1) - \frac{n^2}{1-x^2}\right)y = 0, \tag{8.20}$$

where n is an integer. If $n = 0$, then Equation (8.20) is just Legendre's equation (Equation (8.17)). The term containing n is problematic because its denominator contains a factor $(1-x^2)$, which causes us trouble when $x = 1$. However, we can find power series solutions to this equation, but it takes some effort (see, e.g., Arfken et al., 2013, for more details). These solutions are the associated Legendre functions $P_m^n(x)$ and are related to the Legendre polynomials $P_m(x)$ by

$$P_m^n(x) = (-1)^n(1-x^2)^{n/2}\frac{d^n}{dx^n}P_m(x), \tag{8.21}$$

where for each value of m there are polynomials for integer n values up to m (Table 8.2) and the subscript m is called the *degree* and n is called the *order*. The associated Legendre polynomials have many of the same properties as the Legendre polynomials, but are generally harder to prove. For example, the polynomials for fixed n and different values of m are orthogonal. They are also orthogonal for fixed degree and varying order, but if the order of both polynomials is zero, then the integral in Equation (8.15) is infinite.

The associated Legendre equation appears when we try to solve Laplace's equation $\nabla^2\phi = 0$ in spherical polar coordinates (Chapter 10). When we do this we end up with Equation (8.20) for a function $\Theta(\theta)$,

$$\frac{1}{\sin(\theta)}\frac{d}{d\theta}\left(\sin(\theta)\frac{d\Theta}{d\theta}\right) + n(n+1)\Theta(\theta) = \frac{m^2}{\sin^2(\theta)},$$

and the following equation for a different function $\Phi(\phi)$,

$$\frac{1}{\Phi(\phi)}\frac{d^2\Phi}{d\phi^2} + m^2 = 0,$$

which has solutions $\Phi(\phi) = e^{im\phi}$ and $\Phi(\phi) = e^{-im\phi}$. The angles θ and ϕ are the angular coordinates (e.g., latitude and longitude) on the surface of the sphere.

Table 8.2 Some associated Legendre polynomials

m	n	Legendre polynomial
0	0	$P_0^0(x) = 1$
1	1	$P_1^1(x) = -(1-x^2)^{1/2}$
2	1	$P_2^1(x) = -3x(1-x^2)^{1/2}$
2	2	$P_2^2(x) = 3(1-x^2)$
3	1	$P_3^1(x) = -\frac{3}{2}(5x^2-1)(1-x^2)^{1/2}$
3	2	$P_3^2(x) = 15x(1-x^2)$
3	3	$P_3^3(x) = -15(1-x^2)^{3/2}$

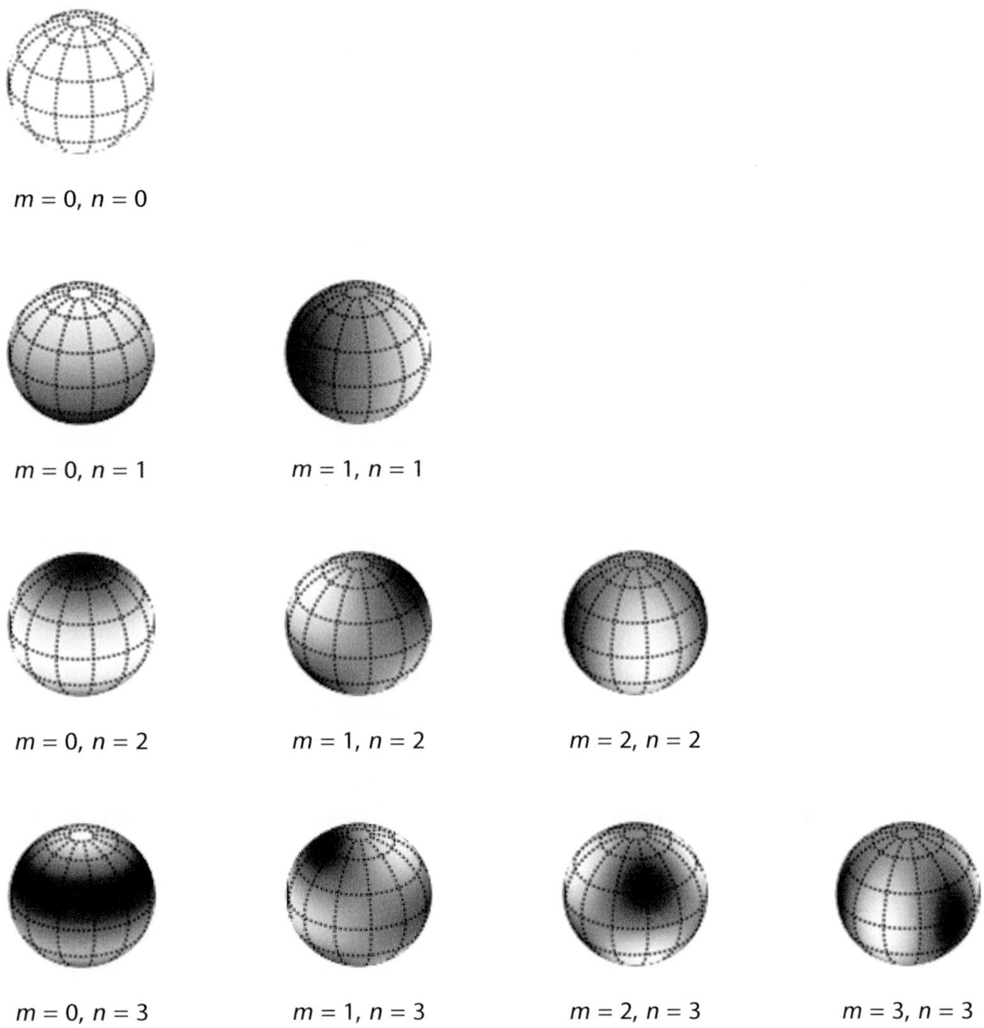

Figure 8.6 Plots of spherical harmonics ($Y_m^n(\theta, \phi)$) for various values of m and n. The grayscale shows regions where $Y_m^n(\theta, \phi)$ has different signs, with black being negative, white being positive, and gray being zero.

Exercise 8.5.4 Show that the functions $\Phi_+(\phi) = \mathrm{e}^{im\phi}$ and $\Phi_-(\phi) = \mathrm{e}^{-im\phi}$ are orthogonal.

The solution to Laplace's equation contains products of the functions $\Theta(\theta)$ and $\Phi(\phi)$, which can be written as (see, e.g., Arfken et al., 2013, for details)

$$Y_m^n(\theta, \phi) = (-1)^n \sqrt{\frac{2m+1}{2}\frac{(m-n)!}{(m+n)!}} P_m^n(\cos(\theta))\mathrm{e}^{im\phi}, \tag{8.22}$$

and these functions are called *spherical harmonics* (Figure 8.6). Spherical harmonics can be used as a basis for functions that vary over the surface of a sphere, so they find many applications in the geosciences. For example, the equation for the gravitational field of the

Earth at a point (r, θ, ϕ) given by spherical coordinates can be solved to give a gravitational potential of

$$\Phi(r,\theta,\phi) = \frac{GM}{R}\left[\frac{R}{r} - \sum_{n=2}^{\infty}\sum_{m=0}^{n}\left(\frac{R}{r}\right)^{n+1}\left(A_{nm}Y^c mn(\theta,\phi) + B_{nm}Y^s mn(\theta,\phi)\right)\right],$$

where M is the mass of the Earth, R is the equatorial radius of the Earth, and θ and ϕ are angular coordinates. The functions $Y^c mn(\theta, \phi)$ and $Y^s mn(\theta, \phi)$ are the real components of Equation (8.22), so that

$$Y^c mn(\theta,\phi) = P_n^m(\cos(\theta))\cos(m\phi), \qquad Y^s mn(\theta,\phi) = P_n^m(\cos(\theta))\sin(m\phi).$$

The values of the constants A_{nm} and B_{nm} can be found by comparing the results of this equation with satellite data, thereby obtaining an accurate model of the Earth's gravitational field (Tapley et al., 2005). Each spherical harmonic gives a different mode of variation across a surface. For example, for $m = 0$ there is no variation in longitude, only a latitudinal variation, and these are called *zonal harmonics*, with each value of n giving more oscillations between the north and south poles (Figure 8.6). If $m = n$, then the harmonics are called sectorial harmonics. Other harmonics are called tesseral harmonics.

8.6 Bessel Functions

Bessel functions, like the Legendre polynomials, are functions that are the solutions of a form of an ODE that appears very often when we are solving partial differential equations (Chapter 10) that involve oscillations in spherical or cylindrical coordinates. Bessel functions are solutions of *Bessel's equation*,[5] which in standard form is

$$x^2\frac{d^2y}{dx^2} + x\frac{dy}{dx} + (x^2 - p^2)y = 0, \tag{8.23}$$

where p is a constant (not necessarily an integer) and gives the *order of the Bessel function*. Note that

$$x(xy')' = x^2y'' + xy',$$

so we can write Bessel's equation in a slightly more compact form as

$$(xy')' + \left(x - \frac{p^2}{x}\right)y = 0,$$

which is in Sturm–Liouville form, and we expect that the solutions to the Bessel equation are orthogonal and form a complete set. Just as with Legendre's equation, the solutions to Equation (8.23) are power series, and in fact we have already met one of these (Equation (6.90)) for positive integer values of p (Figure 8.7),

$$J_p(x) = \sum_{j=0}^{\infty}(-1)^j\frac{1}{j!(j+p)!}\left(\frac{x}{2}\right)^{2j+p}, \tag{8.24}$$

[5] Named after the German scientist Friedrich Wilhelm Bessel (1784–1846), who was also the first scientist to use the method of parallax to find the distance from the Earth of a star other than the Sun.

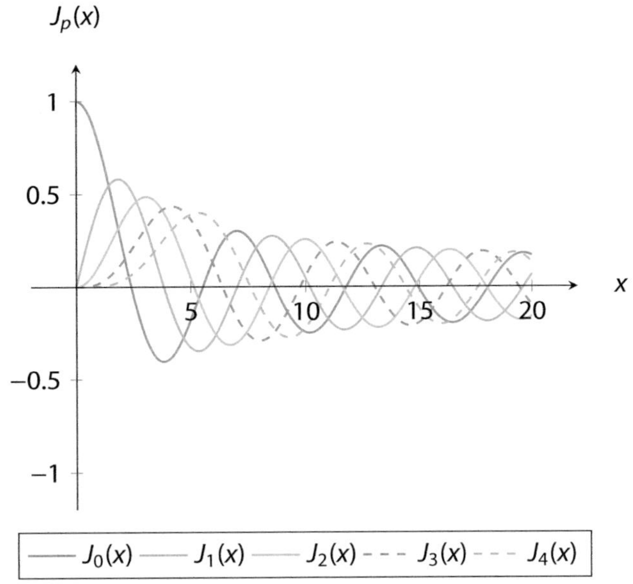

Figure 8.7 Plots of some Bessel functions of the first kind ($J_p(x)$) for positive integral order.

which are called *Bessel functions of the first kind*. Equation (8.24) can be extended to include noninteger values of p

$$J_p(x) = \sum_{j=0}^{\infty}(-1)^j \frac{1}{j!\,\Gamma(j+p+1)}\left(\frac{x}{2}\right)^{2j+p}, \qquad j \neq -1,-2,-3,\ldots \tag{8.25}$$

but this still provides only one solution for a given value of p, so we have to look for a second, linearly independent solution. For the Bessel equation, this is provided by the solution when p is negative,

$$J_{-p}(x) = \sum_{j=0}^{\infty}(-1)^j \frac{1}{j!\,\Gamma(j-p+1)}\left(\frac{x}{2}\right)^{2j-p}. \tag{8.26}$$

The general solution of Equation (8.23) is then $y(x) = AJ_p(x) + BJ_{-p}(x)$, where A and B are constants that need to be determined from the boundary conditions. However, when p is a negative integer, we can write the solution in the form

$$J_{-p}(x) = \sum_{j=0}^{\infty} \frac{(-1)^j}{j!\,(j-p)!}\left(\frac{x}{2}\right)^{2j-p},$$

and if we make the substitution $j = j + p$ in the infinite summation, this becomes

$$J_{-p}(x) = \sum_{j=0}^{\infty} \frac{(-1)^{j+p}}{(j+p)!\,j!}\left(\frac{x}{2}\right)^{2j+p} = (-1)^p J_p(x)$$

so that the two solutions are no longer independent. The standard form for the second solution in this case is called the *Bessel function of the second kind* or the *Neumann function* (Figure 8.8),

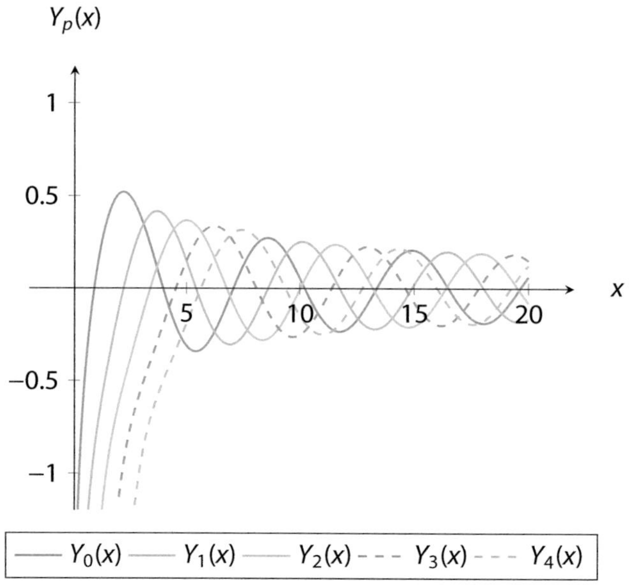

Figure 8.8 Plots of some Bessel functions of the second kind ($Y_p(x)$) for integral order.

$$Y_p(x) = \frac{\cos(p\pi)J_p(x) - J_{-p}(x)}{\sin(p\pi)}, \tag{8.27}$$

which also provides a solution for noninteger values of p because it is a linear combination of the solutions $J_p(x)$ and $J_{-p}(x)$.

8.7 Further Reading

In this chapter we have had a brief introduction to some of the special functions that are in common use. There are many more! Some find use in solving specialized problems, and others are more widely used. The *Handbook of Mathematical Functions with Formulas, Graphs, and Mathematical Tables* by Abramowitz and Stegun (1972) is an extremely useful book to have ready access to because it provides a comprehensive summary of many special functions and their properties. However, to learn more about how certain special functions are used one has to turn elsewhere. Books on mathematical methods in physics and engineering are good places to start, though the examples they give may not necessarily be relevant. Some good examples to look at include *Mathematical Methods for Physicists* by Arfken et al. (2013), *A Guided Tour of Mathematical Methods for the Physical Sciences* by Snieder and van Wijk (2015), and *Mathematical Methods in the Physical Sciences* by Boas (2006).

Problems

8.1 Consider the function

$$y(x) = \begin{cases} 0 & x < 0 \\ 1 & x \geq 1 \end{cases}$$

and make sketches of the following from $x = -2$ to $x = 2$:

$$1.\, u(x), \quad 2.\, u(x+1), \quad 3.\, x^2 u(x), \quad 4.\, u(x) - u(x-1).$$

8.2 What is the value of the following expressions?

1. $x\delta(x)$
2. $\displaystyle\int_{-\infty}^{\infty} f(x-y)\delta(y)\, dy$
3. $\displaystyle\int_{-\infty}^{\infty} f(y)\delta(x-y)\, dy$
4. $\displaystyle\int_{-\pi/2}^{\pi/2} \cos(x)\delta(x-\pi)\, dx$

8.3 By making a suitable change of variables, show that

$$\delta(ax) = \frac{1}{|\, a\,|}\delta(x).$$

8.4 Solve the differential equation

$$\frac{du}{dt} = \delta(t-1), \quad u(t=0) = 0.$$

8.5 Show that the change of variable $x = \cos(\theta)$ turns the following equation into Legendre's equation:

$$\frac{d^2 y}{d\theta^2} + \cot(\theta)\frac{dy}{d\theta} + m(m+1)y = 0, \quad 0 < \theta < \pi.$$

8.6 Show that $P_{2m}(x)$ is an even function and P_{2m+1} is an odd function, where $P_m(x)$ is a Legendre polynomial.

8.7 Use a change of variables to show that

$$\Gamma(x) = \int_0^1 (-\ln(t))(x-1)\, dt.$$

8.8 Show that

$$\frac{d}{dx}\operatorname{erf}(x) = \sqrt{2}\sqrt{\pi}\mathrm{e}^{-x^2}.$$

8.9 The error function is closely tied to the Gaussian distribution. Show that the cumulative probability function for a standardized Gaussian distribution is

$$\frac{1}{\sqrt{2\pi}}\int_{-\infty}^{x} \mathrm{e}^{-z^2/2}\, dz = \frac{1}{2} + \frac{1}{2}\operatorname{erf}\left(\frac{x}{\sqrt{2}}\right).$$

8.10 Show that

$$J_{1/2}(x) = \sqrt{\frac{2}{x\pi}}\sin(x) \qquad \text{and} \qquad J_{-1/2}(x) = \sqrt{\frac{2}{x\pi}}\cos(x).$$

8.11 Show that

$$\int_{-1}^{1} P_n(x)\,dx = 0 \qquad \text{for } n > 0.$$

8.12 Consider the eigenvalue problem

$$x\frac{d^2y}{dx^2} + (1-x)\frac{dy}{dx} = \lambda y(x).$$

Show that this equation is not in self-adjoint form. Write the equation in self-adjoint form by finding the function $w(x)$ that, when multiplying the equation, converts it into self-adjoint form.

8.13 If $P_m(x)$ is a Legendre polynomial, use Rodrigues' formula to show that

$$\int_{-1}^{1} x^n P_m(x)\,dx = 0.$$

8.14 We can take the analogy between orthogonal functions and vectors one step further and use the Gram–Schmidt orthogonalization procedure (Section 4.3.1.1) to construct a basis for polynomials. Consider polynomials defined on the interval $-1 \leq x \leq 1$. The simplest polynomial is $p_0(x) = 1$.

1. Show that the magnitude of $p_0(x)$ is 2 and normalize $p_0(x)$ to have length 1. This will be the analogue of a unit vector.
2. Show that the next simplest polynomial ($p_1(x) = x$) is orthogonal to $p_0(x)$ on the interval $-1 \leq x \leq 1$, and form a normalized function $p_1(x)$ with magnitude 1.
3. Show that the polynomial $p_2(x) = x^2$ is orthogonal to $p_1(x)$, and use an analogue of the Gram–Schmidt process to show that $p_2(x) = x^2 - 1/3$.
4. Repeat this procedure to show that $p_3(x) = \sqrt{7/8}(5x^3 - 3x)$.
5. Compare the polynomial basis you have constructed with the Legendre polynomials.

9 Fourier Series and Integral Transforms

We have seen that orthogonal functions can be used as a basis for developing series representations of more complicated functions. Two functions that are very useful in this regard are the sine and cosine function. Many phenomena in the natural world display periodic, or almost periodic behavior, and these can be analyzed using series made from these functions. For example, changes in sea level caused by tidal effects (Figure 9.1a) can be decomposed into a series of sines and cosines having different frequencies and amplitudes. Other natural phenomena produce complicated signals that do not appear to have any periodicity at all. For example, the Southern Oscillation Index (SOI) is a measure of the difference in atmospheric pressure between the eastern and western tropical Pacific Ocean and is used as an indicator of El Niño and La Niña phenomena that affect many aspects of the climate of the tropics and subtropics. The SOI shows a very complicated behavior over time (Figure 9.1b) and, unlike the tidal signal, it would be hard to detect any periodicity in the signal by eye. It turns out that El Niño and La Niña are *quasiperiodic*, which means that they do not occur with a regular periodicity but they do occur with a period between two and seven years.

Fourier techniques and integral transforms provide the foundation for techniques that can be used to analyze and find periodicities in the type of data shown in Figure 9.1. Once we know that a signal has a periodic component, we can look for processes that can cause the signal and that have the same periodicity.

9.1 Fourier Series

Sine and cosine functions will take center stage throughout most of this chapter, so it is worth spending a little time reminding ourselves of some terminology (Figure 9.2). If t represents time, then the two functions $S(t) = A\sin(\omega t)$ and $C(t) = B\cos(\omega t)$ represent an oscillation (or wave) with time. The constants A and B are the *amplitudes* of the oscillation and ω is the *angular velocity*—i.e., the number of complete cycles in a unit of time—with units of rad s^{-1}. The *period* of the oscillation is $T = 2\pi/\omega$ with dimensions of [T], and the *frequency* of the oscillation is $1/T$ with dimensions of $[T]^{-1}$. We can always write $\cos(\omega t) = \sin(\omega t + \pi/2)$, where the angle $\pi/2$ is called the *phase* between the two signals—the sine and cosine functions are said to be $\pi/2$ out of phase with each other. The phase need not be $\pi/2$, so two sine functions, $\sin(\omega t)$ and $\sin(\omega t + \theta_0)$, are out of phase by θ_0. The meaning of the phase is clear if we put $t = 0$ in both signals—it represents how much further one signal is in the oscillation when the other signal starts its oscillation.

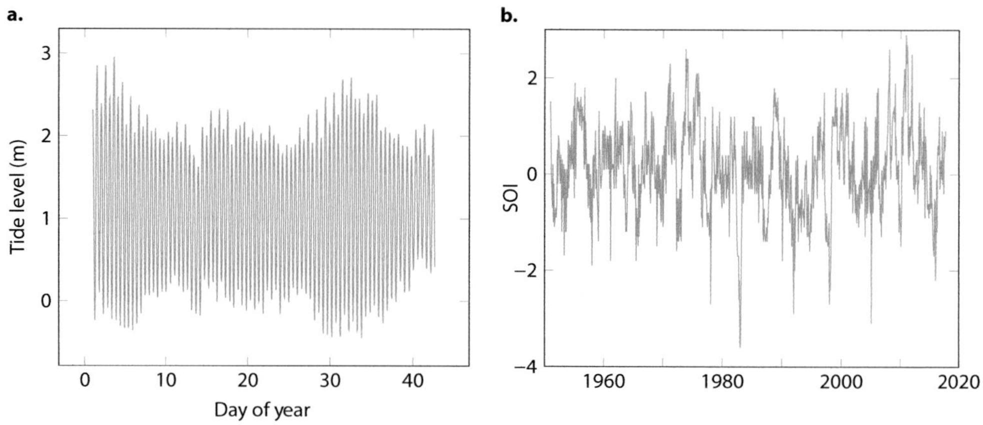

Figure 9.1 Examples of periodic and quasiperiodic signals: (a.) the tidal height at Fort Pulaski, Georgia, USA, (b.) a time series of the Southern Oscillation Index. Both data sets are publicly available from the National Oceanic and Atmospheric Administration.

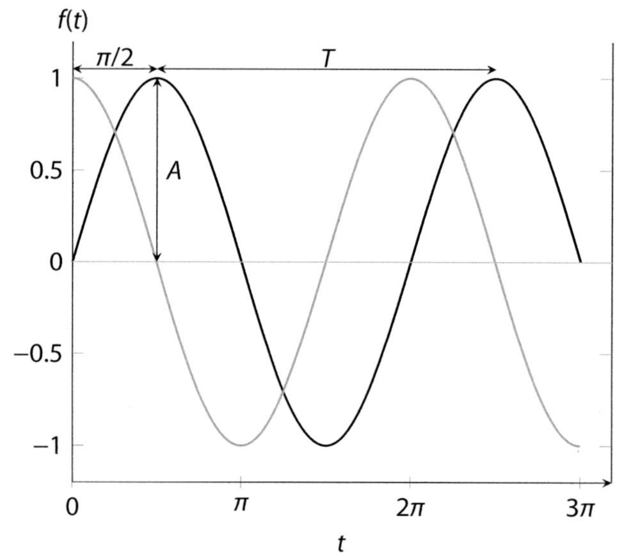

Figure 9.2 The sine and cosine functions, showing the amplitude A, period T, and phase lag $\pi/2$.

Terminology and notation traditionally changes slightly if we are looking at oscillations with respect to space so that $S(x) = A\sin(kx)$ and $C(x) = B\cos(kx)$. In this case, k is called the *angular wave number* and is the number of complete oscillations in 2π, and the *wavelength* is $\lambda = 2\pi/k$. You will often see wave numbers referred to as frequencies, or spatial frequencies; and if the dimensions of the variables in the trigonometric functions are not given, authors will often call k a frequency. It is important for what follows to be able to switch between thinking in terms of the period of oscillation and thinking in terms of frequency, or wavelength and wave number if we have oscillations in space.

Now, let us consider the periodic functions $g(x) = \sin^3(x)$ and $f(x) = \sin^4(x)$. These can be written as a sum of trigonometric terms with different frequencies using equations in Appendix B:

$$\sin^3(x) = \frac{1}{4}(3\sin(x) - \sin(3x)), \quad \sin^4(x) = \frac{1}{8}(3 - 4\cos(2x) + \cos(4x)). \tag{9.1}$$

It appears that as the exponent increases from 3 to 4, the right-hand side of the equations become more complicated and include sines and cosines of higher frequencies. This suggests that we might be able to approximate an arbitrary periodic function $f(x)$ defined on an interval $0 \leq x \leq 2L$ as an infinite series of sines and cosines of increasing frequency. This insight leads us to the definition of the *Fourier series*,[1]

$$f(x) = \frac{a_0}{2} + \sum_{n=1}^{\infty}\left[a_n \cos\left(\frac{n\pi x}{L}\right) + b_n \sin\left(\frac{n\pi x}{L}\right)\right], \tag{9.2}$$

where a_n and b_n are constant coefficients we have yet to determine and the factor of 1/2 that multiplies the a_0 term is to simplify the mathematics later on. You will sometimes see Equation (9.2) written slightly differently as

$$f(x) = \sum_{n=0}^{\infty}\left[a_n \cos\left(\frac{n\pi x}{L}\right) + b_n \sin\left(\frac{n\pi x}{L}\right)\right],$$

where the a_0 term has been incorporated into the cosine term.

If we know the function $f(x)$, how can we find the values of the coefficients? To make our mathematics simpler, we will consider a function for which $L = \pi$. We will see later that we can use different values of L without affecting our results. Equation (9.2) then becomes

$$f(x) = \frac{a_0}{2} + \sum_{n=1}^{\infty} a_n \cos(nx) + b_n \sin(nx). \tag{9.3}$$

Let us start by integrating the function over a single period:

$$\int_{-\pi}^{\pi} f(x)\,dx = \frac{1}{2}\int_{-\pi}^{\pi} a_0\,dx + \int_{-\pi}^{\pi}\sum_{n=1}^{\infty}(a_n \cos(nx) + b_n \sin(nx))\,dx. \tag{9.4}$$

We can easily evaluate the first term on the right-hand side to give πa_0. Let us assume for the time being that we can integrate the terms in the summation term by term, just as we did for power series. For an arbitrary value of n, this will give two integrals that both evaluate to zero,

$$\int_{-\pi}^{\pi} \cos(nx)\,dx = 0 \quad \text{and} \quad \int_{-\pi}^{\pi} \sin(nx)\,dx = 0,$$

so we are left with

$$a_0 = \frac{1}{\pi}\int_{-\pi}^{\pi} f(x)\,dx. \tag{9.5}$$

[1] These are named after Jean-Baptiste Joseph Fourier (1768–1830), who introduced them as a means of solving a type of partial differential equation called the heat equation.

In fact, the term $a_0/2$ is the mean value of the function $f(x)$ over the interval $0 \le x \le 2\pi$, and this explains why the expansion of $\sin^3(x)$ in Equation (9.1) did not have a constant term but the expansion of $\sin^4(x)$ did.

Exercise 9.1.1 Show that the mean value over the interval $-\pi \le x \le \pi$ of the function in Equation (9.3) is $a_0/2$.

To find the coefficients a_n and b_n, we make use of the orthogonality of the sine and cosine functions. First, we multiply the function $f(x)$ by $\cos(mx)$ and integrate over the interval $-\pi \le x \le \pi$,

$$\int_{-\pi}^{\pi} \cos(mx) f(x)\, dx = \frac{1}{2}\int_{-\pi}^{\pi} a_0 \cos(mx)\, dx + \int_{-\pi}^{\pi} \sum_{n=1}^{\infty} (a_n \cos(nx) + b_n \sin(nx)) \cos(mx)\, dx.$$

The first term on the right-hand side of the equation evaluates to zero, and if we again assume that we can integrate the remaining terms term by term, we see that we have two types of integral:

$$a_n \int_{-\pi}^{\pi} \cos(nx)\cos(mx)\, dx \quad \text{and} \quad b_n \int_{-\pi}^{\pi} \sin(nx)\cos(mx)\, dx.$$

Concentrating on the first integral, the orthogonality of cosines tells us that

$$\int_{-\pi}^{\pi} a_n \cos(nx)\cos(mx)\, dx = \pi\delta_{mn},$$

and similarly, the second integral is

$$b_n \int_{-\pi}^{\pi} \sin(nx)\cos(mx)\, dx = 0.$$

So, putting this all together we find that

$$a_n = \frac{1}{\pi}\int_{-\pi}^{\pi} \cos(nx) f(x)\, dx, \tag{9.6}$$

and similarly,

$$b_n = \frac{1}{\pi}\int_{-\pi}^{\pi} \sin(nx) f(x)\, dx. \tag{9.7}$$

Exercise 9.1.2 Derive Equation (9.7) in the same way that we derived Equation (9.6).

Exercise 9.1.3 Calculate the coefficients in the Fourier series of $\sin^4(x)$ using Equations (9.5)–(9.7) and compare your answer with the expansion given in Equation (9.1).

Now we have to be a little careful. You will often see formulae for a_n and b_n that look slightly different. We have looked at Fourier series for functions that are periodic with periods of 2π on the interval $-\pi < x < \pi$. However, the function $f(x)$ may be periodic over a different interval, say $-L \le u \le L$. To obtain the Fourier series for this function we simply transform variables in Equation (9.2) and Equations (9.5)–(9.7) to get

$$f(u) = \frac{a_0}{2} + \sum_{n=1}^{\infty} \left[a_n \cos\left(\frac{n\pi u}{L}\right) + b_n \sin\left(\frac{n\pi u}{L}\right)\right]$$

$$a_0 = \frac{1}{L}\int_{-L}^{L} f(u)\,du, \quad a_n = \frac{1}{L}\int_{-L}^{L} f(u)\cos\left(\frac{n\pi u}{L}\right) du$$
$$b_n = \frac{1}{L}\int_{-L}^{L} f(u)\sin\left(\frac{n\pi u}{L}\right) du.$$

Exercise 9.1.4 Write down the equations for the Fourier coefficients a_0, a_n, and b_n for a function $f(x)$ that is periodic on the interval $0 \leq x \leq 2L$.

One of the main advantages of Fourier series is that we can use them to represent functions that are not smooth, or even discontinuous. For example, let us calculate the Fourier series for the periodic function (Figure 9.3)

$$g(x) = \begin{cases} 1, & 0 < x < \pi \\ -1, & \pi < x < 2\pi \end{cases}, \tag{9.8}$$

which is basically a discontinuous step function. To evaluate the coefficients of the Fourier series using Equations (9.5)–(9.7) we need to integrate over the period $0 < x < 2\pi$, so we split the integral into two parts: the first with limits from 0 to π and the second with limits from π to 2π. For a_0, Equation (9.5) gives

$$a_0 = \frac{1}{\pi}\int_0^{\pi} g(x)\,dx + \frac{1}{\pi}\int_{\pi}^{2\pi} g(x) = \frac{1}{\pi}\int_0^{\pi} dx - \frac{1}{\pi}\int_{\pi}^{2\pi} dx = 1 - 1 = 0, \tag{9.9}$$

and similarly, we find that $a_n = 0$ for all values of n.

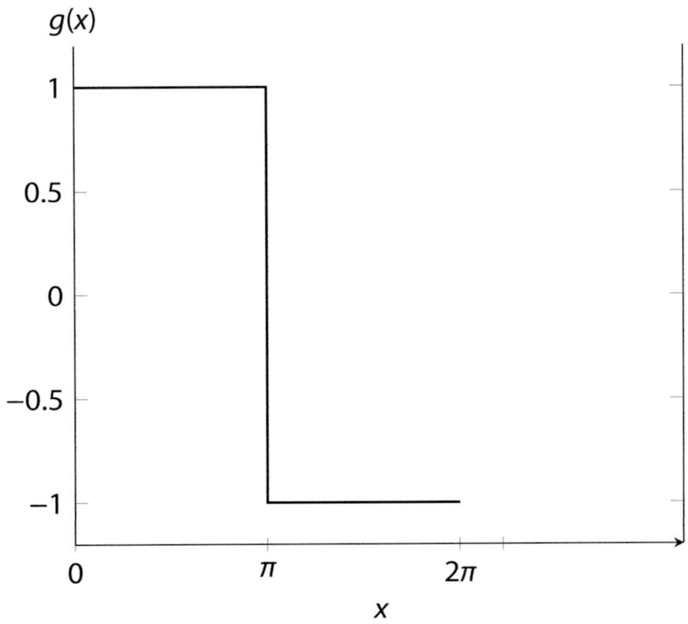

Figure 9.3 The step function (Equation (9.8)).

Exercise 9.1.5 Show that the coefficients $a_n = 0$ in the Fourier series for the function given in Equation (9.8).

Exercise 9.1.6 Show that the coefficients b_n in the Fourier series for the function in Equation (9.8) are given by

$$b_n = \frac{2(1 - \cos(n\pi)}{n\pi}. \tag{9.10}$$

The fact that $a_n = a_0 = 0$ for the function in Equation (9.8) is probably not too surprising in retrospect because Equation (9.8) is an odd function whereas $\cos(x)$ is an even function. So, the Fourier series for the function Equation (9.8) is

$$g(x) = \sum_{n=1}^{\infty} \frac{2(1 - \cos(n\pi)}{n\pi} \sin(nx).$$

We can simplify this a little by recalling that $\cos(n\pi) = (-1)^n$, so when n is even, $(1 - \cos(n\pi)) = 1 - 1 = 0$; and when n is an odd number, $(1 - \cos(n\pi)) = 1 - (-1) = 2$. Therefore, only the odd values of n actually contribute nonzero values to the summation. We can rewrite the series by substituting $n = 2m - 1$ so that n is always an odd number, giving finally

$$g(x) = \frac{4}{\pi} \sum_{m=1}^{\infty} \frac{\sin((2m-1)x)}{(2m-1)}, \tag{9.11}$$

which we can also write as

$$g(x) = \frac{4}{\pi} \sum_{n=1}^{\infty} \frac{\sin(nx)}{n}, \qquad n \text{ is odd},$$

so long as we remember that n takes only odd values.

If we plot the original step function (Equation (9.8)) and its Fourier series (Equation (9.11)) for different values of m, then we see that the Fourier series overshoots the value of the actual function wherever there is a sharp corner (Figure 9.4). We can make this overshoot very small by including many terms in the series in Equation (9.11), but it is always present. This is called the *Gibbs phenomenon*,[2] and it occurs because we are trying to represent a nonsmooth function (the square wave) with a finite number of smooth, continuous functions. If we could actually take the series to an infinite number of terms, then the overshoot would disappear, but since in practice we cannot do this, it is something we have to live with.

Exercise 9.1.7 Calculate the Fourier series of the function

$$f(x) = \begin{cases} 1 & -\pi < x < 0 \\ 2 & 0 < x < \pi \end{cases}.$$

Being able to represent difficult functions like the one in Equation (9.8) in terms of a series of sines and cosines can be very helpful. For example, an earthquake underneath the ocean can cause a sudden vertical displacement in the seafloor that will lead to a sudden

[2] This is named after Josiah Willard Gibbs (1839–1903), an American theoretical physicist who also made significant advances in chemistry and mathematics, and especially in the fields of thermodynamics and statistical physics.

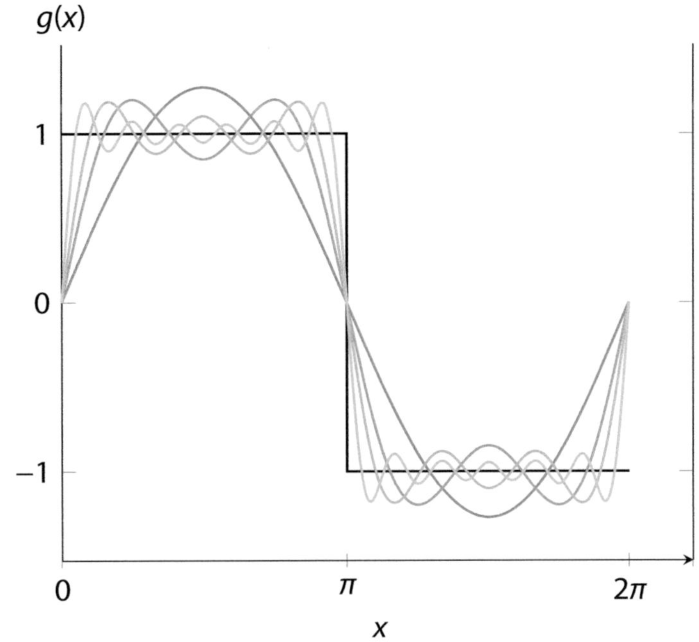

Figure 9.4 The Fourier series of Equation (9.8) showing the Gibbs phenomenon. Each curve shows the effect of including more terms in the Fourier series, with the black curve showing the first term (the sine term), with more terms being used as the colour of the curves gets lighter.

displacement in sea level. One way to model the effects of this on the ocean is to consider the equations of motion for the ocean with a forcing function that is a step function. Let us look at a slightly different example in detail. In Section 6.4.2 we saw that the particular solution of a damped oscillator forced with a cosine term was a sum of sines and cosines (Equation (6.76)) and that the particular solution described how the system behaved once the transients had died down. We can think of the particular solution as being the *quasi steady state solution* of the differential equation because although it depends on time, it is periodic and so repeats itself. What happens if instead of forcing the differential equation with a smooth periodic function, we force it with a function that looks like Equation (9.8)? Let us consider the ODE

$$\ddot{x} + 2b\dot{x} + \omega_0^2 x(t) = Gg(t), \tag{9.12}$$

where G is a constant and $g(t)$ is a repeating, periodic square wave where we have simply repeated the function given by Equation (9.8) over and over (Figure 9.5). We know that we can write the function $g(t)$ as a Fourier series:

$$g(t) = \frac{4}{\pi}\sum_{n=1}^{\infty}\frac{\sin(nt)}{n}.$$

We can then think of Equation (9.12) as being a damped oscillator that is being forced by an infinite number of smooth periodic forcing functions, each with a different frequency

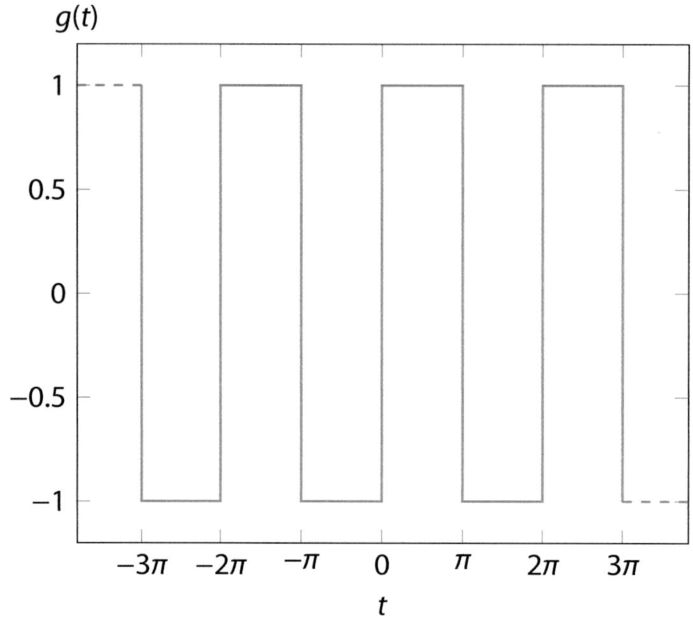

Figure 9.5 The square wave forcing function $g(t)$ in Equation (9.12).

and amplitude. For each of these frequencies, the ODE will have the form of Equation (9.12) but with $g(t) \sim \sin(nt)$, which is something we know how to solve. What is more, because the ODE is a linear equation, solutions at different frequencies do not interact. This means that, for any single frequency (called a *harmonic*), we can write the solution of Equation (9.12) as $x_n(t) = A_n \cos(nt) + B_n \sin(nt)$, where A_n and B_n are constants that will be different for different values of n. Substituting this solution into Equation (9.12) and collecting the sine and cosine terms gives

$$[(\omega_0^2 - n^2)A_n + 2bnB_n]\cos(nt) + [(\omega_0^2 - n^2)B_n - 2bnA_n]\sin(nt) = \frac{4}{n\pi}\sin(nt),$$

which implies that

$$(\omega_0^2 - n^2)A_n + 2bnB_n = 0, \qquad (\omega_0^2 - n^2)B_n - 2bnA_n = \frac{4}{n\pi}.$$

We can solve these equations for A_n and B_n to get the general quasi steady state solution as

$$x(t) = \sum_{n=1}^{\infty} \frac{4(\omega_0^2 - n^2)}{n\pi[(\omega_0^2 - n^2)^2 + 4n^2b^2]}\sin(nt) - \frac{8bn}{n\pi[(\omega_0^2 - n^2)^2 + 4n^2b^2]}\cos(nt), \quad n \text{ odd}.$$

Thus, we can see how we can use Fourier series to solve a forced, linear ODE; we just have to be careful to make sure we equate harmonics and do not mix up different frequencies (or wave numbers).

9.1.1 Complex Fourier Series

We can write the Fourier series (Equation (9.2)) more compactly using complex numbers. We know that we can write the sine and cosine functions in terms of the exponentials of complex numbers (Appendix B),

$$\sin(nx) = \frac{1}{2i}(e^{inx} - e^{-inx}), \quad \cos(nx) = \frac{1}{2}(e^{inx} + e^{-inx}),$$

and this allows us to reformulate the whole Fourier series apparatus in terms of complex exponentials. Doing this can be useful because it is often easier to manipulate exponentials than trigonometric functions. Using these equations in Equation (9.2) we get

$$\begin{aligned} f(x) &= \frac{a_0}{2} + \sum_{n=1}^{\infty}\left(\frac{a_n}{2}(e^{inx} + e^{-inx}) + \frac{b_n}{2i}(e^{inx} - e^{-inx})\right) \\ &= \frac{a_0}{2} + \sum_{n=1}^{\infty}\left[\left(\frac{a_n}{2} + \frac{b_n}{2i}\right)e^{inx} + \left(\frac{a_n}{2} - \frac{b_n}{2i}\right)e^{-inx}\right] = \sum_{-\infty}^{\infty} c_n e^{inx}, \end{aligned}$$

where in the last summation we have extended the sum to $-\infty$, which accounts for both the exponential terms and the constant term (when $n = 0$). To find the values of c_n we employ the same logic as before. Integrating just the function on its own term by term gives

$$\frac{1}{2\pi}\int_0^{2\pi} f(x)\,dx = \frac{c_0}{2\pi}\int_0^{2\pi} dx + \frac{c_1}{2\pi}\int_0^{2\pi} e^{ix}\,dx + \frac{c_{-1}}{2\pi}\int_0^{2\pi} e^{-ix}\,dx + \cdots$$

All the terms containing integrals of the exponential have the form

$$\int_0^{2\pi} e^{\pm inx}\,dx = \pm\left.\frac{e^{\pm inx}}{in}\right|_0^{2\pi} = 0 \implies c_0 = \frac{1}{2\pi}\int_0^{2\pi} f(x)\,dx.$$

To find the other coefficients, we evaluate the integral

$$\frac{1}{2\pi}\int_0^{2\pi} f(x)e^{-imx}\,dx = \frac{c_n}{2\pi}\int_0^{2\pi} e^{-i(n-m)x}\,dx = c_n.$$

Exercise 9.1.8 Work through the details of the calculations for c_0 and c_n.

Exercise 9.1.9 Calculate the complex Fourier series of $f(x) = \sin(x)$.

9.1.2 Even and Odd Functions

Recall from Chapter 2 that the sine function is an odd function (i.e., $\sin(-x) = -\sin(x)$) and the cosine is an even function (i.e., $\cos(-x) = \cos(x)$), and this affects the values of integrals of these functions. The coefficients a_n involve terms that are integrals of $f(x)\cos(nx)$ and the coefficients b_n involve integrals of $f(x)\sin(nx)$. So, if $f(x)$ is an odd function, all the integrals involved in calculating a_n are zero (because an odd function multiplied by an even one is an odd function); whereas if $f(x)$ is an even function, all the integrals involved in calculating b_n are zero. This means that we can write the Fourier series of an odd function using only sines and the Fourier series of an even function using only cosines. These are sometimes called the *Fourier sine series* and the *Fourier cosine series*.

Exercise 9.1.10 Find the Fourier cosine series of the function $f(t) = \sin(t)$ on the interval $0 \leq t \leq \pi$.

Example 9.1 Let us calculate the Fourier series of the function $f(x) = x$, where $-L \leq x \leq +L$. The function $f(x)$ is an odd function, so when we calculate the Fourier series only the terms including the sine function will remain and we will have a Fourier sine series. We need only calculate the coefficients b_n

$$b_n = \frac{1}{L}\int_{-L}^{L} x \sin\left(\frac{n\pi x}{L}\right) dx = \frac{2}{L}\int_{0}^{L} x \sin\left(\frac{n\pi x}{L}\right) dx,$$

where we have used the fact that the integrand is an even function (an odd function multiplied by an odd function) to change the limits of the integral. Evaluating the integral gives us that

$$b_n = \frac{2L}{n^2\pi^2}(\sin(n\pi) - n\pi\cos(n\pi)).$$

We can simplify this equation further using similar arguments to those we used to arrive at Equation (9.11). Thus, we know that the first term in parentheses is zero because n is an integer, and likewise we know that $-\cos(n\pi) = (-1)^{n+1}$. So, we end up with the Fourier sine series

$$x = \frac{2L}{\pi}\sum_{n=1}^{\infty}\frac{(-1)^{n+1}}{n}\sin\left(\frac{n\pi x}{L}\right).$$

You may have noticed something odd about the function in Example (9.1)—apart from the function being an odd function, it is also decidedly not periodic! And yet we went ahead and calculated its Fourier series anyway. How can we do this? The answer is quite simple. We basically imagine that we repeat the function over and over again (Figure 9.6). Once we have calculated the Fourier series we restrict the values of x to once again lie in the interval $-L \leq x \leq +L$.

Exercise 9.1.11 Calculate the Fourier series of $f(x) = x^2$ for $0 < x < 2\pi$.

Exercise 9.1.12 Calculate the Fourier cosine series of $f(x) = \sin(x)$, $0 < x < \pi$.

9.1.3 Dirichlet Conditions

There is an important technical issue concerning Fourier series that we have so far avoided. When we looked at power series approximations for functions we had to examine the convergence of the series to make sure that the series, summing all infinite terms, converged to the actual value of the function. We now need to discuss if we should be concerned about whether or not a Fourier series will converge. Recall from Chapter 2 that for a Taylor series to represent a function series we required that the function was continuous and differentiable. But we have seen that we can calculate the Fourier series of a discontinuous function (Equation (9.8)), and that the Fourier series exhibits the Gibbs

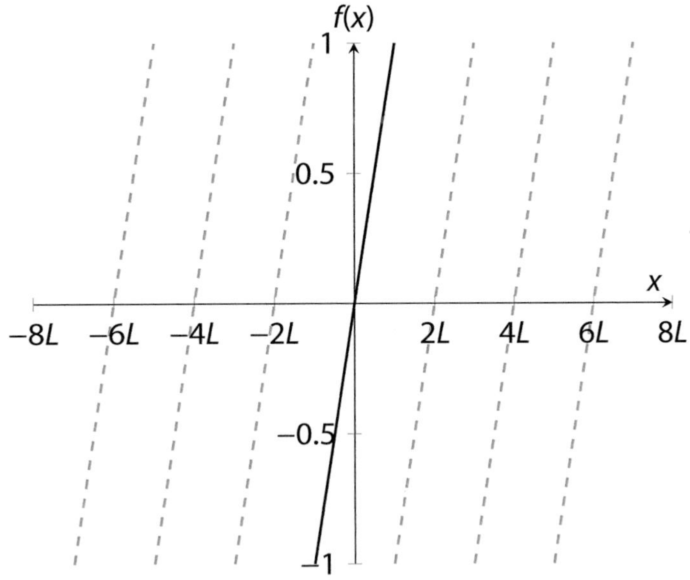

Figure 9.6 Extending the function $f(x) = x$ defined over the interval $-L \leq x \leq +L$ (solid black line) to a periodic function (the dashed gray lines) so that we can calculate its Fourier series.

phenomenon. So it is legitimate to ask if the Fourier series converges to the value of the function. And if it does, what happens at the discontinuities?

These questions are answered by the *Dirichlet conditions*, which state that if the function $f(x)$ is periodic with a period of $2L$ and if for $0 < x < 2L$ the function is single-valued, has a finite number of maxima and minima, only a finite number of finite discontinuities, and if

$$\int_0^{2L} |f(x)|\, dx \quad \text{is finite,} \tag{9.13}$$

then the Fourier series of $f(x)$ evaluated at $x = a$ converges to

$$\frac{1}{2}\left(\lim_{x\to a^+} f(x) + \lim_{x\to a^-} f(x)\right). \tag{9.14}$$

Let us break this statement down and see what each piece means. First, the function $f(x)$ has to be *single-valued*. This means that for each value of x, $f(x)$ has only one value. This restriction rules out the Fourier series of functions like $f(x) = \pm\sqrt{1 - x^2}$ unless we restrict ourselves to either $f(x) = +\sqrt{1 - x^2}$ or $f(x) = -\sqrt{1 - x^2}$.

The second restriction that there are only a finite number of maxima or minima within a single period of the function may seem strange at first, but it rules out the Fourier series of functions like $\sin(1/x)$, which has an infinite number of maxima and minima as $x \to 0$.[3]

[3] We can see this intuitively by noting that as $x \to 0$, $1/x \to \infty$. This means that if we look at the value of $\sin(1/x)$ in a finite region of x that contains $x = 0$, we will have an infinite number of values of $1/x$ that are multiples of π, an infinite number of values of $1/x$ that are multiples of $\pi/2$, and an infinite number of values that are multiples of $3\pi/2$. So, the function $\sin(1/x)$, oscillates an infinite number of times between $+1$, 0, and -1 as $x \to 0$, so it will have an infinite number of maxima and minima in a finite interval of x.

The restriction on the number and type of discontinuities rules out functions like $(1-x)^{-2}$ that have infinite discontinuities.

The Equation (9.13) condition is a statement that stops the function growing infinitely large. For example, if $f(x) = 1/x$, then

$$\int_0^{2L} \frac{1}{|x|}\, dx = \ln(x)|_0^{2L} = \infty,$$

so we cannot take the Fourier series of $1/x$. In reality, we only need to check that $f(x)$ is bounded (i.e., $f(x) \leq \pm M$ for a finite value of M) because, recalling that an integral is an area under the curve, we have

$$\int_0^{2L} |f(x)|\, dx \leq \int_0^{2L} M\, dx = 2LM,$$

which is always finite.

Lastly, Equation (9.14) states that if the function is continuous at $x = a$, then the Fourier series converges to the value $f(a)$. But if the function is discontinuous at $x = a$, then the Fourier series converges to the *average value* of $f(x)$ as we approach $x = a$ from above and below. For the most part, the functions that we meet in environmental sciences and geosciences will not violate any of these conditions and we can happily calculate their Fourier series. However, it is worth keeping these conditions in the back of one's mind, just in case.

Exercise 9.1.13 Show that the function $f(x) = |x|$, $-L < x < L$, satisfies the Dirichlet conditions.

9.1.4 Parseval's Theorem

Parseval's theorem relates the integral of the square of a periodic function to the coefficients of the Fourier series of that function.[4] This seems rather obscure, but many interesting quantities (such as the energy in a wave) are related to the square of a periodic function.

To prove Parseval's theorem, let us start with the Fourier series

$$f(x) = \frac{a_0}{2} + \sum_{n=1}^{\infty}(a_n\cos(nx) + b_n\sin(nx)).$$

The average of $|\ f(x)\ |^2$ is

$$\frac{1}{2\pi}\int_{-\pi}^{\pi} |\ f(x)\ |^2\ dx = \frac{1}{2\pi}\int_{-\pi}^{\pi}\left\{\frac{a_0}{2} + \sum_{n=1}^{\infty}(a_n\cos(nx) + b_n\sin(nx)\right\} \times$$
$$\left\{\frac{a_0}{2} + \sum_{n=1}^{\infty}(a_n\cos(nx) + b_n\sin(nx)\right\}\ dx.$$

Rather than go through the tedium of evaluating this explicitly, we will use what we know about integrals of sines and cosines to evaluate this expression. When we expand out the

[4] This theorem is named after the French mathematician Marc-Antoine Parseval (1755–1836).

integrand we will get a term that is the average of $(a_0/2)^2$ over the interval from $x = -\pi$ to $x = \pi$, which is $a_0^2/4$. Next we will get terms that are the average over the same interval of functions like $(a_n \cos(nx))^2$, each of which will produce a term that looks like $a_n^2/2$. We will also get terms that are the average of $(b_n \sin(nx))^2$ over the interval from $x = -\pi$ to $x = \pi$, each of which will give a term like $b_n^2/2$. Lastly, we will also have three types of cross product, each involving constants multiplied by $\cos(nx)$, $\sin(nx)$, or $\cos(nx)\sin(mx)$, and the average of each of these over the interval from $x = -\pi$ to $x = \pi$ is zero. So, we end up with

$$\frac{1}{2\pi}\int_{-\pi}^{\pi} | f(x) |^2 \, dx = \frac{1}{4}a_0^2 + \frac{1}{2}\sum_{n=1}^{\infty} a_n^2 + \frac{1}{2}\sum_{n=1}^{\infty} b_n^2. \tag{9.15}$$

Exercise 9.1.14 Evaluate the integral of

$$\frac{1}{2\pi}\int_{-\pi}^{\pi}\left(\frac{a_0}{2} + a_1\cos(x) + b_1\sin(x)\right)\left(\frac{a_0}{2} + a_1\cos(x) + b_1\sin(x)\right) dx.$$

Exercise 9.1.15 Show that Equation (9.15) is true, using the complex form of the Fourier series.

What does Parseval's theorem mean and why is it important? We can think of a Fourier series as being the act of decomposing a function into its components for a certain set of basis functions; i.e., the functions 1, $\cos(x)$, $\sin(x)$, $\cos(2x)$, $\sin(2x)$ .... We know that $\cos(nx)$ and $\sin(nx)$ are orthogonal, so we can think of them as forming an infinite (n can go over infinity) set of basis "vectors" but in an abstract "vector space," a space where the vectors are functions. The coefficients of the Fourier series can then be thought of as the components of the function with respect to the basis functions. Now let us extend this analogy with vectors. Recall that if we have a vector $\mathbf{V} = V_x\mathbf{e}_x + V_y\mathbf{e}_y$, where $\mathbf{e}_x$ and $\mathbf{e}_y$ are basis vectors, the magnitude, or length, of $\mathbf{V}$ is $\mathbf{V}\cdot\mathbf{V} = V_x^2 + V_y^2$. If we transform coordinates to a different basis such that $\mathbf{V}' = V_x'\mathbf{e}'_x + V_y'\mathbf{e}'_y$, the length of the vector remains unchanged: $\mathbf{V}'\cdot\mathbf{V}' = (V_x')^2 + (V_y')^2$, even though the components of the vector itself change. In two dimensions this is just Pythagoras' theorem, and in higher dimensions it is an extension of Pythagoras' theorem. We can therefore think of Equation (9.15) as being an extension of Pythagoras' theorem to infinite dimensions. We can think of this in another way by realizing that there are no cross terms in Equation (9.15) (i.e., there are no terms that contain $a_n b_n$). This means that each harmonic (i.e., each value of n) contributes independently to $| f(x) |^2$ and does not interact with any of the others. So, for example, if $| f(x) |^2$ is proportional to the energy in the original signal, then this is the same as the sum of the energies of all the harmonics, which is what Equation (9.15) is telling us.

Parseval's theorem is sometimes used to prove some useful summations of series. For example, the Fourier series of the function $g(x) = x$ on the interval $0 \le x \le 2$ is

$$g(x) = \frac{4}{\pi}\sum_{m=1}^{\infty}\frac{(-1)^{m+1}}{m}\sin\left(\frac{m\pi x}{2}\right).$$

So, by Parseval's theorem

$$\int_0^2 |x|^2\, dx = \frac{8}{3} = \left(\frac{4}{\pi}\right)^2 \sum_{n=1}^{\infty} \frac{1}{n^2},$$

which tells us that

$$\sum_{n=1}^{\infty} \frac{1}{n^2} = \frac{\pi^2}{6}.$$

Exercise 9.1.16 Use the Fourier series of the function $f(x) = x^2$ to calculate $\sum_{n=1}^{\infty} n^{-4}$.

9.1.5 Differentiating and Integrating Fourier Series

If we can represent a function with a Fourier series, is the derivative (or integral) of the Fourier series the same as the Fourier series of the derivative (or integral) of the function? Let us consider differentiation first and look at a familiar example, the Fourier series of $f(x) = x$ on the interval $-L \leq x \leq L$. The Fourier series of this function is

$$f(x) = x = \frac{2L}{\pi} \sum_{n=1}^{\infty} \frac{(-1)^{n+}}{n} \sin\left(\frac{n\pi x}{L}\right), \qquad (9.16)$$

and if we differentiate both sides of this equation, we get

$$\frac{df}{dx} = 1 = 2\sum_{n=1}^{\infty}(-1)^{n+1} \cos\left(\frac{n\pi x}{L}\right).$$

But the Fourier series of the number 1 is simply 1; the only term in the series that is nonzero is the constant term. So, something has gone amiss, and it would appear that we cannot always differentiate a Fourier series term by term and get the Fourier series of the derivative of $f(x)$. If turns out that we can only differentiate the Fourier series of a function $f(x)$ term by term if $f(x)$ is continuous *and* its derivative $f'(x)$ is piecewise smooth. If this is the case, then the derivative of the Fourier series of $f(x)$ is the Fourier series of $f'(x)$.

It turns out that term-by-term integration of Fourier series is a little bit less restrictive. In this case, we only require that the original function $f(x)$ be piecewise smooth; however, the series we obtain for the integral need not be a Fourier series, although it will be an infinite series that will converge to the integral of $f(x)$.

Exercise 9.1.17 Use the Fourier series of $f(x) = x^2$ on the interval $-\pi \leq x \leq \pi$ to calculate the Fourier series of $f(x) = x^3$.

9.2 Fourier Transform

The Fourier series can be a very useful construction, but it has its limitations. For example, although the Fourier series considers an infinite number of frequencies, they are all discrete.

Another limitation is that, while we can calculate the Fourier series of a function that is only defined between say $x = 0$ and $x = L$ by adding duplicates of the function to make it periodic, we cannot deal with a function that is defined from $x = -\infty$ to $x = +\infty$ and that is not periodic. The *Fourier transform* is an integral transform, and it can help us deal with these limitations of the Fourier series.

To see how the Fourier transform comes about let us start with the Fourier series of a function $f(x)$ that is periodic on the interval $x = -L$ to $x = L$,[5] so that

$$f(x) = \frac{a_0}{2} + \sum_{n=1}^{\infty} \left[a_n \cos\left(\frac{n\pi x}{L}\right) + b_n \sin\left(\frac{n\pi x}{L}\right) \right]$$

$$a_0 = \frac{1}{L}\int_{-L}^{L} f(x)dx, \quad a_n = \frac{1}{L}\int_{-L}^{L} f(x)\cos\left(\frac{n\pi x}{L}\right)dx, \quad b_n = \frac{1}{L}\int_{-L}^{L} f(x)\sin\left(\frac{n\pi x}{L}\right)dx.$$

To derive the Fourier transform we are basically going to take the equations for a_0, a_n, and b_n and substitute them into the equation for the Fourier series itself. The set of all frequencies $(n\pi/l)$ forms the *frequency spectrum* of the Fourier series, and the Fourier series involves a summation over all these frequencies. We would like to replace this summation by an integral. This means we need to have a frequency interval that corresponds to the interval Δx in the definition of the Riemann integral. But we do not have a frequency interval that we can let tend to zero as we did for the interval Δx when we defined integration as a limit of a sum in Chapter 2. However, if we let $L \to \infty$, the difference between successive frequencies $0, \pi/L, 2\pi/L, 3\pi/L \ldots$ etc. becomes smaller until eventually the discrete frequency spectrum becomes a continuous, so perhaps we can use that. The difference between two successive frequencies $(\Delta\omega)$ is

$$\Delta\omega = \frac{(n+1)\pi}{L} - \frac{n\pi}{L} = \frac{\pi}{L} \implies L = \frac{\pi}{\Delta\omega},$$

and we can use this as our frequency interval because as $L \to \infty$, $\Delta\omega \to 0$, which is what we want.

Substituting the equations of a_0, a_n, and b_n into the equation for the Fourier series and using the trigonometric identity (see Appendix B) $\cos(\theta)\cos(\phi) + \sin(\theta)\sin(\phi) = \cos(\theta - \phi)$ to simplify the integrands we get

$$f(x) = \frac{1}{2L}\int_{-L}^{L} f(\xi)\,d\xi + \sum_{n=1}^{\infty}\left\{\frac{1}{L}\int_{-L}^{L} f(\xi)\cos\left(\frac{n\pi}{L}(\xi - x)\right)\,d\xi\right\}, \qquad (9.17)$$

where we have used ξ as a dummy variable in the integrals for a_0, a_n, and b_n. As $L \to \infty$, we can constrain the size of the first term on the right-hand side of Equation (9.17),

$$\left|\frac{1}{2L}\int_{-L}^{L} f(\xi)\,d\xi\right| \le \frac{1}{2L}\int_{\infty}^{\infty} \mid f(\xi) \mid \, d\xi.$$

Now, $1/(2L) \to 0$ as $L \to \infty$, and if we also demand that

$$\int_{\infty}^{\infty} \mid f(\xi) \mid \, d\xi \quad \text{is finite as } L \to \infty,$$

[5] What follows is not a rigorous derivation of the Fourier transform, although it may appear so. To make the derivation totally rigorous would take us well beyond the material covered in this book.

then the whole term will tend to zero as $L \to \infty$. We will now use our equation for $\Delta\omega$ to write the summation in Equation (9.17) as

$$\sum_{n=1}^{\infty} \left\{ \frac{\Delta\omega}{\pi} \int_{-L}^{L} f(\xi) \cos(n\Delta\omega(\xi - x))\, d\xi \right\}, \tag{9.18}$$

and letting $\Delta\omega \to 0$ in Equation (9.18) and substituting back into Equation (9.17) we get that

$$f(x) = \frac{1}{\pi} \int_0^{\infty} \left\{ \int_{-\infty}^{\infty} f(\xi) \cos(\omega(\xi - x))\, d\xi \right\} d\omega, \tag{9.19}$$

where in the limit of $\Delta\omega \to 0$ we have written $n\Delta\omega = \omega$. Recall that as the difference between successive frequencies in the frequency spectrum becomes zero, the discrete frequency spectrum becomes continuous.

Exercise 9.2.1 Show that we can write Equation (9.19) as

$$f(x) = \int_0^{\infty} [a(\omega)\cos(\omega x) + b(\omega)\sin(\omega x)]\, d\omega,$$

where $a(\omega)$ and $b(\omega)$ are functions of ω. This equation is often called the Fourier integral.

We can simplify Equation (9.19) further by writing the cosine function in terms of complex exponentials, giving

$$f(x) = \frac{1}{2\pi} \int_0^{\infty} \int_{-\infty}^{\infty} f(\xi) \mathrm{e}^{i\omega(\xi - x)}\, d\xi\, d\omega + \frac{1}{2\pi} \int_0^{\infty} \int_{-\infty}^{\infty} f(\xi) \mathrm{e}^{-i\omega(\xi - x)}\, d\xi\, d\omega.$$

We can make the transformation ω to $-\omega$ in the first integral (it is a dummy variable that is integrated over) and swap the limits on that integral to finally give

$$f(x) = \frac{1}{2\pi} \int_{\infty}^{\infty} \left\{ \int_{-\infty}^{\infty} f(\xi) \mathrm{e}^{-i\omega\xi}\, d\xi \right\} \mathrm{e}^{i\omega x}\, dx,$$

which we can write as

$$f(x) = \int_{-\infty}^{\infty} g(\omega) \mathrm{e}^{i\omega x}\, d\omega, \qquad g(\omega) = \frac{1}{2\pi} \int_{-\infty}^{\infty} f(x) \mathrm{e}^{-i\omega x}\, dx. \tag{9.20}$$

The two functions $f(x)$ and $g(\omega)$ are called a *Fourier transform pair*; $g(\omega)$ is the *Fourier transform* of $f(x)$, and $f(x)$ is the *inverse Fourier transform* of $g(\omega)$. It does not really matter how we apportion the factor of $1/(2\pi)$ between the two integrals, and you will see all possibilities used. The important thing is that when we substitute for $g(\omega)$ in $f(x)$, we end up with an overall factor of $1/(2\pi)$. You will also see the Fourier transform pair defined as

$$f(x) = \int_{-\infty}^{\infty} g(\omega) \mathrm{e}^{2\pi i\omega x}\, d\omega, \qquad g(\omega) = \int_{-\infty}^{\infty} f(x) \mathrm{e}^{-2\pi i\omega x}\, dx,$$

in which the frequency ω has been redefined to involve the factor of 2π. The Fourier transform is an example of an integral transform, and there are many different types of integral transform that vary by the limits of integration and the function (called the *kernel*) used to multiply $f(x)$—the kernel for the Fourier transform is $\mathrm{e}^{-i\gamma x}$.

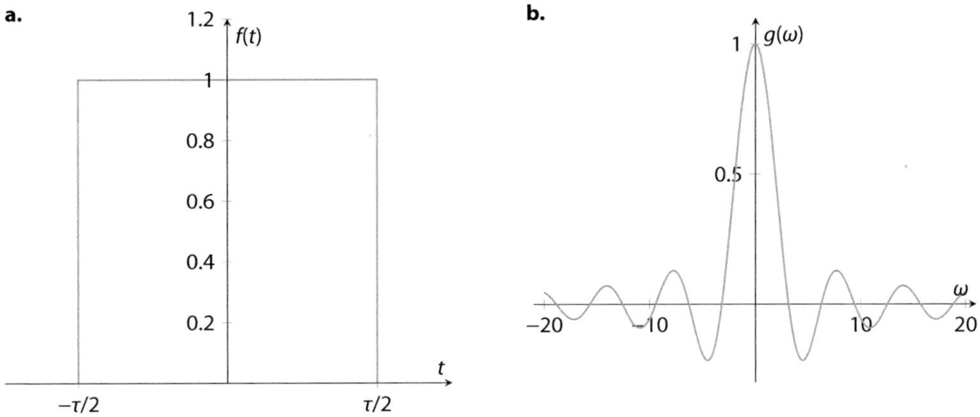

Figure 9.7 The Fourier transform of a rectangular pulse. (a.) shows the rectangular pulse (Equation (9.23) and (b.) shows the function sinc(ω), the Fourier transform of the rectangular pulse.

Example 9.2 To see how this works, let us take the Fourier transform of the delta function

$$g(\omega) = \frac{1}{2\pi}\int_{-\infty}^{\infty} \delta(x)e^{-i\omega x}\,dx = \frac{1}{2\pi} \tag{9.21}$$

and, because a function and its Fourier transform are related by Equation (9.20),

$$f(t) = \frac{1}{2\pi}\int_{-\infty}^{\infty} e^{i\omega x}\,d\omega = \delta(x), \tag{9.22}$$

which provides another representation of the delta function.

Example 9.3 As another example, let us calculate the Fourier transform of a rectangular pulse (Figure 9.7) of duraction τ defined by

$$f(t) = \begin{cases} 0 & |\,t\,| > \frac{\tau}{2} \\ 1 & |\,t\,| \leq \frac{\tau}{2} \end{cases}. \tag{9.23}$$

The Fourier transform is

$$\begin{aligned} g(\omega) &= \frac{1}{2\pi}\int_{-\infty}^{\infty} f(t)e^{-i\omega t}\,dt = \frac{1}{2\pi}\int_{-\tau/2}^{\tau/2} e^{-i\omega t}\,dt = -\frac{1}{2\pi i\omega}(e^{-i\omega\tau/2} - e^{i\omega\tau/2}) \\ &= \frac{1}{2\pi}\tau\frac{\sin(\omega\tau/2)}{\omega\tau/2} = \frac{\tau}{2\pi}\,\text{sinc}\left(\frac{\omega\tau}{2}\right), \end{aligned} \tag{9.24}$$

where the function $\text{sinc}(x) = (1/x)\sin(x)$.

The Fourier series decomposes a signal (i.e., a function) into sines and cosines of different frequencies and amplitudes. A Fourier transform produces a new function (the transform) that is the representation of the original signal in terms of the frequencies of its component parts. To see how this works, let us calculate the Fourier transform of a cosine function $f(t) = \cos(\omega_0 t)$. Writing the cosine function in terms of complex exponentials, we get

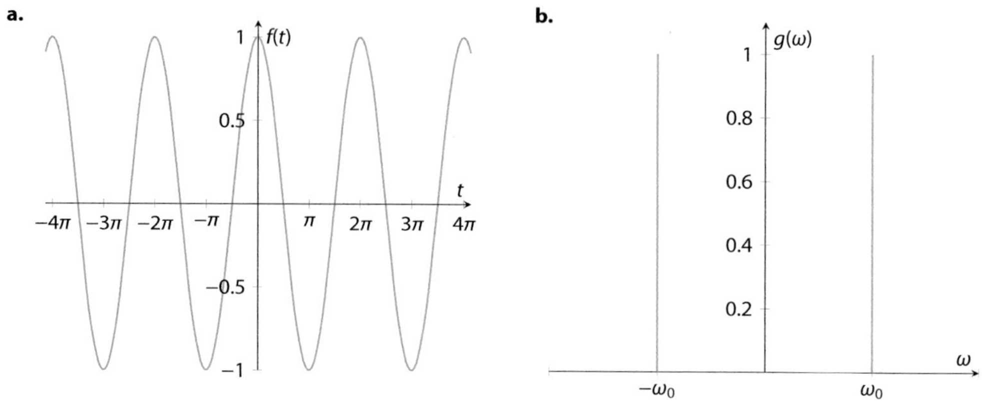

Figure 9.8 The cosine function (a.) and its Fourier transform (b.).

$$g(\omega) = \frac{1}{2\pi}\int_{-\infty}^{\infty}\frac{1}{2}(e^{i\omega_0 t} + e^{-i\omega_0 t})e^{-i\omega t}\,dt = \frac{1}{4\pi}\int_{-\infty}^{\infty} e^{-i(\omega-\omega_0)t}\,dt + \frac{1}{4\pi}\int_{-\infty}^{\infty} e^{-i(\omega+\omega_0)t}\,dt$$
$$= \frac{1}{2}\delta(\omega-\omega_0) + \frac{1}{2}\delta(\omega+\omega_0), \tag{9.25}$$

where we have used Equation (9.22) for the delta function. This shows that the Fourier transform consists of two delta functions (Figure 9.8), one located at $\omega = \omega_0$ and the other at $\omega = -\omega_0$. This seems strange. We started with a function that has a single frequency (ω_0) and ended up with a Fourier transform that has two delta functions, one at the expected frequency ($\omega = \omega_0$) and the other at $\omega = -\omega_0$. Where does the additional frequency come from? Remember that the cosine function is an even function so that $\cos(-\omega_0 x) = \cos(\omega_0 x)$. We can also see this in the complex representation of the cosine function, $\cos(\omega_0 x) = (1/2)(e^{i\omega_0 x} + e^{-i\omega_0 x})$, which indicates that when we use the complex representation we pick up the negative frequency in addition to the positive one. A similar thing occurs for the Fourier transform of the sine function $f(t) = \sin(\omega_0 t)$, which is

$$g(\omega) = -\frac{i}{2}\delta(\omega-\omega_0) + \frac{i}{2}\delta(\omega+\omega_0). \tag{9.26}$$

Exercise 9.2.2 Show that the Fourier transform of the function $f(t) = \sin(\omega_0 t)$ is given by Equation (9.26). Explain why the Fourier transform of the sine function is complex and why one component is negative and the other positive.

Exercise 9.2.3 Calculate the Fourier transform of $f(t) = 2\cos(\omega_0 t) + 4\cos(3\omega_0 t)$ and sketch it.

9.2.1 Sine and Cosine Transforms

We have seen that we could simplify the Fourier series of a function considerably for even and odd functions. A similar simplification occurs with the Fourier transform. We can write the Fourier transform of $f(x)$ as

$$g(k) = \frac{1}{2\pi}\int_{-\infty}^{\infty} f(x)[\cos(kx) - i\sin(kx)]\,dx.$$

If $f(x)$ is an odd function, then $f(x)\cos(kx)$ is odd and its integral will vanish, but the integral of $f(x)\sin(kx)$ will remain and we can write

$$g(k) = -\frac{i}{\pi}\int_0^\infty f(x)\sin(kx)\,dx. \tag{9.27}$$

This transformation is called the *Fourier sine transform* and can be used to find the Fourier transform of odd functions. Similarly, the inverse transform is

$$f(x) = 2i\int_0^\infty g(k)\sin(kx)\,dk. \tag{9.28}$$

Similarly, the Fourier cosine transform pair is given by

$$g(k) = \sqrt{\frac{2}{\pi}}\int_0^\infty f(x)\cos(kx)\,dx, \qquad f(x) = \sqrt{\frac{2}{\pi}}\int_0^\infty g(k)\cos(kx)\,dk. \tag{9.29}$$

Exercise 9.2.4 Find the Fourier sine transform of

$$f(x) = \begin{cases} x & |\,x\,| \le 1 \\ 0 & |\,x\,| > 1 \end{cases}.$$

9.2.2 Properties of the Fourier Transform

Explicitly calculating the Fourier transform and inverse transform of different functions is not always easy. Fortunately, the Fourier transform has many nice properties that can be used to help us. In what follows we will represent the operation of taking the Fourier transform of a function $f(x)$ as $\hat{f}(k) = \mathcal{F}(f(x))$ and the operation of taking the inverse transform as $f(x) = \mathcal{F}^{-1}(\hat{f}(k))$, where $\hat{f}(k)$ represents the actual Fourier transform of the function $f(x)$—this is a common notation, but you will come across others. The first useful property of the Fourier transform and its inverse is that they are linear. That is, if a and b are constants and $u(x)$ and $v(x)$ are functions of x, then

$$\begin{aligned} \mathcal{F}(au(x) + b(v(x)) &= a\mathcal{F}(u(x)) + b\mathcal{F}(v(x)) = a\hat{u}(k) + b\hat{v}(k), \\ \mathcal{F}^{-1}(a\hat{u}(k) + b\hat{v}(k)) &= a\mathcal{F}^{-1}(\hat{u}(x)) + b\mathcal{F}^{-1}(\hat{v}(x)) = au(x) + bv(x). \end{aligned} \tag{9.30}$$

Exercise 9.2.5 Show that Equation (9.30) is true.

Second, the Fourier transform of the derivative of a function is simply related to the Fourier transform of the function itself, given some restrictions on the function. So, if we have a function $f(t)$ such that all its derivatives up to the $(n-1)$th derivative tend to zero as $t \to \pm\infty$ and the integral

$$\int_{-\infty}^{\infty} \left| \frac{d^n f}{dt^n} \right| dt$$

converges for all values of $n = 0, 1, 2\ldots$, then

$$\mathcal{F}\left(\frac{d^n f}{dt^n}\right) = (i\omega)^n \hat{f}(\omega). \tag{9.31}$$

We can prove this using a mixture of induction and integration by parts.[6] If $n = 0$, Equation (9.31) is just the definition of the Fourier transform, so it must be true. Now, using integration by parts we can write (using the notation that $f^{(n)}$ is the nth derivative of the function f, and f^n is the function f raised to the power of n)

$$\begin{aligned}\mathcal{F}(f^{(n+1)}(t)) &= \int_{-\infty}^{\infty} f^{(n+1)}(t)\mathrm{e}^{-i\omega t}\,dt = \left|f^{(n)}(t)\mathrm{e}^{-i\omega t}\right|_{-\infty}^{\infty} + i\omega\int_{-\infty}^{\infty} f^{(n)}\mathrm{e}^{-i\omega t}\,dt \\ &= \left|f^{(n)}(t)\mathrm{e}^{-i\omega t}\right|_{-\infty}^{\infty} + i\omega\left[(i\omega)^n\hat{f}(\omega)\right] = \left|f^{(n)}(t)\mathrm{e}^{-i\omega t}\right|_{-\infty}^{\infty} + (i\omega)^{n+1}\hat{f}(\omega),\end{aligned}$$

where we have assumed that Equation (9.31) holds. We still have to contend with the first term on the right-hand side, the boundary term (so named because it is evaluated at the boundaries $\pm\infty$). To evaluate this, we use the fact that if z_1 and z_2 are two complex numbers, then $\mid z_1 z_2 \mid=\mid z_1 \mid\mid z_2 \mid$ (Appendix B) to write

$$\left|f^{(n)}(t)\mathrm{e}^{-i\omega t}\right| = \left|f^{(n)}(t)\right|\left|\mathrm{e}^{-i\omega t}\right|.$$

The exponential term is a sum of a sine and a cosine, so it will have a modulus of 1, and the condition that we placed on the derivatives of $f(t)$ as $t \to \pm\infty$ ensures that the other term will go to zero at the boundaries. So, the boundary term is zero, leaving us with

$$\mathcal{F}(f^{(n+1)}(t)) = (i\omega)^{n+1}\hat{f}(\omega).$$

This shows that if Equation (9.31) is true for n, then it is also true for $n + 1$, so we have proven Equation (9.31).

Another useful property is the so-called *shift property* or *translation property*. This states that

$$\mathcal{F}(f(x-a)) = \mathrm{e}^{-ia\omega}\hat{f}(\omega), \quad \mathcal{F}^{-1}(\hat{f}(\omega-a)) = \mathrm{e}^{-iat}f(t). \tag{9.32}$$

Equation (9.32) shows that a shift in time or space in the original function is also a shift in frequency (or wave number), which is equivalent to multiplying the function by a complex exponential, so that

$$\mathcal{F}(\mathrm{e}^{i\omega_0 t}f(t)) = \hat{f}(\omega-\omega_0). \tag{9.33}$$

We can see that Equation (9.33) should be true because when we take the Fourier transform of $f(t)$ we multiply it by the complex exponential $\mathrm{e}^{i\omega t}$, which gives us the frequency spectrum of $f(t)$ for ω. But when we multiply $f(t)$ by $\mathrm{e}^{i\omega_0 t}$ and take the Fourier transform, we end up multiplying $f(t)$ by $\mathrm{e}^{i(\omega+\omega_0)t}$, so we will have shifted the frequency spectrum by ω_0.

The last property of the Fourier transform that we want to talk about here relates the Fourier transform of a convolution of two functions to the product of the inverse transforms of the two functions. If we write the convolution of two functions as

$$\int_{-\infty}^{\infty} f(t-\xi)g(\xi)\,d\xi = (f * g)(t),$$

then

$$\mathcal{F}(f * g) = \hat{f}(\omega)\hat{g}(\omega), \qquad \mathcal{F}(\hat{f}(\omega)\hat{g}(\omega)) = f * g. \tag{9.34}$$

[6] As a reminder, proof by the method of induction works by first showing that the statement is true for a given value (e.g., $n = 0$) and then showing that if the statement is true for any value of n, it is also true for $n + 1$.

We can show this is true by first changing the order in which we do the integrations,

$$\mathcal{F}(f*g) = \int_{-\infty}^{\infty}\left(\int_{-\infty}^{\infty} f(t-\xi)g(\xi)\,d\xi\right)\mathrm{e}^{-i\omega t}\,dt = \int_{-\infty}^{\infty} g(\xi)\left(\int_{-\infty}^{\infty} f(t-\xi)\mathrm{e}^{-i\omega t}\,dt\right)\,d\xi,$$

and then defining a new variable $\eta = t - \xi$ and substituting for t so that

$$\begin{aligned}\mathcal{F}(f*g) &= \int_{-\infty}^{\infty} g(\xi)\left(\int_{-\infty}^{\infty} f(\eta)\mathrm{e}^{-i\omega(\eta+\xi)}\,dt\right)\,d\xi \\ &= \left(\int_{-\infty}^{\infty} g(\xi)\mathrm{e}^{-i\omega\xi}\,d\xi\right)\left(\int_{-\infty}^{\infty} f(\eta)\mathrm{e}^{-i\omega\eta}\,d\eta\right) = \hat{g}(\omega)\hat{f}(\omega).\end{aligned}$$

When we looked at Fourier series we found that there was an interesting relationship, Parseval's theorem. That theorem told us that we can think of the coefficients in the Fourier series as somewhat like the components of an infinite dimensional vector, and that we can define the equivalent of a dot product to calculate the "length" of the function. A similar results holds for Fourier transforms,

$$\int_{-\infty}^{\infty} f^*(t)g(t)\,dt = 2\pi\int_{-\infty}^{\infty}\hat{f}^*(\omega)\hat{g}(\omega)\,d\omega, \tag{9.35}$$

where $f^*(t)$ is the complex conjugate of the function $ftx)$. If we let $f(t)$ and $g(t)$ be the same function, then we end up with *Plancherel's theorem*:

$$\int_{-\infty}^{\infty} |\, f(t)\,|^2\ dt = 2\pi\int_{-\infty}^{\infty} |\,\hat{f}(\omega)\,|^2\ d\omega. \tag{9.36}$$

This is important because the left-hand side of Equation (9.36) is often interpreted as an energy. For example, when analyzing signals that vary with time, it is regarded as the total energy in the signal. Parseval's theorem, for both Fourier series and Fourier transforms, then tells us that the energy in the signal is the integral of the energies of all the harmonics.

Calculating the Fourier transform and its inverse by explicitly calculating integrals in Equation (9.20) is not always easy, or even possible. However, we can frequently use the properties of the Fourier transform to derive Fourier transforms from known transforms. For example, consider the function

$$f(x) = 2\mathrm{e}^{-|x|} - 4\mathrm{e}^{-2|x+1|}.$$

The linearity of the Fourier transform gives us that

$$\mathcal{F}(f(x)) = 2\mathcal{F}(\mathrm{e}^{-|x|}) - 4\mathcal{F}(\mathrm{e}^{-2|x+1|}).$$

Using the table of Fourier transforms in Appendix B, the first term on the right-hand side becomes

$$2\mathcal{F}(\mathrm{e}^{-|x|}) = 2\frac{2}{\omega^2+1}.$$

If we write the second transform as $\mathcal{F}(\mathrm{e}^{-|2x+2|})$, then we use the Fourier transform of the function $g(ax+b)$ (see Appendix B) to write

$$\mathcal{F}(\mathrm{e}^{-2|x+1|}) = \frac{\mathrm{e}^{i\omega}}{2}\frac{2}{(\omega/2)^2+1},$$

so that

$$\mathcal{F}(f(x)) = 2\frac{2}{\omega^2 + 1} - 2e^{i\omega}\frac{2}{(\omega/2)^2 + 1}.$$

As another example, let us consider the Fourier transform of a signal that consists of a finite duration cosine. We can think of this as the product of a cosine function and a rectangular box function $h(t)$. For example, if our cosine signal lasts from $t = -20$ to $t = 20$, then our signal can be represented as $f(t) = h(t)\cos(t)$, where $h(t)$ is the box function that has a value 1 between $-10 \leq x \leq 10$ and is zero everywhere else. The transforms of these two functions are

$$\mathcal{F}(\cos(t)) = \frac{1}{2}\delta(\omega - 1) + \frac{1}{2}\delta(\omega + 1), \qquad \mathcal{F}(h(t))\frac{1}{\pi}\frac{\sin(20\omega)}{\omega},$$

so by the convolution property, the Fourier transform of the signal is

$$\mathcal{F}(f(t) = \int_{-\infty}^{\infty} \frac{\sin(20\omega)}{\omega}(\delta(\omega - 1) + \frac{1}{2}\delta(\omega + 1)) = \frac{\sin(20(\omega - 1)}{\omega - 1} + \frac{\sin(20(\omega + 1)}{\omega + 1}.$$

Exercise 9.2.6 Use the convolution properties of the Fourier transform to calculate the transform of $u(t)v(t)$, where $u(t) = \cos(2\pi\nu t)$ and

$$v(t) = \begin{cases} e^{-\beta t} & t \geq 0 \\ 0 & \text{otherwise} \end{cases}.$$

9.2.3 Applications of the Fourier Transform

The Fourier transform has important applications in analyzing digital signals of all kinds. In the geo- and environmental sciences it is often applied to spatially and temporally varying signals because the Fourier transform is able to take a series of observations and extract any periodicity that might be present in the signal. For example, Figure 9.9 shows the result of performing a Fourier transform on the tidal data shown in Figure 9.1a. What is shown in Figure 9.9 is the result of a *discrete Fourier transform*, which works on data sampled at a uniform rate (once per hour, once per year, etc.) rather than continuous functions. The ordinate axis is a measure of the *power* at different frequencies in the signal and is given by the absolute value of the Fourier transform (remember that the Fourier transform of a function can be a complex quantity). In Figure 9.9 we see a very strong signal centered on a frequency of two cycles per day. This represents the two tides per day that are seen at that location, and this is by far the strongest signal indicating that this component is the dominant part of the variability in tidal height. Looking carefully we see that there are some other spikes at both shorter and longer frequencies. Some of these may represent oscillations in the signal at other frequencies. Computer calculations of discrete Fourier transforms usually use an extremely clever algorithm called the *fast Fourier transform* (FFT), which is highly efficient and very fast. You can learn more about the FFT from the books listed in the Further Reading section of this chapter.

We can also use Fourier transforms to help find the solutions of differential equations. This is because Equation (9.31) allows us to use the Fourier transform to convert an ODE

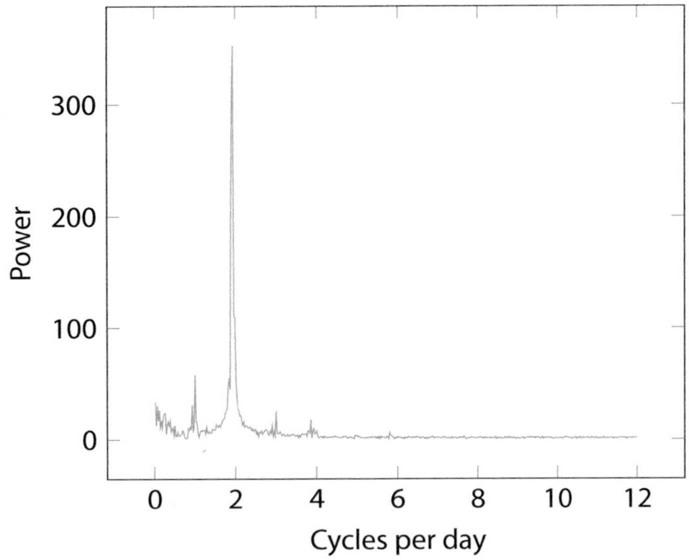

Figure 9.9 The result of performing a Fourier transform on the Fort Pulaski tidal data shown in Figure 9.1a.

into an algebraic one. For example, one of the problems in Chapter 6 concerns the solution of a differential equation that describes the flexure of the oceanic crust under the additional pressure from an island chain. The equation in question is

$$D\frac{d^4 z}{dx^4} + \alpha z = F, \tag{9.37}$$

where D and α are constants. If F is a constant as well, then the equation can be solved using the techniques from Chapter 6. If F is a function of x, however, then solving the equation may need additional techniques. Let us not give any specific form to $F(x)$, but just keep it as a function of position. Taking the Fourier transform of Equation (9.37) and using the property of linearity give us

$$D\mathcal{F}\left(\frac{d^4 z}{dx^4}\right) + \alpha\mathcal{F}(z) = \mathcal{F}(F(x)).$$

We can use Equation (9.31) to write this as (writing the Fourier transform of z as $\hat{z}$):[7]

$$D(i\omega)^4\hat{z} + \alpha\hat{z} = \hat{F},$$

which we can solve for $\hat{z}$. Using the convolution theorem to take the inverse Fourier transform gives us

$$z = \mathcal{F}^{-1}\left(\left(\frac{1}{D\omega^4 + \alpha}\right)\hat{F}\right) = \mathcal{F}^{-1}\left(\frac{1}{D\omega^4 + \alpha}\right) * \mathcal{F}^{-1}(\hat{F}).$$

[7] Recall that in proving Equation (9.31) we had to place restrictions on the function z and its derivatives. So, we have to assume that these are satisfied.

We can now use tables or explicit calculation to evaluate the inverse transform,

$$\mathcal{F}^{-1}\left(\frac{1}{D\omega^4 + \alpha}\right) = \frac{\mu}{\alpha\sqrt{2}} e^{-\mu|x|} \sin(\mu \mid x \mid + \pi/4),$$

where $\mu = (\alpha/(4D))^{1/4}$, and the inverse transform of $\hat{F}$ is just $F(x)$ itself. So, using the convolution property we can write

$$z = \frac{\mu}{\alpha\sqrt{2}} \int_{-\infty}^{\infty} e^{-\mu|x-y|} \sin\left(\mu \mid x - y \mid + \frac{\pi}{4}\right) F(y)\, dy.$$

It might not seem that we have gained anything, but we can now choose functions $F(x)$ that are both physically realistic and that allow us to evaluate this integral, thereby obtaining an explicit solution for the flexure of the plate as a function of distance from the island chain.

Exercise 9.2.7 Use Fourier transforms to find a solution for the ODE $u'' - \beta^2 u(x) = -g(x)$, where $-\infty < x < \infty$ for which $u(x) \to 0$ and $u'(x) \to 0$ as $\mid x \mid \to \infty$ and β is a real constant.

9.3 Laplace Transform

The Laplace transform is another useful integral transform, and it shares many of the properties of the Fourier transform. The Laplace transform $\mathcal{L}(f)$ of a function $f(t)$ is defined by the equation

$$\mathcal{L}(f) = \int_0^{\infty} f(t)e^{-pt}\, dt = F(p). \tag{9.38}$$

Like the Fourier transform, the Laplace transform takes a function of one variable (e.g., t) and returns a different function of another variable (p). In the case of the Fourier transform, it was easy to interpret this new variable because the transform was defined in terms of oscillations using the exponential of a purely imaginary number (i.e., $e^{-i\omega t}$, where t and ω are real). In the case of the Laplace transform, the variable p in Equation (9.38) may contain both real and complex parts, so it is not always easy to interpret.

To see how the Laplace transform works, let us start by calculating the Laplace transform of $f(t) = a =$ constant,

$$\mathcal{L}(a) = \int_0^{\infty} a e^{-pt}\, dt = \frac{a}{p}, \quad \text{for } \mathrm{Re}(p) > 0,$$

where we have specified that this result holds only for the cases when the real part of p is positive because otherwise the integral would be infinite.

Exercise 9.3.1 Show that the Laplace transform of the function $f(t) = e^{-at}$, where a is a real constant, is

$$\mathcal{L}(e^{-at}) = \frac{1}{p - a}, \qquad \mathrm{Re}(p) > a.$$

Exercise 9.3.2 Show that the Laplace transform of the Heaviside step function $H(x-a)$ is

$$\mathcal{L}(H(x-a)) = \frac{1}{p}\mathrm{e}^{-ap}, \quad \mathrm{Re}(p) > 0.$$

In principle, we could continue like this and evaluate the necessary integral whenever we need to calculate the Laplace transform. However, it is often easier to make use of the properties of the Laplace transform. Like the Fourier transform, the Laplace transform is a linear operator. That means that $\mathcal{L}(f(t)+g(t)) = \mathcal{L}(f(t)) + \mathcal{L}(g(t))$. In addition, $\mathcal{L}(af(t)) = a\mathcal{L}(f(t))$, where a is a constant.

Example 9.4 Let us use these properties to evaluate the Laplace transforms of $f(t)=\cos(t)$ and $f(t)=\sin(t)$. We know how to calculate the Laplace transform of an exponential, so we can start with $g(t)=\mathrm{e}^{it}=\cos(t)+i\sin(t)$ and find

$$\mathcal{L}(g(t)) = \mathcal{L}(\mathrm{e}^{it}) = \frac{1}{p-it} = \frac{p+it}{p^2+t^2}, \quad \mathrm{Re}(p-it) > 0.$$

Therefore, splitting this into its real and imaginary parts we get

$$\mathcal{L}(\cos(t)+i\sin(t)) = \mathcal{L}(\cos(t)) + i\mathcal{L}(\sin(t)) = \frac{p}{p^2+t^2} + \frac{it}{p^2+t^2}.$$

If p and t are real, we can equate real and imaginary parts, but we do not know a priori if this is the case. So, to be cautious, we calculate

$$\mathcal{L}(\cos(t)-i\sin(t)) = \mathcal{L}(\cos(t)) - i\mathcal{L}(\sin(t)) = \frac{p}{p^2+t^2} - \frac{it}{p^2+t^2},$$

and by solving these two equations, we find

$$\mathcal{L}(\cos(t)) = \frac{p}{p^2+t^2}, \quad \mathcal{L}(\sin(t)) = \frac{t}{p^2+t^2}.$$

Other useful properties of the Laplace transform include the fact that if we have a function $u(x) = df/dx$, we can calculate the transform of u using integration by parts:

$$\begin{aligned}\mathcal{L}(u(x)) = \mathcal{L}\left(\frac{df}{dx}\right) &= \int_0^\infty \frac{df}{dx}\mathrm{e}^{-px}\,dx = \mathrm{e}^{-px}f(x)\Big|_0^\infty + p\int_0^\infty y(x)\mathrm{e}^{-px}\,dx \\ &= -y(x=0) + p\mathcal{L}(y(x)) = pY - y_0,\end{aligned} \tag{9.39}$$

where we have used the capital letter Y to denote the Laplace transform and $y_0 = y(x=0)$.

Exercise 9.3.3 Show that

$$\mathcal{L}\left(\frac{d^2f}{dx^2}\right) = p^2Y - py_0 - \left.\frac{dy}{dx}\right|_{x=0}$$

Exercise 9.3.4 Write down the Laplace transform of the third derivative of $f(x)$. (Hint: look at the progression in going from the first to the second derivative.)

Exercise 9.3.5 Show that if the Laplace transform $\mathcal{L}(f(t) = F(p)$ exists for $p > c$, then

$$\mathcal{L}(\mathrm{e}^{\alpha t}f(t)) = F(p-\alpha), \qquad p > \alpha + c.$$

So, taking the Laplace transform of derivatives results in algebraic expressions. This can be useful for solving differential equations because if we take the Laplace transform of an ODE, we not only convert the differential equation into an algebraic equation (which is easier to solve), but we also involve the initial conditions. This means that we automatically find the solution to the problem without having to first find the general solution.

Example 9.5 Let us use the Laplace transform to solve a second order linear inhomogeneous equation with constant coefficients. We already know how to solve such an equation, but we will often find such equations with a right-hand side that makes a solution difficult to find. Let us solve

$$\frac{d^2y}{dx^2} + 4\frac{dy}{dx} + 4y(x) = x^2\mathrm{e}^{-2x}, \quad y(0) = y'(0) = 0.$$

Taking the Laplace transform of each term in the equation in turn and using the initial conditions we get

$$p^2Y + 4pY + 4Y = \mathcal{L}(x^2\mathrm{e}^{-2x}) = \frac{2}{(p+2)^3}.$$

We could calculate the Laplace transform of $x^2\mathrm{e}^{-2x}$ explicitly, but it is often easier to use tables of Laplace transforms. Using the table in Appendix B we find that

$$\mathcal{L}(x^2\mathrm{e}^{-2x}) = \frac{2}{(p+2)^3},$$

so our fully transformed equation becomes

$$(p+2)^2Y = \frac{2}{(p+2)^3} \implies Y = \frac{2}{(p+2)^5}.$$

In order to find the actual solution of the ODE, we need to calculate the inverse Laplace transform of Y. This can be tricky to do and generally involves the use of techniques from the analysis of complex variables, which is beyond the scope of this text. In practice, the inverse Laplace transform is often determined by again using tables of Laplace transform and manipulating the function into a form that appears in the tables. This is easy for the equation we are considering here, because we have already used the fact that

$$\mathcal{L}(x^n\mathrm{e}^{\alpha x}) = \frac{n!}{(p-\alpha)^{n+1}}, \quad n = 1, 2, 3\ldots$$

from which we can see that the inverse Laplace transform of $24/(p+2)^5$ is $x^4\mathrm{e}^{-2x}$, so the solution to the ODE is

$$y(x) = \frac{x^4\mathrm{e}^{-2x}}{12}.$$

Exercise 9.3.6 Use partial fractions and the table in Appendix B to calculate the inverse Laplace transform of

$$a.\ \frac{1}{p(p-1)^3}, \qquad b.\ \frac{p^2}{(p^2+1)(p^2+2)}.$$

Exercise 9.3.7 Use Laplace transforms to solve the initial value problem

$$\frac{d^2x}{dt^2} + 4\frac{dx}{dt} + 8x(t) = \cos(2t), \qquad x(t=0) = 2,\ \dot{x}(t=0) = 1.$$

If our ODE is a linear equation with constant coefficients, then we can see something interesting. Let us consider a second order linear inhomogeneous equation with constant coefficients

$$a\frac{d^2y}{dx^2} + b\frac{dy}{dx} + cy = f(x), \qquad y(x=0) = y'(x=0) = 0.$$

Taking the Laplace transform and rearranging the equation we get

$$Y = \left(\frac{1}{ap^2 + bp + c}\right) F(p) = \mathcal{T}(p)F(p),$$

where $F(p) = \mathcal{L}(f(x))$ and the function $\mathcal{T}(p)$ is sometimes called the *transfer function*. We can always factorize the transfer function so that

$$\mathcal{T}(p) = \frac{1}{a(p+\alpha)(p+\beta)},$$

which has the advantage that we know the inverse Laplace transform because (see Appendix B)

$$\mathcal{L}\left(\frac{e^{-\alpha x} - e^{-\beta x}}{\alpha - \beta}\right) = \frac{1}{a(p+\alpha)(p+\beta)}.$$

But, to solve our equation, we need to take the inverse transform of a product of functions $y(x) = \mathcal{L}^{-1}(\mathcal{T}(p)F(p))$, so let us look at the Laplace transform of a product of functions in general. Let $F(p)$ and $G(p)$ be the Laplace transforms of the functions $f(t)$ and $g(t)$. The product of the two transforms is then

$$F(p)G(p) = \left(\int_0^\infty e^{-p\xi} f(\xi)\, d\xi\right)\left(\int_0^\infty e^{-p\eta} g(\eta\, d\eta\right) = \int_0^\infty \int_0^\infty e^{-p(\eta+\xi)} f(\xi)g(\eta)\, d\xi\, d\eta.$$

Now we are going to perform a standard trick and change variables. Recall that in doing so we need to ensure that the same region of the (ξ, η) plane is covered. We are going to change variables in the ξ integral, which means that we treat η as if it were a constant. We define $t = \xi + \eta$ so that, since η is considered as a constant, $d\xi = dt$. Now we also have to worry about getting the limits on the integral correct. When $\xi = 0, t = \eta$, and when $\xi = \infty, t = \infty$. So, we can write the integrals as

$$F(p)G(p) = \int_{\eta=0} \eta = \infty \int_{t=\eta}^\infty e^{-pt} f(t-\eta)g(\eta)\, dt\, d\eta.$$

Now we want to swap the order of the integration so that

$$\begin{aligned} F(p)G(p) &= \int_{t=0}^\infty \int \eta = 0^\eta e^{-pt} f(t-\eta)g(\eta)\, d\eta\, dt \\ &= \int_0^\infty e^{-pt}\left(\int_0^t f(t-\eta)g(\eta)\, d\eta\right) dt = \mathcal{L}\left(\int_0^t f(t-\eta)g(\eta)\, d\eta\right). \end{aligned}$$

We recognize the term in the parentheses as the *convolution*. So, this tells us that the product of two Laplace transforms equals the convolution of the original functions. To see how this can help us, let us solve the initial value problem

$$y'' + 3y'_2 y = \mathrm{e}^{-t}, \quad y(0) = y'(0) = 0.$$

Taking the Laplace transform we find

$$Y = \frac{1}{p^2 + 3p + 2}\mathcal{L}(\mathrm{e}^{-t}).$$

We know that the inverse transform of the transfer function is $(\mathrm{e}^{-t} - \mathrm{e}^{-2t})$, so we have

$$Y = \mathcal{L}(\mathrm{e}^{-t} - \mathrm{e}^{-2t})\mathcal{L}(\mathrm{e}^{-t}) = F(p)G(p).$$

It is often a good tactic to put the $(t - \eta)$ part of the convolution into the easier integral, so with that in mind we get

$$y = \int_0^t f(\eta)g(t - \eta)\, d\eta = \int_0^t (\mathrm{e}^{-t} - \mathrm{e}^{-2t})\mathrm{e}^{-(t-\eta)}\, d\eta = t\mathrm{e}^{-t} + \mathrm{e}^{-2t} - \mathrm{e}^{-t}.$$

9.4 Further Reading

We will see later that Fourier series, Fourier transforms, and Laplace transforms feature heavily in techniques for solving partial differential equations, and any book on mathematical methods will include discussions on these topics. Some of my favorites, which do tend to have a bias toward the physical sciences, include *Mathematical Methods in the Physical Sciences* by Boas (2006), *A Guided Tour of Mathematical Methods for the Physical Sciences* by Snieder and van Wijk (2015), and *Mathematical Methods for Physicists* by Arfken et al. (2013). The first of these has very clear explanations and plenty of examples, and the second book is wonderful for working through on your own (as you might guess from the title). The book by Arfken and colleagues is very comprehensive, but can seem a little intimidating at times. Fourier series and Fourier transforms are a major component of analyzing spatial and temporal signals, and there are many good books that cover this; see, for example, *Data Analyis Methods in Physical Oceanography* by Thomson and Emery (2014).

Problems

9.1 Use the complex form of the Fourier Series of $f(x)$ to show that

$$\frac{1}{2\pi}\int_{-\pi}^{\pi} (f(x))^2\, dx = \sum_{n=-\infty}^{\infty} |c_n|^2.$$

9.2 Calculate the Fourier transform of the following functions:

1. A rectangular pulse of height 1 between $x = -1$ and $x = 1$. Use the Heaviside function to write the equation of the box function.
2. An exponential decay function

$$f(t) = \begin{cases} 0 & x < 0 \\ e^{-\alpha t} & x \geq 0 \end{cases},$$

where $\alpha > 0$.

9.3 Consider the repeating sawtooth pattern given by

$$f(x) = A\frac{x}{X}, \quad f(x + X) = f(x),$$

where A and X are constants.

1. Sketch the shape of the curve from $x = -4X$ to $x = 4X$.
2. Find the Fourier series of $f(x)$ using the complex form of the Fourier series.

9.4 Which of the following functions satisfy the Dirichlet conditions?

$$1.\ f(x) = H(x-1) - H(x), \quad 2.\ f(x) = \frac{1}{1+x^2}, \quad 3.\ f(x) = \frac{1}{1-x^2}.$$

9.5 Find the Fourier series of the function $f(x) =\mid x \mid$ on the interval $-\pi \leq x \leq \pi$.

9.6 Find the Fourier series of the function $f(x) = x^2$ on the interval $-\pi \leq x \leq \pi$.

9.7 Find the Fourier transform of the function $g(x) = 5e^{2t}H(-t) + 4\delta(t-4)$, where $H(x)$ is the Heaviside function.

9.8 Calculate the Fourier transform of the function $h(x) = 10e^{-2|x|}$.

9.9 If $g(\omega)$ is the Fourier transform of the function $f(t) = e^{-t^2}$, show that

$$\frac{dg}{d\omega} + \frac{\omega g(\omega)}{2} = 0.$$

Solve this differential equation assuming that $g(\omega) = 0 = \sqrt{\pi}$ to show that $g(\omega) = \sqrt{\pi}e^{-\omega^2/4}$.

9.10 Fourier analysis is an important tool in understanding the discrete sampling of continuous data. Consider a continuous signal $x(t) = \sin(\omega_0 t)$ that is sampled with a sampling period (the time between samples) of T so that the sampled data is $x_s(n) = \sin(\omega_0 nT)$, where $n = 0, \pm1, \pm2, \ldots$. Now consider a second set of signals,

$$y(t) = \sin\left(\left(\omega_0 + \frac{2\pi k}{T}\right)t\right),$$

where $k = 0, \pm1, \pm2, \ldots$. Show that $y(n) = x_s(n)$. Consider two oscillatory signals, $y_1(t) = \cos(0.4\pi t$ and $y_2(t) = \cos(2.4\pi t)$. Plot both curves on the same plot and show graphically that the two signals are indistinguishable if they are sampled at an interval of $\Delta t = 1$. This phenomenon is called *aliasing* and is important for analyzing time series data.

9.11 Use Laplace transforms to solve the equation

$$\frac{d^2x}{dt^2} - 2\frac{dx}{dt} + x(t) = 0, \qquad x(t=0) = 1, \dot{x}(t=0) = 0.$$

9.12 Use Laplace transforms to solve the equation

$$\frac{d^2x}{dt^2} + x(t) = H(t-1), \qquad x(t=0) = \dot{x}(t=0) = 0,$$

where $H(t-1)$ is the Heaviside function.

9.13 Calculate the Laplace transform of the function

$$f(x) = \begin{cases} 1 & 0 \le x < 3 \\ x-3 & x \ge 3 \end{cases}.$$

10 Partial Differential Equations

Ordinary differential equations (ODEs) are equations that involve the ordinary derivative of an unknown function, and solving them involves finding this function, often with the constraint that it satisfies some given initial or boundary conditions. The unknown function in an ODE is a function of a single variable, but many processes in the natural world vary with space and time, and we need to use partial derivatives to describe how they change. Consequently, partial differential equations (PDEs) are equations that involve partial derivatives of an unknown function of more than one variable. Examples of PDEs appear in the description of how fluids (e.g., air and water) move, how chemicals are transported in the environment, and how energy moves through a system such as the Earth's crust. As we shall see, solving PDEs is harder than solving ODEs. This is largely because of the extra freedom that exists in dealing with functions of multiple variables. In this chapter we will look at the structure of PDEs and what that tells us about the nature of their solutions. We will also investigate some techniques for finding solutions analytically, some of which will involve Fourier series and Sturm–Liouville theory. Finally, we will look at some common numerical techniques that we can use to solve PDEs. As with ODEs, we will find that the technique we use to solve an equation depends on the type of equation, so the classification of PDEs is important.

10.1 Introduction

As with all our mathematical investigations, we will start by looking at some very simple PDEs so that we can develop our intuition about what is involved in solving them, To start with, let us consider what seems to be a very simple PDE,

$$\frac{\partial u(x, y)}{\partial x} = c, \tag{10.1}$$

where c is a constant. We can integrate this equation straight away, but remember that when we take the partial derivative of a function, we differentiate with respect to one variable, treating all the others as if they were constants. This means that when we integrate Equation (10.1), instead of getting an unknown constant of integration as we would for an ODE, we get an unknown function of the other variables (as well as a constant, which is often incorporated into the function). So, integrating Equation (10.1) gives

$$u(x, y) = c \int dx + f(y) + k = cx + f(y) + k = cx + g(y), \tag{10.2}$$

where $f(y)$ is an unknown function, k is a constant, and $g(y) = f(y) + k$. We will need information from the boundary conditions for the specific problem we are interested in if we want to find out what the function $f(y)$ actually is.

Exercise 10.1.1 Find the general solution of

$$\frac{\partial u(x, y, z)}{\partial y} = xz + yz.$$

If the PDE includes higher order derivatives we will accumulate unknown functions for each integration we have to perform.

Example 10.1 Let us find the general solution of

$$\frac{\partial^2 u(x, y)}{\partial x^2} = 2xy.$$

The presence of the second order derivative means that we will have to perform two integrations and obtain a solution that involves two unknown functions. Integrating once, we get

$$\frac{\partial u(x, y)}{\partial x} = 2\int xy\, dx = x^2 y + f(y),$$

where $f(y)$ is an unknown functions; integrating a second time gives

$$u(x, y) = \int (x^2 y + f(y))\, dx = \frac{1}{3}x^3 y + xf(y) + g(y),$$

where $g(y)$ is another unknown function.

Example 10.2 Now let us find the general solution of

$$\frac{\partial u(x, y)}{\partial x \partial y} = 2xy,$$

where the right-hand side of the equation is the same as in Example 10.1, but the left-hand side involves a mixed derivative. Integrating once with respect to y gives

$$\frac{\partial u(x, y)}{\partial x} = \int 2xy\, dy = xy^2 + f(x).$$

But now we have to integrate with respect to x. This is a problem because our solution involves an unknown function of x, $f(x)$. However, we can proceed formally to get

$$u(x, y) = \int (xy^2 + f(x))\, dx.$$

We can integrate the first term in the integrand, but we can only integrate the second term formally to give

$$u(x, y) = \frac{1}{2}x^2 y^2 + F(x) + g(y), \qquad F(x) = \int (f(x)\, dx,$$

where $g(y)$ is an unknown function of y that arises because we have integrated over x.

Exercise 10.1.2 Solve the differential equation in Example 10.2 by integrating with respect to x first, and then integrating with respect to y. Compare your answer with the one obtained in Example 10.2, explaining any differences or similarities.

Now that we have seen some of the issues that arise when we are confronted with a PDE to solve, let us explore these equations in more detail and investigate methods to solve them in a more systematic manner. As with our exploration of ODEs, we will start with the simpler equations first, gradually adding complexity.

10.2 First Order Linear Partial Differential Equations

We will start by looking at first order linear PDEs. In general, these equations have the form in two dimensions

$$a(x,y)\frac{\partial f(x,y)}{\partial x} + b(x,y)\frac{\partial f(x,y)}{\partial y} + c(x,y)f(x,y) = w(x,y), \tag{10.3}$$

where $a(x,y)$, $b(x,y)$, $c(x,y)$, and $w(x,y)$ are known functions of x and y, and $f(x,y)$ is the unknown function we want to find. How do such equations arise? Let us consider the flow of a fluid through a channel (Figure 10.1). A chemical with a concentration $c(x,t)$ flows with the fluid and varies with position along the channel (x) and with time (t). The fluid flows at a velocity v. We would like to know how the concentration of the fluid varies with space and with time. To do this, we will use the principle of conservation of mass. Let us consider two cross sections, located at $x = x_a$ and $x = x_b$, and consider the flow across each surface of area A (Figure 10.1). The volume of fluid in the channel between $x = x_a$ and $x = x_b$ is $A(x_b - x_a) = A\Delta x$, so the total mass of the chemical in that volume is $cV = cA\Delta x$. If there are no sources and sinks of the chemical within V, then in a time interval Δt the amount of chemical in V will change by

$$V\Delta c = (Avc(x_b,t) - Avc(x_a,t))\Delta t,$$

where $Avc(x_a,t)\Delta t$ is the amount of chemical flowing into the volume across the cross section at $x = x_a$, and $Avc(x_b,t)\Delta t$ is the amount flowing out across the cross section at $x = x_b$. Substituting for the volume and rearranging the equation gives

$$\frac{\Delta c}{\Delta t} = -\frac{[c(x_b,t) - c(x_a,t)]}{\Delta x}v,$$

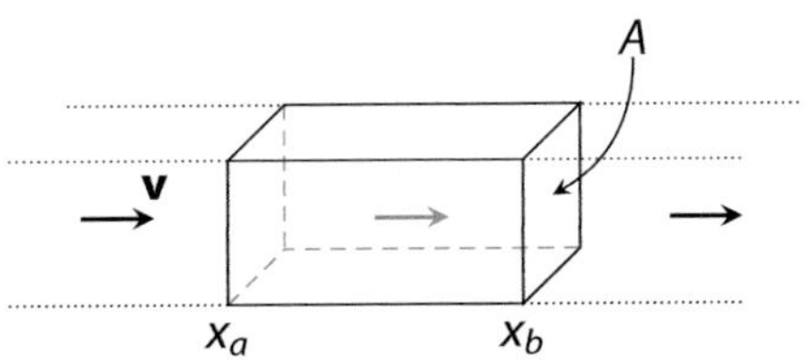

Figure 10.1 Fluid flowing with a velocity **v** through a channel with cross-sectional area A. Considering the differences between flow across the channel at $x = x_a$ and $x = x_b$ leads to the transport equation (Equation (10.4)).

and in the limit as Δt and Δx tend to zero we can replace the finite differences with derivatives to get (assuming all our relevant functions are smooth and differentiable)

$$\frac{\partial c}{\partial t} + v\frac{\partial c}{\partial x} = 0, \tag{10.4}$$

which is known as the *transport equation* or *advection equation* and has the form of Equation (10.3). Notice that if there are sources and/or sinks of the chemical within the volume V, then there will be additional source and sink terms in the equation—we will deal with those a little later.

Equation (10.4) is a little more involved than those we came across in Section 10.1, and it is a little unclear how we can solve it. If we try to integrate with respect to one of the variables, say x, then we have to be able to evaluate the integral of the time derivative of $c(x, t)$, which we do not know.

Let us look in detail at a slightly more general form of Equation (10.4):

$$v\frac{\partial c}{\partial x} + u\frac{\partial c}{\partial t} = 0, \tag{10.5}$$

where v and u are constants and $c = c(x, t)$. We know that we can write the gradient of a scalar function as

$$\nabla c = \frac{\partial c}{\partial x}\hat{\mathbf{i}} + \frac{\partial c}{\partial t}\hat{\mathbf{t}},$$

where $\hat{\mathbf{i}}$ and $\hat{\mathbf{t}}$ are unit vectors in the x and t directions respectively. This may seem a little abstract, as we are not used to thinking of time (t) as being a coordinate with a unit vector, but there is nothing to stop us doing so. We can also combine the two constants u and v into a vector $\mathbf{u} = v\hat{\mathbf{i}} + u\hat{\mathbf{t}}$ so that we can write Equation (10.5) as

$$\mathbf{u} \cdot \nabla c = 0. \tag{10.6}$$

Equation (10.6) is telling us that the gradient of $c(x, t)$ in the direction of the vector $\mathbf{u}$ is zero. In other words, if we were to move along the direction of the vector $\mathbf{u}$ in (x, t) coordinates, the concentration $c(x, t)$ would be constant. The equation of a straight line through the origin with direction $v\hat{\mathbf{i}} + u\hat{\mathbf{t}}$ is

$$\frac{t}{x} = \frac{u}{v} \implies ux - vt = 0,$$

so the equation of any line parallel to $\mathbf{u}$ but not necessarily passing through the origin is $ux - vt = w$, where w is a constant. Because each line has the same direction as $\mathbf{u}$, the function $c(x, t)$ is a constant on each line, but the value of that constant can be different between different lines. These curves on which $c(x, t)$ is constant are called *characteristics*. This implies that the value of $c(x, t)$ on each line depends on the value of w; it is $w = ux - vt$ that distinguishes the different lines (Figure 10.2). So,

$$c(x, t) = f(w) = f(ux - vt), \tag{10.7}$$

where $f(ux - vt)$ is a function that has to be determined from the initial or boundary conditions. We should check that Equation (10.7) is indeed a solution of Equation (10.5):

$$v\frac{\partial c}{\partial x} + u\frac{\partial c}{\partial t} = vu\frac{\partial f}{\partial w} - uc\frac{\partial f}{\partial w} = 0,$$

showing that it is, as claimed, a solution to the PDE.

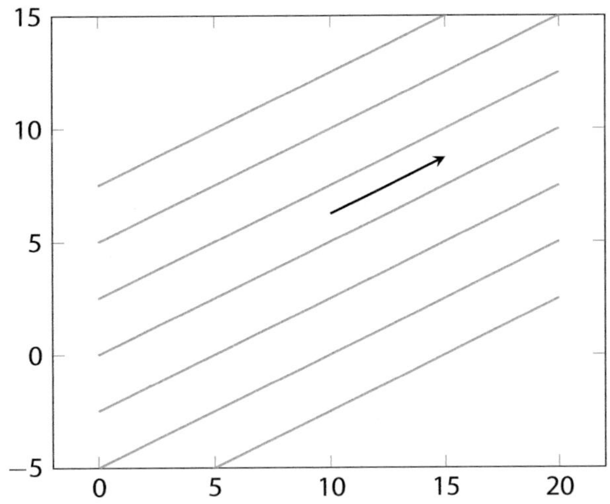

Figure 10.2 The direction of the vector $v\hat{\mathbf{\imath}} + u\hat{\mathbf{t}}$ (shown in black) and the parallel lines $ux - vt = w$ (the characteristics) for different values of w that are in the same direction as the vector.

We have just shown how to construct a solution to Equation (10.5) using geometric arguments. However, we can also solve the equation by changing the PDE into an ODE using a suitable coordinate change. We have seen that the solutions to Equation (10.5) are constant in the direction given by the vector $\mathbf{u} = v\hat{\mathbf{\imath}} + u\hat{\mathbf{t}}$. What we want to do is to rotate the coordinates (x, t) so that one of the new coordinates lines up with direction $\mathbf{u} = v\hat{\mathbf{\imath}} + u\hat{\mathbf{t}}$. Doing this means that there is no variability of c with that coordinate, so all derivatives with respect to it are zero. The new coordinates are simply

$$\xi = vx + ut, \quad \eta = ux - vt, \tag{10.8}$$

so

$$\frac{\partial c}{\partial x} = \frac{\partial c}{\partial \xi}\frac{\partial \xi}{\partial x} + \frac{\partial c}{\partial \eta}\frac{\partial \eta}{\partial x} = v\frac{\partial c}{\partial \xi} + u\frac{\partial c}{\partial \eta}$$
$$\frac{\partial c}{\partial t} = \frac{\partial c}{\partial \xi}\frac{\partial \xi}{\partial t} + \frac{\partial c}{\partial \eta}\frac{\partial \eta}{\partial t} = u\frac{\partial c}{\partial \xi} - v\frac{\partial c}{\partial \eta},$$

and Equation (10.5) becomes

$$v\left(v\frac{\partial c}{\partial \xi} + u\frac{\partial c}{\partial \eta}\right) + u\left(u\frac{\partial c}{\partial \xi} - c\frac{\partial c}{\partial \eta}\right) = (v^2 + u^2)\frac{\partial c}{\partial \xi},$$

and if $u^2 + v^2 \neq 0$, then the transport equation becomes

$$\frac{\partial c}{\partial \xi} = 0,$$

which we can easily solve to give $c(\xi, \eta) = f(\eta)$, where $f(\eta)$ is an unknown function. Transforming back into the original variables, we get $c(x, t) = f(ux - vt)$, which is identical to Equation (10.7). This method of solution is called the *method of characteristics* because we are making use of the characteristics to transform the variables and convert the PDE into an ODE.

Example 10.3 To see how the method of characteristics works in practice, let us solve the following equation:

$$2\frac{\partial c}{\partial x} - \frac{\partial c}{\partial t} = 0, \qquad c(x,0) = \cos(x).$$

By comparing with Equation (10.5) we see that we need to transform to variables $\xi = 2x-t$, $\eta = -x - 2t$, and the solution to the PDE is $c(x,t) = f(-x-2t)$. We can now use the initial conditions to find the form of the function $f(-x-2t)$. Substituting $t = 0$ into the solution and equating the solution with the initial condition gives us

$$c(x,0) = f(-x) = \cos(x),$$

so $c(x,t) = \cos(x + 2t)$, which, by differentiating with respect to x and t, satisfies the original PDE.

Exercise 10.2.1 Use the method of characteristics to solve the initial value problem

$$4\frac{\partial c}{\partial x} - 2\frac{\partial c}{\partial t} = 0, \qquad c(x,0) = e^{-x},$$

and check that the solution you find is a solution to the original PDE.

Equation (10.5) is a homogeneous equation, so it assumes that there are no sources or sinks of the chemical. If there are sources or sinks, then we need to tackle an inhomogeneous PDE. Fortunately, because the equation is linear, we can use our knowledge of ODEs to solve it. Let us consider the equation

$$v\frac{\partial c}{\partial x} + u\frac{\partial c}{\partial t} + wc(x,t) = g(x,t), \tag{10.9}$$

where v, u, and w are all constants and $g(x,t)$ is a given function. The equation is linear, so we know that the solution can be broken down into a sum of two terms, the solution to the corresponding homogeneous equation and a particular solution that depends on the form of $g(x,t)$. First, we transform the equation using the coordinate transformation derived from the characteristics.

Exercise 10.2.2 Show that the transformation Equation (10.8) transforms Equation (10.9) into

$$(v^2 + u^2)\frac{\partial c}{\partial \xi} + wc(\xi,\eta) = g(\xi,\eta). \tag{10.10}$$

To solve Equation (10.10), we first solve the corresponding homogeneous equation

$$(v^2 + u^2)\frac{\partial c}{\partial \xi} + wc(\xi,\eta) = 0$$

to give

$$c_h(\xi,\eta) = \exp\left(-\frac{w\xi}{v^2+u^2}\right) f(\eta),$$

where $f(\eta)$ is an unknown function of η. We can now use a technique such as variation of parameters (Section 6.2.1) to calculate a particular solution:

$$c_p(\xi, \eta) = \exp\left(-\frac{w\xi}{v^2 + u^2}\right) \int \frac{g(\xi, \eta)}{v^2 + n^2} \exp\left(\frac{w\xi}{v^2 + u^2}\right) d\xi,$$

and so long as we can evaluate the integral, we can solve the equation and transform back to the original coordinates. The formal solution to Equation (10.9) is then the sum of the functions c_h and c_p.

Example 10.4 As an example, let us solve the nonhomogeneous equation

$$\frac{\partial c}{\partial x} - \frac{\partial c}{\partial t} + 2c(x, t) = e^{x+t}.$$

The transformed variables are $\xi = x - t$ and $\eta = -x - t$, and $v^2 + u^2 = 2$, so the transformed equation is

$$2\frac{\partial c}{\partial \xi} + 2c(\xi, \eta) = e^{-\eta}.$$

The solution to the homogeneous equation is $c_h(\xi, \eta) = e^{-\xi} f(\eta)$, where $f(\eta)$ is an unknown function. Using the technique of variation of parameters, the particular solution is

$$c_p(\xi, \eta) = \frac{1}{2}e^{-\xi} \int e^{-\eta} e^{\xi}\, d\xi = \frac{1}{2}e^{-\eta}.$$

Adding the two solutions and transforming back to the original variables we get the solution

$$c(x, t) = e^{-(x-t)} f(-x - t) + \frac{1}{2}e^{x+t}.$$

Exercise 10.2.3 Show that the solution obtained in Example 10.4 satisfies the original PDE in that example.

The PDEs we have looked at so far all have constant coefficients. Can we extend the method of characteristics to the general first order linear PDE in Equation (10.3)? Let us first consider the slightly simpler equation

$$v(x, t)\frac{\partial c}{\partial x} + u(x, t)\frac{\partial c}{\partial t} = 0. \tag{10.11}$$

Equation (10.11) is a simple generalization of Equation (10.5), but with nonconstant coefficients, and is a little simpler than Equation (10.3). We can write Equation (10.11) in vector form using the same arguments as we did for Equation (10.5), so that

$$\mathbf{u} \cdot \nabla c = 0, \qquad \mathbf{u} = v(x, t)\hat{\mathbf{i}} + u(x, t)\hat{\mathbf{t}}.$$

When we analyzed Equation (10.5) we found that the characteristics were straight lines. This is because v and u were constants, so the slopes of the characteristics were constant. Now u and v are functions of x and t, so the slopes of the characteristics are curves with changing slopes, and as a result we have to try a more general approach to finding the characteristics of the equation. We will try the well-worn tactic of assuming that we can find new coordinates ξ and ζ as before, and demand that the coordinate transformation allows us to reduce the PDE to an ODE, just as in the case of constant coefficients.

First, let us assume that there is a coordinate transformation such that $\xi = \xi(x,t)$, $\eta = \eta(x,t)$. Equation (10.11) then becomes

$$\frac{\partial c}{\partial \xi}\left(v\frac{\partial \xi}{\partial x} + u\frac{\partial \xi}{\partial t}\right) + \frac{\partial c}{\partial \eta}\left(v\frac{\partial \eta}{\partial x} + u\frac{\partial \eta}{\partial t}\right) = 0.$$

When v and u were constant, we found the transformation that reduced the PDE to an equation with only a derivative with respect to ξ. To do the same thing here we require that the coordinate transformation is such that the second term in the equation vanishes. In other words, we require that

$$v\frac{\partial \eta}{\partial x} + u\frac{\partial \eta}{\partial t} = 0 \implies \frac{dt}{dx} = \frac{u}{v} \tag{10.12}$$

so that the characteristic curves are given by the same equation as before (i.e., the slope of the characteristic curves is u/v), except that now u and v are functions of x and t and not constant. Equation (10.12) also gives us a differential equation for η, which has a solution $\eta(x,t) = a =$ constant. What do we choose for ξ? We are going to simply choose $\xi = x$. We want to make sure that the transformation does not collapse the coordinates giving a zero area, which means that the Jacobian of the transformation must not be zero, i.e.,

$$J = \frac{\partial(\xi,\eta)}{\partial(x,t)} = \begin{vmatrix} \frac{\partial xi}{\partial x} & \frac{\partial xi}{\partial t} \\ \frac{\partial \eta}{\partial x} & \frac{\partial \eta}{\partial t} \end{vmatrix} = \begin{vmatrix} 1 & 0 \\ \frac{\partial \eta}{\partial x} & \frac{\partial \eta}{\partial t} \end{vmatrix} = \frac{\partial \eta}{\partial t}.$$

So, we also require the derivative of η with respect to t to be nonzero. After making this transformation, we can proceed to find the solution to the resulting ODE and hence find the solution to the PDE.

The procedure we have outlined also applies to the more general Equation (10.3). In the new coordinates, Equation (10.3) becomes

$$a(\xi,\eta)\frac{\partial f}{\partial \eta} + c(\xi,\eta)f(\xi\eta) = w(\xi,\eta),$$

which is called the *canonical form*, and any first order linear PDE can be put into this form.

Example 10.5 To see how this more general procedure works in practice, let us find the general solution of the equation

$$x\frac{\partial c}{\partial x} - t\frac{\partial c}{\partial t} + tc(x,t) = t.$$

Using Equation (10.12), the equation for the characteristics is

$$\frac{dt}{dx} = -\frac{t}{x} \implies xt = a = \text{constant},$$

so our new variables are $\xi = x$ and $\eta = xt$. This gives us

$$\frac{\partial c}{\partial t} = \frac{\partial c}{\partial \xi}\frac{\partial \xi}{\partial t} + \frac{\partial c}{\partial \eta}\frac{\partial \eta}{\partial t} = x\frac{\partial c}{\partial \eta}, \quad \frac{\partial c}{\partial x} = \frac{\partial c}{\partial \xi}\frac{\partial \xi}{\partial x} + \frac{\partial c}{\partial \eta}\frac{\partial \eta}{\partial x} = \frac{\partial c}{\partial \xi} + t\frac{\partial c}{\partial \eta}$$

so that the PDE becomes

$$\frac{\partial c}{\partial \xi} + \frac{\eta}{\xi^2}c(\xi, \eta) = \frac{\eta}{\xi^2}.$$

We can solve this equation using an integrating factor to give

$$c(\xi, \eta) = e^{\eta/\xi} f(\eta) + 1,$$

where $f(\eta)$ is a function that we can determine from the initial or boundary conditions. So, in the original coordinates, the solution is

$$c(x, t) = e^t f(xt) + 1.$$

Exercise 10.2.4 Confirm that the solution obtained in Example 10.5 satisfies the PDE from that example.

The method of characteristics provides an elegant technique for solving any first order linear PDE, though the usual caveats apply; e.g., we have to be able to evaluate the integrals that arise in the technique.

10.3 Classification of Second Order Linear Partial Differential Equations

We now turn our attention to second order linear PDEs, where the classification of the equation is important. The most general second order linear PDE for a function $u(x, y)$ in two variables x and y can be written as

$$A(x, y)\frac{\partial^2 u}{\partial x^2} + B(x, y)\frac{\partial^2 u}{\partial x \partial y} + C(x, y)\frac{\partial^2 u}{\partial y^2} + D(x, y)\frac{\partial u}{\partial x} + E(x, y)\frac{\partial u}{\partial y} + F(x, y)u = G(x, y). \tag{10.13}$$

In order to save space (and typing), we will use the following notation for the derivatives

$$\frac{\partial u}{\partial x} = u_x, \quad \frac{\partial^2 u}{\partial x^2} = u_{xx}, \quad \frac{\partial^2 u}{\partial x \partial y} = u_{xy}, \quad \text{etc.}$$

Equation (10.13) is reminiscent of the equation of a general conic

$$ax^2 + bxy + cy^2 + dx + ey + f = 0,$$

where a–f are constants. We know that for a conic we can transform the variables to remove the cross term (bxy) and obtain canonical forms of the equations for an ellipse, parabola, or hyperbola, depending on whether the discriminant $b^2 - 4ac$ is negative, zero, or positive respectively. We can do something very similar with Equation (10.13), but the coefficients A–C are not necessarily constant, so we should expect to see cases where a PDE changes classification as x and y change. Since we are going to concentrate on only the terms involving A, B, and C, we will write Equation (10.13) as

$$\begin{aligned} A(x,y)u_{xx} + B(x,y)u_{xy} + C(x,y)u_{yy} &= G(x,y) - D(x,y)u_x - E(x,y)u_y - F(x,y)u \\ &= H(u_x, u_y, u, x, y). \end{aligned}$$

So, inspired by the similarity of Equation (10.13) to a conic, we are going to try and look for a transformation of variables x and y that will allow us to put the equation in a simpler form. We will assume that we have a coordinate transformation to new coordinates ξ and η so that $\xi = \xi(x,y)$ and $\eta = \eta(x,y)$. In making such a transformation we do not want to inadvertently transform to coordinates that are degenerate (i.e., have zero volume), so we also require that the Jacobian of the transformation is nonzero,

$$J = \frac{\partial(\xi,\eta)}{\partial(x,y)} = \begin{vmatrix} \xi_x & \xi_y \\ \eta_x & \eta_y \end{vmatrix} = \xi_x\eta_y - \xi_y\eta_x \neq 0,$$

and using the chain rule we can get expressions for the derivatives in terms of the new variables.

Exercise 10.3.1 Use the coordinate transformations $\xi = \xi(x,y)$ and $\eta = \eta(x,y)$ with the chain rule to get expressions for u_x, u_y, u_{xx}, u_{yy}, and u_{xy} in terms of the new variables.

Making this coordinate transformation in the PDE yields

$$\tilde{A}(\xi,\eta)u_{\xi\xi} + \tilde{B}(\xi,\eta)u_{\xi\eta} + \tilde{C}(\xi,\eta)u_{\eta\eta} = \tilde{H}(u_\xi, u_\eta, u, \xi, \eta),$$

where

$$\begin{aligned} \tilde{A} &= A(\xi_x)^2 + B\xi_x\xi_y + C(\xi_y)^2, \\ \tilde{B} &= 2A\xi_x\eta_x + B(\xi_x\eta_y + \xi_y\eta_x) + 2C\xi_y\eta_y, \\ \tilde{C} &= A(\eta_x)^2 + B\eta_x\eta_y + C(\eta_y)^2, \end{aligned}$$

and

$$\tilde{B}^2 - 4\tilde{A}\tilde{C} = (\xi_x\eta_y - \xi_y\eta_x)(B^2 - 4AC) = J(B^2 - 4AC).$$

We want to choose the transformation such that $\tilde{A} = \tilde{C} = 0$, i.e.,

$$A(\xi_x)^2 + B\xi_x\xi_y + C(\xi_y)^2 = A(\eta_x)^2 + B\eta_x\eta_y + C(\eta_y)^2 = 0. \tag{10.14}$$

For this to be true, both ξ and η satisfy the same differential equation. The solutions to these equations form a set of curves called *characteristic curves*. With first order PDEs we used the characteristic curves to reduce the equation to an ODE by rotating the original x and y variables so that one of the new variables lay along the characteristic curve. With a second order PDE we want to do something similar, but rather than reduce the equation to an ODE, we want to use the characteristic curves to simply remove some of the terms in the equation. First, we can rewrite Equation (10.14) as

$$A\left(\frac{\xi_x}{\xi_y}\right)^2 + B\left(\frac{\xi_x}{\xi_y}\right) + C = A\left(\frac{\eta_x}{\eta_y}\right)^2 + B\left(\frac{\eta_x}{\eta_y}\right) + C = 0. \tag{10.15}$$

Now, before we move ahead, let us think about what we are trying to achieve. We want to transform the coordinates from x and y to a new set of coordinates ξ and η. This means that we are looking for solutions to these equations that have the form $\xi = \xi(x, y) = \text{constant}$ and $\eta = \eta(x, y) = \text{constant}$ so that

$$d\xi = \frac{\partial \xi}{\partial x}dx + \frac{\partial \xi}{\partial y}dy = 0, \qquad \text{and} \qquad d\eta = \frac{\partial \eta}{\partial x}dx + \frac{\partial \eta}{\partial y}dy = 0,$$

from which we can calculate the slopes (i.e., the derivatives) of the curves $\xi = \text{constant}$ and $\eta = \text{constant}$ in the (x, y) coordinates:

$$\frac{dy}{dx} = -\frac{\xi_x}{\xi_y} \qquad \text{and} \qquad \frac{dy}{dx} = -\frac{\eta_x}{\eta_y}.$$

Substituting these expressions back into Equation (10.15) gives the single ODE

$$A\left(\frac{dy}{dx}\right)^2 - B\left(\frac{dy}{dx}\right) + C = 0. \tag{10.16}$$

This is a quadratic equation with a solution

$$\frac{dy}{dx} = \frac{1}{2A}\left(B \pm \sqrt{B^2 - 4AC}\right),$$

so there will be either two distinct, real solutions if $(B^2 - 4AC) > 0$, a single real solution if $(B^2 - 4AC) = 0$, or a solution that consists of a complex conjugate pair if $(B^2 - 4AC) < 0$.

10.3.1 Hyperbolic Equations

If $(B^2 - 4AC) > 0$, the solutions to Equation (10.16) are

$$y = \frac{1}{2A}\left(B \pm \sqrt{B^2 - 4AC}\right)x + h,$$

where h is a constant of integration. We can choose either ξ or η to be the solution with the plus sign, and we will pick ξ to have the plus sign so that (remembering that ξ and η are constant along the characteristics)

$$\xi(x, y) = \frac{1}{2A}\left(B + \sqrt{B^2 - 4AC}\right)x - y, \qquad \eta(x, y) = \frac{1}{2A}\left(B - \sqrt{B^2 - 4AC}\right)x - y.$$

Substituting this back into the equations for $\tilde{A}$, $\tilde{B}$, and $\tilde{C}$ tells us that in this case $\tilde{A} = \tilde{C} = 0$, so we have been able to achieve our goal of choosing a coordinate transformation that makes these coefficients zero. The coefficient $\tilde{B} = 4C - (B^2/A)$, which is not zero. In these new coordinates, the PDE becomes

$$u_{\xi\eta} = \tilde{H}(u_\xi, u_\eta, u, \xi, \eta) \tag{10.17}$$

and is called a *hyperbolic equation*.

Exercise 10.3.2 Show that the coordinate transformation $z = \xi + \eta$, $t = \xi - \eta$ transforms Equation (10.17) into

$$u_{zz} - u_{tt} = K(u_z, u_t, u, z, t). \tag{10.18}$$

Equation (10.17) and Equation (10.18) are called the *canonical forms* of a hyperbolic equation, and as we will see shortly, these equations are often used to describe wave phenomena.

10.3.2 Parabolic Equations

Things are not quite so straightforward in the case when $(B^2 - 4AC) = 0$ because we will have only a single family of solutions to Equation (10.16), so we will have only one characteristic curve, say $\xi =$ constant. This means that we are free to choose the other coordinate (η), so long as we do not inadvertently choose it such that the coordinate transformation has a zero Jacobian. Following similar arguments to those in Section 10.3.1 we find that

$$\xi = \frac{Bx}{2A} - y,$$

and we are free to choose η. A simple choice is $\eta = x$.

Exercise 10.3.3 Show that the transformation from coordinates (x, y) to (ξ, η) is nondegenerate.

Under this transformation, the PDE becomes

$$u_{\eta\eta} = \tilde{H}(u_\xi, u_\eta, u, \xi, \eta), \tag{10.19}$$

which is the canonical form for a *parabolic equation*. PDEs of this type are used to describe a wide range of phenomena, including diffusion of dissolved chemicals in the environment, and the conduction of heat through the Earth.

10.3.3 Elliptic Equations

The last case to consider is when $(B^2 - 4AC) < 0$ and the solution to Equation (10.16) is a complex conjugate pair:

$$\xi(x, y) = \left(\frac{1}{2A}\left(B \pm i\sqrt{|B^2 - 4AC|}\right)x\right) - y.$$

In this case, we choose our coordinates to be the real and imaginary components of ξ, so that

$$\zeta(x, y) = \text{Re}(\xi) = \frac{Bx}{2A} - y, \qquad \eta(x, y) = \text{Im}(\xi) = \frac{1}{2A}\sqrt{|B^2 - 4AC|})x.$$

Exercise 10.3.4 Show that $\tilde{B} = 0$ and $\tilde{A} = \tilde{C}$ under this coordinate transformation.

The PDE can then be written as

$$u_{\zeta\zeta} + u_{\eta\eta} = \tilde{H}(u_\zeta, u_\eta, u, \zeta, \eta). \tag{10.20}$$

Equation (10.20) is the canonical form of an *elliptic equation* and describes gravitational and electrostatic fields.

In all of the arguments we have used an equation that depends on two variables. Surely we can have PDEs in more than two dimensions? So, how do we deal with them? If we have N coordinates (e.g., we could be looking at the equations describing the motion of the atmosphere, in which case we have three spatial coordinates and time, so $N = 4$), we can write the general second order linear PDE in the form

$$\sum_{i=1}^{N}\sum_{j=1}^{N} A_{ij}\frac{\partial^2 u}{\partial x_i \partial x_j} + \sum_{i=1}^{N} B_i \frac{\partial u}{\partial x_i} + Cu + D = E(x_i),$$

where A_{ij}, B_i, C, and D are all functions of the coordinates x_i. If the matrix A_{ij} is symmetric, then we can perform a similar classification of this equation based on the eigenvalues (λ_i) of the matrix A_{ij}. If all of the eigenvalues are greater than zero or all of them are less than zero, then the equation is an elliptic equation; if one or more of the eigenvalues is zero, the equation is parabolic; and if all of the eigenvalues except one are of the same sign, then the equation is hyperbolic.

10.3.4 Boundary Value Problems

PDEs involve spatial derivatives, and the solution of a PDE will depend on the type of boundary conditions it needs to satisfy. There are several different types of boundary conditions, and we have met some of them before in Section 6.10.3. Boundary conditions are called *Dirichlet boundary conditions* when we specify the value of the function on the boundary. For example, if we have a function $u(x, t)$ that is the solution to a PDE with a boundary at $x = 0$, then the condition $u(0, t) = f(t)$ is a Dirichlet boundary condition.[1] The function $f(t)$ may be a constant, and if $f(t) = 0$, then the boundary condition is a *homogeneous* Dirichlet boundary condition.

We could give the derivative of the function on the boundary instead of the function itself. For example, if the problem has a boundary at $x = 0$, then we could require the solution satisfy the condition

$$\left.\frac{\partial u}{\partial x}\right|_{x=0,t} = f(t).$$

This boundary condition is called a *Neumann boundary condition*.[2] This condition specifies that the solution has to have a certain flow across the boundary. If $f(t) = 0$, then there is no flow across the boundary and the boundary conditions are often called *perfectly insulating* boundary conditions.

A boundary condition can also specify both the value of the function and its derivative at the boundary. For example, with a boundary at $x = 0$, we can specify

$$\alpha u(0, t) + \beta \left.\frac{\partial u}{\partial x}\right|_{x=0,t} = f(t),$$

[1] Named after the German mathematician Johann Peter Gustav Lejeune Dirichlet (1805–1859).
[2] Named after Carl Gottfried Neumann (1832–1925).

where α and β are constants. These are called *Robin boundary conditions*,[3] and they imply that the boundary absorbs some of the material passing through it and transmits the rest.

Another common type of boundary condition is the periodic boundary condition. These are used in cases where we want to solve a PDE over a domain that repeats itself. For example, we may want to solve the heat equation on a domain $x = 0$ to $x = 2\pi$, which is periodic; i.e., as we pass through $x = 2\pi$ we wrap around to $x = 0$ again. Such boundary conditions are used to represent processes on the surface of spheres and cylinders, for example. In such cases, we want the solution and its derivatives to be continuous so that the solution wraps smoothly around the domain—there are no discontinuities in going from $x = 2\pi$ to $x = 0$. This means we need boundary conditions such that $u(0, t) = u(2\pi, t)$ and $u_x(0, t) = u_x(2\pi, t)$.

Now that we have examined a classification of PDEs and their boundary conditions, we can look at some examples and examine techniques for solving them. We will concentrate on linear equations since that covers many of the types of PDE you will encounter, but the field of nonlinear PDEs is an incredibly rich one and has application in understanding solitons (i.e., solitary waves) and rogue waves on the ocean.[4]

10.4 Parabolic Equations: Diffusion and Heat

Diffusion is an important process in the environment. For example, it describes the spread of dissolved substances in water, the motion of particles through a turbulent fluid, and the spread of an invasive species. Diffusion describes the random motions of particles as they move through space. A molecule of substance A will move through the air along a straight line with a certain velocity until it collides with an air molecule. After the collision, the molecule of A will be moving in a different direction. This is the basic mechanism that we want to capture in an equation.

Let us start by deriving an equation to describe the movement of particles (e.g., molecules of a chemical) by diffusion in one spatial dimension (x); i.e., the particles can only move forward or backward along the x direction. In addition, we assume that the distribution of particles in the y and z directions are uniform and constant. Now assume that at time t we know the number of particles ($N(x)$) sitting at each value of x. In a time interval Δt, some of these particles will move a distance Δx to the left and some will move a distance Δx to the right. But how many will move in each direction? Diffusion is a random process, so we would expect that on average half the particles that were at x would move to the left (i.e., to $x - \Delta x$) and half would move to the right (to $x + \Delta x$). This

[3] Named after the French mathematician Victor Gustave Robin (1855–1897).

[4] Solitons are special types of waves that are spatially localized so that they appear as a single pulse that propagates without spreading. These were first documented by John Scott Russell (1808–1882), who observed them propagating along a canal. Solitons describe tidal bores, the waves that can occur within a body of water —so-called internal waves—rather than on the surface, roll clouds in the atmosphere, and rogue waves—also called freak waves—which are very large waves that occur in the open ocean and some have been recorded as high as 29 m (Residori et al., 2017).

means that if we were to count the number of particles traveling in each direction across a cross-sectional area A placed between x and $x + \Delta x$, $N(x)/2$ particles would move from position x to $(x + \Delta x)$, and $N(x + \Delta x)/2$ would move left from $(x + \Delta x)$ to x. So the *net* number of particles moving to the right across A is

$$-\frac{1}{2}\left[N(x + \Delta x) - N(x)\right].$$

The number of particles moving to the right per unit area and per unit time (i.e., the *flux*) is

$$J_x = -\frac{1}{2A\Delta t}\left[N(x + \Delta x) - N(x)\right]. \tag{10.21}$$

The right-hand side of this equation looks like a derivative, but we are missing a Δx in the denominator. So, we will multiply the whole equation by $(\Delta x)/(\Delta x)$ to get

$$J_x = -\frac{\Delta x}{2A\Delta t}\left[\frac{N(x + \Delta x) - N(x)}{\Delta x}\right].$$

We can see that, by taking an appropriate limit, we will have the derivative of N with respect to x on the right-hand side of the equation. But we have seen several times before that while it is good to use numbers of particles or mass to derive an equation (because these are conserved quantities), we are usually interested in concentration, because it is easier to measure. But, to form a volume, we need the area A and another factor of Δx, so we will multiply the equation by $(\Delta x)/(\Delta x)$ again to get

$$J_x = -\frac{(\Delta x)^2}{2\Delta t}\frac{1}{\Delta x}\left[\frac{N(x + \Delta x) - N(x)}{A\Delta x}\right] = -\frac{(\Delta x)^2}{2\Delta t}\left[\frac{C(x + \Delta x) - C(x)}{\Delta x}\right],$$

where $C(x)$ is the concentration, and as we take the limits $\Delta x \to 0$ and $\Delta t \to 0$ we obtain

$$J_x = -\mathcal{D}\frac{\partial C}{\partial x}, \tag{10.22}$$

where $\mathcal{D}$ is the diffusion coefficient that we met back in Chapter 1. Equation (10.22) is called *Fick's first law of diffusion*,[5] and it tells us that the net flux is negatively proportional to the gradient of the concentration. That means that diffusion redistributes material from areas of high concentration to areas of low concentration, with the faster rates occurring when the difference between high and low concentrations is the largest. As the concentration of material becomes smoother, the rate of diffusion gets slower. If material is uniformly distributed so that $\partial C/\partial x = 0$, then the flux $J_x = 0$ and an equilibrium state has been reached.

The same arguments we used to derive Equation (10.22) could be used to obtain equations for the flux in the y direction (J_y) and the z direction (J_z), so we can create a flux vector $\mathbf{J}$ such that

$$\mathbf{J} = J_x\hat{\mathbf{i}} + J_y\hat{\mathbf{j}} + J_z\hat{\mathbf{k}} = -\mathcal{D}\frac{\partial C}{\partial x}\hat{\mathbf{i}} - \mathcal{D}\frac{\partial C}{\partial y}\hat{\mathbf{j}} - \mathcal{D}\frac{\partial C}{\partial z}\hat{\mathbf{k}} = -\mathcal{D}\nabla C. \tag{10.23}$$

Equation (10.23) tells us the flux of molecules at each location in space, but it does not tell us the rate of change of concentration with time at any point. To derive such an

[5] Named after the German physician Adolf Eugen Fick (1829–1901).

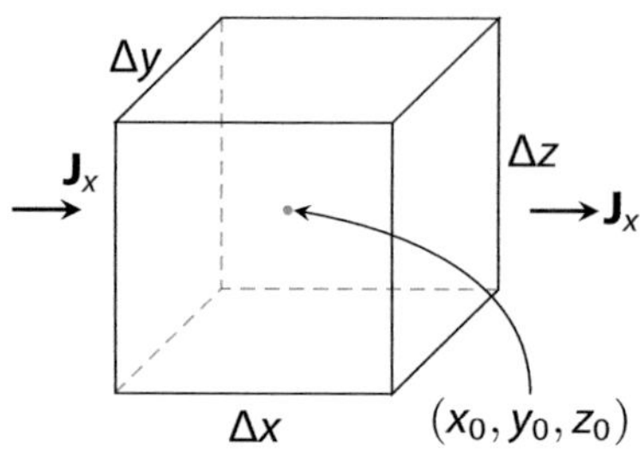

Figure 10.3 The fluxes in the x direction through a cube centered on the point (x_0, y_0, z_0) and with sides Δx, Δy, and Δz.

equation we will use Fick's first law and consider the change to the concentration within a small volume, assuming that there are no sources or sinks within it (Figure 10.3). If we first consider flow in just the x direction, then the net flow of material into the cube along that direction is

$$J_x(x_0 - \Delta x/2, y_0, z_0)\Delta y \Delta z - J_x(x_0 + \Delta x/2, y_0, z_0)\Delta y \Delta z.$$

Now we expand each of these flux terms using Taylor's theorem, keeping only terms up to Δx,

$$J_x(x_0 + \Delta x/2, y_0, z_0)\Delta y \Delta z \approx J_x(x_0, y_0, z_0) + \frac{\Delta x}{2}\frac{\partial J_x}{\partial x},$$
$$J_x(x_0 - \Delta x/2, y_0, z_0)\Delta y \Delta z \approx J_x(x_0, y_0, z_0) - \frac{\Delta x}{2}\frac{\partial J_x}{\partial x},$$

so that we can write the net flow of material out of the cube along the x axis as

$$\text{net flow out of cube along } x \text{ axis} = \frac{\partial J_x}{\partial x}\Delta x \Delta y \Delta z. \tag{10.24}$$

We can do the same thing in the y and z directions so that

$$\begin{aligned} \text{net flow into cube from all directions} &= \left(\frac{\partial J_x}{\partial x} + \frac{\partial J_y}{\partial y} + \frac{\partial J_z}{\partial z}\right)\Delta x \Delta y \Delta z \\ &= (\nabla \cdot \mathbf{J})\,\Delta x \Delta y \Delta z. \end{aligned} \tag{10.25}$$

Because we have assumed that there are neither sources nor sinks within the cube, this net flow must be equal to the rate of change of the chemical within the volume of the cube, i.e.,

$$\frac{\partial C}{\partial t}\Delta x \Delta y \Delta z.$$

Equating these two equations we get

$$\frac{\partial C}{\partial t} = \nabla \cdot \mathbf{J}, \tag{10.26}$$

which is precisely what we expect for a conserved quantity (see Section 7.5.1). We can go further than this because Fick's first law gives us an equation for $\mathbf{J}$ in terms of the gradient of the concentration C, so we should be able to end up with an equation that just involves the concentration. Substituting Equation (10.23) into Equation (10.26) gives

$$\frac{\partial C}{\partial t} = \frac{\partial}{\partial x}\left(\mathcal{D}\frac{\partial C}{\partial x}\right) + \frac{\partial}{\partial y}\left(\mathcal{D}\frac{\partial C}{\partial y}\right) + \frac{\partial}{\partial z}\left(\mathcal{D}\frac{\partial C}{\partial z}\right) = \nabla\cdot(\mathcal{D}\nabla C)$$
$$= \nabla\mathcal{D}\cdot\nabla C + \mathcal{D}\nabla^2 C. \qquad (10.27)$$

If we further assume that $\mathcal{D}$ = constant, then Equation (10.27) simplifies to

$$\frac{\partial C}{\partial t} = \mathcal{D}\nabla^2 C. \qquad (10.28)$$

Equation (10.28) is called, creatively, Fick's second law and together with Equation (10.23) provides a description of diffusion. If we had considered the conduction of heat instead of the diffusion of a chemical, then we would have arrived at a very similar equation,

$$\rho c_p \frac{\partial T}{\partial t} = k\nabla^2 T, \qquad \text{or} \qquad \frac{\partial T}{\partial t} = \kappa\nabla^2 T, \qquad (10.29)$$

where T is the temperature, ρ the density of the material, c_p its specific heat capacity, and k its thermal conductivity, and $\kappa = (k/(\rho c_p))$. You will often see any PDE of the form $\dot{u} = -\alpha\nabla^2 u$, where α is a constant, described as a *heat equation* or *diffusion equation*.

Exercise 10.4.1 Show that the diffusion equation is a parabolic equation.

Equation (10.23) and Equation (10.28) assume that there are no sources of heat or material within the region being considered. This is not always true, so these equations need to be modified if sources and sinks are present. For example, if there is a source or sink of material in the region, then Equation (10.28) becomes

$$\frac{\partial C}{\partial t} = \mathcal{D}\nabla^2 C + \mathcal{R}, \qquad (10.30)$$

where $\mathcal{R}$ represents the rate of production minus the rate of destruction; i.e., the source minus the sink. Equation (10.30) is called a *reaction–diffusion equation*.

10.4.1 Solving the Diffusion Equation

Now that we have derived the diffusion equation, we need to examine ways to solve it. We will follow our usual approach and start simply. So let us start by looking at diffusion in only the x direction so that Equation (10.28) becomes

$$\frac{\partial C}{\partial t} = \mathcal{D}\frac{\partial^2 C}{\partial x^2}. \qquad (10.31)$$

We simplify the problem further by considering the steady state situation so that Equation (10.31) reduces to the ODE

$$\frac{d^2 C}{dx^2} = 0,$$

which has a general solution $C(x) = a + bx$. We need two boundary conditions to obtain a specific solution, so assuming that the concentrations at $x = x_1$ and $x = x_2$ are C_1 and C_2 respectively, the final solution is

$$C(x) = \left(\frac{C_1 x_2 - C_2 x_1}{x_2 - x_1}\right) + \left(\frac{C_2 - C_1}{x_2 - x_1}\right) x.$$

The concentration varies linearly, and since the concentration is in steady state, there must be sources and sinks outside of the region between x_1 and x_2. If there were no sources or sinks, the concentration gradient would gradually decrease to zero.

Let us look at a slightly different situation where we have diffusion to a sphere. We will work in spherical coordinates because the problem has a natural spherical symmetry to it, so that the concentration $C(r, \theta, \phi)$ is a function of the radius r and two angles θ and ϕ. We briefly looked at this problem in Chapter 1, where we used dimensional analysis. The Laplacian in spherical coordinates is given by Equation (7.20) so that

$$\frac{\partial C}{\partial t} = \frac{1}{r^2}\frac{\partial}{\partial r}\left(r^2 \frac{\partial C}{\partial r}\right) + \frac{1}{r^2 \sin\theta}\frac{\partial}{\partial \phi}\left(\sin\theta \frac{\partial C}{\partial \theta}\right) + \frac{1}{r^2 \sin^2\theta}\frac{\partial^2 C}{\partial \theta^2}. \tag{10.32}$$

To simplify this even further, we assume that the system is in steady state and that diffusion is only radial, that is the concentration does not depend on θ or ϕ. This yields another ODE,

$$\frac{\mathcal{D}}{r^2}\left\{\frac{d}{dr}\left(r^2 \frac{dC}{dr}\right)\right\} = \frac{d}{dr}\left(r^2 \frac{dC}{dr}\right) = 0, \tag{10.33}$$

which has a general solution

$$C(r) = k_1 + \frac{k_2}{r}. \tag{10.34}$$

To find values for the constants k_1 and k_2 in Equation (10.34) requires boundary conditions.

Exercises 10.4.2 If the flux (i.e., total transport of material per unit time per unit area) across a sphere of radius r is F, then assuming there are no sources or sinks, the total amount of material crossing the surface of a sphere will be the same for any value of r. Use Fick's first law to show that $k_2 = F/(4\pi\mathcal{D})$.

Exercises 10.4.3 If the concentration of material is constant far away from the sphere (i.e., $C(r = \infty) = C_\infty = \text{constant}$), show that $k_1 = C_\infty$.

The solution of Equation (10.33) is then

$$C(r) = C_\infty + \frac{F}{4\pi\mathcal{D}r}$$

and the concentration at the surface of the sphere for a sphere of radius $r = a$ is

$$F(r) = 4\pi\mathcal{D}a(C_a - C_\infty), \qquad C_a = \frac{F}{4\pi\mathcal{D}a} + C_\infty. \tag{10.35}$$

What is the *maximum* transport to the sphere? This will tell us the *fastest* rate that material can get to the surface of the sphere. The fastest flow rate occurs when $C_a = 0$ (remember that flow to the sphere is negative) and is given by $F_{\text{max}} = -4\pi\mathcal{D}aC_\infty$. It occurs when the concentration of the nutrient at the surface of the sphere is zero. This is what we expect because we know that the rate of transport by diffusion depends on the gradient of the concentration and, with C_∞ fixed, the greatest physical gradient occurs when the concentration at $r = a$ has its smallest value.

Exercise 10.4.4 Derive expressions for $C(r)$ and $F(r = b)$ if the furthest boundary condition is not at $r = \infty$ but if $C = C_b$ at $r = b$.

Exercise 10.4.5 In Chapter 1 we used dimensional analysis to look at the rates of nutrient uptake of phytoplankton cells. Redo the calculation using the exact solutions we have obtained in this chapter.

Exercise 10.4.6 Show by direct substitution that the Gaussian function

$$C(x,t) = \frac{\alpha}{\sqrt{t}} \exp\left(-\frac{x^2}{\gamma t}\right),$$

where α and γ are constants is a solution to the one-dimensional diffusion equation Equation (10.31) for a specific value of γ.

Exercise 10.4.7 Consider two solutions $f(x,t)$ and $g(x,t)$ to the one-dimensional diffusion equation. Show that $h(x,t) = f(x,t) + g(x,t)$ is also a solution, as expected for a linear equation.

Now let us relax the assumption of steady state and solve Equation (10.31) in full. To do this, we will use a method called *separation of variables*, which is a common method for solving a linear homogeneous PDE. The basic idea is to try and convert the PDE into an ODE by making some assumptions about the form of the solution. To see how this works and what is involved, let us consider the diffusion equation in rectangular coordinates for a constant diffusion coefficient,

$$\frac{\partial C}{\partial t} = \mathcal{D}\frac{\partial^2 C}{\partial x^2}, \tag{10.36}$$

and we will save stating the boundary conditions for a bit later. The first thing we are going to do is *nondimensionalize* the equation. One way of doing this is to choose characteristic length and time scales and scale with respect to these. Since the scales will change from problem to problem, we will represent them as a generic length scale L_*, a generic time scale T_*, and a characteristic value for C, C_*. We can then define new, dimensionless coordinates and the function

$$\hat{x} = \frac{x}{L_*}, \quad \hat{t} = \frac{t}{T_*}, \quad \hat{C}(\hat{x},\hat{t}) = \frac{C(x,t)}{C_*}.$$

Making these changes in variables, Equation (10.36) becomes

$$\frac{\partial \hat{C}}{\partial \hat{t}} = \frac{T_*}{L_*^2}\mathcal{D}\frac{\partial^2 \hat{C}}{\partial \hat{x}^2}.$$

If we now choose $L_*^2 = \mathcal{D}T_*$, the equation becomes (dropping the hats on the variables)

$$\frac{\partial C}{\partial t} = \frac{\partial^2 C}{\partial x^2}. \tag{10.37}$$

To solve this equation using separation of variables, we assume that we can find a solution of the form $C(t,x) = T(t)X(x)$. Substituting our hypothesized solution into the equation, dividing by $C(t,x)$, and rearranging we find that

$$\frac{1}{T(t)}\frac{dT(t)}{dt} = \frac{1}{X(x)}\frac{d^2X(x)}{dx^2}. \tag{10.38}$$

We have been able to rearrange the equation such that all terms depending on time t are on one side of the equality, and all terms that depend on the spatial variable x are on the other We would not have been able to make this separation if the equation had been nonlinear. The *only* way that Equation (10.38) can be true is if both sides of the equation are equal to a constant and we can write Equation (10.38) as two ODEs,[6]

$$\frac{1}{T(t)}\frac{dT(t)}{dt} = -\lambda^2, \qquad \frac{1}{X(x)}\frac{d^2X(x)}{dx^2} = -\lambda^2. \tag{10.39}$$

The solutions to these equations depend on the value of λ. For the moment, let us assume that $\lambda > 0$, then the solution to Equation (10.39) is

$$T(t) = A\exp(-\lambda^2 t), \qquad X(t) = B_1\sin(\lambda x) + B_2\cos(\lambda x), \tag{10.40}$$

where A, B_1, and B_2 are constants of integration. We can combine the equations in Equation (10.40) to write the general solution as

$$C(t,x) = \mathrm{e}^{-\lambda^2 t}\left(C_1\sin(\lambda x) + C_2\cos(\lambda x)\right), \tag{10.41}$$

where we have combined the constants so that $C_1 = AB_1$ and $C_2 = AB_2$, just for the sake of convenience.

We have not specified the boundary or initial conditions yet. Let us assume that the spatial domain of our problem extends from $x = 0$ to $x = L$ and impose Dirichlet boundary conditions $C(0,t) = C(L,t) - 0$. We can substitute these conditions into Equation (10.41) and realizing that these boundary conditions hold for all values of t, we get the two conditions $C_2\cos(\lambda 0) = C_2 = 0$ and $C_1\sin(\lambda L) = 0$. To satisfy the second equation we have to have either $C_1 = 0$, in which case $C(t,x) = 0$ everywhere in our domain which is not very interesting, or $\sin(\lambda L) = 0$, which means that $\lambda_n L = n\pi$ where n is an integer, so that the solution of the equation is

$$C(t,x) = \sum_{n=1}^{\infty} C_n \mathrm{e}^{-\lambda_n^2 t}\sin(\lambda_n x), \qquad \lambda_n = \frac{n\pi}{L}, \tag{10.42}$$

where C_n are constants. This should look familiar—the solution is basically a Fourier series, and we have essentially used Sturm–Liouville theory to find a set of eigenfunctions that satisfy the equation. As a result of this we can calculate the values of C_n using the orthogonality property of the eigenfunctions:

$$C_n = \frac{\int_0^L C(0,x)\sin(\lambda_n x)\,dx}{\int_0^L \sin^2(\lambda_n x)\,dx} = \frac{2}{L}\int_0^L C(0,x)\sin(\lambda_n x)\,dx. \tag{10.43}$$

Exercise 10.4.8 What would happen if, in Equation (10.39), we had $\lambda = 0$ or $\lambda < 0$? Do the resulting solutions make physical sense?

[6] You might ask why we have used $-\lambda^2$ as the constant rather than the simpler λ. The reason is that the X equation is a second order equation and we know that the characteristic equation of this will be a quadratic, so we will end up taking the square root of any constant that appears. So it is partly laziness—we do not have to keep writing square root signs everywhere! The sign of the constant will be determined by the boundary and initial conditions.

The problem we have just solved had homogeneous Dirichlet boundary conditions, but what if the boundary conditions are not homogeneous and we have $C(0, t) = \alpha$ and $C(L, t) = \beta$, where α and β are nonzero constants. In cases like this, we make use of the linearity of the PDE and write the solution in the form

$$C(x, t) = v(x, t) + w(x),$$

so that the heat equation becomes

$$v_t = \mathcal{D}(v_{xx} + w_{xx}),$$

which can be satisfied if v satisfies the original heat equation and w satisfies $w_{xx} = 0$. We then set $v(x, t)$ to satisfy the homogeneous boundary conditions $C(0, t) = 0$ and $C(L, t) = 0$ so that we can solve for $v(x, t)$ using separation of variables as above, and w has to satisfy the inhomogeneous boundary conditions $w(0) = \alpha$, $w(L) = \beta$. The initial condition for v is then given by $v(x, 0) = C(x, 0) - w(x)$.

It is important to realize that separation of variables does not necessarily give the only solution to the PDE because we have made a specific assumption about the form of the solution that other solutions might not obey. For example, the solution

$$C(x, t) = \frac{\alpha}{\sqrt{t}} \exp\left(-\frac{x^2}{\gamma t}\right) \tag{10.44}$$

cannot be written as $C(x, t) = X(x)T(t)$, but it is still a solution to the heat equation.

Exercise 10.4.9 Show by direct substitution that Equation (10.44) is a solution to the one-dimensional heat equation in Cartesian coordinates.

Can we apply the same techniques for solving equations in higher dimensions? Yes, though the calculations become more involved. For example, let us apply the method of separation of variables to the diffusion equation in spherical polar coordinates. If $C(t, r, \theta, \phi)$ is a concentration, where r is the radius and θ and ϕ are the usual polar angles, then the diffusion equation becomes

$$\frac{\partial C}{\partial t} = \nabla^2 C = \frac{1}{r^2}\frac{\partial}{\partial r}\left(r^2 \frac{\partial C}{\partial r}\right) + \frac{1}{r^2 \sin(\theta)}\frac{\partial}{\partial \theta}\left(\sin(\theta)\frac{\partial C}{\partial \theta}\right) + \frac{1}{r^2 \sin^2(\theta)}\frac{\partial^2 C}{\partial \phi^2}, \tag{10.45}$$

which appears to be quite a formidable equation. With a deep breath to calm our nerves, we will assume a solution of the form

$$C(t, r, \theta, \phi) = T(t)R(r)\Theta(\theta)\Phi(\phi),$$

and substituting this into Equation (10.45) and rearranging gives

$$\frac{1}{T}\frac{\partial T}{\partial t} = \frac{1}{Rr^2}\frac{\partial}{\partial r}\left(r^2 \frac{\partial R}{\partial r}\right) + \frac{1}{\Theta r^2 \sin(\theta)}\frac{\partial}{\partial \theta}\left(\sin(\theta)\frac{\partial \Theta}{\partial \theta}\right) + \frac{1}{\Phi r^2 \sin^2(\theta)}\frac{\partial^2 \Phi}{\partial \phi^2} = -\lambda^2. \tag{10.46}$$

The left-hand side of the equation is now a function of time, and the right-hand side is a function of r, θ, and ϕ. As before, we argue that the only way this can be true is if both sides of the equation are constant ($-\lambda^2$), so that

$$\frac{1}{T}\frac{\partial T}{\partial t} = -\lambda^2 t, \quad \text{i.e.,} \quad T(t) = e^{-\lambda^2 t}, \tag{10.47}$$

where we have forced the constant to be negative so that we have a decaying exponential in the solution. If $\lambda = 0$, the solution has no time dependence, and if we picked the constant to be $+\lambda^2$, the solution would grow without bound as $t \to \infty$. Substituting this back into Equation (10.46) we can separate out the ϕ-dependent terms to give

$$\sin(\theta)\left\{\frac{1}{R}\frac{\partial}{\partial r}\left(r^2\frac{\partial R}{\partial r}\right) + \frac{1}{\Theta\sin(\theta)}\frac{\partial}{\partial\theta}\left(\sin(\theta)\frac{\partial\Theta}{\partial\theta}\right) + \lambda^2 r^2\right\} = -\frac{1}{\Phi}\frac{\partial^2\Phi}{\partial\phi^2} = m^2,$$

where m is a constant. This gives another eigenvalue equation for Φ, which we can solve to get

$$\Phi(\phi) = A_1\sin(m\phi) + A_2\cos(m\phi), \tag{10.48}$$

where we require m to be an integer.

Exercise 10.4.10 Why must m be an integer in Equation (10.48)? (Use the fact that ϕ is an angular coordinate on the surface of a sphere.)

When we substitute Equation (10.48) back into Equation (10.45) we can separate out the r and θ equations to give

$$\frac{1}{R}\frac{d}{dr}\left(r^2\frac{dR}{dr}\right) + \lambda^2 r^2 = -\frac{1}{\Theta}\frac{1}{\sin(\theta)}\frac{d}{d\theta}\left(\sin(\theta)\frac{d\Theta}{d\theta}\right) + \frac{m^2}{\sin^2(\theta)} = k \tag{10.49}$$

where k is a constant. Let us tackle the θ equation first, which we will rewrite as

$$\frac{d^2\Theta}{d\theta^2} + \frac{\cos(\theta)}{\sin(\theta)}\frac{d\Theta}{d\theta} - \frac{m2}{\sin^2(\theta)}\Theta + k\Theta = 0.$$

We now make the substitution $x = \cos(\theta)$ so that the θ equation becomes

$$(1 - x^2)\frac{d^2\Theta}{d\theta^2} - 2x\frac{d\Theta}{d\theta} + \left(k - \frac{m^2}{1 - x^2}\right)\Theta = 0.$$

This equation is an associated Legendre equation (see Section 8.5.1) and has two linearly independent solutions that are the associated Legendre polynomials. It is convention to write $k = n(n+1)$, and the associated Legendre polynomials are then written as $P_n^m(x)$ and $Q_n^m(x)$, though usually only $P_n^m(x)$ are considered because the polynomials $Q_n^m(x)$ diverge as $x \to \pm 1$. In addition, we have to restrict the value of m because the polynomials diverge if $m > n$. So, our solution for Θ can be written

$$\Theta(\theta) = BP_n^m(\cos(\theta)), \qquad 0 \le m \le n. \tag{10.50}$$

Lastly, we have to deal with the radial equation

$$\frac{d^2R}{dr^2} + \frac{2}{r}\frac{dR}{dr} + \left(\lambda^2 - \frac{n(n+1)}{r^2}\right) = 0.$$

This equation is a generalization of the Bessel equation (Section 8.6) called the *spherical Bessel equation*. It has two linearly independent solutions $j_n(x)$ and $y_n(x)$ that can be written in terms of the Bessel functions $J_n(x)$ and $Y_n(x)$:

$$j_n(x) = \sqrt{\frac{\pi}{2x}}J_{n+1/2}(x), \qquad y_n(x) = \sqrt{\frac{\pi}{2x}}Y_{n+1/2}(x).$$

The solution to the radial equation is then

$$R(r) = C_1 j_n(\lambda r) + C_2 y_n(\lambda r), \tag{10.51}$$

where C_1 and C_2 are constants and n is an integer. We can now write the solution to Equation (10.45) as

$$C(t, r, \theta, \phi) = \sum_{n=0}^{\infty}\sum_{m=0}^{n} \alpha_{mn}(C_1 j_n(\lambda r) + C_2 y_n(\lambda r) P_n^m(\cos(\theta))[A_1 \sin(m\phi) + A_2 \cos(m\phi)]\mathrm{e}^{-\lambda^2 t},$$

where α_{mn} are constants and the constant B has been absorbed into the other constants. The values of these constants can be found from the boundary and initial conditions of the problem. For example, the functions $y_n(x)$ are infinite at $x = 0$, so $C_2 = 0$ for almost all problems we might consider. If the problem we are considering does not depend on the azimuthal angle ϕ, then the solution simplifies to

$$C(t, r, \theta, \phi) = \sum_{n=0}^{\infty}\sum_{m=0}^{n} \alpha_{mn} j_n(\lambda r) P_n^m(\cos(\theta))\mathrm{e}^{-\lambda^2 t}.$$

What happens if we add a source or a sink to the diffusion equation Equation (10.28)? In other words, how do we use separation of variables to solve an inhomogeneous diffusion equation

$$\frac{\partial \phi}{\partial t} - \kappa \nabla^2 \phi = f(t, \mathbf{x}),$$

where $f(t, \mathbf{x})$ is a given function? Recall that when we used separation of variables to find the solution (Equation (10.42)) to the homogeneous equation in Cartesian coordinates, the solution looked suspiciously like a Fourier series. That gives us a clue as to how to proceed. If we can write $f(t, \mathbf{x})$ as a Fourier series, perhaps we will be able to solve the inhomogeneous problem.

Example 10.6 We will use separation of variables to solve the following problem

$$\phi_t - \phi_{xx} = f(x, t), \quad 0 \le x \le L,\ t > 0,$$

with the conditions $\phi(0, t) = \phi(L, t) = 0$ and $\phi(x, 0) = h(x)$. This problem consists of both an inhomogeneous diffusion equation and an inhomogeneous initial condition. We start the solution by making our usual assumption, $\phi(x, t) = X(x)T(t)$. We know that this gives us the following eigenvalue problem for X:

$$\frac{d^2 X}{dx^2} + \lambda X = 0, \quad X(0) = X(L) = 0,$$

with eigenvalues λ and eigenvector $X_n(x)$ given by

$$\lambda = \left(\frac{n\pi}{L}\right)^2, \qquad X_n(x) = \sin\left(\frac{n\pi x}{L}\right).$$

If we now write

$$\phi(x, t) = \sum_{n \ge 1} T_n(t) \sin\left(\frac{n\pi x}{L}\right),$$

we have something that looks like a Fourier series, but we do not know the functions $T_n(t)$. We can differentiate this equation to get

$$\dot{\phi} = \sum_{n\geq 1} \dot{T}_n(t) \sin\left(\frac{n\pi x}{L}\right), \qquad \phi''(x) = -\sum_{n\geq 1} T_n(t) \left(\frac{n\pi}{L}\right)^2) \sin\left(\frac{n\pi x}{L}\right),$$

and substituting into the PDE we get

$$\phi_t - \phi_{xx} = \sum_{n\geq 1} \left\{ \dot{T}_n(t) + T_n(t)\left(\frac{n\pi}{L}\right)^2 \right\} \sin\left(\frac{n\pi x}{L}\right) = f(x,t).$$

Now we expand both $f(x,t)$ and $h(x)$ as the Fourier sine series

$$f(x,t) = \sum_{n\geq 1} f_n(t) \sin\left(\frac{n\pi x}{L}\right), \qquad f_n(t) = \frac{2}{L}\int_0^L f(x,t)\sin\left(\frac{n\pi x}{L}\right) dx$$

and

$$h(x) = \sum_{n\geq 1} h_n \sin\left(\frac{n\pi x}{L}\right), \qquad h_n = \frac{2}{L}\int_0^L h(x)\sin\left(\frac{n\pi x}{L}\right) dx.$$

The inhomogeneous PDE now becomes

$$\sum_{n\geq 1} \left\{ \dot{T}_n(t) + T_n(t)\left(\frac{n\pi}{L}\right)^2 \right\} \sin\left(\frac{n\pi x}{L}\right) = \sum_{n\geq 1} f_n(t) \sin\left(\frac{n\pi x}{L}\right),$$

which implies that

$$\dot{T}_n(t) + T_n(t)\left(\frac{n\pi}{L}\right)^2 = f_n(t), \tag{10.52}$$

and the inhomogeneous initial condition becomes

$$\phi(x,0) = \sum_{n\geq 1} T_n(0) \sin\left(\frac{n\pi x}{L}\right) = \sum_{n\geq 1} h_n \sin\left(\frac{n\pi x}{L}\right),$$

which gives us the initial condition $T_n(0) = h_n$. We can now solve Equation (10.52) by calculating the integrating factor to give

$$T_n(t) = h_n \exp\left[-\left(\frac{n\pi}{L}\right)^2 t\right] + \int_0^t \exp\left[\left(\frac{n\pi}{L}\right)^2 (t-u)\right] f_n(u)\, du,$$

so we have the final solution

$$\phi(x,t) = \sum_{n\geq 1} h_n \exp\left[-\left(\frac{n\pi}{L}\right)^2 t\right] \sin\left(\frac{n\pi x}{L}\right) + \sum_{n\geq 1} \sin\left(\frac{n\pi x}{L}\right) \int_0^t \exp\left[\left(\frac{n\pi}{L}\right)^2 (t-u)\right] f_n(u)\, du.$$

We have seen that we can use the method of separation of variables to solve the heat equation, and in doing so we end up with BVPs that involve Sturm–Liouville problems. The solutions to these equations can be written in terms of Fourier series and polynomial functions. What about the other forms of PDE?

10.5 Hyperbolic Equations: Wave Equation

We claimed in Section 10.3.1 that hyperbolic equations described wave phenomena. We can demonstrate this from first principles. There are many different ways to derive the wave equation depending on what exactly is doing the waving; water waves, waves on a rope, electromagnetic waves. All of these are completely legitimate and use the physics of the particular wave to obtain the wave equation. We are going to take a different, more geometric approach. To start with, let us ask ourselves the question "What is a wave?" Consider the wave shown in Figure 10.4. To make the example a little more concrete, let us assume it represents the oscillation of a water surface as a wave passes. As the wave propagates with a velocity c, the surface of the water undergoes the same oscillation at different locations, only the amplitude of the oscillations at the two locations are shifted in time. This means that the shape of the wave remains the same, it just gets shifted along the x-direction. If λ is the wavelength and ν the frequency of the wave, then $c = \lambda\nu$; and if $h(x,t)$ represents the height of the displacement of the water surface from its mean level, then h must be a function of $x - ct$ so that the wave pattern repeats itself with a wavelength of λ. A small change (Δx) in position is then equivalent to a shift $c\Delta t$ in the function $h(x,t)$, so

$$\frac{\Delta h}{\Delta x} = \frac{1}{c}\frac{\Delta f}{\Delta t} \implies \frac{dh}{dx} = \frac{1}{c}\frac{dh}{dt},$$

which then gives us the wave equation

$$\frac{d^2h}{dt^2} = c^2\frac{d^2h}{dx^2}. \tag{10.53}$$

If we are working in more than one dimension, then the wave equation becomes

$$\frac{d^2h}{dt^2} = c^2\nabla^2 h.$$

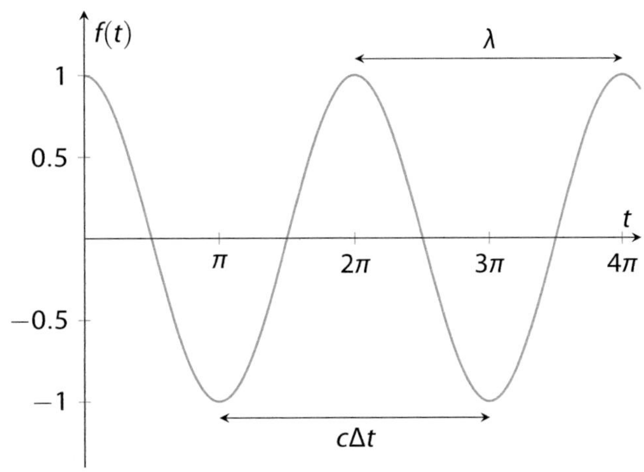

Figure 10.4 A wave with wavelength $\lambda = c\Delta t$, where c is the velocity of the wave and Δt is its period.

Exercise 10.5.1 Show that the wave equation, Equation (10.53), is a hyperbolic PDE.

Exercise 10.5.2 Show that we can nondimensionalize the wave equation to give

$$\frac{d^2h}{dt^2} = \frac{d^2h}{dx^2}.$$

We can solve the wave equation using separation of variables, just as we did with the heat equation. As an example, we will solve the equation in spherical polar coordinates:

$$\frac{d^2h}{dt^2} = c^2\nabla^2 h = \frac{1}{r^2}\frac{\partial}{\partial r}\left(r^2\frac{\partial h}{\partial r}\right) + \frac{1}{r^2\sin(\theta)}\frac{\partial}{\partial\theta}\left(\sin(\theta)\frac{\partial h}{\partial\theta}\right) + \frac{1}{r^2\sin^2(\theta)}\frac{\partial^2 h}{\partial\phi^2}. \quad (10.54)$$

We make the assumption that the solution can be written as $h(t, r, \theta, \phi) = T(t)R(r)\Theta(\theta)\Phi(\phi)$ so that Equation (10.54) becomes

$$\frac{1}{c^2T}\frac{d^2T}{dt^2} = \frac{1}{Rr^2}\frac{d}{dr}\left(r^2\frac{dR}{dr}\right) + \frac{1}{\Theta r^2\sin(\theta)}\frac{d}{d\theta}\left(\sin(\theta)\frac{d\Theta}{d\theta}\right) + \frac{1}{\Phi r^2\sin^2(\theta)}\frac{d^2\Phi}{d\phi^2} = -k^2, \quad (10.55)$$

where k is a constant. The equation for $T(t)$ is the equation of simple harmonic motion,

$$\frac{d^2T}{dt^2} = -c^2k^2T(t) = -\omega_k^2T(t),$$

which has solutions $T(t) = A_k\cos(\omega_k t) + B_k\sin(\omega_k t)$, where A_k and B_k are constants. We can rearrange the rest of Equation (10.55) to get

$$\frac{1}{R}\frac{d}{dr}\left(r^2\frac{dR}{dr}\right) + k^2r^2 = -\frac{1}{\Theta\sin(\theta)}\frac{d}{d\theta}\left(\sin(\theta)\frac{d\Theta}{d\theta}\right) - \frac{1}{\Phi\sin^2(\theta)}\frac{d^2\Phi}{d\phi^2} = \lambda,$$

where λ is a constant. The equation for the r dependence then becomes

$$\frac{d^2R}{dr^2} + \frac{2}{r}\frac{dR}{dr} + \left(k^2 - \frac{\lambda}{r^2}\right)R = 0,$$

which, making the substitution $s = kr$, becomes

$$s^2\frac{d^2R}{ds^2} + 2s\frac{dR}{ds} + (s^2 - \lambda)R = 0,$$

which is the spherical Bessel equation and has solutions given by Equation (10.51).

Now let us look at the angular dependence of the solution:

$$\frac{\sin(\theta)}{\Theta}\frac{d}{d\theta}\left(\sin(\theta)\frac{d\Theta}{d\theta}\right) + \lambda\sin^2(\theta) - \frac{1}{\Phi}\frac{d2}{d\Phi}\phi = m^2. \quad (10.56)$$

The ϕ-equation has solutions that we can write in terms of a complex exponential, $\Phi(\phi) = e^{\pm im\phi}$. But just as with the case of the diffusion equation, ϕ is an angular coordinate and we do not want to have any discontinuities as ϕ goes from 2π to 0. In fact, we would like the solutions to be the same if we change ϕ to $\phi + 2\pi$, otherwise we would see a different solution at the same location each time we completed a full circuit of the sphere. This means that m must be an integer so that $\sin(m\phi) = \sin(m(\phi + 2\pi))$.

Lastly, we will look at the equation for the θ-dependence:

$$\frac{1}{\sin(\theta)}\frac{d}{d\theta}\left(\sin(\theta)\frac{d\Theta}{d\theta}\right) + \left(\lambda - \frac{m^2}{\sin^2(\theta)}\right)\Theta = 0.$$

If we make the substitution $x = \cos(\theta)$, we can transform the equation to the associated Legendre equation if $\lambda = l(l+1)$, where the same conditions on the values of m and l apply as with the diffusion equation. The separable solution to the wave equation is then

$$h = \sum_{l=0}^{\infty}\sum_{m=0}^{l}(A_{kl}j_l(kr)+B_{kl}y_l(kr))C_{ml}P_l^m(\cos(\theta))(D_m e^{im\phi}+E_m e^{-im\phi})(F_k e^{ikct}+G_k e^{-ikct}), \tag{10.57}$$

where A_{kl}, B_{kl}, C_{ml}, D_m, E_m, F_k, and G_k are constants that need to be determined from the boundary and initial conditions.

10.6 Elliptic Equations: Laplace's Equation

Two common examples of elliptic PDEs are *Laplace's equation*,[7]

$$\nabla^2\phi = 0, \tag{10.58}$$

and *Poisson's Equation*,[8]

$$\nabla^2\phi = \rho, \tag{10.59}$$

where ρ is a function of the spatial coordinates.

Exercise 10.6.1 Show that both Laplace's equation and Poisson's equation are elliptic equations.

You will not be surprised to learn that one method of solving Laplace's equation is to use separation of variables.

Example 10.7 Let us solve Laplace's equation for the function $\phi(x, y)$ in two dimensions using Cartesian coordinates (x, y) on a square $0 \le x \le 1$, $0 \le y \le 1$ with the boundary conditions $\phi(x, 0) = \phi(0, y) = \phi(1, y) = 0$ and $\phi(x, 1) = A\sin(n\pi x)$. We make our usual assumption that the solution can be written $\phi(x, y) = X(x)Y(y)$, and substituting this into Equation (10.58) gives us two ODEs:

$$\frac{d^2X}{dx^2} + \kappa^2 X = 0, \qquad \frac{d^2Y}{dy^2} - \kappa^2 Y = 0,$$

which have solutions

$$X(x) = a\sin(\kappa x) + b\cos(\kappa x), \qquad Y(y) = u\sinh(\kappa y) + v\cosh(\kappa y),$$

where a, b. u, and v are constants that have to be determined from the boundary conditions. The boundary condition $\phi(0, y) = 0$ implies that $b = 0$, and the condition $\phi(1, y) = 0$ tells

[7] Named for Pierre-Simon Laplace, who developed (among many other things) a dynamic theory of the tides that gives a better description of the tides than Newton's static theory, as well as the foundations of inductive reasoning that form the foundations of Bayesian probability.

[8] Named for French mathematician, engineer, and physicist Siméon Denis Poisson (1781–1840).

us that $a\sin(\kappa x) = 0$, which, so long as $a \neq 0$, implies that $\kappa = k\pi$, where k is an integer. The boundary condition $\phi(x, 0) = 0$ tells us that $v = 0$, so the solution reduces to

$$\phi(x, y) = a\sin(k\pi x) \times v\sinh(k\pi y).$$

The last boundary condition tells us that $a\sin(k\pi) \times v\sinh(k\pi) = A\sin(n\pi x)$, so $k = n$ and $v\sinh(k\pi) = A$, which tells us v in terms of A. We can write the final solution as

$$\phi(x, y) = \left(\frac{A}{\sinh(n\pi)}\right)\sin(n\pi x)\sinh(n\pi y).$$

10.7 More Laplace Transforms

Another useful technique for solving PDEs is to use Laplace transforms. In Section 9.3 we saw that taking the Laplace transform of an ODE could produce an algebraic equation that was easier to solve. Likewise, taking the Laplace transform of a PDE can lead to an ODE that is easier to solve. For example, let us use this technique to solve the equation

$$\frac{\partial g}{\partial t} + \frac{\partial g}{\partial x} + g(x, t) = 0, \qquad g(x, 0) = \sin(x),\ g(0, t) = 0, \tag{10.60}$$

for $x > 0$ and $t > 0$. If we take the Laplace transform of this equation, and using the fact that the Laplace transform is a linear operation, we get

$$\mathcal{L}\left(\frac{\partial g}{\partial t}\right) + \mathcal{L}\left(\frac{\partial g}{\partial x}\right) + \mathcal{L}(g) = 0, \quad \text{i.e.,} \quad sG(x, s) - g(x, 0) + \mathcal{L}\left(\frac{\partial g}{\partial x}\right) + G(x, s) = 0,$$

where $G(x, s)$ is the Laplace transform of $g(x, t)$. We can evaluate the remaining Laplace transform if we recall the definition of the Laplace transform:

$$\mathcal{L}\left(\frac{\partial g}{\partial x}\right) = \int_0^\infty \left(\frac{\partial g}{\partial x}\right) e^{-st}\, dt = \frac{\partial}{\partial x}\int_0^\infty g(x, t)e^{-st}\, dt = \frac{\partial G}{\partial x},$$

so, using the boundary conditions, we end up with the differential equation

$$\frac{\partial G}{\partial x} + (s + 1)G(x, s) = \sin(x).$$

This is a, first order linear differential equation, so we can solve it by calculating the integrating factor to give

$$G(x, s) = e^{-(s+1)x}\int_0^x e^{-(s+1)v}\sin(v)\, dv + ce^{-(s+1)x}. \tag{10.61}$$

Exercise 10.7.1 Show that the integral in Equation (10.61) is

$$\int_0^x e^{-(s+1)v}\sin(v)\, dv = e^{(s+1)x}\left(\frac{(s + 1)\sin(x) - \cos(x)}{(s + 1)^2 + 1}\right).$$

The Laplace transform of the solution $g(x, t)$ is then

$$G(x, s) = \left(\frac{(s + 1)\sin(x) - \cos(x)}{(s + 1)^2 + 1}\right) + ce^{-(s+1)x},$$

and using the table of Laplace transforms in Appendix B we get

$$g(x, t) = e^{-t}[\sin(x)\cos(t) - \cos(x)\sin(t) + H(t - x)\sin(t - x)] = e^{-t}(1 + H(t - x))\sin(x - t),$$

where $H(t - x)$ is the Heaviside function. This discontinuity in the solution would have caused some problems if we had tried to use separation of variables, and this is where transform techniques can be very helpful. Of course, the use of any transform technique relies on being able to calculate the inverse transforms, which may not always be easy or possible.

We can use Laplace transforms to find some very useful and very general solutions to the heat equation. Let us consider the one-dimensional diffusion equation with $-\infty < x < \infty$ and $t > 0$ (recall that the latter condition is needed for the Laplace transform to be finite):

$$\frac{\partial g}{\partial t} = \frac{\partial^2 g}{\partial x^2}, \tag{10.62}$$

where $g = g(x, t)$. Now, rather than give a specific initial condition, we are going to be very general and set $g(x, 0) = h(x)$, and we are going to also require that the solution is finite for all values of t and x.

If we take the Laplace transform of Equation (10.62), we end up with a second order linear inhomogeneous differential equation,

$$\frac{\partial^2 G}{\partial x^2} - sG(x, s) = -h(x),$$

where $G(x, s)$ is the Laplace transform of $g(x, t)$. We know from Chapter 6 that we can divide the solution to this equation into two parts: one (G_h) that solves the corresponding homogeneous equation, and a particular solution (G_p). The solution to the homogeneous equation is

$$G_h(x, s) = \alpha_1 e^{\sqrt{s}x} + \alpha_2 e^{-\sqrt{s}x},$$

but we have an apparent problem with the particular solution because we have not specified the function $h(x)$. However, we can find a formal solution using the Wronskian in a manner similar to that in Section 6.4.1.2. The solution to the homogeneous equation gives us two linearly independent solutions of the homogeneous equation ($e^{\pm\sqrt{s}x}$). The Wronskian of these solutions is $W = -2\sqrt{s}$, so using Equation (6.68) we can calculate the particular solution as

$$\frac{e^{-\sqrt{2}x}}{2\sqrt{s}} \int_0^x e^{\sqrt{s}v} h(v)\, dv - \frac{e^{\sqrt{2}x}}{2\sqrt{s}} \int_0^x e^{-\sqrt{s}v} h(v)\, dv,$$

although we cannot evaluate the integrals until we know the function $h(x)$. The general solution for $G(x, s)$ is then

$$G(x,s) = G_p(x,s) + G_h(x,s) = e^{\sqrt{s}x}\left\{\alpha 1 - \frac{1}{2\sqrt{s}}\int_0^x e^{-\sqrt{s}v} h(v)\,dv\right\} + e^{-\sqrt{s}x}\left\{\alpha_2 + \frac{1}{2\sqrt{s}}\int_0^x e^{\sqrt{s}v} h(v)\,dv\right\}, \tag{10.63}$$

where α_1 and α_2 are constants. This does not look to be very helpful, but we have not made use of the fact that we want the solutions to be finite for all values of x, irrespective of what $h(x)$ is. This condition implies that the Laplace transform of $g(x, t)$ must also remain finite for all values of x and for all values of $s > 0$. Now, $G(x, s)$ is the sum of two terms, and for $G(x, s)$ to remain finite both of these terms must also remain finite. The first bracketed term in Equation (10.63) is premultiplied by $e^{\sqrt{s}x}$, and so for the whole first term to remain finite, the first bracketed term must tend to zero as $x \to +\infty$. The term α_1 is a constant, so the first term in Equation (10.63) tells us that

$$\lim_{x\to\infty}\left(\alpha 1 - \frac{1}{2\sqrt{s}}\int_0^x e^{-\sqrt{s}v} h(v)\,dv\right) = 0 \implies \alpha_1 = \frac{1}{2\sqrt{s}}\int_0^\infty e^{-\sqrt{s}v} h(v)\,dv,$$

and by a similar argument, the second term in Equation (10.63) gives us the value of α_2 as

$$\alpha_2 = \frac{1}{2\sqrt{s}}\int_{-\infty}^0 e^{\sqrt{s}v} h(v)\,dv.$$

Substituting these values back into $G(x, s)$ and rearranging gives

$$G(x,s) = e^{-\sqrt{s}x}\int_{-\infty}^x \frac{e^{\sqrt{s}v}}{2\sqrt{s}} h(v)\,dv + e^{\sqrt{s}x}\int_x^\infty \frac{e^{-\sqrt{s}v}}{2\sqrt{s}} h(v)\,dv = \int_{-\infty}^\infty \frac{h(v)}{2\sqrt{s}} \exp(-\sqrt{s}|x-v|)\,dv. \tag{10.64}$$

This is rather nice because it tells us that no matter what $h(x)$ is, the Laplace transform of the solution to the heat equation is given by the integral in the Equation (10.64). Can we get any further? We would like to have an equation for the solution $g(x, t)$ rather than its Laplace transform, i.e.,

$$g(x,t) = \mathcal{L}^{-1}(G(x,s)) = \mathcal{L}^{-1}\left(\int_{-\infty}^\infty \frac{h(v)}{2\sqrt{s}} \exp(-\sqrt{s}|x-v|)\,dv\right). \tag{10.65}$$

Let us think a little a bit about Equation (10.65). First, the inverse Laplace transform is going to do its work on s, so a function such as $h(v)$ will be untouched. This means that we do not have to worry about finding the inverse Laplace transform of the unknown function $h(v)$, instead we need

$$\mathcal{L}^{-1}\left(\frac{1}{2\sqrt{s}}\exp(-\sqrt{s}|x-v|)\right) = \frac{1}{\sqrt{4\pi t}}\exp\left(\frac{-|x-v|^2}{4t}\right),$$

where we have used our table of Laplace transforms. So, now we can write down the solution to heat equation as

$$g(x,t) = \frac{1}{\sqrt{4\pi t}}\int_{-\infty}^\infty \frac{h(v)}{\exp}\left(\frac{-|x-v|^2}{4t}\right) dv. \tag{10.66}$$

How does this help us? Equation (10.66) is telling us that if we have a one-dimensional heat equation defined on $-\infty < x < \infty$ with a solution that remains finite over this range and for all $t > 0$, and with an initial condition $g(x, 0) = h(x)$, then to obtain the solution we need to evaluate the integral in Equation (10.66). You will sometimes see the function

$$K(x,t) = \frac{1}{\sqrt{4\pi t}} \exp\left(-\frac{x^2}{4t}\right)$$

referred to as the *kernel* of the heat equation or the *fundamental solution*, and it is essentially the same as the concept of a Green's function (Section 6.12.1). In the limit of $t \to 0$ the heat kernel looks suspiciously like the Gaussian approximation to the Dirac delta function (Equation (8.5)), and indeed it is. As t increases from zero, $K(x, t)$ acts as a Gaussian distribution whose variance increases with time. So, a delta function spike at $t = 0$ spreads out over the x axis from $x = -\infty$ to $x = +\infty$ as t increases—this is just what a diffusive process does. Let us rewrite Equation (10.66) using the heat kernel,

$$g(x,t) = \int_{-\infty}^{\infty} K(|x-v|, t) h(v)\, dv,$$

which we recognize as a *convolution* (Equation (5.53)). So, the solution of the heat equation is the convolution of the initial condition with the heat kernel and describes how that initial condition gets spread out along the x axis as t increases.

10.8 Numerical Methods

Most PDEs cannot be solved analytically, so we have to resort to numerical methods. This is especially true when we consider computer models for the dynamics of the oceans and atmosphere, or of rivers and lakes, where we might have many coupled PDEs; for example, for an ocean model we may have equations for the three components of fluid velocity, temperature, and salinity, as well as equations for dissolved oxygen, carbon dioxide, and so on. As you can see, the numbers of equations such models deal with rapidly becomes very large as new variables are added. However, as with ODEs, we should not jump to the computer straight away, but instead we should try to learn something about our equations—we might be able to use some simplifying assumptions and solve the resulting equations analytically. This is always a useful thing to do because, at the very least, such analytical solutions provide test cases for our computer codes to make sure we have not made any programming errors.

There are many ways of approaching the numerical solution of PDEs, and in this chapter we will focus on one family of methods, *finite difference techniques*. These are simple to implement, but for complicated problems may not be the best techniques. The basic idea is to turn our continuous PDE into a set of discrete equations, where the solution is evaluated on a uniform grid of points (Figure 10.5). Let us start by considering a derivative with respect to t and recall the definition of a partial derivative (Equation (2.20))

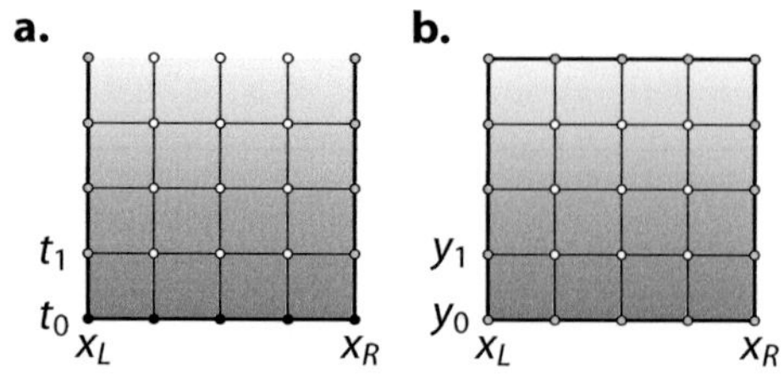

Figure 10.5 Finite difference methods involve finding the solution of the PDE on a set of grid points (white circles). (a.) Equations that involve time and space require initial conditions (black circles) at the grid points, where the initial time $t = t_0$, and boundary conditions (gray circles) for all values of time at the left and right boundaries of the spatial coordinate (x_L and x_R). (b.) Equations that involve only spatial derivatives require boundary conditions on all sides.

$$\frac{\partial u(x,t)}{\partial t} = \lim_{\Delta t \to 0} \frac{u(x,t+\Delta t) - u(x,t)}{\Delta t}.$$

If we discretize our domain into a grid with constant divisions (Δt and Δx) between the grid lines, then we can use this equation to approximate the derivative so that

$$\frac{\partial u(x,t)}{\partial t} \approx \frac{u(x,t+\Delta t) - u(x,t)}{\Delta t}. \tag{10.67}$$

This is called the *forward difference* because, given the value of the function at (x,t), we have to evaluate the function at a time (or location) forward of this, $(t + \Delta t)$ for example. Similarly, we could use a *backward difference* by

$$\frac{\partial u(x,t)}{\partial t} \approx \frac{u(x,t) - u(x,t-\Delta t)}{\Delta t}. \tag{10.68}$$

These finite difference equations are reminiscent of the Euler method for solving ODEs (Section 6.10.1), where we learned that we obtained better accuracy by decreasing the size of Δt. We can see why this should be so by recasting the finite difference formulae in terms of Taylor series:

$$u(x,t+\Delta t) \approx u(x,t) + \Delta t \frac{\partial u(x,t)}{\partial t} + \frac{(\Delta t)^2}{2}\frac{\partial^2 u(x,t)}{\partial t^2} + \frac{(\Delta t)^3}{2}\frac{\partial^3 u(x,t)}{\partial t^3} + O(\Delta t^4),$$
$$u(x,t-\Delta t) \approx u(x,t) - \Delta t \frac{\partial u(x,t)}{\partial t} + \frac{(\Delta t)^2}{2}\frac{\partial^2 u(x,t)}{\partial t^2} - \frac{(\Delta t)^3}{2}\frac{\partial^3 u(x,t)}{\partial t^3} + O(\Delta t^4).$$

By truncating these expressions to get the forward and backward difference equations we see that we introduce an error that is $O(\Delta t^2)$. However, if we subtract these two equations we see that the terms containing even powers of Δt cancel and we are left with

$$u(x,t+\Delta t) - u(x,t-\Delta t) \approx 2\Delta t \frac{\partial u(x,t)}{\partial t} + \frac{(2\Delta t)^3}{2}\frac{\partial^3 u(x,t)}{\partial t^3} + O(\Delta t^5),$$

which, if we use as an approximation for the derivative, incurs an error that is $O(\Delta t^3)$. So, instead of using Equation (10.67) or Equation (10.68) we can take *central differences*:

$$\frac{\partial u(x,t)}{\partial t} \approx \frac{u(x,t+\Delta t) - u(x,t-\Delta t)}{2\Delta t}. \tag{10.69}$$

Second order derivatives can be dealt with in the same way. If we add the two Taylor series and solve for the second derivative, we find

$$\frac{\partial^2 u(x,t)}{\partial t^2} \approx \frac{u(x,t+\Delta t) + u(x,t-\Delta t) + 2u(x,t)}{\Delta t^2}. \tag{10.70}$$

To use these finite difference equations we first have to set up a discrete grid that covers the region we are considering in our problem. For example, if the problem is defined on a spatial domain $0 \leq x \leq L$ and we want to integrate our equations from $t = 0$ to $t = T$, then we divide the domain into a grid with an x interval of $\Delta x = L/M$ and a time interval of $\Delta t = T/N$, where M and N are the number of grid cells in the x and t directions respectively (Figure 10.5). This produces a sequence of discrete values of x $(x_0, x_1, \ldots, x_j, \ldots x_M)$ and t $(t_0, t_1, \ldots, t_n, \ldots, t_N)$. It is conventional to represent the discrete values of the function $u(x,t)$ as u_j^n, where the subscript is the index for the spatial coordinate (x_j) and the superscript the index for the time coordinate (t_n); thus, the function values u_j^n is the value $u(x_j, t_n)$.

We can now use these finite difference schemes to build numerical methods for solving different PDEs. However, in doing so we will find that some methods are ill-advised for certain types of equations.

10.8.1 Advection Equation

We will start by developing finite difference approximations for the one-dimensional advection equation (Equation (10.4))

$$\frac{\partial c}{\partial t} + v\frac{\partial c}{\partial x} = 0, \tag{10.71}$$

where $c(x,t)$ could represent the concentration of a dissolved chemical and v the velocity of water. We have seen that the method of characteristics tells us that any function of the form $c(x,t) = f(x - vt)$ satisfies this equation. We want to find a finite difference approximation that will allow us to calculate the concentration at time t given the concentration at time $(t - \Delta t)$.

For our first method we will approximate the time derivative using forward differences (Equation (10.67)), which involve c^{n+1} and c^n. For the spatial derivative we could choose either the forward difference approximation again or choose the backward difference formula (Equation (10.68)). In this case, there is a natural choice depending on the sign of v in Equation (10.71). If $v > 0$, material that is at the spatial location x_j at $t_n + 1$ will have originated at x_{j-1} at the previous time step t_n (Figure 10.6), so it makes sense to choose the finite difference

$$\frac{\partial c}{\partial x} \approx \frac{c_j^n - c_{j-1}^n}{\Delta x}.$$

If $v < 0$, then material at x_j at t_{n+1} comes from location x_{j+1} at the previous time, so a natural finite difference scheme is

$$\frac{\partial c}{\partial x} \approx \frac{c_{j+1}^n - c_j^n}{\Delta x}.$$

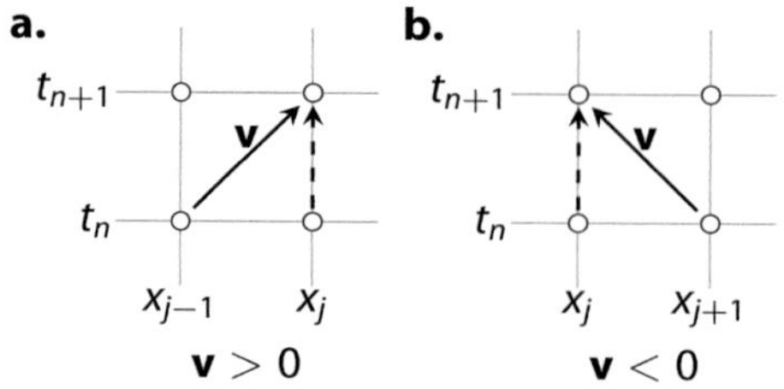

Figure 10.6 The ambiguity in choosing the forward or backward differencing scheme. If the velocity is positive, then it is natural to choose the forward differencing scheme (a.), but if the velocity is negative, then it is natural to choose the backward differencing scheme (b.). This ensures that information about the solution flows from upstream in both cases.

With these schemes, Equation (10.71) becomes

$$\frac{c_j^{n+1} - c_j^n}{\Delta t} + v\left(\frac{c_j^n - c_{j-1}^n}{\Delta x}\right) = 0, \qquad v > 0, \tag{10.72}$$

$$\frac{c_j^{n+1} - c_j^n}{\Delta t} + v\left(\frac{c_{j+1}^n - c_j^n}{\Delta x}\right) = 0, \qquad v < 0. \tag{10.73}$$

These equations allow us to calculate c_j^{n+1} in terms of quantities that we know at the previous time t_n, so long as we have specified initial conditions at $t = 0$. In this scheme, information is always propagating in the direction of the flow irrespective of the sign of v, so the scheme is called an *upwind scheme*. We can rearrange Equation (10.72) and Equation (10.73) to give

$$c_j^{n+1} = c_j^n - \frac{v\Delta t}{\Delta x}(c_j^n - c_{j-1}^n), \qquad v > 0, \tag{10.74}$$

$$c_j^{n+1} = c_j^n - \frac{v\Delta t}{\Delta x}(c_{j+1}^n - c_j^n), \qquad v < 0. \tag{10.75}$$

The upwind scheme is an explicit scheme because the right-hand sides of the equations contain only quantities that we know when we calculate c_j^{n+1}.

We have not specified any initial or boundary conditions yet. The initial conditions are just the specification of c_j^n for $n = 0$ for all values of j. For example, if the initial condition is $c(x, 0) = \sin(x)$, then the discretized initial condition is just the set of values of $\sin(x)$ evaluated at $x = x_j$ for $j = 0, \ldots, M$. The discretized boundary conditions will differ depending on the type of boundary conditions specified for the problem. In this example, we will consider periodic boundary conditions. So, we want the values of the function to be the same as the two spatial boundaries; i.e., we require $c_0^n = c_M^n$. Now, let us write out the equations for calculating the solution at time step t_{n+1} from the solution at time step t_n:

$$\begin{aligned}
c_1^{n+1} &= (1-\alpha)c_1^n + \alpha c_M^n \\
c_2^{n+1} &= (1-\alpha)c_2^n + \alpha c_1^n \\
c_3^{n+1} &= (1-\alpha)c_3^n + \alpha c_2^n \\
\vdots &= \quad \vdots \quad + \quad \vdots \\
c_M^{n+1} &= (1-\alpha)c_M^n + \alpha c_{M-1}^n,
\end{aligned}$$

where we have used the periodic boundary condition in the first line and the constant $\alpha = (v\Delta t)/(\Delta x)$. This is just a large number of linear equations that we can write in matrix form $\mathbf{c}^{n+1} = \mathbf{A}\mathbf{c}^n$, where $\mathbf{A}$ is the constant matrix

$$\mathbf{A} = \begin{pmatrix} (1-\alpha) & 0 & 0 & \cdots & 0 & \alpha \\ \alpha & (1-\alpha) & 0\cdots & 0 & 0 & \\ \vdots & \vdots & \vdots & \vdots & \vdots \\ 0 & 0 & 0 & \cdots & \alpha & (1-\alpha) \end{pmatrix}.$$

This means that we can use what we learned in Chapter 4 to solve these equations. An important thing to recognize about the matrix $\mathbf{A}$ is that many of its entries are zero. Such a matrix is called a *sparse matrix*, and there are many highly efficient numerical programs that can be employed for solving them (see, e.g., Press et al., 1992).

Recall that when we looked at numerical methods for solving ODEs we had to be concerned about the stability of the method. That is, we needed to know if small errors in the numerical solution would decay away or grow. We have the same concerns here, but determining the stability of the method is a bit harder. To examine the stability of the numerical scheme we will look at wavelike solutions of the numerical scheme, and if the amplitude of these waves grows, then we say the method is unstable. This technique produces a local stability criterion and is called a *von Neumann stability analysis*.[9] The basic idea is that we assume the solution can be written as a Fourier series and explore the stability individual Fourier modes. So, let us assume that the solution can be written as

$$c_j^n = \xi^n e^{ikx_j},$$

where k is the spatial wave number of the Fourier mode. Substituting this into Equation (10.74) and writing $x_{j+1} = x_j + \Delta x$ and $x_{j-1} = x_j - \Delta x$ gives

$$\xi^{n+1} e^{ikx_j} = [(1-\alpha) + \alpha e^{-ik\Delta x}]\xi^n e^{ikx_j}.$$

We can see from this equation that the wave amplitude (ξ^{n+1}) at time t_{n+1} is the amplitude (ξ^n) at the previous time step multiplied by an *amplification factor*, $g = [(1-\alpha)+\alpha e^{-ik\Delta x}]$. If $|g| \leq 1$, then the amplitude of the wave will decrease over time and the numerical scheme is said to be stable; but if $|g| > 1$, the amplitude grows over time and the scheme is unstable. We can write the amplification factor as

$$g = 1 - \alpha[1 - \cos(k\Delta x)] - i\alpha \sin(k\Delta x),$$

so the square of its modulus is

$$gg^* = 1 - 2\alpha(1-\alpha)(1 - \cos(k\Delta x)).$$

This implies that for $|g| \leq 1$ we need to have $\alpha(1-\alpha) \geq 0$, so

$$\alpha = \frac{v\Delta t}{\Delta x} \leq 1. \tag{10.76}$$

[9] This technique is named after John von Neumann (1903–1957), who was arguably the greatest mathematician of the early twentieth century, making major contributions to fields as diverse as quantum theory, hydrodynamics, economics, and computing. The technique was originally described by John Crank (1916–2006) and Phyllis Nicolson (1917–1968) and was later explored in more detail by von Neumann and colleagues.

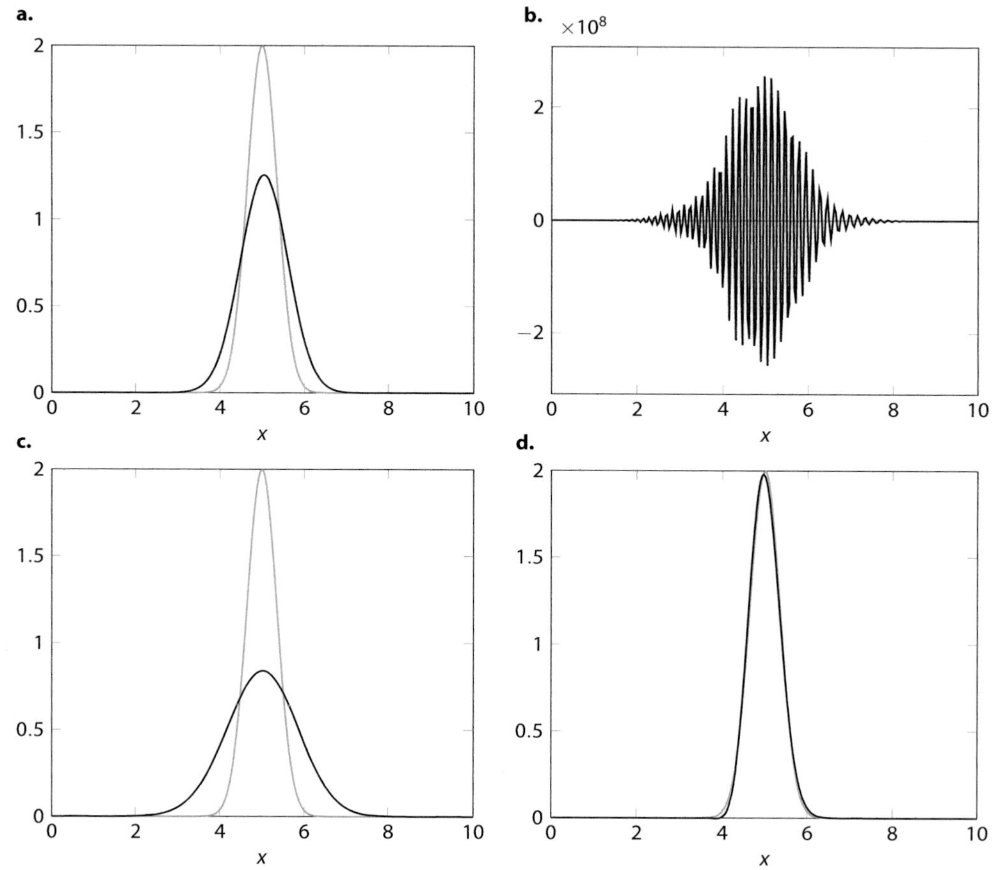

Figure 10.7 A comparison of the results of four different algorithms for solving the advection equation in one dimension. The equation is solved with a Gaussian pulse initial condition and Dirichlet boundary conditions. The initial condition is shown in gray and the final solution in black and shifted in x to make comparison with the initial condition easier: the upwind scheme (a.) the unstable forward time centered space scheme (b.), the Lax–Friedrichs scheme (c.), and the Lax–Wendroff scheme (d.).

Equation (10.76) is called the *Courant–Friedrichs–Lewy condition*, or CFL condition,[10] and it places limits on the sizes of Δx and Δt that we can use. The upwind scheme is *conditionally stable*; that is, it is stable as long as the CFL condition is satisfied. As nice and simple as it may seem, the upwind scheme suffers from a problem. Figure 10.7a shows the result of using the upwind scheme to solve Equation (10.71), where the two curves show the initial and final solution. The upwind scheme ensures that the wave travels at the correct speed (v in Equation (10.71)), but as we can see, the amplitude of the pulse has dramatically decreased over time. However, there is nothing in the advection equation that would lead to such a reduction. The fact that the pulse becomes spread out provides a clue as to what is going on. The culprit is the fact that we truncated the finite difference

[10] Named after the German-American mathematicians Richard Courant (1888–1872), Kurt Otto Friedrichs (1901–1982), and Hans Lewy (1904–1988).

approximation of the spatial derivative at the first order—we linearized the function. As we shall see a bit later, the neglected second order terms act like diffusion, causing the pulse to spread out over time and the amplitude to decrease.

What would happen if we had decided to use central differences (Equation (10.69)) for the spatial derivative instead of forward differences? This might seem like a good idea because, as we described above, central differences incur a smaller error than forward differences. This would give us a scheme

$$c_j^{n+1} = c_j^n - \frac{1}{2}\frac{v\Delta t}{\Delta x}(c_{j+1}^n - c_{j-1}^n). \tag{10.77}$$

This scheme is called the *forward time centered space (FTCS) scheme* and seems as benign as the the upwind scheme. Sadly, it is unconditionally unstable; that is, perturbations will grow no matter how we choose Δt and Δx. This is dramatically shown in Figure 10.7b.[11]

Exercise 10.8.1 Derive Equation (10.77) and use von Neumann stability analysis to show that it is unconditionally unstable.

What happens is that very small errors in the numerical solution get amplified very rapidly, and as can be seen from Figure 10.7b, they can rapidly grow to be very large. These numerical errors are inevitable and are called *round-off errors* or *truncation errors*. They arise because computers have only a fixed finite amount of memory to store a number, so they have to round off numbers that contain more digits than their memory can hold. The accumulation of these truncations over many, many calculations can result in a significant round-off error. However, there is a way out. Let us replace the first term on the right-hand side of Equation (10.77) with the average of the solution at the points either side of it,

$$c_j^n = \frac{1}{2}(c_{j+1}^n + c_{j-1}^n),$$

so that the finite difference equation becomes

$$c_j^{n+1} = \frac{1}{2}(c_{j+1}^n + c_{j-1}^n) - \frac{1}{2}\frac{v\Delta t}{\Delta x}(c_{j+1}^n - c_{j-1}^n), \tag{10.78}$$

which is called the Lax–Friedrichs scheme.[12] The results of a numerical solution using this scheme are shown in Figure 10.7c, and we can see that the scheme is indeed stable.

Exercise 10.8.2 Show that the Lax–Friedrichs scheme is stable as long as the CFL condition, $(v\Delta t)/\Delta x < 1$), is satisfied.

What just happened? How did swapping out the term c_j^n in Equation (10.77) turn the scheme from one that is unconditionally unstable to one that is conditionally stable? The cause is again diffusion, but in this case it has helped prevent the instability. Let us rewrite Equation (10.78) as

$$\frac{c_j^{n+1} - c_j^n}{\Delta t} = -v\left(\frac{c_{j+1}^n - c_{j-1}^n}{2\Delta x}\right) + \frac{1}{2}\left(\frac{c_{j+1}^n + c_{j-1}^n - 2c_j^n}{\Delta x}\right),$$

[11] The online codes provide an animation of the development of this instability that is well worth watching to understand how it arises.

[12] Named after Hungarian-American mathematician Peter Lax (b. 1926) and Kurt Otto Friedrichs.

which, if we interpret each term as a finite difference approximation to a derivative, is the finite difference approximation to the PDE

$$\frac{\partial c}{\partial t} = -v\frac{\partial c}{\partial x} + \frac{1}{2}\frac{(\Delta x)^2}{\Delta t}\frac{\partial^2 c}{\partial x^2}.$$

It appears that we have effectively added a diffusion-like term to the transport equation. This might seem like a very bad thing to do because we have changed the original PDE (the transport equation) we were working with. However, it turns out that this extra term is cleverly constructed such that it diffuses away perturbations before they grow, but decays away itself sufficiently fast that it has minimal effect on the actual solution of the transport equation. The problem with the Lax–Friedrichs scheme is that it still does not reproduce the solution well. The speed of the propagation of the wave is good, but the amplitude of the wave decreases over time. A technique called the *Lax–Wendroff* helps alleviate the amplitude error (Figure 10.7d).[13] To derive this scheme, we will take a slightly different approach (Rood, 1987) than we have in deriving the other finite difference schemes. At each value of x we can expand $c(x, t)$ as a Taylor series so that

$$c(x, t + \Delta t) = c(x, t) + \Delta t\frac{\partial c}{\partial t} + \frac{(\Delta t)^2}{2}\frac{\partial^2 c}{\partial t^2} + \cdots$$

Equation (10.71) furnishes us with an expression for the first derivative of c with respect to t, and if v is constant, we also have

$$\frac{\partial^2 c}{\partial t^2} = -v\frac{\partial}{\partial t}\frac{\partial c}{\partial x} = -\frac{\partial}{\partial x}\frac{\partial c}{\partial t} = v^2\frac{\partial^2 c}{\partial x^2}.$$

In general,

$$\frac{\partial^n c}{\partial t^n} = (-c)^n\frac{\partial^n c}{\partial x^n}$$

so that we can substitute the time derivatives in the Taylor series with spatial derivatives. Then, using central differences for the first derivative we arrive at the scheme

$$c_i^{n+1} = c_i^n - \frac{v\Delta t}{2\Delta x}(c_{i+1}^n - c_{i-1}^n) + \frac{v^2(\Delta t)^2}{2(\Delta x)^2}(c_{i+1}^n + c_{i-1}^n - 2c_i^n). \tag{10.79}$$

This scheme has the added diffusion that makes the Lax–Friedrichs scheme stable, but it also has minimal dissipation and therefore very low amplitude errors (Figure 10.7d).

Now let us consider the diffusion equation in one dimension:

$$\frac{\partial c}{\partial t} = \mathcal{D}\frac{\partial^2 c}{\partial x^2}.$$

We have seen that we can make the FTCS scheme stable by adding a numerical diffusion; well, the diffusion equation has that process already built in. As a result, the FTCS scheme is a stable and reasonably accurate scheme for the diffusion equation; however, the CFL condition must be modified so that

$$\frac{\mathcal{D}\Delta t}{\Delta x^2} < \frac{1}{2}.$$

[13] Named after Peter Lax and American applied mathematician Burton Wendroff (b. 1930).

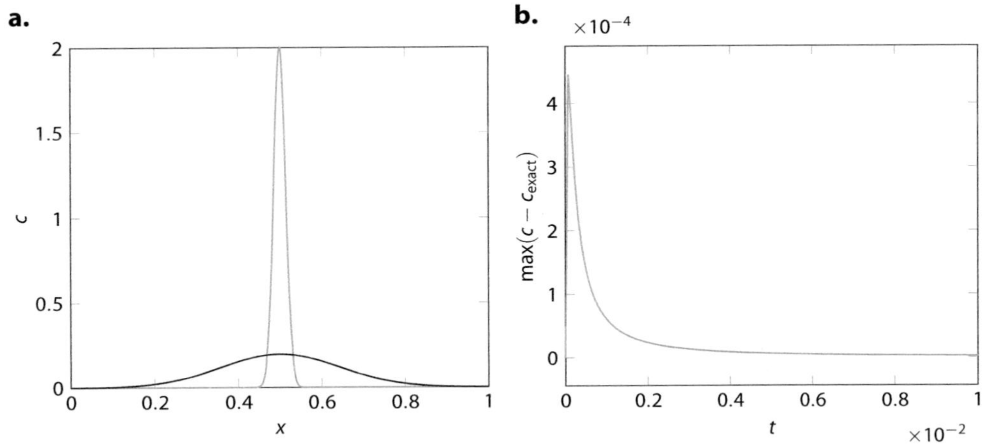

Figure 10.8 Solving the diffusion equation using a FTCS scheme with an initial condition (in gray) of a Gaussian pulse and Dirichlet boundary conditions. After a finite time, the solution shows a Gaussian curve that is more spread out, as expected, but still centered on the same location as the initial pulse (a.). The maximum error (i.e., the difference between the numerical and exact solution) at each time step is small and decreases rapidly over time (b.).

The results of using an FTCS scheme to solve the diffusion equation in one dimension are shown in Figure 10.8. The initial condition is a Gaussian pulse, and we have used Dirichlet boundary conditions—this has the advantage that the analytical solution is a Gaussian that spreads over time, so we can compare the numerical and exact solutions to assess the accuracy of the algorithm. The difference between the numerical solution and the exact solution is shown in Figure 10.8b.

Lastly, let us consider how to use finite differences to solve elliptic equations. As an example, let us consider Poisson's equation in two dimensions,

$$\nabla^2\phi = \frac{\partial^2\phi}{\partial x^2} + \frac{\partial^2\phi}{\partial y^2} = f(x, y), \tag{10.80}$$

on a rectangular domain $0 \leq x \leq a$, $0 \leq y \leq b$. Notice that this equation does not involve time, so we do not need initial conditions, but we do need four boundary conditions, $\phi(x, 0) = u(x)$, $\phi(0, y) = v(x)$, $\phi(x, b) = r(y)$, and $\phi(a, y) = s(y)$. As before, we are going to be calculating the solution at discrete points on a grid that covers the domain, so we divide the interval $0 \leq x \leq a$ into M equally sized intervals with a distance $\Delta x = a/M$ between them. Similarly, we divide the interval $0 \leq y \leq b$ into N intervals with a distance $\Delta y = b/N$ between them. This means that we have $(M + 1)$ points in the x direction and $(N + 1)$ in the y direction, with each point having coordinates $p_{jk} = (j\Delta x, k\Delta y)$, where $j = 0, 1, 2, 3, \ldots, M$, and $k = 0, 1, 2, 3, \ldots, N$. We can now apply the finite difference approximation Equation (10.70) to each of the derivatives in Equation (10.80) to get

$$\frac{1}{(\Delta x)^2}(\phi_{j-1,k} - 2\phi_{j,k} + \phi_{j+1,k}) + \frac{1}{(\Delta y)^2}(\phi_{j,k-1} - 2\phi_{j,k} + \phi_{j,k+1}) = f_{j,k}, \tag{10.81}$$

where $f_{j,k}$ is the value of the function evaluated at the point $(x_j, y_k) = (j\Delta x, k\Delta y)$. If we choose a grid such that $\Delta x = \Delta y = h$, then Equation (10.81) simplifies to

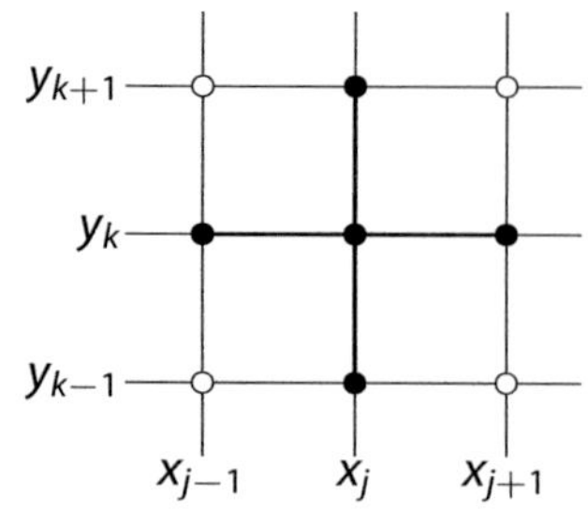

Figure 10.9 The five-point template for numerically solving an elliptic equation. The black dots show the grid points that are used in determining the solution at grid point (x_j, y_k).

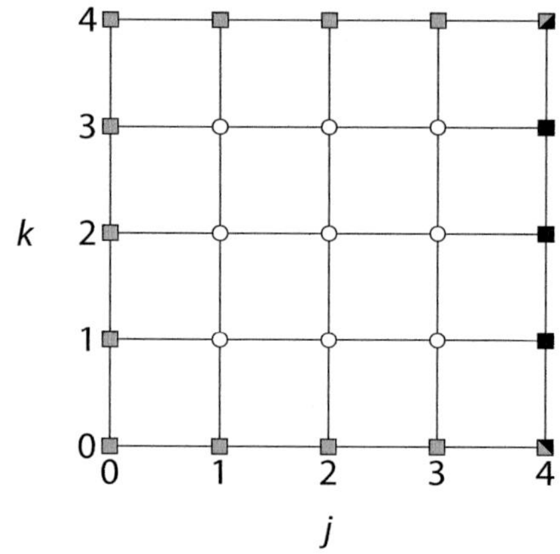

Figure 10.10 The complete (5×5) grid for examining the finite difference solution to the Laplace equation. The open circles represent the interior points of the grid. The grid points making up the boundary conditions are shown as squares: gray squares for where the boundary conditions are zero, and black squares for where they are nonzero. The two corner squares at $j = 4,\ k = 0$ and $j = 4,\ k = 4$ are colored both black and gray to indicate that the boundary conditions must be compatible at these points.

$$\phi_{j-1,k} + \phi_{j+1,k} + \phi_{j,k-1} + \phi_{j,k+1} - 4\phi_{j,k} = h^2 f_{j,k}. \tag{10.82}$$

Equation (10.82) is often called a *five-point rule* because it uses a template of five points to calculate a solution (Figure 10.9). Equation (10.82) is simply a set of linear equations, albeit possibly a very large set, so we can write it as a matrix equation. To make the structure of the matrix equation more transparent, we will look at solving Laplace's equation (i.e., $f_{j,k} = 0$), in which case Equation (10.82) becomes

$$\phi_{j,k} = \frac{1}{4}\left(\phi_{j-1,k} + \phi_{j+1,k} + \phi_{j,k-1} + \phi_{j,k+1}\right). \tag{10.83}$$

To make things more concrete, we will consider a domain $0 \le x \le 1$, $0 \le y \le 1$, and we will assume the boundary conditions are such that $\phi(x, 0) = \phi(x, 1) = \phi(0, y) = 0$ and $\phi(1, y) = g(y)$, with Dirichlet boundary conditions. The grid for the domain is shown in Figure 10.10. Notice that in order for our equation and boundary conditions to be consistent, we must have $g_{4,0} = g_{4,4} = 0$, otherwise we would have a conflict with the boundary conditions at the two corners along the boundary $x = 1$. Let us look at the finite difference equations for the interior points only, and we will start with the row $k = 1$,

$$\begin{aligned}
\phi_{0,1} + \phi_{1,0} + \phi_{2,1} + \phi_{1,2} - 4\phi_{1,1} &= 0 + 0 + \phi_{2,1} + \phi_{1,2} - 4\phi_{1,1} &&= 0,\\
\phi_{1,1} + \phi_{2,0} + \phi_{3,1} + \phi_{2,2} - 4\phi_{2,1} &= \phi_{1,1} + 0 + \phi_{3,1} + \phi_{2,2} - 4\phi_{2,1} &&= 0,\\
\phi_{2,1} + \phi_{3,0} + \phi_{4,1} + \phi_{3,2} - 4\phi_{3,1} &= \phi_{2,1} + 0 + g_{4,1} + \phi_{3,2} - 4\phi_{3,1} &&= 0,
\end{aligned}$$

where $g_{4,1} = g(x_4, y_1)$. If we take each row in turn and do the same thing, we find that we can write the whole system of nine equations for the interior points in matrix form as

$$\left(\begin{array}{ccc|ccc|ccc}
-4 & 1 & 0 & 1 & & & & & \\
1 & -4 & 1 & 0 & 1 & & & & \\
0 & 1 & -4 & 0 & 0 & 1 & & & \\
\hline
1 & 0 & 0 & -4 & 1 & 01 & & & \\
 & 1 & 0 & 1 & -4 & 1 & 0 & 1 & \\
 & & 1 & 0 & 1 & -4 & 0 & 0 & 1 \\
\hline
 & & & 1 & 0 & 0 & -4 & 1 & 0 \\
 & & & & 1 & 0 & 1 & -4 & 1 \\
 & & & & & 1 & 0 & 1 & -4
\end{array}\right)
\left(\begin{array}{c}
\phi_{1,1}\\ \phi_{2,1}\\ \phi_{3,1}\\ \hline \phi_{2,1}\\ \phi_{2,2}\\ \phi_{2,3}\\ \hline \phi_{1,3}\\ \phi_{2,3}\\ \phi_{3,3}
\end{array}\right)
=
\left(\begin{array}{c}
0\\ 0\\ g_{4,1}\\ \hline 0\\ 0\\ g_{4,2}\\ \hline 0\\ 0\\ g_{4,3}
\end{array}\right). \tag{10.84}$$

The missing entries in Equation (10.84) are all zero; we have left them out to make the structure of the matrix clearer and to save typing! We can calculate the values of ϕ on the boundaries using the boundary conditions, so we can calculate the solution at every grid point in the domain.

Example 10.8 As an example, we can solve the Poisson equation

$$\nabla^2\phi = \frac{\partial^2\phi}{\partial x^2} + \frac{\partial^2\phi}{\partial y^2} = -8\pi^2 \sin(2\pi x)\sin(2\pi y) \tag{10.85}$$

on the domain $0 \le x \le 1$, $0 \le y \le 1$, with Dirichlet boundary conditions. The results of the numerical solution using 50 grid cells in each direction is shown in Figure 10.11. We can determine the accuracy of the numerical solution because the solution to the equation is $\phi_{\text{exact}}(x, y) = \sin(2\pi x)\sin(2\pi y)$, as you can check by direct substitution. The accuracy of the solution improves as we decrease the grid spacing; however, the trade off is that the computational time increases, but not linearly. In practical terms this means that there is an optimal grid spacing that gives the best accuracy of the solution in a reasonable time (Table 10.1).

10.9 Further Reading

In this chapter we have covered some of the elementary aspects of PDEs and their solutions. PDEs form the mainstay of many numerical models in the Earth and environmental

Table 10.1 The number of grid cells (N), computational time (τ), and maximum error ϵ for the numerical solution of Equation (10.85).

N	τ (s)	ϵ
10	0.01	3×10^{-2}
20	0.03	8.3×10^{-3}
30	0.04	3.6×10^{-3}
40	0.23	2.1×10^{-3}
50	0.43	1.3×10^{-3}
60	1.02	9.1×10^{-4}
70	2.33	6.7×10^{-4}
80	4.14	5.1×10^{-4}
90	7.83	4.1×10^{-4}
100	14.17	3.2×10^{-4}

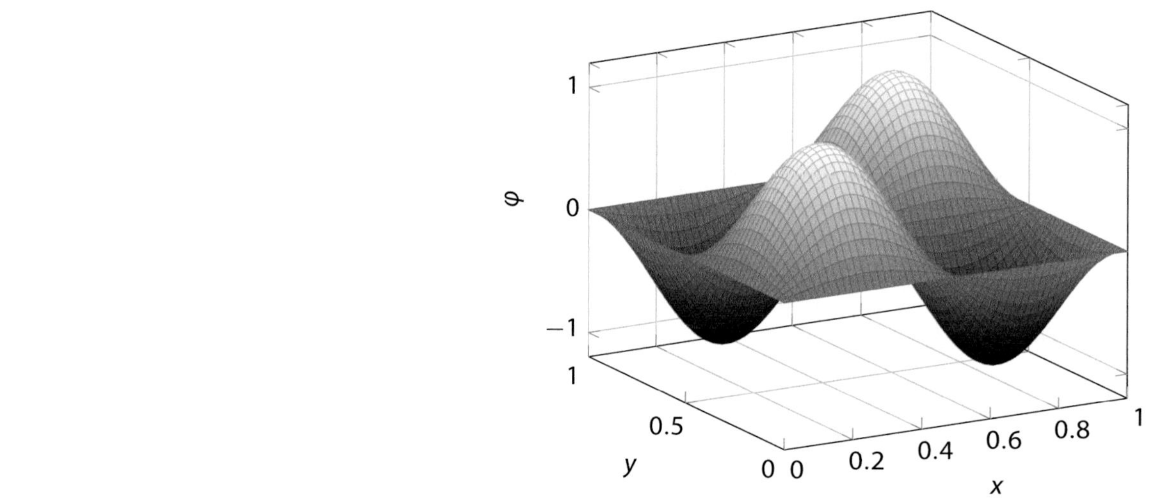

Figure 10.11 The numerical solution of Equation (10.85), using finite differences.

sciences, and there is a rich literature dealing with their properties and numerical techniques for solving them. For example, a good guide to the practical use and solution of PDEs in marine science can be found in the later chapters of *Modeling Methods for Marine Science* by Glover et al. (2011). Many textbooks that cover mathematical methods in physics or engineering cover PDEs to some degree; for example, the relevant chapters in *Mathematical Methods in the Physical Sciences* by Boas (2006) are very good and approachable. Two other books that are at a level that is very approachable are *An Introduction to Partial Differential Equations with* MATLAB by Coleman (2005) and *Partial Differential Equations* by Gockenbach (2002). As we have seen, we often have to choose different methods for solving different types of PDEs. A good book for learning the practical aspects of solving PDEs numerically is *Numerical Recipes* by Press et al. (1992);

the book is available in multiple versions, each using a different programming language, but the text is the same. There are also books that focus on only one type of equation. For example, the classic text on solutions to the heat equation is *Conduction of Heat in Solids* by Carslaw and Jaeger (1959). The book contains analytic solutions to the heat equation for many different types of problems and geometries and also covers some numerical methods, though it does not cover modern numerical methods. Another specialist book is *Numerical Methods for Wave Equations in Geophysical Fluid Dynamics* by Durran (1999), which covers advanced numerical techniques for solving wave equations in contexts relevant to the geosciences; a somewhat more general, updated text is *Numerical Methods for Fluid Dynamics*, also by Durran (2010). A good text that deals with the mathematical analysis of numerical methods, but at quite a more advanced, technical level than this one, is the book *A First Course in the Numerical Analysis of Differential Equations* by Iserles (2008).

Problems

10.1 Use a von Neumann stability analysis to determine the conditions under which the FTCS scheme is a stable scheme to use for solving a one-dimensional diffusion equation.

10.2 The temperature within soils with depth will vary according to the amount of sunlight striking the surface. This can be modeled as a one-dimensional conduction problem:

$$\frac{\partial T(t,z)}{\partial t} = \mathcal{K}\frac{\partial^2 T(t,z)}{\partial z^2},$$

where z is the depth in the soil, t is time, and $\mathcal{K}$ is the thermal diffusivity of the soil

$$\mathcal{K} = \frac{\kappa}{\rho c},$$

where k is the thermal conductivity, ρ the density, and c the heat capacity of the soil. Solve the equation with the following boundary conditions:

$$T(t,0) = T_0 + A\sin(\omega t + \phi), \qquad T(t,\infty) = T_0,$$

to show that the solution is

$$T(t,z) = T_0 + Ae^{-z/\mathcal{D}}\sin\left(\omega t - \frac{z}{D} + \phi\right), \qquad D = \sqrt{\frac{2K}{\omega}}.$$

10.3 The shape of the side of a hill depends on the rate of uplift, pushing the ground upward, and erosion processes such as rainfall, which move material down the hill. Let us assume that we can model the sum of all the erosion processes as a diffusive process with a diffusion coefficient κ. Let $h(x,t)$ represent the height of the ground above some set level, where x is the distance from the hill top, and t is time, and h obeys a diffusion equation.

1. If the rate of uplift (u) is constant, then the steady state diffusion equation becomes $\kappa h_{xx} + u = 0$. Solve this equation with a no-flux boundary condition at $x = 0$ and $h(x = L) = 0$ to show that

$$h(x) = \frac{u(L^2 - x^2)}{2\kappa}.$$

What does the no-flux boundary condition at the top of the hill mean physically? What is the shape of the resulting hillside?

2. To calculate how long the hill slope takes to come to steady, consider the equation $h_t - \kappa h_{xx} = u$, which represents the net change in height h from uplift and diffusive processes removing material and where the uplift u is constant. Solve the equation with an initial condition $h(x, 0) = 0$ and boundary conditions as before. Plot the quantity $(h\kappa)/(uL^2)$ against x for a hillside that has a length L of 100 m and for values of κt ranging from 100 m^2 to 1000 m^2. A useful trick to solve the equation is to define a new variable,

$$w(x, t) = h(x, t) - \frac{u(L^2 - x^2)}{2\kappa}.$$

3. Use dimensional analysis to estimate the time to achieve a steady shape for a hill, assuming $\kappa = 1$ m^2 kyr^{-1} for $L = 20, 50, 100$ m.

10.4 Glacial moraines form when glaciers scrape the surface of the land, building up a large pile of debris in front of them. When the glaciers retreat they leave behind them a triangular-shaped formation, called a moraine, that gradually erodes, and which can be modeled as a diffusive process with a diffusion coefficient κ. If the highest point of the moraine is at $x = 0$ and the height of the base of the moraine at $x = L$, then there is a no-flux boundary condition at $x = 0$. Assuming that the initial shape of the moraine is triangular, with the height of the peak being $h_0 L$, solve the diffusion equation for the height of the moraine as a function of x and t.

10.5 Even great scientists can make mistakes, as this problem illustrates. There was considerable debate in the nineteenth century about the age of the Earth. Charles Darwin had estimated an age of at least 300 million years, but the physicist William Thomson (who later became Lord Kelvin) thought it much younger. Thomson knew that the temperature increases as one goes deeper into the Earth, so it seemed to him as if the Earth was cooling down from a molten state at the time of its formation. Consequently, he based his estimate on the time it takes a uniform sphere to cool.

1. Solve the diffusion equation $u_t = \kappa\nabla^2 u$ for $u(r, \theta, \phi)$ for a homogeneous sphere of radius R with an initial condition $u(r, 0) = u_0 =$ constant for $r < R$ and boundary condition $u(R, t) = 0$ for $t > 0$.
2. Thomson used measured temperature gradients in deep mine shafts to constrain his numbers. From your solution, calculate the gradient of u with respect to the radius and find an expression for the gradient when $r = R$.

3. Use the fact that if $\kappa t \ll R^2$, then

$$\sum_{n=-\infty}^{\infty} \exp\left(-\frac{n^2\pi^2\kappa t}{R^2}\right) \approx \int_{-\infty}^{\infty} \exp\left(-\frac{n^2\pi^2\kappa t}{R^2}\right)\, dx$$

to show that the time it takes to achieve a temperature gradient of du/dr is

$$\tau \approx \frac{u_0^2}{\pi\kappa}\left(\frac{du}{dr}\right)^{-2}.$$

4. Assuming $u_0 = 1000°\text{C}$, $\kappa \approx 0.2 \times 10^{-6}\ \text{m}^2\ \text{yr}^{-1}$, and a surface temperature gradient of $\approx 25°\text{C km}^{-1}$, calculate the age of the Earth.
5. We know that the age of the Earth is approximately 4.5 billion years, so why was Kelvin so wrong?

10.6 The PDE

$$\frac{\partial C}{\partial t} = \mathcal{D}\frac{\partial^2 C}{\partial x^2} - u\frac{\partial C}{\partial x}$$

is used to describe the simultaneous diffusion and advection of a substance whose concentration is $C(x,t)$. Use Laplace transforms to solve this equation with boundary and initial conditions

$$C(0,t) = C_0, \quad C(x,0) = C_1, \quad \left.\frac{\partial C}{\partial x}\right|_{x\to\infty} = 0,$$

where C_0 and C_1 are constants. You might find it useful to use a new variable $c(x,t) = C(x,t) - C_1$.

10.7 Material in sediments is gradually moved deeper and deeper as more material arrives at the top the sediment. This can be described as an advective process. If that material is also consumed, by microbial activity for example, then the concentration will also decay over time. The combined process can be described by an advection–reaction equation. Solve the advection–reaction equation for $C(z,t)$,

$$\frac{\partial C}{\partial t} + u\frac{\partial C}{\partial z} = -kC,$$

with initial condition $C(z,0) = 0$ and $C(0,t) = 1$.

10.8 Some PDEs have solutions, that are called *scaling solutions*, and one of these is the diffusion equation. Consider diffusion of a substance along an infinite (ie. $-\infty < x < \infty$) pipe of uniform cross-sectional area A. At $t = 0$, a mass M_i of material is placed centered on the origin (i.e., from $-\delta x$ to $+\delta x$, where δx is a small distance) such that the concentration is $C(x, t = 0) = M_i A \delta x$. This problem can be described by a one-dimensional diffusion equation $C_t = \mathcal{D}C_{xx}$ with boundary conditions $C(\pm\infty, t) = 0$ and initial condition $C(0,t) = (M_i/A)\delta x$.

1. Use the Buckingham pi theorem to show that there are two dimensionless ratios and that the concentration can be written as

$$C(x,t) = \frac{M_i}{A(\mathcal{D}t)^{1/2}} f\left(\frac{x}{(\mathcal{D}t)^{1/2}}\right). \tag{10.86}$$

2. Show that Equation (10.86) combined with the new variable $y = x/(\mathcal{D}t)^{1/2}$ reduces the diffusion equation to

$$\frac{d^2 f}{dy^2} + \frac{y}{2}\frac{df}{dy} + \frac{1}{2}f(y) = \frac{d}{dy}\left(\frac{df}{dy} + \frac{fy}{2}\right) = 0. \tag{10.87}$$

3. Convert the boundary and initial conditions to conditions on the function $f(y,t)$.
4. Solve Equation (10.87) with the appropriate boundary and initial conditions, and use the fact that

$$\int_{-\infty}^{\infty} f(y)\, dy = 1$$

to write the final solution as

$$C(x,t) = \frac{M_i}{A(4\pi\mathcal{D}t)^{1/2}} \exp\left(-\frac{x}{4\mathcal{D}t}\right).$$

10.9 Consider the first order PDE

$$\frac{\partial u}{\partial t} + \frac{\partial}{\partial x}(f(u)) = 0,$$

where $f(u)$ is a function of u and is sometimes called the *flux function*. This equation describes a family of equations, among which is the Buckley–Leverett equation that is used to describe flow in a medium consisting of two phases (e.g., solid and liquid, liquid and gas) and is used to determine the flow of fluids through rock. Use the method of characteristics to show that

$$x = a(u_0)(t - t_0) + x_0,$$

where $a(u) = f'(u)$, u_0 is the value of u along the characteristic, and (x_0, t_0) is a point on the characteristic.

10.10 The equation for the flow of a homogeneous fluid on a rotating Earth is

$$\frac{\partial \mathbf{u}}{\partial t} + (\mathbf{u}\cdot\nabla)\mathbf{u} + 2\boldsymbol{\omega}\times\mathbf{u} = -\frac{1}{\rho}\nabla p + \mathbf{g}, \tag{10.88}$$

where $\mathbf{u}$ is the fluid velocity, t is time, $\boldsymbol{\omega}$ is the angular velocity of the Earth, ρ is the density of the fluid, p is the pressure of the fluid, and $\mathbf{g}$ is the gravitational acceleration written as a vector.

1. Choose typical scales of U for the velocity, L for distance, and P for pressure. Show that the typical time scale is $T = L/U$.
2. Nondimensionalize the variables in Equation (10.88) and write down the nondimensional version of Equation (10.88). The nondimensional coefficient of the rotation term is called the Rossby number, $\mathrm{Ro} = U/(2\omega L)$.

10.11 Write a computer code to solve the one-dimensional heat equation on the interval $0 \le x \le L$, with initial conditions $u(x,0) = \sin(\pi x/L)$ and boundary conditions $u(0,t) = u(L,t) = 0$. Use a range of Δx and Δt values, and compare your results with the exact solution,

$$u(x,t) = \exp\left(\frac{-\mathcal{D}\pi^2 t}{L^2}\right)\sin\left(\frac{\pi x}{L}\right).$$

11 Tensors

We have already seen that physical quantities can be described mathematically by scalars or vectors. A scalar describes the situation where there is a single number at each point of space—a temperature, or a concentration of a chemical, for example. Vectors require both a magnitude and a direction in order to specify them—a velocity, a force, etc. There are still other physical objects that cannot be described by these two types of objects.

As an example, let us consider how a body deforms under both stress and pressure. This is a question that is important in the study of geophysical fluid dynamics, where one is interested in the motions of packets of air or water, and also in geology and geophysics, where one wants to understand how rocks will deform under the forces acting on them. Let us look at a cubical volume in a fluid and think about the forces acting on a point in the upper face (Figure 11.1). There can be a force parallel to the z axis that gives rise to a normal stress σ_{zz}, and there are also tangential forces that deform the surface in the x or y directions τ_{zx} and τ_{zy} respectively. So, each face of the cube can have three forces acting on it. If we shrink the cube down to an infinitesimal point, then we can write the forces acting at that point as a matrix,

$$\sigma_{ij} = \begin{pmatrix} \sigma_{xx} & \tau_{xy} & \tau_{xz} \\ \tau_{yx} & \sigma_{yy} & \tau_{yz} \\ \tau_{zx} & \tau_{zy} & \sigma_{zz} \end{pmatrix}, \tag{11.1}$$

called the *stress tensor*. This matrix has nine entries, so we need something other than a vector (which in three dimensions has only three components) to represent it—we need a tensor.

11.1 Covariant and Contravariant Vectors

To better understand tensors we need to go back and take another look at vectors. This is because we have not told the whole story concerning vectors. In particular, we want to look again at how vectors transform under coordinate changes. Recall that although the components of a vector may change when we change coordinates, the magnitude of the vector does not. Let us consider the position vector of a point P in three-dimensional Cartesian coordinates (x, y, z). We can write that position vector as $\mathbf{r} = x\hat{\mathbf{\imath}} + y\hat{\mathbf{\jmath}} + z\hat{\mathbf{k}}$. We now transform to a new set of coordinates (α, β, γ) so that the old coordinates are functions of the new ones $x = x(\alpha, \beta, \gamma)$, $y = y(\alpha, \beta, \gamma)$, $z = z(\alpha, \beta, \gamma)$, and vice versa. For example,

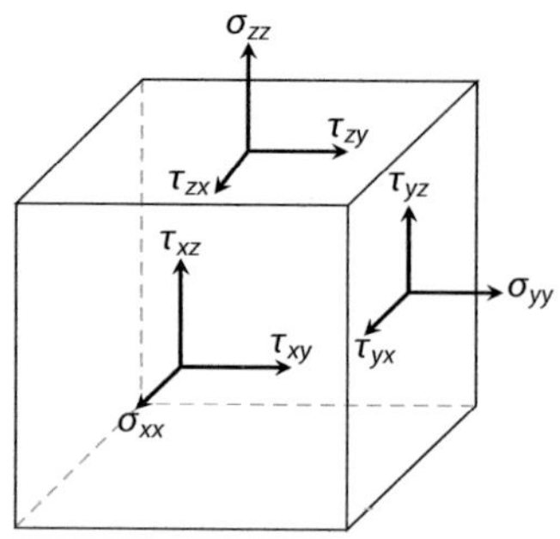

Figure 11.1 The forces acting on a cube. The force acting on each face of the cube can be decomposed into a force perpendicular to the surface (σ) that squeezes and stretches the cube, and forces (τ) that deform the cube in directions tangent to the surface.

our new coordinates might be spherical polar coordinates (r, ϕ, θ) (Figure 7.11), in which case,

$$x = r\cos(\phi)\sin(\theta), \quad y = r\sin(\phi)\sin(\theta), \quad z = r\cos(\theta), \quad 0 \le \phi \le 2\pi, \quad -\frac{\pi}{2} \le \theta \le \frac{\pi}{2}.$$

Converting the coordinates does not change the position of the point in space, only the way we label that point. For example, the position vector of P in spherical coordinates will have the form $\mathbf{r} = r\cos(\phi)\sin(\theta)\hat{\mathbf{i}} + r\sin(\phi)\sin(\theta)\hat{\mathbf{j}} + r\cos(\theta)\hat{\mathbf{k}}$, so the point represented by $x = y = z = 1$ is the same point as $r = \sqrt{3}$, $\theta = \cos^{-1}(1/\sqrt{3})$, $\phi = \pi/4$ so that $r\cos(\phi)\sin(\theta) = r\sin(\phi)\sin(\theta) = r\sin(\phi)\sin(\theta) = 1$.

We saw in Figure 7.10 that a point (x_P, y_P, z_P) is the point of intersection of three planes, each of which has one of the coordinates held constant while the other two vary forming the plane. Similarly, a point in the (α, β, γ) coordinate system will be defined by the intersection of three *coordinate surfaces*, but these need not be planes. To use the example of spherical coordinates again, the surfaces will be a surface of constant r—i.e., the surface of a sphere—a plane from the z axis making a constant angle ϕ with the x axis, and the base of a cone whose axis is the z axis and which has an angle of θ with the z axis (Figure 11.2). We have also seen in Chapter 7 that we can parameterize paths, so let us choose a set of very special parameterizations that will represent the different coordinate lines passing through the point P. For example, if we let $r = r_P$ and $\phi = \phi_P$ and allow θ to vary, we have a parameterized curve given by the position vector

$$\mathbf{r} = r_P\cos(\phi_P)\sin(\theta)\hat{\mathbf{i}} + r_P\sin(\phi_P)\sin(\theta)\hat{\mathbf{j}} + r_P\cos(\theta)\hat{\mathbf{k}},$$

which describes a meridian of constant ϕ on the surface of radius r_P and passing through the point P. Similarly, we can form two other parameterized curves by either holding r and θ constant or by holding θ and ϕ constant.

Exercise 11.1.1 Sketch the parameter curves we would get by holding r and θ constant and by holding θ and ϕ constant.

Exercise 11.1.2 What would the paths be if we used Cartesian (x, y, z) coordinates and held x and z constant, x and y constant, and y and z constant.

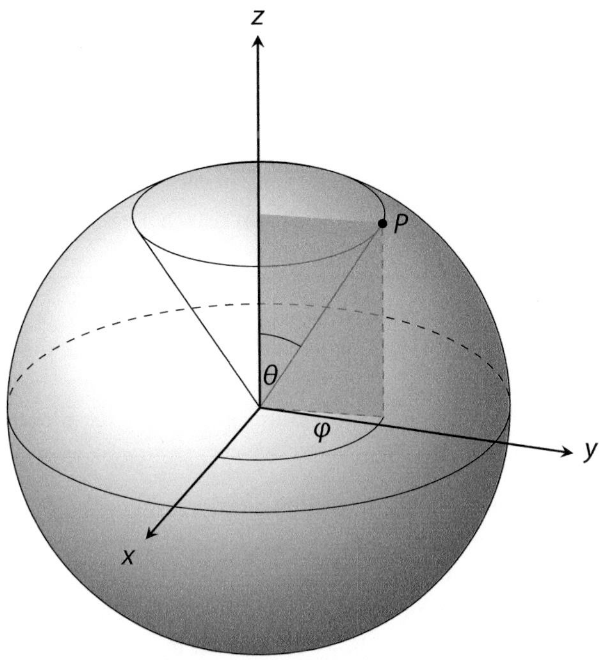

Figure 11.2 The three coordinate surfaces defining a point P in spherical polar coordinates, and the three basis vectors $\hat{\mathbf{e}}_r$, $\hat{\mathbf{e}}_\theta$, and $\hat{\mathbf{e}}_\phi$ at that point.

The curves we form in this way are simply the coordinate curves in the coordinate system we are using. So, in general we could write these curves in the form

$$\begin{aligned}\mathbf{r}(\alpha) &= x(\alpha, \beta_P, \gamma_P)\hat{\mathbf{\imath}} + y(\alpha, \beta_P, \gamma_P)\hat{\mathbf{\jmath}} + z(\alpha, \beta_P, \gamma_P)\hat{\mathbf{k}},\\ \mathbf{r}(\beta) &= x(\alpha_P, \beta, \gamma_P)\hat{\mathbf{\imath}} + y(\alpha_P, \beta, \gamma_P)\hat{\mathbf{\jmath}} + z(\alpha_P, \beta, \gamma_P)\hat{\mathbf{k}},\\ \mathbf{r}(\gamma) &= x(\alpha_P, \beta_P, \gamma)\hat{\mathbf{\imath}} + y(\alpha_P, \beta_P, \gamma)\hat{\mathbf{\jmath}} + z(\alpha_P, \beta_P, \gamma)\hat{\mathbf{k}}.\end{aligned}$$

The derivatives of each of these curves with respect to the varying parameter is the tangent to that curve, i.e., the tangent to the coordinate curve, and we can use these to form a set of basis vectors,

$$\mathbf{e}_\alpha = \frac{\partial}{\partial \alpha}\mathbf{r}(\alpha), \quad \mathbf{e}_\beta = \frac{\partial}{\partial \beta}\mathbf{r}(\beta), \quad \mathbf{e}_\gamma = \frac{\partial}{\partial \gamma}\mathbf{r}(\gamma). \tag{11.2}$$

This set of basis vectors forms what is called the *natural basis*.

Example 11.1 To find the natural basis in spherical polar coordinates, we first write a general position vector as

$$\mathbf{r} = r\cos(\phi)\sin(\theta)\hat{\mathbf{\imath}} + r\sin(\phi)\sin(\theta)\hat{\mathbf{\jmath}} + r\cos(\theta)\hat{\mathbf{k}}$$

and differentiate it with respect to each of the coordinates r, θ, and ϕ:

$$\mathbf{e}_r = \frac{\partial \mathbf{r}}{\partial r} = \cos(\phi)\sin(\theta)\hat{\mathbf{i}} + \sin(\phi)\sin(\theta)\hat{\mathbf{j}} + \cos(\theta)\hat{\mathbf{k}},$$
$$\mathbf{e}_\phi = \frac{\partial \mathbf{r}}{\partial r} = -r\sin(\phi)\sin(\theta)\hat{\mathbf{i}} + \cos(\phi)\sin(\theta)\hat{\mathbf{j}},$$
$$\mathbf{e}_\theta = \frac{\partial \mathbf{r}}{\partial \theta} = r\cos(\phi)\cos(\theta)\hat{\mathbf{i}} + r\sin(\phi)\cos(\theta)\hat{\mathbf{j}} - r\sin(\theta)\hat{\mathbf{k}}.$$

Exercise 11.1.3 Show that the natural basis in spherical polar coordinates is orthogonal. Find the magnitudes of the basis vectors and form a set of orthonormal vectors.

Exercise 11.1.4 Calculate the natural basis in three-dimensional cylindrical polar coordinates.

Using the tangents to the coordinate lines might be the natural way to form a basis, but are there others? Yes, and in fact one of the more useful ways is to form the basis by looking at the normal gradient to the coordinate lines. We know from Chapter 7 that the gradient of a function is orthogonal to the level curves of the function. So, if the coordinate systems (x, y, z) and (α, β, γ) are related by $x = x(\alpha, \beta, \gamma)$, $y = y(\alpha, \beta, \gamma)$, $z = z(\alpha, \beta, \gamma)$ and we can form the inverse of these functions so that $\alpha = \alpha(x, y, z)$, $\beta = \beta(x, y, z)$, and $\gamma = \gamma(x, y, z)$ (this will be true for almost all the coordinate systems that we come across in the Earth and environmental sciences), then we can define the basis

$$\mathbf{e}^\alpha = \nabla\alpha = \frac{\partial \alpha}{\partial x}\hat{\mathbf{i}} + \frac{\partial \alpha}{\partial y}\hat{\mathbf{j}} + \frac{\partial \alpha}{\partial z}\hat{\mathbf{k}},$$
$$\mathbf{e}^\beta = \nabla\beta = \frac{\partial \beta}{\partial x}\hat{\mathbf{i}} + \frac{\partial \beta}{\partial y}\hat{\mathbf{j}} + \frac{\partial \beta}{\partial z}\hat{\mathbf{k}},$$
$$\mathbf{e}^\gamma = \nabla\gamma = \frac{\partial \gamma}{\partial x}\hat{\mathbf{i}} + \frac{\partial \gamma}{\partial y}\hat{\mathbf{j}} + \frac{\partial \gamma}{\partial z}\hat{\mathbf{k}}.$$

This basis is called the *dual basis*, sometimes also called the *reciprocal basis*, and these vectors are written with superscripts instead of subscripts.

Example 11.2 To calculate the dual basis in spherical polar coordinates, we again start with the position vector

$$\mathbf{r} = r\cos(\phi)\sin(\theta)\hat{\mathbf{i}} + r\sin(\phi)\sin(\theta)\hat{\mathbf{j}} + r\cos(\theta)\hat{\mathbf{k}},$$

where

$$r^2 = x^2 + y^2 + z^2, \quad \theta = \cos^{-1}\left(\frac{z}{\sqrt{x^2 + y^2 + z^2}}\right), \quad \phi = \tan^{-1}\left(\frac{y}{x}\right).$$

Then,

$$\mathbf{e}^r = \nabla r = \frac{\partial r}{\partial x}\hat{\mathbf{i}} + \frac{\partial r}{\partial y}\hat{\mathbf{j}} + \frac{\partial r}{\partial z}\hat{\mathbf{k}}.$$

The easiest way to evaluate the partial derivatives is to note that, for example,

$$\frac{\partial r^2}{\partial x} = 2r\frac{\partial r}{\partial x} = 2x,$$

with similar results for the other derivatives of r, giving us

$$\mathbf{e}^r = \frac{x}{r}\hat{\mathbf{i}} + \frac{y}{r}\hat{\mathbf{j}} + \frac{z}{r}\hat{\mathbf{k}} = \sin(\theta)\cos(\phi)\hat{\mathbf{i}} + \sin(\theta)\sin(\phi)\hat{\mathbf{j}} + \cos(\theta)\hat{\mathbf{k}}.$$

This vector has a magnitude of 1 and looks very much like the natural base vector $\mathbf{e}_r$ from Example 11.1. Let us move on and see if this pattern continues. Using the derivative of $\cos^{-1}(\theta)$ from Appendix B with the chain rule we get that

$$\mathbf{e}^\theta = \nabla\theta = \frac{1}{r}\cos(\theta)\cos(\phi)\hat{\mathbf{i}} + \frac{1}{r}\cos(\theta)\sin(\phi)\hat{\mathbf{j}} - \frac{1}{r}\hat{\mathbf{k}}.$$

This vector looks similar to the corresponding natural basis vector from Example 11.1 but with an extra factor of $1/r$. The magnitude of $\mathbf{e}^\theta$ is r, so the unit vector $\hat{\mathbf{e}}^\theta$ in the dual basis is the same as the unit vector $\hat{\mathbf{e}}_\theta$ in the natural basis.

Exercise 11.1.5 Calculate the remaining basis vector $\mathbf{e}^\phi$ in the dual basis for spherical polar coordinates from Example 11.2 and compare it with the corresponding vector in the natural basis from Example 11.1.

Exercise 11.1.6 Show that the basis vectors in the dual vector basis are orthogonal to each other.

These examples seem to indicate that the dual basis vectors and the natural basis vectors point in the same direction, but can have different magnitudes. This is true only for *orthogonal* coordinate systems, where the coordinate surfaces intersect each other at right angles, and this is what makes such coordinate systems nice to work with. However, in special circumstances we might want to use coordinates that are not orthogonal. For example, some large-scale ocean and atmosphere computer models make use of nonorthogonal coordinate systems, especially when looking at the vertical coordinate (Griffies, 2004). That having been said, most of the coordinate systems you will encounter will be orthogonal, but the methods we explore in this chapter can be applied to both.

The natural and the dual sets of vectors form equally valid sets of basis vectors, so we can write any arbitrary vector as a linear combination of either the natural basis vectors or the dual basis vectors, $\mathbf{u} = u^i\mathbf{e}_i = u_i\mathbf{e}^i$. The components u^i of the vector are called the *contravariant components of the vector* and the components u_i are called the *covariant components of the vector*. How do these different type of components transform under changes of coordinates? Let us recall what happens to a vector under a coordinate transformation. To make things concrete, Let us consider a particle moving in two dimensions (we choose two dimensions so as to make our mathematics a little easier, these results will hold in three dimensions as well). The position vector of the particle is $\mathbf{s} = x(t)\hat{\mathbf{i}} + y(t)\hat{\mathbf{j}}$, where t is a parameter (e.g., time) along the particle's trajectory. In a time

δt, the particle will move to a new position $\delta\mathbf{s}$ that has components $(x + \delta x, y + \delta y)$ and the velocity of the particle is given by the vector $\mathbf{v}$ that has components

$$v^x = \frac{\delta x}{\delta t}, \quad v^y = \frac{\delta y}{\delta t}. \tag{11.3}$$

We now convert from Cartesian coordinates to polar coordinates (r, θ). The displacements corresponding to δx and δy are

$$\delta r = \frac{\partial r}{\partial x}\delta x + \frac{\partial r}{\partial y}\delta y, \quad \delta\theta = \frac{\partial \theta}{\partial x}\delta x + \frac{\partial \theta}{\partial y}\delta y,$$

so the components of the velocity are

$$v^r = \frac{\partial r}{\partial x}v_x + \frac{\partial r}{\partial y}v_y, \quad v^\theta = \frac{\partial \theta}{\partial x}v_x + \frac{\partial \theta}{\partial y}v_y. \tag{11.4}$$

The components of a velocity vector are *contravariant components* and transform like the components of a natural basis. Other vectors that transform in the same way include the position vector and acceleration.

Let us look at a different vector, the gradient of a scalar field $g(x, y)$. In Cartesian coordinates, the components of the gradient are

$$(\nabla g)_x = \frac{\partial g}{\partial x}, \quad (\nabla g)_y = \frac{\partial g}{\partial y}$$

and in polar coordinates they are

$$(\nabla g)_r = \frac{\partial g}{\partial r}, \quad (\nabla g)_\theta = \frac{\partial g}{\partial \theta}.$$

Using the chain rule, we get

$$(\nabla g)_r = \frac{\partial g}{\partial x}\frac{\partial x}{\partial r} + \frac{\partial g}{\partial y}\frac{\partial y}{\partial r} = (\nabla g)_x\frac{\partial x}{\partial r} + (\nabla g)_y\frac{\partial y}{\partial r} \tag{11.5a}$$

$$(\nabla g)_\theta = \frac{\partial g}{\partial x}\frac{\partial x}{\partial \theta} + \frac{\partial g}{\partial y}\frac{\partial y}{\partial \theta} = (\nabla g)_x\frac{\partial x}{\partial \theta} + (\nabla g)_y\frac{\partial y}{\partial \theta}. \tag{11.5b}$$

The components of the gradient of a scalar are *covariant components* and transform like the components of a dual basis. When we transformed the velocity vector, we needed derivatives of the form $\partial r/\partial x$, but for the gradient we required $\partial x/\partial r$. So, we have two different types of vector defined by the way their components transform under a change of coordinates. In general, if we have two coordinate systems, an unprimed coordinate system (x, y, z) and a primed system (x', y', z'), then

$$(V')^i = \sum_j \frac{\partial (x')^i}{\partial x^j}V^j \quad \text{(contravariant vector)}, \tag{11.6}$$

$$(V')_i = \sum_j \frac{\partial x^j}{\partial (x')^i}V_j \quad \text{(covariant vector)}. \tag{11.7}$$

By convention, the components of contravariant vectors are written with superscript indices, and the components of covariant vectors with indices as subscripts.

We can now start to look at tensors. As you might guess from our discussion of the shear tensor at the start of this chapter, a tensor is a generalization of the concepts of a scalar and a vector. In fact, scalar quantities are *tensors of rank* 0 and vectors are *tensors of rank* 1. A tensor of rank n has components that have n indices, each of which can take D values, where D is the dimension of the space the tensor is defined in; in most cases we will come across, the dimension of the space will be 3. A scalar quantity does not have any indices, so it is a rank-0 tensor, whereas the components of vectors have a single index, making them rank-1 tensors. The way in which the components of the tensor transform under coordinate transformations determines the type of tensor; so, for example, we can define three types of rank-2(second rank) tensor according to

$$(T')^{ij} = \sum_{k,n} \frac{\partial (x')^i}{\partial x^k} \frac{\partial (x')^j}{\partial x^n} T^{kn} \quad \text{(contravariant tensor),} \tag{11.8}$$

$$(T')_{ij} = \sum_{k,n} \frac{\partial x^k}{\partial (x')^i} \frac{\partial x^n}{\partial (x')^i} T_{kn} \quad \text{(covariant vector),} \tag{11.9}$$

$$(T')^i_j = \sum_{k,n} \frac{\partial (x')^i}{\partial x^k} \frac{\partial x^n}{\partial (x')^i} T^k_n \quad \text{(mixed tensor).} \tag{11.10}$$

Example 11.3 We first met the Kronecker delta in Chapter 4, where we just treated it as a simple mathematical object. However, we can show that it is in fact a second-rank tensor. For example, Let us consider δ^i_j. We will use the fact that the derivative of one coordinate variable with respect to another is either 1 (if the two coordinate variables are the same) or 0 (if they are different). Then, under a coordinate transformation from unprimed to primed coordinates, we find

$$(\delta')^i_j = \frac{\partial (x')^i}{\partial x^k} \frac{\partial x^n}{\partial (x')^j} \delta^k_n = \frac{\partial (x')^i}{\partial x^k} \frac{\partial x^k}{\partial (x')^j} = \frac{\partial (x')^i}{\partial (x')^j} = (\delta')^i_j,$$

therefore, the Kronecker delta is a mixed second-rank tensor (Equation (11.10)).

You will see many different notations used to represent tensors, and many of the tensors that we come across can be represented as matrices. For example, Let us look at Equation (11.8):

$$(T')^{ij} = \sum_{k,n} \frac{\partial (x')^i}{\partial x^k} \frac{\partial (x')^j}{\partial x^n} T^{kn}.$$

The quantities $\partial (x')^i / \partial x^k = M_{ik}$ are matrices, so we can write this equation as a matrix equation,

$$t'_{ij} = \sum_{k,n} M_{ik} M_{jn} t_{kn} = \sum_{k,n} M_{ik} t_{kn} (M_{nj})^T = \mathbf{MtM}^T,$$

where we have used t_{ij} to represent the matrix form of the tensor T_{ij}. We used a matrix to represent the stress tensor at the beginning of this chapter.

A common tool for simplifying the way we write tensor equations is to use the Einstein summation convention that we met in Section 4.6. Recall that this is a compact way of writing a summation over repeated indices, so that

$$v_i u^i = \sum_i v_i u^i.$$

When tensors are involved, this is often called a *contraction* over the index i.

11.2 Metric Tensors

We know that, for a given set of coordinates we can define a natural basis $\mathbf{e}_i$ such that any vector $\mathbf{U}$ can be written as $\mathbf{U} = U^i \mathbf{e}_i$. The dot product between vectors $\mathbf{U}$ and $\mathbf{V}$ is then

$$\mathbf{U} \cdot \mathbf{V} = (U^i \mathbf{e}_i) \cdot (V^j \mathbf{e}_j) = \mathbf{e}_i \cdot \mathbf{e}_i U^i V^j = g_{ij} U^i V^j,$$

where $g_{ij} = \mathbf{e}_i \cdot \mathbf{e}_i$ is the *metric tensor* that we have met before (Section 7.4). In exactly the same way we can also define the contravariant components of the metric tensor by

$$\mathbf{U} \cdot \mathbf{V} = (U_i \mathbf{e}^i) \cdot (V_j \mathbf{e}^j) = \mathbf{e}^i \cdot \mathbf{e}^i U_i V_j = g^{ij} U_i V_j.$$

Exercise 11.2.1 Show that $g^{ij} g_{jk} = \delta^i_k$.

You will often see the metric tensor used to raise or lower indices on tensors, so that $U_i = g_{ij} U^j$. This converts the covariant components of a tensor to the contravariant components of the same tensor; in other words we switch between the natural and dual bases. We can relate the covariant and contravariant components of the metric in a different way by representing the metric as a matrix. If we use a matrix $\mathbf{G}$ to represent the contravariant components of the metric and the matrix $\tilde{\mathbf{G}}$ to represent the covariant components, then we can write the equation $g^{ij} g_{jk} = \delta^i_k$ as $\mathbf{G}\tilde{\mathbf{G}} = \mathbf{I}$, where $\mathbf{I}$ is the identity matrix. So, the matrices representing the contravariant and covariant forms of the metric tensor are inverses of each other. The metric tensor plays an important role in that it defines distances.[1] If we have a position vector $\mathbf{r}$ in a general coordinate system (u^i), then we can write $d\mathbf{r} = du^i \mathbf{e}_i$, and the square of the length of this increment in position vector is

$$ds^2 = d\mathbf{r} \cdot d\mathbf{r} = \mathbf{e}_i \cdot \mathbf{e}_j du^i du^j = g_{ij}\, du^i\, du^j.$$

In Cartesian coordinates, this is just a statement of Pythagoras' theorem.

Example 11.4 If we have a set of basis vectors for a given coordinate system, we can easily find the metric by taking the dot products of the basis vectors. For example, in spherical polar coordinates we have the covariant basis vectors

[1] In mathematics, a metric is a generalization of the concept of a distance, and it is not necessarily related to the metric system of units.

$$\mathbf{e}_r = \cos(\phi)\sin(\theta)\hat{\mathbf{\imath}} + \sin(\phi)\sin(\theta)\hat{\mathbf{\jmath}} + \cos(\theta)\hat{\mathbf{k}},$$
$$\mathbf{e}_\phi = -r\sin(\phi)\sin(\theta)\hat{\mathbf{\imath}} + r\cos(\phi)\sin(\theta)\hat{\mathbf{\jmath}},$$
$$\mathbf{e}_\theta = r\cos(\phi)\cos(\theta)\hat{\mathbf{\imath}} + r\sin(\phi)\cos(\theta)\hat{\mathbf{\jmath}} - r\sin(\theta)\hat{\mathbf{k}}.$$

We know that these vectors are orthogonal, so only the dot products of the basis vectors with themselves will be nonzero, and

$$g_{rr} = \cos^2(\phi)\sin^2(\theta) + \sin(\phi)\sin(\theta) + \cos^2(\theta) = 1,$$
$$g_{\phi\phi} = r^2\sin^2(\phi)\sin^2(\theta) + r^2\cos^2(\phi)\sin^2(\theta) = r^2\sin^2(\theta),$$
$$g_{\theta\theta} = r^2\cos^2(\phi)\cos^2(\theta) + r^2\sin^2(\phi)\cos^2(\theta) + r^2\sin^2(\theta) = r^2,$$

so

$$ds^2 = dr^2 + r^2 d\phi^2 + r^2\sin^2(\theta)d\theta^2,$$

and in matrix form

$$g_{ij} = \begin{pmatrix} 1 & 0 & 0 \\ 0 & r^2 & 0 \\ 0 & 0 & r^2\sin^2(\theta) \end{pmatrix}.$$

Exercise 11.2.2 Calculate the metric tensor for cylindrical coordinates in three dimensions.

11.3 Manipulating Tensors

Because we can represent a tensor in three dimensions as a matrix, we can manipulate tensors in many of the same ways we manipulate matrices. We can add and subtract tensors; i.e., we add and subtract the components, which implies that we can only add and subtract tensors that have the same shape. So, if U^{ij} and V^{ij} have the same shape, then $T^{ij} = U^{ij} + V^{ij}$.

Tensors, like matrices, can have certain symmetries that make calculations easier. For example, a general second-rank tensor in three dimensions (the type of tensor we meet most often in the geosciences) can be represented as a (3×3) matrix, which in general has nine independent components; that is, we can pick the nine slots in the matrix independently of each other. This means that an equation involving such a tensor is actually not one equation, but nine—just as a vector equation is a compact way of writing three equations, one for each component of the vectors. A *symmetric* second-rank tensor has components such that $T^{ij} = T^{ji}$, which is basically the same as a symmetric matrix. In this case, the three diagonal elements can be picked independently, as can the three elements above or below the diagonal; giving values to the three elements above the diagonal automatically gives values for the three elements below the diagonal. So, a symmetric second-rank tensor in three dimensions has six independent components. An *antisymmetric* second-rank tensor is defined such that its components satisfy $T^{ij} = -T^{ji}$.

Exercise 11.3.1 Show that an antisymmetric second-rank tensor in three dimensions has only three independent components.

Exercise 11.3.2 Show that the metric tensor is a symmetric tensor.

In general we can decompose any tensor into a symmetric and antisymmetric part:

$$T^{ij} = \frac{1}{2}(T^{ij} + T^{ji}) + \frac{1}{2}(T^{ij} - T^{ji}). \tag{11.11}$$

The relationship between tensors and matrices runs deeper, and we can do things like diagonalize a tensor just as we can a matrix.

We have already seen that we can take a scalar product of vectors using a contraction of indices, and this process generalizes easily to other tensors. However, if we multiply the components of two tensors, we create a new tensor that has the same rank as the sum of the ranks of the two initial tensors. This is called a *direct product*. That is, if we multiply the components of two vectors (i.e., rank 1-tensors) together, we get a rank-2 tensor. For example, if **U** and **V** are covariant and contravariant vectors, then their product $u_i v^j$ is a mixed second-rank tensor T_i^j. We can show this by considering the transformation

$$(T')_i^j = (a')^j (b')_i = \left(\frac{\partial (x')^j}{\partial x^k} a^k\right)\left(\frac{\partial x^n}{\partial (x')^j} b_n\right) = \frac{\partial (x')^j}{\partial x^k}\frac{\partial x^n}{\partial (x')^j} a^k b_n = \frac{\partial (x')^j}{\partial x^k}\frac{\partial x^n}{\partial (x')^j} T_n^k,$$

This is why the gradient of a vector ($\nabla \mathbf{v}$) is a tensor, not a vector.

11.4 Derivatives of Tensors

Vectors are simply tensors, so we can revisit the derivative of a vector using the machinery of tensors. Doing this may seem tedious at first, but it has the advantage of showing us explicitly why taking the derivative of a vector is more complicated than taking the derivative of a scalar function. It also shows us how we can take derivatives of tensors in general.

We know that we can write a vector in terms of the basis vectors of the coordinate system being used. If we differentiate that vector, we also have to differentiate the basis vectors as well. For Cartesian (x, y, z) coordinates the basis vectors are all constants, so their derivatives are zero. But in other coordinate systems, such as spherical coordinates, this is not the case. The other problem we have to deal with is that we define vectors and tensors according to how they transform under a coordinate transformation. We would like to have the situation where the derivative of a vector transforms as a tensor—the derivative of a vector is a two-index object, so it should transform as a tensor. So, Let us see if it does. Consider a vector **V** with contravariant components V^i in the coordinate system (x^1, x^2, x^3) and components $\tilde{V}^i$ in the coordinates $(\tilde{x}^1, \tilde{x}^2, \tilde{x}^3)$, so that

$$\tilde{V}^i = \frac{\partial x^i}{\partial \tilde{x}_k} V^k.$$

We can differentiate this expression with respect to $\tilde{x}^j$ to see how the derivative of the vector transforms,

$$\begin{aligned}\frac{\partial \tilde{V}^i}{\partial \tilde{x}^j} &= \frac{\partial x^i}{\partial \tilde{x}_k}\frac{\partial V^k}{\partial \tilde{x}^j} + \frac{\partial^2 x^i}{\partial \tilde{x}^j \partial \tilde{x}^k} V^k \\ &= \frac{\partial V^k}{\partial \tilde{x}^j}\mathbf{e}_k + V^k\left(\frac{\partial \mathbf{e}_k}{\partial \tilde{x}^j}\right). \end{aligned} \tag{11.12}$$

We know that we must be able to write each of the terms on the right-hand side of Equation (11.12) in terms of the basis vectors, so we can write the second term on the right-hand side of the equation as

$$V^k\left(\frac{\partial \mathbf{e}_k}{\partial \tilde{x}^j}\right) = V^k \Gamma^m_{jk}\mathbf{e}_m,$$

where Γ^m_{jk} is a set of, possibly coordinate dependent, coefficients that we have to determine. We define these coefficients in such a way that the derivative of the vector transforms as a tensor under coordinate transformations. A somewhat lengthy but simple calculation then shows that we can define the *covariant derivative* of a vector such that

$$V^i_{,k} = \frac{\partial V^i}{\partial x^k} + V^m \Gamma^i_{mk} \quad \text{with} \quad \Gamma^i_{mk} = \frac{1}{2}g^{in}\left(\frac{\partial g_{mn}}{\partial x^k} + \frac{\partial g_{kn}}{\partial x^m} - \frac{\partial g_{mk}}{\partial x^n}\right), \tag{11.13}$$

where Γ^i_{mk} are the *Christoffel symbols of the second kind.*[2] We have also used a common notation for the covariant derivative by representing it using a comma.

Example 11.5 In spherical coordinates $x = r\sin(\theta)\cos(\phi)$, $y = r\sin(\theta)\cos(\phi)$, $z = r\cos(\theta)$, the square of the line element is

$$ds^2 = dr^2 + r^2 d\theta^2 + r^2\sin^2(\theta)d\phi^2,$$

so that the components of the metric tensor are

$$g_{rr} = 1, \quad g_{\theta\theta} = r^2, \quad g_{\phi\phi} = r^2\sin^2(\theta),$$

with all other values being zero. Using Equation (11.13) we can methodically calculate the Christoffel symbols for the different combinations of indices. For example,

$$\Gamma^r_{\theta\theta} = \frac{1}{2}g^{rk}\left(\frac{\partial g_{\theta k}}{\partial \theta} + \frac{\partial g_{\theta k}}{\partial \theta} - \frac{\partial g_{\theta\theta}}{\partial x^k}\right).$$

The two derivatives with respect to θ are both zero, leaving only the three terms obtained from the contraction (i.e., summation) over the index k in the last term. However, we know that the metric is diagonal, so the only term that is nonzero is

$$-\frac{1}{2}g^{rr}\left(\frac{\partial g_{\theta\theta}}{\partial r}\right) = -r.$$

Similarly, we can find the remaining nonzero Christoffel symbols.

---

[2] This is named after Elwin Bruno Christoffel (1829–1900), whose work helped develop differential geometry and tensor calculus. There is a Christoffel symbol of the first kind, but they are not as commonly used.

Exercise 11.4.1 Calculate the remaining Christoffel symbols for spherical polar coordinates.

Exercise 11.4.2 Use Equation (11.13) to calculate the divergence of a vector **F** in spherical polar coordinates.

11.5 Further Reading

Our discussion of tensors has been quite brief. A more detailed and very accessible introduction can be found in *A Guided Tour of Mathematical Methods for the Physical Sciences* by Snieder and van Wijk (2015). Two other good texts that go into more detail are *A Student's Guide to Vectors and Tensors* by Fleisch (2012) and *Vector and Tensor Analysis with Applications* by Borisenko and Tarapov (1968). A very clear discussion of the use of tensors in geology can be found in the books *Fundamentals of Structural Geology* by Pollard and Fletcher (2005) and *Structural Geology Algorithms* by Allmendinger et al. (2012). Tensors really belong to a branch of mathematics called *differential geometry*, and an excellent, modern exposition of this can be found in the book *Geometrical Methods of Mathematical Physics* by Schutz (1980), though the approach is very different from those in the other books listed above.

Problems

11.1 Consider two coordinate systems, the usual orthogonal Cartesian coordinate system (x, y, z) and the coordinate system (α, β, γ) such that a position vector is

$$\mathbf{r}(\alpha, \beta, \gamma) = (\alpha - \beta)\hat{\mathbf{\imath}} - (\alpha + \beta)\hat{\mathbf{\jmath}} + (\alpha\beta - \gamma)\hat{\mathbf{k}}.$$

Calculate the natural and dual basis vectors for the coordinate system (α, β, γ) and determine if the coordinate system is orthogonal or not.

11.2 Show that the eigenvalues of the stress tensor (Equation (11.1)) satisfy the cubic equation

$$\lambda^3 - I_1\lambda^2 - I_2\lambda - I_3 = 0,$$

and give expressions for I_1, I_2, and I_3 in terms of the components of the stress tensor. The quantities I_1, I_2, and I_3 are called *invariants* of the tensor and have important physical interpretations.

11.3 Consider the Levi-Civita symbol (Equation 4.83). If a coordinate transformation from a coordinate system $\{x^i\}$ to a coordinate system $\{x'^i\}$ is written as a matrix, i.e.,

$$\frac{\partial (x')^j}{\partial x^k} = M^j_k,$$

then show that $\epsilon'_{i_1 i_2 i_3} = \pm\det(M)\epsilon_{i_1 i_2 i_3}$.

11.4 An antisymmetric tensor changes sign when you interchange two indices (i.e., $T_{ij} = -T_{ji}$). If A_{ij} is an antisymmetric tensor and S^{ij} is a symmetric tensor (i.e., $S^{ij} = T^{ji}$), show that $A_{ij}S^{ij} = 0$.

11.5 What are the values of the following quantities? 1. δ_{kk}, 2. $\delta_{ij}\delta_{ij}$.

11.6 Consider an arbitrary second-rank tensor. Show that it can be written as the sum of three tensors: a symmetric tensor with a zero trace, a diagonal tensor, and an antisymmetric tensor. What are the values of the elements in the diagonal tensor?

11.7 Consider a general second-rank tensor that has covariant components T_{ij} in a basis $\{\hat{\mathbf{e}}_1, \hat{\mathbf{e}}_2, \hat{\mathbf{e}}_3\}$. The coordinates are transformed to new coordinates $\{\hat{\mathbf{e}}'_1, \hat{\mathbf{e}}'_2, \hat{\mathbf{e}}'_3\}$ that are formed by a rotation of $\pi/2$ about the axis in the direction of $\hat{\mathbf{e}}_3$.

1. Write down the equations transforming $\{\hat{\mathbf{e}}_1, \hat{\mathbf{e}}_2, \hat{\mathbf{e}}_3\}$ into $\{\hat{\mathbf{e}}'_1, \hat{\mathbf{e}}'_2, \hat{\mathbf{e}}'_3\}$.
2. Write down the rotation matrix that transforms $\{\hat{\mathbf{e}}_1, \hat{\mathbf{e}}_2, \hat{\mathbf{e}}_3\}$ into $\{\hat{\mathbf{e}}'_1, \hat{\mathbf{e}}'_2, \hat{\mathbf{e}}'_3\}$.
3. Use the matrix form of the tensor transformation law to show that

$$\begin{pmatrix} T'_{11} & T'_{12} & T'_{13} \\ T'_{21} & T'_{22} & T'_{23} \\ T'_{31} & T'_{32} & T'_{33} \end{pmatrix} = \begin{pmatrix} T_{22} & -T_{21} & T_{23} \\ -T_{12} & T_{11} & -T'_{13} \\ T_{32} & -T_{31} & T_{33} \end{pmatrix}.$$

4. Now assume that the components of the tensor are unchanged by the transformation (i.e., the tensor is *isotropic*, so any rotation of coordinates leaves the components of the tensor unchanged) so that $T'_{ij} = T_{ij}$. Show that the tensor must have the form $a\delta_{ij}$, where a is a constant.

11.8 Diagonalize the tensor

$$T_{ij} = \begin{pmatrix} 2 & -1 & 0 \\ -1 & 2 & 0 \\ 0 & 0 & 4 \end{pmatrix}.$$

11.9 Show that an antisymmetric tensor A_{ij} can be written as

$$A_{ij} = \frac{1}{2}A_{mn}\left(\delta_{im}\delta_{jn} - \delta_{jm}\delta_{in}\right) = -\frac{1}{2}A_{mn}\epsilon_{ijk}\epsilon_{nmk}.$$

11.10 Show that for a second-rank tensor T_{ij}, $T'_{ii} = T_{jj}$ under a coordinate transformation.

Appendix A Units and Dimensions

All measurable, physical quantities have associated with them a numerical value (determined by the units we are using) and a dimension. The dimension of a quantity defines what that quantity is: a velocity always has dimensions of a length per unit time ($[L][T]^{-1}$).[1] The units that we use to measure length and time determine the numerical value we assign to that measurement.

As an example, consider a measurement of water velocity yielding a value of $10.3\ \text{cm s}^{-1}$. This is the same as $0.103\ \text{m s}^{-1}$ or 0.200 knots.[2] Although the numerical values are all different, these are all the same speed. We know that each of these numbers represents a speed because in each case the units consist of a length divided by a time; length and time are the *dimensions* that define the variable *velocity*, and the *units* (e.g., miles, centimeters, hours) used to measure the length and time determine the actual numerical value of the velocity—even though all the velocities given here have the same magnitude.

There are many standard systems of units that have been designed (e.g., cgs, SI), and for any consistent system of units, one first requires a set of dimensions that are considered to be fundamental. This set of dimensions is then used to derive the dimensions of all other quantities (e.g., length and time can be combined to define velocity and acceleration). The choice of fundamental dimensions is, to some degree, arbitrary. For example, we could choose a system with length ([L]), mass ([M]), and time ([T]) as fundamental dimensions, or we could choose length ([L]), force ([F]), and time ([T]) (Barenblatt, 1996); in the former system, density has the familiar dimensions of mass per length cubed ($[M][L]^{-3}$), but in the latter it has dimensions of force multiplied by time squared per length to the fourth power ($[F][T]^2[L]^{-4}$).

Once we have chosen a consistent set of dimensions we must choose standard magnitudes for each of them. These are the units that we choose to use. There is nothing in the structure of nature that leads us to choose one set of units over another; but for the system to work, there must be general agreement on what to use. Unfortunately, systems of units have changed over time, and convention is frequently violated. This means that as practicing scientists we must be able to reliably convert between different systems of units.

[1] Dimensions are usually represented by a single capital letter surrounded by square brackets. For example, length is represented as [L], time as [T], mass as [M], etc.

[2] A knot is 1 nautical mile per hour, and a nautical mile is 6076.1 ft, 1852 m, or 1.1508 miles.

A.1 International System of Units

The system of units in use in the world today is the International System of Units (or SI system), established in 1960. The SI system is derived from seven basic dimensions, with seven corresponding base units (Table A.1). All other units within the SI system are derived from these seven fundamental ones. These *derived units* are made up of combinations of the fundamental units raised to some power (see Table A.2); one can show mathematically that this is the *only* consistent way to combine dimensions and units (Barenblatt, 1996). For the sake of convenience 22 derived units have been given special names and symbols (Table A.3).

The SI system also specifies 20 *prefixes* for dealing with decimal multiples and submultiples of the units defined in the SI system (Table A.4).[3] These apply only to decimal multiples (i.e., powers of ten). Multiple prefixes cannot be used, so when talking about mass, the prefixes listed in Table A.4 are used with "gram" and not "kilogram"—for example, we do not talk about 1 μkg, but instead refer to 1 mg.

There are some units that are in common use that are not part of the SI system. Some are accepted and can be used along with SI units, others are not; a partial list of those that can be officially used is given in Table A.5. There are others–such as the nautical mile (1852 m),

Table A.1 Base dimensions in the SI system

Base dimension	Symbol	Unit	Unit symbol
Mass	[M]	kilogram	kg
Length	[L]	meter	m
Time	[T]	second	s
Electric current	[I]	ampere	A
Thermodynamic temperature	[θ]	kelvin	K
Amount of substance	[N]	mole	mol
Luminous intensity	[J]	candela	cd

Table A.2 Derived quantities in the SI system

Quantity name	Dimension	Unit
Area	$[L]^2$	m^2
Volume	$[L]^3$	m^3
Velocity	$[L][T]^{-1}$	$m\ s^{-1}$
Acceleration	$[L][T]^{-2}$	$m\ s^{-2}$
Mass density	$[M][L]^{-3}$	$kg\ m^{-3}$
Specific volume	$[L]^3[M]^{-1}$	$m^3\ kg^{-1}$
Amount-of-substance concentration	$[N][L]^{-3}$	$mol\ m^{-3}$

[3] Note that the kilogram is the only SI unit that starts with a prefix already.

Table A.3 Special derived quantities within the SI system

Derived quantity	Name (symbol)		Base units
Plane angle	radian (rad)		$m\ m^{-1} = 1$
Solid angle	steradian (sr)		$m^2\ m^{-2} = 1$
Frequency	hertz (Hz)		s^{-1}
Force	newton (N)		$m\ kg\ s^{-2}$
Pressure, stress	pascal (Pa)	$N\ m^{-2}$	$m^{-1}\ kg\ s^{-2}$
Energy, work	joule (J)	N m	$m^2\ kg\ s^{-2}$
Power, radiant flux	watt (W)	$J\ s^{-1}$	$m^2\ kg\ s^{-3}$
Electric charge	coulomb (C)		s A
Potential difference	volt (V)	$W\ A^{-1}$	$m^2\ kg\ s^{-3}\ A^{-1}$
Capacitance	farad (F)	$C\ V^{-1}$	$m^{-2}\ kg^{-1}\ s^4\ A^2$
Electric resistance	ohm (Ω)	$V\ A^{-1}$	$m^2\ kg\ s^{-3}\ A^{-2}$
Electric conductance	siemens (S)	$A\ V^{-1}$	$m^{-2}\ kg^{-1}\ s^3\ A^2$
Magnetic flux	weber (Wb)	V s	$m^2\ kg\ s^{-2}\ A^{-2}$
Magnetic flux density	tesla (T)	$Wb\ m^{-2}$	$kg\ s^{-2}\ A^{-1}$
Inductance	henry (H)	$Wb\ A^{-1}$	$m^2\ kg\ s^{-2}\ A^{-2}$
Celsius temperature	celsius (°C)		K
Luminous flux	lumen (lm)	cd sr	$m^2\ m^{-2}\ cd$
Illuminance	lux (lx)	$lm\ m^{-2}$	$m^{-2}\ cd$
Radionuclude activity	becquerel (Bq)		s^{-1}
Absorbed dose	gray (Gy)	$J\ kg^{-1}$	$m^2\ s^{-2}$
Dose equivalent	sievert (Sv)	$J\ kg^{-1}$	$m^2\ s^{-2}$
Catalytic activity	katal (kat)		$s^{-1}\ mol$

Table A.4 Prefixes in the SI system

Factor	Name	Symbol	Factor	Name	Symbol
10^{24}	yotta	Y	10^{-1}	deci	d
10^{21}	zetta	Z	10^{-2}	centi	c
10^{18}	exa	E	10^{-3}	milli	m
10^{15}	peta	P	10^{-6}	micro	μ
10^{12}	tera	T	10^{-9}	nano	n
10^{9}	giga	G	10^{-12}	pico	p
10^{6}	mega	M	10^{-15}	femto	f
10^{3}	kilo	k	10^{-18}	atto	a
10^{2}	hecto	h	10^{-21}	zepto	z
10^{1}	deka	da	10^{-24}	yocto	y

the knot (1 nautical mile per hour), the hectare (10^4 m), the bar (10^5 Pa)—that also lie outside of the SI system but can be used if they are defined in the documents where they are used. There are other units that are not part of the SI system but are generally restricted to use within a specialized field. These generally arise because to use the SI system would

Table A.5 Units outside the SI system that can officially be used

Name	Symbol	Value in SI units
minute (time)	min	1 min = 60 s
hour	h	1 h = 60 min = 3 600 s
day	d	1 d = 24 h = 86 400 s
degree (angle)	°	$1° = (\pi/180)$ rad
minute (angle)	′	$1' = (1/60)° = (\pi/10\,800)$ rad
second (angle)	″	$1'' = (1/60)' = (\pi/648\,000)$ rad
liter	L	$1\text{ L} = 1\text{ dm}^3 = 10^{-3}\text{ m}^3$
metric ton (tonne)	t	$1\text{ t} = 10^3$ kg

require carrying around multiple factors of 10. One such unit used by oceanographers is the Sverdrup (Sv); this is used to express volume transport in the ocean and is defined such that $1\text{ Sv} = 10^6\text{ m}^3\text{ s}^{-1}$.

A.2 Converting between Units

Being able to convert between different sets of units is a critical skill for a scientist to have. But why? Although most scientists today try to use SI units, different units have been used in the past. So, if you are comparing your results with those obtained in the past, you will likely have to convert between system of units. Another reason for having to convert units is that different measurements of the same quantity may have different units. For example, primary production is a measure of the photosynthetic productivity of a biological system, and it can be measured in terms of changes in oxygen or changes in carbon, and normalized per unit chlorophyll or biomass.

Different units are frequently used for reporting measurements of concentration, and so converting between them is an essential skill. As an example, how do we calculate the molar concentration of a solution of 1 mmol m^{-3} of calcium carbonate? The molar concentration will have units of moles per liter. So, putting in all the steps,

$$1\,\frac{\text{mmol}}{\text{m}^3} = 10^{-3}\,\frac{\text{mol}}{\text{m}^3} = 10^{-3}\,\frac{\text{mol}}{10^3\text{ L}} = 10^{-6}\,\frac{\text{mol}}{\text{L}} = 10^{-6}\text{ M} = 1\ \mu\text{M}.$$

There is an easy way to remember how to convert from one set of units to another—you multiply by 1, arranging the factors such that you get the units you need. So, in the previous example we needed to convert from mmol m^{-3} to mol L^{-1}, so, rewriting the previous calculation to make this more explicit gives

$$1\,\frac{\text{mmol}}{\text{m}^3} = 1\,\frac{\text{mmol}}{\text{m}^3} \times \frac{1\text{ mol}}{10^3\text{ mmol}} \times \frac{1\text{ m}^3}{10^3\text{ L}} = 10^{-6}\,\frac{\text{mol}}{\text{L}} = 1\ \mu\text{M}.$$

Each of the factors in the second part of the expression is equal to one, and you treat the units as arithmetical quantities and cancel them to get the units you want.

It is well worthwhile to get into the habit of checking the units of the results of all your calculations. This is because we all make mistakes, and it is all too easy to make a simple slip and, say, copy down m^{-2} from one line as m^{-3} in another line of a calculation.

Appendix B Tables of Useful Formulae

This appendix contains summaries of some useful relationships, tables of series, derivatives, and integrals

B.1 Properties of Basic Functions

B.1.1 Trigonometric Functions

$$\sin^2(\theta) + \cos^2(\theta) = 1$$
$$1 + \tan^2(\theta) = \sec^2(\theta)$$
$$1 + \cot^2(\theta) = \mathrm{cosec}^2(\theta)$$
$$1 - \cos(2\theta) = 2\sin^2(\theta)$$
$$\sin(2\theta) = 2\sin(\theta)\cos(\theta)$$
$$\cos(2\theta) = \cos^2(\theta) - \sin^2(\theta)$$
$$\tan(2\theta) = \frac{2\tan(\theta)}{1 + \tan^2(\theta)}$$
$$1 + \cos(2\theta) = 2\cos^2(\theta)$$
$$\tan(\theta + \phi) = \frac{\tan(\theta) + \tan(\phi)}{1 - \tan(\theta)\tan(\phi)}$$
$$\tan(\theta - \phi) = \frac{\tan(\theta) - \tan(\phi)}{1 + \tan(\theta)\tan(\phi)}$$

$$\sin(\theta + \phi) = \sin(\theta)\cos(\phi) + \cos(\theta)\sin(\phi)$$
$$\sin(\theta - \phi) = \sin(\theta)\cos(\phi) - \cos(\theta)\sin(\phi)$$
$$\cos(\theta + \phi) = \cos(\theta)\cos(\phi) - \sin(\theta)\sin(\phi)$$
$$\cos(\theta - \phi) = \cos(\theta)\cos(\phi) + \sin(\theta)\sin(\phi)$$
$$\sin(\theta)\sin(\phi) = \frac{1}{2}(\cos(\theta - \phi) - \cos(\theta + \phi))$$
$$\cos(\theta)\cos(\phi) = \frac{1}{2}(\cos(\theta - \phi) + \cos(\theta + \phi))$$
$$\sin(\theta)\cos(\phi) = \frac{1}{2}(\sin(\theta + \phi) + \sin(\theta - \phi))$$
$$\sin(\theta) + \sin(\phi) = 2\sin\left(\frac{\theta + \phi}{2}\right)\cos\left(\frac{\theta - \phi}{2}\right)$$
$$\sin(\theta) - \sin(\phi) = 2\cos\left(\frac{\theta + \phi}{2}\right)\sin\left(\frac{\theta - \phi}{2}\right)$$
$$\cos(\theta) + \cos(\phi) = 2\cos\left(\frac{\theta + \phi}{2}\right)\cos\left(\frac{\theta - \phi}{2}\right)$$
$$\cos(\theta) - \cos(\phi) = -2\sin\left(\frac{\theta + \phi}{2}\right)\sin\left(\frac{\theta - \phi}{2}\right)$$

$$a^2 = b^2 + c^2 - 2bc\cos(A) \quad \text{Law of cosines (see Figure B.1)} \tag{B.1}$$

$$\frac{\sin(A)}{a} = \frac{\sin(B)}{b} = \frac{\sin(C)}{c} \quad \text{Law of sines (see Figure B.1)} \tag{B.2}$$

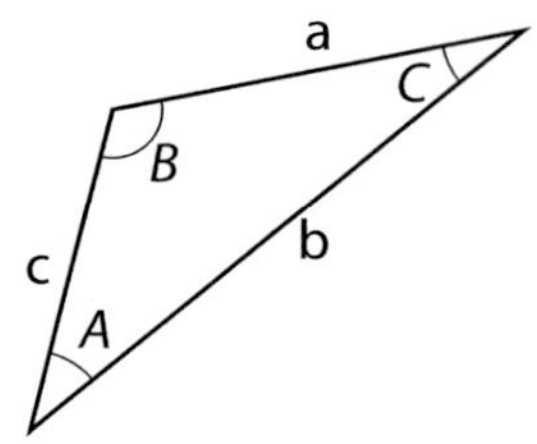

Figure B.1 A triangle showing the labelling of the triangle sides and angles used in the law of sines and law of cosines.

B.1.2 Logarithms and Exponentials

$$\log(xy) = \log(x) + \log(y) \qquad \ln(e^x) = x$$
$$\log(x/y) = \log(x) - \log(y) \qquad e^x e^y = e^{x+y}$$
$$\log(x^n) = n\log(x) \qquad e^x/e^y = e^{x-y}$$
$$\log_a(x) = (\log_b(x))/(\log_b(a)) \qquad (e^x)^n = e^{nx}$$

B.1.3 Hyperbolic Functions

Hyperbolic functions are less common, but they do have their uses. They are frequently found in functions representing photosynthesis, as well as in approximations for the density of seawater as a function of depth, for example.

$$\sinh(x) = \frac{1}{2}(e^x - e^{-x}) \qquad \cosh^2(x) - \sinh^2(x) = 1$$
$$\cosh(x) = \frac{1}{2}(e^x + e^{-x}) \qquad \cosh(x+y) = \cosh(x)\cosh(y) + \sinh(x)\sinh(y)$$
$$\tanh(x) = \sinh(x)/\cosh(x) \qquad \sinh(x+y) = \sinh(x)\cosh(y) + \cosh(x)\sinh(y)$$
$$\sinh(x) = -i\sin(ix) \qquad \sinh(2x) = 2\sinh(x)\cosh(x)$$
$$\cosh(x) = \cos(ix) \qquad \cosh(2x) = 2\cosh^2(x) - 1$$

B.2 Some Important Series

$$\exp(x) = \sum_{n=0}^{\infty} \frac{x^n}{n!} \qquad -\infty < x < \infty$$
$$\sin(x) = \sum_{n=0}^{\infty} \frac{(-1)^n x^{2n+1}}{(2n+1)!} = x - \frac{x^3}{3!} + \frac{x^5}{5!} \cdots \qquad -\infty < x < \infty$$

$$\cos(x) = \sum_{n=0}^{\infty} \frac{(-1)^n x^{2n}}{(2n)!} = 1 - \frac{x^2}{2!} + \frac{x^4}{4!} \cdots \qquad -\infty < x < \infty$$

$$\sinh(x) = \sum_{n=0}^{\infty} \frac{x^{2n+1}}{(2n+1)!} = x + \frac{x^3}{3!} + \frac{x^5}{5!} \cdots \qquad -\infty < x < \infty$$

$$\cosh(x) = \sum_{n=0}^{\infty} \frac{x^{2n}}{(2n)!} = 1 + \frac{x^2}{2!} + \frac{x^4}{4!} \cdots \qquad -\infty < x < \infty$$

$$\frac{1}{(1-x)} = \sum_{n=0}^{\infty} x^n = 1 + x + x^2 + x^3 + \cdots \qquad -1 \le x \le 1$$

$$\ln(1+x) = \sum_{n=1}^{\infty} \frac{(-1)^{n-1} x^n}{n} = x - \frac{x^2}{2} + \frac{x^3}{3} \qquad -1 < x \le 1$$

$$(1+x)^p = \sum_{n=0}^{\infty} \binom{p}{n} x^n = \sum_{n=0}^{\infty} \frac{(p-n+1)_n}{n!} x^n \qquad -1 < x < 1$$

B.3 Some Common Derivatives

$$\frac{d}{dx} x^n = n x^{n-1}$$

$$\frac{d}{dx} e^x = e^x$$

$$\frac{d}{dx} \ln(x) = \frac{1}{x}$$

$$\frac{d}{dx} \log_a(x) = \frac{1}{x \ln(a)}$$

$$\frac{d}{dx} a^x = a^x \ln(a)$$

$$\frac{d}{dx} \ln(f(x)) = \frac{f'(x)}{f(x)}$$

$$\frac{d}{dx} \sin(x) = \cos(x)$$

$$\frac{d}{dx} \cos(x) = -\sin(x)$$

$$\frac{d}{dx} \tan(x) = \sec^2(x)$$

$$\frac{d}{dx} \csc(x) = -\csc(x)\cot(x)$$

$$\frac{d}{dx} \sec(x) = \sec(x)\tan(x)$$

$$\frac{d}{dx} \cot(x) = -\csc^2(x)$$

$$\frac{d}{dx} \sin^{-1}(x) = \frac{1}{\sqrt{1-x^2}}$$

$$\frac{d}{dx} \cos^{-1}(x) = -\frac{1}{\sqrt{1-x^2}}$$

$$\frac{d}{dx} \tan^{-1}(x) = \frac{1}{1+x^2}$$

$$\frac{d}{dx} \csc^{-1}(x) = -\frac{1}{x\sqrt{1-x^2}}$$

$$\frac{d}{dx} \sec^{-1}(x) = \frac{1}{x\sqrt{1-x^2}}$$

$$\frac{d}{dx} \tan^{-1}(x) = -\frac{1}{1+x^2}$$

$$\frac{d}{dx} \sinh(x) = \cosh(x)$$

$$\frac{d}{dx} \cosh(x) = \sinh(x)$$

$$\frac{d}{dx} \tanh(x) = \operatorname{sech}^2(x)$$

$$\frac{d}{dx} \operatorname{sech}(x) = -\operatorname{sech}(x)\tanh(x)$$

$$\frac{d}{dx} \operatorname{csch}(x) = -\operatorname{csch}(x)\coth(x)$$

$$\frac{d}{dx} \coth(x) = -\operatorname{csch}^2(x)$$

B.4 Some Common Integrals

$$\int x^n\,dx = \frac{1}{n+1}x^{n+1} + c \qquad \int \frac{1}{a^2+x^2}\,dx = \frac{1}{a}\tan^{-1}\left(\frac{x}{a}\right) + c$$

$$\int \frac{1}{x}\,dx = \ln\mid x\mid + c \qquad \int \frac{x^2}{a^2+x^2}\,dx = x - a\tan^{-1}\left(\frac{x}{a}\right) + c$$

$$\int e^{ax}\,dx = \frac{1}{a}e^{ax} + c \qquad \int \frac{1}{\sqrt{a^2-x^2}}\,dx = \sin\left(\frac{x}{a}\right) + c$$

$$\int \sin(ax)\,dx = -\frac{1}{a}\cos(ax) + c \qquad \int xe^{ax}\,dx = \left(\frac{x}{a} - \frac{1}{a^2}\right)e^{ax} + c$$

$$\int \cos(ax)\,dx = \frac{1}{a}\sin(ax) + c \qquad \int \sin^2(ax)\,dx = \frac{x}{2} - \frac{\sin(2ax)}{4a} + c$$

$$\int \tan(ax)\,dx = -\frac{1}{a}\ln\mid\cos(ax)\mid + c \qquad \int \cos^2(ax)\,dx = \frac{x}{2} + \frac{\sin(2ax)}{4a} + c$$

$$\int \frac{1}{ax+b}\,dx = \frac{1}{a}\ln\mid ax+b\mid + c \qquad \int \tan^2(ax)\,dx = -x + \frac{1}{a}\tan(ax) + c$$

$$\int x\cos(ax)\,dx = \frac{1}{a^2}\cos(ax) + \frac{x}{a}\sin(ax) + c \qquad \int x\sin(ax)\,dx = \frac{1}{a^2}\sin(ax) - \frac{x}{a}\cos(ax) + c$$

B.5 Fourier and Laplace Transforms

Tables B.1 and B.2 contain some common Fourier and Laplace transform pairs.

Table B.1 Some common Fourier transforms

Function $f(t)$	Fourier transform $F(\omega)$	Function $f(t)$	Fourier transform $F(\omega)$
$\delta(t)$	1	$e^{i\omega_0 t}$	$\delta(\omega-\omega_0)$
$\cos(\omega_0 t)$	$\frac{1}{2}(\delta(\omega-\omega_0) + \frac{1}{2}(\delta(\omega+\omega_0)$	$\sin(\omega_0 t)$	$\frac{1}{2i}(-\delta(\omega-\omega_0) + \frac{1}{2i}(\delta(\omega+\omega_0)$
$u(t)$	$\pi\delta(\omega) + \frac{1}{i\omega}$	$u(t)e^{-\alpha t}$	$\frac{1}{\alpha+i\omega}$
$f(t-t_0)$	$F(\omega)e^{-i\omega t_0}$	$f(at)$	$\frac{1}{\mid a\mid}F\left(\frac{\omega}{a}\right)$

Table B.2 Some common Laplace transforms

Function $f(x)$	Laplace transform $F(s)$	Function $f(x)$	Laplace transform $F(s)$
1	$\frac{1}{s}$	$x^n,\ n = 0, 1, 2\cdots$	$\frac{n!}{s^{n+1}}$
$\sin(ax)$	$\frac{a}{s^2+a^2}$	$x\sin(ax)$	$\frac{2as}{(s^2+a^2)^2}$
$\cos(ax)$	$\frac{s}{s^2+a^2}$	$x\cos(ax)$	$\frac{s^2-a^2}{(s^2+a^2)^2}$
e^{ax}	$\frac{1}{s-a}$	$x^n\mathrm{e}^{ax}$	$\frac{n!}{(s-a)^{n+1}}$
$f(ax)$	$\frac{1}{a}F\left(\frac{a}{s}\right)$	$\delta(x-a)$	e^{-as}
$H(x-a)$	$\frac{1}{s}\mathrm{e}^{-as}$	$H(x-a)f(x-a)$	$\mathrm{e}^{-as}F(s)$
$\frac{1}{\sqrt{\pi t}}\mathrm{e}^{-a^2/4t}$	$\frac{1}{\sqrt{s}}\mathrm{e}^{-a\sqrt{s}}$	$\frac{a}{\sqrt{\pi t^3}}\mathrm{e}^{-a^2/4t}$	$\mathrm{e}^{-a\sqrt{s}}$

B.6 Further Reading

There are a great many books that cover elementary functions and mathematics. One good example is *Used Math for the First Two Years of College Science* by Swartz (1993). There are also two venerable and invaluable books that are worth seeking out. The first is *Handbook of Mathematical Functions with Formulas, Graphs, and Mathematical Tables* by Abramowitz and Stegun (1972), which contains a wealth of information about functions, integrals, integral transforms, and a whole lot more; an online successor, the *Digital Library of Mathematical Functions*, can be found at https://dlmf.nist.gov. The second is *Table of Integrals, Series, and Products* by Gradshteyn and Ryzhik (1980), with later editions containing over ten thousand entries, which is a wonderful resource for evaluating integrals and integral transforms.

Appendix C Complex Numbers

This appendix contains a brief reminder of some essential aspects of complex numbers. Complex numbers provide a means to simplifying mathematical expressions and manipulations and arise when we try to take the square root of a negative number. We shall see that, remarkably, functions that describe oscillations and periodic phenomenon (waves, tides, etc.) can be very compactly and usefully written in the form of complex numbers.[1]

C.1 Making Things Complex

The square root of minus one is usually written as

$$i = \sqrt{-1},$$

though some authors use $j = \sqrt{-1}$. A complex number is written $z = x \pm iy$, and it has a real component ($\text{Re}(z) = x$) and an imaginary component ($\text{Im}(z) = y$). It is important to realize that the imaginary component y of a complex number is itself a real number. The complex conjugate $\bar{z}$ of a complex number is the complex number with the sign of the imaginary component reversed. For example, if $z = 3 + 2i$, then $\bar{z} = 3 - 2i$.

Adding and subtracting complex numbers involves adding and subtracting the real and imaginary components of the two numbers separately, i.e.,

$$(a + ib) + (x + iy) = (a + x) + i(b + y).$$

Multiplication of two complex numbers involves cross multiplication of real and imaginary components, and we have to keep track of the powers of i that this generates. So, for example,

$$(a + ib)(x + iy) = ax + i^2 by + i(bx + ay) = (ax - by) + i(bx + ay).$$

If we multiply a complex number by its complex conjugate, then we obtain a real number: $(a + ib)(a - ib) = a^2 + b^2$. Division by a complex number involves regularizing the denominator such that it becomes a real number. This is achieved by multiplying and dividing by the complex conjugate of the denominator:

$$\frac{a + ib}{x + iy} = \frac{a + ib}{x + iy}\frac{x - iy}{x - iy} = \frac{ax + by) + i(bx - ay)}{x^2 + y^2}.$$

[1] The fact that complex numbers play such a significant role in the mathematics we use to describe the natural phenomena has led some to wonder about the role that mathematics plays in nature (Wigner, 1960).

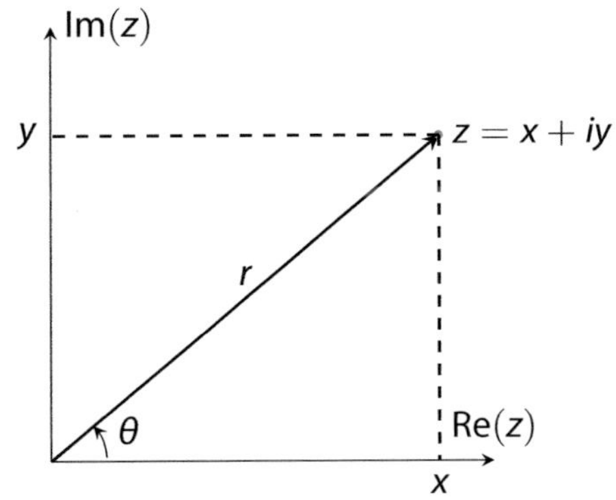

Figure C.1 An Argand diagram representing a complex number $z = x + iy$.

C.2 Complex Plane

We can represent complex numbers graphically using an Argand diagram (Figure C.1). This is basically an (x, y) plane in which the x-coordinate represents the real component of the complex number and the y-coordinate represents the imaginary component. We can represent the point $z = x + iy$ in the Argand diagram using polar coordinates so that

$$x + iy = r\cos(\theta) + ir\sin(\theta) = r(\cos(\theta) + i\sin(\theta)),$$

where $r = \sqrt{x^2 + y^2} =\mid z \mid= (z\bar{z})^{1/2}$.

Example C.1 As an example, let us write $z = -1 - i$ in polar form. We have that $\mathrm{Re}(z) = x = -1$ and $\mathrm{Im}(z) = y = -1$ so that $r = \sqrt{2}$. The angle θ has an infinite number of values

$$\theta = \tan^{-1}\left(\frac{\mathrm{Im}(z)}{\mathrm{Re}(z)}\right) = \frac{5\pi}{4} + 2n\pi,$$

where n is a positive or negative integer. The convention is to take $n = 0$, and this is called the *principal angle*.

C.3 Series

Infinite series of complex numbers involve separate series of the real and imaginary components, so that, for example, the partial sums of an infinite series of a complex number consist of two partial sums, one for the real part and one for the imaginary part: $S_n = X_n + iY_n$. Convergence of series of complex numbers is defined in a similar manner to that of real numbers, but if S_n approaches a limit $S = X + iY$ as $n \to \infty$, then both X_n and Y_n must also converge. For example, consider the two series:

$$f(z) = 1 - z + \frac{z^2}{2} - \frac{z^3}{3} + \frac{z^4}{4} - \cdots \tag{C.1}$$

$$g(z) = 1 + iz + \frac{(iz)^2}{2!} + \frac{(iz)^3}{3!} \cdots = 1 + iz - \frac{z^2}{2!} + \frac{iz^3}{3!} \cdots \tag{C.2}$$

We can examine the convergence of these series by looking at the Ratio Test (Equation (3.22)). For the series $f(z)$ we have

$$\lim_{n\to\infty} \left| \frac{nz}{n+1} \right| = \mid z \mid,$$

so the series converges as long as $\mid z \mid < 1$; in other words, $\sqrt{x^2 + y^2} < 1$. So, the series converges as long as z lies inside a circle of radius 1 centered on the origin in the complex plane. This interior of a circle, called the *circle of convergence*, replaces the interval of convergence for real numbers and gives us a *radius of convergence*.

For the function $g(z)$, we have

$$\lim_{n\to\infty} \left| \frac{(iz)^{n+1}}{(n+1)!} \times \frac{n!}{(iz)^n} \right| = \lim_{n\to\infty} \left| \frac{iz}{n+1} \right| = 0,$$

so this series converges for all values of z.

Functions of complex numbers are often defined using power series expansions and have the same properties as their real counterparts. For example:

$$e^z = 1 + \frac{z}{1!} + \frac{z^2}{2!} \cdots \tag{C.3}$$

has the same properties as the exponential of a real number:

$$e^{z_1} e^{z_2} = e^{z_1 + z_2}.$$

C.4 Euler's Formula

One of the most important properties of complex numbers is that there is a relationship between complex numbers and trigonometric functions and exponential functions. To see this, we start with Equation (C.3):

$$\begin{aligned} e^{i\theta} &= 1 + (i\theta) + \frac{(i\theta)^2}{2!} + \frac{(i\theta)^3}{3!} + \frac{(i\theta)^4}{4!} + \frac{(i\theta)^5}{5!} \cdots \\ &= 1 + i\theta - \frac{\theta^2}{2!} - \frac{i\theta^3}{3!} + \frac{\theta^4}{4!} + \frac{i\theta^5}{5!} \cdots \\ &= \left(1 - \frac{\theta^2}{2!} + \frac{\theta^4}{4!} + \cdots\right) - i\left(\theta\frac{\theta^3}{3!} + \frac{\theta^5}{5!} \cdots\right) \\ &= \cos(\theta) + i\sin(\theta), \end{aligned}$$

where we assume that θ is a real number and we have used the power series expansions

$$\sin(\theta) = \theta - \frac{\theta^3}{3!} + \frac{\theta^5}{5!} \cdots \qquad \cos(\theta) = 1 - \frac{\theta^2}{2!} + \frac{\theta^4}{4!} \cdots$$

This means that we can write any complex number as

$$z = x + iy + r(\cos(\theta) + i\sin(\theta)) = re^{i\theta},$$

which is called *Euler's formula*.[2] This formula is very important because manipulating exponential functions is easier than manipulating trigonometric functions, so complex numbers are frequently written as exponentials.

C.5 De Moivre's Theorem

Another important theorem is De Moivre's theorem,[3] which states that

$$(\cos(\theta) + i\sin(\theta))^n = \cos(n\theta) + i\sin(n\theta).$$

We can see that this must be true by using Euler's formula:

$$z^n = (re^{i\theta})^n = r^n e^{in\theta} = r^n(\cos(n\theta) + i\sin(n\theta)).$$

We can use De Moirvre's theorem to calculate the nth root of a complex number

$$z^{1/n} = (re^{i\theta})^{1/n} = r^{1/n}e^{i\theta/n} = r^{1/n}\left(\cos\left(\frac{\theta}{n}\right) + i\sin\left(\frac{\theta}{n}\right)\right)$$

so that the nth root of a complex number has multiple roots.

[2] Named after the Swiss mathematician Leonhard Euler (1707–1783).
[3] Named after the French mathematician Abraham de Moivre (1667–1754).

References

Abramowitz, Milton, and Stegun, Irene A. 1972. *Handbook of Mathematical Functions with Formulas, Graphs, and Mathematical Tables*. National Bureau of Standard Applied Mathematics Series 55. US Government Printing Office, Washington DC.

Acton, Forman S. 1990. *Numerical Methods That (Usually) Work*. Mathematical Association of America Washington DC.

Acton, Forman S. 1996. *Real Computing Made Real: Preventing Errors in Scientific and Engineering Calculations*. Princeton University Press, Princeton, NJ.

Adam, J. A. 2003. *Mathematics in Nature: Modeling Patterns in the Natural World*. Princeton University Press, Princeton, NJ.

Albarède, Francis. 1995. *Introduction to Geochemical Modeling*. Cambridge University Press, Cambridge.

Allmendinger, Richard W., Cardozo, Nestor, and Fisher, Donald M. 2012. *Structural Geology Algorithms: Vectors and Tensors*. Cambridge University Press, Cambridge.

Althoen, Steven C., and McLaughlin, Renate. 1987. Gauss–Jordan reduction: A brief history. *American Mathematical Monthly*, **94**(2), 130–142.

Arfken, George B., Weber, Hans J., and Harris, Frank E. 2013. *Mathematical Methods for Physicists: A Comprehensive Guide*, 7th edn. Academic Press, New York, NY.

Arnold, V. I. 1978. *Ordinary Differential Equations*. MIT Press, Cambridge, MA.

Barenblatt, G. I. 1996. *Scaling, Self-Similarity, and Intermediate Asymptotics*. Cambridge University Press, Cambridge.

Berg, Howard C. 1993. *Random Walks in Biology*. Princeton University Press, Princeton NJ.

Bertsch McGrayne, Sharon. 2011. *The Theory That Would Not Die: How Bayes' Rule Cracked the Enigma Code, Hunted Down Russian Submarines, and Emerged Triumphant from Two Centuries of Controversy*. Yale University Press, New Haven, CT.

Boas, Mary L. 2006. *Mathematical Methods in the Physical Sciences*. Wiley, Hoboken, NJ.

Borisenko, A. I., and Tarapov, I. E. 1968. *Vector and Tensor Analysis with Applications*. Dover, New York.

Boyce, William E., and DiPrima, Richard C. 2012. *Elementary Differential Equations and Boundary Value Problems*. Wiley, Hoboken, NJ.

Butikov, Eugene I. 2002. A dynamical picture of the oceanic tides. *American Journal of Physics*, **70**, 1001–1011.

Carry, Benoit. 2012. Density of asteroids. *Planetary and Space Science*, **73**, 98–118.

Carslaw, H. S., and Jaeger, J. C. 1959. *Conduction of Heat in Solids*. 2nd edn. Clarendon Press, Oxford.

Cartwright, Julyan H. E., Eguíluz, V. M., Hernández-García, E., and Piro, O. 1999. Dynamics of elastic excitable media. *International Journal of Bifurcation and Chaos*, **9**, 2197–2202.

Chaytor, Jason D., ten Brink, Uri S., Solow, Andrew R., and Andrews, Brian D. 2009. Size distribution of submarine landslides along the US Atlantic margin. *Marine Geology*, **264**, 16–27.

Chillingworth, D. R. J., and Holmes, P. J. 1980. Dynamical systems and models for reversals of the Earth's magnetic field. *Journal of the International Association for Mathematical Geology*, **12**, 41–59.

Coleman, Matthew P. 2005. *An Introduction to Partial Differential Equations with MATLAB*. Chapman and Hall, Boca Raton, FL.

Coles, Peter. 2006. *From Cosmos to Chaos: The Science of Unpredictability*. Oxford University Press, Oxford.

Cooper, Necia Grant (ed.). 1989. *From Cardinals to Chaos: Reflections on the Life and Legacy of Stanislaw Ulam*. Cambridge University Press, Cambridge.

Cortini, M., and Barton, C. C. 1994. Chaos in geomagnetic reversal records: A comparison between Earth's magnetic field data and model disk dynamo data. *Journal of Geophysical Research*, **99B**, 18021–18033.

Crowe, M. J. 1994. *A History of Vector Analysis: The Evolution of the Idea of a Vectorial System*. Dover, New York.

Darrigol, Olivier. 2005. *Worlds of Flow: A History of Hydrodynamics from the Bernoullis to Prandtl*. Oxford University Press, Oxford.

Denny, Mark W. 1993. *Air and Water: The Biology and Physics of Life's Media*. Princeton University Press, Princeton, NJ.

Denny, Mark W. 2008. *How the Ocean Works: An Introduction to Oceanography*. Princeton University Press, Princeton, NJ.

Denny, Mark, and Gaines, Steven. 2002. *Chance in Biology: Using Probability to Explore Nature*. Princeton University Press, Princeton, NJ.

Dodds, Peter Sheridan, and Rothman, Daniel H. 2000. Scaling, universality, and geomorphology. *Annual Review of Earth and Planetary Science*, **28**, 571–610.

Duren, Peter. 2009. Changing faces: The mistaken portrait of Legendre. *Notices of the American Mathematical Society*, **56**, 1440–1443.

Durran, Dale R. 1999. *Numerical Methods for Wave Equations in Geophysical Fluid Dynamics*. Springer, New York.

Durran, Dale R. 2010. *Numerical Methods for Fluid Dynamics: With Applications to Geophysics*. 2nd edn. Springer, New York.

Dziewonski, Adam M., and Anderson, Don L. 1981. Preliminary reference Earth model. *Physics of Earth and Planetary Interiors*, **25**, 297–356.

Falkner, K. Kenison, and Edmond, J. M. 1990. Gold in seawater. *Earth and Planetary Science Letters*, **98**, 208–221.

Feulner, Georg. 2012. The faint young Sun problem. *Reviews of Geophysics*, **50**, RG2006.

Fleisch, Daniel A. 2012. *A Student's Guide to Vectors and Tensors*. Cambridge University Press, Cambridge.

Follows, Michael J, and Dutkiewicz, Stephanie. 2011. Modeling diverse communities of marine microbes. *Annual Review of Marine Science*, **3**, 427–451.

Glover, David M., Jenkins, William J., and Doney, Scott C. 2011. *Modeling Methods for Marine Science*. Cambridge University Press, Cambridge.

Gockenbach, Mark S. 2002. *Partial Differential Equations: Analytical and Numerical Methods*. Society for Industrial and Applied Mathematics, Philadelphia.

Gradshteyn, I. S., and Ryzhik, I. M. 1980. *Table of Integrals, Series, and Products*. Academic Press, New York.

Griffies, Stephen M. 2004. *Fundamentals of Ocean Climate Models*. Princeton University Press, Princeton, NJ.

Guckenheimer, John, and Holmes, P. J. 1983. *Nonlinear Oscillations, Dynamical Systems, and Bifurcations of Vector Fields*. Springer, New York.

Hack, J. T. 1957. *Studies of Longituindal Profiles in Virginia and Maryland*. Tech. rept. US Geological Survey Professional Paper 294B.

Hafemeister, D. 2007. *Physics of Societal Issues: Calculations on National Security, Environment, and Energy*. Springer, New York.

Haldane, J. B. S. 1945. On Being the Right Size. In: *Possible Worlds and Other Essays* (pp. 18–26). Chatto and Windus, London.

Harte, John. 1988. *Consider a Spherical Cow: A Course in Environmental Problem Solving*. University Science Books, Sausalito, CA.

Henderson, Paul, and Henderson, Gideon M. 2009. *The Cambridge Handbook of Earth Science Data*. Cambridge University Press, Cambridge.

Heuschele, Jan, and Selander, Erik. 2014. The chemical ecology of copepods. *Journal of Plankton Research*, **36**(4), 895–913.

Hilborn, Ray, and Mangel, Marc. 1997. *The Ecological Detective: Confronting Models with Data*. Princeton University Press, Princeton, NJ.

Holsapple, K. A., and Schmidt, R. M. 1982. On the scaling of crater dimensions: 2. Impact processes. *Journal of Geophysical Research*, **87**, 1849–1870.

Huang, J., and Turcotte, D. L. 1990. Evidence for chaotic fault interactions in the seismicity of the San Andreas fault and Nankai trough. *Nature*, **348**, 234–236.

Hunt, Bruce. 2012. Oliver Heaviside: A first-rate oddity. *Physics Today*, **65**, 48–54.

Iserles, Arieh. 2008. *A First Course in the Numerical Analysis of Differential Equations*. 2nd edn. Cambridge University Press, Cambridge.

Ito, K. 1980. Chaos in the Rikitake two-disc dynamic system. *Earth and Planetary Science Letters*, **51**, 451–456.

Jevrejeva, S., Moore, J. C., Grinsted, A., Woodworth, P. L. (2008). Recent global sea level acceleration started over 200 years ago? *Geophysical Research Letters*, **35**, L08715, doi://10.1029/2008GL033611.

Julian, Bruce R., Foulger, Gillian R., Hatfield, Oliver, Jackson, Samuel E, Simpson, Emma, Einbeck, Jochen, and Moore, Andrew. 2015. Hotspots in Hindsight. In: *The Interdisciplinary Earth: A Volume in Honor of Don L. Anderson*, edited by G. R. Foulger, M. Lustrino, and S. D. King, 105Ű201. Geological Society of America Boulder, CO.

Kaper, Hans, and Engler, Hans. 2013. *Mathematics and Climate*. Society for Industrial and Applied Mathematics, Philadelphia, PA.

Karamouz, M., Nazif, S., and Falahi, M. 2012. *Hydrology and Hydroclimatology: Principles and Applications*. CRC Press, Boca Raton, FL.

Kiørboe, Thomas. 2008. *A Mechanistic Approach to Plankton Ecology*. Princeton University Press, Princeton, NJ.

Kline, Morris. 1977. *Calculus: An Intuitive and Physical Approach*. Dover, New York.

Kring, David A. 2007. The Chicxulub impact event and its environmental consequences at the Cretaceous–Tertiary boundary. *Paleogeography, Paleoclimatology, Paleoecology*, **255**, 4–21.

Krumbein, William Christian, and Aberdeen, Esther Jane. 1937. The sediments of Barataria Bay (Louisiana). *Journal of Sedimentary Research*, **7**(1), 3–17.

Legendre, Pierre, and Legendre, Louis. 2012. *Numerical Ecology*. 3rd edn. Elsevier, London.

Le Quéré, Corinne, Andrew, Robbie M., Candadell, Josep G., et al. 2016. Global Carbon Budget 2016. *Earth System Science Data*, **8**, 605–649.

Lorenz, Edward N. 1963. Deterministic nonperiodic flow. *Journal of the Atmospheric Sciences*, **20**, 130–141.

Ludwig, D., Jones, D. D., and Holling, C. S. 1978. Qualitative analysis of insect breakout systems: The spruce budworm and forest. *Journal of Animal Ecology*, **47**, 315–332.

Mackey, M. D., Mackey, D. J., Higgins, H. W., and Wright, S. W. 1996. CHEMTAX—A program for estimating class abundances from chemical markers: Application to HPLC measurements of phytoplankton. *Marine Ecology Progress Series*, **144**, 265–283.

Mahajan, Sanjoy. 2010. *Street-Fighting Mathematics: The Art of Educated Guessing and Opportunistic Problem Solving*. MIT Press, Cambridge, MA.

Marshall, John, and Plumb, R. Alan. 2008. *Atmosphere, Ocean, and Climate Dynamics: An Introductory Text*. Academic Press, New York.

Martin-Silverstone, Elizabeth, Vincze, Orsolya, McCann, Ria, Jonsson, Carl H. W., Palmer, Colin, Kaiser, Gary, and Dyke, Gareth. 2015. Exploring the relationship between skeletal mass and total body mass in birds. *PLoS One*, **10**, e0141794.

May, R. M., Wishart, D. M. G., Bray, J., and Smith, R. L. 1987. Chaos and the dynamics of biological populations. *Proceedings of the Royal Society Series A*, **413**, 27–44.

Miller, Robert N. 2007. *Numerical Modeling of Ocean Circulation*. Cambridge University Press, Cambridge.

Mobley, Curtis D. 1994. *Light and Water: Radiative Transfer in Natural Waters*. Academic Press, New York.

Monson, Russell, and Baldocchi, Dennis. 2014. *Terrestrial Biosphere–Atmosphere Fluxes*. Cambridge University Press, Cambridge.

Murphy, G. M. 1960. *Ordinary Differential Equations and Their Solutions*. Van Nostrand, London.

Nahin, Paul J. 2014. *Inside Interesting Integrals*. Springer, New York.

Niklas, Karl J. 1994. *Plant Allometry: The Scaling of Form and Process*. University of Chicago Press, Chicago, IL.

North, Gerald R. 1975. Theory of Energy-Balance Climate Models. *Journal of the Atmospheric Sciences*, **32**, 2033–2043.

North, Gerald R., and Kim, Kwang-Yul. 2017. *Energy Balance Climate Models*. Wiley-VCH, Weinheim, Germany.

Peters, Robert Henry. 1983. *The Ecological Implications of Body Size*. Cambridge University Press, Cambridge.

Pollard, David D., and Fletcher, Raymond C. 2005. *Fundamentals of Structural Geology*. Cambridge University Press, Cambridge, UK.

Press, William H, Teukolsky, Saul A, Vetterling, William T, and Flannery, Brian P. 1992. *Numerical Recipes in C: The Art of Scientific Computing*. 2nd edn. Cambridge University Press, Cambridge.

Prothero, John William. 2015. *The Design of Mammals: A Scaling Approach*. Cambridge University Press, Cambridge.

Pruppacher, H. R., and Klett, J. D. 2010. *Microphysics of Clouds and Precipitation*. Springer, New York.

Residori, S., Onorato, M., Bortolozzo, U., and Arecchi, F. T. 2017. Rogue waves: A unique approach to multidisciplinary physics. *Contemporary Physics*, **58**, 53–69.

Risch, R. H. 1969. The problem of integration in finite terms. *Transactions of the American Mathematical Society*, **139**, 167–189.

Risch, R. H. 1970. The solution of the problem of integration in finite terms. *Bulletin of the American Mathematical Society*, **76**, 605–608.

Rood, R. B. 1987. Numerical advection algorithms and their role in atmospheric transport and chemistry. *Reviews of Geophysics*, **25**, 71–100.

Schey, H. M. 2004. *Div, Grad, Curl, And All That: An Informal Text on Vector Calculus*. 4th edn. W. W. Norton, New York.

Schmidt-Nielsen, Knut. 1984. *Scaling: Why Is Animal Size So Important?* Cambridge University Press, Cambridge.

Schutz, Bernard. 1980. *Geometrical Methods of Mathematical Physics*. Cambridge University Press, Cambridge.

Shampine, Lawrence F. 1994. *Numerical Solution of Ordinary Differential Equations*. Chapman and Hall, Boca Raton, FL.

Snieder, Roel, and van Wijk, Kasper. 2015. *A Guided Tour of Mathematical Methods for the Physical Sciences*. 3rd edn. Cambridge University Press, Cambridge.

Solé, Ricard V., and Bascompte, Jordi. 2006. *Self-Organization in Complex Ecosystems*. Princeton University Press, Princeton, NJ.

Spivak, Michael. 2008. *Calculus*. 4th edn. Publish or Perish Press, Houston, TX.

Stensrud, David J. 2009. *Parameterization Schemes: Keys to Understanding Numerical Weather Prediction Models*. Cambridge University Press, Cambridge.

Stommel, Henry M., and Moore, Dennis W. 1989. *An Introduction to the Coriolis Force*. Columbia University Press, New York.

Strang, Gilbert. 2006. *Linear Algebra and Its Applications*. 4th edn. Thomson Learning, Belmont, CA.

Strang, Gilbert. 2017. *Calculus*. 3rd edn. Wellesley-Cambridge Press, Wellesley, MA.

Strogatz, Steven H. 2001. *Nonlinear Dynamics and Chaos: With Applications to Physics, Biology, Chemistry, and Engineering*. Westview Press, Boulder, CO.

Swartz, Clifford E. 1993. *Used Math for the First Two Years of College Science*. American Association of Physics Teachers, College Park, MD.

Szirtes, T. 2007. *Applied Dimensional Analysis and Modeling*. Butterworth-Heinemann, New York.

Tapley, B., Ries, J., Bettadpur, S., Chambers, D., Cheng, M., Condi, F., Gunter, B., Kang, Z., Nagel, P., Pastor, R., Pekker, T., Poole, S., and Wang, F. 2005. GGM02—An improved Earth gravity field model from GRACE. *Journal of Geodesy*, **79**(8), 467–478.

Tennekes, Henk, and Lumley, John L. 1972. *A First Course in Turbulence*. MIT Press, Cambridge, MA.

Thomson, Richard E., and Emery, William. 2014. *Data Analysis Methods in Physical Oceanography*. 3rd edn. Elsevier, New York.

Toon, Owen B., Zahnle, Kevin, Morrison, David, Turco, Richard P., and Covey, Curt. 1997. Environmental perturbations caused by the impacts of asteroids and comets. *Reviews of Geophysics*, **35**, 41–78.

Turcotte, Donald L. 1997. *Fractals and Chaos in Geology and Geophysics*. 2nd edn. Cambridge University Press, Cambridge.

Turcotte, Donald, and Schubert, Gerald. 2014. *Geodynamics*. Cambridge University Press, Cambridge.

Vallis, Geoffrey K. 2017. *Atmospheric and Oceanic Fluid Dynamics: Fundamentals and Large-Scale Circulation*. 2nd edn. Cambridge University Press, Cambridge.

Weinstein, L., and Adam, J. A. 2008. *Guesstimation: Solving the World's Problems on the Back of a Cocktail Napkin*. Princeton University Press, Princeton, NJ.

Wentworth, C. K. 1922. A scale of grade and class terms for clastic sediments. *Journal of Geology*, **30**(5), 377–392.

Wigner, E. 1960. The unreasonable effectiveness of mathematics in the natural sciences. *Communications in Pure and Applied Mathematics*, **13**, 1–14.

Williams, Richard G., and Follows, Michael J. 2011. *Ocean Dynamics and the Carbon Cycle: Principles and Mechanisms*. Cambridge University Press, Cambridge.

Zhou, Hua-Wei. 2014. *Practical Seismic Data Analysis*. Cambridge University Press, Cambridge.

Zwillinger, Daniel. 1992. *Handbook of Integration*. Jones and Bartlett, Boca Raton, FL.

Zwillinger, Daniel. 1997. *Handbook of Differential Equations*. Academic Press, New York.

Index

Printed in the United Kingdom by TJ Clays Ltd.